Basic Neurochemistry

BASIC NEUROCHEMISTRY

Edited by

R. Wayne Albers, Ph.D.
National Institute of Neurological
Diseases and Stroke, Bethesda

George J. Siegel, M.D.
Mount Sinai School of Medicine
of the City University of New York

Robert Katzman, M.D.
Albert Einstein College of Medicine
of Yeshiva University, Bronx

Bernard W. Agranoff, M.D.
University of Michigan Medical School,
Ann Arbor

Little, Brown and Company
Boston

First Edition

Library of Congress catalog card No. 71-189721

ISBN 0-316-02050

Published in Great Britain by Churchill/Livingstone, Edinburgh and London

Printed in the United States of America

Preface

NEUROCHEMISTRY has emerged in the past half-century as a distinct, although hybrid, discipline. Its central, unifying objective is the elucidation of biochemical phenomena that subserve the characteristic activity of the nervous system or are associated with neurological diseases. This objective generates certain subsidiary and prerequisite goals common to all biological investigative disciplines: (1) isolation and identification of components; (2) analysis of their organization, i.e., the effects of the components upon each other; and (3) a description of the temporal and spatial relations of these components and their interactions to the activity of the intact organ.

A comprehensive description of the nervous system function ideally should be continuous from the molecular level to the most complex level of integration. Presumably such a description would provide intellectually satisfying and socially useful explanations of nervous system interactions with the environment. These interactions may be observed as mentation, behavior, physiological responses, and pathological changes.

The validity and vitality of neurochemistry as a distinct discipline are proportionate to the successes of its practitioners in providing biochemical explanations for neural processes. Information as to which biochemical phenomena subserve some identifiable neural function cannot be derived from chemical analyses alone, because, in the course of analyses, organization and function are destroyed. Those who would gather such information must be equipped to study different levels of complexity in both structure and function.

Accordingly, the areas of juncture between the "horizontal" field of biochemistry and the "vertically" structured fields of neurobiology, neurology, and the behavioral sciences constitute the scope of neurochemistry. Neurochemists are crucially dependent upon data from these diverse subjects to formulate functionally meaningful molecular hypotheses.

As a result of this diversity, the student of neurochemistry must become familiar with concepts and information which are widely dispersed in the scientific literature.

It is often difficult to evaluate data and conclusions that are far outside one's own experience and training. A number of neurochemistry courses have been organized in medical and graduate schools within recent years, and the organizers have become acutely aware of the difficulties in selecting the most significant material and in placing it within the perspective of brain function. As a source of aid, few books are at once comprehensive and sufficiently concise to be practical as texts in neurochemistry.

In order to consider these problems, the Conference on Neurochemistry Curriculum was held June 19 and 20, 1969, in New York City under the sponsorship of the National Institute of Neurological Diseases and Stroke. This conference brought together individuals engaged in teaching and investigation in the basic neurobiological and clinical neurological sciences in institutions across the United States and in Canada. This interdisciplinary group constructed a syllabus which has served to delineate a consensus of the proper scope for the subject matter and which has subsequently been developed into this textbook on basic neurochemistry.

The book embodies three broad sections: General Neurochemistry (edited by R. Wayne Albers); Medical Neurochemistry (edited by Robert Katzman); and Behavioral Neurochemistry (edited by Bernard W. Agranoff). George J. Siegel served as Coordinating Editor. Substantial cross-indexing among these three parts helps to provide a cohesive unit. We believe that this format can be used in the basic science training of both M.D. and Ph.D. candidates and in postgraduate clinical science training as well, offering to each training program an opportunity for interdisciplinary exposure.

Despite the wide input into the initial outlines and the preparation of this book, the editors are aware that the treatment of some important topics is necessarily curtailed. This is the result of limitations both of space and of time. An example is the biochemistry and pharmacology of seizures. The editors anticipate that the experience gained through the construction and use of this text will initiate a continuing reappraisal of the scientific and educational aspects of neurochemistry which will be reflected in later editions.

The references have been divided into two categories: general references, including books and reviews, and lists of original sources, which, in some cases, incorporate reviews of special topics. There is no attempt to cite all relevant literature or to identify all the sources for the text. Most of these sources and many bibliographies can be found in the general references listed after each chapter.

For those interested in the historical development of neurochemistry, several essays are available, in particular, D. L. Drabkin: *Thudicum, Chemist of the Brain* (Philadelphia, University of Pennsylvania Press, 1958); I. H. Page: Chemistry of the brain – past imperfect, present indicative, and – future perfect? (*Science* 125:721, 1957); D. B. Tower: Origins and development of neurochemistry (*Neurology* 8 [Suppl. 1]:3, 1957).

We would like to thank all contributors for their cooperation and also those who, by their work at the Conference, contributed to the planning of this volume: Julius Axelrod, Abel Lajtha, John Menkes, John O'Brien, Jerome Posner, Eric Shooter, Wallace Tourtellotte, and Myron Winick.

The editors are very pleased to acknowledge the most professional and meticulous editorial assistance of Mrs. Helene Jordan Waddell. Her own devotion to the task of making the text coherent and comprehensible is responsible in large measure for achieving a volume presentable for publication.

R. W. A.
G. J. S.
R. K.
B. W. A.

Acknowledgment

The Conference on Neurochemistry Curriculum, held at Greystone Conference Center, Riverdale, New York, June 19–20, 1969, was supported by the United States Public Health Service, National Institutes of Health, Grant 1T14NB05671-01.

Contents

Contributing Authors

LEO G. ABOOD, Ph.D.
Professor of Biochemistry and Brain Research, University of Rochester School of Medicine and Dentistry, Rochester, New York
Chapter 3

BERNARD W. AGRANOFF, M.D.
Professor of Biological Chemistry and Research Biochemist, Mental Health Research Institute, The University of Michigan Medical School, Ann Arbor
Chapter 31

R. WAYNE ALBERS, Ph.D.
Chief, Section on Enzyme Chemistry, Laboratory of Neurochemistry, National Institute of Neurological Diseases and Stroke, National Institutes of Health, Bethesda, Maryland
Chapters 1 and 4

STANLEY H. APPEL, M.D.
Professor and Chief, Division of Neurology, and Associate Professor of Biochemistry, Duke University School of Medicine, Durham, North Carolina
Chapter 21

SAMUEL H. BARONDES, M.D.
Head, Division of Neurobiology, and Professor of Psychiatry, University of California, San Diego, School of Medicine, La Jolla
Chapter 12

JOYCE A. BENJAMINS, Ph.D.
Assistant Professor of Neurology, The Johns Hopkins University School of Medicine, Baltimore
Chapter 14

MURRAY B. BORNSTEIN, M.D.
Professor of Neurology, Albert Einstein College of Medicine of Yeshiva University, Bronx, New York
Chapter 24

ROSCOE O. BRADY, M.D.
Chief, Developmental and Metabolic Neurology Branch, National Institute of Neurological Diseases and Stroke, National Institutes of Health, Bethesda, Maryland
Chapter 23

MAYNARD M. COHEN, M.D., Ph.D.
Professor and Chairman, Department of Neurological Sciences, Rush Medical College; Chairman, Department of Neurological Sciences, Presbyterian-St. Luke's Hospital, Chicago
Chapter 27

EUGENE D. DAY, Ph.D.
Professor of Immunology, Duke University School of Medicine, Durham, North Carolina
Chapter 21

PIERRE M. DREYFUS, M.D.
Professor and Chairman, Department of Neurology, University of California, Davis, School of Medicine; Director, Department of Neurology, Sacramento Medical Center, Sacramento
Chapter 25

GARY R. DUTTON, Ph.D.
Assistant Research Biochemist, Department of Psychiatry, University of California, San Diego, School of Medicine, La Jolla
Chapter 12

JOSEPH S. EISENMAN, Ph.D.
Associate Professor of Physiology, Mount Sinai School of Medicine of the City University of New York
Chapter 17

STANLEY E. GEEL, Ph.D.
Assistant Professor of Neurochemistry, Department of Neurology, University of California, Davis, School of Medicine
Chapter 25

J. GERGELY, M.D., Ph.D.
Associate Professor of Biological Chemistry, Harvard Medical School; Director, Department of Muscle Research, Boston Biomedical Research Institute, Boston
Chapter 19

GORDON GUROFF, Ph.D.
Chief, Section on Intermediary Metabolism, Laboratory of Biomedical Sciences, National Institute of Child Health and Human Development, National Institutes of Health, Bethesda, Maryland
Chapter 10

RICHARD HAMMERSCHLAG, Ph.D.
Associate Research Scientist, Division of Neurosciences, City of Hope Medical Center, Duarte, California
Chapters 5 and 8

BOYD K. HARTMAN, M.D.
Assistant Professor of Psychiatry, Washington University School of Medicine; Assistant Psychiatrist, Barnes and Renard Hospitals, St. Louis
Chapter 30

FRANCIS C. G. HOSKIN, Ph.D.
Professor of Biology, Illinois Institute of Technology, Chicago; Senior Investigator, Marine Biological Laboratory, Woods Hole, Massachusetts
Chapter 7

Y. EDWARD HSIA, B.M., B.Ch., M.R.C.P., D.C.H.
Associate Professor, Division of Medical Genetics, Departments of Pediatrics and Medicine, Yale University School of Medicine, New Haven, Connecticut
Chapter 22

ROBERT KATZMAN, M.D.
Professor and Chairman, Department of Neurology, Albert Einstein College of Medicine of Yeshiva University, Bronx; Director, Neurological Services, Bronx Municipal Hospital Center, Bronx, New York
Chapter 16

GERARD M. LEHRER, M.D.
Professor of Neurology and Director, Division of Neurochemistry, Mount Sinai School of Medicine of the City University of New York; Attending Neurologist, The Mount Sinai Hospital, New York
Chapter 9

HENRY R. MAHLER, Ph.D.
Research Professor of Chemistry, Indiana University, Bloomington
Chapter 13

HOWARD S. MAKER, M.D.
Assistant Professor of Neurology, Mount Sinai School of Medicine of the City University of New York; Assistant Attending Neurologist, The Mount Sinai Hospital, New York
Chapter 9

DAVID B. McDOUGAL, JR., M.D.
Professor of Pharmacology, Washington University School of Medicine, St. Louis
Chapter 20

GUY M. McKHANN, M.D.
Kennedy Professor of Neurology and Director, Department of Neurology, The Johns Hopkins University School of Medicine; Neurologist-in-Chief, The Johns Hopkins Hospital, Baltimore
Chapter 14

DON D. MICKEY, Ph.D.
Research Associate, Department of Microbiology and Immunology, Duke University School of Medicine, Durham, North Carolina
Chapter 21

PIERRE MORELL, Ph.D.
Assistant Professor of Neurology and Biochemistry, Albert Einstein College of Medicine of Yeshiva University, Bronx, New York
Chapter 24

WILLIAM T. NORTON, Ph.D.
Professor of Neurology, Albert Einstein College of Medicine of Yeshiva University, Bronx, New York
Chapters 18 and 24

EUGENE ROBERTS, Ph.D.
Director, Division of Neurosciences, City of Hope Medical Center, Duarte, California; Adjunct Professor of Biochemistry, University of Southern California at Los Angeles
Chapters 5 and 8

ELI ROBINS, M.D.
Wallace Renard Professor and Head, Department of Psychiatry, Washington University School of Medicine; Psychiatrist-in-Chief, Barnes and Renard Hospitals, St. Louis
Chapter 30

GEORGE J. SIEGEL, M.D.
Associate Professor of Neurology and Physiology, Mount Sinai School of Medicine of the City University of New York; Associate Attending Neurologist, The Mount Sinai Hospital, New York
Chapter 17

SOLOMON H. SNYDER, M.D.
Professor of Pharmacology and Psychiatry, The Johns Hopkins University School of Medicine; Psychiatrist, The Johns Hopkins Hospital, Baltimore
Chapter 6

LOUIS SOKOLOFF, M.D.
Chief, Laboratory of Cerebral Metabolism, National Institute of Mental Health, Bethesda, Maryland
Chapter 15

THEODORE L. SOURKES, Ph.D.
Professor of Biochemistry, Departments of Psychiatry and Biochemistry, McGill University Faculty of Medicine; Associated Scientist, Royal Victoria Hospital and Allan Memorial Institute of Psychiatry, Montreal, Quebec, Canada
Chapters 28 and 29

WILLIAM L. STAHL, Ph.D.
Associate Professor of Physiology and Biophysics and Medicine (Neurology), University of Washington School of Medicine; Head, Neurochemistry Laboratory, Veterans Administration Hospital, Seattle
Chapter 2

KUNIHIKO SUZUKI, M.D.
Professor of Neurology, Rose F. Kennedy Center for Research in Mental Retardation and Human Development, Albert Einstein College of Medicine of Yeshiva University, Bronx, New York. Formerly Associate Professor of Neurology, University of Pennsylvania School of Medicine, Philadelphia
Chapter 11

PHILLIP D. SWANSON, M.D., Ph.D.
Associate Professor of Medicine and Head, Division of Neurology, University of Washington School of Medicine, Seattle
Chapter 2

DONALD B. TOWER, M.D., Ph.D.
Chief, Laboratory of Neurochemistry, National Institute of Neurological Diseases and Stroke, National Institutes of Health, Bethesda, Maryland
Chapter 26

Basic Neurochemistry

GENERAL NEUROCHEMISTRY

SECTION I

Structure and Function of Neural Membranes

1

Biochemistry of Cell Membranes

R. Wayne Albers

THE INTEGRATING FUNCTION of the nervous system depends upon the ability of combinations of neurons to generate differential responses to various stimuli. A single neuron is typically characterized by a number of receptor specializations. Receptors are usually confined to the dendritic or somal regions of neurons. A wide range of specialization occurs in different neurons. There are receptors for various physical stimuli in sensory neurons and for specific chemical neurotransmitters in multineuronal circuits. By releasing specific neurotransmitters, axonal propagation elicits further neuronal responses which finally impinge upon the external environment via muscles or glands.

Most of this activity involves primary interactions with the outer cell membrane (plasmalemma) of neurons. This chapter will consider some biological membrane features which appear to be either general principles or to be particularly relevant to neural functions. A number of recent reviews of membrane biochemistry are available [1-4, 6].

Membranes are predominantly composed of proteins and complex lipids (Table 1-1). The lipids are well characterized [40], but only a few individual membrane proteins have been sufficiently purified to yield reliable chemical information.

PHYSICAL PROPERTIES OF AMPHIPHILIC LIPIDS

The structural organization of membranes is closely related to the physicochemical properties of membrane lipids because the forces of molecular interactions in membranes primarily involve the formation of noncovalent bonds [41]. These may be classified as ionic, hydrogen, and lyophobic bonds (or, in aqueous solutions, hydrophobic bonds). Such interactions are weak relative to covalent bonds, but a particular complementary orientation between molecules, which maximizes the number of these interactions, may produce a stable or favored association.

Molecules are soluble if their interactions with the solvent are stronger than their

TABLE 1-1. Composition of Cell Membranes

	Liver			Neural		
	Microsomes [1, 4, 41]	Golgi [1, 41]	Plasma [3, 41]	Axolemma [5]	Myelin [18, 33]	Microsomes [18]
			% Solids			
Protein	70	60	68	37	25	
Cholesterol	1.5	5	8	9	18	10
Phospholipids	28	21	15	29	32	44
Glycolipids	0	....	....	....	20	7
			% Total Phospholipids			
Phosphatidylcholine	65	45	35	....	22	47
Phosphatidylethanolamine	20	17	22	....	36	26
Phosphatidylserine	4	4	4	....	22	12
Phosphatidylinositol	12	9	8	....	2	3
Sphingomyelin	4	12	20	....	11	11
			% Unsaturation			
Phosphatidylcholine	49	41	30	....	....	....
Phosphatidylethanolamine	34	34	39	....	....	....
Phosphatidylserine	75	55	14	....	....	....
Sphingomyelin	30	10	11	....	....	....

interactions with other solute molecules. Ionic and polar groups of a molecule will become hydrated, i.e., they will interact with hydrogen or oxygen dipoles of water. If such polar constituents predominate, the molecule will dissolve. Regions of large molecules which lack such polar groups will associate with nonpolar regions of other molecules to form *micelles.* Within such micelles, molecules assume an orientation that minimizes the exposure of nonpolar groups to water or, conversely, maximizes the interactions of polar groups with water.

Both complex lipids and proteins may be classified as *amphiphilic molecules,* i.e., molecules that contain both polar and nonpolar moieties. Amphiphilic molecules in aqueous solution either will form aggregates or, in the case of macromolecules, will fold in such a way as to sequester the hydrophobic regions in minimal volumes. Hydrophobic bonding is now recognized as a major factor in determining the stability of associations of both proteins and lipids.

FORMATION OF LIPID BILAYERS

The complex lipid constituents of membranes are characterized by distinct segregation of polar and nonpolar moieties within a single molecule. Thus, phosphatides can be represented as having a polar "head" composed of a glycerophosphoryl ester moiety and a "tail" composed of two fatty acid chains. Aqueous emulsions of phospholipids consist of a few to several hundred molecules that form more or less spherical aggregates (micelles), wherein the polar heads contact water and sequester the hydrocarbon tails in the interior. With increasing concentration, these micelles will coalesce farther to form large, multilayered complexes that sometimes are called *liposomes.* Similar structures may be formed by adding small amounts of water to solid phospholipids. These have been called *myelin figures.* Their basic structural unit is a bilayer sheet of phospholipid molecules oriented with polar heads outward on either side in contact with water, and with the hydrocarbon tails directed inward, the hydrocarbon chains being more or less perpendicular to the plane of the bilayer.

Single bilayer membranes may be formed from various mixtures of amphiphiles by a process of spontaneous thinning which takes place after a solution in organic solvent is painted across a small aperture immersed in aqueous solution. A single bilayer membrane is too thin to diffract light and is often called a *black lipid membrane* [2].

The experimental use of lysosomes and black lipid membranes has been providing extensive correlations between their chemical and physical properties. Their permeability to small, uncharged molecules is increased by increasing the degree of hydrocarbon chain unsaturation, by decreasing the chain length, or by decreasing the cholesterol content.

PHYSICAL STATE OF LIPIDS IN BIOLOGICAL MEMBRANES

The existence of some form of lipid bilayer in natural membrane as a barrier to the entrance of small nonelectrolytes was recognized by Overton in 1895 [33]. In

1925, Fricke deduced that the hydrocarbon barrier should have a thickness of about 33 Å from electrical impedance measurements [18]. In the same year, Gorter and Grendel estimated the surface area of red cells and, from the lipid content, deduced that there was precisely enough lipid to cover the surface with a bilayer [20]. More recently, the existence of bilayers in myelin [8] and the plasma membranes of *Mycoplasma laidlawii* [15] has been established by x-ray diffraction.

MEMBRANE PROTEINS

TERTIARY AND QUATERNARY PROTEIN STRUCTURE

Current reviews of this subject are readily available [27, 38]. The chief point to be emphasized here is that hydrophobic forces are major determinants of higher order structure in proteins as well as in lipids. Proteins often associate to form polymeric structures. These range from simple dimerization of similar proteins to the formation of highly ordered complexes of several unlike subunits. The one case in which the interfacial contacts between subunits are known is hemoglobin. Here the contacts between unlike subunits ($\alpha\beta$) are chiefly nonpolar, while $\alpha\alpha$ and $\beta\beta$ contacts are polar [36, 38].

PHYSICAL STATE OF PROTEINS IN BIOLOGICAL MEMBRANES

Early workers felt that the low surface tension and elasticity of plasmalemma could not be properties of a lipid film. Danielli [14] found that proteins were effective in lowering surface tension at an oil-water interface and proposed that a layer of "unrolled" protein might be adsorbed to the polar surfaces of the lipid bilayer. This would lead to the expectations that cell membranes should have fairly constant ratios of protein to lipid and that much of the "extended" protein may exhibit β secondary structure [38]. Neither expectation has been upheld. Protein-lipid ratios seem to vary over a range of nearly tenfold from cell to cell [28]. Proteins containing peptide chains in β conformation have a characteristic infrared absorption band at 1630 cm^{-1}. This has not been detected in plasma membrane proteins. On the other hand, optical rotatory dispersion and circular dichroism measurements are compatible with as much as 60 percent α-helical content, which is higher than that of many globular proteins [1].

CHEMICAL COMPOSITION OF MEMBRANE PROTEINS

As procedures are developed for solubilizing and resolving membrane proteins, it becomes clear that many different proteins are involved. Amino acid analyses are available for membranes from a variety of cells and they have been discussed with respect to the content of nonpolar amino acids. These are, in fact, not much higher than in many soluble proteins; nevertheless, these proteins have a marked tendency to form large aggregates. From studies of red-cell membrane proteins, Maddy and

Kelly [29] have concluded that at least three classes of proteins can be separated on the basis of their dominant aggregative forces: (1) bivalent cation bridges, (2) direct electrostatic bridges, and (3) hydrophobic bonding.

LIPID-PROTEIN INTERACTIONS IN MEMBRANES

There are few covalent bonds between membrane proteins and lipids. Hypothetical models for membrane structure differ with respect to the relative importance of polar and hydrophobic interactions. Models that suppose that membranes contain a continuous core of lipid bilayer must, explicitly or by exclusion, rely chiefly on polar lipid-protein bonding. However, membrane proteins are not commonly dissociated from lipids at high ionic strength, which seems to eliminate electrostatic binding as a dominant force. The most effective solvents for membrane proteins are certain organic solvents, and solubilization is promoted by low ionic strength, low pH, and divalent metal chelating agents. Anionic and neutral detergents are often effective in bringing membrane proteins into aqueous solutions. A number of "chaotropic" anions, such as SCN^- and ClO_4^-, have strong solubilizing effects which appear to arise from their ability to disrupt the hydrogen-bonded structure of water and thereby to reduce the relative stability of hydrophobic associations [24].

This type of information indicates that hydrophobic forces are generally dominant in binding membrane components together. Some recent work has suggested that membrane proteins may exist as globular units, or arrays, which "float" in the lipid bilayer [22] (Fig. 1-1). The composition of the hydrocarbon part of the bilayer of biological membranes is such that individual lipid molecules have a high degree of mobility within the plane of the bilayer [15].

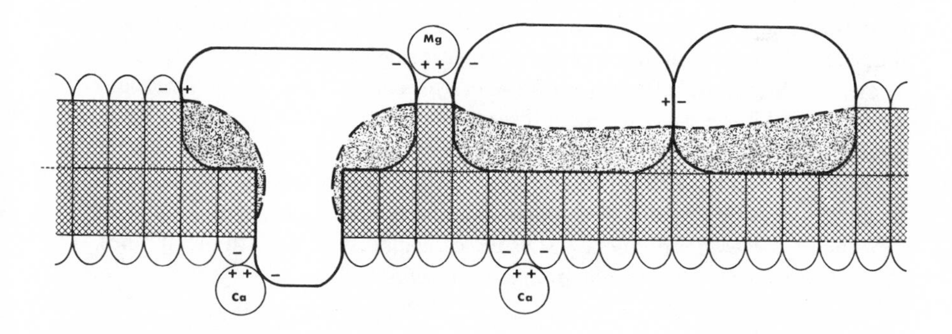

FIGURE 1-1. Binding forces in cell membranes. Some current concepts of structure are illustrated. The lipid bilayer is interrupted by proteins inserted primarily (but not exclusively) into the cytoplasmic half of the bilayer, and, in some cases, extending through the bilayer. Hydrophobic bonding is a dominant force (indicated by adjacent shaded areas). This is supplemented by electrostatic bonds directly between lipids and proteins as well as by divalent cation bridges.

Functional requirements dictate that certain proteins, e.g., those of transport systems and the external-receptor–internal-enzyme systems such as adenyl cyclase, must extend through the lipid bilayer. There has been limited confirmation of this by use of the freeze-fracture method of preparing membranes for electron microscopy. However, this technique shows that most of the globular protein units appear to be associated with the inner half of the lipid bilayer [37, 42].

The examples of freeze-etched membrane topography that have been published to date reveal a rather low density of such globular subunits, far too few to fulfill a structural role. Other techniques of electron microscopy have supported quite a different concept. The usual trilaminar image of thin-sectioned membranes stained with heavy metals is still evident in mitochondria after about 95 percent of the lipid has been extracted [17], indicating that protein is a major determinant of this image. The negative-staining technique presents an image of closely spread globular subunits attached to the membrane surface in the mitochondrial inner membrane [34] and sarcoplasmic reticulum [21]. Subunit structure has been seen within membranes in special cases, such as retinal photoreceptor discs (see Chap. 4) and Mauthner cell synapses [39].

Thus, although the existence of lipid bilayer structure in native membranes has been receiving increasing experimental confirmation, few generalizations can be made about the physical properties of membrane proteins. Evidence about the modes of association between these proteins and lipids indicates considerable diversity [29]. It seems likely that the molecular organization of membranes is at least as variable as the range of associations of soluble proteins. If this is true, satisfactory models can be derived only from the intensive study of specific cases.

The practice of recent reviews of membrane structure has been to present the Danielli model and the subunit model as opposing concepts. In fact, most biologists in this research area do not hold such fixed ideas, and one could find few unequivocal defenders of either. Those most involved with the development of lipid bilayer models are ready to admit a host of modifications to allow for what has been called the "inability to explain most of the biological functions of membranes" [6]. Those who find most difficulty with the bilayer concept are, by and large, those whose central interest is to explain these biological functions. However, dissatisfaction with the bilayer theory does not lead to a clear alternative. The current arguments are reminiscent of the early attempt to find geometric principles underlying the structure of globular proteins. Simple synthetic polypeptides can, in fact, show highly ordered structures, but as the case appears to be with membranes, these structures are less prominent in real proteins.

MEMBRANE BIOSYNTHESIS

The composite nature of membranes introduces a number of interdependent questions relating to the synthesis of individual components and their assembly into a functional structure. How are newly synthesized lipids and proteins assembled to form new membranes? Are membranes synthesized only de novo, by addition of supramolecular lipoprotein subunits, or by the insertion of new molecules into old membranes? What controls the type of membrane formed? Is there genetic control at the supramolecular level, or is this a "spontaneous" assembly determined by intrinsic characteristics of individual protein and lipid components?

Ribosomes in combination with messenger RNA, i.e., polyribosomes, are found in close association with endoplasmic reticulum (ER). In some cases they appear to be closely adherent to these membranes. Correlations have been made between the appearance and structure of this rough endoplasmic reticulum and the rate of synthesis of certain membrane-bound enzymes. The evidence suggests that, at least in some instances, nascent proteins may move directly from ribosomes into the membranes of the endoplasmic reticulum. Such studies must be extended to more membrane proteins before the generality of this mechanism can be established. If most membranes originate as part of the endoplasmic reticulum, other questions arise. How are they transformed into other types of membranes? How do the structural and functional differences of "mature" membranes arise?

Almost all of the complex lipids of cells are associated with membranes. The enzymes for phospholipid synthesis are largely bound to endoplasmic reticulum [45]. Little is known, however, about the relation between the lipid biosynthetic process and the assembly of lipids into membranes.

Different hypotheses of membrane biosynthesis are current. One might be termed the sequential transformation hypothesis [23]. It is possible to view the nuclear, endoplasmic reticulum, Golgi, and plasma membranes as forming a continuum, and indeed instances of physical continuity have been demonstrated between nuclear and rough endoplasmic membranes, between endoplasmic reticulum (rough and smooth) and Golgi, and between smooth endoplasmic reticulum (SER) and plasma membrane. One may envision the polyribosomes assembling at the nuclear membrane to form rough endoplasmic reticulum (RER), which will generate new membrane proteins, including those which catalyze phospholipid synthesis. The Golgi apparatus may be a region into which these newly synthesized membranes flow and which, to a large extent, mediates the transition from rough to smooth endoplasmic reticulum. This Golgi region is also a site of protein modification reactions, in particular, the addition of carbohydrates to form glycoproteins. Furthermore, in secretory cells, elements of SER appear to convey secretory precursor material into the interiors of the Golgi membranes preparatory to the formation of membrane-bound secretory granules [12].

A functional continuity is postulated between the Golgi membranes and the plasma membrane [11], perhaps via a type of smooth endoplasmic reticulum [9]. In certain cases, Golgi membranes can be shown to have enzyme profiles that are intermediate between those of isolated endoplasmic reticulum and plasma membranes. This has been interpreted as evidence for the sequential transformation of ER to Golgi to plasma membrane [11].

A somewhat different, but not incompatible, view arises from the work of Palade, Siekevitz, and co-workers, who have studied the development of membranes in rat liver cells during a period of rapid development [32, 44]. Nuclear, ER, and plasma membrane proteins appear to be synthesized at about the same average rate. On the other hand, there are differences in the rates of synthesis of particular enzymes. Even within the ER, lipids turn over more rapidly than do proteins and, within the lipids, the glycerol moiety turns over more rapidly than do the fatty acids [32]. They proposed that ER membranes may be "mosaics of functionally different patches, each bearing a characteristic enzyme or enzyme set . . . assembled in a single-step operation from structural protein, phosphatides, and corresponding enzymes, but various types [of patches] produced in different proportions at different times" [13].

Both of the preceding hypotheses envisage the RER as the primary site of membrane protein synthesis. However, Glick and Warren have reported that plasma membranes isolated from mouse fibroblasts contain 60 to 80 μg RNA per milligram of protein, most of which appeared to be ribosomal. These membranes incorporate amino acids into protein at a rate comparable to that of microsomes from the same cells, and these authors propose that "association of the protein-synthesizing machinery with the surface membranes could facilitate production of proteins for membrane synthesis or for export" [19]. Some mechanism of protein incorporation into membranes, other than into RER, must take place because the enzyme content of plasma membrane differs markedly from that of RER [11, 31].

Neuronal membrane synthesis poses several unique questions. Because ribosomal protein synthesis is restricted to the perikaryon (Chap. 12), the problems related to assembly and transport into the axon are especially obvious. Not only must membrane proteins be transported the length of the axon, but specialized synaptic structures [43] must be assembled. Moreover, the axolemma itself must contain an exceedingly complex array of specific proteins.

Mahler and co-workers have studied the rate of in vivo protein turnover in subcellular brain fractions from rats [5]. The average rates of turnover of proteins were all very similar in the following fractions: soluble, synaptic vesicle, mitochondrial, and plasma membrane components of synaptosomes. The initial rate of incorporation of radioactive amino acid was much more rapid into the microsome fraction than into synaptic plasma membranes. This finding is consistent with a temporal separation of the processes of membrane protein synthesis and synaptic membrane assembly; possibly this delay corresponds to the time required for axoplasmic transport (Chap. 12). A large fraction of the radioactive protein that actively moves down axons is insoluble. It is not known how much of this is in the form of membranes or precursors of membranes.

Membrane biosynthesis is also an important aspect of the process of quantal release of neurotransmitters (Chap. 8). The origin and disposition of the synaptic vesicle membranes is speculative. Mahler and co-workers find that the qualitative electrophoretic pattern of vesicle membranes is distinct from that of synaptic plasma membranes, although their average turnover rates are similar [2]. This information is not consistent with theories that vesicles either arise from or merge with synaptic plasma membrane.

Several workers have investigated the possibility that there may be protein synthesis within the axon, despite the general absence of ribosomes. These studies are extremely difficult because all axons are closely invested with glial or Schwann cells, which do contain ribosomes. There are suggestions that the axolemma may contain bound RNA and that some axonal and synaptic membrane protein might be synthesized locally. This would require some transport of genetic information to those sites rather than the transport of precursor protein [2]. The interchange of macromolecules between glia and neurons is also a possibility that has been discussed but not demonstrated.

Regardless of whether or not there may be some potential for local synthesis, there is no doubt that a large increase in protein synthesis takes place within the neuronal perikaryon during the process of axonal regeneration [1]. During the outgrowth of axons, their tips become enlarged and project long, membranous extensions. Although the axon proper contains neurofilaments and neurotubules, only a type of SER extends into this "growth cone" region where most of the new membrane synthesis appears to occur [9].

COOPERATIVE EFFECTS OF MEMBRANE PROTEINS

The nerve impulse is a process localized in the plasma membrane and seems to involve several temporally and spatially coordinated changes in membrane structure. These changes are characterized by abrupt transitions, e.g., with respect to cation permeabilities as a function of transmembrane potential. As discussed elsewhere (Chap. 3), the molecular basis of these effects is still speculative.

From a biochemical viewpoint, it is attractive to consider whether these changes may be consistent with the known ability of proteins to mediate interactions between diverse molecules and ions. In a fundamental sense, this role of proteins is simply a molecular example of Newton's third law. If molecules A and B each react with a protein, P, then to some extent P will mediate an interaction that will depend largely upon the properties of P.

If A and B bind to distinct sites on P, some physical effects are necessarily propagated between the sites. These may be sufficient to cause changes in the protein folding (conformation). Detailed studies of these changes are available for hemoglobin, oxygen, and 2,3-diphosphoglycerate [35]. Oxygen-free hemoglobin contains six cationic groups between the two β-subunits that form a site complementary to the diphosphoglycerate anion. The oxygen-free conformation of hemoglobin can be established by combination with this "effector" molecule. Upon combination with oxygen, electrostatic bonds between hemoglobin subunits are broken, H^+ is absorbed, and the binding site for diphosphoglycerate is distorted. Thus, diphosphoglycerate acts to hinder the binding of oxygen and, conversely, it binds only weakly to oxyhemoglobin.

Figure 1-2 is a qualitative illustration of effects that may accrue from the interaction of multiple binding sites for the same effector. Such effects are common

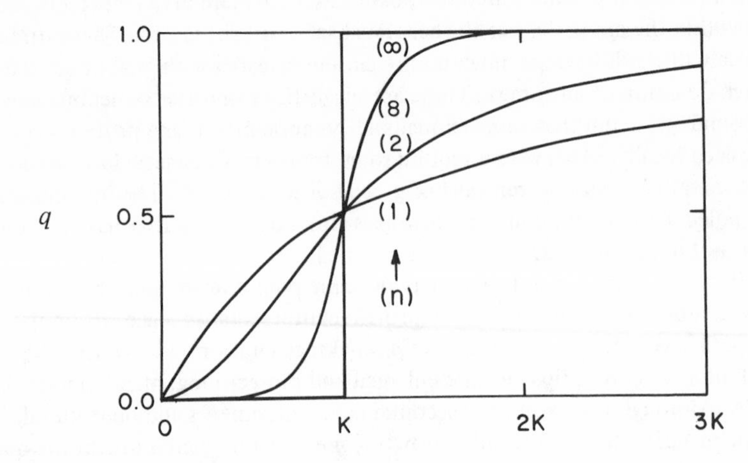

FIGURE 1-2. The influence of an effector molecule on the conformational or functional state of a protein as depicted by the Hill approximation (see text).

occurrences, usually resulting from the interaction of several identical or similar protein subunits, each with one effector binding site. If the protein can exist in only two states, P and Q, and the fraction, q, in state Q is a function of the concentration of free effector molecule, m, then

$$q = [(K/m)^n + 1]^{-1}$$

which is an approximation proposed by Hill [25] to describe this function. When $n = 1$, this becomes the Michaelis equation (Chap. 7, Eq. [11]). As n increases, the response becomes increasingly sigmoidal, and as n approaches ∞, there is an increasingly abrupt transition between states P and Q. In this form of the equation, K is the concentration of m at which the protein molecules are equally distributed between states P and Q. The physical interpretation of n is ambiguous, since it reflects the number of binding sites per protein molecule, the number of interacting protein molecules, and the strength of the interactions between binding sites [7]. However, n can never be greater than the total number of strongly interacting binding sites. Usually there will be only one binding site per protein subunit, so an abrupt transition between two states requires a large array of strongly interacting proteins. In the Hodgkin-Huxley description (Chap. 3), the factors controlling Na^+ and K^+ permeabilities are related to these permeabilities by exponents of 3 and 4, respectively, which might reflect high values of n in the Hill formulation.

Although a number of membrane-bound enzymes exhibit marked cooperative effects, the extent to which these concepts can be extended to such phenomena as impulse conduction, transmitter release, or neuroreceptor actions is uncertain. All of these processes involve highly specific interactions and nonlinear, or threshold, responses. In impulse conduction, tetrodotoxin and DDT appear to interact specifically with membrane Na^+ channels at two distinct sites [26]; DDT and veratrine prolong the action potential by interfering with the closure of Na^+ channels. The "cooperative" nature of DDT action is indicated by the fact that the Na^+ channels must be in the open state before DDT binding can occur.

Membranes could accommodate an "infinite" two-dimensional array of interacting subunits [10]. Morphological evidence of large arrays of globular subunits is seen in mitochondrial inner membranes, chloroplasts, certain elements of sarcoplasmic reticulum, photoreceptor discs, and some junctional areas of plasma membranes. In general, however, freeze-etching and negative-staining techniques demonstrate rather widely spaced units or small clusters of globular units in plasma membranes [37, 42].

Calculations for the node of Ranvier in frog indicate a density of Na^+ channels of about 160 per square micron. If each channel is associated with a protein of 100,000 molecular weight, only about 10^{-4} of the membrane area would be occupied by that protein [26]. Thus, even such a specialized area does not require a high density of specialized protein molecules.

Coupling in membranes is undoubtedly mediated to a large extent by the membrane potential. This is a large force (about 100,000 volts per centimeter), and the effect of membrane potentials on membrane proteins has not been studied intensively. The effect of adding various proteins to black lipid membranes has been reported in terms of changes in membrane conductances [2], and more studies are now appearing concerning enzymes artificially bound to membranes, but the field still is relatively unexplored.

Another possible factor in the propagation of effects through membranes is the lipid bilayer. Bilayers may undergo state transitions from one packing geometry to another [30]. This could lead to protein transformations and the change of properties of the bilayer itself, as well.

PREPARATION OF PURIFIED PLASMA MEMBRANE FRACTIONS

The general aspects of subcellular fractionation (Chap. 21) and the preparation of synaptosomes and myelin (Chap. 18) are discussed elsewhere in this volume. Biochemical studies of membranes are almost totally contingent upon the adequacy of these methods.

Several of the more useful methods depend on a strategic choice of starting material. Mammalian erythrocyte membranes are probably the most readily prepared and the most intensively studied. The red cells are easily separated from serum and white cells by low-speed centrifugation in isotonic media. Cytoplasmic proteins, predominantly hemoglobin, are then removed in a second stage of washing in hypotonic media. Because these cells are essentially devoid of internal organelles, such preparations can be pure. Even in this simple case, however, there are some uncertainties when individual components are studied: some artifactual adsorption of soluble protein may occur or some "loosely" associated components may be removed.

Myelin is another relatively rich source of plasma membranes, and purification strategies have been based upon the possibility of preparing multilaminar vesicles of low density and large size (Chaps. 18 and 21).

Heterogeneity of more complex tissues is a major problem at both the histological and subcellular levels. In some cases pure cell lines can be grown in culture, but the amount of material is usually limited and the properties of the plasma membrane structure of cultured cells may differ importantly from that of normal cells. Accordingly, considerable effort has been directed to preparing plasma membrane fractions from complex tissues.

The strategies may involve preliminary dissociation of the tissue and fractionation of cell types; mild disruptive techniques designed to obtain large, readily sedimenting fragments of membranes; the use of low ionic strength media and chelating agents to minimize aggregation artifacts; and the use of density gradients to separate plasma membranes from the somewhat more dense membranes of nuclei and mitochondria. These methods require certain criteria for purity in terms of morphology, chemical composition, enzymatic activity, and so on. Adenyl cyclase, (Na^+, K^+)-ATPase, and 5'-nucleotidase are frequently used as "marker enzymes" for plasma membranes.

Brain and electric organ synaptosomes are advantageous starting points, and several subfractionation procedures have been published (see Chap. 4) with the goal of investigating synaptic receptors.

Preliminary work has appeared on the isolation of axolemma [16]. The strategy here is the converse of that involved in preparing myelin. The large, unmyelinated, retinal nerve bundles of squid were chosen as starting tissue because up to 10 g of nerve fibers per squid is available. These membranes have a high protein-lipid ratio relative to myelin. However, no satisfactory way of distinguishing axolemma from Schwann cell membranes has yet been applied.

REFERENCES

BOOKS AND REVIEWS

1. Chapman, D. (Ed.). *Biological Membranes, Physical Fact and Function.* New York: Academic, 1968. (This is a comprehensive series of essays on membranes and covers biophysical, chemical, and physiological aspects. Chapter 4 reviews physical studies of membrane structure.)
2. Henn, F. A., and Thompson, T. E. Synthetic lipid bilayer membranes. *Annu. Rev. Biochem.* 38:241, 1969.
3. Korn, E. D. Current concepts of membrane structure and function. *Fed. Proc.* 28:6, 1969. (A review which raises questions about much of the classic evidence in support of the bilayer membrane model.)

4. Korn, E. D. Cell membranes: Structure and synthesis. *Annu. Rev. Biochem.* 38:263, 1969.
5. Lajtha, A. (Ed.). *Protein Metabolism of the Nervous System.* New York: Plenum, 1970. (Chapters of particular relevance are those by Austin et al. [Protein biosynthesis in axons], Koenig [membrane protein synthesis in axons], and Mahler and Cotman [proteins in synaptic plasma membranes].)
6. Stoeckenius, W., and Engelman, D. M. Current models for the structure of biological membranes. *J. Cell Biol.* 42:613, 1969.

ORIGINAL SOURCES AND SPECIAL TOPICS

7. Atkinson, D. E. Regulation of enzyme activity. *Annu. Rev. Biochem.* 35:85, 1966.
8. Blaurock, A. J. Structure of the nerve myelin membrane: Proof of the low-resolution profile. *J. Mol. Biol.* 56:35, 1971.
9. Bunge, M. B., and Bray, D. Fine structure of growth cones from cultured sympathetic neurons. *J. Cell Biol.* 47:241a, 1970.
10. Changeaux, J., Thiery, J., Tung, Y., and Kittel, C. On the cooperativity of biological membranes. *Proc. Natl. Acad. Sci. U.S.A.* 57:335, 1967.
11. Cheetham, R. D., Morré, D. J., and Yunghans, W. N. Isolation of a Golgi apparatus – Rich fraction from rat liver. II. Enzymatic characterization and comparison with other cell fractions. *J. Cell Biol.* 44:492, 1970.
12. Claude, A. Growth and differentiation of cytoplasmic membranes on the course of lipoprotein granule synthesis in the hepatic cell. I. Elaboration of elements of the Golgi complex. *J. Cell Biol.* 47:745, 1970.
13. Dallner, G., Siekevitz, P., and Palade, G. E. Biogenesis of ER membranes. II. Synthesis of constitute microsomal enzymes in developing rat hepatocyte. *J. Cell Biol.* 30:97, 1966.
14. Danielli, J. F. Protein films at the oil-water interface. *Quant. Biol.* 6:190, 1938.
15. Engelman, D. M. Structure in the membrane of *Mycoplasma laidlawii. J. Mol. Biol.* 58:153, 1971.
16. Fischer, S., Cellino, M., Zambrano, F., Zampighi, G., Tellez-Nagel, M., Marcus, D., and Canessa-Fischer, M. The molecular organization of nerve membranes. I. Isolation and characterization of plasma membranes from retinal axons of the squid: An axolemma-rich preparation. *Arch. Biochem. Biophys.* 138:1, 1970.
17. Fleischer, S., Fleischer, F., and Stoeckenius, W. Fine structure of whole and fragmented mitochondria after lipid depletion. *Fed. Proc.* 24:296, 1965.
18. Fricke, H. A mathematical treatment of the electric conductivity and capacity of disperse systems. II. The capacity of a suspension of conducting spheroids surrounded by a non-conducting membrane for a current of low frequency. *Physical Rev.* 26:678, 1925.

19. Glick, M. C., and Warren, L. Membranes of animal cells. III. Amino acid incorporation by isolated surface membranes. *Proc. Natl. Acad. Sci. U.S.A.* 63:563, 1969.
20. Gorter, E., and Grendel, F. On bimolecular layers of lipoids on the chromocytes of the blood. *J. Exp. Med.* 41:439, 1928.
21. Greaser, M. L., Cassens, R. J., Hoekstra, W. G., and Briskey, E. J. Purification and ultrastructural properties of the calcium accumulating membranes in isolated sarcoplasmic reticulum preparations from skeletal muscle. *J. Cell Physiol.* 74:37, 1969.
22. Green, N. M. Possible modes of organization of protein molecules in membranes. *Biochem. J.* 122:37P, 1971.
23. Grove, S. N., Bracker, C. E., and Morré, D. J. Cytomembrane differentiation in the endoplasmic reticulum–Golgi apparatus–vesicle complex. *Science* 161:171, 1968.
24. Hatefi, Y., and Hanstein, W. G. Solubilization of particulate proteins and nonelectrolyte by chaotropic agents. *Proc. Natl. Acad. Sci. U.S.A.* 62:1129, 1969.
25. Hill, A. V. The combinations of haemoglobin with oxygen and with carbon monoxide. *Biochem. J.* 7:471, 1913.
26. Hille, B. Pharmacological modifications of the sodium channels of frog nerve. *J. Gen. Physiol.* 51:199, 1968.
27. Klotz, I. M., Langerman, N. R., and Darnell, D. W. Quaternary structure of proteins. *Annu. Rev. Biochem.* 39:25, 1970.
28. Korn, E. D. Structure of biological membranes. *Science* 153:1491, 1966. (Contains a reevaluation of evidence for the unit membrane theory.)
29. Maddy, A. H., and Kelly, P. G. Factors affecting the interactions of proteins isolated from erythrocyte membranes. *Biochem. J.* 122:62P, 1971.
30. McFarland, B. G., and McConnell, H. M. Bent fatty acid chains in lecithin bilayers. *Proc. Natl. Acad. Sci. U.S.A.* 68:1274, 1971.
31. Meldolesi, J., Jamieson, J. D., and Palade, G. E. Composition of cellular membranes in the pancreas of the guinea pig. III. Enzymatic activities. *J. Cell Biol.* 49:150, 1971.
32. Omura, T., Siekevitz, P., and Palade, G. E. Turnover of constituents of the endoplasmic reticulum membranes of rat hepatocyte. *J. Biol. Chem.* 242:2389, 1967.
33. Overton, E. Ueber die osmotischen eigenschaften der lebenden pflanzen und tiergelle. *Vjschr. Naturforsch. Ges. Zurich* 40:159, 1895.
34. Parsons, D. F. Mitochondrial structure: Two types of subunits on negatively stained mitochondrial membranes. *Science* 140:985, 1963.
35. Perutz, M. F. Stereochemistry of cooperative effects in hemoglobin. *Nature (Lond.)* 228:726, 1970.
36. Perutz, M. F., and Lehmann, H. Molecular pathology of human haemoglobin. *Nature (Lond.)* 219:902, 1968.

37. Pinto da Silva, P., and Branton, D. Membrane splitting in freeze-etching covalently bound territin as a membrane marker. *J. Cell Biol.* 45:598, 1970.
38. Rich, A. Molecular Configuration of Synthetic and Biological Polymers. In Oncley, J. L. (Ed.), *Biophysical Science: A Study Program.* New York: Wiley, 1959.
39. Robertson, J. D. The occurrence of a subunit pattern on the unit membranes of club endings in Mauthner cell synapses in goldfish brains. *J. Cell Biol.* 19:201, 1963.
40. Rouser, G., and Yamamoto, A. Lipids. In Lajtha, A. (Ed.), *Handbook of Neurochemistry.* Vol. 1. New York: Plenum, 1969, p. 121.
41. Scheraga, H. A. Intramolecular Bonds in Proteins. II. Noncovalent Bonds. In Neurath, H. (Ed.), *The Proteins.* Vol. 1 (2nd ed.). New York: Academic, 1963.
42. Tillack, T. W., and Marchesi, V. T. Demonstration of the outer surface of freeze-etched red blood cell membranes. *J. Cell Biol.* 45:649, 1970.
43. Whittaker, V. P. The Synaptosome. In Lajtha, A. (Ed.), *Handbook of Neurochemistry.* Vol. 2. New York: Plenum, 1969, p. 327.
44. Widnell, C. C., and Siekevitz, P. The turnover of the constituents of various rat liver membranes. *J. Cell Biol.* 35:142A, 1967.
45. Willgram, G. F., and Kennedy, E. P. Intracellular distribution of some enzyme catalyzing reactions in the biosynthesis of complex lipids. *J. Biol. Chem.* 238:2615, 1963.

2

Ion Transport

Phillip D. Swanson and
William L. Stahl

THIS CHAPTER deals with mechanisms by which the neuronal elements maintain intracellular concentrations of ions that differ from the concentrations in the surrounding fluid. The first section contains information about general ways in which ions and other molecules are able to move across membranes, using examples of nonneuronal membranes, including bacterial and artificial membranes. The second section deals with transport processes that are believed to be important in neural tissues.

GENERAL CONSIDERATIONS

Different cellular and subcellular compartments of the body are separated from each other and from the external environment by a series of membranes. In order to maintain the compartmental concentrations of essential nutrients and ions at levels necessary for normal cellular activity, membrane transport processes are essential. Each compartment has a reasonably characteristic composition that usually differs from the composition in neighboring compartments, so steep chemical and electrical gradients often exist across the membranes. Also, transported materials often participate in cellular function. For example, sodium and potassium ions are involved in the generation and propagation of the action potential in nerve and muscle tissues.

Movement of molecules through membranes can, in some instances, be described in terms of relatively simple forces that are sufficient to bring about movement without the addition of metabolic energy. In these *passive transport* processes, a net flux of material arises due to dissipation of the total free energy of the system. The energy of the system decreases until thermodynamic equilibrium is reached and all net fluxes are zero. In other instances, movements of molecules across membranes require expenditure of energy *(active transport).* Neither passive nor active transport is understood fully in precise molecular terms.

In general, the rate of passage of a molecule through a membrane depends on the magnitude of forces responsible for the movement (concentration or electrical potential gradients) and the relative ease with which the molecule passes through the membrane, the latter being expressed in terms of permeability of the membrane. Mechanisms of movement remain largely unproved, but may involve direct movement through the membrane, passage through membrane pores, and, possibly, pinocytosis.

In an incisive monograph, Stein [6] has carefully examined data on the permeation of a large number of substances through membranes. The model of a simple bimolecular leaflet with substances directly diffusing through the hydrophobic lipid phase accounts for much of the assembled permeability data in several cell types. This is especially true for nonelectrolytes as based on considerations of the permeant's size, the number of potential hydrogen-bonding groups, and, in some cases, the number of bare methylene groups in relation to the oil-water partition coefficient. However, this model does not account for the anomalous permeability of water, urea, and ions in most natural membrane systems. Alternatively, a model with pores in the membrane has been considered. In this model, movement of the permeant takes place within channels penetrating through the hydrophobic environment of the membrane. Here diffusion might be determined exclusively by the size of the penetrating species. This view is not wholly adequate because no single value for the pore radius can account completely for the available data. Penetration of many molecules may occur through nonrigid lattice spaces between the lipoprotein chains of the membrane matrix. There is probably a range of such pore sizes in the membrane.

It is well known that a number of molecules – glucose and glycerol, for example – pass through cell membranes faster than is predicted on the basis of the above considerations. This accelerated transport allows specific molecules to pass through membranes by specific transport systems and imparts greater flexibility to the membrane. Systems controlling the rate at which specific molecules enter the cell down a preexisting concentration gradient have been termed *facilitated diffusion* systems. These systems do not require metabolic energy. The kinetics of such systems have been successfully interpreted in terms of membrane carriers. Such a model assumes reversible binding of a permeating substance P with a membrane component C, designated as the carrier (Fig. 2-1). The complex then moves from one side of the membrane to the other, where the complex dissociates, thereby releasing the permeant into the cell. The carrier then returns to the opposite face of the membrane to complete the cycle.

As a result of their cyclic nature, such processes can effect movements of large amounts of material. As in enzyme catalysis, they can be described as utilizing Michaelis-Menten kinetics. Assuming the presence of carriers enables one to explain many of the known kinetic characteristics of membrane transport processes, including saturation, specificity, and competition. Although the mechanism shown in Figure 2-1 indicates movement of the carrier through the membrane, the carrier molecule could be viewed as rotating or changing in conformation in order to permit movement of the permeant through the membrane (see Fig. 2-4).

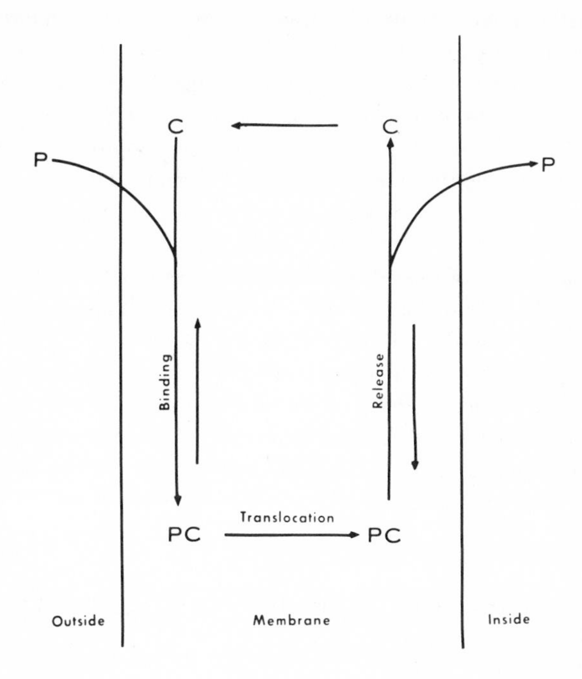

FIGURE 2-1. Carrier-mediated transport. P, permeant; C, carrier.

Cells are able to maintain a steady state for many substances not in equilibrium by supplying energy derived from metabolic processes. Therefore, active transport of a substance is transport against its electrochemical gradient, requiring a source of energy to bring about vectorial movement of particles through a membrane. Energy might be supplied at one or more of the steps shown in Figure 2-1 in such a way that transfer of permeant is more effective in one direction than in the other. A useful example is the movement of sodium ions. During the initial phase of a nerve action potential, sodium moves rapidly into the cell by passive diffusion due to forces that result from its concentration and electrical potential gradients. During the next phase, however, sodium moves out of the cell against these gradients. Thus, the transport during the recovery phase is an "active" process requiring energy for the transport.

Active transport of sodium out of the cell is important for maintenance and restoration of ion gradients that are necessary for initiation of an action potential, as well as for regulation of intracellular cation concentrations and for maintenance of a constant cell volume. The concentrations of Na^+ and K^+ within the cell undoubtedly influence the rates of a variety of enzyme reactions. Complex systems, such as mitochondrial oxidative phosphorylation, are inhibited by high Na^+ or low K^+ concentrations. (This homeostatic function of active transport is discussed in Chap. 3.)

The necessity of active transport for cell volume stabilization has been discussed in theoretical terms by Post and Jolly [21] and by Stein [6]. A cell is surrounded by a water-permeable membrane that must be nonrigid because the cell volume can vary if the cell is placed into media of varying tonicity. The steady-state volume will change if the cell is prevented from metabolizing appropriate substrates. The membrane is permeable to some solutes (e.g., inorganic ions) and not to others (e.g., certain proteins). If there were no active transport, the permeable constituents would become distributed in equilibrium with equal internal and external concentrations or, in the case of ions, distributed according to a Donnan equilibrium ratio (see Chap. 3). Because of the presence of impermeable substances within the cell, water would move into the cell, thus increasing the volume. As developed by Post and Jolly [21], the three factors that control the volume of a cell without a rigid cell wall are: (1) the amount of fixed intracellular material, (2) the concentration of external permeable material, and (3) the ratio of the rate constants of the "leak" to the "pump." Increasing leakiness or decreasing pump rates will bring about an increased cell volume.

ISOLATION AND IDENTIFICATION OF TRANSPORT SYSTEMS

TRANSPORT PROTEINS

Although many of the properties of transport systems have been described in great detail, little is known about these systems at the molecular level. A major goal has been to isolate and to characterize transport components and, hopefully, to reconstitute activity in model systems by reassociation of membrane components, as in the synthetic membranes described below. Study of transport proteins has been hampered because the methods used to isolate the transport species necessarily lead to dissolution of the cell membrane with concomitant loss of the transport function. The properties of the transporting system and the isolated transport protein must be compared to establish a relationship between the two. For example, in both the bacterial sulfate-transporting system and a sulfate-binding protein discussed below, they: (1) are identically inhibited by a series of anions and a series of protein reagents; (2) are lost from cells to a similar degree by osmotic shock; (3) are simultaneously lost by mutation, are regained by reversion or transduction, and are lacking in certain transport-negative mutants; and (4) show a similarity in affinity constants for binding and transport.

Membrane-transport proteins have been successfully isolated in pure form, mainly from microorganisms. The only exception at present is the calcium-binding protein of chick duodenum. Bacterial systems offer several advantages. Many transport proteins are liberated easily by osmotic shock, which simplifies further purification, and specific transport function can often be induced by growing the organisms in the presence of substrate.

Beta-Galactoside Transport Protein

In the now-classic experiments of Fox and Kennedy [20], evidence was presented for the involvement of a protein (M protein) in the facilitated entrance of β-galactosides into *E. coli.* This component is distinct from previously characterized proteins of the lactose system, is localized in the membrane fraction, and has a high affinity for certain β-galactosides. Organisms that were induced or noninduced for the transport system were used. *N*-Ethylmaleimide (NEM), which combines irreversibly with the sulfhydryl groups (–SH) of proteins, blocks transport of β-galactosides. The bacteria were first treated with nonradioactive NEM in the presence of thiodigalactoside. This improved the specificity of isotopic labeling in a subsequent step by first permitting all –SH groups not protected by thiodigalactoside to react with unlabeled NEM. The excess NEM and the thiodigalactoside were then removed. The induced organisms were treated with ^{14}C-NEM and the uninduced organisms with ^{3}H-NEM. The labeled bacteria were then mixed and the proteins were extracted and fractionated. It was hoped that the transport species (induced) would be selectively labeled with ^{14}C-NEM. Indeed, a membrane fraction that had a higher ratio of ^{14}C to ^{3}H was isolated and the component was purified to yield a protein with a molecular weight of 30,000.

Sulfate-Binding Protein

Transport of sulfate into *Salmonella typhimurium* occurs against a concentration gradient and requires energy. Utilizing substrate recognition (binding of $^{35}SO_4{}^{2-}$) as a means of testing for the transport protein, Pardee and his associates [20] have isolated and crystallized a protein with a molecular weight of 32,000. One sulfate ion appears to be bound per protein molecule, with a dissociation constant of about 0.1 μM. Thus far, attempts to reconstitute the transport system by adding pure sulfate-binding protein to shocked organisms (which are unable to transport sulfate) have been largely unsuccessful. It is likely that both the sulfate-binding protein and M protein are involved in the initial binding step of the transport process (Fig. 2-1).

Phospho Transferase System

Roseman and his colleagues have elucidated the details of a sugar transport system in microorganisms [5]. They have related the transport of nine sugars to a phospho transferase system (Fig. 2-2). The net process

$$\text{sugar} + \text{phosphoenolpyruvate (PEP)} \rightarrow \text{sugar-phosphate} + \text{pyruvate}$$

involves three proteins and results in release of the phosphorylated derivative inside the cell. The sugar-phosphate cannot pass back through the membrane, thereby achieving an active transport of the sugar. The energy donor for this system is PEP,

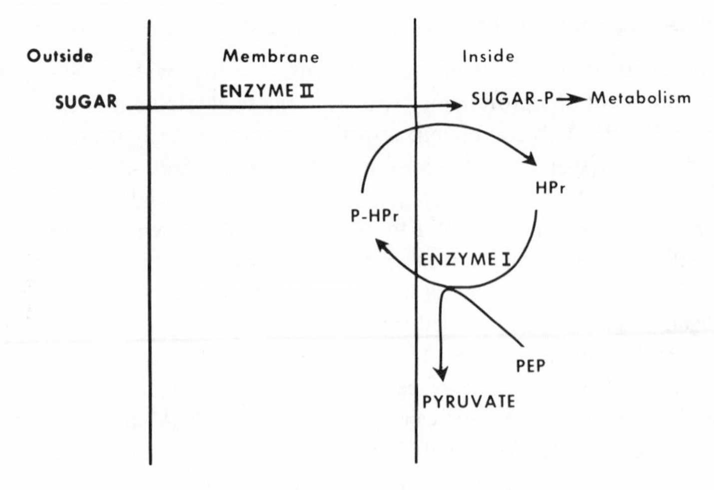

FIGURE 2-2. Sugar transport by the phospho transferase system.

making the phospho transferase system different from that of other known sugar kinases. A cytoplasmic enzyme (Enzyme I) mediates phosphorylation of a cytoplasmic energy-donating protein (HPr, a protein of molecular weight 9400) on a histidine residue to form phospho-HPr:

$$\text{PEP} + \text{HPr} \underset{\substack{Mg^{2+} \\ \text{(cytoplasm)}}}{\overset{\text{Enzyme I}}{\rightleftarrows}} \text{phospho-HPr} + \text{pyruvate}$$

This is followed by transfer of phosphate to the sugar, thereby regenerating dephospho-HPr:

$$\text{phospho-HPr} + \text{sugar} \xrightarrow[]{\substack{\text{(Membrane)} \\ \text{Enzyme II}}} \text{sugar-P} + \text{HPr}$$

The specificity of the system resides in the membrane-bound Enzyme II. Apparently, a specific Enzyme II exists for each sugar. Bacterial mutants that lack HPr or Enzyme I, which eliminates phospho-HPr as an energy donor, are not able to accomplish active transport of sugars. Also, there is little passive transport, suggesting that energy is required at the translocation step in this system.

MACROCYCLIC COMPOUNDS AS CARRIER MODELS

A large number of organic compounds that possess the ability to transport cations across membranes are now known. The earliest compounds described were macrocyclic compounds synthesized by microorganisms and possessing antibiotic activity. More recently, structurally simpler cyclic polyethers have also been shown to exhibit transport properties. These substances have certain common features: they are low molecular-weight compounds (MW <2000) with covalently linked or potential cyclic structures having lipophilic external surfaces. These compounds will sequester cations selectively. They are often able to bring about ion transport into mitochondria in vitro, including alkali-metal cation movement against concentration gradients by an energy-requiring process. The accumulation in mitochondria is presumed to be a form of active transport, as interrupting the energy supply reverses the direction of ion movement.

These active compounds also have demonstrable electrical effects on artificial bilayers. The factors underlying the ion selectivity of these compounds are discussed here briefly because the capacity of biological membranes to distinguish sodium and potassium ions underlies such important fundamental processes as excitation and conduction in the nervous system and active sodium and potassium transport in general. The detailed structures of these compounds are known, so the study of systems employing them as ion carriers offers an important alternate approach (as opposed to isolating carriers from the membrane) to the study of the molecular basis of membrane transport.

Table 2-1 lists the classes of ion carriers that have been studied. The valinomycin and cyclic polyether groups form positively charged ion pairs. The nigericin group forms neutral ion pairs. Alamethicin, a pure peptide containing one free carboxyl group, can form neutral or positively charged ion pairs. These compounds contain either covalently bonded closed rings or linear chains, which can assume a ring configuration by cyclizing noncovalently.

Many of these active compounds show a dramatic ability to discriminate between monovalent cations. For example, when added to mitochondrial systems, valinomycin has an apparent complexing preference of K^+ over Na^+ as high as 10,000 to 1. Complex formation in these compounds occurs by induced ion-dipole interactions with oxygen atoms of the ring. Polar groups face inward and hydrophobic groups face toward the exterior of the molecule. The number of ring atoms varies for the active polycyclic compounds. Those with fewer than 18 atoms, however, will not select for K^+. This is especially clear in the case of the cyclic polyethers, where there is excellent correlation between the ionic diameters of Li (1.20 Å), Na (1.90 Å), and K (2.66 Å) and the "holes" of the cyclic polyethers. The 4-oxygen, 5-oxygen, and 6-oxygen polyethers (Table 2-1) have holes approximately 1.8, 2.7, and 4.0 Å in diameter to accommodate ions, which probably accounts for their indicated ion preferences. Complexes are not formed if the ion is too large to lie in the hole of the polyether ring.

Interaction between the active compounds and the transported ions involves the unhydrated cation at the center of the active compound interacting with oxygen portions of the molecule. Hydrogen bonding is important in the formation of stable complexes in such open-chain compounds as monensin. In the cyclic polyethers, an increase in the number of oxygen atoms increases the stability of the

TABLE 2-1. Macrocyclic Carrier Models

Active Compounds	No. Ring Atoms	Ion Selectivity[1]
Valinomycin group (positively charged ion pairs formed)		
Nonactin, monactin, dinactin (macrotetrolides)	32	Potassium[2,3]
Enniatins A and B (depsipeptides)	18	Potassium[3]
Gramicidins A, B, and C (neutral linear polypeptides)	–	Potassium[3,4,5]
Valinomycin (depsipeptide)	36	Potassium[2,3,5,6]
Nigericin group (neutral ion pairs formed)		
Nigericin (macrocyclic monocarboxylic acid)	–	Potassium[2,5]
Monensin (macrocyclic monocarboxylic acid)	–	Sodium[2]
Cyclic polyethers (positively charged ion pairs formed)		
4-Oxygen polyethers (e.g., bis-butylcyclohexyl 14-crown-4)	14	Lithium[2]
5-Oxygen polyethers (e.g., butylcyclohexyl 15-crown-5)	15	Sodium[2,4]
6-Oxygen polyethers (e.g., dicyclohexyl 18-crown-6)	18	Potassium[2,5,6]
Alamethicin (neutral or positively charged ion pairs formed) (cyclic polypeptide)	53	Poor[2]

[1]Refers primarily to selectivity of potassium vs. sodium ions.
[2]Based on ion complexing power.
[3]Based on transport through artificial membranes.
[4]Based on cation permeability of red cells.
[5]Based on transport in mitochondria.
[6]Based on ion permeability of phospholipid micelles.

complex, provided the oxygen atoms are coplanar and symmetrically distributed in the polyether ring.

Alamethicin, an extracellular cyclopeptide from *Trichoderma viride* (Table 2-1), is able to induce energy-linked K^+ accumulation by mitochondria [2] but shows little ion selectivity, probably due to the large ring size, which should be flexible enough to assume many ion-selective conformations. This compound differs from the other compounds discussed above in that unselective cation transport can be induced as a function of time and membrane current in experimental bilayers. When such basic proteins as histone, spermine, or protamine are present in the bilayer system, however, alamethicin can induce action potentials that are sensitive to the ionic gradient or applied voltage. In this system, therefore, conditions can be varied to permit electrical activity that resembles an action potential in nerve.

These membrane characteristics can also be induced by a still-uncharacterized material, "excitability inducing material" (EIM), which is very likely a peptide or protein and which was isolated from *Enterobacter cloacae* by Mueller and Rudin [18]. These substances work, although with varying sensitivities, in membrane systems made from mixed lipids, phospholipids, or "oxidized" cholesterol. It has been proposed from electrokinetic and chemical data that six or more molecules form channels through the synthetic membrane, thereby permitting ion flow; assembly and disassembly of these aggregates by voltage and chemical parameters would regulate membrane conduction [19].

Carrier Mechanisms

These studies have given rise to schemes for two main molecular mechanisms by which cations might be transported across membranes through the mediation of macrocyclic compounds. In the first, *channel formation,* several active molecules would stack together and form a channel through the membrane (Fig. 2-3A). The tunnel or channel would be hydrophilic in nature so that anhydrous or partially hydrated ions could pass through; intermediary complexes could be formed with the active molecules that form the channel. In the process of channel formation, valinomycin molecules, for example, would stack together to form a channel lined with oxygen atoms. In this situation, water molecules would form a bridge between the K^+ and carbonyl group so that the partially hydrated ion could transit the membrane. Ion selectivity is presumed to arise from differences in hydrated ion size. Mueller and Rudin [19] favor channel formation in explaining the action of

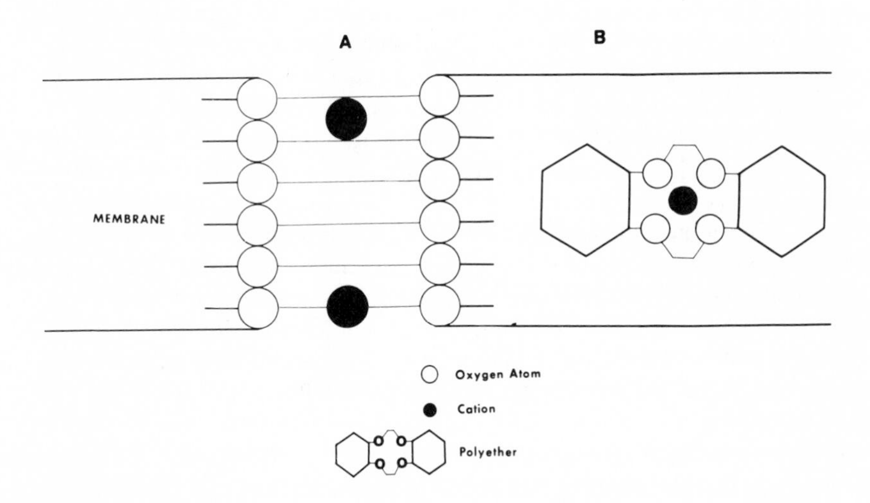

FIGURE 2-3. Possible mechanisms for movement of cations through biological membranes in the presence of a macrocyclic compound. *A*, channel formation; *B*, mobile carrier. (Adapted from [2].)

alamethicin and EIM. This is based on observations that the maximum change in steady-state conductance in synthetic membranes varies exponentially with the sixth power of the potential for a wide range of alamethicin concentrations. Also, as in nerve, the apparent kinetic order of the reaction is six.

The highly nonlinear cooperative effects shown by alamethicin and EIM would appear to exclude a monovalent carrier process and would favor a carrier or channel composed of six or more active molecules. The alternative to a channel is a *mobile carrier.* In this case the active complexing molecule would displace the water of hydration from the cation and form a lipid-soluble complex that could diffuse through the membrane and release the cation at the opposite interface of the membrane (Fig. 2-3B). In this mechanism, ion specificity can best be explained by a combination of unhydrated ion with the active molecule to form the mobile lipid-soluble complex. This mechanism is favored by evidence gathered from x-ray crystallographic and nuclear magnetic resonance (NMR) analyses for valinomycin, monactin, and the enniatins. For example, NMR data [14] indicate that transfer of cations between valinomycin molecules in nonpolar media is unlikely, thus favoring a mobile carrier mechanism.

EVIDENCE FOR ACTIVE TRANSPORT OF Na^+ AND K^+

There is ample evidence that sodium and potassium ions are actively transported in a large number of cell types. Na^+, for example, is transported from the cytoplasm of cells against an electrochemical gradient, and cells are able to maintain low internal Na^+ and high internal K^+ concentrations. The active transport of Na^+ depends on the presence of K^+ in the extracellular fluid, indicating a coupling between the active outward transport of Na^+ and inward transport of K^+. Na^+ extrusion and K^+ assimilation require metabolic energy because they are greatly reduced by 2,4-dinitrophenol, azide, and cyanide. The relationship between the adenosine triphosphate (ATP) supply and transport was clearly demonstrated by Caldwell et al. [12]. They injected ATP, arginine phosphate, or phosphoenolpyruvate into axons of *Loligo* poisoned with 2,4-dinitrophenol or cyanide. Na^+ efflux was restored and the number of Na ions extruded was proportional to the number of high-energy phosphate molecules injected. In other experiments, Hoffman [3] depleted human red cells of metabolic energy reserves and labeled them internally with ^{24}Na. After transfer to sodium-free media, only adenosine (and not inosine) stimulated ATP formation and sodium transport. These and other experiments indicate a close link between energy in the form of ATP and transport of Na^+ and K^+.

In contrast to the energy-requiring active transport of Na^+ and K^+, the shifts of these cations that occur during conduction of a nerve impulse take place without addition of an energy source as long as the necessary cation gradients are present [9]. Active transport is necessary for the maintenance of the cation gradients, and hence for the membrane potential. Development of the membrane potential also depends upon the relatively greater permeability of the neuronal membrane to K^+ than to Na^+ [7].

SPECIFIC MECHANISMS INVOLVING ACTIVE TRANSPORT OF CATIONS

TRANSPORT OF Na^+ AND K^+: (Na^+, K^+)-STIMULATED ADENOSINE TRIPHOSPHATASE

Extrusion of Na^+ ions from the interior of a cell or cell process and the coupled accumulation of K^+ ions require the presence of a mechanism by which energy can be utilized for cation pumping. Because the direct source of energy is likely to be ATP, a system which consumes ATP (an adenosine triphosphatase) and which has properties differentiating it from other ATP-consuming systems was sought in membrane-derived subcellular fractions. In 1948 adenosine triphosphatase (ATPase) activity was demonstrated in the sheath of the giant axon of the squid *Loligo pealii.* The activity of this bound enzyme was 19 to 100 times greater than the activity in the axoplasm. Appreciable magnesium-activated ATPase activity was also found in membrane-derived subcellular fractions prepared from rat peripheral nerve. Particular interest was generated by studies of Skou [24], which demonstrated in microsomal membranes from crab nerve the presence of ATPase activity that was markedly stimulated by the presence of Na^+ plus K^+ in the incubation medium. Either cation alone was much less effective than both together. The requirement for K^+ was less specific than that for Na^+ because Rb^+ could substitute for K^+. Since this first detailed study, a great deal of circumstantial evidence has suggested involvement of this enzyme system in active Na^+ and K^+ transport.

The properties of the (Na^+, K^+)-ATPase system that suggest its role in active Na^+ and K^+ transport are:

1. It is located in the cell membrane.
2. Cation requirements are similar to those of the active transport system: there is a higher affinity for Na^+ than K^+ on the intracellular side of cell membrane and opposite affinity on the outside of cell membrane.
3. Substrate requirements are similar to those of the active transport system: ATP is utilized as the energy source for movement of cations.
4. The enzyme system hydrolyzes ATP at a rate dependent on concentration of Na^+ inside and K^+ outside the cell.
5. The system is found in all cells that have active, linked transport of Na^+ and K^+.
6. The system is inhibited by cardiac glycosides.
7. There is good correlation between enzyme activity and rates of cation flux in different tissues.

Tissue Localization

The enzyme is found in many tissues, with very high activity in electrically excitable tissues such as eel electric organ, brain, nerve, and muscle. Appreciable activity also is present in secretory organs, including kidney, choroid plexus, and the ciliary body. Bonting and Caravaggio [11] correlated (Na^+, K^+)-ATPase activities with rates of active cation fluxes (Na^+ efflux, K^+ influx, or their average) in six tissues in which flux data were available. Recognizing that the conditions of enzyme assay and those of flux determinations cannot exactly coincide, the ratios of enzyme and flux activity remain remarkably constant over a 22,000-fold range in magnitude (Table 2-2).

TABLE 2-2. (Na^+, K^+)-ATPase Activity and Cation Flux in Tissues

Tissue	Temp. (°C)	A Cation Flux (10^{-14}moles/ sq cm/sec)	B (Na^+, K^+)-ATPase[1] (10^{-14}moles/ sq cm/sec)	(A)/(B)
Human erythrocytes	37	3.87	1.38 ± 0.36	2.80
Frog toe muscle	17	985	530 ± 94	1.86
Squid giant axon	19	1200	400 ± 79	3.00
Frog skin	20	19,700	6640 ± 1100	2.97
Toad bladder	27	43,700	17,600 ± 1640	2.48
Electric eel, noninnervated membranes of Sachs' organ	23	86,100	38,800 ± 4160	2.22
Average	....	....	....	2.56 ± 0.19

[1]ATPase activities were determined by standard methods on portions of lyophilized and reconstituted tissue homogenates or microdissected samples of frozen-dried tissue. (Data from Bonting and Caravaggio [11].)

Localization in Subcellular Fractions

In every tissue with which subcellular fractionations have been carried out, (Na^+, K^+)-ATPase activity has been found in fractions that contain particulate material derived from cell membranes. In dispersions prepared in isotonic sucrose from mammalian brain, the highest specific activity is found in the primary microsomal fraction, consisting morphologically of membrane fragments, and in the synaptosomal fraction that is prepared by centrifuging the primary "mitochondrial" fraction on a sucrose density gradient [15]. Lowest specific activities are found in the supernatant and purified mitochondrial fractions. When synaptosomes are disrupted by dispersion in water, (Na^+, K^+)-ATPase activity is absent from the synaptic vesicle subfraction and is concentrated in fractions that contain membrane fragments derived from external synaptosomal membranes.

Other Similarities Between Active Cation Transport Requirements and (Na^+, K^+)-ATPase Activity Requirements

Comparisons between the activity of an enzyme system and the physiological function of that system are necessarily limited by the conditions under which each system is assayed. Most enzyme assays are carried out on purified or partially purified material that has been separated, as far as possible, from other constituents of the tissue from which it is derived. In contrast, measurement of cation transport requires tissue that is as intact as possible in order to allow measurement of cation movements across membranes. Therefore, it is rather remarkable that, in tissues in which comparisons have been made, close correlations are found in the requirements for the two processes. The erythrocyte has been particularly useful for such comparisons, and Post et al. [22] carefully documented the most important features of both systems. Both systems utilize adenosine triphosphate rather than inosine triphosphate as substrate. Both require Na^+ and K^+ together, rather than either cation alone, and NH_4^+ can substitute for K^+ but not for Na^+ in each system. In both systems, high concentrations of Na^+ competitively inhibit K^+ activation. In both systems, ouabain is inhibitory. The concentrations at which Na^+, K^+, ouabain, and NH_4^+ show half of their maximal effects are the same in both systems.

Vectorial Stimulation of (Na^+, K^+)-ATPase

A technique has been developed with erythrocytes that makes possible the production of "ghosts," whose internal composition can be reconstituted with known concentrations of cations and other substances. In one series of experiments carried out by Whittam and Ager [29], the procedure used was the following:

> Cells were hemolyzed in 5 to 60 volumes of a hypotonic solution containing 0.25 to 4 mM $MgCl_2$ and ATP. To the suspension, enough NaCl or other salt was added to raise the osmotic pressure to about 0.3 osmolar. The suspension was then incubated at 37°C for 30 min, during which time the cells regained their low permeability to Na^+ and K^+ and to ATP. The reconstituted ghosts were washed with 0.15 M NaCl or choline chloride, and they contained the cation in high concentration that had been added to the hemolysate. Determination of the amount of inorganic phosphate produced upon further incubation was used as a measure of ATP hydrolysis.

When cells were incubated in a medium containing 10 mM KCl, the rate of ATP hydrolysis was greatest when the internal cation was Na^+. In addition, incubation in K^+-free medium or in medium containing ouabain decreased ATP hydrolysis. In contrast, external Na^+ failed to activate ATP hydrolysis. These and similar experiments have suggested that (Na^+, K^+)-ATPase is stimulated by internal Na^+ and by external K^+, and that the orthophosphate produced from ATP hydrolysis is liberated inside the cell.

Lipid Requirements of the (Na^+, K^+)-ATPase System

This enzyme system has been found to be bound particularly tightly to other membrane constituents and has not been isolated in pure form. For optimal activity, the enzyme appears to require the presence of lipid, and alteration in enzyme activity occurs upon treatment with detergents or phospholipases:

1. Activity is markedly stimulated by addition of low concentrations of detergents [28].
2. Phospholipases inhibit enzyme activity. Addition of phospholipid may activate enzyme that has become inactivated by diverse means. Whether this is a specific requirement for enzyme activity has not been settled.
3. However, cholesterol can be removed by preparation of acetone powders without loss of activity.

Purification of (Na^+, K^+)-ATPase

Separation of enzyme activity from other membrane components has not been achieved in spite of a great deal of effort. Material with high specific activity has been obtained by treatment of membrane fractions with a concentrated solution of sodium iodide. Detergent treatment can also convert the enzyme system into a form that is not sedimented by high-speed centrifugation [28]. The "solubilized" enzyme system still contains considerable lipid and probably represents membrane fragments that are incorporated into detergent micelles. When electrophoresed, such material contains many protein components [17].

Component Parts of the (Na^+, K^+)-ATPase Reaction

Three reactions have been suggested to represent partial reactions of the (Na^+, K^+)-ATPase system: (1) Na^+-stimulated incorporation of ^{32}P from ATP into membrane phosphoprotein, (2) Na^+-dependent ADP-ATP exchange reaction, and (3) K^+-stimulated phosphatase activity.

Phosphorylation of (Na^+, K^+)-ATPase Preparations. Membrane preparations from kidney, brain, electrophorus electric organ, and erythrocytes incorporate labeled phosphate from γ-^{32}P-ATP after brief incubation. Incorporation is increased in the presence of Na^+ and is diminished by adding K^+ or NH_4^+. Cardiac glycosides slow the turnover of the "phosphoenzyme" and diminish its sensitivity to K^+ [23]. The rate of Na^+-dependent incorporation is consistent with its presumed role as part of the (Na^+, K^+)-ATPase system. The labeled material appears to be an acyl phosphate, as it is stable in acid, released by incubation with an acyl phosphatase preparation, and labile in hydroxylamine. Peptic digests of membrane fractions have contained labeled peptides, and thus the labeled "intermediate" is very likely bound to protein.

ADP-ATP Exchange Reaction. Formation of labeled ATP occurs when membrane preparations are incubated in the presence of radioactively labeled adenosine diphosphate (ADP). Much of this ADP-ATP exchange activity is easily separated without loss of (Na^+, K^+)-ATPase activity and is thus due to nonrelated reactions. A component that remains associated with (Na^+, K^+)-ATPase activity is stimulated by Na^+ under carefully specified conditions [13, 25]. This Na^+-stimulated exchange reaction has substrate requirements similar to those of the overall (Na^+, K^+)-ATPase activity and requires incubation at very low Mg^{2+} concentrations for its demonstration.

K^+-Stimulated Phosphatase Activity. Using either *p*-nitrophenyl phosphate or acetyl phosphate as substrate, (Na^+, K^+)-ATPase preparations will form inorganic phosphate. This reaction is stimulated by K^+ and inhibited by ouabain, and it has therefore been regarded as resembling the dephosphorylation component of the overall (Na^+, K^+)-ATPase reaction [1].

Relationship of (Na^+, K^+)-ATPase to Active Cation Transport

The similarities of some of the properties of the (Na^+, K^+)-ATPase to those of active Na^+ and K^+ transport lend credence to the contention that the enzyme activity is involved in transport of these cations. Precise definition of the nature of the relationship is not complete. Models are revised as new facts become known. Two such models are illustrated in Figure 2-4. The first model (Fig. 2-4A) [4] pictures the (Na^+, K^+)-ATPase as a single enzyme molecule situated on the inner surface of the cell membrane at the mouth of an ion-selective pore. ATP is in the form of a complex with Mg^{2+}, which is assumed to be bound to two oxygen atoms of the phosphoric anhydride chain rather than to the adenine portion of the molecule. Magnesium-ATP becomes associated with the enzyme through its adenosine group.

A negative group at the pore with a high affinity for Na^+ combines with Na^+ and at the same time allows a basic group on the enzyme to be occupied by the phosphate moiety from ATP. When the phosphate group is cleaved from ATP, new negatively charged groups repel each other, allowing the basic group of the enzyme to recombine with the acidic group at the pore and to displace Na^+ toward the cell exterior. This mechanism takes into account the association of the enzyme with the membrane and postulates the presence of membrane groups with selective affinity for Na^+. It does not require the presence of a carrier molecule other than the enzyme itself.

A second model (Fig. 2-4B) incorporates more details about the partial reactions of the (Na^+, K^+)-ATPase system and about the way in which inhibitors interfere with enzyme activity [1, 8]. The reaction is pictured as taking place in several steps:

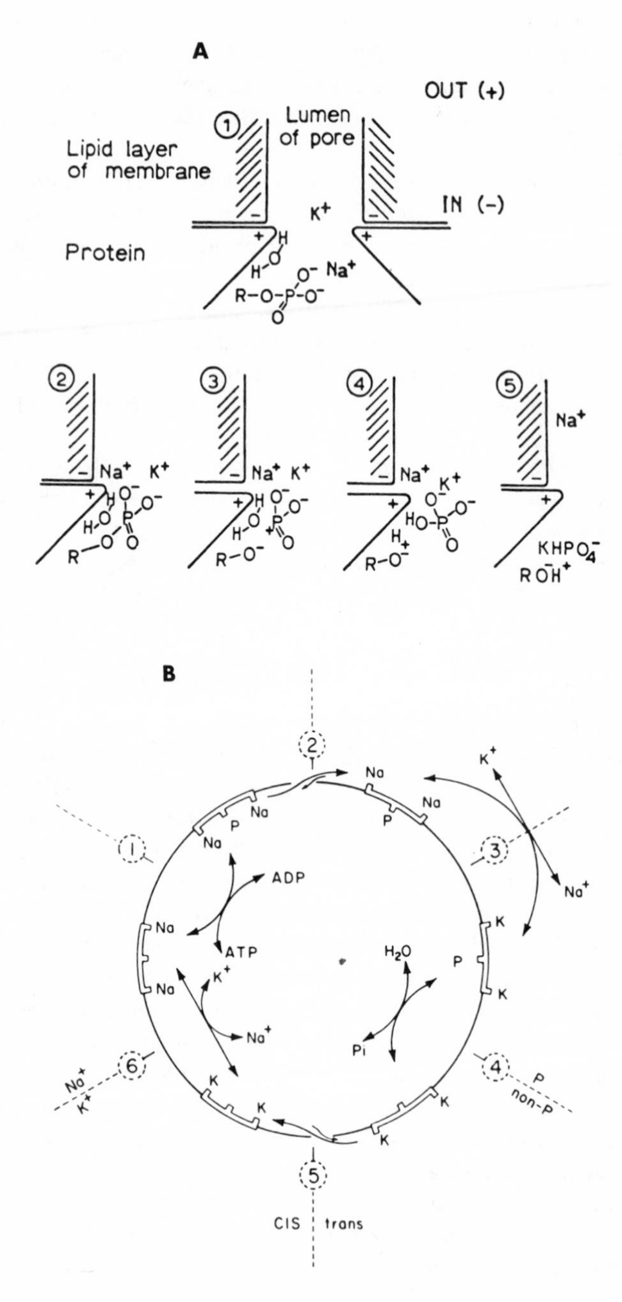

FIGURE 2-4. Two proposed mechanisms for active Na^+ and K^+ transport. (A) The (Na^+, K^+)-ATPase is located at the mouth of an ion-selective pore. (Reproduced with permission [4].) (B) The (Na^+, K^+)-ATPase brings about active transport through allosteric transition. (Reproduced with permission [8].) For details see text.

$$E_1 + MgATP \overset{Na^+}{\rightleftharpoons} E_1 \sim P + MgADP \qquad (1, 6)$$

$$E_1 \sim P \xrightarrow{Mg^{2+}} E_2 - P \qquad (2)$$

$$E_2 - P \xrightarrow{K^+} E_2 + P \qquad (3, 4)$$

$$E_2 \rightleftharpoons E_1 \qquad (5)$$

Reaction (1), which requires Na^+, phosphorylates the enzyme reversibly and can be measured as an ADP-ATP exchange reaction and as a Na^+-dependent membrane phosphorylation. In reaction (2), the phosphorylated enzyme is converted to a second (trans) form of the enzyme, which can now react with K^+ (reaction 3) to cleave orthophosphate from the phosphorylated enzyme (reaction 4). In this scheme, inwardly directed sites have high affinity for Na^+, and outwardly directed sites have high affinity for K^+. In reactions (2) and (5), the orientation of the enzyme is altered by allosteric conversions and provides the vectorial force of transport. The cycle is completed when Na^+ displaces K^+ from the cis enzyme (reaction 6).

CALCIUM TRANSPORT

Mechanisms by which Ca^{2+} are excluded from the interior of neural cells are not well understood (see Chap. 19). The concentration of ionized Ca^{2+} within nerve cells has been estimated to be 10^{-5}M, whereas a concentration of 1 M would be expected if this cation were distributed according to a Donnan ratio.

In various tissues, several mechanisms exist for regulating the intracellular concentration of Ca^{2+}. In muscle, sarcoplasmic reticulum contains an ATP-dependent mechanism that sequesters Ca^{2+} and that plays an important role in the coupling between excitation of the muscle membrane and contraction of the myofibrils. Erythrocytes appear to extrude Ca^{2+} actively in the presence of ATP. Mitochondria from various tissues possess an energy-requiring mechanism for Ca^{2+} uptake.

In neural tissues the evidence for an energy-dependent Ca^{2+} transport system is less firm. The most detailed studies have been carried out on squid giant axons [10]. In this tissue, efflux of $^{45}Ca^{2+}$ was reduced when external Na^+ was replaced by Li^+, choline, or dextrose. From this and from other data, it was suggested that part of the calcium efflux is coupled to sodium entry by means of a mechanism for exchange between internal Ca^{2+} and external Na^+.

In brain slices, Ca^{2+} uptake is increased by ouabain, a cardiac glycoside that inhibits active Na^+ and K^+ transport, as well as by omission of glucose or by addition of metabolic inhibitors [26]. Such results have been interpreted in terms of a Ca^{2+}-Na^+ exchange mechanism similar to that proposed for the squid axon [27]. Energy-dependent calcium uptake has also been demonstrated in synaptosomes isolated from brain homogenates, and it is similar in a number of characteristics to uptake by cerebral mitochondria [16].

ACKNOWLEDGMENTS

This publication was supported in part by U.S. Public Health Service Grant NB05424 from the National Institute of Neurological Diseases and Stroke and a grant (01/8099.01/170-03) from the Veterans Administration.

REFERENCES

BOOKS AND REVIEWS

1. Albers, R. W. Biochemical aspects of active transport. *Annu. Rev. Biochem.* 36:727, 1967.
2. Eisenman, G. (Chairman). Symposium on biological and artificial membranes. *Fed. Proc.* 27:1249, 1968.
3. Hoffman, J. F. Molecular Mechanism of Active Cation Transport. In Shanes, A. M. (Ed.), *Biophysics of Physiological and Pharmacological Action.* Washington: American Association for the Advancement of Science, #69, 1961.
4. McIlwain, H. *Chemical Exploration of the Brain. A Study of Cerebral Excitability and Ion Movement.* Amsterdam: Elsevier, 1963.
5. Roseman, S. The Transport of Carbohydrates by a Bacterial Phosphotransferase System in Membrane Proteins. In *Membrane Proteins.* (Proceedings of a Symposium Sponsored by the New York Heart Association.) Boston: Little, Brown, 1969, p. 138.
6. Stein, W. D. *The Movement of Molecules Across Cell Membranes.* New York: Academic, 1967.
7. Woodbury, J. W. The Cell Membrane: Ionic and Potential Gradients and Active Transport. In Ruch, T. C., and Patton, H. D. (Eds.), *Physiology and Biophysics.* Philadelphia: Saunders, 1965.

ORIGINAL SOURCES

8. Albers, R. W., Koval, G. J., and Siegel, G. J. Studies on the interaction of ouabain and other cardioactive steroids with sodium-potassium-activated adenosine triphosphatase. *Mol. Pharmacol.* 4:324, 1968.
9. Baker, P. F., Hodgkin, A. L., and Shaw, T. I. The effects of changes in internal ionic concentrations on the electrical properties of perfused giant axons. *J. Physiol. (Lond.)* 164:355, 1962.
10. Blaustein, M. P., and Hodgkin, A. L. The effect of cyanide on the efflux of calcium from squid axons. *J. Physiol. (Lond.)* 200:497, 1969.

11. Bonting, S. L., and Caravaggio, L. L. Studies on sodium-potassium-activated adenosine triphosphatase. V. Correlation of enzyme activity with cation flux in six tissues. *Arch. Biochem. Biophys.* 101:37, 1963.
12. Caldwell, P. D., Hodgkin, A. L., Keynes, R. D., and Shaw, T. I. The effects of injecting "energy-rich" phosphate compounds on the active transport of ions in the giant axons of *Loligo. J. Physiol. (Lond.)* 152:561, 1960.
13. Fahn, S., Koval, G. J., and Albers, R. W. Sodium-potassium-activated adenosine triphosphatase of electrophorus electric organ. I. An associated sodium-activated transphosphorylation. *J. Biol. Chem.* 241:1882, 1966.
14. Haynes, D. H., Kowalsky, A., and Pressman, B. C. Application of nuclear magnetic resonance to the conformational charges in valinomycin during complexation. *J. Biol. Chem.* 244:502, 1969.
15. Hosie, R. J. A. The localization of adenosine triphosphatase in morphologically characterized subcellular fractions of guinea pig brain. *Biochem. J.* 96:404, 1965.
16. Lust, W. D., and Robinson, J. D. Calcium accumulation by isolated nerve ending particles from brain. I. The site of energy-dependent accumulation. *J. Neurobiol.* 1:303, 1970.
17. Medzihradsky, F., Kline, M. H., and Hokin, L. E. Studies on the characterization of the sodium-potassium transport adenosine triphosphatase. I. Solubilization, stabilization and estimation of apparent molecular weight. *Arch. Biochem. Biophys.* 121:311, 1967.
18. Mueller, P., and Rudin, D. O. Action potential phenomena in experimental lipid membranes. *Nature (Lond.)* 213:603, 1967.
19. Mueller, P., and Rudin, D. O. Action potentials induced in bimolecular lipid membranes. *Nature (Lond.)* 217:713, 1968.
20. Pardee, A. B. Membrane transport proteins. *Science* 162:632, 1968.
21. Post, R. L., and Jolly, P. C. The linkage of sodium, potassium, and ammonium active transport across the human erythrocyte membrane. *Biochim. Biophys. Acta* 25:118, 1957.
22. Post, R. L., Merritt, G. R., Kinsolving, C. R., and Albright, C. D. Membrane adenosinetriphosphatase as a participant in the active transport of sodium and potassium in the human erythrocyte. *J. Biol. Chem.* 235:1796, 1960.
23. Sen, A. K., Tobin, T., and Post, R. L. A cycle for ouabain inhibition of sodium- and potassium-dependent adenosine triphosphatase. *J. Biol. Chem.* 244:6596, 1969.
24. Skou, J. C. The influence of some cations on an adenosine triphosphatase from peripheral nerves. *Biochim. Biophys. Acta* 23:394, 1957.
25. Stahl, W. L. Sodium stimulated ^{14}C-adenosine diphosphate-adenosine triphosphate exchange activity in brain microsomes. *J. Neurochem.* 15:511, 1968.
26. Stahl, W. L., and Swanson, P. D. Uptake of calcium by subcellular fractions isolated from ouabain-treated cerebral tissues. *J. Neurochem.* 16:1553, 1969.
27. Stahl, W. L., and Swanson, P. D. Movements of calcium and other cations in isolated cerebral tissues. *J. Neurochem.* 18:415, 1971.

28. Swanson, P. D., Bradford, H. F., and McIlwain, H. Stimulation and solubilization of the sodium ion-activated adenosine triphosphatase of cerebral microsomes by surface-active agents, especially polyoxyethylene ethers: Actions of phospholipases and a neuraminidase. *Biochem. J.* 92:235, 1964.
29. Whittam, R., and Ager, M. E. Vectorial aspects of adenosine triphosphatase activity in erythrocyte membranes. *Biochem. J.* 93:337, 1964.

Excitation and Conduction in the Neuron

Leo G. Abood

THE PROBLEM OF NERVE CONDUCTION and its fundamental significance for neural function first emerged in 1850 when Helmholtz conducted his measurements on the velocity of the nerve impulse in frogs. The present concept of the membrane potential originated with Ostwald in 1890 with the demonstration that colloidal membranes, by virtue of their ion selectivity, can promote the development of electrochemical forces of considerable magnitude. During the next fifty years that preceded the definitive work of Bernstein, the fundamental theories of electricity and electrochemistry were developed by Arrhenius, Nernst, Planck, and many others; these theories were to be of primary importance for the subsequent development of hypotheses for neural excitation [4, 5, 17, 22].

EXCITABILITY

In its broadest sense the term *excitability* can be defined as the ability of a cell or an organelle to respond to a stimulus. Unless this definition is narrowed, however, it would include virtually all processes of the cell – including even metabolic ones – that interact with the cell's immediate environment. Historically the term has been restricted to a description of the transient, rapid responsiveness of cells to a physicochemical stimulus in a way that involves a reversible change in the cellular membrane potential. Theoretically, all cells manifesting an electrochemical gradient would exhibit a transient change in the membrane potential, accompanied by a transient change in membrane permeability. Normally nonexcitable tissues, such as skin, erythrocytes, glandular cells, and even plant cells, elicit an action potential when they are hyperpolarized electrically. Furthermore, as will be discussed later, a number of artificial membrane preparations, both organic and inorganic, exhibit excitatory properties. Although the physicochemical events of such systems are not unlike those that occur in physiologically excitable tissues, the term *excitation* is primarily used to describe bioelectrical processes in muscle (including smooth, skeletal, and heart muscle) and nerve cells. Nevertheless, the fact that most living cells manifest a membrane potential leaves open the possibility that bioelectrical phenomena may somehow participate in a variety of other cellular functions (e.g., secretion, mitosis, and permeability).

There are generally two aspects of excitability as applied to tissues: one is the initiation of a nerve impulse; the other, its propagation or transmission. The initiation of the nerve impulse takes place within the neuron and involves the summation of the activities of innumerable synapses. Propagation refers to the cablelike transmission of an action potential along an axon or through any conductive medium of excitable tissues.

Originally the phenomenon of excitation was assumed to be the result of an explosivelike chemical reaction that led to an instantaneous release of energy. With an increasing understanding of bioenergetics and membrane biophysics, the excitatory membrane has become regarded as a dynamic metastable unit capable of manifesting several states of equilibrium. One of the crucial questions concerning excitation is whether the rhythmic instability, characteristic of excitatory tissues, is caused by a biochemical reaction or by a purely physical one. Is the sudden change in membrane permeability triggered by a chemical event, such as the splitting of ATP, or is it the result of the interplay of such physical forces as electrochemical potential, fixed charges, and osmotic gradients? Much of the discussion in this chapter will focus on these and related questions.

POSSIBLE MECHANISMS FOR MODIFYING MEMBRANE STRUCTURE

If it is assumed that the initial event in excitation is an alteration of the configuration and physicochemical characteristics of the excitatory membrane, it is appropriate to consider the ways in which membrane structure can be reversibly altered electrically, mechanically, or chemically. Among the more apparent mechanisms are the following: (1) those causing the alteration of the conformation (i.e., tertiary or quaternary structure) of the structural protein by affecting electrical charge distribution or through combination with a divalent cation or other mobile substance, (2) those modifying the state (phase transition) or charge of membranous lipid, (3) those affecting the overall equivalent pore radius of the membrane, (4) those affecting the configuration or charge characteristics of a sialopolysaccharide, (5) those initiating an enzymic or chemical reaction within the membrane, or (6) those triggering a purely electrochemical process controlling ion conductance. The various theories of excitation are based on one or more of these mechanisms, and only the more important ones will be discussed.

Initially, it may be helpful to justify and qualify the use of the term *membrane* as it is used here. Although the term refers to such cellular constituents as the axolemma, sarcolemma, and synaptic membranes which are familiar to the ultrastructuralist, it is equally applicable to the interface between any macromolecular matrix and its surrounding fluid. Primarily, the term is meant to describe the interfacial regions that constitute an energy barrier to the diffusion of ions and other mobile substances – a barrier, in the case of the excitatory membrane, that is responsive to specific stimuli.

NATURE OF THE MEMBRANE POTENTIAL

The membrane potential of a living cell is an electrical potential difference between two electrolyte systems of varying concentration which are separated by the semipermeable cellular membrane. Because all systems seek to maintain electroneutrality, the membrane potential is essentially an equilibrium potential that balances the net forces resulting from all ionic fluxes. The tendency of the two solutions to mix provides the free energy of the system, which, in the case of an ideally permselective membrane, is totally conserved or utilizable. A permselective membrane is an ion-exchange membrane that is selective only for counter ions, whereas a semipermeable membrane is one that is permeable to one type of ion or molecule in preference to another. Among the factors that determine the magnitude of ionic flux across the semipermeable cellular membrane are the electrolyte concentration difference, the electrical potential difference, the permeability characteristics of the membrane barrier, and energy-linked ion transport systems.

When colloid science developed near the turn of the century, it was recognized that polyvalent colloids, including proteins, in contact with an electrolyte will establish a Donnan equilibrium potential even in the absence of a semipermeable membrane. The potential arises because the electrical charge of the colloidal matrix is fixed (immobile) and the diffusible ions distribute themselves unequally in order to maintain electrical neutrality. The Donnan potential does not successfully account for the resting potential, so a theory was proposed based on the concept that a charged membrane, which may behave as a colloidal matrix, establishes a separate equilibrium system with a counter ion. This results in oppositely charged and unequal potentials at each membrane interface [19, 45, 50]. Within the interior of the membrane, a diffusion potential develops. The membrane potential, then, is the sum total of the two Donnan potentials and the diffusion potential (Fig. 3-1).

In recent years considerable attention has focused on the mechanisms responsible for maintenance of the electrochemical balance of excitable tissues because this balance is crucial for excitation. It should be stressed, however, that the ionic movements associated with excitation take place independently of those involved in active ion transport (see Chap. 2). Among other differences, the ionic movements that take place during depolarization are from zones of higher to those of lower concentration gradients. They require no energy and involve considerably greater fluxes than do active transport processes [27].

Although different mechanisms are obviously involved, the possibility remains that similar membranous components participate in both types of ionic processes. For example, the ATP-dependent systems responsible for active transport may be

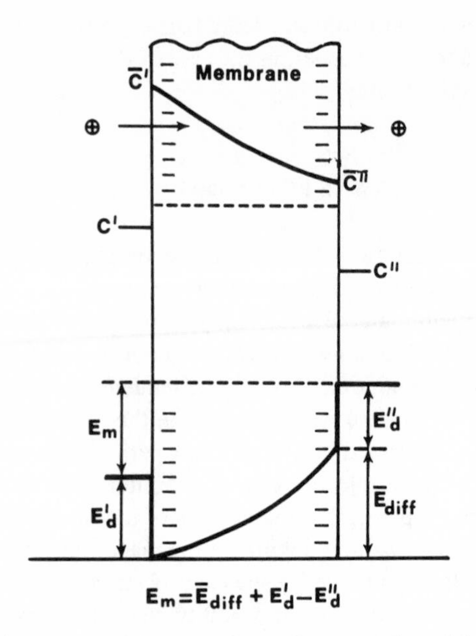

FIGURE 3-1. Profile of a cation exchange membrane demonstrating origin of the membrane potential, E_m. E_m is essentially a concentration potential resulting from the unequal distribution of a cation (e.g., K^+). Because the membrane is more permeable to cations than to anions, the cations diffuse across more readily; the momentary excess cation transfer creates an electrical potential. A steady state is attained in which the potential enforces equivalence of cation and anion fluxes, thus inducing a positive potential (E_d'') in the more dilute (external) solution. E_d' = potential in the internal solution, $\bar{E}_{diff}$ = diffusion potential, E_d = Donnan equilibrium potential, C = concentration of cation, ($\bar{\ }$) indicates membrane, and (') and ('') internal and external components. E_m is the sum total of the diffusion and the two Donnan potentials.

identical with or closely related to those regulating the ionic fluxes of excitation. Although in some tissues, such as the electric organ of the eel, the two systems are believed to be distinct, this system is functionally unique and may not be representative of neuronal systems. Excitation may be regarded as a transient interruption of the active transport mechanisms, leading to a partial reversal of specific physical and chemical reactions that reside in the membrane. Nevertheless, the ionic fluxes associated with the action potential are peculiar to excitable tissues and have provided the basis for the major theory of excitation.

HODGKIN-HUXLEY HYPOTHESIS

In the resting state, an axon or nerve cell has a resting potential (E_R) of about –90 mv, which results from the unequal distribution of ions across its functional membrane. Originally, it was thought that the resting potential was a K^+ equilibrium potential and describable by the Nernst equation

$$E_K = (RT/F) \ln([K^+]_i/[K^+]_o)$$

where R is the gas constant (8.31 joules/mole/deg.)

T = degrees K

F = Faraday's constant (96,500 coulombs/mole)

Although this equation does yield a reasonably correct value for E, it does not take into account the differential Na^+ (and Cl^-) distribution. If the ratio $[Na^+]_i/[Na^+]_o$ is incorporated in the equation, the value for the potential would be 0. The addition of the Cl^- ratio does not rectify the problem. In order to circumvent this difficulty, Hodgkin and Huxley [23, 24] proposed a steady-state model in which the permeability (P) differences of the various ions were considered in the squid axon. It was determined experimentally that the relative permeability of K^+, Na^+, and Cl^- in the resting state was 1.0 to 0.04 to 0.45. The revised equation for E_R was then

$$E_R = (RT/F) \ln \frac{P_K [K^+]_i + P_{Na} [Na^+]_i + P_{Cl} [Cl^-]_o}{P_K [K^+]_o + P_{Na} [Na^+]_o + P_{Cl} [Cl^-]_i}$$

or

$$E_R = 58 \log_{10} \frac{[K^+]_i + (P_{Na}/P_K) [Na^+]_i + (P_{Cl}/P_K) [Cl^-]_o}{[K^+]_o + (P_{Na}/P_K) [Na^+]_o + (P_{Cl}/P_K) [Cl^-]_i}$$

at 20°C. Substitution of the P ratios and $[Na^+]_o$ = 120 mM, $[Na^+]_i$ = 4 mM, $[K^+]_o$ = 2.5 mM, $[K^+]_i$ = 140 mM, and $[Cl^-]_o$ = 120 mM yields the correct value for E, −70 mv.

If a driving force (electrical or chemical) that increases P_{Na} is then applied, the ionic permeabilities are altered so that at the peak of the action potential the ratio P_K to P_{Na} to P_{Cl} is 1.0 to 0.20 to 0.45. The inward Na^+ current (I_{Na}) could then be expressed as a function of the electrochemical potential difference $E - E_{Na}$ by a variant of the Goldman equation [18]

$$I_{Na} = P_{Na} \frac{F^2 E}{RT} [Na^+]_o \frac{e^{(E - E_{Na}) F/RT} - 1}{e^{EF/RT} - 1}$$

Based on the supposition that the initial unit event in nerve conduction involves a sudden change in P_{Na}, a series of empirical differential equations were formulated. These described in quantitative terms the permeability changes that presumably were responsible for the action potential.

Briefly, the Hodgkin-Huxley ionic hypothesis can be described as follows. At rest, the excitatory membrane exhibits some permeability to K^+ and Cl^-, but it is virtually impermeable to Na^+. When activated, either chemically or electrically, the membrane potential decreases to a critical threshold, where a sudden, marked increase occurs in Na^+ permeability. The transient, inward movement of Na^+ down an electrochemical gradient results in an inward current, and the magnitude and rate of depolarization is a function of the influx of Na^+. With a diminution of the transmembrane potential, the driving (electrochemical) force of Na^+ influx slows and eventually ceases. At the same time, there is a compensatory outward current, which is the result of K^+ outflux (and Cl^- influx), so that at the height of the action potential, the outward and inward currents are balanced. The increased Na^+ influx is only transient, so the transmembrane potential gradually decreases until the resting equilibrium potential is attained. The net result of the single action potential is the exchange of equal quantities of Na^+ for K^+, and both the shape and magnitude of the action potential are described mainly in terms of the time course of the Na^+ and K^+ conductance changes.

By employing the ingenious technique of the voltage clamp, which permits the maintenance of the transmembrane potential at a fixed value, Hodgkin and Huxley demonstrated that the time course of the mechanisms that determine ionic fluxes across the membrane is expressible by rate constants dependent only upon voltage and temperature. They were thus able to solve the following formulations:

$$I = C(dV/dt) + P_K (V - V_K) + P_{Na} (V - V_{Na}) + P_\lambda (V - V_\lambda) \tag{1}$$

$$P_{Na} = \bar{g}_{Na} m^3 h \tag{2}$$

$$P_K = \bar{g}_K n^4 \tag{3}$$

$$dm/dt = \alpha_m(1-m) - \beta_m m \tag{4}$$

$$dh/dt = \alpha_h(1-h) - \beta_h h \tag{5}$$

$$dn/dt = \alpha_n(1-n) - \beta_n n \tag{6}$$

where I = total ionic current density, C = capacitance/cm^2, and V = transmembrane potential. P_{Na} and P_K are the permeability coefficients for Na^+ and K^+; V_{Na} and V_K are the resting potentials for Na^+ and K^+. The subscript λ refers to similar parameters for other ions involved. The ionic permeabilities are expressed in terms of $\bar{g}_{Na}$ and $\bar{g}_K$, the conductance due to Na^+ and K^+, respectively. The letters m, h, and n represent the mole fractions of unspecified factors that regulate ionic permeabilities, and α and β the empirical rate constants. By fixing the voltage (i.e., V = constant), equation (1) becomes

$$I = P_{Na} (V - V_{Na}) + P_K (V - V_K) + P_\lambda (V - V_\lambda) \tag{7}$$

or

$$I = I_{Na} + I_K + I_\lambda \tag{8}$$

where I_{Na} is the inward Na^+ current and $I_K + I_\lambda$ the outward currents due to K^+ and other ions. Under these conditions, the total current is only a function of the combined ionic permeabilities. The rate constants are monotonic functions of the transmembrane potential, and their values are obtained over the range V = resting potential to V = Na equilibrium potential

$$V_{Na} = RT/F \ln [Na^+]_o/[Na^+]_i = -50 \text{ mv}$$

After determining the values for the constants α and β, the four differential equations (first order, nonlinear) are solved simultaneously for the boundary condition V = constant. A comparison is then made between the calculated current values and the data from the voltage-clamp experiments to determine the fit of the α and β functions and the necessity for making adjustments. A plot of the data obtained from the solution of the equations reveals excellent agreement between the computed and experimental curves for the action potential with respect to size and shape, as well as to threshold and subthreshold responses to the stimuli.

The overall equation describing the action potential is obtained by substituting equations (2) and (3) into (1)

$$I = C(dV/dt) + \bar{g}_{Na}m^3h(V - V_{Na}) + \bar{g}_K n^4(V - V_K) + \bar{g}_\lambda(V - V_\lambda) \qquad (9)$$

where the first term refers to the current contribution of the membrane capacitance and the subsequent terms to the current contributions made by Na^+, K^+, and non-specified conductances.

From the standpoint of neurochemistry, one of the most challenging and important problems is to define the chemical nature of the dimensionless parameters m, h, and n which regulate ionic permeability during activity. The molecular nature of the parameters is not known, although it could conceivably represent the mole fraction of a depolarizing transmitter, such as acetylcholine. At the end plate, acetylcholine contributes to the increased Na^+ permeability. Conceivably, the parameters m and n involve certain metastable structural components of the membrane, such as phospholipids and protein, which undergo reversible rearrangements affecting both the spatial and charge distribution of the components. Because the rate constants of these substances change dramatically with slight changes in the membrane potential, a cooperative effect probably prevails, so that a rearrangement of one molecule is transferred to a number of neighboring molecules.

It is apparent, therefore, that a comprehension of the quantitative aspects of nerve conduction must await an understanding of the physical and chemical nature of the excitable membrane. Apart from describing their rate constants, the ionic hypothesis provides no hint as to their nature or if one or more mechanisms are involved. As will be discussed later, one plausible mechanism is the association of Ca^{2+} and ATP.

Na^+ CONDUCTANCES AND ION CHANNELS

Since the early observation of Lorente de Nó [34] that quaternary ammonium compounds can be substituted for Na^+ in the excitatory process, the question has risen as to the dispensability of Na^+. Although there is a wide variety of Na^+ substitutes, including guanidine, hydrazine, and hydroxylamine, only hydrazine yields action potentials similar to those obtained with Na^+ [28, 29]. More recently, it has been shown that in some tissues, such as the molluscan giant nerve and the crustacean muscle, the inward current is produced by Ca^{2+} and not by Na^+. Such observations have led to the hypothesis that within the excitatory membrane are unspecific channels that are opened (or activated) during the initial phase of the action potential. Presumably, the various ionic channels are distinct, since the transient currents of inward and outward Na^+, but not of Ca^{2+} and K^+, are blocked by low concentrations of tetrodotoxin [56]. Insofar as tetrodotoxin (as well as saxitoxin and batrachotoxin) blocks the action potential when hydrazine (but not tetraethylammonium) is substituted for Na^+, it might be inferred that hydrazine has a high affinity for the Na^+ channel. In some species of fish Na^+ conductance is not blocked by the toxin, although Ca^{2+} conductance is blocked, so it is premature to accept the notion of operationally separate channels for Na^+ and other ions. A recent observation that tetrodotoxin interacts with cholesterol and not with other membranous lipids has led to the suggestion that cholesterol may be a constituent of the Na^+ channel [55]. Apart from the question of specificity, agents such as tetrodotoxin should prove valuable in elucidating the biochemical nature of the excitable membrane.

MECHANISMS FOR ION SELECTIVITY AND MAINTENANCE OF ELECTROCHEMICAL GRADIENTS

Because excitability is dependent upon the maintenance of an electrochemical potential, the intracellular mechanisms involved in the regulation of the ionic balance are of fundamental importance. Attention has focused mainly on two problems: the mechanisms of K^+ selectivity (influx) and Na^+ extrusion (outflux). Several theories have been proposed to account for the K^+ selectivity, and they can be classified into three categories. One is based on the notion that the membrane operationally consists of "pores" that permit the entry of K^+ over Na^+, as the latter has a greater hydrated diameter. The second assumes that a metabolically driven "pump" is responsible for Na^+ extrusion, and the third attributes selectivity to the permselective nature of the cytoplasm and its ability to act as an ion exchange system. For a detailed discussion of the theories, the reader is referred elsewhere [33, 44], and only those features that relate directly to the problem of excitability will be discussed.

The need for an energy-dependent mechanism for intracellular K^+ accumulation and Na^+ extrusion derives from the fact that both processes must occur against a large electrochemical gradient. Although the "Na^+ pump theory" postulates that K^+ accumulates by exchanging for the extruded Na^+, a specific mechanism for K^+ accumulation is still required because of the inward K^+ gradient and high $[Na^+]_o$. It has been well established that respiratory metabolism per se is not essential for excitation because excitable tissues can retain excitability after complete inhibition of respiration and depletion of phosphocreatine [1, 33]. When, however, the [ATP] falls below 70 to 75 percent of normal, the membrane potential diminishes to less than –60 mv (frog ganglion or muscle cells) and excitability is no longer possible unless the cell is activated by a hyperpolarizing current.

On the other hand, it has been reported that Na^+ extrusion in frog sartorius muscle is not affected by depletion of the ATP + phosphocreatine supply (DNP inhibition), but is completely stopped after glycolysis is inhibited by iodoacetate [10]. Such observations have been cited in support of the hypothesis that ion transport – at least in skeletal muscle – depends upon a metabolic redox potential, i.e., an electromotive driving force which provides the energy for ion transport. The assumption is that the muscle contains two metal redox systems, one a donor and the other an acceptor of electrons [10].

Na^+ transport is conceived as a process whereby the reduced form of the donor combines with Na^+, which is then extruded upon transfer of the electron to the oxidized form of the acceptor. The total energy change within the muscle can be expressed as

$$dG/dn = F(E_a - E_d) + RT \ln[\mathrm{Na}]_i$$

where $F(E_a - E_d)$ is the available free energy per equivalent of electrons donated, and E_a and E_d the potentials of the acceptor and donor systems. When Na^+ is extruded, the total energy change, which remains the same, is given by

$$dG/dn = RT \ln[\mathrm{Na}^+]_o - EF + F(E_a - E_d')$$

$F(E_a - E_d')$ reflects the free energy that results from the increase in the potential of the donor system as $[Na^+]_i$ decreases, and E is the membrane potential. Combination of the two equations yields

$$F(E_a - E_d) = RT \ln([\mathrm{Na}^+]_o/[\mathrm{Na}^+]_i) - EF + F(E_a - E_d')$$

where the first two terms on the right represent the energy barrier to Na^+ extrusion. The energy barrier reaches a maximal (critical) value when $F(E_a - E_d')$ approaches zero, at which point the Na^+ pump is presumed to become inactive.

No specific enzymic redox system has been proposed for the scheme, but components of the electron transport scheme linked to various dehydrogenases could qualify. Although there are numerous shortcomings in the redox pump theory [7, 33], particularly when applied to neural tissue, it is at least an attempt to relate a specific metabolic system to electrochemical potentials.

ASSOCIATION THEORY FOR ION SELECTIVITY

There are essentially two theories for ion selectivity in cells: the so-called membrane theory, which adheres to the concept of a permselective membrane, and the "association," or "sorption," theory, which emphasizes the role of the protoplasmic matrix. The major difference between the association theory and the membrane theory is based on their assumptions concerning the physical state of intracellular water and ions. According to the association theory, a major fraction of the intracellular cations must be associated with fixed anionic sites provided by cytoplasmic proteins, while the intracellular water must exist in both a free and a bound state. Upon depolarization, changes take place in the degree of association between the intracellular cations and fixed ionogenic sites. The resting potential of the cell results from the dissociation of the protein-electrolyte complex induced by the recording microelectrode.

The membrane theory, on the other hand, requires that no differences exist in the state of intracellular and extracellular ions and water. Whatever differences occur during depolarization are a reflection of permeability changes that result in an alteration of the equilibrium conditions between the intracellular and extracellular ions.

Some support for the association theory derives from the lowered mean activities of the intracellular Na^+ and K^+ of some cells, such as skeletal muscle; however, measurements in squid axons have disclosed no difference. Only a very small fraction of the total ions are involved in excitatory phenomena, so activity measurements on whole cells are of doubtful value in settling the issue.

If either the cytoplasmic or membranous matrix is regarded as a cation exchange system, it is reasonable to assume that part of the water is free and the remainder is associated with the cations (i.e., as water of hydration). With the use of nuclear magnetic resonance (NMR), the presence of bound water has been detected in frog sartorius muscle and brain tissue (see [13] for reviews). When the less hydrated K^+ exists within the matrix, more free water is present than would be present when the more hydrated Na^+ is present. This difference in free water content is explained on the basis that the larger hydrated Na^+ increases the swelling pressure and thereby forces out free water. According to this model, the physical state of water can be influenced by altering either the ionic milieu or the molecular configuration and electrical charge of the ionogenic matrix.

The importance of the electrical field strength in determining cation selectivity stems from the fact that a cation is desolvated as it associates with the polyvalent anionic matrix. With a series of monovalent cations it could be demonstrated that the order of cationic selectivity by an anionic source (e.g., a glass surface) is dependent primarily on the electrostatic field strength [14]. The rule for cationic selectivity is based on the assumption that the least hydrated cation is more readily desolvated as the negative field strength increases. Consequently the sequential order of increasing selectivity for a series of monovalent cations would be Cs^+, Rb^+, K^+, Na^+, Li^+ at low anionic field strength and the exact reverse at high anionic field strength.

The theory has been subjected to a thermodynamic treatment based on the free-energy changes in the transitional states as the unsolvated cation interacts with water and the anion exchanger. It can be shown that selectivity of the alkali cation requires an optimal field strength that is comparable to those of phosphate and carboxyl groups, which are abundant in biological membranes. In addition to the magnitude of the charge, the state and geometry of the anionic charge matrix is an important consideration in determining cation selectivity, a fact that is well documented with ion exchange systems [19].

ROLE OF Ca^{2+} IN EXCITABILITY

Since the work of Ringer in 1890, the importance of Ca^{2+} in the maintenance of excitability has been recognized (see [2, 28, 35] for reviews). Ca^{2+} is believed to be responsible for the organization of the membranous components directly involved in the ion permeability changes accompanying the action potential. In addition to regulating the Na^+ and K^+ conductance during excitation, Ca^{2+} serves as a charge carrier in some tissues and, possibly, in others in which Na^+ and K^+ are mainly involved [28]. Still another role of Ca^+ is its participation in the storage and release of neurotransmitters at synaptic junctions [8, 11, 25]. In the ensuing discussion, repeated reference will be made to this important divalent cation and the various hypotheses concerning its functional mechanisms.

"TWO-STABLE-STATES" THEORY

Another theory of excitation is based on the notion that the action potential results from a reversible conformational change in the macromolecular complex that comprises the excitable membrane [49]. The action potential results not from the influx and outflux of inorganic ions, but rather from intermolecular or intramolecular rearrangements triggered by the exchange of ions. In the resting (stable) state, the negatively charged membrane (due to the charge of phospholipids and proteins at physiological pH) is occupied largely by Ca^{2+}. When the axon is excited, the inward positive current electrophoretically displaces the Ca^{2+} from the fixed negative sites, which, through ion exchange, are replaced by Na^+ (or any monovalent cation substituted for Na^+). Tasaki [49], who originally proposed this "two-stable-states" theory of excitation, conceived of the ion exchange process as a cooperative phenomenon triggering a "phase transition or a discontinuous conformational change of the membrane macromolecules to the excited state." The transient changes in membrane impedance and potential that accompany excitation are the combined effect of the flux of ions and water and fluctuations in the fixed charges of the membrane.

Until the introduction of the perfused axon preparation, the theory had little experimental support other than the observation that Na^+ was not indispensable.

After the axoplasm has been removed from an excised squid axon by proteolytic digestion (pronase), it can be perfused with various media while electrical and other measurements are being made. When the axon is immersed in artificial sea water (300 to 450 mM NaCl, 50 to 200 mM $CaCl_2$) and perfused with potassium fluoride or other appropriate potassium salts, excitability can be maintained for hours. The ability of the various potassium salts to maintain excitability conforms to the relative lyotropic number of the anionic species. Fluoride is the most favorable and thiocyanate (SCN^-) is the least. Because the affinity of anions for the $-NH_4^+$ groups of proteins also conforms to the lyotropic series but in the opposite direction (i.e., SCN^- is greater in affinity than F^-), proteins are believed to be the major constituents involved in excitation. Presumably, anions with the greatest affinity for ammonium would more readily disrupt salt linkages in proteins, thereby disrupting the appropriate protein conformations. Just as a variety of monovalent cations could substitute for Na_o in an intact axon, it is possible to make substitutions in the internal perfusate. The order for maintaining excitability is Cs > Rb > K > NH_4 > Na, and is in agreement with that found for the affinity to phosphoryl groups of various colloids [6].

The theory attempts to explain nerve conduction as well as excitation. As a region of the membrane is converted to the excited state, a local current is established between the excited and adjacent resting areas. This current, which flows inward in the excited area and outward in the resting area, drives the internal cations into the resting membrane and thus establishes a self-propagating reaction along the axon. Meanwhile, immediately behind the excited region, the reversible Ca^{2+} exchanges for the monovalent ion, restoring the membrane to the resting or stable state.

Changes in turbidity, birefringence, and fluorescence (employing 8-anilinonaphthalene-1-sulfonate) after tetanization of axons all have been cited in support of the theory; however, such changes could be attributed to the transfer of water and mobile ions as well as to rearrangements in macromolecular structure.

MOLECULAR MODEL FOR THE EXCITATORY MEMBRANE BASED ON A Ca^{2+}-ATP MEMBRANOUS COMPLEX

A molecular model of an excitatory membrane has been proposed in which the functional unit is a macromolecular complex of Ca-ATP-phospholipid-protein [1, 2]. The Ca^{2+} is linked to ATP and the phosphate group of the phospholipid (e.g., phosphatidylserine), which in turn interacts hydrophobically with a structural protein moiety. In addition to having a high affinity for ATP, the protein may have certain catalytic functions, such as ATP-ADP exchange and ATP hydrolysis.

The initial chemical event in excitation is believed to be the displacement (electrophoretically) of Ca^{2+} followed by the release of ATP and its eventual hydrolysis. Associated with the chemical events are the conformational changes in membrane structure responsible for the increased Na^+ permeability. Restoration of the membranous structure requires the resynthesis and recombination of ATP, which in turn promotes the reuptake of Ca^{2+}. (For a detailed discussion of the kinetic and chemical aspects of the model, see [1].)

In addition to interfacial and other studies demonstrating the interaction of Ca^{2+}, ATP, and phospholipids, the model derives some support from studies on synaptic and other neural membranes that demonstrate the presence of high concentrations and combinations of the various constituents. Such membranes contain a high concentration of ATP which is bound directly to the protein, whereas the Ca^{2+} seems to be associated largely with phospholipids. According to the model, the state of membranous Ca^{2+} is dependent upon ATP, which, by virtue of its chelating and sequestering roles as well as its catalytic role, determines the conformational state of the membrane. Although the model provides an explanation for the change in membrane permeability, it also conforms to the notion that excitation results from fluctuations in the fixed charges of the membrane.

Because Ca^{2+} evidently plays a critical role in regulating ion permeability during excitation, some consideration should be given to possible mechanisms involved in its regulation. If the surface layer of the excitatory membrane behaves like a surface (cationic) film, the equilibrium concentration of Ca^{2+} within the surface can be expressed as

$$a' = Kae^{(-2F\psi/RT)}$$

where a' is the surface concentration of Ca^{2+}, a the concentration in the external solution, ψ the phase boundary (surface) potential, K a constant, and the other symbols have their customary designation.

Electrical depolarization may produce a change in the dipole orientation of phospholipid groups (e.g., $CH_2{-}O{-}P$, $P{=}O$, and $P{-}O{-}CH_2CH_2N$) attached to a Ca^{2+}. As a result, a change takes place in ψ followed by a change in the surface concentration of Ca^{2+}. Depolarization may also result in the hydrolysis of a complex of Ca^{2+} with ATP by promoting hydrolysis of the Ca-phosphate bond. Although such mechanisms are intended to be no more than examples, there is some experimental evidence in support of them.

PHYSICAL CHANGES IN MOLECULAR STRUCTURE

The availability of such techniques as resonance spectroscopy and optical rotary dispersion has stimulated interest in measuring molecular changes in neural tissue during excitation. With the use of nuclear magnetic resonance spectroscopy, it is possible to detect an emission of energy in the μ wavelength from an excited axon, which is believed to originate from the membrane surface and to be the result of such events as ion movements [15]. Studies on muscles and nerves through use of the proton magnetic resonance spectrometer reveal that the relaxation time of water is greatly reduced upon depolarization [16]. Presumably the change is a reflection of the rate of exchange of "bound" water with the free intracellular water, and not the exchange of intracellular and extracellular water. Other studies on natural and artificial membranes employing NMR spectroscopy have shown that the immobilized water in the vicinity of phospholipid molecules is released in the presence of Ca^{2+}, while the mobility of the phospholipid molecules is increased. Despite the importance of resonance spectroscopy in detecting molecular changes in pure materials, the problem of interpretation in complex systems poses considerable difficulties.

PHYSICOCHEMICAL MODELS FOR EXCITATION

In attempting to explain bioelectrical phenomena, mention should be made of some purely physicochemical models. To account for the fact that ice has greater conductivity than water, it has been proposed that protons (or electrons) may jump from one molecule of water to another [53]. Such a mechanism requires that water molecules be arranged in a hexagonal lattice. It is possible to conceive of such arrangements of water molecules hydrogen-bonded to proteins that have the conformation of β-keratin. The mode and degree of hydrogen bonding of the water molecules would be determined by the amino acid sequence as well as by the overall conformation of proteins, so specific proteins with such properties would have to exist in the excitatory membrane. In addition to the peculiar conductive properties of crystalline water, the changes in state accompanying the association and dissociation of water, lipoproteins, ions, and other membranous components may be related to bioelectrical phenomena. Such changes in state involve significant entropy changes, which directly or indirectly determine electrochemical potentials.

Another physical property of proteins and lipids that is relevant to the problem of excitability is their ability to undergo changes in dipole orientation when they are spread as insoluble monolayers and subjected to an electrostatic field. This distortion polarizability, which is especially high for zwitterions, may serve as a "triggering" phenomenon for the membrane conformational change required for the increased ionic permeability during the action potential [43]. Polarization distortion is associated with entropy changes and could also account for some of the heat changes associated with the action potential.

Such physicochemical studies have just begun to be carried out on excitable tissues as a result of the advent of such techniques as differential thermal analysis and NMR spectroscopy. As will be discussed in relation to the Wien dissociation effect, polarization can also account for certain pH changes of possible significance for excitation.

EXCITATION BASED ON THE WIEN DISSOCIATION EFFECT

A theory of excitation proposed recently is based upon the so-called Wien dissociation effect [3]. It derives from the notion that a positive increment of 0.1 to

0.2 pH units (either externally or internally in perfused axons) can produce depolarization and that this pH increment can occur as a result of the depolarizing electrical field on the dissociation constant of a weak electrolyte (i.e., depolarization increases the association of protons with fixed anionic sites). By employing Onsager's equation, which relates the dissociation in the presence of an electrical field to the dielectric constant of the medium (assumed to be about 8 within the high-impedance membrane), the critical potential needed to initiate the action potential (20 mv) can be provided for by the positive pH increment. By utilizing differential equations describing the diffusional proton transfer across the membrane and assuming that the relaxation time for the association of the protons is much smaller than the diffusional relaxation time of H^+ (1 msec), the kinetics of the pH changes are shown to be consistent with the time course of the action potential. The internal chemical equilibrium of the proton association-dissociation reaction (possibly involving the phosphate groups of phospholipids) is then effectively reached before neutralization by the external protons, a process that restores the membrane to the resting state.

Although this theory is interesting and appears to have a sound physicochemical basis, it suffers from numerous shortcomings. To begin with, the depolarizing effect of a small increase in pH is only demonstrable internally (perfused axons) and not at the external surface of the membrane. In addition, the dielectric constant of the membrane may be considerably greater than 8 because of the presence of significant amounts of proteins within the membrane. Furthermore, because of the amphoteric nature and strong buffering capacity of proteins, pH changes may be compensated for internally by intramolecular rearrangements or transient protonic jumps.

PROTOCHEMICAL THEORY FOR EXCITATION

The suggestion has been made that bioelectrical potentials may be derived from the free energy of a proton-generating system analogous to that of a galvanic cell [48]. Although metabolic redox reactions are the source for protons, the immediate excitatory events are presumed to be the result of proton transfer reactions. More recently, a chemiosmotic or proton gradient scheme has been advanced to explain ion transport in mitochondria [37]. According to this scheme, energy coupling promotes the flow of protons across the membrane, which, by creating a potential gradient, induces the flow of cations in the opposite direction. If such a scheme is valid and is applicable to membrane phenomena in general, any transient change in the proton gradient would influence cationic flow.

OSCILLATORY WAVE THEORY FOR NERVE CONDUCTION

Some of the earliest theories of excitation held that the action potential resulted from oscillatory phenomena associated with electrochemical reactions at the membrane interface. Early studies included the work of Lillie [32], whose "iron nerve" model generated a propagating, oscillatory wave of polarization and depolarization along an iron wire coated with a film of iron oxide as it underwent oxidation and reduction.

More recently, a nonmetallic model has been developed by Teorell [51]. This consists of a porous membrane with fixed charges (e.g., a fritted glass disc) that separates two stirred solutions of different electrical conductance. When a constant electrical current is passed through the membrane, electrophoresis or electroendosmosis occurs, which tends to increase the fluid volume in the compartment with higher conductance. The electroosmotic flow establishes a hydrostatic pressure difference, p, across the membrane and regulates the velocity of flow, v, so that

$$v = q(dp/dt)$$

where q is a "geometry" coefficient. With a variation in v (which is a function of the applied voltage and p), a corresponding variation takes place in the concentration profile of the electrolyte across the membrane. This, in turn, affects the electrical resistance, R, of the membrane. As v is varied, a time lag is required for R to reach a steady state, and the rate of change of the membrane resistance is found to be directly proportional to the deviation of the resistance from the steady state, R^o. Therefore

$$dR/dt = k(R - R^o)$$

where k is a rate constant dependent upon the nature of the membrane and electrolyte. A set of differential equations is then derived expressing the oscillatory, time-dependent variation for the fluid velocity v and the drive electrical potential E (from Ohm's law). The solution of the equations yields integral curves, which may be damped, undamped, or "growing" sinusoidal oscillations, depending on the "damping-factor" in the equations. For a given fixed-charge porous membrane and electrolyte, the model can describe accurately the cyclical changes in R (or E) induced by electroosmosis. This experimental model incorporates many electrochemical and physical concepts relevant to bioelectrical phenomena, and so serves as an excellent teaching device. Like many such physical models, however, the chemical structural nature of the model membrane (fritted glass) bears little resemblance to a biological system.

LIPID MODELS FOR EXCITATION

Since the observation of Langmuir about fifty years ago that a monolayer of lipids assumes a particular molecular orientation, the interactions of all surface-active substances at water-lipid interfaces have received considerable attention by both physical chemists and biologists. Such interactions, which can be described for the most part in terms of Gibbsian adsorption (i.e., as a lowering of surface free energy), are found in biological systems as well as in pure lipid-water systems. Through the use of the techniques of surface chemistry, x-ray diffraction, electron microscopy, and polarization optics, much is known about the physical chemistry and ultra-structure of such lipid systems.

Among the unique characteristics of lipid-water systems is their ability to undergo reversible phase transformations, ranging from a solid form to a liquid crystalline (anisotropic) form to various lipid isotropic forms [26]. Of interest to the problems of membrane phenomena is that such phase transformations can occur readily, even under physiological conditions, and may account for such characteristics as ionic selectivity, hydrophilic-lipophilic shifts (i.e., dielectric changes), permeability changes, variations in equivalent pore size, and, possibly, even excitability. A number of dynamic membrane models have been proposed based on the phase transitions of lipids between micellar and lamellar forms [26, 42, 52]. Many drugs and detergents have been shown to increase conductivity and selectivity to cations and anions respectively, presumably by promoting lamellar-to-micellar transitions in the lipid bilayer membrane.

The lipid bilayer preparation of Rudin and Mueller is another model system that has been employed for the study of membrane phenomena associated with excitability [40]. In addition to having the molecular configuration and dimensions of biological membranes, the lipid bilayer has comparable electrical and permeability characteristics and exhibits excitability characteristics. A variety of lipophilic substances, including an antibiotic polypeptide (alamethicin) and certain enzymes and antigens, can produce transient decreases in the membrane resistance of the bilayer by inducing ionic conduction pores. By introducing certain heterocylic antibiotics into the membrane, the membranes can be made to become K^+ selective [47].

One of the shortcomings of the lipid bilayer is that, in addition to phospholipids, it requires the presence of relatively high concentrations of lipid solvents (e.g., *n*-decane) and cholesterol-like materials to maintain the stability and fluidity needed for carrying out experiments. It has been estimated that the ratio of *n*-decane to phospholipid in the membrane preparation may be as high as ten to one [20]. Consequently, many of the electrical and permeability characteristics of the membrane may be largely attributable to the hydrocarbon. A further difficulty in utilizing it as a membrane model is that it does not account for the role of proteins in determining the structural and functional characteristics of membranes, which is felt by many investigators to be the major one.

Apart from its questionable resemblance to natural excitatory phenomena, the modification of the electrical properties of the lipid bilayer by alamethicin and the peptide isolated from *Enterobacter cloacae* is of interest from the standpoint of the role of proteins in regulating ion transport. Presumably, the peptides develop cation-selective channels within the membrane, while the addition of the protamine, a basic peptide, renders some of these channels anion-selective. Under certain critical conditions, the transmembrane potential can be made to open and close either type of channel, causing an oscillatory change in ion conductance [39].

Such cation-selective agents – including many cycloethers and cyclopeptides which facilitate ion transport and agents such as tetrodotoxin which specifically block ion transport – are of considerable importance in elucidating the structural and functional role of the excitatory membrane.

Another simple artificial lipid model for excitable membranes has been developed [38]. This consists of a thin, flexible membrane (1 μ thick) formed by spreading a drop of an unsaturated glyceride (e.g., linseed oil) on a solution of 0.1 percent $KMnO_4$. A film of this type exhibits such characteristics as cation selectivity, biionic potentials, electrical parameters comparable to biological membranes, and "excitatory" responses. After discounting the Wien dissociation effect as responsible for the oscillatory transient responses elicited by an applied constant current, a mechanism was proposed based on the notion that cation transfer requires the breaking of the hydrogen bonds between water and the hydroxyl groups of lipids [38]. Conceivably, a depolarizing electrical field could initiate the cleavage of hydrogen bonds and thereby promote cation transfer.

METABOLISM AND EXCITATION

The question of the manner in which respiratory metabolism participates in excitatory activity has preoccupied neurochemists for almost a half century. If respiratory metabolism is accelerated during neuronal excitation (this point remains problematical), it is not known whether the metabolic increment is a cause or a consequence of the activity or if it is associated entirely with the recovery phase. It is conceivable that the generalized increase in membrane permeability during activity accelerates the exchange of substrates, cofactors, and metabolites and thus leads to increased metabolism. Until neurochemists are able to understand fully the biophysical events of excitation in terms of energetics, it is futile to attempt to relate energetic metabolism to neuronal activity. At present it is more appropriate to consider the energetic requirements of such functions as Na^+ extrusion, neurotransmitter metabolism and storage, and the maintenance of membrane configuration.

One approach to the problem of relating metabolic activity to excitability is to examine the chemical changes occurring in excitable tissues under the influence of stimulating and depressant drugs. Although the respiratory metabolism of the brain in vivo is diminished during surgical anesthesia, the content of ATP, phosphocreatine, and other energy-rich metabolites is essentially unchanged. It is not yet known whether the reduced metabolism is entirely the result of a specific suppression of brain centers that control cerebral blood flow and respiratory rate or whether the anesthetics affect metabolic activity that is more directly associated with excitation [36]. In the isolated ganglion preparation of the cat, a direct relationship was found between the inhibition of oxidative metabolism by various anesthetics and the degree of synaptic transmission, particularly postsynaptic [31].

The major difficulty with this approach is that anesthetics, particularly at the concentrations used, could affect the physicochemical characteristics of the excitatory membrane in a variety of ways, and the diminished respiratory metabolism may not necessarily be responsible for diminished excitatory activity. Anesthetics and other lipophilic drugs could alter the hydrophilic-lipophilic balance, electrical charge distribution, and molecular configuration of the membrane, all of which are critical for excitation. The reader is referred elsewhere for discussions of this problem [46].

An impressive body of evidence indicates that nerve conduction is not necessarily accompanied by an increase in respiratory metabolism, insofar as it proceeds in the absence of oxygen or in the presence of oxidative and glycolytic inhibitors [1, 33, 46]. In such tissues as frog spinal ganglia, spinal cord, and sciatic nerve, there is no increase in the rate of incorporation of $^{32}P_i$ into ATP during excitation. There is a stage, however, after metabolic inhibition (e.g., by inhibitors of oxidative phosphorylation), when excitation begins to fail; at this point the cellular concentration of ATP falls below 70 percent of its normal value. At this stage the membrane potential has fallen below a critical value (about –60 mv). Upon hyperpolarization, however, the nerves and ganglia are once again excitable. Such studies suggest that the concentration of ATP is a critical factor for the maintenance of the membrane potential above a critical level necessary for excitability [1, 7].

HEAT CHANGES DURING EXCITATION

Attempts to measure the heat production of stimulated nerve extend back to Helmholtz's work in 1848. For many years they were aimed largely at attempting to resolve the issue of whether the nerve impulse is a purely physical wave in which the total energy for propagation remains constant after the initial burst. Subsequent studies on various axons by Hill [21] had shown that the heat change is diphasic, consisting of an initial rise in temperature associated with the initial impulse, followed by a fall that lasts a few tenths of a second afterward. Although one is tempted to attribute the heat exchange to such phenomena as a $Na^+ = K^+$ interchange (a charging and recharging of the condenserlike surface membrane) such an explanation is unsatisfactory. It is more plausible that the changes are due either to specific metabolic reactions or to phase changes in the macromolecular conformation of the excitatory membrane. The initial heat production could conceivably be due to a hydrolytic (exergonic) reaction, involving such substances as acetylcholine, ATP, or other high-energy intermediates (phosphorylated or acylated).

SYNAPTIC TRANSMISSION

In the foregoing discussion, conduction (transmission) and excitation were treated as if they involved similar physicochemical events and mechanisms. Although there is some justification for this approach, it is essential to distinguish synaptic transmission from axonal transmission and conduction in muscle fibers. Transmission at the neuromuscular junction (end plate) and at neuronal synapses is mediated, for the most part, by a chemical transmitter (or neurotransmitter) which is stored within the synapse. As impulses (presynaptic) reach the synapse, the transmitter is released in quanta to produce a postsynaptic potential which, in turn, causes the neuron (or muscle) to respond. No attempt will be made here to discuss neurotransmitters in detail. Instead, we will discuss briefly certain aspects related to their possible mechanisms of action (see Chaps. 4 to 7).

The release of the neurotransmitter into the synaptic cleft is believed to be under the influence of a membranous complex of Ca^{2+}. One possibility is that a Ca^{2+}-anionic complex (e.g., a Ca^{2+}-lipoprotein-ATP complex), by virtue of having a net negative charge, may serve as a carrier for the positive charged transmitter [11]. With the catelectrotonus spread induced by the action potential, the entire complex with the amine is driven to the internal membrane surface, where the transmitter is released. Another hypothesis, which attempts to account for the increased latency between depolarization and transmitter release, suggests that depolarization retards the inward movement of Ca^{2+} or a positively charged Ca^{2+} complex [25].

Apart from its role in synaptic transmission, acetylcholine (ACh) has been proposed as the depolarizing agent in nerve conduction [41]. According to this theory, the combination of ACh with a specific receptor site on the excitatory membrane results in the molecular rearrangement leading to increased Na^+ conductance. Whether the receptor site is acetylcholinesterase or a protein with similar binding characteristics has not been resolved. The major difficulty with this theory is that a variety of agents which block the action of ACh at synaptic junctions (e.g., cholinesterase inhibitors and bis-quaternary ammonium compounds) fail to block nerve conduction at sufficiently low concentrations. Furthermore, a variety of noncholinergic agents, such as local anesthetics, antihistamines, and phenothiazines, also block nerve conduction. It has been proposed that such agents, including the cholinergic ones, interact with a receptor of the conducting membrane that is distinct from the synaptic cholinergic receptor and may be a phospholipoprotein [12].

OTHER ENDOGENOUS METABOLITES

A unique characteristic of the excitatory cell is its extreme sensitivity to an endless variety of metabolites and other chemical agents. Within the neuron itself, a variety of metabolites are produced in addition to the chemical transmitters. These metabolites are capable of either depolarizing or hyperpolarizing the membrane. Included among the depolarizing metabolites are ATP, glutamate, aspartate, citrate, and other intermediates of the tricarboxylic acid cycle; examples of hyperpolarizing metabolites are γ-aminobutyric acid, glutamine, and glycine. It is conceivable that neuronal excitability is regulated by the concentration profile of specific derivatives from a number of metabolic schemes. Although considerable progress is being made in elucidating the mechanisms of a few known chemical transmitters, very little attention has been given to the bioelectrical effects of other metabolites.

ROLE OF PROTEINS IN BIOELECTRICAL PHENOMENA

Until recently the structural role of membranous proteins had not been emphasized, so their functional role in membrane phenomena has remained largely obscure. The emphasis in the past has been mainly on the structural and functional role of the lipids, particularly the phospholipids. It has now become recognized that the membranous proteins play an important role in both structure and function of membranes. In addition to their catalytic, conformational, contractile, and ion exchange properties, they possess a strong hydrophobic character that permits them to interact with one another as well as with lipids. Recent studies with structural proteins – whether derived from mitochondria, erythrocytes, or excitatory membranes – indicate that they are readily capable of undergoing reversible polymerization, contain sialopolysaccharides, bind ATP, and may be equivalent to the (Na^+, K^+)-ATPase normally found in many membranes. One of the unique characteristics of proteins is their ability to undergo cooperativity, a process whereby an interaction at one point of a macromolecule (e.g., with a substrate, ATP, or a metal) can induce a change (e.g., proton transfer or other energy change) elsewhere [57]. It is highly probable that such macromolecular characteristics, which involve intramolecular energetic changes, are of prime significance to the problem of excitation.

ROLE OF GLIA IN NERVE EXCITATION

Among the most poorly understood, yet most important, problems for the neuroscientist is the role of glia in neuronal function. Apart from their apparent role in transporting substances between capillaries and neurons, they are believed to play a role in the regulation of the ionic environment of the neuron [30]. The Schwann cell of the squid giant axon is known to have a membrane potential of –40 mv (a K^+ equilibrium potential), which remains unchanged during axonal conduction [54]. Insofar as the Schwann cell contains six times $[Na^+]_o$, it has been suggested that the cell may be responsible for maintaining the ionic composition of the axolemma-Schwann cell space. A ouabain-sensitive K^+ transport system may be involved in the maintenance of the $[K^+]_i$ at the level of the membrane potential. In the absence of a K^+ transport system, the K^+ equilibrium potential would be about –80 mv.

CONCLUSION

In spite of all the available biophysical and biochemical data concerning bioelectrical phenomena, the mechanisms underlying excitation are still largely obscure.

Although the ionic hypothesis appears to be the most adequate theory to date, it has not provided any insight into the biochemical nature or physicochemical characteristics of the excitatory membrane. One of the most important areas for neurochemical investigation is that concerned with membrane biochemistry, particularly the dynamic aspects associated with the action potential. Apart from the need to determine if specific lipids or proteins play a primary role, it is necessary to determine how a depolarizing stimulus or chemical agent can initiate the change in ionic permeability or whatever other physicochemical events may be responsible for excitatory potentials. As Cole [9] so aptly stated in the conclusion of a classic treatise on the ionic hypothesis: "It is most unfortunate, yet most challenging, that a possible and reasonable explanation of the most powerful parts of ion permeabilities may be completely wrong."

REFERENCES

BOOKS AND REVIEWS

1. Abood, L. G. Interrelationships between phosphates and calcium in bioelectric phenomena. *Int. Rev. Neurobiol.* 9:223, 1966.
2. Abood, L. G. Calcium-ATP-Lipid Interactions and Their Significance in the Excitatory Membrane. In Ehrenpreis, S., and Solnitzky, O. C. (Eds.), *Neurosciences Research.* New York: Academic, 1969.
3. Bass, L., and Moore, W. J. A Model of Nervous Excitation Based on the Wien Dissociation Effect. In Rich, A., and Davidson, N. (Eds.), *Structural Chemistry and Molecular Biology.* San Francisco: W. H. Freeman, 1968.
4. Bernstein, J. *Electrobiologie.* Braunschweig: Wieweg und Sohn, 1912.
5. Bull, H. B. *An Introduction to Physical Biochemistry.* Philadelphia: Davis, 1964.
6. Bungenberg de Jong, H. G. Reversal of Charge Phenomena, Equivalent Weight and Specific Properties of the Ionized Groups. In Kruyt, H. (Ed.), *Colloid Science.* Vol. 2. New York: Elsevier, 1949, p. 259.
7. Caldwell, P. C. Factors governing movement and distribution of inorganic ions in nerve and muscle. *Physiol. Rev.* 48:1, 1968.
8. Cavallito, E. J. Some speculations on the chemical nature of post-junctional membrane receptors. *Fed. Proc.* 26:1647, 1967.
9. Cole, K. S. *Membranes, Ions, and Impulses.* Berkeley: University of California Press, 1968.
10. Conway, E. J., and Mullaney, M. Anaerobic Secretion of Sodium Ions from Skeletal Muscle. In Kleinzeller, A., and Kotyk, A. (Eds.), *Membrane Transport and Metabolism.* New York: Academic, 1961, p. 117.
11. Dodge, J. A., Jr., and Raminoff, R. Cooperative action of calcium ions in transmitter release at the neuromuscular junction. *J. Physiol. (Lond.)* 193:419, 1967.
12. Ehrenpreis, S. Acetylcholine and nerve activity. *Nature (Lond.)* 201:887, 1964.

13. Eisenberg, D., and Kauzmann, W. *The Structure and Properties of Water.* New York: Oxford, 1969.
14. Eisenman, G. Some Elementary Factors Involved in Specific Ion Permeation. In Noble, D. (Ed.), *Lectures and Symposia of the XXIII International Congress of Physiological Sciences, Tokyo, September, 1965.* Amsterdam: Excerpta Medica Foundation, 1965, p. 489.
15. Fraser, A., and Frey, A. H. Electromagnetic emission at micron wavelengths from active nerves. *Biophys. J.* 8:731, 1967.
16. Fritz, O. G., and Swift, T. J. The state of water in polarized and depolarized frog nerves. *Biophys. J.* 7:675, 1967.
17. Glasstone, S. *Textbook of Physical Chemistry.* New York: Van Nostrand, 1948.
18. Goldman, D. E. Potential, impedence, and rectification in membranes. *J. Gen. Physiol.* 27:37, 1943.
19. Helfferich, F. *Ion Exchange.* New York: McGraw-Hill, 1962, p. 106.
20. Henn, F. A., and Thompson, T. E. Synthetic lipid bilayer membranes. *Annu. Rev. Biochem.* 38:693, 1969.
21. Hill, A. V., and Howarth, J. V. The initial heat production of stimulated nerve. *Proc. R. Soc. Lond. [Biol.]* 149:167, 1958.
22. Höber, R. *Physical Chemistry of Cells and Tissues.* Philadelphia: Blakiston, 1945.
23. Hodgkin, A. L., and Huxley, A. F. Currents carried by sodium and potassium through the membrane of the giant axon of *Lodigo. J. Physiol. (Lond.)* 116:449, 1952.
24. Hodgkin, A. L., and Huxley, A. F. A quantitative description of membrane current and its application to conduction. *J. Physiol. (Lond.)* 117:500, 1952.
25. Katz, B., and Miledi, R. The release of acetylcholine from nerve endings by graded electrical pulses. *Proc. R. Soc. Lond. [Biol.]* 167:23, 1967.
26. Kavanau, J. K. *Structure and Function in Biological Membranes.* San Francisco: Holden-Day, 1965.
27. Keynes, R. D. Ion transport in excitable cells. *Protoplasma* 63:13, 1967.
28. Koketsu, K. Calcium and the Excitable Cell Membrane. In Ehrenpreis, S., and Solnitzky, O. C. (Eds.), *Neurosciences Research.* New York: Academic, 1969.
29. Koketsu, K., and Nishi, S. Effects of tetrodotoxin on the action potential in Na-free media. *Life Sci.* 5:2341, 1966.
30. Kuffler, S. W., and Nicholls, J. G. Physiology of neuroglia cells. *Ergeb. Physiol.* 57:1, 1966.
31. Larrabee, M. G., and Posternak, J. M. Selective actions of anesthetics on synapses and axons in mammalian sympathetic ganglia. *J. Neurophysiol.* 15:91, 1952.
32. Lillie, R. S. The passive iron wire model of protoplasmic and nervous transmission and its physiological analogies. *Biol. Rev.* 11:181, 1936.
33. Ling, G. N. *A Physical Theory of the Living State.* New York: Blaisdell, 1962.

ORIGINAL SOURCES

34. Lorente de Nó, R. On the effect of certain quaternary ammonium ions upon frog nerve. *J. Cell. Comp. Physiol.* 33:(Suppl. 1), 1949.
35. Manery, J. F. Effect of calcium on membranes. *Fed. Proc.* 25:1804, 1966.
36. McIlwain, H. *Biochemistry and the Central Nervous System* (3rd ed.). Boston: Little, Brown, 1966.
37. Mitchell, P. *Chemiosmotic Coupling and Energy Transduction.* Bodmin, England: Glynn Research Laboratories, 1968.
38. Monnier, A. M. Experimental and theoretical data on excitable artificial lipidic membranes. *J. Gen. Physiol.* 51:265, 1968.
39. Mueller, P., and Rudin, D. O. Resting and action potentials in experimental bimolecular lipid membranes. *J. Theoret. Biol.* 18:222, 1968.
40. Mueller, P., and Rudin, D. O. Bimolecular Lipid Membranes: Techniques of Formation, Study of Electrical Properties and Induction of Ionic Gating Phenomena. In Passow, H., and Stämpfli, R. (Eds.), *Laboratory Techniques in Membrane Biophysics.* Berlin: Springer-Verlag, 1969, p. 141.
41. Nachmansohn, D. *Chemical and Molecular Basis of Nerve Activity.* New York: Academic, 1959.
42. Parsons, D. F. Ultrastructural and Molecular Aspects of Cell Membranes. *Canadian Cancer Conference.* Vol. 7. New York: Pergamon, 1966, p. 193.
43. Pethica, B. A. Structure and physical chemistry of membranes. *Protoplasma* 63:147, 1967.
44. Rothstein, A. Membrane phenomena. *Annu. Rev. Physiol.* 30:15, 1968.
45. Schlögl, R. Elektrodiffusion in freier Losung und geladenen Membranen. *Z. Physik. Chem.* 1:305, 1954.
46. Seeman, P. M. Membrane stabilization by drugs, tranquilizers, steroids and anesthetics. *Int. Rev. Neurobiol.* 9:145, 1966.
47. Seufer, W. D. Induced permeability changes in reconstituted cell membrane structure. *Nature (Lond.)* 207:174, 1964.
48. Shedlovsky, T. Electromotive force from proton transfer reactions. *Cold Spring Harbor Symp.* 17:97, 1952.
49. Tasaki, I. *Nerve Excitation: A Macromolecular Approach.* Springfield, Ill.: Thomas, 1968.
50. Teorell, T. Transport Processes and Electrical Phenomena in Ionic Membranes. In Butler, J. A. V., and Randall, J. T. (Eds.), *Progress in Biophysics and Biophysical Chemistry.* Vol. 3. London: Academic, 1953, p. 305.
51. Teorell, T. Oscillatory Phenomena in a Porous, Fixed Charge Membrane. In Passow, H., and Stämpfli, R. (Eds.), *Laboratory Techniques in Membrane Biophysics.* Berlin: Springer-Verlag, 1969, p. 130.
52. Torch, W. The Effect of Sodium, Calcium, and Aluminum Chlorides on the Rate of Formation, Stability, and Electrical Resistance of Black Lipid Membranes. Ph.D. thesis, University of Rochester, 1968.
53. Vanderheuvel, F. A. Structural role of water in lipoprotein systems. *Protoplasma* 63:188, 1967.

54. Villegas, J. Transport of electrolytes in the Schwann cell and location of sodium. *J. Gen. Physiol.* 51:615, 1968.
55. Villegas, R., Barnola, F. V., and Camejo, G. Ionic channels and nerve membrane lipids. *J. Gen. Physiol.* 55:548, 1970.
56. Watanabe, A., Tasaki, I., Singer, I., and Lerman, L. Effects of tetrodotoxin on excitability of squid giant axons in sodium-free media. *Science* 155:95, 1967.
57. Wyman, J. Regulation in macromolecules as illustrated by hemoglobin. *Q. Rev. Biophys.* 1:35, 1968.

Molecular Biology of Neural Receptors

R. Wayne Albers

THE PHYSIOLOGICAL CONCEPT of *receptors* encompasses the processes of recognition and response. A particular receptor is defined in terms of a set of stimuli that elicit qualitatively identical responses. The stimuli may be either a set of molecules or a range of energy. The observable response may represent a complex chain of events that culminates in muscle contraction, nerve impulses, or glandular secretion (Fig. 4-1).

At the molecular level, the study of receptors immediately encounters a specific difficulty: most analytical techniques require a degree of disruption of cellular organization. Such disruption destroys many or all of the physiological responses used to identify that response mechanism.

Two approaches have been employed to meet this problem. Most frequently, one attempts to identify receptors by some characteristic that is independent of the response mechanism, e.g., the optical absorption of visual pigments or the drug-binding characteristics of cholinergic receptors. The other possibility is to obtain receptor-deficient cells that may respond to the addition of exogenous receptors. This tactic has been successful in certain bacterial systems in which receptor-deficient mutants have been identified [22].

MOLECULAR RECOGNITION BY RECEPTORS

Pioneering pharmacologists such as Emil Fischer and Paul Ehrlich emphasized that systematic modifications in the structure of drugs can produce systematic changes in the toxic or therapeutic effects of those drugs. Studies of peripheral cholinergic receptors are particularly extensive ([7]; see also Chap. 7).

Figure 4-2 summarizes the results of two recent studies on the topography of acetylcholine receptors [11, 17]. *Nicotinic* receptors are predominant at skeletal neuromuscular junctions and autonomic ganglia. These receptors are subject to stimulation by nicotine and to blockade by curare. *Muscarinic* receptors predominate

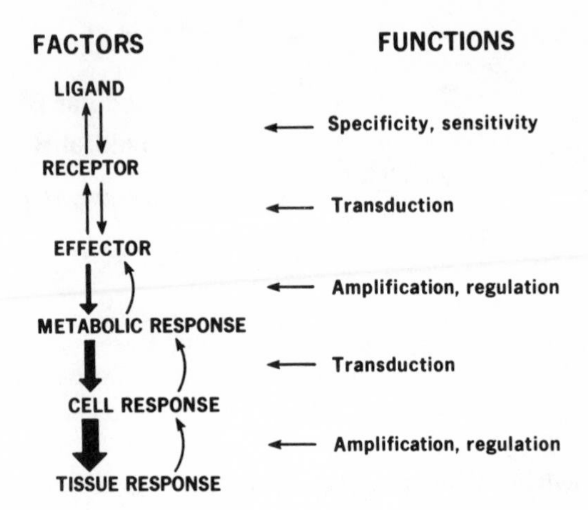

FIGURE 4-1. Receptor systems. Generalized representation of the complex chain of events between the interaction of an agonist (ligand) and the observed physiological response.

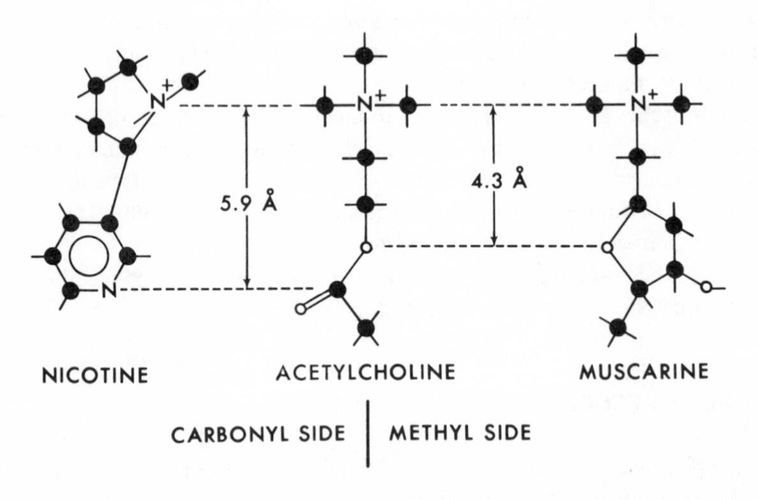

FIGURE 4-2. Recognition factors that differentiate muscarinic and nicotinic receptors.

at visceral autonomic target cells. They are subject to stimulation by muscarine and to blockade by atropine. Because acetylcholine is effective at both types of receptor, one must conclude either that different parts of the molecule are involved or that acetylcholine assumes different conformations as it interacts with different receptors. The latter explanation is unlikely because it appears that the conformations of the relevant parts of nicotinic and muscarinic molecules are similar.

Chothia [17] postulates that muscarinic activity is associated with an aspect of the acetylcholine molecule defined as the "methyl side" and nicotinic activity with the "carbonyl side" (Fig. 4-2). Beers and Reich [11] propose that in both cases the effector part of the molecule is the positively charged, hydrophobic nitrogen because, in sufficient concentration, the simple organic cation trimethylammonium stimulates both types of receptor. In their view, muscarinic receptors can form a hydrogen bond with the bridge oxygens of acetylcholine or muscarine at a distance of 4.3 Å from the positive nitrogen. Nicotinic receptors, in contrast, require a hydrogen bond acceptor at 5.9 Å from the positive nitrogen, i.e., the carbonyl oxygen of acetylcholine or the pyridine nitrogen of nicotine. It follows from either hypothesis that (1) certain parts of the molecule interact directly with the receptor to initiate a response, (2) certain parts may interact further to increase the stability of the drug-receptor complex, and (3) certain parts of the molecule act to support the interacting groups in an effective steric conformation.

The topography of adrenergic receptors is less well defined. Pharmacological criteria have led to the definition of two general classes of receptors [8] (Fig. 4-3). The α-receptors produce vasoconstriction, smooth muscle and myometrial contraction, pupil dilation, and excitation of exocrine glands. They are directly stimulated by phenylephrine and blocked by phenoxybenzamine. The β-receptors produce vasodilation, relax bronchial smooth muscle, and increase the rate and force of the heartbeat, among other effects. They are directly stimulated by isoproterenol and blocked by propanolol. The more potent activators acting at α- and β-receptors appear to have dissociation constants of the order of 10^{-7} to 10^{-8} M. Introduction into catecholamine of *N*-alkyl groups of increasing size generally increases β affinity and decreases α activity. The phenolic hydroxyls are essential for β activators and their removal or replacement may produce β antagonists [10].

PROCESS OF RECEPTOR ACTIVATION

Some generalizations about receptor mechanisms have been established. Drugs acting on receptors are classified as *agonists* and *antagonists.* Agonists activate receptors, while antagonists produce no observable effect by themselves, but reduce the effect of agonists (Fig. 4-4). Agonists are characterized in terms of "affinity" and "efficacy," which are approximately analogous to the K_s and V parameters of enzyme-substrate interactions, where K_s is the substrate concentration producing half-maximal reaction velocity and V is the maximum velocity of an enzyme reaction. Antagonists are characterized primarily in terms of their mode of

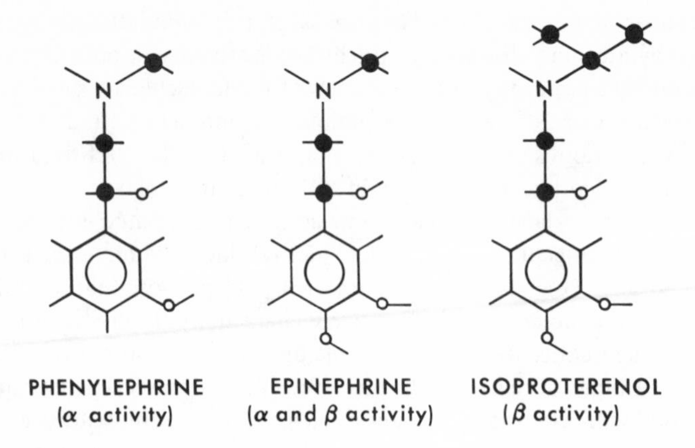

FIGURE 4-3. Adrenergic α and β agonists.

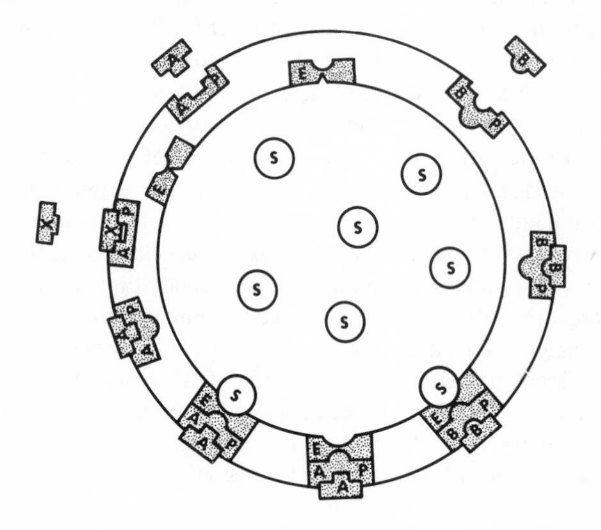

FIGURE 4-4. Hypothetical scheme to explain several aspects of receptor activation. The circle represents a plasma membrane. Two different receptor proteins, *AP* and *BP*, are specific for the agonists (drugs, hormones, neurotransmitters, etc.), *A* and *B* respectively. *X*, which combines with *AP* but does not activate it, is a competitive inhibitor of *A*. Either *AP* or *BP*, when activated, can combine with the effector protein *E*. This protein might be an enzyme such as adenyl cyclase; it might equally well be a regulator of membrane permeability (cholinergic systems), a factor in active transport, or in a contractile process. In some cases, *E* might be the final link in the process (a gate in an ion channel) or it might act on other factors, *S*, which might be either cytoplasmic (ATP in the cyclase reaction) or membrane-bound. Note that there may be unequal amounts of receptor and effector proteins, e.g., spare receptors. The range of specificity is determined by the number of specificity of the receptor proteins (binding proteins or permeases), while the range of effects is determined by the number and function of the effector proteins.

antagonism and, in the case of reversible antagonists, affinity for the receptor can be measured. In cases of irreversible antagonists, the rate of interaction with the receptor is the relevant parameter.

THE "SPARE-RECEPTOR" THEORY

Receptor activation cannot always be described in terms of simple occupancy of sites by agonist molecules. For example, a series of alkyl trimethylammonium homologs acting on the muscarinic receptors of intestinal muscle differ markedly in their effects. The lower homologs, C_1 to C_5, are primarily agonists, whereas higher homologs have antagonistic effects similar to those of atropine. Moreover, the maximum attainable contractile response differs for different members of this series, as does the concentration of drug necessary for half-maximal response. These and related observations led Stephenson [33] to propose that, in many cases, receptors are present in a large excess over those necessary to elicit a maximal response by combination with a potent agonist. With less effective agonists, more sites must be occupied to produce the same effect, whereas antagonists occupy sites but produce little or no effect.

THE RATE THEORY

Another common observation is that, for the same receptors, the onset and offset kinetics of antagonists are usually much slower than those of agonists. Paton has suggested [29] that the activation of receptors may occur transiently during the association or dissociation of the drug-receptor complex. If such is the case, drugs that associate and dissociate rapidly will have large positive effects (i.e., will be agonists); those that have slow kinetics will activate infrequently and will act chiefly as antagonists. Since affinity is a reflection of a ratio of rate constants, two drugs may have similar affinities and quite different absolute interaction rates.

The differences between the Stephenson and Paton theories are more subtle than might at first appear. Both state that different drugs which act at similar receptors may form complexes that differ in their effectiveness. Stephenson's proposal was originally interpreted in terms of structural differences in the drug-receptor interaction. In more current terms, conformational differences of the complexes were equated with efficacy. Paton's kinetic theory states that the important differences lie in the frequencies with which the complexes form and dissociate. Experimental tests of these conceptual differences have been inconclusive.

Receptor activation has been studied in the electroplax of the electric eel by a combination of microelectrode and pharmacological techniques that gives some insight into the molecular events ([24]; see also Chap. 7). Activation of this cholinergic receptor appears to take place when the drug-receptor complex can assume a conformation that is characterized by a 9 Å separation of the cationic binding site from an adjacent sulfhydryl-disulfide pair. Such pairs are normally in the hydrophobic oxidized states (in the disulfide form). Acetylcholine interacts more readily with the reduced receptor (in the sulfhydryl form). Other drugs act as antagonists by binding at the same sites and forming inactive conformations in which the cationic site is farther than 9 Å from the sulfhydryl-disulfide pair.

EFFECTOR MECHANISMS

REGULATION OF ION CHANNELS BY RECEPTORS

Conduction in nerve and muscle consists of a propagated sequence of regenerative changes in the ionic conductances of their plasma membranes. The various ions involved probably flow through physically distinct channels [30]. Acetylcholine and the amino acid transmitters act at specialized areas of these membranes by increasing transiently the conductance of one or more of these channels. In skeletal muscle, the sensitivity to acetylcholine is confined effectively to the neuromuscular junction, and ion channels are opened selectively to initiate impulses. In embryonic or denervated skeletal muscle, the entire length of the sarcolemma may be sensitive to acetylcholine [21]. The formation of a neuromuscular contact somehow affects the whole sarcolemma. Suppression of the acetylcholine-receptor interaction, the receptor-channel interaction, or the biosynthesis of receptors are all possible explanations.

SECOND-MESSENGER HYPOTHESIS

Hormones and neurotransmitters acting at plasma membrane receptors can regulate intracellular and even intranuclear reactions. One mechanism for mediating the response of membrane receptors is through regulation of the level of intracellular cyclic AMP [5] by linkage to adenyl cyclase, another component of the plasma membrane (Fig. 4-4). In the central nervous system, catecholamines, serotonin, and histamine are adenyl cyclase activators. The most general role of cyclic AMP appears to be as a factor in the activation of certain protein phosphokinases. These enzymes mediate the phosphorylation of other enzymes and regulatory proteins. Thus, the nature of the humoral response of such cells depends upon the presence of membrane receptors, their linkage to adenyl cyclase, the specificity of the phosphokinases, and the presence of the phosphoryl-accepting proteins. In all documented instances, a seryl residue is phosphorylated by a protein kinase that is dependent on cyclic AMP.

Although this phosphorylation pathway is the only one for which the second-messenger hypothesis is well founded, other parallel systems may exist. There is evidence that, in brain, cyclic GMP (guanosine monophosphate) has a different role, perhaps involving cholinergic activation [20]. There are numerous other ways in which proteins are modified subsequent to their synthesis. These include acetylation, methylation, amidation, hydroxylation, and glycosylation. Little is known of the regulation of these modifications or of the extent that second messengers may be involved. The neural regulation of muscle properties, in which protein methylation may be a factor, is discussed in Chapter 19.

In the instance of hormonal regulation, some cells respond to several hormones which act upon adenyl cyclase. For instance, adenyl cyclase in fat cells is sensitive to catecholamines, glucagon, ACTH, TSH, LH, and prolactin [12]. Each hormone acts on a specific receptor and each molecule of adenyl cyclase is influenced by multiple receptors.

These observations demonstrate clearly that the receptors and the effectors (cyclase) are physically distinct. It is difficult to imagine how a single enzyme could be linked simultaneously with five different receptors, so it may be that the association occurs only with activated receptors (Fig. 4-4). This proposal requires that the interacting membrane proteins have a degree of mobility within the lipid bilayer matrix (see Chap. 1).

SENSORY RECEPTOR SYSTEMS

PHOTORECEPTORS

The photoreceptor in vertebrate retinal rods is a protein (opsin) conjugated with the compound 11-*cis*-retinal ([2, 3] and Fig. 4-5). The conjugate (rhodopsin) has a broad optical absorption maximum at about 500 nm. Upon exposure to light, a series of transient chemical intermediates are formed, leading ultimately to a completely dissociated mixture of opsin and all *trans*-retinal [6]. If heat, rather than light, is used to effect the dissociation, i.e., by denaturing the opsin, the 11-*cis*-retinal is released without isomerization.

Many amphibians and fish contain a different class of photoreceptor pigments. These are conjugates of opsins with 3-dehydroretinal and characteristically have absorption maxima around 523 nm. Free 11-*cis*-retinal and 3-dehydroretinal have absorption maxima at 380 and 400 nm respectively. Evidently the retinal-opsin linkage confers additional electron delocalization. It has been proposed that the covalent part of this linkage may be a thiohemialdimine involving cysteine and lysine residues of the opsin [23]. This in itself cannot account for the absorption maxima of the different visual pigments because different opsins combined with the same kind of retinal have different maxima. For example, chicken opsin and retinal form iodopsin with a 562 nm maximum; chicken opsin plus 3-dehydroretinal form cyanopsin with a 620 nm maximum. In primates, color vision probably depends upon the existence of three different classes of cone cells that contain three types of opsin, all conjugated with retinal [6]. This shift of the pigment spectrum toward red, relative to that of free retinal or retinal oxime, must result from factors in the environment of the retinal binding state, such as "stacking" interactions with aromatic amino acid residues [23]. Opsins have an unusually high phenylalanine content [32].

The light activation of visual pigments also appears to involve mutual interaction of retinal and opsin to produce a specific conformation at the binding site. In the dark, the 11-cis isomer of retinal will combine spontaneously with opsin to regenerate rhodopsin. The primary action of light on rhodopsin is to isomerize the bound

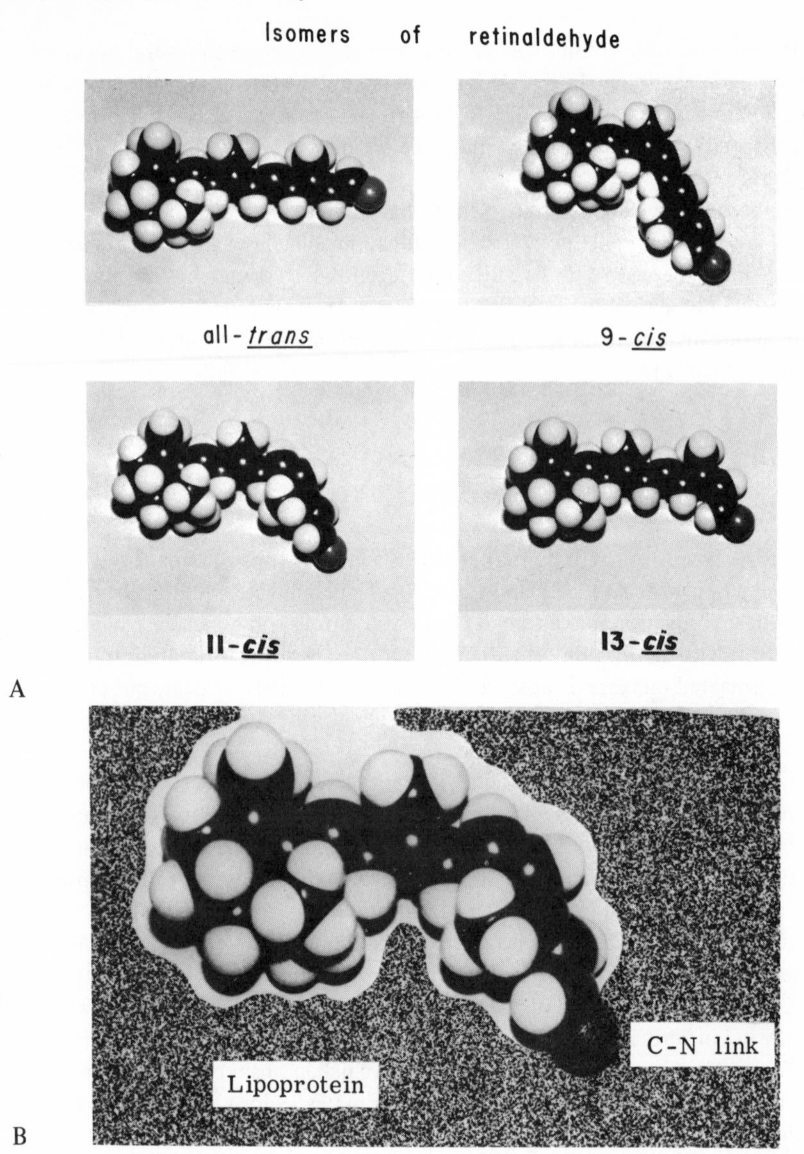

FIGURE 4-5. (A) Models of four retinaldehyde isomers. (B) Representation of the role of geometric configuration at the chromophoric site in rhodopsin. (From Bridges [2]; reproduced with permission.)

11-*cis*-retinal to the all trans isomer which dissociates from the protein. An intermediate complex, called prelumirhodopsin, can be observed at low temperatures. This intermediate has an absorption spectrum that is shifted even farther to the red end of the spectrum than is rhodopsin, indicating some increased electronic interaction between *trans*-retinal and protein. The subsequent intermediates are all associated with spectral changes to the shorter wavelengths, corresponding to progressive dissociation of the *trans*-retinal. These later changes are much too slow to be related directly to visual stimulation. Thus the transient formation of prelumirhodopsin, with its increased retinal-protein interaction, must reflect a change in the shape of the opsin-active site which is closely related to, if not in fact, the activated receptor. Many physical studies indicate large changes in the conformation of opsin as a result of bleaching, but these are not all temporally relevant to the initial activation process.

The regeneration of a photoreceptor requires isomerization of partially or completely dissociated retinal. Free *trans*-retinal is isomerized by visible light to all possible isomers; however, only the 11-cis isomer is somehow selected out for recombination with opsin. Although the 9-cis isomer can be shown to combine with opsins experimentally, it does not do so physiologically. Spectral evidence for the regeneration cycle can be obtained in both the living eye and the isolated retina, but many aspects of the cycle are still unknown [18].

COUPLING OF PHOTORECEPTOR ACTIVATION TO EXCITATION

Vertebrate photoreceptor pigments are organized within the intracellular membranes of the photoreceptor discs. From several hundred to a few thousand of these discs are stacked within the outer segments of the receptor cells. In cone cells of some species, continuity between the mature discs and the plasma membrane has been demonstrated. Such continuity in rod cells is in doubt. If continuity is a general feature, however, one might consider that visual excitation ensues from receptor regulation of ion permeability, as in synaptic excitation, because the internal fluid of the disc might be functionally extracellular. The possibility that no continuity exists makes it necessary to examine other hypotheses.

At high resolution, electron micrographs of the discs in frog show triple-layered membranes of 65 to 70 Å in thickness [28]. There are, however, strong indications of globular 50 Å subunits. Similar subunit organization is seen in plasma membranes of outer segments of rod. Other types of subunits have been reported for other species and by other electron microscopic techniques. In any case, optical studies indicate that rhodopsin exists in an ordered structure in these membranes: the axis of the chromophore is parallel to the plane of the membranes.

Illumination of the retina with a flash of high-intensity light produces a biphasic electrical response with essentially no latency (the early receptor potential). This response seems to arise as a result of the displacements of electrons within the oriented chromophores. Conceivably, a solid-state charge could be conducted tangentially through these membranes, but if so, some additional conduction path to the plasma membranes is required.

Another hypothesis that shares this difficulty is that lipid bilayers in the membrane may be involved in some type of phase change that could be transmitted through the membrane. Under some conditions, a Schiff's base of retinal with phosphatidylethanolamine can be extracted from rod outer segments. However, this Schiff's base does not appear to be the natural linkage of retinal in the dark-adapted retina [9].

Wald [34] proposed that the conformational change ensuing from the isomerization of retinal may unmask some intrinsic enzyme activity of the opsin molecule. Such activation could, in turn, activate other molecules, analogous to the cascade of factors in blood clotting, thereby achieving a large amplification of effect.

Recently it has been reported that outer segments of frog retinal rods contain adenyl cyclase activity that is inhibited more than sixfold by exposure to light [13]. Microelectrode studies have shown that vertebrate photoreceptors respond to light by a decreased conductance of Na^+. Therefore it is proposed that cyclic AMP is formed at the disc membranes and that regulation of plasma membrane conductance takes place by means of the diffusion of reaction products through the micron or so that exists between the disc and the plasma membrane. This implies that Na^+ conductance is regulated by some protein phosphokinase that is dependent on cyclic AMP.

TASTE AND OLFACTION

It is generally agreed that taste is based upon four distinct qualities: acid, salt, sweet, and bitter [1]. By contrast, investigators do not agree on theories of olfaction. Estimates of the number of "primary odors" have ranged from 4 to 44, with later theories tending toward the higher numbers. The molecular bases for distinguishing between salt and acid appear to be self-evident. Sweetness, however, is elicited by a variety of apparently unrelated compounds: sugars, cyclamates, saccharin, some amino acids and peptides, and lead acetate.

Most theories of taste and olfaction accept the formation of a stimulus-receptor complex. Amoore and his co-workers [4] have correlated the molecular shape of odorant molecules with their subjectively rated similarity in odor to certain primary odorants. It seems that certain aspects of quite different molecules may have similar effects, depending on the topography of a specific type of olfactory receptor.

CHEMORECEPTORS IN LOWER ORGANISMS

Motile bacteria will move toward concentrations of certain chemicals. The chemotactic response in *E. coli* has been defined in terms of eight or more distinct chemoreceptors [22]. A "galactose-receptor" mutant can be isolated that does not respond to galactose but does respond normally to ribose and amino acids. The component of the galactose receptor involved in recognition is distinct from the transport system because some mutants that are defective in galactose transport have normal chemotactic responses. Evidently, the various chemoreceptors share a common pathway that links them to the flagellar response because some mutants that do not respond to chemical stimuli in general are otherwise fully motile.

The galactose-binding protein is thought to be a "pericytoplasmic" protein, i.e., it is present between the cell wall and the plasma membrane proper. When this protein is purified, it undergoes a conformational change in vitro when it binds galactose [14]. This change may reflect the mechanism by which the presence of galactose affects the membrane. In turn, the plasma membrane is attached to the base of the flagellum and therefore may transmit the effect of chemotactic agents from receptor to flagellum, thereby producing a change in activity.

More complex responses occur in simple multicellular organisms such as hydra [26]. In this organism, reduced glutathione can elicit a highly specific feeding response. Along with more complex responses, many higher organisms have developed highly selective recognition systems, particularly with respect to regulation of sexual activity. For example, the antennae of male silkworm moths have an extensive olfactory system that is selective for hexadeca-10-*trans*,12-*cis*-dien-1-ol, which is the female sexual pheromone [31].

ISOLATION AND PURIFICATION OF RECEPTOR PROTEINS

Many questions about the nature of receptors remain unanswered because of the absence of pure receptor material. Early attempts to isolate cholinergic receptors are discussed in Chapter 7. Some recent reports indicate that their successful isolation may be possible, although there are still discrepancies among laboratories [15, 16, 19, 27]. In general it seems that lipoproteins of less than 100,000 daltons can be isolated and still retain the characteristic drug-binding properties of the cholinergic receptor.

A particulate fraction from heart has been characterized as containing β-adrenergic receptors by similar criteria [25]. Other purified receptor proteins include the visual pigment proteins (opsins) and the bacterial galactoside-binding proteins [22].

REFERENCES

BOOKS AND REVIEWS

1. Beidler, L. M., and Reichart, W. E. Sensory transduction. *Neurosci. Res. Prog. Bull.* 8:459, 1970.
2. Graymore, C. N. (Ed.). *Biochemistry of the Eye.* New York: Academic, 1970.
3. Morton, R. A., and Pitt, G. A. J. Aspects of visual pigment research. *Adv. Enzymol.* 32:97, 1969.
4. Pfaffmann, C. (Ed.). Olfaction and Taste. *Proceedings of the Third International Symposium.* New York: Rockefeller University Press, 1969.
5. Rall, T. W., and Gilman, A. G. The role of cyclic AMP in the nervous system. *Neurosci. Res. Prog. Bull.* 8:221, 1970.

6. Wald, G. Molecular basis of visual excitation. *Science* 162:230, 1968.
7. Waud, D. R. Pharmacological receptors. *Pharmacol. Rev.* 20:49, 1968.

ORIGINAL SOURCES AND SPECIAL TOPICS

8. Ahlquist, R. P. Agents which block adrenergic β-receptors. *Annu. Rev. Pharmacol.* 8:259, 1968.
9. Anderson, R. E., Hoffman, R. T., and Hall, M. O. Linkage of retinal to opsin. *Nature New Biol.* 229:249, 1971.
10. Ariens, E. J. The structure-activity relationships of beta adrenergic drugs and beta adrenergic blocking drugs. *Ann. N.Y. Acad. Sci.* 139:606, 1967.
11. Beers, W. H., and Reich, E. Structure and activity of acetylcholine. *Nature (Lond.)* 228:917, 1970.
12. Birnbaumer, L., and Rodbell, M. Adenyl cyclase in fat cells. II. Hormone receptors. *J. Biol. Chem.* 244:3477, 1969.
13. Bitensky, M. W., Gorman, R. E., and Miller, W. H. Adenyl cyclase as a link between photon capture and changes in membrane permeability of Feg photoreceptors. *Proc. Natl. Acad. Sci. U.S.A.* 68:561, 1971.
14. Boos, W., Gordon, A., and Hall, R. E. Substrate-induced conformational change of the galactose-binding protein of *E. coli.* *Fed. Proc.* 30:1062, 1971.
15. Changeux, J.-P., Kasai, M., Huchet, M., and Meunier, J. C. Extraction a partir du tissu electrique de gymnote d'une protein presentant plusieurs proprietes characteristiques du recepteur physiologique de l-acetylcholine. *C. R. Acad. Sci. [D] (Paris)* 270:2864, 1970.
16. Changeux, J.-P., Kasai, M., and Lee, C. H. Use of a snake venom toxin to characterize the cholinergic receptor protein. *Proc. Natl. Acad. Sci. U.S.A.* 67:1241, 1970.
17. Chothia, C. Interaction of acetylcholine with different cholinergic nerve receptors. *Nature (Lond.)* 225:36, 1970.
18. Cone, R. A., and Brown, P. K. Spontaneous regeneration of rhodopsin in the isolated rat retina. *Nature (Lond.)* 221:818, 1969.
19. De Robertis, E. Molecular biology of synaptic receptors. *Science* 171:963, 1971.
20. Ferrendelli, J. A., Steiner, A. L., McDougal, D. B., Jr., and Kipnis, D. M. The effect of oxotremorine and atropine on c-GMP and c-AMP levels in mouse cerebral cortex and cerebellum. *Biochem. Biophys. Res. Commun.* 41:1061, 1970.
21. Guth, L. Trophic influences of nerve on muscle. *Physiol. Rev.* 48:645, 1968.
22. Hazelbauer, G. L., and Adler, J. Role of the galactose binding protein in chemotaxis of *E. coli* toward galactose. *Nature New Biol.* 230:101, 1971.
23. Heller, J. Structure of visual pigments. II. Binding of retinal and conformational changes on light exposure in bovine visual pigment. *Biochemistry* 7:2914, 1968.

24. Karlin, A. Chemical modification of the active site of the acetyl choline receptor. *J. Gen. Physiol.* 54:245, 1969.
25. Lefkowitz, R. J., and Haber, E. A fraction of the ventricular myocardium that has the specificity of the cardiac beta-adrenergic receptor. *Proc. Natl. Acad. Sci. U.S.A.* 68:1773, 1971.
26. Lenhof, H. M. Behavior, hormones and hydra. *Science* 161:434, 1968.
27. Miledi, R., Malinoff, P., and Potter, L. J. Isolation of the cholinergic receptor protein of *Torpedo* electric tissue. *Nature (Lond.)* 229:554, 1971.
28. Nilsson, S. E. G. Receptor cell outer segment development and ultrastructure of the disk membranes in the retina of the tadpole *(Rana pipiens). J. Ultrastruct. Res.* 11:581, 1964.
29. Paton, W. D. M. A theory of drug action based on the rate of drug receptor combination. *Proc. R. Soc., Ser. B. (Lond.)* 154:21, 1961.
30. Rojas, E., and Armstrong, C. Sodium conductance activation without inactivation in pronase-perfused axons. *Nature (Lond.)* 229:177, 1971.
31. Schneider, D. Insect olfaction: Deciphering system for chemical messages. *Science* 163:1031, 1969.
32. Shichi, H., and Lewis, M. S. Biochemistry of visual pigments. I. Purification and properties of bovine rhodopsin. *J. Biol. Chem.* 244:529, 1969.
33. Stephenson, R. P. A modification of receptor theory. *Bri. J. Pharmacol.* 11:379, 1956.
34. Wald, G. Visual excitation and blood clotting. *Science* 150:1028, 1965.

SECTION II

Biochemistry of Synaptic Function

5

An Overview of Transmission

Eugene Roberts and
Richard Hammerschlag

ALL NORMAL OR ADAPTIVE ACTIVITY in nervous systems (information processing) is a result of the coordination of a dynamic interplay of excitation and inhibition within and between neuronal subsystems in a particular organism. Most communication that takes place between receptor and neuron, neuron and neuron, and neuron and effector probably occurs via the presynaptic liberation of substances that have either excitatory or inhibitory effects on specialized regions of membranes (presynaptic axonal endings, dendrites, or soma) of neurons or membranes of muscle or gland cells. Although completely conclusive evidence of such mediator action by a particular substance at a particular junction or synapse is lacking in all but a few instances, the existence of such mechanisms is not questioned by most neuroscientists because, in general, postsynaptic membranes are not electrically excitable.

Chemical transmission in the nervous system is a well-documented possibility: morphological studies of neuronal junctions at an electron microscopic level have revealed presynaptic vesicles and a characteristic synaptic cleft; physiological studies have detected a synaptic time-delay that is characteristic of diffusion and postsynaptic potential changes in discrete quantal steps; and pharmacological studies have demonstrated that at least a dozen chemicals normally found in nervous tissue can excite or inhibit neurons. A key factor that remains is the actual identification of the various chemical transmitter substances in relation to their specific neuronal sites.

From physiological studies, it has been deduced that contact by substances that have excitatory effects with the outer surfaces of neuronal membranes produces a physical change in the state of the membranes that is accompanied by a general increase in permeability to small cations. Experimental evidence and theories suggest that the excitable membrane may be thought of as a cation exchanger, whose affinity for divalent ions (calcium) is decreased by the excitatory transmitter. Displacement of Ca^{2+} causes a conformational change in the membrane which permits Na^{+} to enter and K^{+} to leave the cell. The excitatory effects usually are attributed to a depolarization of the membrane produced by inward movement of Na^+. Inhibitory substances produce an increase in K^+ or Cl^- conductances of the membranes upon which they impinge, or they might activate an electrogenic Na^+ ion pump. In these ways they accelerate the rate of return to the resting potential of all depolarized membrane segments which they would contact, and they would stabilize (i.e., decrease sensitivity to stimulation) undepolarized membrane segments.

In many neural systems that have been studied, it has been demonstrated that, in addition to the so-called mainline neurons, there exist many types of interneurons which have their axonal endings on the presynaptic (axonal) or postsynaptic (dendritic or somatic) membranes of other neurons. (The reader is referred to an excellent book [5] and a symposium [3] on the subject of interneurons.) In such systems the interneurons may liberate substances or transmitters that have inhibitory or excitatory effects on neuronal membranes. The possibility also should be left open that, immediately after depolarization and permeabilization of a postsynaptic site, substances that could act as feedback inhibitors on the presynaptic membrane of a synaptic contact might be liberated from the postsynaptic neuronal site into the extraneuronal synaptic environment. If the latter type of inhibition should exist, it could play an important role in neural function because such a mechanism could exist at all synapses regardless of the details of the neuronal circuits in which the synapses are found. During neural activity, glia, which invest the synaptic regions, could liberate membrane-active substances into the synaptic area or remove them. Increased metabolism consequent to synaptic activity may result in local increases in pCO_2 and decreases in pH and pO_2, which in turn could cause a dilation and permeabilization of capillaries in the synaptic region and enhanced rates of exchanges of substances between the extracellular compartments and the blood, with resultant modification of synaptic activity.

For illustrative purposes, some potential neuronal relations are schematized in a highly oversimplified fashion in Figure 5-1. Synaptic relations and the time sequence of their action are shown for an excitatory presynaptic ending *(A)*, which can excite major excitatory neuron *(B)* and an inhibitory interneuron *(C)*. Neuron *B* in turn can excite another excitatory neuron *(E)* by liberation of transmitter from its presynaptic ending *(D)* and also an inhibitory interneuron *(F)*. Finally, as a direct or indirect result of the activity of the neural circuit of which neurons *A, B,* and *E* are a part, inhibitory interneuron *H* is activated by transmitter released from the presynaptic ending of excitatory neuron *G*. Assuming that each neuron liberates only one effective transmitter, either inhibitory or excitatory, the above scheme shows three presynaptic sites of release of three possibly different excitatory substances onto dendritic endings, and nine possible sites of presynaptic release of three possibly different inhibitory substances on presynaptic axonal sites and on postsynaptic dendritic and somatic membranes.

The above scheme is a gross underestimation of the complexity of the situation that must exist in reality, for there can be hundreds or even thousands of excitatory and inhibitory synapses on the dendrites and somata of the excitatory and inhibitory neurons, such as those shown in Figure 5-1, and there must be a multiplicity

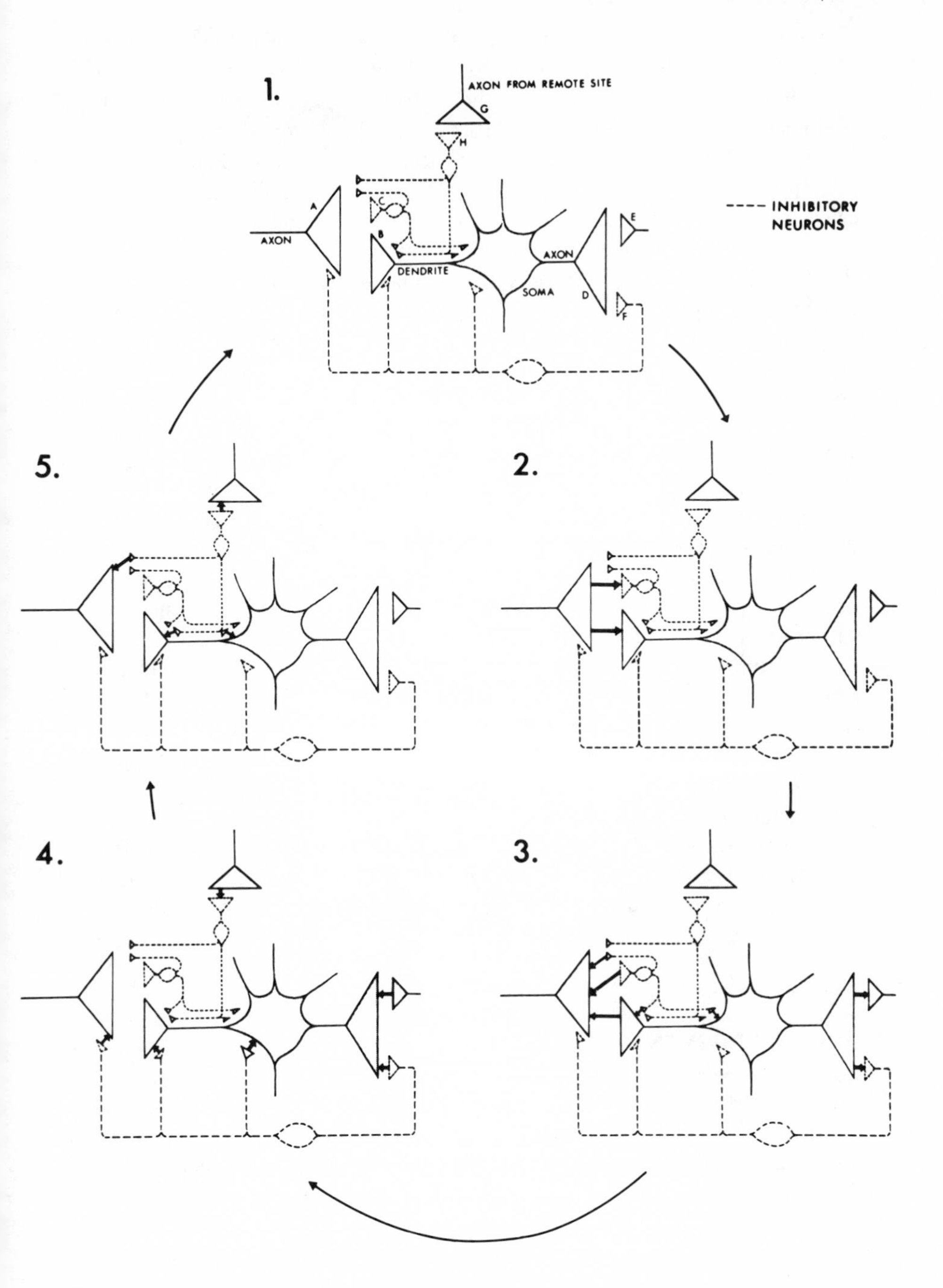

FIGURE 5-1. Schematic presentation of sequential liberation of excitatory and inhibitory transmitters acting in a simple model neuron environment (see text).

of transmitters of various presynaptic origins acting on a given cell at any time. When one deals with whole neuronal systems in intact organisms, the situation becomes even more complex because the overall effect of inhibition at the synaptic level may be either activation or inhibition. Thus, inhibition of inhibitory neurons may lead to disinhibition or excitation of the system as a whole, inhibition of excitatory neurons may lead to inhibition of the system, inhibition of excitatory neurons that act on inhibitory neurons may lead to disinhibition or excitation, and so on. It is therefore understandable why so little definitive information is available about the nature of the transmitters that are operative at particular synapses in vertebrate nervous systems and why many neuroscientists have chosen to search for simpler, invertebrate preparations in which to study basic synaptic properties.

There are data, ranging from suggestive to convincing, that implicate several known, naturally occurring amino acids as potential excitatory or inhibitory transmitters. γ-Aminobutyric acid (GABA) and glycine may be inhibitory transmitters and glutamic and aspartic acids may be excitatory transmitters. The physiological and pharmacological properties of these substances have been studied in a variety of biological systems. Knowledge of the properties and distributions of the enzymes that form and degrade some of these substances and of the neural circuits in which they exist is becoming available. The transport mechanisms which have been identified in neuronal membranes, and which in most instances may be the chief mechanisms for removal of the active substances from synapses, also are becoming known.

The ultimate goals in the study of transmitters are: (1) to be able to identify the neurons that may employ the substances as transmitters, (2) to localize at an ultrastructural level those synaptic junctions at which these substances are released and act, (3) to elucidate the modes of action of the transmitters on neural membranes at a molecular level, and (4) to define the roles of such neurons in the neural systems of functioning organisms.

In Chapter 8 we will consider data about the aforementioned amino acids as potential transmitters which may be pertinent to the attainment of these goals. The reader may consult a number of recent reviews for more thorough coverage of various aspects of the literature [1, 2, 4, 6, 7].

ACKNOWLEDGMENT

This investigation was supported in part by Grant NB-01615 from the National Institute of Neurological Diseases and Stroke, National Institutes of Health, and was also made possible in part by a supporting fund established in the name of Robert F. Kennedy.

REFERENCES

1. Aprison, M. H., Davidoff, R. A., and Werman, R. Glycine: Its Metabolic and Possible Transmitter Roles in Nervous Tissue. In Lajtha, A. (Ed.), *Handbook of Neurochemistry.* Vol. 3. New York: Plenum, 1970.
2. Baxter, C. F. The Nature of γ-Aminobutyric Acid. In Lajtha, A. (Ed.), *Handbook of Neurochemistry.* Vol. 3. New York: Plenum, 1970.
3. Brazier, M. A. B. (Ed.). *The Interneuron.* Berkeley and Los Angeles: University of California Press, 1969.
4. Curtis, D. R., and Johnston, G. A. R. Amino Acid Transmitters. In Lajtha, A. (Ed.), *Handbook of Neurochemistry.* Vol. 4. New York: Plenum, 1970.
5. Horridge, G. A. *Interneurons.* London and San Francisco: W. H. Freeman, 1968.
6. Kravitz, E. A. Acetylcholine, γ-Aminobutyric Acid, and Glutamic Acid: Physiological and Chemical Studies Related to Their Roles as Neurotransmitter Agents. In Quarton, G. C., Melnechuk, T., and Schmitt, F. O. (Eds.), *The Neurosciences.* New York: Rockefeller University Press, 1967.
7. Roberts, E., and Kuriyama, K. Biochemical-physiological correlations in studies of the γ-aminobutyric acid system. *Brain Res.* 8:1, 1968.

Catecholamines and Serotonin

Solomon H. Snyder

THE CATECHOLAMINES, norepinephrine and dopamine, as well as the indoleamine, serotonin, are putative neurotransmitters in certain neuronal tracts in the brain. They occupy a uniquely important place in neurobiology because they are the only putative neurotransmitters whose localization in particular brain tracts has been established and whose relationship to specific animal or human behaviors has, at least in part, been worked out. Hence, a discussion of the neurochemistry of these compounds can integrate findings derived from electron microscopy, histochemistry, neurophysiology, enzymology, pharmacology, psychology, and even clinical psychiatry.

A great deal of information about catecholamines and serotonin had accumulated prior to knowledge of which neuronal tracts contained them or even whether they were localized in neurons at all. Nonetheless, for clarity's sake, it might be best to begin by presenting the histochemical observations that have delineated the neuronal systems which contain these compounds in the brain.

HISTOCHEMICAL MAPPING OF MONOAMINE NEURONS IN THE BRAIN

The histochemical fluorescence method for the identification of catecholamines and serotonin, developed by a group of Swedish workers [2], incorporates the condensation of these amines in tissue sections with formaldehyde in a humid environment to form intensely fluorescent isoquinolines. In this way, cell bodies, axons, and nerve terminals of the monoamine-containing neurons have been mapped throughout the brain. Then, by making selective brain lesions and following the anterograde loss or retrograde build-up of amine, the course of the tracts has been clarified [10].

In the histochemical fluorescence method, serotonin can be distinguished from catecholamines by the wavelength of fluorescence, which varies in such a way that serotonin appears bright yellow and the catecholamines appear bright green.

Although both dopamine and norepinephrine fluoresce green, they can be differentiated by their response to drugs.

What are the pathways of the monoamine neuronal systems? The cell bodies of all of them are in one or another location within the brain stem. Axons ascend or descend throughout the brain and into the spinal cord. For each of the amines there are several separate and distinct tracts (Fig. 6-1).

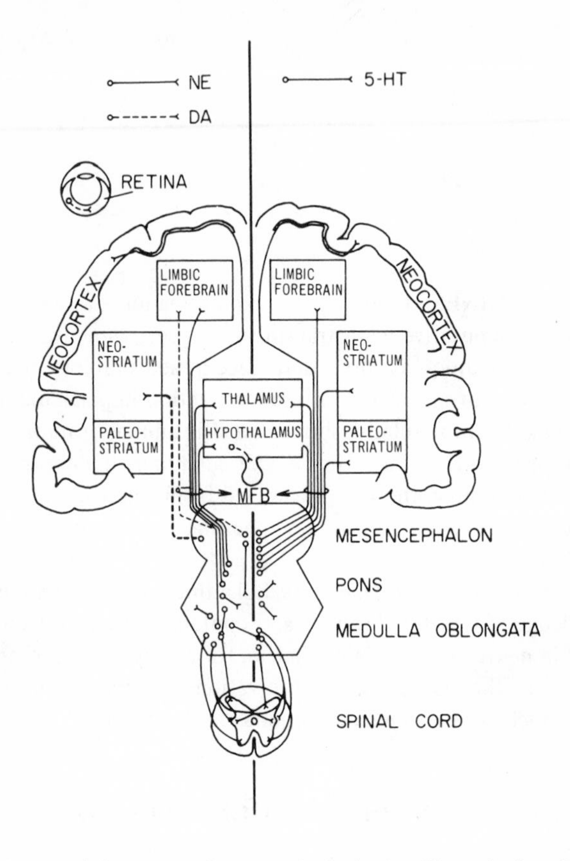

FIGURE 6-1. Pathways of the monoamine tracts in the brain. (From Anden et al. [10].)

NOREPINEPHRINE

The cell bodies of the norepinephrinergic neurons occur in ten or more clusters in the medulla oblongata, pons, and midbrain, mostly scattered throughout the reticular formation. In the locus coeruleus, however, norepinephrinergic cell bodies are so densely packed that they may be the only ones in that nucleus. Axons from some of the norepinephrinergic cells in the medulla oblongata descend in the lateral sympathetic columns of the spinal cord and terminate at various levels. Others ascend primarily in the medial forebrain bundle, giving off terminals in all areas of the brain; the highest density is in the hypothalamus, a lesser density is in the limbic system, and the lowest frequency is in the cerebral cortex and cerebellum [2, 10].

SEROTONIN

The cell bodies of the serotonin-containing neurons are localized in a series of nuclei in the lower midbrain and upper pons that are called the *raphe* nuclei. These are a group of old structures, phylogenetically, whose function, until the discovery that they contained serotonin, was obscure. As with norepinephrine and the locus coeruleus, serotonin is so highly concentrated in the raphe nuclei that probably all of its cells are serotoninergic. Serotoninergic axons, like those containing norepinephrine, ascend primarily in the medial forebrain bundle and also give off terminals in all brain regions, with the major proportion in the hypothalamus and the least in the cerebellum and cerebral cortex.

DOPAMINE

Dopaminergic neuronal tracts are more circumscribed than those of norepinephrine or serotonin. The largest dopaminergic tract originates in the zona compacta of the substantia nigra, with terminals in the caudate nucleus and putamen of the corpus striatum. In Parkinson's disease, this nigrostriatal tract degenerates, with an attendant depletion of brain dopamine. That the dopamine deficiency accounts for the symptoms of this disease is attested to by the therapeutic efficacy of L-DOPA, the amino acid precursor of dopamine, in treating Parkinson's disease (see also Chap. 28).

Another dopamine tract arises from cell bodies in the midbrain near the interpeduncular nucleus and terminates principally in the nucleus accumbens of the septal region. Yet a third dopamine system begins in the arcuate and anterior periventricular nuclei of the hypothalamus and passes downward to terminate in the primary capillary plexus that supplies the pituitary gland. It is via this capillary plexus that hypothalamic releasing factors reach the pituitary gland to regulate discharge of the anterior pituitary tropic hormones (see Chap. 17). Dopamine can stimulate the release of some of these hormones [16]. The retina also contains isolated dopamine fibers, the function of which is obscure.

CATECHOLAMINES

BIOSYNTHESIS

While histochemical studies have delineated the tracts containing biogenic amines, investigation of their metabolism and turnover and the effects of drugs has contributed greatly to an understanding of their function.

Tyrosine is the dietary amino acid precursor of dopamine and norepinephrine (Fig. 6-2). The rate-limiting step in the biosynthesis of both these catecholamines

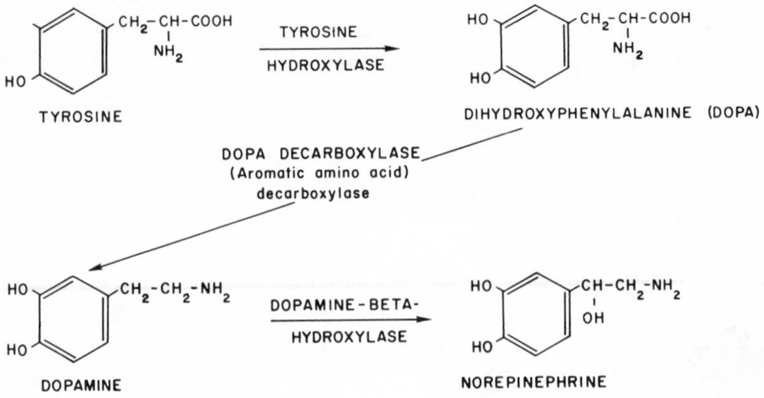

TYROSINE → TYROSINE HYDROXYLASE → DIHYDROXYPHENYLALANINE (DOPA)

DOPA DECARBOXYLASE (Aromatic amino acid) decarboxylase

DOPAMINE → DOPAMINE-BETA-HYDROXYLASE → NOREPINEPHRINE

FIGURE 6-2. Pathways of catecholamine synthesis. (From Snyder [6].)

is the hydroxylation of tyrosine to form DOPA by the enzyme tyrosine hydroxylase. The activity of this enzyme is relatively feeble, so it can be detected only radioisotopically. It requires a biopterin as a cofactor; biopterin is a pteridine compound that also enhances the activity of several other hydroxylating enzymes [15].

Researchers have been eager to define the localization of tyrosine hydroxylase within neurons. For instance, if this enzyme, whose activity determines the rate of catecholamine synthesis, were primarily confined to neuronal cell bodies, it would support the concept that catecholamines are primarily synthesized there and then transported down the axon to the nerve terminal. If the enzyme were mostly in the synaptic vesicles, one would conclude that those vesicles are factories for synthesizing, as well as for storing and releasing, the amines. Although these questions are not altogether resolved, it appears likely that, while some tyrosine hydroxylase can be located in cell bodies, the majority is confined to catecholamine nerve terminals. Within the nerve terminals, the enzyme is largely cytoplasmic, so that at least the initial step of catecholamine synthesis probably does not take place within the synaptic vesicles.

Several tyrosine analogs, such as α-methyltyrosine, can inhibit tyrosine hydroxylase. These drugs deplete dopamine and norepinephrine in the brain. When the synthesis of a neurotransmitter is inhibited, its levels can be expected to fall at a rate determined by the rate at which the neurons fire and by the subsequent catabolism of the transmitter molecule. Thus, the rate of decline of brain catecholamine in animals that have been treated with α-methyltyrosine provides a reflection of the turnover rate of catecholamines [5a]. Catecholamine turnover measured in this way corresponds well with measurements of the rate of catecholamine synthesis from tyrosine, and the two types of measures are affected similarly by drugs or altered physiological conditions.

BIOSYNTHESIS AND NEURONAL ACTIVITY

Tyrosine hydroxylase activity also can be inhibited by the catecholamines, suggesting that normally they may exert a type of "feedback" inhibition [5a, 15]. This notion would imply that catecholamine release produced by nerve stimulation or drugs would induce enhanced catecholamine synthesis (to replace that which was discharged) by stimulating the hydroxylation of tyrosine without actually increasing levels of the tyrosine hydroxylase itself. In support of this are findings that stimulation of peripheral sympathetic nerves (which seems to regulate norepinephrine metabolism as it does in the brain) produces a rapid acceleration of norepinephrine synthesis from tyrosine [9]. In such a situation, the formation of norepinephrine from administered DOPA is not enhanced, showing that hydroxylation of tyrosine was the "stimulated" step in the pathway. Cold stress, as well as a variety of drugs, increase tyrosine hydroxylation in the intact brain without increasing the activity of this enzyme when subsequently the tissues are assayed in vitro. If catecholaminergic nerve activity is enhanced for several days, the levels of tyrosine hydroxylase become elevated. Thus, after brief stresses, tyrosine hydroxylation can be stimulated by apparent release from feedback inhibition without an increase in enzyme levels; prolonged stimuli provoke the synthesis of new tyrosine hydroxylase molecules.

Variations in the activity of norepinephrine synapses alter the synthesis of this amine in a predictable way. For instance, drugs that block adrenergic receptors accelerate the formation of norepinephrine from tyrosine. Presumably a message from the postsynaptic neuron, saying "I am not getting enough norepinephrine; send me more," is transmitted via neuronal feedback to the presynaptic neurons, which then speed both their firing rates and the synthesis of norepinephrine. Drugs such as haloperidol, which block dopamine receptors, and those such as apomorphine, which stimulate the same receptors, respectively enhance and slow the synthesis of dopamine (see Chap. 29, under chlorpromazine).

FORMATION OF DOPAMINE AND NOREPINEPHRINE

DOPA is decarboxylated to dopamine by an enzyme, DOPA decarboxylase. The same enzyme can decarboxylate 5-hydroxytryptophan, the amino acid precursor of serotonin, as well as several other aromatic amino acids. Accordingly, it has also been called "aromatic amino acid decarboxylase." Like tyrosine hydroxylase, DOPA decarboxylase seems to be contained in the soluble portion of catecholamine nerve terminals in the brain. The activity of DOPA decarboxylase is not rate limiting in the synthesis of the catecholamines, and hence is not a regulating factor in their formation. Normally, there seems to be a great excess of DOPA decarboxylase present in the brain, so that drugs that inhibit its activity by as much as 95 percent fail to lower brain levels of the catecholamines.

The hydroxylation of dopamine at the beta carbon to form norepinephrine is mediated by the enzyme dopamine β-hydroxylase. This enzyme requires copper for its optimal functioning. The enzyme is difficult to assay in crude tissue preparations because of the presence of unidentified inhibitory substances. In the adrenal medulla, where it participates in the synthesis of epinephrine, dopamine β-hydroxylase is localized, at least in part, in the wall of the epinephrine-storing granules. It is not released by the adrenal gland at times that epinephrine is released, and this has been taken as evidence that the secretion of epinephrine involved exocytosis, a process by which the granule or storage vesicle fuses with the cell membrane, discharges its contents, and returns to the cell interior. One would expect that only soluble constituents of the vesicles are released during exocytosis, and that chemicals in the vesicle wall would stay inside. Although there is no direct evidence, it is reasonable to suppose that the process of catecholamine release in the brain might be similar to or the same as that in the adrenal gland [5b].

As might be expected, dopamine β-hydroxylase is lacking in those parts of the brain, such as the caudate nucleus, in which dopamine is the predominant catecholamine and which have only negligible concentrations of norepinephrine.

CATECHOLAMINE CATABOLISM

Two enzymes are primarily responsible for the degradation of catecholamines. One of these is monoamine oxidase (MAO) (see also Chap. 28), which oxidatively deaminates dopamine or norepinephrine to the corresponding aldehydes (Fig. 6-3).

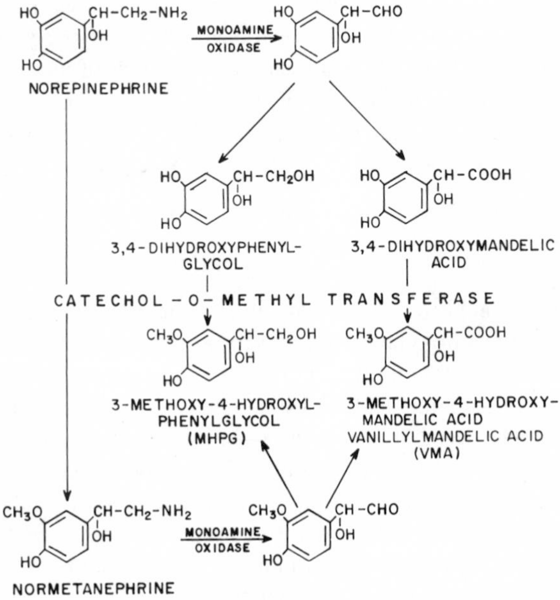

FIGURE 6-3. Pathways of catecholamine degradation.

These, in turn, can be converted by aldehyde dehydrogenase to analogous acids. The aldehydes may also be reduced to form alcohols. Dietary factors, such as the ingestion of ethanol, can determine the relative amounts of catechol acids or alcohols formed from the catecholamines because ethanol also competes for aldehyde dehydrogenase.

Catecholamines can be methylated by the enzyme catechol-*O*-methyl transferase (COMT), which transfers the methyl group of *S*-adenosylmethionine to the meta (3 position) hydroxyl of the catecholamines [1]. To a very limited extent, COMT can also methylate the para hydroxyl grouping. COMT will act on any catechol compound, including the aldehydes and acids formed from the action of MAO on the catecholamines. When norepinephrine is methylated by this enzyme, the product is called normetanephrine. There is no corresponding name for the methylated derivative of dopamine, which is simply referred to as 3-*O*-methyldopamine.

Like COMT, MAO is relatively nonspecific and will act on any monoamine, including normetanephrine and 3-*O*-methyldopamine, converting these first into their respective aldehydes and then into acids or alcohols. The formation of the acid is probably mediated by aldehyde oxidase and the alcohol by alcohol dehydrogenase. Thus an *O*-methylated alcohol or acid will result from the combined actions of MAO and COMT. Measurement of the levels in tissues or body fluids of *O*-methylated acids or alcohols, of course, will give no indication of which enzyme acted first. To find this out, one must measure the *O*-methylated amines or the catechol acids. Determining if released catecholamines were first acted on by MAO or by COMT conveys useful information about neuronal function. In the peripheral sympathetic nervous system, the extent of oxidation of the aldehyde product exceeds the reductive pathway so that the *O*-methylated acid product of norepinephrine degradation, called vanillylmandelic acid (VMA), is the final breakdown product of norepinephrine. Quantitatively, VMA is the major metabolite of norepinephrine in the periphery and is readily detectable in the urine, so its levels are measured in clinical laboratories as an index of sympathetic nervous function and to diagnose tumors that produce norepinephrine or epinephrine, such as pheochromocytomas and neuroblastomas.

In the brain, reduction of the aldehyde formed from the action of MAO on norepinephrine or normetanephrine predominates, so that the major norepinephrine metabolite in the brain is an alcohol derivative called 3-methoxy-4-hydroxylphenylglycol (MHPG). The MHPG formed in the brain is conjugated to sulfate. Because MHPG can diffuse from brain to the general circulation, estimates of its levels in the urine might be thought to reflect directly the activity of norepinephrinergic neurons in the brain. However, although MHPG is proportionately only a minor metabolite of norepinephrine in the peripheral sympathetic nervous system, quantitatively most of it in the urine still derives from the periphery and to only a minor extent (about 20 percent) from the brain.

Whether the aldehyde formed from dopamine or 3-*O*-methyl dopamine is primarily oxidized or reduced is not altogether certain. Most evidence suggests that it is largely oxidized to an acid, 4-hydroxy-3-methoxy-phenylacetic acid, more commonly known as homovanillic acid (HVA). HVA levels are often measured in the brain or cerebrospinal fluid, and those levels are taken as reflections of the activity of dopaminergic neurons, especially of those with endings in the caudate nucleus and putamen. HVA levels are markedly lower in the spinal fluid of parkinsonian patients (see also Chap. 28).

Epinephrine differs chemically from norepinephrine only in the addition of a methyl group to the amine nitrogen. It is the predominant catecholamine in the adrenal glands of most species. Only extremely small amounts of epinephrine are found in sympathetic nerves or in the brain. The synthesis of epinephrine occurs by the same steps as does norepinephrine, with the addition of an *N*-methylating enzyme which transforms norepinephrine to epinephrine. That enzyme is designated phenylethanolamine-*N*-methyl transferase (PNMT). PNMT is almost wholly confined to the adrenal gland, although small amounts are in the brain. The activity of this enzyme and, accordingly, the synthesis of epinephrine, is regulated by hormones of the adrenal cortex. Thus, removal of the pituitary gland results in a profound lowering of adrenal PNMT, which can be restored to normal levels either by ACTH or by large doses of adrenal glucocorticoids. This provides a means whereby the adrenal cortex can regulate the functions of the adrenal medulla. Epinephrine is catabolized by the same enzymes that act on the other catecholamines, and its major end-metabolite is VMA.

The subcellular localization of MAO and COMT in brain can provide important information about how these enzymes function in vivo. MAO is present in the outer membrane of mitochondria in almost every tissue of the body and, in the brain, in glia as well as in neurons. However, the portion of MAO that is primarily concerned with the deamination of norepinephrine and dopamine is localized in mitochondria within the nerve terminals of the catecholaminergic neurons, where it deaminates the surplus catecholamines that leak out of synaptic vesicles [1]. Accordingly, levels of catechol acids, which arise when MAO is the first enzyme to act on these compounds, reflect catecholamines that leak out of vesicles within nerve terminals and are metabolized before leaving them. Hence, these catecholamines cannot reach the synaptic cleft or act on postsynaptic receptors.

There are several isozymes (distinct proteins that catalyze the same chemical reaction) of MAO in the brain. In the caudate nucleus, the area of the brain richest in dopamine, one of the MAO isozymes is extraordinarily active in deaminating dopamine, with much less effect on other monoamines. This suggests that different classes of aminergic neurons may have their own "tailor-made" MAO isozymes.

COMT also occurs in a wide variety of tissues. The COMT moiety that has first access to catecholamines in the brain, however, seems to be located largely outside the catecholamine nerve terminals. Hence, levels of normetanephrine or 3-*O*-methyldopamine reflect catecholamines that are released outside the nerve terminals into the synaptic cleft with access to postsynaptic receptors before being metabolized by COMT [1].

REUPTAKE INACTIVATION OF SYNAPTICALLY RELEASED CATECHOLAMINES

After discharge at synapses, acetylcholine is inactivated via hydrolysis by the enzyme acetylcholinesterase. Two enzymes can degrade the catecholamines. Surprisingly, neither of them is the primary mode of catecholamine inactivation. Instead, synaptically discharged catecholamines are inactivated by reuptake into the nerve terminals that had released them (Fig. 6-4). Proof that this mechanism terminates the actions of released catecholamines was obtained largely from the peripheral sympathetic nervous system, but the process is probably the same in the brain [3].

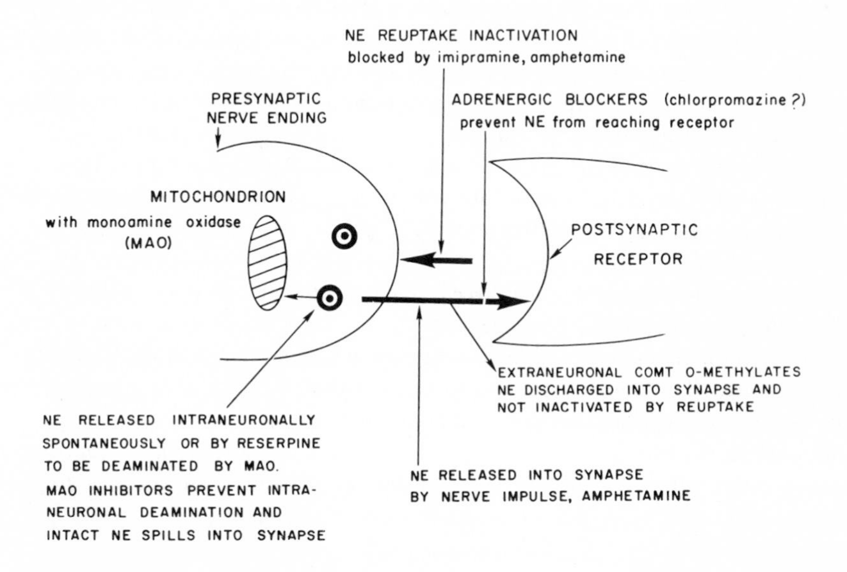

FIGURE 6-4. Model of postulated events at catecholamine synapses. (From Snyder [6].)

Catecholamine reuptake was discovered by Axelrod and his co-workers [1]. They found that when radioactive norepinephrine is injected intravenously, it accumulates primarily in tissues that have a dense sympathetic innervation. Cutting the sympathetic nerves to these organs abolishes the accumulation of the radioactive amine, so its uptake must be into sympathetic nerves. When these tissues are homogenized and centrifuged on sucrose density gradients, the accumulated norepinephrine is localized in a fraction of particles about the size of microsomes and with sedimentation properties similar to those of synaptic vesicles (see Chap. 21). Drugs that inhibit this uptake process increase the quantity of norepinephrine released after sympathetic nerves are stimulated and also potentiate the effects of sympathetic nerve stimulation. Inhibitors of MAO or of COMT do not potentiate the effects of sympathetic nerve stimulation or injected catecholamines.

There are no simple ways to measure the effects of nerve stimulation in the brain, so the physiological role of catecholamine reuptake cannot be readily assessed [5b]. However, brain tissue is useful for characterizing the uptake process itself. Catecholamine uptake in the brain can be measured after radioactive amine has been injected intraventricularly or if brain slices or even synaptosomes (pinched-off nerve terminals, see Chap. 21) are incubated with the catecholamine. The uptake process is energy dependent, since it is depressed at 0°C and by a variety of metabolic inhibitors. It is saturable, obeys Michaelis-Menten kinetics, and requires sodium and potassium, functioning optimally at the physiological concentrations of these ions. Catecholamine uptake is inhibited by ouabain, an inhibitor of (Na^+, K^+)-ATPase. Its relationship to the cell's sodium pump probably reflects a dependence of catecholamine uptake on the relative concentrations of sodium within and without the neuron [18]. The norepinephrine-containing neurons in the brain can accumulate dopamine and the dopamine neurons can take up norepinephrine. Interestingly, while the norepinephrine nerve terminals have four times the affinity for the physiologically occurring *l*-norepinephrine than for *d*-norepinephrine, the dopaminergic neurons of the caudate nucleus have equal affinities for the two isomers. Because dopamine is optically symmetrical and has no stereoisomers, perhaps its neuronal membrane has no physiological need to distinguish optical isomers of the catecholamines [11].

Catecholamine uptake can take place at the neuronal membrane of cell bodies, axons, and nerve terminals. It can be distinguished from the process whereby catecholamines are retained within synaptic vesicles, or storage granules inside the nerve terminal. The granular storage process is disrupted by the drug reserpine, which does not affect the neuronal membrane transport system [3, 6]. Other drugs, such as the tricyclic antidepressant imipramine, are selective inhibitors of the neuronal membrane transport [3, 6]. (Sourkes discusses these tricyclic drugs in Chap. 29.)

SEROTONIN

METABOLISM

Tryptophan, the dietary amino acid precursor of serotonin, is hydroxylated by the enzyme tryptophan hydroxylase to form 5-hydroxytryptophan (Fig. 6-5). The enzymology of tryptophan hydroxylase is not worked out as well as is that of tyrosine hydroxylase [12], and it is not yet clear whether tryptophan hydroxylase is "inducible" when the activity of nerves that contain serotonin is altered. There are some conditions under which the conversion of tryptophan to serotonin in the intact animal is changed, but whether this is accompanied by variations in tryptophan hydroxylase activity is not known. *d*-Lysergic acid diethylamide (LSD) slows the turnover rate of serotonin by retarding its formation from tryptophan [17]. It is thought that LSD mimics serotonin at its postsynaptic receptors, and that the postsynaptic neuron transmits the message "too much serotonin" via a neuronal feedback to the presynaptic neuron, whose rate of firing then slows [8]. Presumably, the slowed discharge rate of the neuron results in a decrease in the

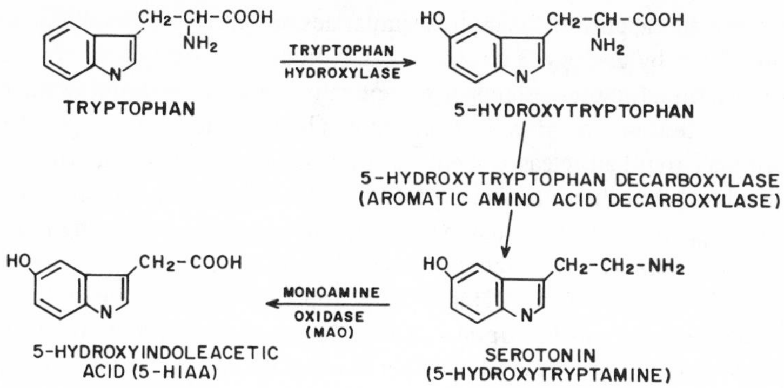

TRYPTOPHAN → (TRYPTOPHAN HYDROXYLASE) → 5-HYDROXYTRYPTOPHAN → (5-HYDROXYTRYPTOPHAN DECARBOXYLASE (AROMATIC AMINO ACID DECARBOXYLASE)) → SEROTONIN (5-HYDROXYTRYPTAMINE) → (MONOAMINE OXIDASE (MAO)) → 5-HYDROXYINDOLEACETIC ACID (5-HIAA)

FIGURE 6-5. Metabolism of serotonin.

synthesis of new serotonin, as there is less need for the synthetic machinery to keep up with the loss of serotonin by neuronal discharge. Although this picture may not reflect the true events accurately, direct neurophysiological recordings of the raphe nuclei have shown that LSD slows their firing rate [8] and biochemical studies have shown a slowing of serotonin turnover after LSD [5a, 17]. (Hallucinogens are discussed in Chap. 29.)

5-Hydroxytryptophan is decarboxylated to serotonin by 5-hydroxytryptophan decarboxylase, an enzyme whose range of substrate preference is the same as DOPA decarboxylase, so that it also can be termed an aromatic amino acid decarboxylase. Serotonin itself is metabolized by only one enzyme, MAO (Fig. 6-5).

SEROTONIN TRANSPORT INTO NEURONS

Serotonin is not established as a neurotransmitter in any major system of peripheral nerves, so there is no simple way to determine what mechanism accounts for its synaptic inactivation. As with norepinephrine, serotonin can be accumulated by serotoninergic neurons via a highly specific uptake system operating at the level of the neuronal membrane. Like the catecholamine process, the serotonin uptake system is energy dependent, requires sodium, is saturable, obeys Michaelis-Menten kinetics, and can be demonstrated both by intraventricular injection of radioactive serotonin and by incubation of the radioactive amine with brain slices or synaptosomes. Serotonin also appears to be stored in vesicles within the nerve endings by a process which can be disrupted by reserpine. Although both the serotonin and the catecholamine uptake systems can be inhibited by some of the same drugs, especially the tricyclic antidepressants, there are differences in the relative affinities of various antidepressants for the neurons that contain serotonin and catecholamine. It is commonly thought that the mechanism of action of the tricyclic antidepressants (see Chap. 29) involves inhibition of norepinephrine uptake, with resultant potentiation of amine released into the synaptic cleft [3, 6]. These drugs might also owe their therapeutic efficacy to a similar action on the neuronal uptake of serotonin. Thus, by analogy, a good case can be made for a major role of neuronal reuptake in terminating the effects of synaptically released serotonin in the brain.

The notion that neurotransmitter actions can be terminated by reuptake into the nerve terminals that had released them was first thought to be unique for the catecholamines. As other putative neurotransmitters have been investigated, it begins to appear that reuptake may be a universal means of terminating transmitter action, and that enzymatic degradation, as with acetylcholine, may be the exception rather than the rule. Besides norepinephrine, dopamine, and serotonin, there is strong evidence for a neuronal uptake system that accumulates γ-aminobutyric acid (GABA) and glycine. (See also Chap. 8 for discussion on transmitter uptake.)

PHYSIOLOGICAL ROLE OF SEROTONIN IN THE BRAIN

Histochemical identification of the raphe nuclei as a repository of cell bodies of serotoninergic neurons opened the way for studies that have elucidated, in part, the function of these neurons. Jouvet, the French neurophysiologist, lesioned the raphe nuclei of cats and examined their subsequent sleeping patterns [4]. Serotoninergic neurons degenerate when their cell bodies are destroyed, so serotonin levels in the brain are depleted proportionally to the extent of the lesion. With 90 percent destruction of the raphe nuclei, cats became totally insomniac. Jouvet then administered *p*-chlorophenylalanine to cats. This drug profoundly inhibits tryptophan hydroxylase activity in intact animals, although in test-tube systems it is only a weak inhibitor of the enzyme. After drug treatment, the serotonin content of the brain is depleted, and in Jouvet's experiments the cats became markedly insomniac. Of course, *p*-chlorophenylalanine, an amino acid analog, might have many effects besides inhibition of serotonin synthesis.

To establish the specificity of its action, Jouvet administered 5-hydroxytryptophan, which can be converted into serotonin even when tryptophan hydroxylase is inhibited because it bypasses the enzymatic block. Treatment with 5-hydroxytryptophan reversed the *p*-chlorophenylalanine-induced insomnia and put the cats to sleep. This pharmacological paradigm has been repeated in other animal species. Moreover, stimulating the raphe nuclei of rats at a physiological frequency makes the animals somnolent [14]. Interestingly, stimulation of these nuclei at a frequency five to ten times the physiological rate produces a state of hyperalertness similar to that which follows LSD administration. This accords well with the fact, cited above, that LSD seems to mimic the excessive synaptic liberation of serotonin.

SEROTONIN AND ITS METABOLITES IN THE PINEAL GLAND

The pineal gland, although situated in the middle of the head, does not appear to be part of the brain in most species. In the rat, in which the best neuroanatomical studies have been performed, the pineal gland is attached to the brain only by some connective tissue and receives its major or sole innervation from postganglionic sympathetic nerve fibers, the cell bodies of which originate in the superior cervical ganglion [7]. The pineal gland contains the highest serotonin concentration of any known animal tissue. In the rat, serotonin concentration is 200 times greater in the pineal gland than in the brain. The levels of serotonin in the pineal gland undergo marked diurnal fluctuations; at noon, levels are ten times higher than they are at midnight. This biological rhythm persists in the absence of light-dark changes in illumination and in blinded animals. It is not synchronized within the pineal gland but receives regulatory information from the central nervous system via the sympathetic nerves. Almost every chemical that has been examined in the pineal gland exhibits rhythmicity, although sometimes with altered phases.

The pineal gland also has unique pathways for the metabolism of serotonin (Fig. 6-6). The aldehyde formed from the action of MAO can be oxidized to 5-hydroxyindoleacetic acid or reduced to 5-hydroxytryptophol. 5-Hydroxytryptophol can then be methylated on the 5-hydroxyl group of the indole ring to form

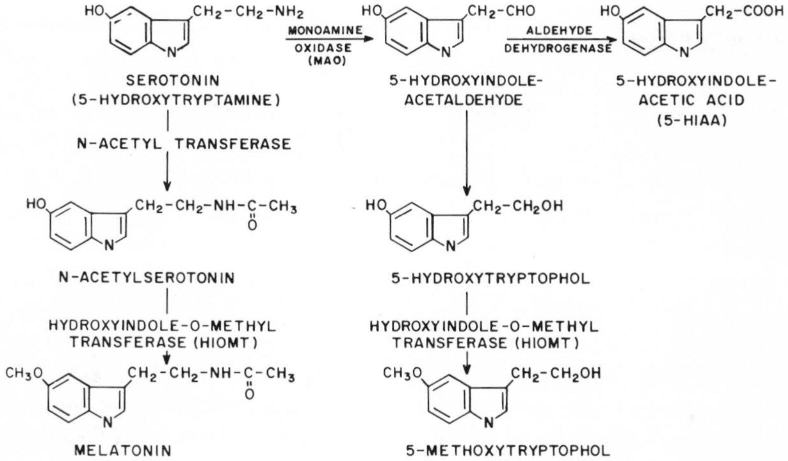

FIGURE 6-6. Unique pathways of serotonin metabolism in the pineal gland.

5-methoxytryptophol. This reaction is mediated by an enzyme, hydroxyindole-*O*-methyl transferase (HIOMT), which is found only in the pineal gland. Serotonin can be acetylated by *N*-acetyl transferase to form *N*-acetylserotonin, which can be acted on by HIOMT to form a compound known as melatonin. Both melatonin and 5-methoxytryptophol are produced solely in the pineal gland.

These pineal-specific metabolic pathways may have important physiological significance [7]. The pineal glands of rats maintained in constant darkness have HIOMT activity ten times higher than those of rats maintained in constant light. Such rats also show a decreased incidence of vaginal estrus. Removing the innervation of the pineal gland by ablating the superior cervical ganglia abolishes the ability of different lighting conditions to affect either vaginal estrus or pineal HIOMT activity. Thus, lighting seems to affect the gonads via the pineal gland. Because exposure to constant darkness is associated with increased HIOMT activity and a decreased incidence of vaginal estrus, one would predict that the compounds produced by this enzyme, melatonin or 5-methoxytryptophol, ought to inhibit estrus, which is, in fact, the case when either is administered to rats.

Another enzymatic step in the synthesis of melatonin, the *N*-acetylation of serotonin, also appears to be inducible. The activity of *N*-acetyl transferase in pineal glands in organ culture is increased 15 to 20 times by the addition of norepinephrine to the cultures, an effect which appears to be mediated by cyclic 3', 5'-AMP [13]. Norepinephrine is the neurotransmitter in the sympathetic nerves that innervates the pineal gland, so these effects in organ culture may reflect the means whereby sympathetic nervous activity regulates the enzymology of serotonin metabolism in the pineal glands of intact animals.

ACKNOWLEDGMENTS

This study was supported by U.S. Public Health Service Grants MH 18501, 1-RO1-NB07265, and 1-PO1-GM-16492. The author is a recipient of National Institute of Mental Health Research Scientist Development Award, K3-MH-33128.

REFERENCES

BOOKS AND REVIEWS

1. Axelrod, J. The metabolism, storage, and release of catecholamines. *Recent Progr. Horm. Res.* 21:597, 1965. (A review, summarizing a decade of work in Dr. Axelrod's laboratory, which reads like a history of the field of catecholamines, since so many of the major breakthroughs can be attributed to his innovative discoveries.)
2. Hillarp, N. A., Fuxe, K., and Dahlström, A. Demonstration and mapping of central neurons containing dopamine, noradrenaline, and 5-hydroxytryptamine and their reactions to psychopharmaca. *Pharmacol. Rev.* 18:727, 1966. (A summary of the epochal mapping by this Swedish group of the monoamine neurons in the brain.)
3. Iversen, L. L. *The Uptake and Storage of Noradrenaline in Sympathetic Nerves.* London: Cambridge University Press, 1967. (The best general reference on all aspects of catecholamines.)
4. Jouvet, M. Biogenic amines and the states of sleep. *Science* 163:32, 1969. (A review of the research relating serotonin to sleep mechanisms.)
5. Lajtha, A. (Ed.). *Handbook of Neurochemistry.* Vol. 4. New York: Plenum, 1970. See in particular:

5a. Costa, E., and Neff, N. H. Estimation of Turnover Rates to Study the Metabolic Regulation of the Steady-State Level of Neuronal Monoamines, p. 45.
5b. Glowinski, J. Storage and Release of Monoamines in the Central Nervous System, p. 91.
6. Snyder, S. H. New developments in brain chemistry. Catecholamine metabolism and its relationship to the mechanism of action of psychotropic drugs. *Am. J. Orthopsychiatry* 37:864, 1967. (A review of catecholamine metabolism and its relationship to the mechanism of the action of psychotropic drugs.)
7. Wurtman, R. J., Axelrod, J., and Kelly, D. E. *The Pineal.* New York: Academic, 1968. (The standard reference on the pineal gland.)

ORIGINAL SOURCES

8. Aghajanian, G. K., Foote, W. E., and Sheard, M. H. Action of psychotogenic drugs on single midbrain raphe neurons. *J. Pharmacol. Exp. Ther.* 171:178, 1970. (An important work, the first to tie changes in brain amine disposition to the firing rate of neurons known to contain the specific amine. Also, the major contribution to date to an understanding of how LSD exerts its effects in man.)
9. Alousi, A., and Weiner, N. The regulation of norepinephrine synthesis in sympathetic nerves: Effect of nerve stimulation, cocaine, and catecholamine-releasing agents. *Proc. Natl. Acad. Sci. U.S.A.* 56:1491, 1966. (A simple and elegant, and possibly the first, demonstration that an increase in the firing rate of catecholamine nerves enhances their biosynthesis.)
10. Anden, N. E., Dahlström, A., Fuxe, K., Larsson, K., Olson, L., and Ungerstedt, U. Ascending monoamine neurons to the telencephalon and diencephalon. *Acta Physiol. Scand.* 67:313, 1966. (Summarizes evidence with lesions and histochemistry detailing the pathways of amine tracts in the brain.)
11. Coyle, J. T., and Snyder, S. H. Catecholamine uptake by synaptosomes in homogenates of rat brain: Stereospecificity in different areas. *J. Pharmacol. Exp. Ther.* 170:221, 1969. (Shows differences between the catecholamine transport system by norepinephrine and dopamine neurons in the brain in terms of stereospecificity.)
12. Ichiyama, A., Nakamura, S., Nishizuka, Y., and Hayaishi, O. Enzymic studies on the biosynthesis of serotonin in mammalian brain. *J. Biol. Chem.* 245:1699, 1970. (A thorough study of tryptophan hydroxylase, the rate-limiting enzyme in serotonin synthesis, which, like tyrosine hydroxylase, is notoriously elusive.)
13. Klein, D. C., Berg, G. R., and Weller, J. Melatonin synthesis: Adenosine 3'5' monophosphate and norepinephrine stimulate N-acetyltransferase. *Science* 168:979, 1970. (Acetylation of serotonin may mediate the effects of sympathetic nerve impulses on pineal function.)
14. Kostowski, W., and Giacolone, E. Stimulation of various forebrain structures and brain 5-HT, 5-HIAA and behavior in rats. *Eur. J. Pharmacol.* 7:176, 1969. (Stimulation of raphe nuclei at slow, physiological rates produces somnolence.)

15. Nagatsu, T., Levitt, M., and Udenfriend, S. Tyrosine hydroxylase. The initial step in norepinephrine biosynthesis. *J. Biol. Chem.* 239:2910, 1964. (A major paper providing the first definitive description of tyrosine hydroxylase, which had for years eluded characterization.)
16. Schneider, H. P. G., and McCann, S. M. Possible role of dopamine as a transmitter to promote discharge of LH-releasing factor. *Endocrinology* 85:121, 1969. (Solid evidence that dopamine may regulate LH release; there is comparable evidence for a role of dopamine in FSH release.)
17. Schubert, J., Nyback, H., and Sedvall, G. Accumulation and disappearance of ^{3}H-5-hydroxytryptamine formed from ^{3}H-tryptophan in mouse brain; effect of LSD-25. *Eur. J. Pharmacol.* 10:215, 1970. (Shows how LSD slows serotonin synthesis from tryptophan.)
18. Tissari, A. H., Schonhofer, P. S., Bogdanski, D. F., and Brodie, B. B. Mechanism of biogenic amine transport. II. Relationship between sodium and the mechanism of ouabain blockade and the accumulation of serotonin and norepinephrine by synaptosomes. *Mol. Pharmacol.* 5:593, 1969. (Demonstrates how amine transport and the Na-K stimulated ATPase are linked via the regulation of intracellular sodium concentrations.)

7

Acetylcholine

Francis C. G. Hoskin

THE METABOLISM of acetylcholine (ACh) could be developed into such an all-embracing topic – sources of acetate, choline, and related 2-carbon entities, transmethylation, lipid synthesis, the citric acid cycle, and oxidative phosphorylation – that the more specific relations of ACh metabolism to nerve function might be largely overlooked. All of these many topics are important in all cells; some have special pertinence to nerve cells. For example, the synthesis and assembly of lipids in myelin is related to the inability of ACh to reach some parts of nerve cells; the production of energy is necessary for the extrusion of Na^+ following the permeability changes that result from ACh interaction with a receptor. In this chapter, however, *ACh metabolism* refers specifically to its relationship to the primary functions of nerve cells: the transmission of signals across synapses and their conduction along axons. Topics included are: (1) ACh synthesis, transport, penetration, storage, and release; (2) its removal, usually by hydrolysis; and (3) what is presently the most interesting aspect of ACh metabolism, its interactions with the postulated receptor.

This chapter should provide the reader with an introduction to the literature on ACh and with laboratory directions that may serve as the starting point for the design of new experiments. With regard to the literature citations, this is not a review, although some features of a review are evident. References chosen are often reviews or, if not, they themselves contain pertinent references. Precedence and priority are frequent and sometimes proper topics in scientific circles, but they have been overlooked here in order to provide the reader with a manageable bibliography.

ACh SYNTHESIS, TRANSPORT, STORAGE, AND RELEASE

Although the presence of an ACh-synthesizing enzyme, choline acetylase (ChAc), was foreseen a number of years ago, Nachmansohn [17] and Lipmann [16] may be credited with understanding the significance of their findings with respect both to nerve function and to the broader aspects of intermediary metabolism. Nachmansohn recognized that the equation

$$\text{acetate} + \text{choline} \longrightarrow \text{ACh} \quad (1)$$

only expressed an overall reaction. The speculation that the "active acetate" might be acetyl phosphate was not unreasonably wide of the mark, and in certain microorganisms the individual steps appear to be

$$\text{ATP} + \text{acetate} \rightleftharpoons \text{acetyl phosphate} + \text{ADP} \quad (2)$$

$$\text{acetyl phosphate} + \text{CoA} \rightleftharpoons \text{acetyl-CoA} + \text{P} \quad (3)$$

The observation [53] that a heat-stable dialyzable cofactor was involved and the elucidation [16] of the structure of CoA made possible a more carefully defined set of reactions for animal tissues, yeast, plants, and most microorganisms [26]

$$\text{ATP} + \text{acetate} \rightleftharpoons \text{adenyl acetate} + \text{pyrophosphate} \quad (4)$$

$$\text{adenyl acetate} + \text{CoA} \rightleftharpoons \text{acetyl-CoA} + \text{AMP} \quad (5)$$

Although the name "choline acetylase" was originally applied to an enzyme thought to catalyze reaction (1), it has since been limited to a final step [18]

$$\text{acetyl-CoA} + \text{choline} \rightleftharpoons \text{ACh} + \text{CoA} \quad (6)$$

Thus, the overall reaction is

$$\text{acetate} + \text{ATP} + \text{choline} \longrightarrow \text{ACh} + \text{AMP} + \text{pyrophosphate} \quad (7)$$

in which the driving force is a free-energy change, the favorability of which may be appreciated by considering the large overall standard free energy, ΔF^{o}, of -4.4 kcal. Inasmuch as CoA is often written CoA-SH, the reader should consult a biochemistry text for the complete structure. He should also consider the significance of the sulfhydryl group and the reasons for certain inhibition tests originally performed [18].

Early studies indicated a general inhibition at rather high concentrations by certain keto acids and quinones. As might be expected, choline analogs afford some inhibition. A group of ChAc inhibitors described recently appears to hold the promise of potency and specificity [5]. Thus, although the compound shown in Formula 7-A shows marked inhibition but little specificity, both 4-(1-naphthylvinyl)pyridine (Formula 7-B) and its *N*-methyl quaternary derivative (Formula 7-C) show inhibition and specificity, especially between ChAc and AChE. They are inhibitory in the trans configuration but are isomerized by light to the noninhibitory cis form. The inhibitory concentrations found in the literature (in the order

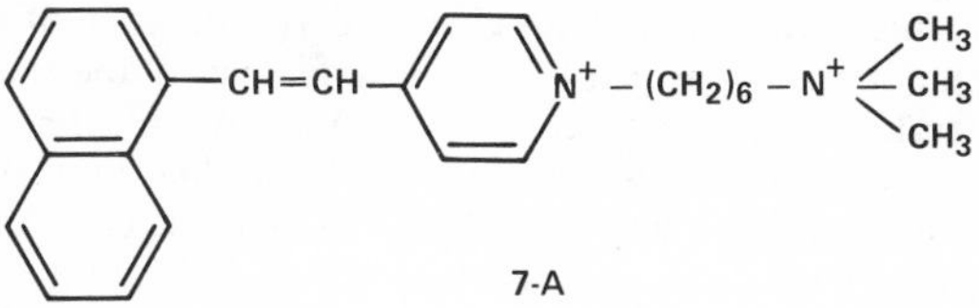

7-A

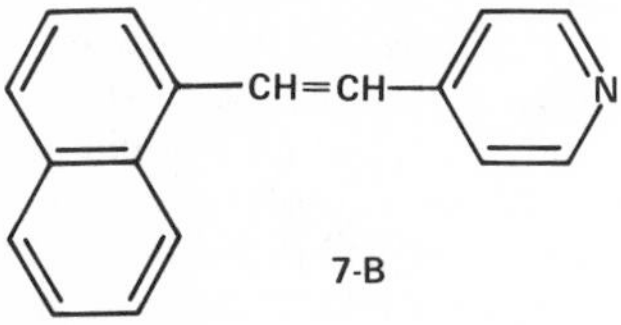

7-B

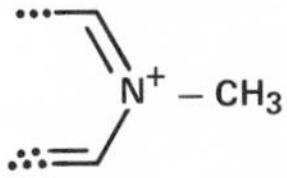

7-C

of 10^{-5} M) may imply to the uninitiated some simple relationship (that of the K_i, for example) in which at a somewhat higher concentration ChAc may be almost completely inhibited. Much work remains to be done in this area – work that should be within the capacity of the well-directed graduate student. The results may be relevant to the use of these inhibitors for relating metabolism to function in intact tissues. As to substrate specificity, one rather unexpected finding is worth noting. L(+)-Acetyl-β-methylcholine reacts specifically with AChE (as will be discussed later in this chapter) and even to a considerable extent with the ACh receptor, but L(+)-β-methylcholine will not serve as a substrate for ChAc. The D(–) isomer is a substrate, but a much poorer one than is choline. These findings [39] may provide a clue to the nature of the active site of ChAc.

The distribution and localization of ChAc has been reviewed from somewhat different points of view [7, 18]. Intracellularly, ChAc appears to be cytoplasmic, with the possibility of a loose binding to membranes; the binding is dependent, perhaps, on pH and ionic environment. Recently, evidence that ChAc may be associated with ribosomes, neurofilaments, and synaptic vesicles [47, 57] raises the possibility of a transitory but functionally significant association in situ with the neuronal membrane.

The determination of ChAc activity is essentially the determination of ACh. The Hestrin colorimetric assay can be taken directly from the literature [40] as has been done frequently by students in the author's laboratory. The leech muscle bioassay was used in a neurosciences course given at the Marine Biological Laboratory, Woods Hole, Massachusetts, in 1970, and is described herewith [60].

A leech is pinned in a stretched attitude, ventral side up, on a white-surfaced dissecting board. The ventral wall is removed by cutting laterally under a dissecting microscope for about 2 cm back from the mouth. The slimy exudate may be removed, and the preparation must be kept moist by frequent additions and removals of mammalian Locke's solution. The inner surface of the dorsal wall is cleaned of viscera, organs, and connective tissue. A longitudinal 10 mm strip of muscle is isolated by using two scalpel blades fastened together and spaced about 0.25 mm apart. A single nylon fiber (obtained by unraveling an ordinary nylon thread and uncrimping it over a warm surface, such as a hot water pipe) is stitched in near each end of the strip with the finest obtainable sewing needle. The strip is cut free, stretched lightly in a drop of Locke's solution, and cleaned of skin and fragments. The muscle is mounted in a clear plastic organ bath, as shown in Figure 7-1, and the upper end is attached to a strain gauge, the output from which goes to a recorder. The muscle is allowed to relax for at least 1 hr in a slow drip of medium made with 8 volumes of water to 10 volumes of mammalian Locke's solution and containing eserine sulfate to a final concentration of 2×10^{-5} M. At the time of assay the drip is discontinued and the medium is allowed to run down to *A* (see Fig. 7-1) by aspirating at *B*. A known concentration of ACh is added in 0.2 ml of the eserinized medium, and at that instant a stopwatch is started. At the end of 1 min, the contraction is noted on the recorder. Conditions should be adjusted so that the threshold is at about 5×10^{-10} M ACh and the full scale at about 10^{-8} M. The strip is washed in dripping buffer for 2 to 3 min before the next dose is applied. A standard curve of contraction versus concentration may be made. In practice, the concentration of ACh in an unknown sample should fall between the values of two known standards.

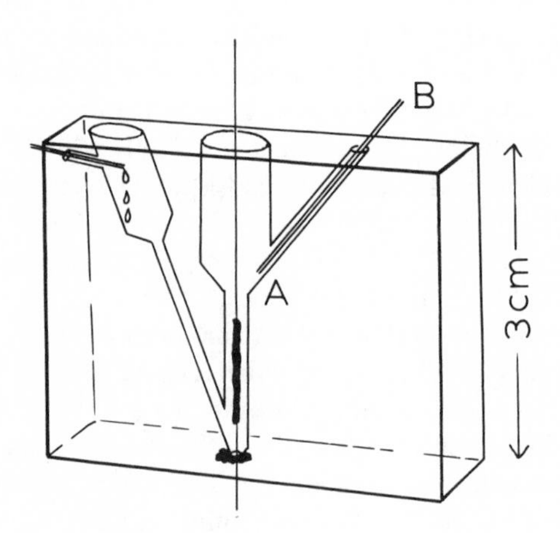

FIGURE 7-1. Clear plastic organ bath for mounting leech muscle for bioassay of ACh.

The leech muscle bioassay is extremely sensitive, but other compounds may interfere or mimic the ACh response. The Hestrin assay is reliable and highly reproducible, but it is limited in sensitivity. Several other extremely sensitive ACh assays have been summarized by Jenden [12], who is himself the contributor of one of the most sensitive, accurate, and expensive assay systems – that of gas chromatography combined with mass spectrometry. With the availability of 1-^{14}C-acetyl-CoA, accurate and sensitive direct assays of ChAc have been developed. One of these is given here in detail,* and has been used in the author's laboratory.

The incubation system consists of (1) 30 μl aqueous medium, suitably buffered, pH 7.8, containing as a part of this volume intact or homogenized tissue (e.g., squid axon), and the following chemicals at the concentrations shown: choline, 10^{-2} M; eserine, 6×10^{-4} M; and ChAc inhibitors or other agents being investigated, as desired; (2) 20 μl buffer containing 1-^{14}C-acetyl-CoA, 1.8×10^{-3} M, with a total radioactivity of approximately 3×10^5 counts per minute; (3) an additional 10 μl of buffer to be used for rinsing the homogenizer pestle. This provides the following final concentrations: choline, 5×10^{-3} M; eserine, 3×10^{-4} M; and acetyl-CoA, 6×10^{-4} M; the inhibitors or other agents being investigated have half the concentration used in (1). The system is incubated for 90 min at 21 to 22°C (the times and temperatures are variable within reasonable limits). Incubation is carried out in the dark if compounds such as the trans inhibitors (see above) are used. The incubation is terminated by dilution to 1 ml with distilled water and immersion in dry ice. The thawed 1 ml sample is then applied to a 0.1 ml column of Dowex-1-X8 acetate resin (100 to 200 mesh), and the effluent is collected in scintillation-counting vials in portions appropriate to the aqueous capacity of Bray's or a similar scintillation fluid. The column is washed with two 1 ml aliquots of distilled water, the radioactivity of which is also measured. The remaining 1-^{14}C-acetyl-CoA is eluted with acidified buffer. Except for the scale, this analytical procedure is essentially that which was previously described [42].

The possibility that ACh is synthesized in one part of a cell and localized at a higher concentration in another part raises questions of transport. ACh penetrates into cells, e.g., the squid giant axon, very poorly [43]; it should be emphasized that in this context penetration means from an external medium of about 5×10^{-3} M into a cell that is surrounded by a Schwann cell, basement layer, and connective tissue. Lower concentrations will undoubtedly show higher relative penetrations. In brain slices and at external concentrations of about 4×10^{-5} M, there appears to be significant uptake. The evidence does not necessarily indicate either an "uphill" process or metabolic mediation. Based on results obtained from studying fluxes of atropine or choline in brain slices and synaptosomes [34, 38], a general indication might be that quaternary ammonium compounds of functional importance enter neurons or neuron substructures by carrier mechanisms. The indication might also be that this entry can be inhibited competitively by certain analogs, notably hemicholinium-3 (Formula 7-D), but that entry or uptake is not significantly blocked by the usual metabolic inhibitors or by anaerobiosis.

$$HO-CH_2CH_2-N^+(CH_3)_2-CH_2C(=O)-C_6H_4-C_6H_4-C(=O)CH_2-{}^+N(CH_3)_2-CH_2CH_2-OH$$

7-D

*Courtesy of Prof. L. T. Kremzner, Columbia University, New York.

The idea of storage of ACh in an inactive form, presumably bound molecularly, has been an integral feature in at least one theory of nerve function [17]. The earlier concepts of "bound" ACh generally overlooked or dismissed the molecular binding in favor of vesicular storage [7]. At least on paper, if one arranges the charged portions of ACh and, e.g., lecithin, in a head-to-tail manner, and then considers the intimate association of proteins and phospholipids (among other components) in membranes and subcellular organelles, molecular binding and vesicular storage seem far from incompatible.

The almost explosive interest in nerve ending particles, as well as the concept of quantal release of ACh as an explanation for the miniature end-plate potentials [14], will probably dominate our thinking in this area for some time to come. The two names most usually associated with these particles are Eduardo de Robertis and V. P. Whittaker. The reader may consult Chapter 21 and a recent review [11] for further references. In brief, nerve endings may be prepared by brain homogenization and density-gradient centrifugation. These synaptosomes may be shocked osmotically to release synaptic vesicles of about 300 Å diameter. These are probably identical with the vesicles found presynaptically in brain and neuromuscular sections when examined in the electron microscope. The preparation of synaptosomes and of synaptic vesicles can be adapted directly from the literature, as has been done by graduate students in the author's laboratory.

A recent modification of the methods, presented below, will be of considerable interest to the researcher [2]. The preparation of synaptic vesicles by zonal rotor centrifugation is likely to give more material of a better quality; in addition, the separation of vesicles from the highly specialized electric organs of electric fish may disclose evidence of a more direct nature concerning the relation of structure and metabolism to function.

The electric organ of *Torpedo nobeliata,* obtainable in the vicinity of the Marine Biological Laboratory, Woods Hole, Massachusetts, is disintegrated in a Waring Blendor in 0.2 M sucrose and 0.3 M NaCl (1:1, w/v) for 1 min. The suspension is filtered through cheesecloth, and the filtrate is homogenized in the usual way and centrifuged for 30 min at 10,000 X *g*. The supernatant is separated by zonal rotor centrifugation on a gradient in which the starting solution is 0.2 M sucrose and 0.3 M NaCl, and the limiting solution is 50% (w/w) sucrose in 0.2 M NaCl.

The gradient is described by the equation

$$C = C_o + (C_m - C_o)\sqrt{v/v_m} \tag{8}$$

where C_o is the concentration of the light component
C_m that of the heavy component
C the concentration after a volume v has been delivered
and v_m the volume of the gradient.

More elegant equations have recently been described, but in practice they do not differ too markedly from this gradient [22]. Certain advantages have also been gained from employing one gradient above another, the two curves of which are discontinuous relative to one another. The presence of about 2 percent Ficoll may improve the preparation. The experimenter should consult handbooks issued by centrifuge manufacturers and should be prepared to vary conditions. For example, *Torpedo* is an elasmobranch, whereas *Electrophorus*, another electric fish, is a teleost, with concomitant differences in body fluids. Some uptake and metabolic studies have been performed with synaptosomes, but as yet little has been done along these lines with the vesicles. In view of the large amount of material that may be processed by zonal rotor centrifugation, this situation will probably change in the near future.

Among the significant features of vesicles are that they are rich in ACh, and that they, or at any rate their counterparts in situ, appear plentifully and almost exclusively on the presynaptic side of synapses. Although electron photomicrographs purporting to show vesicles discharging their contents into the synaptic cleft are likely to be artifactual, the miniature end-plate potentials probably are not. In addition to the amplitude distribution of these end-plate potentials as an indication of the quantal release of ACh at synapses, there are the observations that botulinum toxin abolishes both the miniature and the usual end-plate potentials by preventing the release of ACh, and that end-plate potentials may be produced by the micro-iontophoretic application of ACh near neuromuscular junctions [13].

ACh ENZYMATIC HYDROLYSIS

The molecular mechanism for the release of ACh is, as yet, only poorly portrayed. Indeed, its release specifically from presynaptic sites and the action of the same molecules specifically at postsynaptic sites may still be open to some reexamination [20, 64]. One requirement, however, is unquestioned – a means to remove ACh rapidly after its interaction with a receptor is necessary. It seems doubtful that there is reabsorption of released ACh analogous to the situation at adrenergic synapses. There may be some indirect evidence for diffusion away from the scene of interaction and binding to other components in which dimensions of a few hundred Ångströms and a few microseconds are involved. That such a process is considered nonspecific may only mean that it is not understood. However, a rapid and fairly specific means of removal of ACh is known: its hydrolysis by acetylcholinesterase (AChE)

$$\text{AChE} + \text{ACh} \rightleftharpoons \text{AChE-ACh} \tag{9}$$

$$\text{AChE-ACh} \rightleftharpoons \text{acetyl-AChE} + \text{choline} \tag{10}$$

$$\text{acetyl-AChE} + H_2O \rightleftharpoons \text{AChE} + \text{acetate} \tag{11}$$

The role of the ACh system in nerve function, whether axonal or synaptic or both, is shown in Figure 7-2; however, this is only a schematic representation and

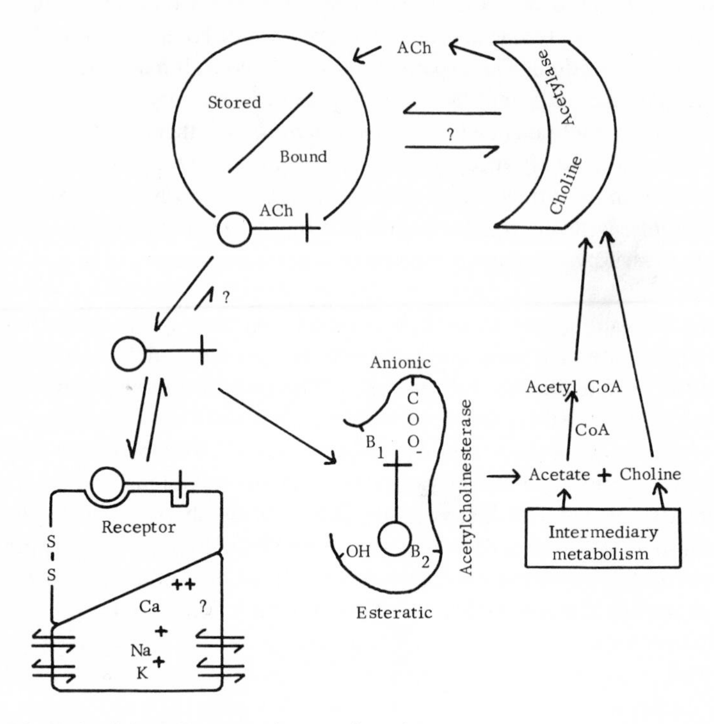

FIGURE 7-2. Role of the ACh system in nerve function.

is undoubtedly a gross oversimplification. The receptor is depicted to combine the findings and speculations of Karlin and Changeux [30, 46]. AChE is depicted to combine the now-classic picture of Wilson and Nachmansohn with Krupka's more recent evidence [17, 50]. The underlying relationship of the vesicle to the stored or bound form of ACh is still an open question, as is the possibility of some highly labile association between ChAc and the vesicle; hence the representation in the upper part of Figure 7-2.

There are several reviews (e.g., [6] and [18]) that deal primarily with the properties of AChE and, to a large extent, with the other cholinesterases and even some related esterases. In brief, virtually all tissues hydrolyze ACh. Neural tissues and washed red blood cells show a V_{max} when ACh is approximately 5×10^{-3} M and a marked decline in the hydrolysis rate at higher concentrations. Other sources, notably blood serum, show no evidence of this decline, although enzyme saturation is approached at about 0.1 M. This property of substrate inhibition has been used to differentiate AChE (also called specific or "true" cholinesterase) from ChE, the serum esterase (also called nonspecific or "pseudo" cholinesterase). ChE hydrolyzes butyrylcholine whereas AChE does not; conversely, only AChE hydrolyzes acetyl-β-methylcholine, or mecholyl (MeCh).

The configuration of MeCh deserves attention, especially since substrate configuration may be an important clue to the nature of the enzyme surface [31]. MeCh is normally a racemic compound consisting of the L(+) and D(–) isomers. Only the former is hydrolyzed by AChE and, contrary to what is sometimes implied, marked substrate inhibition has been consistently found in the author's laboratory. The D(–) isomer is a weak inhibitor of AChE. While this does not detract from the usefulness of (D,L)-MeCh, commercial availability of the L(+) isomer would increase sensitivity of AChE determinations by about twofold.

Among the esterases found in such prolific sources as serum, liver, and kidney are enzymes that have been termed A-esterases, B-esterases, aryl esterases, aliesterases, and so on [1]. These are probably descriptive terms, rather than strictly definitive names, for mixtures of isoenzymes that show subtle differences which result, overall, in a seemingly endless variation of enzymatic properties between species, organs, and, genetically, even between members of the same species.

One such genetic variation may be of special interest to the clinician. Succinyldicholine is a neuromuscular blocking agent which is relatively short-lived because it is hydrolyzed by a serum esterase. It appears that about two persons per thousand in the normal human population may have a homozygous, atypical serum esterase which either fails to hydrolyze succinyldicholine or hydrolyzes it very slowly. Added to this is a small but significant (and increasing) number of persons who may have been exposed to ChE inhibitors, e.g., the organophosphorus insecticides. In all of these cases, the possibility exists of dangerous, unusual, or, at best, prolonged effects on administration of succinyldicholine [1].

In the case of AChE, the chemical nature of the substrate and the phenomenon of substrate inhibition indicate a catalytic center made up of two subsites: one that contains a displaceable proton, termed *esteratic site;* and another, the *anionic site,* that has a residual negative charge (probably an unshared pair of electrons on a nitrogen-containing protein side chain). The use of the tertiary analog of ACh, dimethylaminoethyl acetate, and of other nonhydrolyzable quaternary ammonium compounds and their tertiary amine counterparts, as well as studies of hydrolysis and inhibition as functions of pH (bearing in mind the effects on substrate and on enzyme), have indicated that the anionic site has a pK_a of about 6.5, which seems to implicate the imidazole portion of histidine. Substrate inhibition occurs when, at high ACh concentration, a significant number of ACh molecules bind only through their quaternary ammonium portion to the anionic site; they are not thereby hydrolyzed, but they hinder the approach of another substrate molecule to the esteratic site [6, 17]. The use of organophosphorus enzyme inhibitors of the irreversible type, especially radioactive diisopropylphosphorofluoridate (DFP), on either AChE or its model, chymotrypsin, has consistently led to the finding of labeled phosphorylserine. The integration of such labeling data with a wealth of kinetic data has led to pictures of the active site of AChE in terms of its primary and, to some extent, secondary protein structure [50]. The most recent model (but, of course, subject to revision) involves a carboxyl site for binding the ACh quaternary nitrogen, a serine to account for the DFP labeling, and two imidazoles, one to direct the acetylation onto the serine (or the phosphorylation when DFP is used) and another to direct the attack of water onto the acetylated serine (the attack of water onto phosphorylated serine is generally unsuccessful) to complete the last step in the enzymatic hydrolysis of ACh [50].

AChE is a membrane-bound enzyme that can be solubilized with varying degrees of success. The crystallization of AChE from *Electrophorus electricus* has recently been made possible by the use of benzylated DEAE-cellulose for column chromatography [20]. The subunit nature of AChE may have considerable significance, either as a means of regulating enzyme function or as a means for assembly of the enzyme in membranes by an ordered aggregation [36]. The physiological role for AChE has been pictured in Figure 7-2 in a general way. The specific role has been argued vigorously and frequently [25]. From time to time it has even been speculated that AChE itself may be the ACh receptor. Experimental evidence has usually turned on whether the inhibition of esterase and the inhibition of receptor are separate or identical events. One problem is that esterase activity is measured on intact or homogenized material (the latter, preferably) by methods of solution chemistry, whereas the receptor only reveals itself unequivocally, as yet, as a function measurable by electronic methods on intact and specially arranged cells. The most recent speculation raises the possibility that esterase and receptor may have either overlapping sites or separate sites on the same macromolecule [30]. This may provide excellent properties for allosteric regulation, but such regulation may be achieved in the same general way by other specific combinations of active sites on macromolecules.

AChE activity may be determined by several methods, the choice being governed largely by how small in mass, or low in activity, the enzyme source is. Automatic titration equipment is expensive, but it is generally the most satisfactory. Manufacturers' handbooks are explicit and often use a ChE as their example. The instruments operate by adding small aliquots of alkali to a reaction mixture in which an acid is being produced (in the case of an esterase) in order to hold the pH constant at some preset level. The amount of alkali is recorded automatically as a function of time. The reaction mixture must be essentially unbuffered, although salts may be added. There must be efficient stirring without frothing. Extraneous acidity, notably CO_2, must be excluded by flooding the surface of the reaction mixture with nitrogen. Enzyme and alkali concentrations must be adjusted by trial determination to result in small but discrete additions of alkali. The enzyme activity should be adjusted by dilution or by smaller additions so that not more than 20 percent of the substrate is hydrolyzed in a minimum of one minute. This condition may be partially overcome by more elaborate arrangements, wherein both alkali and replacement aliquots of substrate are added. The nonenzymatic hydrolysis of substrate may be determined on the same sample by first titrating the complete reaction mixture for a few minutes without enzyme. This usually results in a small increment of slope above the time base. Then one adds enzyme in a very small volume. If inhibition is being studied, a larger volume of enzyme is incubated first with inhibitor in a separate vessel and, at some predetermined time or at a series of times, a small volume of enzyme is added to the main reaction vessel. Depending on the inhibitor, this will often serve to stop but not to reverse the inhibition reaction by reducing the inhibitor concentration and by protecting the remaining uninhibited sites with substrate. Exceptions to these suggestions are frequent and may provide useful information [51].

This "pH-stat" method allows the operator to obtain initial rates over a wide range of constant pHs. The older manometric method, which also depends on the formation of an acid, measures the release of CO_2 from a bicarbonate buffer; it is limited in its pH range, it does not actually function at a completely constant pH, and it is not dependable for the most initial rates. Nevertheless, the method has some usefulness in a related determination, which will be discussed shortly.

Two colorimetric methods are available and can be used or adapted directly from the literature [35, 40]. The researcher should also consult Worthington Biochemical Corporation's catalogs and data sheets, noting especially the use of gelatin to stabilize the purified enzyme in solution. In Hestrin's method especially, adaptations that depend on a single point in time are often described. These have usually been developed for rapid clinical or field tests and are used in a "cookbook" manner. They are not dependable, and are certainly unsuitable for research purposes.

Two methods are also available for AChE determinations on very small samples. One is a magnetically controlled Cartesian diver method [27], which requires, in this author's opinion, an unusual amount of dexterity, patience, and investment in special rooms, baths, and ultraconstant temperature equipment. The other method has been described in detail [42], and is not unlike the isotopic method given earlier in this chapter for the determination of ChAc activity.

All of these methods have been used by the author, by colleagues, or by graduate students or technicians in the author's laboratory. They should be amenable to wide modifications, depending on individual research conditions and requirements. Greater, and sometimes almost insurmountable, difficulties arise in attempts to quantitate AChE activity in intact tissue, intact cells, parts of cells, or even in homogenized tissue where subcellular permeability barriers appear to remain. The use of detergents and solubilizing enzymes raises other problems. These matters have frequently been discussed by Nachmansohn and his associates [19, 58].

The preceding several paragraphs have dealt with the laboratory technique of measuring AChE activity. The numbers expressing such activity may vary by orders of magnitude even for the same enzyme source, depending on time, concentrations, temperature, and pressure. The latter two, although usually held to rather narrow limits in biological systems, are within the scope of thermodynamics. Recently, nonequilibrium thermodynamics has begun to give profound insights into the physiological functioning of enzymes. Even to specify rigorously the other two variables (note that for concentration this includes enzyme, substrates, cofactors, salts, H^+, and inhibitors) for each enzyme determination would be cumbersome and would obscure an understanding either of the mechanism of AChE catalysis, or of the role which the enzyme may serve in the functioning cell.

In clinical and field tests, these variables may have been specified in advance by the commonly accepted test procedure. For research purposes, however, conditions may, indeed must, be varied considerably from what is commonly considered "optimal," bearing in mind especially the short times involved in nerve signaling and the unusual concentrations or localizations of all reactants which occur, sometimes temporarily, in or near the excitable membranes of nerves. These subjects are more properly understood by an application of kinetics. Although the enzyme discussed for the next several paragraphs is AChE, much the same treatment applies to all enzymes, including ChAc and the ACh receptor, as is discussed in the final section of this chapter.

In studying the kinetics of an enzymatic reaction, consideration must be given to the time and manner of adding substrates and inhibitors to enzymes. In this technical sense, substrate inhibition offers few problems. Tetraalkylammonium compounds may have complex effects if one considers that a cetyl or a dodecyl substituent confers detergent properties. Thus, addition of substrate may protect the active site of the enzyme, but noncompetitive inhibition may be caused by effects at distant sites. Without these complex effects, quaternary ammonium compounds are weak AChE inhibitors; their inhibition is easily reversed on dilution. Eserine and neostigmine, once regarded in this class, are now recognized as agents which carbamylate the active site of AChE by the formation of a covalent bond. This is a time-dependent reaction and only appears to attain equilibrium in that the rate of carbamylation may come to equal that of decarbamylation [6, 62]. Reversal of inhibition occurs on dilution by favoring the decarbamylation, but not as rapidly as with the simple quaternary ammonium compounds.

Although AChE inhibition by both of these types of inhibitors has been thoroughly studied, it may be worthwhile to review some simple kinetic treatment of data. The familiar reciprocal plot of $1/[s]$ on the abscissa against $1/v$, velocity being in any suitable arbitrary term, is derived from

$$1/v = (K_m/V_{max} \cdot 1/[s]) + 1/V_{max} \tag{12}$$

which is in the form $y = ax + b$. In the equation

$$E + S \underset{k_{-1}}{\overset{k_1}{\rightleftharpoons}} ES \xrightarrow{k_2} E + P \tag{13}$$

if $k_{-1} \gg k_2$, then $K_m = k_{-1}/k_1$ and is the thermodynamic equilibrium constant for the reversible dissociation of ES. If one now considers reversible inhibition as

$$E + I \rightleftharpoons EI \tag{14}$$

and a constant K_I as analogous to K_m one can, by suitable algebraic manipulation, arrive at

$$1/v = \left(\frac{K_m(1 + [I]/K_I)}{V_{max}} \cdot 1/[s]\right) + 1/V_{max} \tag{15}$$

again in the form $y = ax + b$.

For the specific case of irreversible inhibition, such as with DFP, where the situation may be depicted

$$E + I \longrightarrow EI \tag{16}$$

the treatment that gives a most useful rate constant combining both time and concentration is

$$k = \frac{1}{t[I]} \ln \left(\frac{\text{initial activity}}{\text{remaining activity}} \right) \tag{17}$$

In practice, enzyme activity is determined after various concentrations of inhibitor have been incubated with enzyme for various times, and the data compared with enzyme activity after incubations for various times without inhibitor.

For the general case of "acid-transfering" inhibitors where the reactions are

$$E + I \underset{}{\overset{K_I}{\rightleftharpoons}} EI \xrightarrow{k_3} E' \tag{18}$$

when $[I] \gg K_I$, then $k_3 = k_{apparent}\,(1 + K_I/[I])$; in the limiting case when $K_I/[I]$ approaches zero, $k_3 = k_{apparent}$. By suitable manipulation, one obtains

$$1/k_{app} = (K_I/k_3 \cdot 1/[I]) + 1/k_3 \tag{19}$$

which is in the form $y = ax + b$ and can be plotted in the usual way. The one value not apparent is $k_{apparent}$; this is a pseudo-first-order rate constant that is determined for each of several [I]s by the equation

$$k_{app} = \frac{1}{t} \ln \left(\frac{\text{initial activity}}{\text{remaining activity}} \right) \tag{20}$$

In practice, enzyme and inhibitor are mixed and, at time intervals, initial remaining enzyme activity is measured on small aliquots which are withdrawn from the enzyme inhibition mixture. One can use equation (20) or determine the half-time for inhibition by plotting, using the relation $k_{apparent} = 0.693/t_{1/2}$. When $[I] \ll K_I$ then

$$E + I \xrightarrow{k'_3} E! \longrightarrow E$$

A k_4 is determined by greatly diluting a solution of inhibited enzyme and immediately measuring the enzyme activity. The activity with no inhibitor minus the activity measured immediately on diluting is set equal to E'_o; the activity with no inhibitor minus the activity at a certain time, t, after diluting is set equal to E'; then

$$\ln (E'/E'_o) = -k_4 t \tag{21}$$

and

$$I_{50(ss)} = k_4/k'_3 \tag{22}$$

The term $I_{50(ss)}$ refers to the concentration of an inhibitor that causes 50 percent inhibition in the steady state [48, 61]. With regard to the inhibition of AChE by neostigmine, for example, it should be noted that k_4 applies not to the inhibition by the neostigmine but rather to the decomposition of the dimethylcarbamyl enzyme, whether this was originally formed from neostigmine or some other compound. The equation

$$k_{app} = (k'_3/K_I)[I] \tag{23}$$

which is in the form $y = ax$, will permit a determination of K_I.

In what appears to combine both the reversible and irreversible inhibition properties of the organophosphorus inhibitors, Main and Iverson [51] arrive at

$$[I](t_2 - t_1)/\ln(v_1/v_2) = (1/k_p)[I] + 1/k_i \tag{24}$$

which, being in the form $y = ax + b$. may be plotted in the usual way. Another constant, K_a, may be determined graphically, or by the relation $K_a = k_p/k_i$. In the equation

$$E + I \overset{K_a}{\rightleftharpoons} EI_{rev} \longrightarrow EI_{irrev} \tag{25}$$

K_a is the affinity, or initial binding, constant.

The terms I_{50} or pI_{50} are sometimes encountered. The former, as already indicated, means the concentration of inhibitor causing 50 percent inhibition, and the latter is the corresponding negative logarithm. Either is a convenient term because, since the single dimension is concentration, and the reader can immediately appreciate the potency of the inhibitor so described. They are not meaningful terms, however, unless time is specified, whereas the second-order rate constant in equation (17) includes a time parameter. For example, this author [42] has reported a pI_{50} for selenophos (Formula 7-E) of 9.1 after incubation of AChE with inhibitor for 30 min. The pI_{50} was used for a limited purpose, and the inhibition vs. the

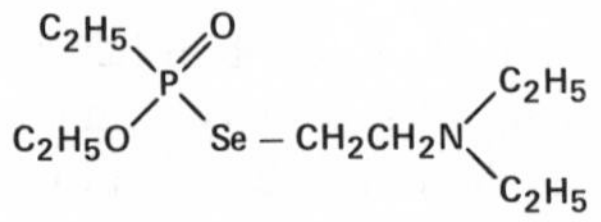

$$(C_2H_5)(C_2H_5O)P(=O)-Se-CH_2CH_2N(C_2H_5)_2$$

7-E

concentration plot shown was a means to an end. Since 50 percent inhibition was attained in 30 min at a selenophos concentration of 8×10^{-10} M, from equation (17)

$$k = \frac{2.303 \log 2}{30 \times 8 \times 10^{-10}} \approx 3 \times 10^{7} \text{ liters/mole/min}$$

This latter value would allow the calculation of the degree of inhibition to be expected by various combinations of concentrations and times for this one compound.

For irreversible inhibition of AChE, one thinks mainly of the organophosphorus compounds and, to some extent, of their sulfur analogs such as methanesulfonyl fluoride. Except for the organophosphates and their sulfur analogs that have been used more recently, Homstedt's review [10] is still a major source of information.

Many of the organophosphorus inhibitors of AChE are both nerve gases and insecticides. The organophosphorus inhibitors are generally of the structure shown in Formula 7-F, where J and K may vary widely, L is usually oxygen or sometimes sulfur, and P–X has some acid anhydride quality. J, K, or X may also have some

J L
P
K X

7-F

directing properties to permit orientation of the inhibitor molecule with respect to the anionic site of AChE so as to promote the cleavage of the P–X bond and the subsequent covalent bonding of the phosphorus to serine in the active site. The inhibition of AChE by this mechanism is both irreversible and competitive, the latter meaning that the presence of high concentrations of substrate will protect the enzyme even in the presence of inhibitor. The organophosphorus inhibitors have two properties which recommend them as insecticides: (1) they are generally not persistent in their toxic form, being subject to weather, soil, and so on, with half-lives of days or weeks, and (2) many of them are metabolized to innocuous substances, especially by mammalian liver. As nerve gases, they have virtually the same properties. The choice between the two is governed by vapor pressure, storage properties, reactions to heat and pressure, and species variations. Nerve gases being what they are, species variation dictates organophosphorus inhibitors which are less readily metabolized by man and which may interact more specifically with the AChE of man. In addition to the practical or diabolical aspects, the more potent and deadly of these compounds are also among the most potent chemical probes for exploring and understanding cellular function at the molecular level.

AChE inhibited by organophosphorus compounds can be reactivated by certain reagents, notably pyridine-2-aldoxime methiodide (2-PAM) and *N, N*-trimethylene-bis-(pyridinium-4-aldoxime)dibromide (TMB-4) (Formulas 7-G and 7-H, respectively) [8]. The functional group for reactivation is the oxime, and the quaternary nitrogen provides orientation on the enzyme surface. To illustrate reactivation, the following experiment has been adapted in the author's laboratory from the literature.

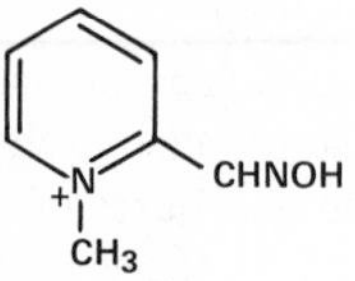

CHNOH CHNOH

+N CHNOH +N N+

CH_3 $CH_2 - CH_2 - CH_2$

7-G 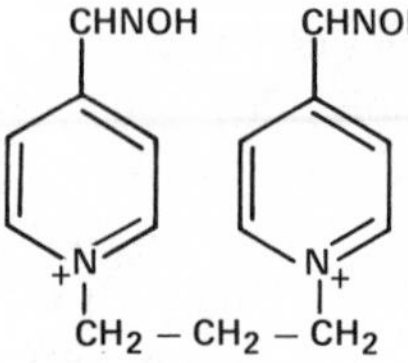 7-H

For inhibition, 10 mg AChE (Worthington, 100 units/mg) in 1 ml salt solution (0.15 M NaCl, 0.02 M $MgCl_2$, 0.006% gelatin) is added to 1 ml of 2×10^{-5} M diisopropylphosphorofluoridate (DFP) in water and allowed to stand at room temperature for 2 hr, at which time 18 ml water is added. The remainder of the experiment is now performed on a pH-stat in a final volume of 5.0 ml in small titration vessels. To each of four vessels is added 0.2 ml of the diluted inhibited enzyme, 2.0 ml of the salt solution, and a sufficient volume of 2-PAM (Eastman) in water or sufficient additional water to give final molar concentrations of 2-PAM of 10^{-3}, 5×10^{-4}, 10^{-4}, and 0. Each reaction vessel is incubated under nitrogen for 30 min, at the end of which time 0.06 ml of 0.25 M ACh is added and the subsequent reaction is followed on the pH-stat. The percentages of enzyme activity, compared to an uninhibited AChE solution, for the three concentrations of 2-PAM and the vessel without 2-PAM, were respectively 90%, 80%, 70%, and zero. If, after incubation of AChE and DFP for 2 hr and addition of the 18 ml of water, this inhibited enzyme solution is dialyzed against salt solution for 48 hr and the balance of the experiment then is performed, the percent reactivation is, in the same order, 30%, 20%, 10%, and zero when compared with a sample of AChE carried through the entire procedure except for treatment with DFP. (These values are approximate, but the same general level and order will be found consistently.)

It should be noted that the degree of reactivation possible is inversely related to the time during which irreversibly inhibited AChE remains in that state. This phenomenon of a decreasing ability to be reactivated is termed *aging.* It is probably caused by the loss of J or K groups in the structural formula for the organophosphorus compounds (Formula 7-F). In sarin inhibition of AChE, such a loss results in the structure shown in Formula 7-I, thus interfering with the nucleophilic attack of, e.g., 2-PAM.

The investigator who undertakes these or similar experiments should be prepared to vary inhibition time, reactivation time, reactivation concentration, and, to a more limited degree, inhibitor concentration in order to illustrate any points

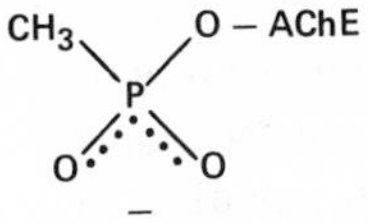

$$CH_3\text{—}P(\text{—}O\text{—}AChE)(\cdots O^-)(\cdots O)$$

7-I

he wishes to explore. The experiment may also be planned so as to run the inhibited enzyme solution through a gel filtration column to remove excess inhibitor, or even to do this after reactivation to remove both excess inhibitor and reactivator.

Spontaneous reactivation of inhibited AChE, especially in intact tissue, may be much more rapid than would be expected from experiments with purified enzyme. Such reactivation may have important consequences on nerve function after organophosphorus treatment [33]. There is a question of whether function can be restored either by spontaneous or by chemical reactivation of AChE after treatment with organophosphorus compounds at sufficiently high concentrations to cause blockade, especially when the reactivators are themselves ChE inhibitors and are not particularly good tissue penetrators by virtue of being quaternary nitrogen compounds. This has not yet been settled fully, although claims and counterclaims have appeared [63].

In subjecting either intact or homogenized whole-tissue preparations to chemical agents, various problems must be considered. Permeability properties have already been mentioned. Furthermore, some of the agents may themselves be metabolized. The surprising finding of a high level of so-called DFPase in the squid giant axon is such a case [45].* Although it is not the purpose of this chapter to discuss DFPase, some consideration must be given to this enzyme in relating the AChE inhibitors to effects on function. The fact that the DFPase is found predominantly in the axoplasm [45] raises the question of whether DFP, when applied to an intact squid axon at external concentrations which do not block conduction (for example, at less than 5×10^{-3} M), crosses the excitable membrane as DFP or as the hydrolysis product. Furthermore, DFP blocks conduction at about 5×10^{-3} M in axons of squid, lobster, and spider crab and in the electroplax of the electric eel, although the approximate relative levels of DFPase in these four preparations are, respectively, 100, 10, 1, and undetectable. Other organophosphorus compounds that are even more potent ChE inhibitors than DFP are not detoxified. They penetrate into the squid axon in their inhibitory form and block conduction, if at all, only at external (and now internal) concentrations of 10^{-3} M or higher [42, 44].

---

*DFP, sarin, tabun, and similar nerve gases present serious hazards in the laboratory. In determining their enzymatic hydrolysis, the older manometric method has a distinct advantage. The Warburg vessels form a closed system, the effluent from gassing the vessels may be passed through one or more alkaline wash-bottles, and at the end of the experiment a drop of strong alkali may be introduced through a side-arm port to destroy any remaining nerve gas. In addition, each mole of hydrolyzed DFP releases two moles of CO_2 from a bicarbonate buffer. The reader should consider the results of hydrolyzing such substances as tabun or paraoxon.

It has been known for some time that there is a parallel between toxicity and potency in inhibiting AChE. It appears that such a parallel has not been found between inhibition and effects on nerve function. These matters have been discussed in a recent exchange of technical comments which the neurobiologist may find interesting [41]. Although the findings, which are generally not in dispute, were again reviewed in great detail in a recent handbook article [21], the questions raised were not dealt with and remain unresolved.

While it is true that "a successful dissociation of electrical and enzyme activity after exposure to organophosphates" [20] has not been accomplished, it seems premature to conclude that the failure to demonstrate such a dissociation of electrical and ChE activities, especially for reasons of technical inadequacy, is proof that the two activities are directly associated. This is not to say that it is not accepted a priori that bioelectrical activity is controlled by macromolecules whose properties are expressed in terms of enzyme kinetics. It may be that an essential role of AChE in nerve function [25] is more likely to be found in the central nervous system, where ACh cannot be removed rapidly enough by diffusion. The possibility that the effects on function of AChE inhibitors that are finally observable at sufficiently high concentrations may be effects on the ACh receptor will be more properly discussed in that section.

An obvious practical interest in the rapid reactivation of organophosphorus-inhibited AChE is the possibility that it may be an antidote to nerve gas or insecticide poisoning. The discovery of 2-PAM and the underlying biochemical explanation of why it and similar reactivators have their remarkable effects on decreasing the toxicity of the organophosphorus inhibitors only in the presence of atropine (the older and more commonly known symptomatic antidote) have been reviewed [17, 24]. The clinically oriented reader may wish to redetermine mouse (or other small animal) toxicities (LD_{50}) for DFP, Parathion, or one of a host of other organophosphorus compounds in combination with either prophylactic or therapeutic doses of reactivators and adjuvants. (Trade names of the compounds can be found on the labels of insecticide dispensers, and small quantities of the pure compounds for research purposes can often be obtained as gifts from the manufacturers.)

The localization of AChE has been reviewed from two rather different points of view [15, 19]. At the molecular level, it is recognized that AChE is a membrane-associated enzyme; at the cellular level, it has been shown in increasingly finer detail that AChE is localized in the synaptic apparatus [3, 4, 37, 49]. It now appears that, after suitable pretreatment either with detergents or with enzymes sufficiently suitable to overcome permeability barriers but not to destroy fine structure, AChE is found generally at or near excitable membranes, i.e., it is not restricted to the synaptic region. At present, histochemistry seems too blunt an instrument with which to examine the most minute details of structure and function. Quantitation is still a major obstacle.

A most interesting research problem, but one fraught with difficulties, would be the electron microscopic examination of histochemically prepared nerves after treatment with penetrating, nonmetabolizable, organophosphorus inhibitors – for example, selenophos (see Formula 7-E) – and an attempt to overcome permeability barriers sufficiently to allow a substrate to reach all parts of the membrane without causing a possibly artifactual spread of inhibitor. To correlate such methods with functional studies would provide significant information concerning either a general role for ACh in the permeability cycle of excitable membranes or the more commonly assumed role of AChE as a transmitter inactivator at cholinergic synapses.

ACh INTERACTIONS WITH RECEPTOR

What has been discussed so far about ACh metabolism has been, in a sense, either prelude (synthesis, storage, and release) or postscript (hydrolysis). The physiological role for ACh, or the pharmacologically demonstrable effects of applied ACh analogs (in the broadest sense), seems to require a receptor (see Fig. 7-2). The history of this concept can be traced back to about the turn of the century through references that are cited in this section. The physiologist-anatomist, especially in the older literature, probably conceived of the receptor in terms of an organelle, a patch on the surface of a receptive cell, or even the cell itself. More recently, the biochemist has thought of the receptor in molecular terms. Our most recent views, strengthened by electron microscopic revelations of increasingly complex substructures within the cell, down to and including molecular components, are that receptors (now limiting our discussion to the ACh receptor) are protein macromolecules built into the excitable membranes of cells in a highly ordered manner. It is believed that they may be further specialized in their nature or arrangement in postsynaptic areas and, when they interact with ACh or a similar receptor effector, something happens that reveals itself by the initiation or alteration of cellular function. If the first part of this description is valid, the biochemist should be able to isolate and purify the receptor by the usual methods; if the last part is valid, the assay for purified receptor may be extremely difficult because cellular function is obviously lost at almost the first approach to purification.

In these circumstances, two approaches have been used. One is to study the functional responses of the postulated receptor in situ to various pharmacological agents. This has been the classic role of pharmacology in receptor investigation. Some compounds, including ACh, act predominantly at synapses, block electrical activity, and depolarize the cell. These are referred to as *receptor activators.* Other compounds, often analogs of ACh, block without depolarization; many of them are tertiary analogs of receptor activators and, since they are therefore able to penetrate cells, they often act on both synaptic and axonal function. These compounds are termed *receptor inhibitors.* The two terms should be regarded as largely descriptive and not as definitive as the same terms are in enzymology. Perhaps *receptor effectors* would be a more general name to apply to both classes. If the receptor responses are studied both qualitatively and quantitatively, compounds can be found that are assumed, on the basis of the functional studies, to bind strongly and fairly specifically to the receptor. The second approach is now obvious. A source of receptor, notably nerve tissue, is treated with such a compound and the tissue is homogenized or otherwise disrupted. The order of these two procedures may be reversed, and an attempt may be made to isolate a fraction to which the compound is bound, bearing in mind that the binding may easily be reversible.

Two attempted isolations [54] have been unsuccessful, in large part, due to two fundamental features. One is that the best receptor effectors are, invariably, charged entities – that is, quaternary ammonium ions – and because of this they show strong binding to various tissue components, such as phospholipids, nucleic acids, mucopolysaccharides, and various proteins. The other is that because the receptor and AChE must be situated in the same tissue and because they must have many features in common (note that ACh interacts with both), it is difficult to know if the binding is to the receptor and not to the AChE as well.

A recent isolation [54, 55] has made use of the idea that AChE is a sort of model for the receptor. Muscarone (Formula 7-J) is considered as a nonhydrolyzable analog of ACh (Formula 7-K). It has been shown that well-authenticated receptor effectors compete with muscarone for binding by a tissue fraction, whereas the organophosphorus compounds such as paraoxon do not markedly reduce the

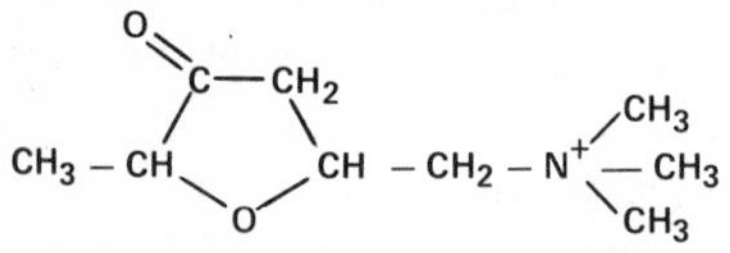

$CH_3-CH\langle O\rangle CH-CH_2-N^+(CH_3)_3$ (ring: CH_3-CH, $C{=}O$, CH_2, CH, O)

7-J

$CH_3-\overset{O}{\overset{\|}{C}}-O-CH_2-CH_2-N^+(CH_3)_3$

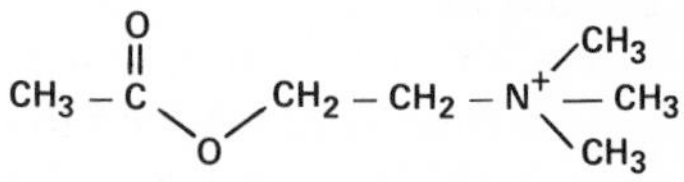

7-K

muscarone binding although they inhibit any remaining AChE in the fraction. The results appear promising, but these experiments, dealing with highly heterogeneous fractions rather than with pure substances, are fraught with possibilities for misinterpretation. This caution is perhaps especially applicable to the recent work of de Robertis et al. [32, 56], in which dimethyl-*d*-tubocurarine has again been used.

What seems to this writer to be the most promising, as well as the most recent, receptor isolations (perhaps "characterization" is a more conservative term) are based on the effects of a snake venom component, α-bungarotoxin, on electrical function and its binding to a cell component separable by column chromatography and ultracentrifugation [28, 52]. The reader should note especially that what may appear to be a single protein peak is separable into two distinguishable but overlapping peaks, one containing α-bungarotoxin bound to protein and the other containing AChE activity [52], and he should now refer back to the unsuccessful isolation attempts. While there is little point in attempting to repeat the earlier isolations, the availability of electric tissue from the torpedo ray in many parts of the world and the possibility of using parts of mammalian brain should assure an interest in O'Brien's experiments [55].

The characterization of the receptor by some novel applications of enzyme kinetics to nerve function has proceeded in advance of receptor isolation, however. For example, equation (12) may be rewritten as

$$v/(V_{\max} - v) = [s]^n/K_m \tag{26}$$

and

$$\log [v/(V_{\max} - v)] = n \log [s] - \log K_m$$

which is in the familiar form $y = ax + b$. The slope of y plotted against x is the interaction coefficient, or the Hill coefficient, n_H. The extent to which n_H is greater than one is a measure of the sigmoidicity of a simple plot of substrate concentration vs. reaction velocity (Chap. 1). Its meaning, however, is much more complex, encompassing the number of interacting substrate-binding sites per enzyme molecule and the strength of this interaction. In an interesting adaptation of this treatment, Changeux and associates [29, 30] have substituted measurement of the change in resting potential of a single electroplax cell of *Electrophorus electricus* treated with a receptor activator, phenyltrimethylammonium, for measurement of the velocity, v, and have substituted the maximum possible change in the presence of a high concentration of the activator for $V_{\max}$. The results were foreshadowed, to some extent, in an earlier work [59].

The reader should carefully think through both the rationale and the experimental tactics in Karlin's [46] receptor study. It may be helpful, as a guide, to point out that in this work (1) the maximum depolarization of electric eel cells by carbamylcholine was noted; (2) after washing and then treatment with the reducing agent dithiothreitol, the cell was not depolarizable; (3) the cell was then treated with a quaternary derivative of *N*-ethylmaleimide for a noted time and concentration; (4) exposed –SH groups, which had not been trapped by the *N*-ethylmaleimide derivative, were then reoxidized with dithio-bis-(2-nitrobenzoate); (5) the depolarization attainable with carbamylcholine was again noted; (6) the initial and final depolarization data were used in an adaptation of equation (17).

The gist of all these experiments seems to be that there are auxiliary sites on both AChE and on the ACh receptor, that there are disulfide bonds at or near both of these auxiliary sites, and that the sites may serve a regulatory function in an allosteric manner. Furthermore, while the evidence indicates the untenability of the view that the AChE and the ACh receptor might be identical, the possibility exists that the regulatory sites on the AChE molecule might serve a physiological function as regulatory sites for the ACh receptor. Although the interpretations are extremely tenuous at this stage, some evidence has been accumulated concerning the nature of subsites in the active center of the receptor. Many of the tentative conclusions may be rejected or may at least be modified in time, but at present they have the stimulating property of suggesting chemical means of examining nerve function.

ACKNOWLEDGMENT

This chapter incorporates research that was supported in part by U.S. Public Health Service grants NB-07055 and NS-09090.

REFERENCES

BOOKS AND REVIEWS

1. Augustinsson, K.-B. Classification and Comparative Enzymology of the Cholinesterases, and Methods for Their Determination. In Koelle, G. B. (Ed.), *Handbuch der experimentellen Pharmakologie; Cholinesterases and Anticholinesterase Agents.* Berlin: Springer-Verlag, 1963.
2. Barker, L. A., Dowdall, M., Essman, W. B., and Whittaker, V. P. The Compartmentation of Acetylcholine in Cholinergic Nerve Terminals. In Heilbronn, E. (Ed.), *Drugs Affecting Cholinergic Systems in the CNS.* Stockholm: Försvarets Forskningsanstalt (Almquist and Wiksell, agents), 1971.
3. Barnard, E. A., and Rogers, A. W. Determination of the number, distribution and some *in situ* properties of cholinesterase molecules in the motor end plate, using labeled inhibitor methods. *Ann. N.Y. Acad. Sci.* 144:584, 1967.
4. Bloom, F. E., and Barrnett, R. J. The fine structural localization of cholinesterases in nervous tissue. *Ann. N.Y. Acad. Sci.* 144:626, 1967.
5. Cavallito, C. J., White, H. L., Yun, H. S., and Foldes, F. F. Inhibitors of Choline Acetyltransferase. In Heilbronn, E. (Ed.), *Drugs Affecting Cholinergic Systems in the CNS.* Stockholm: Försvarets Forskningsanstalt (Almquist and Wiksell, agents), 1971.
6. Cohen, J. A., and Oosterbaan, R. A. The Active Site of Acetylcholinesterase and Related Esterases and Its Reactivity Toward Substrates and Inhibitors. In Koelle, G. B. (Ed.), *Handbuch der experimentellen Pharmakologie; Cholinesterases and Anticholinesterase Agents.* Berlin: Springer-Verlag, 1963.
7. Hebb, C. Formation, Storage and Liberation of Acetylcholine. In Koelle, G. B. (Ed.), *Handbuch der experimentellen Pharmakologie; Cholinesterases and Anticholinesterase Agents.* Berlin: Springer-Verlag, 1963.
8. Heilbronn-Wikstrom, E. Phosphorylated cholinesterases. *Svensk Kemisk Tidskrift* 77:11, 1965.
9. Hodgkin, A. L. *The Conduction of the Nervous Impulse.* Springfield, Ill.: Thomas, 1964.
10. Holmstedt, B. Structure-Activity Relationships of the Organophosphorus Anticholinesterase Agents. In Koelle, G. B. (Ed.), *Handbuch der experimentellen Pharmakologie; Cholinesterases and Anticholinesterase Agents.* Berlin: Springer-Verlag, 1963.
11. Hubbard, J. I. Mechanism of transmitter release. *Progr. Biophys. Mol. Biol.* 21:33, 1970.

12. Jenden, D. J. Recent Developments in the Determination of Acetylcholine. In Heilbronn, E. (Ed.), *Drugs Affecting Cholinergic Systems in the CNS.* Stockholm: Försvarets Forskningsanstalt (Almquist and Wiksell, agents), 1971.
13. Katz, B. *Nerve, Muscle, and Synapse.* New York: McGraw-Hill, 1966.
14. Katz, B. *The Release of Neural Transmitter Substances.* Springfield, Ill.: Thomas, 1969.
15. Koelle, G. B. Significance of acetylcholinesterase in central synaptic transmission. *Fed. Proc.* 28:95, 1969.
16. Lipman, F. Biosynthetic Mechanisms. *Harvey Lect.* 44:99, 1950.
17. Nachmansohn, D. *Chemical and Molecular Basis of Nerve Activity.* New York: Academic, 1959.
18. Nachmansohn, D. Choline Acetylase. In Koelle, G. B. (Ed.), *Handbuch der experimentellen Pharmakologie; Cholinesterases and Anticholinesterase Agents.* Berlin: Springer-Verlag, 1963.
19. Nachmansohn, D. Proteins of excitable membranes. *J. Gen. Physiol.* 54 (No. 1, part 2):187, 1969.
20. Nachmansohn, D. Proteins in excitable membranes. *Science* 168:1059, 1970.
21. Nachmansohn, D. Proteins in Bioelectricity. Acetylcholine-Esterase and Receptor. In Lowenstein, W. R. (Ed.), *Handbook of Sensory Physiology.* Vol. I. *Principles of Receptor Physiology.* Berlin: Springer-Verlag, 1971.
22. Price, C. A. Zonal Centrifugation. In Umbreit, W. W., Burris, R. H., and Stauffer, J. F. (Eds.), *Manometric Techniques.* 5th ed. Minneapolis: Burgess, 1971.
23. Tasaki, I. Nerve Excitation. Springfield, Ill.: Thomas, 1968.
24. Wilson, I. B. Molecular complementary and antidotes for alkylphosphate poisoning. *Fed. Proc.* 18:752, 1959.
25. (For the substance and flavor of this debate, see [7] and [18].) Also compare: Nachmansohn, D. Actions on Axons, and Evidence for the Role of Acetylcholine in Axonal Conduction. In Koelle, G. B. (Ed.), *Handbuch der experimentellen Pharmakologie; Cholinesterases and Anticholinesterase Agents.* Berlin: Springer-Verlag, 1963, with: Koelle, G. B. Cytological Distribution and Physiological Function of Cholinesterases in the same volume. Also compare the presentations and discussions of these two authors in Rodahl, K., and Issekutz, B. (Eds.), *Nerve as a Tissue.* New York: Harper and Row, 1966.

ORIGINAL SOURCES

26. Berg, P. Acyl adenylates. *J. Biol. Chem.* 222:991, 1015, 1956.
27. Brzin, M., and Zeuthen, E. Notes on the possible use of the magnetic diver for respiration measurements (error 10^{-7} μl./hour). *Compt. Rend. Trav. Lab. Carlsberg* 34:427, 1964.
28. Changeux, J.-P., Kasai, M., and Lee, C.-Y. Use of a snake venom toxin to characterize the cholinergic receptor protein. *Proc. Natl. Acad. Sci. U.S.A.* 67:1241, 1970.

29. Changeux, J.-P., and Podleski, T. R. On the excitability and cooperativity of the electroplax membrane. *Proc. Natl. Acad. Sci. U.S.A.* 59:944, 1968.
30. Changeux, J.-P., Podleski, T., and Meunier, J.-C. On some structural analogies between acetylcholinesterase and the macromolecular receptor of acetylcholine. *J. Gen. Physiol.* 54 (No. 1, part 2):225, 1969.
31. Chothia, C., and Pauling, P. Conformation of cholinergic molecules relevant to acetylcholinesterase. *Nature (Lond.)* 223:919, 1969.
31a. Denburg, J. L., Eldefrawi, M. E., and O'Brien, R. D. Macromolecules from lobster axon membranes that bind cholinergic ligands and local anesthetics. *Proc. Natl. Acad. Sci. U.S.A.* 69:177, 1972.
32. De Robertis, E. Molecular biology of synaptic receptors. *Science* 171:963, 1971.
33. Dettbarn, W.-D., Bartels, E., Hoskin, F. C. G., and Welsch, F. Spontaneous reactivation of organophosphorus inhibited electroplax cholinesterase in relation to acetylcholine induced depolarization. *Biochem. Pharmacol.* 19:2949, 1970.
34. Diamond, I., and Kennedy, E. P. Carrier-mediated transport of choline into synaptic nerve endings. *J. Biol. Chem.* 244:3258, 1969.
35. Ellman, G. L., Courtney, K. D., Andres, V., Jr., and Featherstone, R. M. A new and rapid colorimetric determination of acetylcholinesterase activity. *Biochem. Pharmacol.* 7:88, 1961.
36. Froede, H. C., and Wilson, I. B. On the subunit structure of acetylcholinesterase. *Israel J. Med. Sci.* 6:179, 1970.
37. Giacobini, E. Cholinergic and adrenergic cells in sympathetic ganglia. *Ann. N.Y. Acad. Sci.* 144:646, 1967.
38. Heilbronn, E. Further experiments on the uptake of acetylcholine and atropine and the release of acetylcholine from mouse brain cortex slices after treatment with physostigmine. *J. Neurochem.* 17:381, 1970.
39. Hemsworth, B. A., and Smith, J. C. Enzymic acetylation of the stereoisomers of α- and β-methyl choline. *Biochem. Pharmacol.* 19:2925, 1970.
40. Hestrin, S. The reaction of acetylcholine and other carboxylic acid derivatives with hydroxylamine, and its analytical application. *J. Biol. Chem.* 180:249, 1949.
41. Hoskin, F. C. G. Proteins in excitable membranes. *Science* 170:1228, 1970.
42. Hoskin, F. C. G., Kremzner, L. T., and Rosenberg, P. Effects of some cholinesterase inhibitors on the squid giant axon. *Biochem. Pharmacol.* 18:1727, 1969.
43. Hoskin, F. C. G., and Rosenberg, P. Alteration of acetylcholine penetration into, and effects on, venom-treated squid axons by physostigmine and related compounds. *J. Gen. Physiol.* 47:1117, 1964.
44. Hoskin, F. C. G., and Rosenberg, P. Penetration of an organophosphorus compound into squid axon and its effects on metabolism and function. *Science* 156:966, 1967.

45. Hoskin, F. C. G., Rosenberg, P., and Brzin, M. Re-examination of the effect of DFP on electrical and cholinesterase activity of squid giant axon. *Proc. Natl. Acad. Sci. U.S.A.* 55:1231, 1966.
46. Karlin, A. Chemical modification of the active site of the acetylcholine receptor. *J. Gen. Physiol.* 54 (No. 1, part 2):245, 1969.
47. Kása, P., Mann, S. P., and Hebb, C. Localization of choline acetyltransferase. *Nature (Lond.)* 226:812, 814, 1970.
48. Kitz, R. J., Ginsburg, S., and Wilson, I. B. The reaction of acetylcholinesterase with O-dimethylcarbamyl esters of quaternary quinolinium compounds. *Biochem. Pharmacol.* 16:2201, 1967.
49. Koelle, G. B., Davis, R., and Gromadzki, C. G. Electron microscopic localization of cholinesterases by means of gold salts. *Ann. N.Y. Acad. Sci.* 144:613, 1967.
50. Krupka, M. R. Evidence for an intermediate in the acetylation reaction of acetylcholinesterase. *Biochemistry* 6:1183, 1967.
51. Main, A. R., and Iverson, F. Measurement of the affinity and phosphorylation constants governing irreversible inhibition of cholinesterases by diisopropyl phosphorofluoridate. *Biochem. J.* 100:525, 1966.
52. Miledi, R., Molinoff, P., and Potter, L. T. Isolation of the cholinergic receptor protein of *Torpedo* electric tissue. *Nature (Lond.)* 229:554, 1971.
53. Nachmansohn, D., and Berman, M. Studies on choline acetylase. *J. Biol. Chem.* 165:551, 1946.
54. O'Brien, R. D., and Gilmour, L. P. A muscarone-binding material in electroplax and its relation to the acetylcholine receptor. I. Centrifugal assay. *Proc. Natl. Acad. Sci. U.S.A.* 63:496, 1969. [See also references 2-5 in that paper (O'Brien and Gilmour).] See also Trams, E. G., Irwin, R. L., Lauter, C. J., and Hein, M. M. Properties of electroplax protein. *Biochim. Biophys. Acta* 58:602, 1962, and Beychok, S. On the problem of isolation of the specific acetylcholine receptor. *Biochem. Pharmacol.* 14:1249, 1965.
55. O'Brien, R. D., Gilmour, L. P., and Eldefrawi, M. E. A muscarone-binding material in electroplax and its relation to the acetylcholine receptor. II. Dialysis assay. *Proc. Natl. Acad. Sci. U.S.A.* 65:438, 1970.
56. Parisi, M., Rivas, E., and De Robertis, E. Conductance changes produced by acetylcholine in lipidic membranes containing a proteolipid from *Electrophorus*. *Science* 172:56, 1971.
57. Ritchie, A. K., and Goldberg, A. M. Vesicular and synaptoplasmic synthesis of acetylcholine. *Science* 169:489, 1970.
58. Rosenberg, P., and Dettbarn, W.-D. Ester-splitting activity of the electroplax. *Biochim. Biophys. Acta* 69:103, 1963. [See also reference 10 in that paper (Rosenberg and Dettbarn).]
59. Webb, G. D. Affinity of benzoquinonium and ambenonium derivatives for the acetylcholine receptor, tested on the electroplax, and for acetylcholinesterase in solution. *Biochim. Biophys. Acta* 102:172, 1965.
60. Whittaker, V. P., Michaelson, I. A., and Kirkland, R. J. A. The separation of

synaptic vesicles from disrupted nerve-ending particles ("synaptosomes"). *Biochem. J.* 90:293, 1964.

61. Wilson, I. B., Harrison, M. A., and Ginsburg, S. Carbamyl derivatives of acetylcholinesterase. *J. Biol. Chem.* 236:1498, 1961.
62. Wilson, I. B., Hatch, M. A., and Ginsburg, S. Carbamylation of acetylcholinesterase. *J. Biol. Chem.* 235:2312, 1960.
63. Compare: Dettbarn, W.-D., Rosenberg, P., and Nachmansohn, D. Restoration by a specific chemical reaction of "irreversibly" blocked axonal electrical activity. *Life Sci.* 3:55, 1964, with: Ehrenpreis, S., Munson, H. R., Jr., Bigo-Gullino, M., and Avery, M. A. Lack of correlation between cholinesterase activity and responsiveness of nerve (lobster) and smooth muscle (rabbit aorta). *Life Sci.* 3:809, 1964.

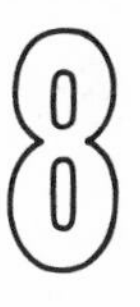

Amino Acid Transmitters

Eugene Roberts and
Richard Hammerschlag

GAMMA-AMINOBUTYRIC ACID (GABA)

The first report of the presence of γ-aminobutyric acid (GABA) in uniquely large concentrations in the vertebrate central nervous system was made in 1950. For several years this finding remained a biochemical curiosity; then definitive evidence for an inhibitory function for GABA at synapses came from studies of the effects of crude brain extracts on the function of the crayfish stretch receptor system. The establishment of GABA as the major factor in brain extracts responsible for the inhibitory action on the stretch receptor system accelerated physiological-to-biochemical correlative efforts. As a result of work in a number of laboratories, GABA now appears established as an inhibitory transmitter at the crustacean neuromuscular junction. Many lines of evidence also make it seem probable that GABA is a major inhibitory transmitter in the vertebrate central nervous system [2, 4, 17, 21, 22].

FORMATION AND METABOLISM OF GABA IN THE NERVOUS SYSTEM

An outline of the principal known reactions of GABA is shown in Figure 8-1, and those of major concern to the present discussion are emphasized (see also Chap. 10). GABA is formed in the CNS of vertebrate organisms to a large extent, if not entirely, from L-glutamic acid. The reaction is catalyzed by an L-glutamic acid decarboxylase (GAD), an enzyme found in mammalian organisms only in the CNS, largely in gray matter. This enzyme is inhibited by anions and carbonyl-trapping agents. (The α-decarboxylation of L-glutamic acid now has been shown to occur in kidney and several other nonneuronal tissues, including glial cells and cerebral blood vessels [67, 68, 85]. The enzyme catalyzing this reaction requires high concentrations of anions for maximal activity and is activated by carbonyl-trapping agents.) The reversible transamination of GABA with α-ketoglutarate is catalyzed by an amino transferase, GABA-T, which is found chiefly in the gray matter in the

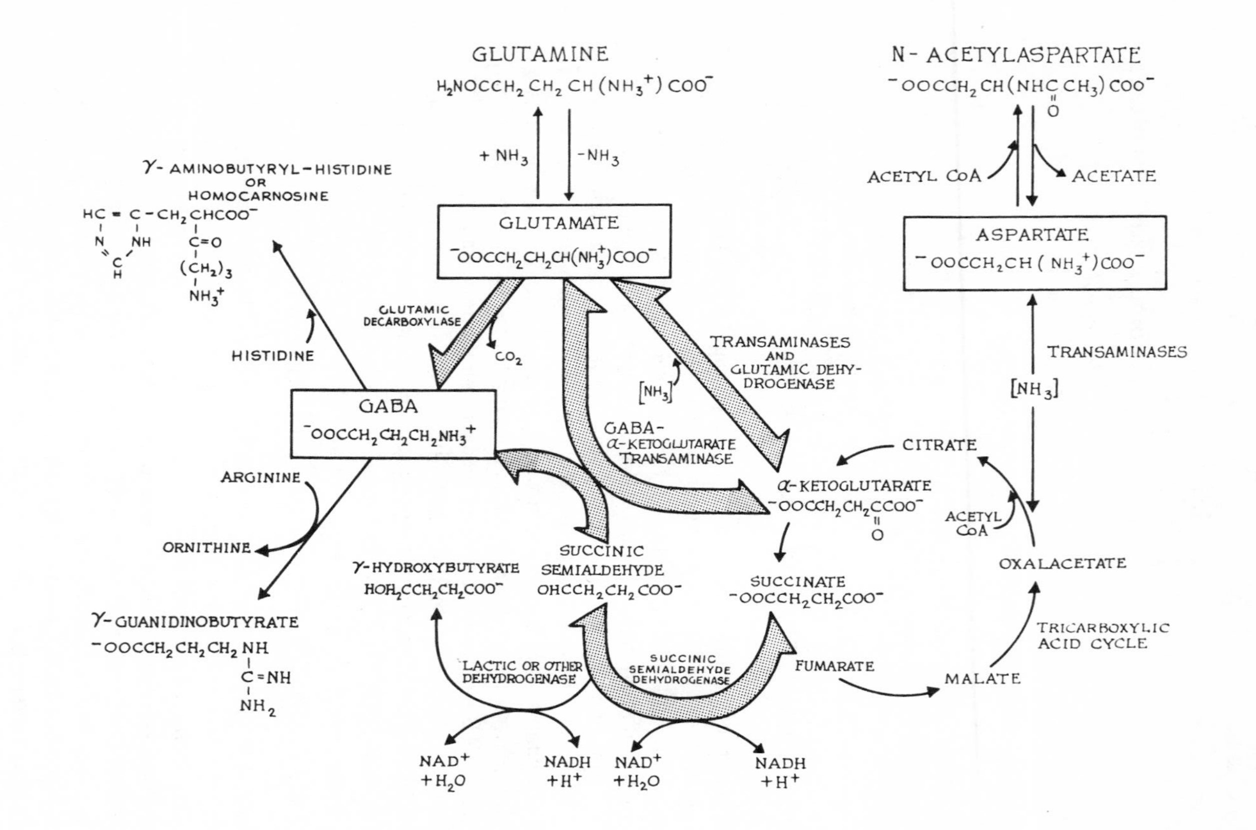

FIGURE 8-1. Principal known reactions of GABA, glutamate, and aspartate in the nervous system. The reactions pertinent to GABA metabolism are emphasized by the large arrows.

CNS but is also found in other tissues. Both GAD and GABA-T are B_6 enzymes that require pyridoxal phosphate as coenzyme (see Chap. 25). The products of the transaminase reaction are succinic semialdehyde and glutamic acid. If a ready metabolic source of succinic semialdehyde were available, GABA could be formed by the reversal of the reaction. To date, however, no convincing evidence has been adduced for the formation of significant amounts of GABA by reactions other than the decarboxylation of L-glutamic acid.

It has been shown that γ-hydroxybutyric acid can be formed enzymatically by reduction of succinic semialdehyde. It also has been shown that rat brain in vivo is capable of converting intracisternally administered ^{3}H-GABA to γ-hydroxybutyrate to a small extent. There is controversy, however, about whether γ-hydroxybutyrate occurs in significant amounts in nervous tissue.

Brain also contains a dehydrogenase which catalyzes the oxidation of succinic semialdehyde to succinic acid, which in turn can be oxidized via the reactions of the tricarboxylic acid cycle. Similar enzymes (GAD, GABA-T, and succinic semialdehyde dehydrogenase) are present in both peripheral and central nervous systems of the lobster [17, 70].

Consistent with these relationships, glutamic acid, GABA, and succinic semialdehyde can be oxidized by various brain preparations and can support oxidative phosphorylation. Therefore, in both vertebrate and invertebrate nervous systems as a whole, a metabolic shunt (termed the "GABA shunt"; see also Chap. 10) can exist around the α-ketoglutarate oxidase system of the tricarboxylic acid cycle, the operation of which depends on the unique presence of GAD and GABA, a formulation consistent with all in vitro and in vivo studies performed to date. The occurrence and metabolism of homocarnosine, guanidinobutyric acid, and other possible metabolites of GABA are reviewed elsewhere [2]. The most recent assessment of the quantitative significance of the GABA shunt pathway in guinea pig cortical slices in vitro showed the GABA flux to be about 8 percent of the total tricarboxylic acid flux [29]. It is not clear what this kind of data might mean in terms of information processing in the nervous system because the components of the GABA system are not homogeneously distributed among all of the diverse cellular structures in the CNS, nor probably even among dendrites, somata, axons, and synaptosomes of individual neurons. It is doubtful that one can talk meaningfully about overall GABA metabolism.

Virtually all the data in the literature are consistent with the interpretation that the steady-state concentrations of GABA in various brain areas normally are governed by GAD activity and not by GABA-T [2, 21, 22]. In the lobster nervous system, the presence of 100 times more GABA in inhibitory than in excitatory axons also appears to be attributable to the presence of 100 times more GAD activity in the inhibitory axons than in the excitatory ones [70].

Pharmacological studies produced the first evidence that the above two enzymes involved in the GABA shunt are not present in the same location in the vertebrate CNS [22]. When hydroxylamine or aminooxyacetic acid, substances which are potent inhibitors of both GAD and GABA-T in vitro, are administered to animals, only the GABA-T is inhibited, and there are marked elevations of GABA content in the brains of animals so treated. One of the simplest possibilities is that the two enzymes are present in different cell types or in different intracellular sites in the same cells, and that the inhibitors penetrate to the regions containing the transaminase but not to those in which the decarboxylase is located. This is consistent with the findings from all cell fractionation studies.

Study of the distribution of the two chief enzymes of GABA metabolism indicates that both GAD and GABA-T are associated with particulate fractions. Work with subtle cell fractionation procedures, employing electron microscopic monitoring, has shown GAD to be particularly rich in presynaptic nerve ending fractions, whereas GABA-T is found largely in the free mitochondria. The free mitochondria, probably a mixture from neuronal perikarya and dendrites and from glial and endothelial cells, are relatively much richer in GABA-T than are those prepared from isolated presynaptic nerve endings, so it is reasonable to propose that the metabolism of GABA takes place chiefly at postsynaptic intracellular neuronal sites and at extraneuronal intracellular sites. In the case of GAD, only approximately 40 percent of the total enzyme activity in brain homogenates has been recovered in fractions containing the presynaptic endings. These findings strongly suggest the possibility that GAD also is present at other neuronal sites, as well as in presynaptic endings.

Study of the subcellular distribution of such simple molecules as GABA is more difficult than that of enzyme proteins because the possibility of diffusion during the fractionation procedures is greater than it is with macromolecular constituents. Careful studies of guinea pig, rat, and mouse brain homogenates do not show highly selective localization of GABA in the dense presynaptic endings, but rather show the distribution to be similar to that of enzymes used as cytoplasmic markers. The amino acid content of the particulate fractions seems to reflect the occlusion of cytoplasm with organized cell fragments (mainly nerve endings) in these fractions [93]. However, quantitative estimates of the amounts of GABA per synaptosome and the minimal amounts of GABA required to inhibit individual neurons led to the conclusion that the results obtained could not eliminate the possibility that GABA is a presynaptically liberated inhibitory transmitter. In inhibitory nerves in the lobster, both the axons and cell bodies contain GABA in similarly high concentrations [17].

From the above, one may derive the tentative picture that, at least in some inhibitory nerves, both GAD and GABA are present and are distributed throughout the neuron, GAD being somewhat more highly concentrated in the presynaptic endings than elsewhere. GABA-T is contained in mitochondria of all neuronal regions, but it seems to be richer in the mitochondria of those neuronal sites onto which GABA might be liberated. Such regions would be expected to exist in perikarya and dendrites that receive inhibitory inputs and possibly in the glial and endothelial cells that are in the vicinity of inhibitory synapses.

EVIDENCE FOR PRESYNAPTIC RELEASE OF GABA

Stimulation of axons of several nerves that inhibit different lobster muscles has been shown to result in the release of GABA in amounts related to the extent of stimulation, while stimulation of the excitatory nerve does not produce GABA release [105]. GABA is not released in response to inhibitory nerve stimulation in a low calcium medium, a condition under which the stimulation also fails to produce inhibitory junctional potentials in the muscle fibers. On returning to a normal calcium medium, inhibitory junctional potentials and GABA release can again be demonstrated in response to inhibitory nerve stimulation.

Data showing the liberation of GABA on stimulation of specific inhibitory neurons in the vertebrate nervous system are extremely difficult to obtain, and we must content ourselves at the present time with successive approaches to this problem. More GABA and less glutamic acid are liberated from the perforated pial surface of the cortex of cats during sleep than in the aroused state [74], and GABA also has been shown to be released specifically from the surface of the posterior lateral gyri of cats during cortical inhibition produced either by stimulation of the ipsilateral geniculate or by direct stimulation of the cortex [95]. In cats pretreated with aminooxyacetic acid to block GABA metabolism, the rate of GABA release in a perfusate of the fourth ventricle increased threefold over the control level during stimulation of the cerebellar cortex [103]. It was concluded that "GABA in the perfusates is derived from the cerebellar subcortical nuclei which occupy a large part of tissues facing the fourth ventricle and where abundant Purkinje cell axons terminate forming inhibitory synapses" [103].

Considerable available circumstantial physiological evidence (to be mentioned subsequently) suggests that GABA may be the inhibitory transmitter liberated when the Purkinje cells are stimulated. Depolarization of cortical slices of rat brain (preloaded with ^{3}H-GABA in the presence of aminooxyacetic acid) by high K^+ concentrations or electrical stimulation leads to an enhanced release of GABA, which, in the case of the high K^+ concentration, is dependent on the Ca^{2+} in the medium [115]. Finally, it has been shown that a preparation of synaptosomes from rat brain released amino acids into the medium during application of electrical pulses of alternating polarity; glutamate, aspartate, and GABA are released in much greater proportion than are other amino acids [32].

MECHANISM OF ACTION OF GABA ON MEMBRANES

Good evidence has been obtained for an ionic basis of the inhibitory effect of GABA on the postsynaptic regions of vertebrate and invertebrate neurons [21, 22]. Applied GABA alters the membrane conductance to Cl^-, and the membrane potential stays near the resting level. GABA also has a presynaptic inhibitory action at the crayfish neuromuscular junction, imitating the action of the natural inhibitory transmitter by increasing permeability to Cl^- [48], thus decreasing the release of excitatory transmitter substances [120].

If GABA produces increased conductance to Cl^- ions in neuronal membranes, it should facilitate the exchange of extraneuronal and intraneuronal Cl^-. It has not been technically feasible to perform direct measurements on the available invertebrate preparations. Assuming that a similar phenomenon might occur in vertebrate neurons, however, the Cl^- exchange has been studied in a preparation rich in isolated, sealed-off presynaptic terminals made from mouse brain homogenates, using ^{36}Cl as tracer [87]. Surprisingly, there was no Cl^- binding or exchange whatsoever in the nerve ending particles under any conditions tested at physiological pH, while several other anions (I^-, Br^-, NO_3^-, succinate, pyruvate, glutamate, and aspartate) were taken up readily. It seems likely that if GABA exerts a presynaptic inhibitory action in mouse brain, it does so by a mechanism that is different from that in the crayfish neuromuscular system mentioned above. The interesting possibility exists that there is a fundamental difference between presynaptic and postsynaptic mammalian neuronal membranes with regard to Cl^- permeability because physiological observations made with intracellular electrodes show unequivocally that Cl^- ions can move through the membranes of perikarya of vertebrate neurons and that this movement is facilitated by GABA.

Experiments performed with neuronal perikarya isolated from mouse brain similar to those above may give a suitable in vitro system for study of the mechanism of GABA action. Prior to attempting to determine the molecular mechanism of action of GABA on neuronal membranes, it is necessary to obtain a manipulable unicellular or membrane-bounded subcellular preparation in which it can be demonstrated that GABA exerts an accurately measurable chemical effect that is directly related to its known physiological effect in intact neuronal systems.

TRANSPORT OF GABA INTO INTRACELLULAR SITES

The first step in any action of a substance upon cellular structures must be through some physical or chemical association of the substance with those structures. The cessation of action of such a substance could be brought about by the removal of the substance from the sensitive sites by destruction, by transport, or by diffusion. In the case of GABA, it is unlikely that the catabolic mechanism is important in this regard because GABA-T, the only known major degradative enzyme for GABA, is associated almost entirely with mitochondria, which are intracellular, and not with neuronal membrane components.

A detailed analysis of the GABA-binding phenomenon under various conditions in crude mixtures of subcellular particulates and in centrifugally separated and electron-microscopically identified microsomal, synaptosomal, and mitochondrial fractions has provided data which are consistent with a carrier-mediated transfer model for the accumulation of GABA in mouse brain (see [22] for discussion and references). A reasonable picture of what happens to free extraneuronal GABA that is liberated into the extracellular region of the synapse is as follows: The membranes (presynaptic, postsynaptic, and possibly glial) contain highly mobile binding sites for GABA that have an absolute Na^+ requirement for their activity and need a high Na^+ concentration (0.1 M) for maximal activation. Thus, GABA binds to membranes to a greater extent in a high Na^+ environment than in a low

Na^+ concentration. The GABA bound on the outer surface of the membranes equilibrates rapidly with the GABA in solution in the extracellular medium. The GABA bound on the inside of the membranes equilibrates rapidly with GABA in solution intracellularly. The binding sites are partially restricted to one or the other side of the membrane by a barrier. The frequency with which the binding sites traverse the barrier may be dependent upon various asymmetries on the two sides of the membranes (such as redox; degree of phosphorylation; Na^+, K^+, and Cl^- concentrations; levels of sugars, amino acids, nucleotides). Under metabolic conditions, because the concentrations of Na^+ present intraneuronally are lower than those extraneuronally, GABA tends to dissociate from the carrier on the intraneuronal side and to become available for mitochondrial metabolism. Thus, the asymmetrical concentration of Na^+ sets up the conditions for a rapid removal of GABA from the extraneuronal synaptic environment into the intraneuronal environment and a rapid metabolism of the GABA therein.

The energy for operation of carrier-mediated transport systems for amino acids (and possibly many other metabolites) is furnished by the Na^+ gradient that animal cells usually maintain. There is no requirement for the linkage of metabolically generated ATP to each transport carrier, but a utilization of ATP for the operation of the Na^+ ion pump (see Chap. 2) which maintains the gradient could presumably enable the many Na^+-requiring membrane carriers to carry out active transport.

GABA binding takes place in the inhibitory synaptic area of crayfish stretch receptors [113] and in several crustacean neuromuscular preparations [73, 96]. The GABA transport system is Na^+-dependent in the latter preparations and it appears, as in the mammalian system, that the uptake serves to terminate the physiological actions of GABA by rapidly removing it from its sites of action in synaptic clefts.

As mentioned previously, there is no hint to date from the work done with GABA about the manner in which its interaction with membranes produces increases in the Cl^- conductance of the membranes. The inhibitory action of GABA at the crustacean neuromuscular junction probably is not directly related to its Na^+-dependent binding at this structure because substitution of choline for Na^+ does not eliminate or even modify the typical inhibitory effects of GABA [121], although in the absence of added Na^+ the total binding is greatly decreased [73, 113]. Imipramine and desmethylimipramine inhibit the binding of GABA to membranous subcellular particles in brain [111], but these substances do not antagonize the inhibitory action of iontophoretically applied GABA on the synaptic responses of spinal neurons [7]. Hydroxylamine and aminooxyacetic acid greatly increase the intracellular content of GABA by blocking its destruction by GABA-T, but these substances fail to potentiate synaptic inhibition or the inhibitory action of GABA in cortical neurons [83]. Obviously, different strategies must be employed in studying the molecular mechanism of the binding of GABA which is a general property of neuronal membranes, from those used to examine conformational changes resulting when GABA binds to specific, active membrane sites that subserve special functions.

ANTAGONISTS OF THE DEPRESSANT ACTION OF GABA

Specific pharmacological antagonists of putative neurotransmitters are extremely useful tools in helping to establish neurotransmitter roles. Until recently, of all the many substances tested for possible antagonism to the inhibitory action of GABA, picrotoxin was found to be active in more of the test systems than was any other substance. In crustacean muscle, picrotoxin specifically blocks the action of the natural inhibitory transmitter and, similarly, that of GABA [109, 122]. Picrotoxin-GABA antagonism also has been shown in cells of the cuneate nucleus [59] and in oculomotor neurons [100] in the vertebrate CNS. In general, however, the degree of specificity of picrotoxin action is not high enough to establish it as the antagonist of choice to be used in analyses of particular inhibitory pathways for which GABA could be the transmitter. Only recently, bicuculline, a phthalide-isoquinoline alkaloid, has been shown to be a relatively more specific inhibitor of GABA action in neurons in several regions of the mammalian CNS [31, 38, 51, 56, 94] and in the crayfish stretch receptor system [36]. Both picrotoxin and bicuculline are currently being employed in many studies.

GABA SYSTEM IN CEREBELLUM

Extensive correlative neuroanatomical and neurophysiological analyses have been made of the cerebellum. Several excellent books are available on the subject [9, 10, 12, 18]. A schematic presentation of neuronal relationships in the mammalian cerebellar cortex is shown in Figure 8-2.

The overall function of the cerebellum probably is entirely inhibitory. The only output cells of the cerebellar cortex – the Purkinje cells – inhibit monosynaptically in Deiters' and intracerebellar nuclei (see Fig. 8-3). The basket, stellate, and Golgi cells are believed to play inhibitory roles within the cerebellum. The basket cells make numerous powerful inhibitory synapses on the lower region of the somata of the Purkinje cells and on their basal processes, or "preaxons." The superficial stellate cells form inhibitory synapses on the dendrites of Purkinje cells. The Golgi cells make inhibitory synapses on the dendrites of the granule cells. Afferent excitatory inputs reach the cerebellum via the climbing and mossy fibers, which excite the dendrites of the Purkinje and granule cells, respectively. The latter are believed to be the only cells that lie entirely within the cerebellum that have an excitatory function.

From the foregoing, and because of its conveniently layered structure, it appears that, in the vertebrate nervous system, the cerebellum is a favorable site for chemical investigation of the possible substance (or substances) which may mediate the activity of neurons with inhibitory functions. After hand-dissection of frozen dried sections, GABA content and GAD activity have been determined in the molecular layer and the Purkinje and granular cell layers of the gray matter and in the subjacent white matter of the rabbit cerebellum [86]. Also, the GABA-transaminase–succinic-semialdehyde route (GABA-T-S) – the metabolic pathway by which GABA is converted to succinic acid – has been visualized histochemically. Measurements

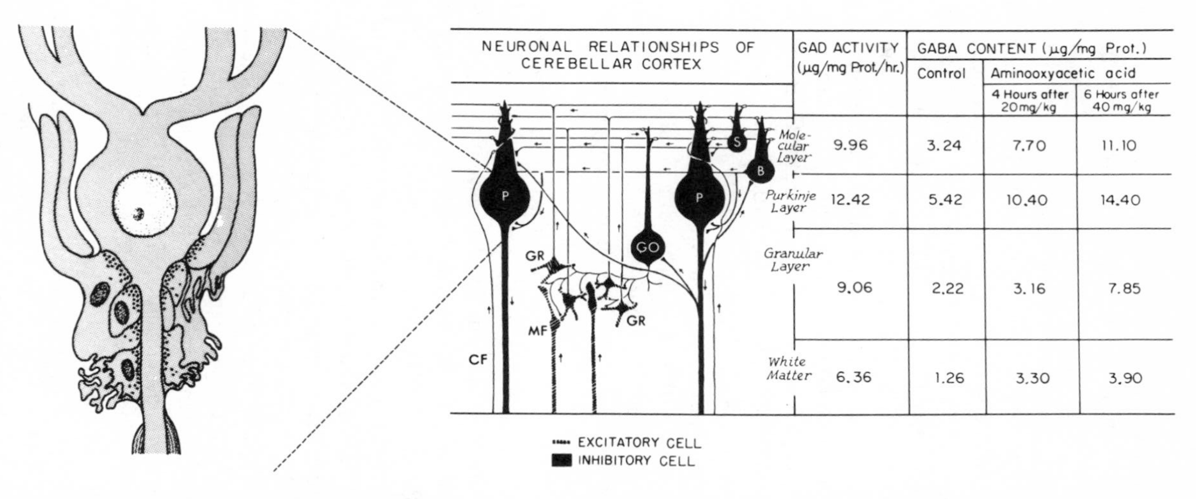

Neuronal relationships of cerebellar cortex	GAD activity (μg/mg Prot/hr.)	GABA content (μg/mg Prot.)		
		Control	Aminooxyacetic acid	
			4 Hours after 20 mg/kg	6 Hours after 40 mg/kg
Molecular Layer	9.96	3.24	7.70	11.10
Purkinje Layer	12.42	5.42	10.40	14.40
Granular Layer	9.06	2.22	3.16	7.85
White Matter	6.36	1.26	3.30	3.90

FIGURE 8-2. Neuronal relationships in the mammalian cerebellar cortex (after Fig. 120 of [9]) and GAD activities and GABA content [86]. *B*, basket cell; *CF*, climbing fiber; *GO*, Golgi cell; *GR*, granule cell; *MF*, mossy fiber; *P*, Purkinje cell; *S*, stellate cell.

of GABA content and the histochemical procedure also have been performed on cerebellar tissue from animals that had been injected previously with aminooxyacetic acid, which in vivo blocks GABA-T, but not GAD, activity and causes elevation of GABA levels.

Both the GAD activity and the GABA content of the layer that contains the Purkinje cells are higher than those found in whole cerebellum or in the other cerebellar cortical regions (see Fig. 8-2). The high values could be attributable to the Purkinje cell bodies themselves, to the numerous presynaptic endings from the basket cells on the somata of the Purkinje cells, or to both. The granular layer is significantly lower in both GABA and GAD. The extensive dendritic arborizations of the Purkinje cells are found largely in the molecular layer, and the axons of the Purkinje cells lie in the granular layer and below. GAD might be concentrated in the presynaptic endings of the basket cell axons since cell fractionation studies have shown that the enzyme is found in the synaptosomes in the cerebellum as well as in other neural structures. As has been pointed out, it is likely that GAD also is found throughout neurons. A high level of the GABA-T-S, the GABA-metabolizing system, is seen uniformly in the cytoplasm of the Purkinje cell. The nuclei of the cells and the axons always are negative. A spatial separation of GAD and GABA-T activities is consistent with the findings, previously cited for whole brain, that the latter enzyme is associated largely with free mitochondria in subcellular fractionation studies, and that the mitochondria within nerve endings have lower GABA-T activity than do those in a mixed mitochondrial population from neuronal perikarya and glial cells.

The effects of the differential inhibition of the transaminase in vivo are shown in those instances in which aminooxyacetic acid was administered (see Fig. 8-2). The inhibition was verified histochemically. The levels of GABA were raised in all areas studied, the extent of elevation being related to the control GAD level and being greatest in the Purkinje cell layer (see Fig. 8-2).

Analyses of various types of individually dissected neurons from cat central nervous system have shown the Purkinje cells to have the highest concentration of GABA [102]. Since such dissected cells still have the large presynaptic basket cell terminals adhering to them, clear-cut assignment of GABA and GAD on the basis of such determinations could not be made either to the Purkinje cells or to the basket cell terminals. Recent experiments, however, with intracellular recordings from Purkinje cells, have shown that both the basket cell inhibition of these cells and their inhibition by iontophoretically applied GABA are blocked by bicuculline, a GABA antagonist [36]. The evidence is strong, therefore, that basket cell inhibition of Purkinje cells is mediated by GABA.

A reasonable interpretation of the above data, combined with data from the study of transport of GABA by subcellular particles, is that in the cerebellum GABA is formed and stored in the presynaptic endings of the basket cells and, upon stimulation, is released from them onto membranes of Purkinje cells. A part of the released GABA could be transported into the presynaptic endings, where most of it would be retained for subsequent release, or could be taken up by the Purkinje cell bodies, where it could be transported down the axons or metabolized by the GABA-T-S pathway. The possibility of glial uptake and metabolism also must be considered.

The granular layer has lower GAD and GABA levels than do the other layers of the cerebellar gray matter and shows less elevation of GABA content in animals treated with aminooxyacetic acid (see Fig. 8-2). The granule cells are completely negative for GABA-T-S, but the Golgi cells and the endings of the mossy fibers are strongly positive. There are numerous, possibly inhibitory axosomatic synapses on the Golgi cell bodies; some of the axons arise from the Purkinje cells. The GAD-GABA system probably is situated at the presynaptic endings at these sites, since the Purkinje cells probably are GABA neurons (see below). The GABA-T-S activity in the Golgi cell bodies may serve the same function as that postulated for the Purkinje cells. Similarly, the Golgi cells may liberate GABA at the mossy-fiber–granule-cell synapse and the GABA-T-S system visualized at the postsynaptic sites would then serve an intracellular degradative function. Golgi axon terminals probably have endings both on the dendrites of the granule cells and on the mossy fiber terminals. The GAD and GABA levels of the molecular layer and the elevations of GABA observed after treatment of rabbits with aminooxyacetic acid are intermediate between those found for the Purkinje and granular cell layers. A study of numerous preparations showed that the GABA-T-S system is distributed throughout the dense axodendritic networks, with only an occasional basket cell appearing positive; the stellate cells and most of the basket cell bodies did not show distinctive formazan deposits characteristic of GABA-T-S activity. Clear visualization of whole dendritic branchings was never achieved by the histochemical procedure.

The inhibitory actions of the recurrent Purkinje axon collaterals probably are also mediated by GABA. It might be expected that the postsynaptic receptor sites for inhibition on the dendrites of the Purkinje cells would have GABA-T-S activity. Such regions probably could only be identified by a combination of histochemical and electron microscopic procedures. Recent data from studies with picrotoxin and bicuculline suggest that GABA may be the transmitter that mediates inhibition at the stellate–Purkinje cell [130] and at Golgi–granule cell synapses [31], as well as at the basket cell–Purkinje junctions in the cerebellar cortex.

GAD activity and GABA were found at the lowest levels in the cerebellar white matter (see Fig. 8-2); the GABA-T-S system was below the level of detection of the histochemical procedure. The principal constituents of this region are the efferent fibers of the Purkinje cells, the axons of the afferent climbing and mossy fibers, glial cells, and a small number of granule cells. The GABA system usually is not found in excitatory cells, so it seems likely that, in this region, GAD and GABA are present in the axons of the Purkinje cells. They may be in a state of transport from the cell body to the presynaptic endings of the Purkinje cells on Deiters' neurons in the vestibular nucleus and the cells of the intracerebellar nuclei, where the Purkinje cells inhibit monosynaptically. A schematic representation of some of the latter relationships is shown in Figure 8-3 (see pp. 227–261 of [9] for detailed discussion).

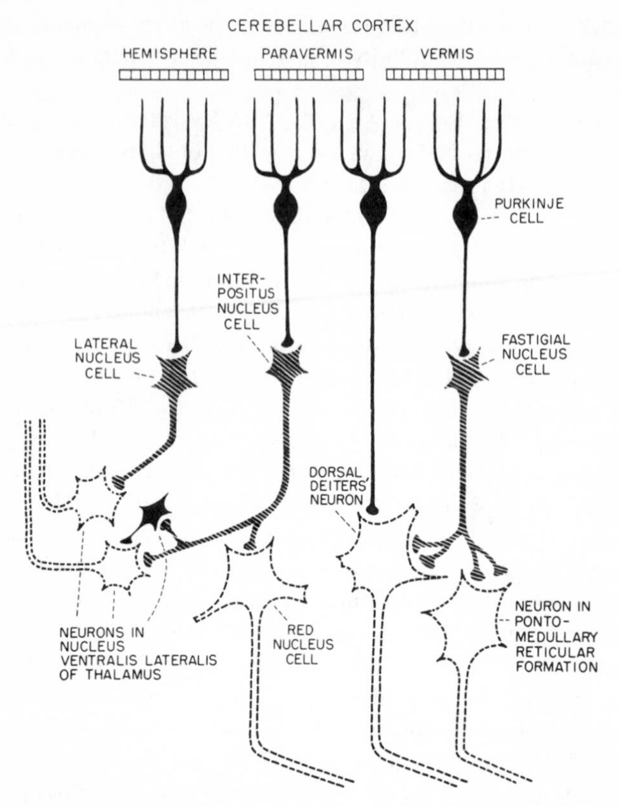

FIGURE 8-3. Neuronal relationships in the cerebellar efferent system (after Fig. 124 of [9]). The black cells are inhibitory, the others excitatory. Climbing and mossy fibers are not shown in this diagram.

The Purkinje cell input to neurons in Deiters' nucleus and the surrounding reticular formation produces a strong depressant effect involving an increase of membrane conductance and hyperpolarization of these neurons. This action is mimicked by the iontophoretic application of GABA to these neurons. Both natural inhibition and GABA were found to produce an increase in the Cl^- conductance of the postsynaptic membranes in such cells [101]. The action of both natural inhibition and GABA on Deiters' neurons is blocked by bicuculline [36]. The above data make a strong physiological case for GABA being the transmitter that is released by the Purkinje cells onto the postsynaptic membranes of recipient cells.

A similar conclusion was reached from combined neuroanatomical and biochemical approaches [58]. GAD activity was found to be 2 1/2 times greater in the dorsal than in the ventral part of Deiters' nucleus; Purkinje cell axons terminate in the dorsal but not in the ventral part. Destruction of the Purkinje cells in the anterior cerebellar vermis results in a loss of approximately two-thirds of the GAD activity in the dorsal region of Deiters' nucleus and no change in the ventral portion.

Destruction of those Purkinje cells in the left cerebellar hemisphere that send their axons to the left nucleus interpositus results in loss of GAD similar to that above, while there is no decrease from the normal level in the right nucleus interpositus. Concomitant determinations of choline acetyltransferase activity show no changes to have taken place in any of the regions as a consequence of the experimental procedures. Taken together, the various anatomical and chemical observations made in the latter study suggest that there is a high concentration of GAD, and consequently GABA, within the Purkinje axon terminals. As would be expected from cells that receive a GABA input, cells in Deiters' and intracerebellar nuclei stain heavily for GABA-T-S activity [86].

In experiments with cats with chronically deafferented cerebella, it was shown that a single effective antidromic stimulus in Purkinje cell axons caused an inhibitory silence [52]. In individual rhythmically discharging Purkinje cells, stronger stimulus caused longer silence. This inhibitory action was attributed to the activity of axon collaterals of other Purkinje cells upon the cells being monitored (see Fig. 8-2). The inhibitory action of GABA liberated in this manner lasted until the GABA was removed from extracellular to intracellular sites by its specific Na^+-dependent binding and transport system which is present on the external surfaces of membranes at the site. It would be expected that a transport mechanism for the inactivation of inhibitory transmitter may be relatively slow (by comparison with the rate of destruction of acetylcholine by cholinesterase), and that the Purkinje axons could be stimulated to such an extent that GABA would accumulate at synapses, resulting in prolonged inhibition.

The results of a detailed morphological and biochemical correlative study of the developing chick embryo cerebellum are consistent with the interpretation that the development of all of the components of the GABA system is better correlated with the development and increase in recognizable synaptic structures than with the accretion of the total mass of the cerebellum [89]. The subcellular fractionation data at all stages of development show GAD to be more highly concentrated in presynaptic endings than elsewhere and the GABA-T to be particularly high in the free mitochondria. The results fit the suggestion that, in the chick cerebellum, GABA is formed largely at presynaptic sites and metabolized at the postsynaptic sites onto which it is liberated.

GABA may be involved not only as an inhibitory transmitter within the cerebellum and as the substance mediating virtually all signals that leave the cerebellum; it also may take part in information processing beyond the Purkinje cell synapses. Rabbit oculomotor neurons are inhibited by neurons found in the vestibular nuclear complex, and this inhibition is blocked by intravenously administered picrotoxin [100]. Picrotoxin administered electrophoretically onto oculomotor neurons blocks the inhibition from secondary vestibular neurons and also blocks the inhibition produced by the application of GABA [100]. Although glycine also inhibits the neurons tested, the inhibition by this amino acid is not affected by picrotoxin. On the other hand, strychnine blocks the action of glycine, but not that of GABA on the oculomotor neurons.

GABA IN OTHER VERTEBRATE NEURONAL SYSTEMS

Biochemical analytical data have shown the presence of GAD and GABA in many regions of the vertebrate CNS. Therefore, it appears likely that GABA plays an inhibitory role in many neuronal systems. In no instance, however, have the functional relationships been worked out to the same extent as they have in the cerebellum.

Retina

The retina is a layered structure of the central nervous system and analyses of the components of the GABA systems in various layers have been performed. Analyses were carried out on four hand-dissected layers of rabbit retina, as shown in Figure 8-4 [88]. The cellular layers all contained GABA and GAD activity, higher contents of GABA being accompanied by higher levels of GAD activity. The highest

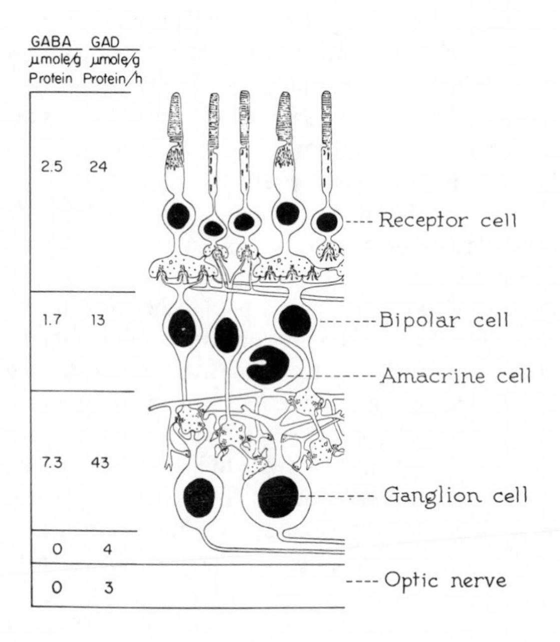

FIGURE 8-4. GABA content and GAD activity in retinal layers [88] (after Fig. 23 of [44]).

values for GABA and GAD were found in the third layer – the ganglion cell layer – and the next highest in the receptor-containing layer. The region of the optic fibers showed no GABA and only traces of GAD. Therefore the neurons containing GABA probably are not the retinal ganglion cells because their axons furnish the bulk of the fibers of the optic nerve.

The morphology of the outer synaptic layer, which is the region of synapses between receptors and bipolar cells, has been studied extensively, but its transmitter chemistry is not known. In this region there are horizontal retinal neurons that form synaptic contact with dendrites and somata of bipolar cells and probably with other horizontal cells, possibly receiving an input from the receptor cells. The inner synaptic layer, the region of synapses of bipolar with ganglion cells, also has been studied extensively [8, 45, 61]. This layer is known to contain various types of horizontal intraretinal cells. The inner nuclear layer probably is a region in which many types of neurons, employing different transmitters, are involved in information processing. Three catecholamine-containing regions within this layer have been identified in rabbit retina, the major catecholamine being dopamine. This layer also appears to be the major, if not the only, site of cholinergic endings in the retina (see [19] for details). The reader is referred to several recent and pertinent morphological and physiological studies of the retina [44, 46, 78, 97, 104, 127, 128].

Some interesting physiological observations are available with regard to the action of GABA and glutamic acid in the visual system. In young chicks with an incomplete blood-brain barrier to GABA, intravenously administered GABA can block all conducted impulses from the retina to the optic tectum. Probably the first effects of GABA could be localized at the bipolar cells or peripheral to them, perhaps at the synapses between the receptors and bipolar cells (see [22] for discussion). GABA applied onto the bullfrog retina inhibits the spike potential of retinal ganglion cells [81]. Recordings were made from ganglion cells in isolated rabbit retinas maintained in vitro [25] in which the effects of a variety of substances added to the bathing fluid were studied. The evidence obtained was consistent with the existence in the retina of cholinergic and adrenergic synapses, and both GABA and glycine were inhibitory at relatively high concentrations. Only small effects of high concentrations of glutamate were observed, although a previous study has indicated that, in vivo, GABA opposed the excitatory action of glutamic acid in the retina [20].

Determinations have been made of GABA and glutamate contents and of GAD and GABA-T activities in frog retina during adaptation to light and darkness [64]. In dark-adapted retinas the GABA and GAD levels are approximately one-half of those found in the light adapted ones, while the glutamate and GABA-T activities are the same in both. These data suggest that the decrease in GABA in the dark is attributable to a decrease in GAD activity and that the GABA system may, indeed, be important in regulation of retinal sensitivity to light. The fascinating problem of how nerve activity influences GAD activity in the retina remains to be solved.

Hippocampus

The hippocampal formation is another structure suitable for laminar analysis. Hippocampal pyramidal cells are excited through axodendritic synapses. The discharges of impulses down the axons of the pyramidal cells are controlled by inhibitory synapses on their somata. The presynaptic inhibitory elements originate from basket cell interneurons indigenous to the same regions [26]. A microanalytical laminar study of the superior region of the rat hippocampus (CA I) showed GAD activity to reach a peak around the cell bodies of the pyramidal cells and to decline gradually on either side, a distribution parallel to that of the basket cell system [57]. Transection of the fimbria, the source of afferent input to the pyramidal cells, did not change the localization of GAD, a result consistent with the occurrence of GABA in intrahippocampal interneurons.

The above data, together with the recent finding that naturally produced inhibition in the pyramidal cells through the basket cells and that caused by iontophoretic application of GABA to the pyramidal cells are sensitive to bicuculline [36], suggest strongly that GABA plays an important regulatory role in the activity of the hippocampus.

Cortex

There is convincing physiological evidence that inhibition in the mammalian cortex may be mediated by GABA neurons [47]. There are remarkable similarities between the intracellularly recorded inhibition in neurons of the cat pericruciate cortex that is produced by electrical stimulation of the cortex and that produced by the iontophoretic application of GABA. Both produce hyperpolarization and a great rise in membrane permeability, as shown by a large increase in electrical conductance, and both show closely similar reversal potentials. The mechanism of the inhibition is probably through an increase in Cl^- ion permeability of the postsynaptic membrane.

The action of GABA or of natural inhibition differs greatly from that of glycine or acetylcholine, both of which have some depressant effect on cortical neurons. Bicuculline depresses both the synaptic inhibition of cortical neurons and the effect of GABA on them, while neither picrotoxin nor strychnine has this effect [36]. Physiological [66] and morphological [117] studies suggest that inhibition in the cortex may be attributable almost exclusively to intracortical interneurons, and from the above it is likely that at least some of these are GABA neurons.

AMINO ACIDS IN SPINAL CORD

A SIMPLIFIED SCHEME FOR SYNAPTIC CONNECTIVITY IN THE SPINAL CORD

The spinal cord is one of the few regions of the vertebrate nervous system in which one can tentatively assign transmitter substances to several different functional types of synapses. Although most of the evidence can be described as suggestive at best, a working model can be set up as shown in Figure 8-5. Glutamate might be the chief postsynaptic excitatory transmitter liberated by the presynaptic endings of the dorsal root fibers. Most of these fibers terminate in the dorsal regions of the cord at the beginning of polysynaptic pathways that end on motor neurons (*2,* Fig. 8-5), and only some of them continue to form monosynaptic contacts with the motor neurons (*1,* Fig. 8-5). Therefore, the content of such a transmitter would be expected to be higher in the dorsal than in the ventral portion of gray matter in the cord. The dorsal portion of the spinal gray matter is a region composed mainly of small interneurons arranged in laminae. Most of these interneurons probably are inhibitory, some of them possibly GABA neurons (*3,* Fig. 8-5), and they play important roles in information processing at the point of entry of somatic sensory information into the cord. It is possible that these interneurons may even be involved in a primitive kind of learning at the spinal cord level [13]. The morphological complexity at the cellular level is very great in this region [112]. Certain interneurons then depolarize motor neurons by releasing the excitatory transmitter aspartic acid (*4,* Fig. 8-5), while others may inhibit motor neurons by release of glycine (*5,* Fig. 8-5). Both GABA and GAD as well as glycine are present at all levels of the gray matter of the cord, so GABA and glycine interneurons may

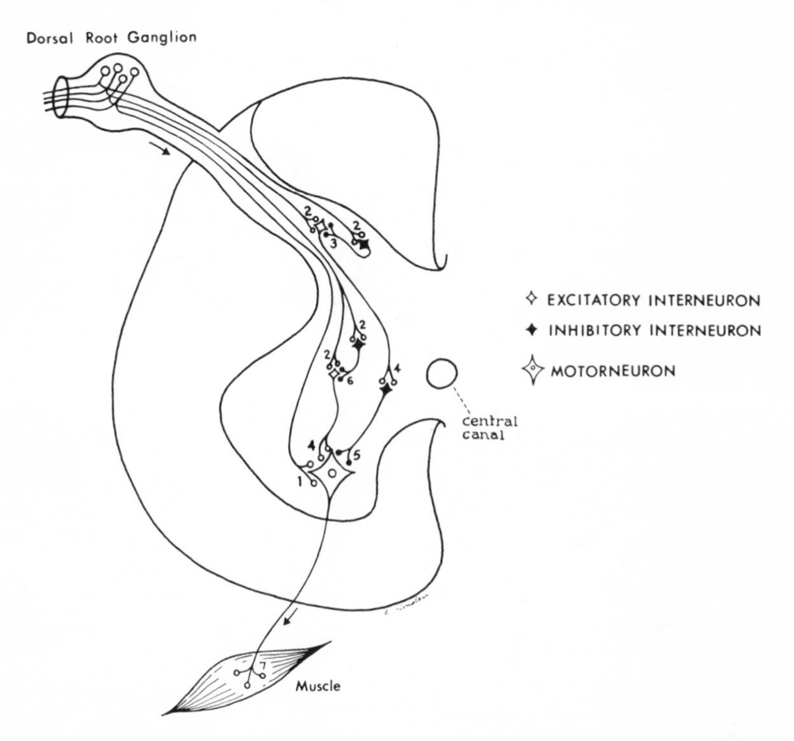

FIGURE 8-5. Suggested loci of action of some putative transmitter substances: *1,* Glutamic acid; *2,* Glutamic acid; *3,* GABA; *4,* Aspartic acid; *5,* Glycine; *6,* Glycine or GABA; *7,* Acetylcholine.

participate in modulatory activities in all regions of the cord, possibly acting as inhibitors in the polysynaptic excitatory pathways (*6,* Fig. 8-5). Finally, it is generally accepted that acetylcholine functions as an excitatory transmitter in motor pathways which may terminate with a neuromuscular junction (*7,* Fig. 8-5).

This scheme of assigning amino acids as transmitters at specific functional synapses is based mainly on two types of studies: the regional distribution of these chemicals in spinal cord, and the electrophysiologically monitored effects of the substances applied iontophoretically directly onto spinal neurons.

DATA SUPPORTING THE SCHEME

In the spinal cord of the monkey [24, 110], the chief enzymes of GABA metabolism, GAD and GABA-T, are significantly higher in the dorsal than in the ventral gray matter (Table 8-1). There is a progressive decrease of the level of both enzymes in the dorsoventral direction, the gradient of GAD being considerably greater than that of GABA-T. Only low levels of both enzymes are present in the dorsal and

TABLE 8-1. GAD and GABA-T Activities in Monkey Spinal Cord

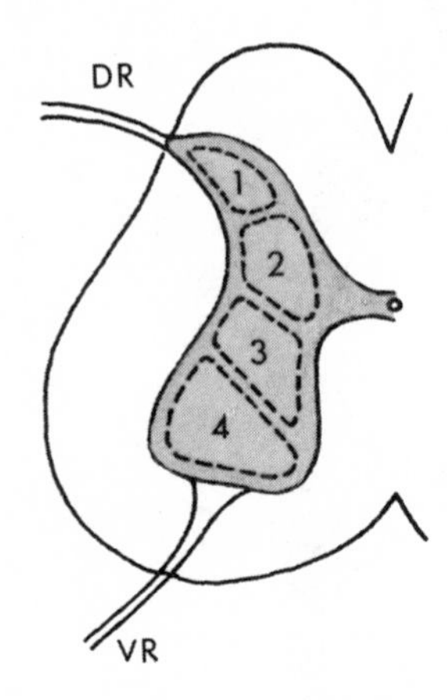

Spinal Cord Areas	GAD (mmoles/kg dry wt/hr)	GABA-T (mmoles/kg dry wt/hr)
Dorsal roots (DR)	1.6	0
Lumbar gray		
1	25.6	315
2	16.9	246
3	11.5	202
4	6.8	197
Ventral roots (VR)	2.5	0

Source: From [24] and [110].

ventral roots. In the cat spinal cord [63, 65], the GABA and GAD levels are approximately twice as high in the dorsal gray matter as in the ventral gray

TABLE 8-2. Amino Acids and Related Enzymes in Spinal Cord[a]

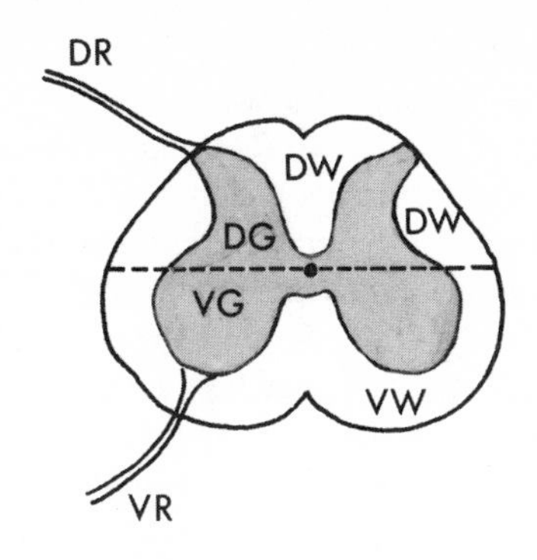

Spinal Cord Areas

Amino Acid-Enzyme System	Dorsal Roots (*DR*)	Dorsal Gray (*DG*)	Ventral Gray (*VG*)	Ventral Roots (*VR*)	Dorsal White (*DW*)	Ventral White (*VW*)
GABA system						
GABA	0.06	2.23(2.12)	1.07(1.33)	0.08	0.43(0.43)	0.44(0.32)
GAD	4	13	7	3	0	0
GABA-T	1	14	13	1	3	3
Glycine system						
Glycine	0.64	5.65(4.05)[b]	7.08(4.95)[b]	0.64	3.04(2.25)	4.39(3.32)[b]
D-Amino acid oxidase (glycine oxidase)	–	2	2	–	0.8	0.4
Glycine amino transferase	–	10	10	–	7	7
Glutamic-aspartic system						
Glutamate	4.61	6.48(4.72)[b]	5.39(4.48)[b]	3.14	4.80(4.18)[c]	3.89(3.41)[c]
Glutamine	1.88	5.30(5.09)	5.35(5.01)	1.90	3.59(3.60)	3.81(3.71)
Glutaminase	12	397	331	13	67	66
Glutamine synthetase	7	67	67	4	20	19
Glutamate dehydrogenase	17	446	448	16	149	139
Aspartate amino transferase	545	5412	5418	537	2454	2480
Aspartate	1.02	2.05(1.54)[c]	3.05(2.41)[c]	1.27	1.11(1.19)	1.29(1.27)

[a]Amino acids: μmoles/g fresh wt tissue; enzymes: μmoles/g fresh wt/hr.
[b]Significant at $P < 0.005$.
[c]Significant at $P < 0.05$.
Sources: Values were obtained from [63] and [65], with the exception of those for D-amino acid oxidase and glycine amino transferase, which came from [43] and [76], respectively. The values in parentheses were obtained 11 to 35 days after aortic occlusion [41].

matter (Table 8-2). It is particularly interesting that the level of GAD is below detection in the dorsal and ventral white matter of cat spinal cord. The above results indicate that the capacity to form GABA in the cord is present almost entirely in interneurons indigenous to the cord and that these interneurons have a higher density in the dorsal gray matter.

Concentrations of glutamate in dorsal roots and in dorsal gray matter are higher than in the corresponding ventral regions [49, 65, 71], suggesting a transmitter role for glutamate in primary afferent neurons. In contrast to glutamate and GABA, the glycine and aspartate contents of cat spinal cord are higher in the ventral than in the dorsal gray matter of the cord [65] (see Table 8-2). The distribution of glycine suggests that it may be the chief, if not the only, transmitter of interneurons that are active on the motor neurons themselves. The distribution of aspartate conforms to the distribution that would be expected for an excitatory transmitter of polysynaptic pathways in the cord. Decreases in aspartate levels are found to correlate well with the decrease in the number of interneurons that follows a period of anoxia produced by aortic occlusion; the decreases in glutamate levels do not correlate with the loss of interneurons [41] (see Table 8-2). Although GABA content does not diminish significantly whereas that of glycine does, it remains possible that GABA is in interneurons that are less sensitive to anoxia. Neither the clinical tests showing loss of inhibitory reflexes nor the histological methods employed to date [41] have been able to distinguish types of interneurons clearly.

GLYCINE

The data discussed in the preceding section, together with a large body of pharmacological and electrophysiological evidence, strongly support the role of glycine as an inhibitory transmitter in the spinal cord (see [1] and [6] for extended bibliographies).

METABOLISM OF GLYCINE IN NERVOUS TISSUE

Although glycine metabolism has been studied extensively in various species and tissues, our knowledge of the metabolism of this amino acid in nervous tissue is rudimentary [1]. It is not yet known whether biosynthetic pathways are of any importance in the glycine economy of the spinal cord or of other regions of the CNS, or what contribution to glycine turnover is made by transport of extracellular glycine into neurons and the transport of synaptically liberated glycine into intracellular presynaptic and postsynaptic neuronal and glial sites.

Glycine may be formed in nervous tissue from serine by a reversible 5,10-methylene tetrahydrofolate–tetrahydrofolate-dependent transformation catalyzed by the enzyme, serine *trans*-hydroxymethylase (Fig. 8-6). Serine can be formed in nervous tissue from glucose via a pathway with 3-phosphoglycerate and 3-phosphoserine as intermediates [33]. Another possible biosynthetic pathway for glycine might be via glyoxylate (see Fig. 8-6). Glyoxylate may be formed by the action of isocitrate lyase [76] and can be transaminated by glutamate to form glycine. Glycine also can be converted to glyoxylate by the action of D-amino acid oxidase; glycine is poorly oxidized by this enzyme but it is a definite substrate [43, 99].

The activities of glycine amino transferase [76] and D-amino acid oxidase [43] are higher in the gray matter of the cord than in the white matter (see Table 8-2), but there are no significant dorsoventral differences. Most of the glycine amino transferase activity of spinal cord homogenates from cat is found in the nonparticulate, soluble, cytoplasmic fraction and not in the synaptosomal fraction. The iontophoretic injection of glycine into the somata of spinal motor neurons in cat, followed by autoradiography, shows that glycine probably is incorporated into

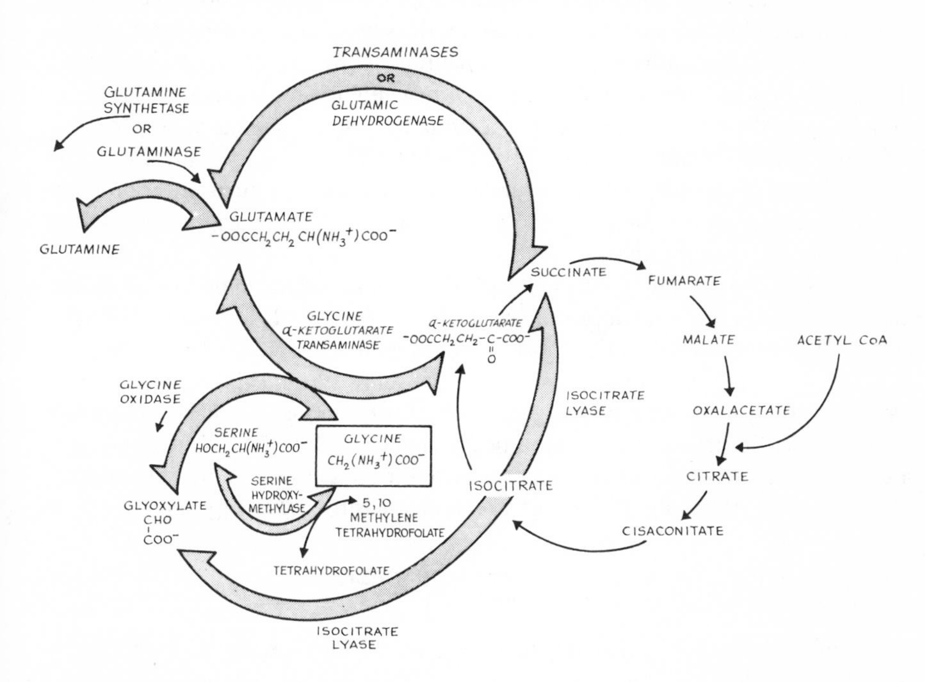

FIGURE 8-6. Principal known reactions of glycine that may be operative in nervous tissue.

protein in the cytoplasm and that the bound glycine spreads rapidly throughout the dendrites [62]. There also may be transfer of bound glycine from the axonal contents of one neuron to another.

EVIDENCE FOR PRESYNAPTIC RELEASE OF GLYCINE

Attempts are being made to show the specific release of glycine from the spinal cord during stimulation of the dorsal roots. Preliminary results with perfused, isolated toad spinal cord prelabeled with ^{14}C-glycine seem to indicate a hopeful approach. The best glycine release appears to occur when the dorsal roots are stimulated at very slow rates [27].

EVIDENCE FOR GLYCINE AS A POSTSYNAPTICALLY ACTIVE INHIBITORY TRANSMITTER IN SPINAL CORD

The finding of a correlation between glycine content and the number of inhibitory interneurons involved in mediating polysynaptic inhibitory reflexes in the cord [41] and the observations that glycine, applied iontophoretically, inhibits spinal interneurons [1, 6] have led to the suggestion that glycine might be the transmitter for a population of interneurons in the spinal cord [1]. A large series of experiments has shown that both glycine and stimulating inhibitory inputs to motor neurons (the synapses of which are located mainly on the soma and large dendrites) act identically on the synaptic membranes of neurons [1, 6]. In all conditions tested, the effects on ionic fluxes are identical, and the equilibrium potentials for both processes are the same before and after altering the internal anionic concentration of motor neurons by intracellular injection of anions that can move through the membrane. Both glycine and the naturally released inhibitory transmitter probably act on the recipient membranes by producing conformational changes that result in increases in Cl^- conductance.

A convulsant, strychnine, acts by blocking the postsynaptic inhibition of motor neurons, thus leading to an incoordination of their activity. It is presumed to interact with the receptor sites on the postsynaptic membranes and to prevent access of inhibitory transmitter to these sites. Strychnine and strychninelike substances reversibly antagonize the action of glycine on spinal neurons, but such substances do not affect the action of iontophoretically applied GABA [39]. The latter observation is strong pharmacological evidence for the supposition that glycine is the transmitter ordinarily liberated at strychnine-sensitive inhibitory spinal synapses.

EVIDENCE FOR CELLULAR UPTAKE AS A MECHANISM TO TERMINATE THE ACTION OF GLYCINE

The synaptic action of glycine may be terminated by the transport of the amino acid out of the synaptic gap, as in the case of GABA; furthermore, the transport sites on the membranes may be distinct from the receptor sites because the synaptic antagonist strychnine fails to affect glycine transport in slices of cortex and spinal cord [37, 39]. A high-affinity accumulation process for glycine has been found in rat and rabbit spinal cord [98].

GLUTAMATE AND ASPARTATE

One of the earliest studies to demonstrate that glutamate and aspartate might affect neural activity involved a search for a substance in brain that triggered the phenomenon of spreading cortical depression [125]. In the course of this study it was found that glutamate is remarkably effective in causing contractions of crustacean muscles. At present, as has been discussed, the case for an excitatory transmitter role for glutamate at primary sensory nerve endings in the spinal cord rests almost entirely on the greater concentration of glutamate in dorsal over ventral regions of spinal cord and on the excitatory effect of glutamate when applied directly to spinal interneurons and motor neurons. Similarly, aspartate is being considered as the excitatory transmitter released from interneurons because of the predominance of this amino acid in ventral cord areas, its association with interneurons, and its physiological excitatory effects on spinal neurons.

The general difficulties involved in assessing whether glutamate and aspartate may be spinal neurotransmitters can be examined by briefly reviewing the generally accepted criteria for transmitter identification; we find that none of these criteria has been satisfied to a significant extent. Because of this current paucity of information, we tend toward speculation and will therefore discuss possible strategies for future research in this area.

SYNTHESIS

Because glutamate and aspartate are common intermediary metabolites of neural tissue, the means for their synthesis is present in all nerve cells, rather than being restricted to specific functional types of neurons, as are acetylcholine and GABA. Glutamate synthesizing enzymes – aspartic amino transferase, glutamic dehydrogenase, and glutaminase – have been assayed at comparable activities in dorsal and ventral spinal roots and cord regions [63] in spite of the apparent dorsal excess of glutamate (see Table 8-2). A degree of enzymatic specificity, however, may exist in the dorsal root ganglion because the concentration gradient of glutamate detected from cell body (ganglion) to axon (dorsal root) does not exist for other equally ubiquitous amino acids, e.g., aspartate and glycine [50, 75]. Thus, glutamate may be synthesized in the cell body and transported to the nerve endings. If glutamate is a transmitter, it is probable that those cells from which it is released contain a mechanism that enables the synthesis of transmitter to be controlled independently of the glutamate required for normal cellular metabolism. Although a unique glutamate synthetic mechanism may exist in primary afferent fibers, it seems more likely that several metabolic pools of glutamate exist in these neurons (similar to the multiple compartments of glutamate demonstrated in brain tissue [3]) and that the stores of glutamate required for transmitter function may be tapped selectively from one particular pool.

STORAGE

A storage of glutamate in sensory nerve endings may be reflected in the concentration excess of this amino acid in dorsal over ventral gray matter in the spinal cord. Presynaptic sensory nerve endings have been shown by electron microscopy to contain synaptic vesicles [107]; these particles may be the morphological basis of the quantal response detected at spinal monosynaptic sensory synapses [84]. It remains to be demonstrated whether glutamate is selectively concentrated in the synaptic vesicles of cord regions that are rich in primary sensory endings.

RELEASE

A most important criterion to be tested is whether glutamate is released by electrical stimulation from vertebrate sensory nerve endings. This has not yet been reported, but a stimulus-related release of glutamate from frog sciatic nerve has been demonstrated [42, 129]. These latter results are probably unrelated to the question of transmission because the glutamate apparently is released from along the length of the axons. The results suggest, however, that the possible release of transmitter glutamate from nerve endings may be difficult to detect above the high background release of glutamate that may be consequent to metabolic events underlying nerve conduction (see also [28]).

POSTSYNAPTIC ACTION

It also will be difficult to satisfy the criterion that application of the amino acid transmitter candidates to neurons should mimic the effects of the natural transmitter. Glutamate and a series of acidic amino acid analogs have been applied iontophoretically to a wide variety of sites in the central nervous system [7, 82]. All neurons thus far tested were similarly excited by both L-glutamate and L-aspartate. These studies have included such primary afferent target sites as spinal interneurons and motor neurons [40] and the pool of secondary sensory neurons comprising the cuneate and gracilis nuclei of the medulla oblongata [60].

It would appear that a receptor site triggering excitation in response to glutamate and aspartate may be a functional and nonspecific part of all nerve cell membranes. It may be necessary to consider the possibility that presynaptic events determine the specificity of amino acid-mediated transmission; i.e., it may be more pertinent to focus on what is released by the presynaptic neurons.

A more rigorous test of this criterion is to determine whether the postsynaptic effects of the applied transmitter candidate quantitatively reproduce the effects of the natural transmitter released by nerve stimulation. To this end, the equilibrium potential (that potential at which the transmitter will neither depolarize nor hyperpolarize the postsynaptic membrane) that was measured for glutamate excitation of spinal motor neurons was found to be at a somewhat more polarized level than that of the natural transmitter [5]. This type of determination is difficult technically because of the complex geometry of central neurons; an inherent uncertainty exists as to whether the membrane potential is changed as much by injected ions at the actual sites of transmitter action as it is at the sites of the recording pipets. The results of the motor neuron studies should not, then, be considered contradictory to a proposed transmitter role for glutamate.

INACTIVATION

The final criterion is concerned with a specific mechanism for the termination of transmitter action. One line of evidence for an enzymatic inactivation of synaptically released acetylcholine is that pretreatment of a cholinergic synapse with an inhibitor of acetylcholinesterase will prolong the action of iontophoretically applied acetylcholine [15]. By contrast, pretreatment with inhibitors of enzymes that metabolize glutamate has no effect on the latency of glutamate-induced excitation of spinal neurons [40]. The similarities in the time course of action of L- and D-glutamate as neural excitants [35] also suggest that extracellular enzymatic inactivation is not an important means of removing glutamate from the synaptic cleft.

An alternate mechanism for termination of transmitter action may be via active transport into cellular elements, such as presynaptic endings (as proposed for norepinephrine in vertebrate sympathetic nerves [14] and GABA at the crustacean neuromuscular junction [73] and the vertebrate CNS [22]). Such a reuptake route would at least be possible for glutamate because spinal cord [72], like a wide variety of brain regions [92], possesses an active uptake mechanism for glutamate. To satisfy this criterion for transmitter identification, it remains to be shown that glutamate uptake can be directly related to the electrical stimulation of primary afferent neurons.

In summary, most of the studies of glutamate and aspartate in spinal cord have added to our knowledge of the general properties of these amino acids in neural tissue such as their nonspecific depolarization of nerve cells, their transport through neural membranes, their key roles as intermediary metabolites, and the possibility, at least for glutamate, of nonspecific release from neural tissue as a consequence of metabolic activity. These phenomena add complexity to the problem of defining an additional highly specific role for glutamate and aspartate as transmitter substances of specific neural pathways.

EVIDENCE FOR GLUTAMATE AS THE EXCITATORY TRANSMITTER AT THE ARTHROPOD NEUROMUSCULAR JUNCTION

The case for glutamate as a vertebrate transmitter is strengthened by the evidence that glutamate may function as an excitatory motor transmitter at the neuromuscular junction of crustaceans and insects. This hypothesis follows from the neurochemical curiosity that, while vertebrate motor neurons are cholinergic and sensory neurons are not, the reverse pattern appears to exist in arthropods, in which acetylcholine is the putative sensory transmitter and glutamate the possible motor transmitter [11]. A release of glutamate during motor nerve stimulation has been demonstrated in the perfusate from the neuromuscular junction of locust [124] and several species of crustaceans [79]. The concentration of glutamate released in the insect study was shown to be proportional to both the frequency of stimulation and the concentration of Ca^{2+} in the bathing fluid. Apparently, Ca^{2+} directly facilitates transmitter release [16], which suggests that the glutamate released is from the nerve endings.

Of a large number of compounds tested, glutamate was found to be the most effective in causing muscle contraction in crustaceans [108, 118, 126] and insects [30, 54, 80, 123]. The action of iontophoretically applied glutamate has further been shown to be localized to specific patches of postsynaptic membrane of crayfish [118]; prolonged administration of glutamate leads to desensitization of the muscle to both applied glutamate and the natural transmitter, without affecting the depressant action of GABA, the probable inhibitory transmitter [119]. Thus, the identity of action of glutamate and the natural transmitter at the arthropod neuromuscular junction has been fairly well established.

Active transport of glutamate has been demonstrated in a crustacean neuromuscular preparation [73]. More recently, the technique of electron microscopic autoradiography has been utilized to demonstrate glutamate uptake into neuromuscular preparations of cockroach [55]. Uptake is greater at the neuromuscular junction than at other regions of the tissue, and it is stimulated approximately twofold by nerve stimulation. This study offers suggestive evidence for a transport-mediated termination of glutamate transmitter action.

Interestingly, the concentrations of glutamate in the cell bodies and axons of excitatory motor fibers in crustaceans are not significantly different from the concentrations in inhibitory motor fibers [106]. Transmitter substances are usually found to be concentrated in the neurons from which they are released, so the discovery of the mechanism by which transmitter glutamate may be sequestered in the excitatory fibers will be an important finding in terms of the function of glutamate at the neuromuscular junction.

Comparative studies of glutamate action at vertebrate sensory synapses and at arthropod neuromuscular junctions will help to clarify whether the properties of a transmitter substance are similar at functionally different types of synapses.

SUGGESTIONS FOR FUTURE RESEARCH

Well-defined and readily accessible synapses, in some ways comparable to the neuromuscular junction, are required for future studies of both glutamate and aspartate as vertebrate transmitters. The cuneate nucleus of the medulla oblongata comprises a relatively homogeneous and superficially localized population of primary afferent nerve endings; as such, it may provide a site for studies concerned with the storage, the electrophysiological effects, and the inactivation of glutamate as a neurotransmitter. The cell bodies of these endings are packaged in dorsal root ganglia, which offer a site at which to study the metabolism of glutamate in cells that may also utilize this amino acid as a transmitter.

In future studies in vertebrates and invertebrates of the mechanisms by which a common metabolite may also be utilized as a neurotransmitter, the several criteria for transmitter identification may have to be redefined to provide a more rigorous framework for investigation of this class of amino acid transmitters.

REGULATION OF TRANSMITTER BIOSYNTHESIS

Information processing in the nervous system is dynamic. Little could be learned about the actual participation of amino acids in this process even if it were possible to identify those specifically concerned and to assay their steady-state concentrations or the maximal capacities of the pertinent transport and enzyme systems. A more pertinent measurement would be to determine at individual synapses the portion of the total turnover of these substances that can be attributed to the presynaptic release of postsynaptically effective amounts. It is obvious, however, that suitable experimental approaches are not yet available to measure such complex biophysical and biochemical phenomena. A realization of the above may help prevent the reader from accepting too readily causal interpretations of correlations between increases or decreases in total concentrations, either in whole brains or in specific brain areas, of the potential amino acid transmitters and the alterations of physiological or behavioral states that are produced by environmental manipulations such as surgery, drugs, or those that take place during development. The biochemist cannot stand alone; for his results to be meaningful, they must be imbedded in a matrix of appropriate morphological, physiological, and behavioral information.

It is possible that, in the CNS, enzymes responsible for the synthesis of putative neurotransmitter substances may be regulated by levels of their immediate product or by the end-product of the synthetic sequence. Two distinct types of regulatory mechanisms involving negative feedback may exist: inhibition of the enzyme activity itself and repression of enzyme level. Regulatory mechanisms involving inhibition of enzyme activity have been demonstrated in several neurotransmitter systems (see also Chap. 6). Pulse-labeling techniques in vivo have shown that elevated levels of norepinephrine in rat brain decrease the synthesis of norepinephrine by end-product inhibition of the rate-limiting enzyme, tyrosine hydroxylase, in catecholamine synthesis [114]. Choline acetyltransferase is inhibited by acetylcholine in vitro, suggesting that synthesis of acetylcholine in synaptic endings of cholinergic nerves may be controlled by the accumulation of the transmitter in synaptic vesicles [77]. Similarly, the inhibitory effect of serotonin on the decarboxylation of 5-hydroxytryptophan in vitro has been suggested as a possible feedback mechanism controlling the synthesis of serotonin [34].

The second type of regulatory mechanism, the repression of enzyme levels, has not been extensively studied to date. In the brain of developing chick embryo, the hydroxylation of tryptophan may be repressed by serotonin [53].

The amount of GABA in nervous tissue certainly is highly correlated with GAD activity. Therefore, it is of interest to attempt to gain an insight into how GAD activity is regulated. Only recently it was demonstrated that, in chick embryo and mouse brains, elevation of endogenous levels of GABA by administration of GABA or by blockade of its metabolic degradation with aminooxyacetic acid is associated with a concomitant reduction of the steady-state levels of GAD, the enzyme that synthesizes GABA [69, 116]. These findings suggest that brain GAD may be regulated by a feedback mechanism involving product repression of enzyme synthesis. The genetically induced high level of acetylcholine in the brains of *ep* mice, an inbred strain known to be susceptible to convulsive seizures, is paralleled by a proportionately increased level of GABA [90, 91]. This may very well imply that the synthesis of GABA is regulated by levels of acetylcholine. Thus, it appears possible that biochemical cross regulation may exist between an inhibitory and an excitatory transmitter system (see [23] for discussion of such possibilities). Because the biosynthetic routes of synaptically liberated glycine, glutamate, and aspartate are not yet known, meaningful studies of relevant enzymes cannot be made at the present time.

ACKNOWLEDGMENTS

This investigation was supported in part by grant NB-01615 from the National Institute of Neurological Diseases and Stroke, National Institutes of Health, and was also made possible in part by a supporting fund established in the name of Robert F. Kennedy.

REFERENCES

BOOKS AND REVIEWS

1. Aprison, M. H., Davidoff, R. A., and Werman, R. Glycine: Its Metabolic and Possible Transmitter Roles in Nervous Tissue. In Lajtha, A. (Ed.), *Handbook of Neurochemistry*. Vol. 3. New York: Plenum, 1970.
2. Baxter, C. F. The Nature of γ-Aminobutyric Acid. In Lajtha, A. (Ed.), *Handbook of Neurochemistry*. Vol. 3. New York: Plenum, 1970.
3. Berl, S., and Clarke, D. D. Compartmentation of Amino Acid Metabolism. In Lajtha, A. (Ed.), *Handbook of Neurochemistry*. Vol. 2. New York: Plenum, 1969, p. 447.
4. Brazier, M. A. B. (Ed.). *The Interneuron*. Berkeley and Los Angeles: University of California Press, 1969.
5. Curtis, D. R. The Actions of Amino Acids Upon Mammalian Neurons. In Curtis, D. R., and McIntyre, A. K. (Eds.), *Studies in Physiology*. Heidelberg: Springer-Verlag, 1965, p. 34.
6. Curtis, D. R., and Johnston, G. A. R. Amino Acid Transmitters. In Lajtha, A. (Ed.), *Handbook of Neurochemistry*. Vol. 4. New York: Plenum, 1970.
7. Curtis, D. R., and Watkins, J. C. The pharmacology of amino acids related to gamma-aminobutyric acid. *Pharmacol. Rev.* 17:347, 1965.
8. Dowling, J. E. The site of visual adaptation. *Science* 155:273, 1967.
9. Eccles, J. C., Ito, M., and Szentágothai, J. *The Cerebellum as a Neuronal Machine*. Berlin: Springer-Verlag, 1967, pp. 227–261.
10. Fields, W. S., and Willis, W. D., Jr. (Eds.). *The Cerebellum in Health and Disease*. St. Louis: Warren H. Green, 1970.

11. Florey, E. Neurotransmitters and modulators in the animal kingdom. *Fed. Proc.* 26:1164, 1967.
12. Fox, C. A., and Snider, R. S. (Eds.). *The Cerebellum.* Amsterdam: Elsevier, 1967.
13. Horridge, G. A. *Interneurons.* London and San Francisco: Freeman, 1968.
14. Iversen, L. L. *The Uptake and Storage of Noradrenaline in Sympathetic Nerves.* Cambridge: Cambridge University Press, 1967, p. 136.
15. Katz, B. *Nerve, Muscle and Synapse.* New York: McGraw-Hill, 1966, p. 106.
16. Katz, B. *The Release of Neural Transmitter Substances.* Springfield, Ill.: Thomas, 1969, p. 34.
17. Kravitz, E. A. Acetylcholine, γ-Aminobutyric Acid, and Glutamic Acid: Physiological and Chemical Studies Related to Their Roles as Neurotransmitter Agents. In Quarton, G. C., Melnechuk, T., and Schmitt, F. O. (Eds.), *The Neurosciences.* New York: Rockefeller University Press, 1967.
18. Llinás, R. (Ed.). *Neurobiology of Cerebellar Evolution and Development.* Chicago: American Medical Association, 1969.
19. Lolley, R. N. Metabolic and Anatomical Specialization Within the Retina. In Lajtha, A. (Ed.), *Handbook of Neurochemistry.* Vol. 2. New York: Plenum, 1969.
20. Noell, W. K. The visual cell: Electric and metabolic manifestations of its life processes. *Am. J. Ophthalmol.* 48:347, 1959.
21. Roberts, E., and Eidelberg, E. Metabolic and neurophysiological roles of γ-aminobutyric acid. *Int. Rev. Neurobiol.* 2:279, 1960.
22. Roberts, E., and Kuriyama, K. Biochemical-physiological correlations in studies of the γ-aminobutyric acid system. *Brain Res.* 8:1, 1968.
23. Roberts, E., and Matthysse, S. Neurochemistry: At the crossroads of neurobiology. *Anno. Rev. Biochem.* 39:777, 1970.

ORIGINAL SOURCES

24. Albers, R. W., and Brady, R. O. The distribution of glutamic decarboxylase in the nervous system of the rhesus monkey. *J. Biol. Chem.* 234:926, 1959.
25. Ames, A., III, and Pollen, D. A. Neurotransmission in central nervous tissue: A study of isolated rabbit retina. *J. Neurophysiol.* 32:424, 1969.
26. Andersen, P., Eccles, J. C., and Loyning, Y. Location of postsynaptic inhibitory synapses on hippocampal pyramids. *J. Neurophysiol.* 27:592, 1964.
27. Aprison, M. H. Studies on the release of glycine in the isolated spinal cord of the toad. *Trans. Am. Soc. Neurochem.* 1:25, 1970.
28. Baker, P. F. An efflux of ninhydrin-positive material associated with the operation of the Na^+ pump in intact crab nerve immersed in Na^+-free solutions. *Biochem. Biophys. Acta* 88:458, 1964.
29. Balasz, R., Machiyama, Y., Hammond, B. J., Julian, T., and Richter, D. The operation of the γ-aminobutyrate bypath of the tricarboxylic acid cycle in brain tissue *in vitro. Biochem. J.* 116:445, 1970.

30. Beranek, R., and Miller, P. L. The action of iontophoretically applied glutamate on insect muscle fibers. *J. Exp. Biol.* 49:83, 1968.
31. Bisti, S., Iosif, G., and Strata, P. Suppression of inhibition in the cerebellar cortex by picrotoxin and bicuculline. *Brain Res.* 28:591, 1971.
32. Bradford, H. F. Metabolic response of synaptosomes to electrical stimulation: Release of amino acids. *Brain Res.* 19:239, 1970.
33. Bridgers, W. F. The biosynthesis of serine in mouse brain extracts. *J. Biol. Chem.* 240:4591, 1965.
34. Contractor, S. F., and Jeacock, M. K. A possible feedback mechanism controlling the biosynthesis of 5-hydroxytryptamine. *Biochem. Pharmacol.* 16:1981, 1967.
35. Crawford, J. M., and Curtis, D. R. The excitation and depression of mammalian cortical neurons by amino acids. *Br. J. Pharmacol.* 23:313, 1964.
36. Curtis, D. R., Duggan, A. W., Felix, D., and Johnston, G. A. R. GABA, bicuculline and central inhibition. *Nature (Lond.)* 226:1222, 1970; and personal communication from D. R. Curtis, 1971.
37. Curtis, D. R., Duggan, A. W., and Johnston, G. A. R. The inactivation of extracellularly administered amino acids in the feline spinal cord. *Exp. Brain Res.* 10:447, 1970.
38. Curtis, D. R., Felix, D., and McLennan, H. GABA and hippocampal inhibition. *Br. J. Pharmacol.* 40:881, 1970.
39. Curtis, D. R., Hosli, L., Johnston, G. A. R., and Johnston, I. H. The hyperpolarization of spinal motoneurones by glycine and related amino acids. *Exp. Brain Res.* 5:235, 1968.
40. Curtis, D. R., Phillis, J. W., and Watkins, J. C. The chemical excitation of spinal neurons by certain acidic amino acids. *J. Physiol. (Lond.)* 150:656, 1960.
41. Davidoff, R. A., Graham, L. T., Jr., Shank, R. P., Werman, R., and Aprison, M. H. Changes in amino acid concentrations associated with loss of spinal interneurons. *J. Neurochem.* 14:1025, 1967.
42. De Feudis, F. V. Glutamate fluxes in peripheral nerves. Second Meeting, Internat. Soc. Neurochem., Milan, 1969 (Abstract).
43. De Marchi, W. J., and Johnston, G. A. R. The oxidation of glycine by D-amino acid oxidase in extracts of mammalian central nervous tissue. *J. Neurochem.* 16:355, 1969.
44. Dowling, J. E., and Boycott, B. B. Organization of the primate retina; electron microscopy. *Proc. R. Soc. Lond. (Biol.)* 166:80, 1966.
45. Dowling, J. E., Brown, J. E., and Major, D. Synapses of horizontal cells in rabbit and rat retina. *Science* 153:1639, 1966.
46. Dowling, J. E., and Werblin, F. S. Organization of retina of the mudpuppy, *Necturus maculosus.* I. Synaptic structure. *J. Neurophysiol.* 32:315, 1969.
47. Dreifuss, J. J., Kelly, J. S., and Krnjevic, K. Cortical inhibition and γ-aminobutyric acid. *Exp. Brain Res.* 9:137, 1969.
48. Dudel, J., and Kuffler, S. W. Presynaptic inhibition at the crayfish neuromuscular junction. *J. Physiol. (Lond.)* 155:543, 1961.

49. Duggan, A. W., and Johnston, G. A. R. Glutamate and related amino acids in cat, dog, and rat spinal roots. *Comp. Gen. Pharmacol.* 1:127, 1970.
50. Duggan, A. W., and Johnston, G. A. R. Glutamate and related amino acids in cat spinal roots, dorsal root ganglia and peripheral nerves. *J. Neurochem.* 17:1205, 1970.
51. Duggan, A. W., and McLennan, H. Bicuculline and inhibition in the thalamus. *Brain Res.* 25:188, 1971.
52. Eccles, J. C., Llinás, R., and Sasaki, K. The action of antidromic impulses on the cerebellar Purkinje cells. *J. Physiol. (Lond.)* 182:316, 1966.
53. Eiduson, S. 5-Hydroxytryptamine in the developing chick brain. Its normal and altered development and possible control by product repression. *J. Neurochem.* 13:923, 1966.
54. Faeder, I. R., and O'Brien, R. D. Responses of perfused isolated leg preparations of the cockroach, *Gromphadorhina portentosa,* to L-glutamate, GABA, picrotoxin, strychnine, and chlorpromazine. *J. Exp. Zool.* 173:203, 1970.
55. Faeder, I. R., and Salpeter, M. M. Glutamate uptake by a stimulated insect nerve muscle preparation. *J. Cell Biol.* 46:300, 1970.
56. Felix, D., and McLennan, H. The effect of bicuculline on the inhibition of mitral cells of the olfactory bulb. *Brain Res.* 25:661, 1971.
57. Fonnum, F., and Storm-Mathisen, J. GABA synthesis in rat hippocampus correlated to the distribution of inhibitory neurones. *Acta Physiol. Scand.* 76:35A, 1969.
58. Fonnum, F., Storm-Mathisen, J., and Walberg, F. Glutamate decarboxylase in inhibitory neurons. A study of the enzyme in Purkinje cell axons and boutons in the cat. *Brain Res.* 20:259, 1970.
59. Galindo, A. GABA-picrotoxin interaction in the mammalian central nervous system. *Brain Res.* 14:763, 1969.
60. Galindo, A., Krnjevic, K., and Schwartz, S. Micro-iontophoretic studies on neurons in the cuneate nucleus. *J. Physiol. (Lond.)* 192:359, 1967.
61. Gallego, A., and Cruz, J. Mammalian retina: Associational nerve cells in ganglion cell layer. *Science* 150:1313, 1965.
62. Globus, A., Lux, H. D., and Schubert, P. Somadendritic spread of intracellularly injected tritiated glycine in cat spinal motoneurons. *Brain Res.* 11:440, 1968.
63. Graham, L. T., Jr., and Aprison, M. H. Distribution of some enzymes associated with the metabolism of glutamate, aspartate, γ-aminobutyrate and glutamine in cat spinal cord. *J. Neurochem.* 16:559, 1969.
64. Graham, L. T., Jr., Baxter, C. F., and Lolley, R. N. *In vivo* influence of light or darkness on the GABA system of the frog (Rana pipiens). *Brain Res.* 20:379, 1970.
65. Graham, L. T., Jr., Shank, R. P., Werman, R., and Aprison, M. H. Distribution of some synaptic transmitter suspects in cat spinal cord. *J. Neurochem.* 14:465, 1967.

66. Grampp, W., and Oscarsson, O. Inhibitory Neurons in the Group I Projection Area of the Cat's Cerebral Cortex. In von Euler, C., Skoglund, S., and Soderberg, U. (Eds.), *Structure and Function of Inhibitory Neuronal Mechanisms.* Oxford: Pergamon, 1968.
67. Haber, B., Kuriyama, K., and Roberts, E. An anion stimulated L-glutamic acid decarboxylase in non-neural tissues: Occurrence and subcellular localization in mouse kidney and developing chick embryo brain. *Biochem. Pharmacol.* 19:1119, 1970.
68. Haber, B., Kuriyama, K., and Roberts, E. L-Glutamic acid decarboxylase: A new type in glial cells and human brain gliomas. *Science* 168:598, 1970.
69. Haber, B., Sze, P. Y., Kuriyama, K., and Roberts, E. GABA as a repressor of L-glutamic decarboxylase (GAD) in developing chick embryo optic lobes. *Brain Res.* 18:545, 1970.
70. Hall, Z. W., Bownds, M. D., and Kravitz, E. A. The metabolism of gamma aminobutyric acid in the lobster nervous system. *J. Cell Biol.* 46:290, 1970.
71. Hammerschlag, R., and Potter, L. T. Amino compounds in bovine dorsal and ventral spinal roots. Submitted for publication.
72. Hammerschlag, R., Potter, L. T., and Vinci, J. Uptake of ^{14}C-glutamate by slices of rat spinal cord. Submitted for publication.
73. Iversen, L. L., and Kravitz, E. A. The metabolism of GABA in the lobster nervous system – uptake of GABA in nerve-muscle preparations. *J. Neurochem.* 15:609, 1968.
74. Jasper, H. H., Khan, R. T., and Elliott, K. A. C. Amino acids released from the cerebral cortex in relation to its state of activation. *Science* 147:1448, 1965.
75. Johnson, J. L., and Aprison, M. H. The distribution of glutamic acid, a transmitter candidate and other amino acids in the dorsal sensory neuron of the cat. *Brain Res.* 24:285, 1971.
76. Johnston, G. A. R., Vitali, M. V., and Alexander, H. M. Regional and subcellular distribution studies on glycine: 2-oxoglutarate transaminase activity in cat spinal cord. *Brain Res.* 20:361, 1970.
77. Kaita, A. A., and Goldberg, A. M. Control of acetylcholine synthesis – the inhibition of choline acetyltransferase by acetylcholine. *J. Neurochem.* 16:1185, 1969.
78. Kaneko, A. Physiological and morphological identification of horizontal, bipolar, and amacrine cells in goldfish retina. *J. Physiol. (Lond.)* 207:623, 1970.
79. Kerkut, G. A., Leake, L. D., Shapira, A., Cowan, S., and Walker, R. J. The presence of glutamate in nerve-muscle perfusates of *Helix, Carcinus,* and *Periplaneta. Comp. Biochem. Physiol.* 15:485, 1965.
80. Kerkut, G. A., and Walker, R. J. The effect of iontophoretic injection of glutamic acid and γ-amino-n-butyric acid on the miniature end plate potentials and contractures of the coxal muscles of the cockroach, *Periplaneta americana. Comp. Biochem. Physiol.* 20:999, 1967.

81. Kishida, K., and Naka, K. I. Amino acids and the spikes from the retinal ganglion cells. *Science* 156:648, 1967.
82. Krnjevic, K. Actions of drugs on single neurons in the cerebral cortex. *Br. Med. Bull.* 21:10, 1965.
83. Krnjevic, K., and Schwartz, S. The Inhibitory Transmitter in the Cerebral Cortex. In von Euler, C., Skoglund, S., and Soderberg, U. (Eds.), *Structure and Function of Inhibitory Neuronal Mechanisms.* Oxford: Pergamon, 1968.
84. Kuno, M. Quantal components of excitatory synaptic potentials in spinal motoneurons. *J. Physiol. (Lond.)* 175:81, 1964.
85. Kuriyama, K., Haber, B., and Roberts, E. Occurrence of a new L-glutamic acid decarboxylase in several blood vessels of the rabbit. *Brain Res.* 23:121, 1970.
86. Kuriyama, K., Haber, B., Sisken, B., and Roberts, E. The γ-aminobutyric acid system in rabbit cerebellum. *Proc. Natl. Acad. Sci. U.S.A.* 55:846, 1966.
87. Kuriyama, K., and Roberts, E. Association of organic and inorganic anions with synaptosomes from mouse brain. *Brain Res.* 26:105, 1971.
88. Kuriyama, K., Sisken, B., Haber, B., and Roberts, E. The gamma-aminobutyric acid system in rabbit retina. *Brain Res.* 9:165, 1968.
89. Kuriyama, K., Sisken, B., Ito, J., Simonsen, D. G., Haber, B., and Roberts, E. The γ-aminobutyric acid system in the developing chick embryo cerebellum. *Brain Res.* 11:412, 1968.
90. Kurokawa, M., Kato, M., and Machiyama, Y. Choline acetylase activity in a convulsive strain of mouse. *Biochim. Biophys. Acta* 50:385, 1961.
91. Kurokawa, M., Machiyama, Y., and Kato, M. Distribution of acetylcholine in the brain during various states of activity. *J. Neurochem.* 10:341, 1963.
92. Levi, G., and Lajtha, A. Cerebral amino acid transport *in vitro.* II. Regional differences in amino acid uptake by slices from central nervous system of the rat. *J. Neurochem.* 12:639, 1965.
93. Mangan, J. L., and Whittaker, V. P. The distribution of free amino acids in subcellular fractions of guinea-pig brain. *Biochem. J.* 98:128, 1966.
94. McLennan, H. Bicuculline and inhibition of crayfish stretch receptor neurones. *Nature (Lond.)* 228:674, 1970.
95. Mitchell, J. F., and Srinivasan, V. Release of ^{3}H-γ-aminobutyric acid from the brain during synaptic inhibition. *Nature (Lond.)* 224:663, 1969.
96. Morin, W. A., and Atwood, H. L. A comparative study of gamma-aminobutyric acid uptake in crustacean nerve-muscle preparations. *Comp. Biochem. Physiol.* 30:577, 1969.
97. Naka, K., and Nye, P. W. Receptive-field organization of the catfish retina: Are at least two lateral mechanisms involved? *J. Neurophysiol.* 33:625, 1970.
98. Neal, M. J., and Pickles, H. G. Uptake of ^{14}C-glycine by rat spinal cord. *Nature (Lond.)* 222:679, 1969.
99. Neims, A. H., and Hellerman, L. Specificity of the D-amino acid oxidase in relation to glycine oxidase activity. *J. Biol. Chem.* 237:C976, 1962.

100. Obata, K., and Highstein, S. M. Blocking by picrotoxin of both vestibular inhibition and GABA action on rabbit oculomotor neurones. *Brain Res.* 18:538, 1970.
101. Obata, K., Ito, M., Ochi, R., and Sato, N. Pharmacological properties of the postsynaptic inhibition by Purkinje cell axons and the action of γ-aminobutyric acid on Deiters' neurones. *Exp. Brain Res.* 4:43, 1967.
102. Obata, K., Otsuka, M., and Tanaka, Y. Determination of gamma-aminobutyric acid in single nerve cells of cat central nervous system. *J. Neurochem.* 17:697, 1970.
103. Obata, K., and Takeda, K. Release of γ-aminobutyric acid into the fourth ventricle induced by stimulation of the cat's cerebellum. *J. Neurochem.* 16:1043, 1969.
104. Ogden, T. E., and Brown, K. T. Intraretinal responses of the cynomolgus monkey to electrical stimulation of the optic nerve and retina. *J. Physiol. (Lond.)* 27:682, 1964.
105. Otsuka, M., Iversen, L. L., Hall, Z. W., and Kravitz, E. A. Release of gamma-aminobutyric acid from inhibitory nerves of lobster. *Proc. Natl. Acad. Sci. U.S.A.* 56:1110, 1966.
106. Otsuka, M., Kravitz, E. A., and Potter, D. D. Physiology and chemical architecture of a lobster ganglion with particular reference to gamma-aminobutyrate and glutamate. *J. Neurophysiol.* 30:725, 1967.
107. Ralston, H. J. Dorsal root projections to dorsal horn neurons in the cat spinal cord. *J. Comp. Neurol.* 132:303, 1968.
108. Robbins, J. The excitation and inhibition of crustacean muscle by amino acids. *J. Physiol. (Lond.)* 148:39, 1959.
109. Robbins, J., and Van der Kloot, W. G. The effect of picrotoxin on peripheral inhibition in the crayfish. *J. Physiol. (Lond.)* 143:541, 1958.
110. Salvador, R. A., and Albers, R. W. The distribution of glutamic-γ-aminobutyric transaminase in the nervous system of the rhesus monkey. *J. Biol. Chem.* 234:922, 1959.
111. Sano, K., and Roberts, E. Binding of γ-aminobutyric acid by mouse brain preparations. *Biochem. Pharmacol.* 12:489, 1963.
112. Scheibel, M. E., and Scheibel, A. B. A Structural Analysis of Spinal Interneurons and Renshaw Cells. In Brazier, M. A. B. (Ed.), *The Interneuron.* Berkeley and Los Angeles: University of California Press, 1969.
113. Sisken, B., and Roberts, E. Radioautographic studies of the binding of γ-aminobutyric acid to the abdominal stretch receptors of the crayfish. *Biochem. Pharmacol.* 13:95, 1964.
114. Spector, S., Gordon, R., Sjoerdsna, A., and Udenfriend, S. End-product inhibition of tyrosine hydroxylase as a possible mechanism for regulation of norepinephrine synthesis. *Mol. Pharmacol.* 3:549, 1967.
115. Srinivasan, V., Neal, M. J., and Mitchell, J. F. The effect of electrical stimulation and high potassium concentrations of the efflux of [^{3}H] γ-aminobutyric acid from brain slices. *J. Neurochem.* 16:1235, 1969.

116. Sze, P. Y. Possible repression of L-glutamic acid decarboxylase by gamma-aminobutyric acid in developing mouse brain. *Brain Res.* 19:322, 1970.
117. Szentágothai, J. The Synapses of Short Local Neurons in the Cerebral Cortex. In Szentágothai, J. (Ed.), *Modern Trends in Neuromorphology* (Symp. Biol. Hung. 5:251, 1965). Budapest: Akademiai Kiado, 1965.
118. Takeuchi, A., and Takeuchi, N. The effect on crayfish muscle of iontophoretically applied glutamate. *J. Physiol. (Lond.)* 170:296, 1964.
119. Takeuchi, A., and Takeuchi, N. Localized action of gamma-aminobutyric acid on the crayfish muscle. *J. Physiol. (Lond.)* 177:225, 1965.
120. Takeuchi, A., and Takeuchi, N. On the permeability of the presynaptic terminal of the crayfish neuromuscular junction during synaptic inhibition and the action of γ-aminobutyric acid. *J. Physiol. (Lond.)* 183:433, 1966.
121. Takeuchi, A., and Takeuchi, N. Unpublished data, 1967.
122. Takeuchi, A., and Takeuchi, N. A study of the action of picrotoxin on the inhibitory neuromuscular junction of the crayfish. *J. Physiol. (Lond.)* 205:377, 1969.
123. Usherwood, P. N. R., and Machili, P. Chemical transmission at the insect excitatory neuromuscular synapse. *Nature (Lond.)* 210:634, 1966.
124. Usherwood, P. N. R., Machili, P., and Leaf, G. L-Glutamate at insect excitatory nerve-muscle synapses. *Nature (Lond.)* 219:1169, 1968.
125. Van Harreveld, A. Compounds in brain extracts causing spreading depression of cerebral cortical activity and contraction of crustacean muscle. *J. Neurochem.* 3:300, 1959.
126. Van Harreveld, A., and Mendelson, M. Glutamate-induced contractions in crustacean muscle. *J. Cell Comp. Physiol.* 54:85, 1959.
127. Werblin, F. S. Response of retinal cells to moving spots: Intracellular recording in *Necturus maculosus. J. Neurophysiol.* 33:342, 1970.
128. Werblin, F. S., and Dowling, J. E. Organization of the retina of the mudpuppy. II. Intracellular recording. *J. Neurophysiol.* 32:339, 1969.
129. Wheeler, D. D., Boyarsky, L. L., and Brooks, W. H. Release of amino acids from nerve during stimulation. *J. Cell Physiol.* 67:141, 1966.
130. Woodward, D. J., Rushmer, D., Hoffer, B. J., Siggins, G. R., Oliver, A. P., and Armstrong, C. Evidence for the presence of stellate cell inhibition in frog cerebellum and for mediation of this inhibition by gamma aminobutyric acid. *Fed. Proc.* 30:318 (Abstract 700), 1971.

SECTION III

Intracellular Metabolism of the Nervous System

Carbohydrate Chemistry of Brain

Howard S. Maker and
Gerard M. Lehrer

THIS CHAPTER outlines the adaptations of carbohydrate metabolism that are specific for brain. Alterations in energy and carbohydrate metabolism produced by varying functional demands of the nervous system under both normal and pathological conditions also are described.

Like the blind men examining an elephant, our methods limit our interpretation of carbohydrate metabolism. Only by assembling data obtained by different techniques can a valid understanding of brain chemistry be obtained. Measurements of blood gases and metabolites give a picture of overall metabolism, but they give no indication of precise localization or intermediate metabolic steps. (The regulation of blood flow is discussed in detail in Chap. 15.) In vitro studies are limited by tissue disconnection and damage as well as by altered access of oxygen and metabolites to the tissue. By utilizing very thin slices and allowing sufficient time at optimal conditions for rebuilding substrates depleted during tissue preparation, levels of intermediate metabolites approximating those in vivo can be attained in vitro. Although electrolyte gradients are maintained and synaptic activity can be recorded, metabolic rates are well below those in situ, probably due to disruption of the tissue. Although electrical stimulation or the addition of potassium or glutamate can elevate these rates to levels closer to those in situ, in vitro conditions are far from physiological. Maximal velocities of brain enzymes can be estimated in various tissue preparations in vitro, but, in the living brain, enzyme and substrate compartmentation, as well as steady-state conditions, produce significant departures from assay conditions. These problems are discussed in detail elsewhere (see Chap. 20).

ENERGY METABOLISM

Oxidative steps of carbohydrate metabolism normally contribute 36 of the 38 high-energy phosphate bonds (~P) generated in the aerobic metabolism of a single glucose molecule. In the basal state, approximately 15 percent of brain glucose is converted to lactate and does not enter the citric acid cycle. Therefore the total net gain of ~P is only 33 equivalents per mole of glucose utilized. The steady-state level of ATP is high and represents the sum of a very rapid synthesis and utilization. Half of the terminal phosphate groups of ATP turn over in about 3 sec, on the average, and probably in considerably shorter periods in certain regions [13]. The level of ~P is kept constant by regulation of ADP phosphorylation in relation to ATP hydrolysis. The very active adenylate kinase reaction, which forms equal amounts of ATP and AMP from ADP, prevents any great accumulation of ADP. A small amount of AMP is present under steady-state conditions; consequently, a relatively small percentage decrease in ATP may lead to a relatively large percentage increase in AMP. AMP is a positive modulator of several reactions leading to increased ATP synthesis, so such an amplification factor provides a sensitive control for maintenance of ATP levels [19].

The level of creatine phosphate in brain is even higher than that of ATP, and creatine phosphokinase is extremely active. The creatine phosphate level is exquisitely sensitive to changes in oxygenation, providing ~P for ADP phosphorylation, thus maintaining ATP levels. The creatine phosphokinase system may also function in regulating mitochondrial activity (see below). The isoenzyme of creatine phosphokinase in the soluble fraction of brain differs from that in muscle; however, there appears to be more than one isoenzyme present in the particulate fraction of brain (see [1, 3–7]).

GLUCOSE TRANSPORT

Under ordinary conditions, the basic substrate for brain metabolism is glucose. The brain depends on glucose both for energy and as the major carbon source for a wide variety of simple and complex molecules. Even a transient decline in the oxidative metabolism of glucose leads to an abrupt disruption of brain function. Despite this dependence on glucose, the brain at rest extracts only about 10 percent of the glucose from the blood, or about 5 mg (28 μmoles) glucose per 100 gm brain per minute. If the flow of blood is slowed, a relatively greater fraction of both the oxygen and the glucose of the blood is taken up by the brain, as expected from simple diffusion. The mechanism of glucose uptake, however, may be more complex (see Chap. 16). Glucose entry into the cell may be a rate-limiting step in glycolysis.

Although the glucose concentration of brain slices does not appear to be related to the glucose level of the medium, both glucose uptake from the blood and glucose concentration in brain vary directly with the glucose concentration in the blood (within physiological limits). Insulin influences glucose uptake in many tissues [23], but such an effect cannot be demonstrated easily in brain (see [2, 7]).

GLYCOLYSIS

The terms "aerobic" and "anaerobic" glycolysis may be misleading. Since the time when metabolism was studied in the Warburg apparatus, workers have usually referred to the production of lactate under conditions of "adequate" oxygen as aerobic glycolysis or under anoxia as anaerobic glycolysis. Aerobic glycolysis, thus defined, would measure only a small portion of total glycolysis. *Glycolytic flux* is defined indirectly as the rate at which glucose must be utilized in order to produce observed rates of ADP phosphorylation.

Figure 9-1 outlines the flow of glycolytic substrates in brain. Brain glucose apparently is readily available for metabolic needs, disappearing rapidly during ischemia [20]. The hexokinase reaction is far to the right of equilibrium and is essentially irreversible. Brain hexokinase phosphorylates other sugars, such as fructose, but at lesser rates than it does glucose. Although fructose is utilized by brain, it cannot sustain normal metabolic rates. Mannose, however, may be an adequate substrate [25]. The electrophoretically slow-moving (type I) isoenzyme of hexokinase is characteristic of brain. Up to 90 percent of brain hexokinase is attached

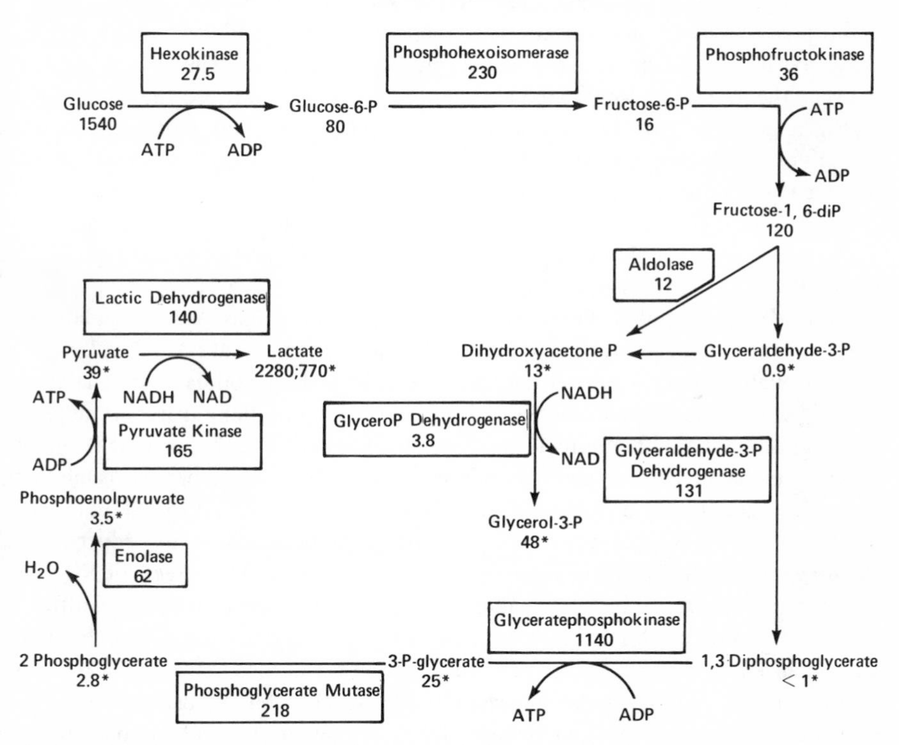

FIGURE 9-1. Glycolysis in brain. Enzyme data from mouse brain homogenates. The figures under each enzyme represent V_{max} at 38°C in mmoles/kg wet wt/min [7]. Metabolite levels from quick-frozen adult mouse brain are indicated in μmoles/kg wet wt. Asterisk indicates 10-day-old mouse brain [6].

to mitochondria after cell fractionation and much of this is firmly bound (see below under "Mitochondria"). Although other glycolytic enzymes may adhere to mitochondria, their association probably is not significant. Mg-ATP^{2-} is the other substrate of the hexokinase reaction, and the ratio of magnesium to ATP may have a regulatory action. Intracellular free magnesium levels in cortex have been estimated to be from 0.5 to 1.5 mM. The measured glycolytic flux, when compared with the maximal velocity of the hexokinase present in brain, indicates that, in the steady state, the hexokinase reaction is 97 percent inhibited. This is an important control point in brain metabolism. Brain hexokinase is inhibited by its product, glucose 6-phosphate, to a lesser extent by ADP, and allosterically by 3-phosphoglycerate and several nucleoside phosphates, including cyclic AMP and free ATP^{4-} Other probable controlling factors are the compartmentation of enzymes and substrates and glucose transport. Glucokinase, a major component of the liver hexokinases, has not been found in brain.

Glucose 6-phosphate represents a branch point in metabolism because it is a common substrate for enzymes involved in glycolytic, pentose phosphate shunt and glycogen-forming pathways. There is also slight, but detectable, glucose 6-phosphatase activity in brain, the significance of which is not clear. In liver, this enzyme is necessary for the conversion of glycogen to glucose. The differences between the liver and brain hexokinase and the differences between the modes of glycogen metabolism of those tissues can be related to the function of liver as a carbohydrate storehouse for the body; brain metabolism is adapted to rapid carbohydrate utilization for energy needs.

In glycolysis, glucose 6-phosphate is the substrate of phosphohexose isomerase. This is a reversible reaction (small free-energy change), whose five to one equilibrium ratio in brain favors glucose 6-phosphate.

Fructose 6-phosphate is the substrate of phosphofructokinase, a key regulatory enzyme controlling glycolysis [19, 22]. The other substrate is Mg-ATP^{2-}. Like other regulatory reactions, it is essentially irreversible. It is modulated by a large number of metabolites and cofactors whose concentrations under different metabolic conditions have great effects on glycolytic flux (Table 9-1). Prominent among these are availability of ~P and citrate levels. Brain phosphofructokinase is inhibited by ATP, magnesium, and citrate and is stimulated by NH_4^+, K^+, P_i, 5'-AMP, cyclic 3',5'-AMP, ADP, and fructose 1,6-diphosphate. Possible inhibition by reduced pyridine nucleotides can be overcome by an increase in pH. The effects of NH_4^+, inorganic phosphate, and AMP are additive, and there is evidence for several different allosteric sites on this complex enzyme. In general, activity is inhibited by ATP and enhanced by substances that accumulate under conditions in which ~P reserves are diminished. Activity is also increased by 3',5'-cyclic AMP, whose presence ultimately leads to increased energy use. Inhibition by citrate would amount to an end-product inhibition if acetate, its precursor, is considered the final product of glycolysis. The concentration of NH_4^+, a product of amino acid and amine metabolism, is controlled by transamination reactions coupled to citric acid cycle intermediates, as well as by glutamic dehydrogenase and glutamine synthetase. Increased

TABLE 9-1. Substances Affecting Phosphofructokinase Activity

Substance	Effect on PFK Activity and on Glycolytic Rate[a]
ATP	–
Citrate	–
Mg^{2+}	–
NH_4^+	+
K^+	+
P_i	+
5′-AMP	+
3′,5′-AMP (cyclic AMP)	+
ADP	+
Fructose diphosphate	+

[a]Inhibitory: –; stimulatory: +.
Source: After [22].

glycolysis, by raising pyruvate and α-ketoglutarate levels, would tend to reduce NH_4^+ levels. The stimulation by NH_4^+ is unaffected by high ATP levels, allowing acceleration of glycolysis to increase availability of α-ketoglutarate and pyruvate for transamination, even when ~P levels are adequate.

When oxygen is admitted to cells metabolizing anaerobically, there is an increased utilization of O_2 and a drop in utilization of glucose and lactate production (Pasteur effect). Modulation of the phosphofructokinase reaction can account directly for the Pasteur effect. In the steady state, ATP and citrate levels in brain are apparently sufficient to keep phosphofructokinase relatively inhibited as long as the level of positive modulators (or disinhibitors) is low. Activation of this enzyme when the steady state is disturbed produces an increase in glycolytic flux that takes place almost as fast as the events changing the internal milieu.

Fructose 1,6-diphosphate is split by brain aldolase to glyceraldehyde 3-phosphate and dihydroxyacetone phosphate. Dihydroxyacetone phosphate is the common substrate for both glycerophosphate dehydrogenase, an enzyme active in NADH oxidation and lipid pathways (see p. 180), and triose phosphate isomerase, which maintains an equilibrium between dihydroxyacetone phosphate and glyceraldehyde phosphate, strongly favoring dihydroxyacetone phosphate.

Following glyceraldehyde phosphate dehydrogenase, brain glycolysis proceeds through the usual steps. It is of interest that brain phosphoenolpyruvate kinase controls an essentially irreversible reaction, which requires not only Mg^{2+} (as do several other glycolytic enzymes) but also K^+ or Na^+. This step may also be regulatory.

Brain tissue, even when it is at rest and well oxygenated, produces a small amount of lactate which is removed in the venous blood. This represents about 13 percent of the pyruvate produced by glycolysis. The measured brain lactate level depends on the success of halting brain metabolism rapidly during tissue processing (see Chap. 20). Five isoenzymes of lactate dehydrogenase are present in adult brain. The one that moves electrophoretically most rapidly toward the anode (band I) predominates. This isoenzyme is generally higher in tissues which are more dependent on aerobic processes for energy; the slower moving isoenzymes are relatively higher in tissue such as white skeletal muscle, which is better adapted to function at suboptimal oxygen levels. The activities of lactate dehydrogenases and their distribution in various brain regions, layers of the retina, brain neoplasms, brain tissue cultures, and during development indicate that the synthesis of these isoenzymes might be controlled by tissue oxygen levels. Lactate dehydrogenase functions in the cytoplasm as a means of oxidizing NADH which accumulates as the result of glyceraldehyde phosphate dehydrogenase action in glycolysis. It thus permits glycolytic ATP production to continue under anaerobic conditions. Lactate dehydrogenase also functions under aerobic conditions because NADH cannot easily penetrate the mitochondrial membrane. The oxidation of NADH in the cytoplasm depends on this reaction and "shuttle mechanisms" to transfer reducing equivalents to the mitochondria (see p. 179).

Another enzyme indirectly associated with glycolysis also participates in the cytoplasmic oxidation of NADH. Glycerol phosphate dehydrogenase reduces dihydroxyacetone phosphate to glycerol 3-phosphate, oxidizing NADH in the process. Under hypoxic conditions, the levels of α-glycerophosphate and lactate initially increase at comparable relative rates, although the amount of lactate production greatly exceeds the former. The relative levels of the oxidized and reduced substrates of these reactions indicate much higher local levels of NADH in brain than are found by gross measurements. In fact, the relative proportions of oxidized and reduced substrates of the reactions linked to the pyridine nucleotides may be a better indicator of local oxidation-reduction states (NAD^+/NADH) in brain than is provided by the direct measurement of the pyridine nucleotides themselves (see [2–4, 6, 7]).

CITRIC ACID CYCLE

The energy output and oxygen consumption in adult brain are associated with high levels of enzyme activity in the citric acid cycle [15]. The actual flux through the citric acid cycle depends on glycolysis and active acetate production (see Chap. 25) which can "push" the cycle, the possible control at several enzyme steps of the cycle, and the local ADP level which is known to be a prime activator of the mitochondrial respiration to which the citric acid cycle is linked. The steady-state level of citrate (Fig. 9-2) in brain is about one-fifth that of glucose. This is relatively high compared with levels of glycolytic intermediates or with that of isocitrate.

As in other tissues, there are two isocitrate dehydrogenases in brain. One is active primarily in the cytoplasm and requires NADP as cofactor; the other, bound

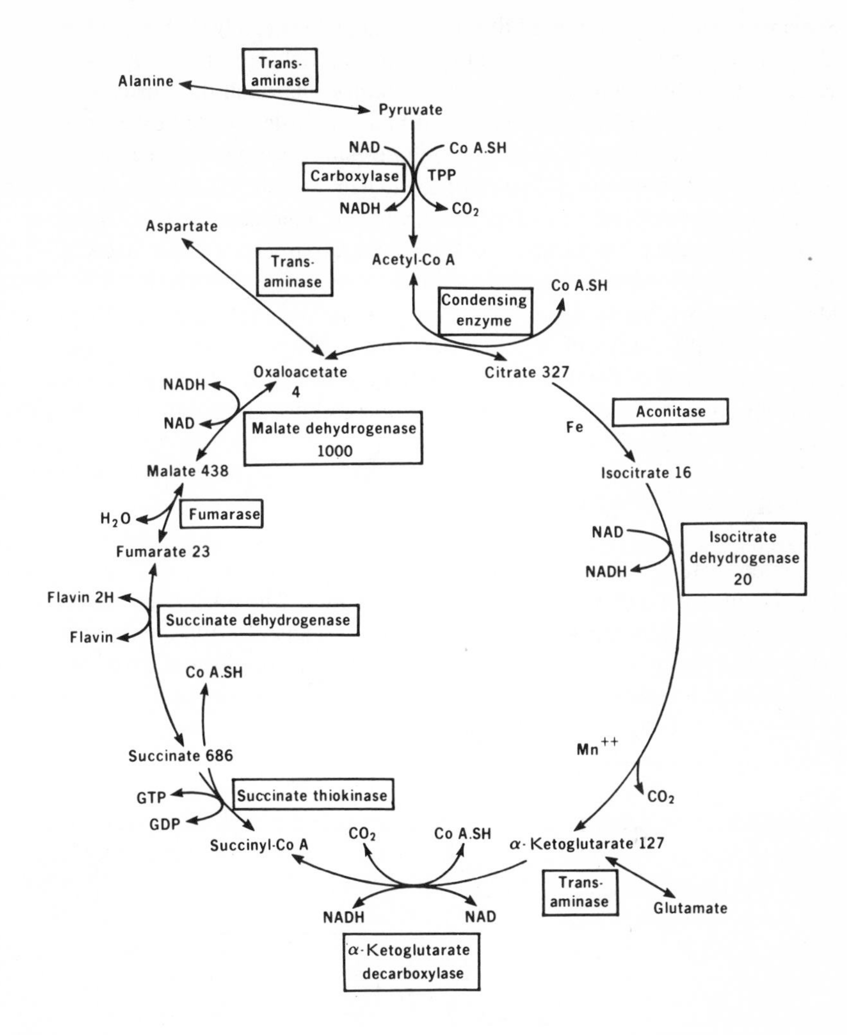

FIGURE 9-2. Citric acid cycle in brain. Data expressed as in Figure 9-1. Metabolite levels from [8].

to mitochondria and requiring NAD, is the enzyme participating in the citric acid cycle. The NAD-linked enzyme catalyzes an essentially irreversible reaction, has allosteric properties, is inhibited by ATP and NADH, and may be stimulated by ADP. The function of cytoplasmic NADP-isocitrate dehydrogenase is uncertain, but it has been postulated that it supplies reduced NADPH that is necessary for many reductive synthetic reactions. The relatively high activity of this enzyme in immature brain and white matter is consistent with such a role. α-Ketoglutarate dehydrogenase, which oxidatively decarboxylates α-ketoglutarate, requires the same cofactors as the pyruvate decarboxylation step (see Chap. 25).

Succinate dehydrogenase of adult mouse brain, which catalyzes the oxidation of succinate to fumarate, is tightly bound to mitochondrial membrane. In brain, succinate dehydrogenase may also have a regulatory role when the steady state is disturbed. Isocitrate and succinate levels of brain are little affected by change in the citric acid cycle flux, as long as an adequate glucose supply is available. The highly unfavorable free-energy change of the malate dehydrogenase reaction is overcome by the rapid removal of oxaloacetate, which is maintained at low concentrations under steady-state conditions by the condensation reaction with acetyl-CoA (see [1–4, 7]).

Malic dehydrogenase is one of several enzymes in the citric acid cycle that is present both in the cytoplasm and in mitochondria. The function of the cytoplasmic components of these enzyme activities is not known, but they may function in the transfer of hydrogen from the cytoplasm into mitochondria (see p. 179).

GLYCOGEN

Glycogen content in brain (Fig. 9-3) is related to the glucose concentration in blood and can be influenced by steroids and other hormones. The bulk of the glycogen is stored in the cytoplasmic granules which can be seen in electron micrographs of suitably prepared sections. The structure of isolated glycogen varies with the

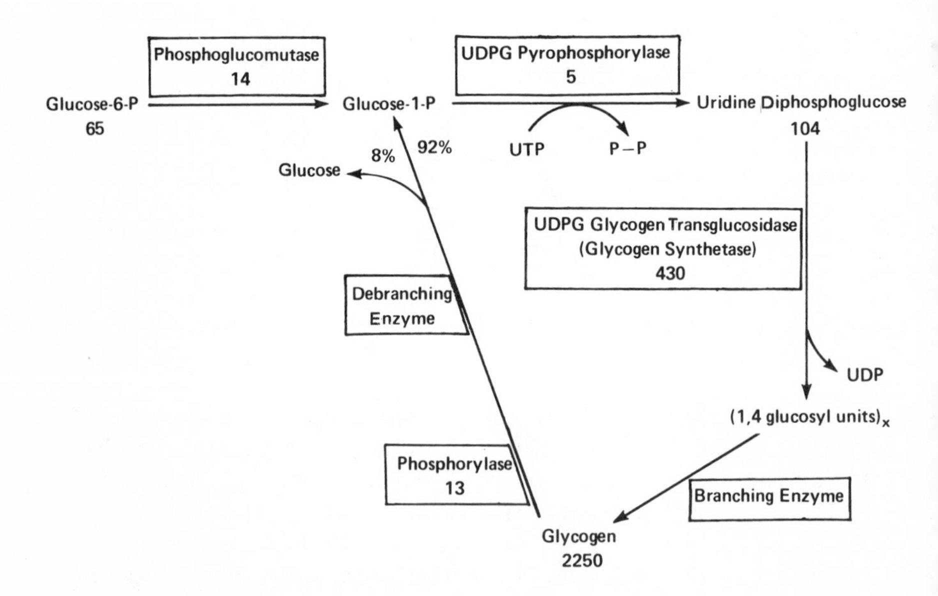

FIGURE 9-3. Glycogen metabolism in brain. Enzyme and metabolite data expressed as in Figure 9-1. Metabolite data from Passonneau et al., *J. Biol. Chem.* 244:902, 1969. Enzyme data from Breckenridge and Gatfield, *J. Neurochem.* 3:234, 1961.

methods used to extract it; gentle methods, which extract incompletely, yield a larger molecule. The accepted role of glycogen is that of a carbohydrate reserve utilized when glucose supply falls below need. There is, however, a rapid, continual breakdown and synthesis of glycogen, subject to elaborate control mechanisms, suggesting that even under steady-state conditions local carbohydrate reserves are important for brain function.

Separate systems for the synthesis and degradation of glycogen allow a greater degree of control than if glycogen were degraded simply by reversal of the synthetic steps. The level of glucose 6-phosphate, the initial synthetic substrate, usually varies inversely with brain glycolytic flux because of greater facilitation of the phosphofructokinase step relative to glucose transport and phosphorylation. Consequently, the decline in glucose 6-phosphate decreases glycogen formation at times of energy need. Phosphoglucomutase requires only catalytic amounts of glucose 1,6-diphosphate as a cofactor. Yet, the concentration of the cofactor in brain exceeds that of the substrate.

The glucosyl group of uridine diphosphate (UDP) glucose is transferred to the terminal glucose of the nonreducing end of an amylose chain in an α-1,4-glycosidic linkage. This reaction, catalyzed by glycogen synthetase, is evidently the rate-controlling reaction in glycogen synthesis [14]. In brain, as in other tissues, glycogen synthetase occurs both in a phosphorylated (D) form, which is dependent for activity on glucose 6-phosphate as a positive modulator, and in a dephosphorylated, independent (I) form. Although in brain the latter form requires no stimulator, it has a relatively high K_M for UDP-glucose. Under circumstances in which the brain glycolytic flux is increased to meet energy needs, there is a change not only from the D form to the I form, but also there seems to be the formation of an I form with even lower affinity for the substrate [14]. The result is inhibition of glycogen synthesis and increased availability of glucose 6-phosphate for glycolysis. The phosphorylation of the I form to the more active (and regulated) D form requires ATP, i.e., adequate energy reserves.

In the brain there is no relative increase in the I form under conditions in which glycogen formation is increased. Glucose 6-phosphate levels are low and the I form is associated with inhibition of glycogen synthesis. In the liver, on the other hand, where glycogen formation and breakdown are major activities, the I form is associated with glycogen formation. The two tissues apparently use the same biochemical apparatus for different purposes in relation to differences in overall metabolic patterns. The regulation of the D form may be the main factor in reducing glycogen formation in brain to about 5 percent of its potential rate.

The breakdown of glycogen depends on the activity of phosphorylase. Under steady-state conditions, it is probable that in brain less than 10 percent of phosphorylase is in the unphosphorylated *b* form (requiring AMP) which is inactive at the very low AMP concentrations present normally. Under some circumstances when the steady state is disturbed (e.g., by reduction of oxygen levels), there is an extremely rapid conversion to the *a* form, which is active at low AMP levels (see p. 178). This conversion depends on the activation of phosphorylase *b* kinase.

The product of phosphorylase is glucose 1-phosphate. The 1-6 linkages of glycogen are broken by a debrancher enzyme which also is active in brain, releasing about one glucose molecule for every eleven of glucose 1-phosphate to the intracellular compartment. α-Glucosidase (acid maltase) is a lysosomal enzyme whose precise function in glycogen metabolism is unknown. In its hereditary absence, Pompe's disease (see Chap. 22), glycogen accumulates in lysosomes in brain as well as elsewhere in the body (see [3, 4, 7]).

INTRACELLULAR MESSENGERS

The action of many hormones and some neural transmitters is indirect. These substances act at membranes of specific receptor cells to cause the intracellular release of an active molecule ("second messenger") which, in turn, modifies cellular metabolism and function. Ionic calcium in muscle is such a substance. 3',5'-Cyclic AMP is another [8].

In sensitive cells, epinephrine activates the membrane-associated adenylate cyclase system, which converts ATP into 3',5'-cyclic AMP. This substance, unlike 5'-AMP, retains a high-energy phosphate bond and can activate various protein kinases, among them phosphorylase *b* kinase kinase. This enzyme catalyzes the phosphorylation of phosphorylase *b* kinase, which, in turn, stimulates the conversion of phosphorylase *b* to the *a* form and thus facilitates the breakdown of glycogen. 3',5'-Cyclic AMP also activates the kinase converting glycogen synthetase to the regulated D form. Relatively high levels of 3',5'-cyclic AMP are found in brain and may function postsynaptically in interneural transmission, particularly at adrenergic synapses [8]. Preganglionic sympathetic stimulation increases cyclic AMP in the postsynaptic cell in proportion to the rate of stimulation.

PENTOSE PHOSPHATE SHUNT

This pathway has a relatively high activity in developing brain, reaching a peak during myelination. Its main contribution probably is to produce the NADPH required for reductive reactions in lipid synthesis. It also provides pentose for nucleotide synthesis; however, only a small fraction of the activity of this pathway would be required to meet such a need. In adult brain, under basal conditions, this pathway has been estimated to be 3 to 8 percent as active as glycolysis. Like glycogen synthesis, turnover in the pentose phosphate pathway decreases under conditions of increased energy need (e.g., high rates of stimulation) [21]. Pentose phosphate flux apparently is regulated by the concentrations of glucose 6-phosphate, NADP, glyceraldehyde 3-phosphate, and fructose 6-phosphate [17]. One of the enzymes in this pathway, transketolase, requires thiamine pyrophosphate as a cofactor (see Chap. 25). Poor myelin maintenance in thiamine deficiency may be due to the failure of this pathway to provide sufficient NADPH for lipid synthesis (see [3, 4, 7]).

COMPARTMENTATION

When the steady state is disturbed, the levels of several carbohydrate metabolites and cofactors in brain change in ways which indicate that enzymes and substrates are not in homogeneous solution. Measurements of the rates of incorporation of labeled carbon into various metabolites also indicate segregation of metabolite pools. Aside from obvious compartmentation into such subcellular organelles as mitochondria, lysosomes, or synaptic vesicles, the locations of other compartments are not clear. These may be intracellular or may represent differences in metabolism among neurons, glia, or synaptic structures [1–4].

MITOCHONDRIA

As in other tissue, brain mitochondria are apparently self-replicating bodies, although most of their active enzymes depend on the cell's chromosomal-ribosomal apparatus for their synthesis. Probably not all mitochondria are identical in enzyme complement or function. Because current methods of separating mitochondria from other cell components depend on sedimentation from homogenates (Chap. 21), differences related to function are difficult to define. Mitochondrial size and function change during maturation. Functional heterogeneity among brain mitochondria may be related to their location in perikarya, synaptosomes, or various glial cells.

The mitochondrial membranes and matrix form metabolic compartments separate from the cytosol. Entry and egress of metabolites and ions are selective. This allows a degree of metabolic control not otherwise possible. Mitochondrial membranes are not freely permeable to pyridine nucleotides. The brain contains several enzymes which could function in postulated shuttle systems that transfer hydrogen generated in the cytosol (e.g., NADH) to the mitochondria to be oxidized by the electron transport system. These enzymes include mitochondrial and cytoplasmic malate dehydrogenases (NAD) as well as cytoplasmic (NAD) and mitochondrial (flavoprotein) glycerol phosphate dehydrogenases.

The sine qua non of brain metabolism is its high rate of respiration. In the coupled, controlled state, the level of mitochondrial function depends on local concentrations of ADP. The entry of ADP into mitochondria is restricted insofar as it must exchange with intramitochondrial ATP. The high steady-state mitochondrial respiration in brain is related to local availability of substrates and the ratio of ADP to ATP. This may not be reflected in the average ratio of ADP to ATP in brain, which is quite low. Brain mitochondria differ from those of other tissues in that they contain higher concentrations of certain "nonmitochondrial" enzymes. Hexokinase, creatine kinase, and perhaps lactic dehydrogenase-1 are partially mitochondrial. Indeed, most of brain hexokinase, a glycolytic enzyme, is bound to mitochondria. Hexokinase and creatine kinase may aid in the maintenance of local levels of ADP by transferring ~P from ATP to creatine or glucose [1, 3, 4, 7].

RELATION OF CARBOHYDRATES TO LIPID METABOLISM

The principal source of lipid carbon in brain is blood glucose. Carbohydrate intermediates and related metabolites, such as acetate (fatty acid and cholesterol), dihydroxyacetone phosphate (glycerol phosphate), mannosamine and pyruvate (neuraminic acid), glucose 6-phosphate (inositol), galactose, and glucosamine supply the building blocks of the complex lipids (Chap. 11). NADPH is also necessary for their reductive synthesis. When a few polyunsaturated fatty acids and sulfate are supplied, immature brain slices readily form all the lipids of myelin and cell membranes, utilizing glucose as the only substrate. Active energy production (as ATP) by carbohydrate pathways is also needed as a source of the nucleoside phosphates required for lipid assembly [4, 7].

AMINO ACID METABOLISM AND CARBOHYDRATES

When isotopically labeled glucose is metabolized by adult brain, a large portion of the carbon appears rapidly not in CO_2 but in amino acids and their derivatives, notably glutamic acid, glutamine, γ-aminobutyric acid, and aspartic acid; whereas amino acid carbon appears in CO_2. This is due to extremely active transamination reactions, in which amino groups are exchanged between keto acids, both from the glycolytic chain (pyruvate) and the citric acid cycle (α-ketoglutarate, oxaloacetate), and various amino acids (Chaps. 8 and 10). If an amino group is exchanged between labeled α-ketoglutarate derived from glucose and unlabeled glutamate, the result is an unlabeled α-ketoglutarate molecule that can be the substrate for α-ketoglutarate dehydrogenase. Adult brain has only a small pool of citric acid cycle intermediates which turns over extremely rapidly and is linked by active transamination with large pools of amino acids, so the labeled carbon of glucose is siphoned off to label the large amino acid pool. In the exchange, however, the energy of the carbon-carbon bonds is returned to the citric acid cycle. Thus, the net source of respiratory energy is glucose, although the CO_2 produced is derived only partially from administered glucose. In the maturing brain, the shunting of glucose carbon into amino acids increases concomitantly with citric acid cycle activity. Transamination reactions with carbohydrate-derived keto acids and the glutamate dehydrogenase reaction involving the reductive amination of α-ketoglutarate with NH_4^+ may be important means for disposal of ammonia in the brain [4, 7].

CARBOHYDRATES IN NEURONAL FUNCTION

The high functional requirement of nervous tissue for $\sim$P is probably related to energy demands for transmitter synthesis, packaging, secretion, uptake, and sequestration; for ion pumping to maintain ionic gradients; for intracellular transport; and for synthesis of complex lipids and macromolecules in both neurons and glial cells. The mode of coupling metabolism to function is derived easily from the discussion of the regulatory mechanisms that control carbohydrate metabolism. This can be illustrated by (Na^+, K^+)-ATPase, which functions in the Na^+, K^+ exchange reaction that is essential for maintaining electrolyte gradients (see also Chap. 2). This enzyme is particularly active in regions with high concentrations of synaptic membranes, e.g., gray matter and subcellular fractions containing nerve endings (synaptosomes). This membrane-associated, topographically oriented enzyme is stimulated by extracellular K^+ and intracellular Na^+, so its activity is increased by the ionic changes that accompany depolarization. The ADP that is released intracellularly is an activating modulator for mitochondria and several rate-limiting reactions: glycolytic (hexokinase, phosphofructokinase), Krebs cycle (NAD-dependent isocitric dehydrogenase), and glycogenolytic (phosphorylase). The changes accompanying the consequent increased metabolic flux lead to inhibition of other pathways, such as glycogen synthesis and the pentose phosphate shunt, and the lowered ~P levels and NADPH levels inhibit various synthetic reactions. Because the products of reactions that use ~P are accelerators of reactions leading to ~P formation, energy production tends to keep pace with utilization.

REGIONAL DIFFERENCES

It is assumed that lipid represents a relatively inert metabolic compartment (although many enzymes are membrane-associated and even lipid-dependent). Therefore, comparisons of different areas of brain are often based on lipid-free dry weight. It is hoped that in this way it will be possible to gain a better estimate of the actively metabolizing compartments in regions of widely varying lipid content. In general, those regions of brain with higher metabolic requirements have higher activity of enzymes in the glycolytic series and the citric acid cycle and higher levels of respiration. Several glycolytic and mitochondrial enzymes are more active in regions with large numbers of synaptic endings (neuropil) than in areas rich in neuronal cell bodies. On the other hand, glucose 6-phosphate dehydrogenase (a rate-limiting enzyme of the pentose phosphate pathway) is high in myelinated fibers and tends to vary in activity with the degree of myelination. Phosphofructokinase and phosphorylase are distributed in a relatively constant ratio in different brain regions, suggesting a relation between glycolysis and glycogenolysis. Hexokinase distribution is more closely linked to mitochondrial enzymes than to glycolytic enzymes.

Blood flow to gray matter is greater than to white, and many in vitro studies indicate that, even when corrected for lipid content, the metabolic activity of white matter is less than that of gray. However, ~P and glycolytic flux may be as high in white matter as it is in gray. The depressing effect of anesthetic agents on these parameters of metabolism may also be greater in white than in gray matter [13]. The neuropil-rich molecular layer of cerebellum has only slightly higher ~P flux than has white matter. High metabolic rates in white matter might be related to axonal transport mechanisms and the maintenance of myelin. The oligodendroglial cell that maintains myelin in the CNS probably ranks, along with the neuron and its processes, as one of the cells with the highest known metabolic requirements (see [5, 6, 27]).

METABOLISM OF THE RETINA

The rabbit retina is avascular and depends almost entirely on diffusion from choroidal capillaries. The primate retina, however, is vascularized from the vitreal surface as far as the bipolar cell layer. The rabbit retina shows high rates of glucose and oxygen consumption and of lactate formation. The high rate of aerobic lactate formation might be due to segregation of glycolytic and oxidative processes and to the adaptation of the poorly vascularized inner layers to a relatively anoxic existence. The rod inner-segment layer, which contains packed mitochondria, has high levels of all mitochondrial enzymes and hexokinase. This is the region closest to the choroidal nutrient supply.

In the vascularized inner layers of the monkey retina, hexokinase activity is almost twice that in the homologous layers of the rabbit retina. Several glycolytic enzymes, including phosphofructokinase and glyceraldehyde phosphate dehydrogenase as well as lactic dehydrogenase and glycogen phosphorylase, tend to be higher toward the vitreous surface in rabbit than in monkey retina. The total energy reserves (especially high glycogen levels) and lactate levels increase as the avascular vitreous surface of the rabbit retina is approached. The citric acid cycle enzymes, malate dehydrogenase and NAD-dependent isocitrate dehydrogenase, vary with the relative density of mitochondria, which is high in the layer of rod inner segments and in the synaptic layers. Data such as these suggest adaptive changes, dependent on the local availability of substrates and oxygen, in carbohydrate enzymes and metabolism. As in brain, however, continued electrical responsiveness of the retina depends upon oxidative metabolism. The response to light is dependent on metabolic processes and control mechanisms similar to those described for the maintenance of electrical activity in brain [6].

PERIPHERAL NERVE

The metabolic rate of peripheral nerve is about 7 percent of that in mouse brain. As in brain, carbohydrate reserves are low. In both the sympathetic system and in brain, glucose is the major metabolite of both nerve and ganglion. Unlike brain, however, there is no apparent barrier to glucose uptake in ganglion, and glucose levels are close to those in blood. Glucose levels in peripheral nerve are intermediate between those of brain and ganglion. The patterns of substrate distribution and utilization of stimulated nerve differs from those of nerve at rest. Depletion of energy reserves leads to the failure of synaptic transmission before failure of conduction. Transmission in sympathetic ganglia can fail after depletion of carbohydrate reserves (glucose and glycogen), despite maintenance of high levels of ~P. This situation is similar to that in hypoglycemic brain, and similar mechanisms may be involved [27].

GLIAL TUMORS

Glial tumors might serve as models for glial metabolism. The metabolic rates in these tumors (with the possible exception of oligodendrogliomas) are relatively low; the reason may at least in part be due to poor vascularization of large areas of the tumors. Both the architecture and the blood supply of tumors are complex, and the overall metabolism of the tumor may represent an adaptation to its environment, rather than the intrinsic metabolic capacity of the parent (glial) tissue or neoplastic change. Metabolic fluxes are higher in areas well supplied with nutrients than in areas more distant from the blood supply. Cellular proliferation is restricted to the well-supplied regions [5].

DEVELOPMENT AND AGING

Brain development and maturation can be divided into overlapping stages (see Chap. 14). These stages change at different rates in different species, but generally follow similar patterns. In animals such as the guinea pig, the brain is already far advanced at birth; in man and rodents, maturation occurs primarily after birth. During the earliest stage – cell proliferation and migration – energy requirements and oxygen consumption are low. At this time the brain is poorly vascularized. During the following stages of cell growth and differentiation, and establishment of synaptic connections, there is a large increase in energy production and utilization that is coincidental with the development of structures and functions with high energy requirements. A major increase in both glycolytic enzymes and glycolytic flux takes place during this period. The lesser sensitivity of immature brain to anoxia or ischemia, as compared with adult brain, is related to a lower energy requirement and thus a decreased dependence on oxygen.

Overshadowing the increase in glycolytic capacity during brain development, and satisfying the bulk of the greatly increased energy requirement, are large increases in mitochondrial enzymes and functions. Several of the carbohydrate enzymes appear to be associated with myelination, which begins after neuronal outgrowth has begun and continues longer. Glucose 6-phosphate dehydrogenase, NADP isocitrate dehydrogenase, β-glucuronidase, and β-galactosidase either reach a peak or may be decreased during myelination, relative to other enzymes. Relative changes in enzyme activities during maturation probably reflect differences in modes of metabolism. For instance, the great increase in the mitochondrial (NAD-dependent) isocitrate dehydrogenase during development is accompanied by a reciprocal decline in the activity of the cytoplasmic isocitrate dehydrogenase (NADP).

A decline in oxygen consumption in brain during old age has been found in humans, but this might be related to changes in cerebral blood flow (Chap. 15). There may also be a decline in metabolic rates in the brains of aged animals. The causes of these metabolic changes are unknown [3, 5, 7, 9, 27].

CONDITIONS AFFECTING THE STEADY STATE

HORMONES

Insulin

Insulin can increase the uptake of glucose by brain slices in vitro and by brain in vivo after intracisternal injection. The effect of circulating insulin, however, remains in doubt. Lower activities of hexokinase and higher activities of several enzymes associated with glycogen metabolism have been found in the cerebral cortex of rats made diabetic by administration of alloxan. Glycogen levels are also higher. In the diabetic state, sorbitol and fructose are high in brain and peripheral nerve. The sorbitol pathway provides an alternate route for the conversion of glucose to fructose [23].

Thyroid

Unlike most tissues, metabolism in adult brain is not demonstrably affected by thyroid hormone. Thyroxin is necessary, however, for the development and maturation of brain. Administration of thyroxin to developing rats can speed the rate of increase in brain oxygen consumption associated with maturation. Yet, no direct effect on brain carbohydrate metabolism has been demonstrated [26].

Steroids

Cortisol administered to immature mice leads to an increase in total energy reserves, mainly by increasing brain glucose levels. Similar effects have been found in adrenalectomized adult animals. This may indicate that steroids facilitate the transfer of glucose into the brain. Administration of cortisol also leads to increased glycerophosphate dehydrogenase activity in the brain. Neither hypophysectomy nor adrenal insufficiency reduce brain oxygen consumption. Steroids may improve function clinically in brain that has been disturbed by diverse pathological conditions, including neoplasm, vascular occlusion, or certain demyelinating conditions. The mechanism of this action is not known [4, 12].

ISCHEMIA

Interference with blood flow deprives an organ of metabolic substrates and oxygen simultaneously. When the blood flow to the brain is less than optimal, changes occur in carbohydrate metabolism without at first affecting energy production or neuronal function: glucose and O_2 extraction increase, oxidative phosphorylation decreases, and glycolysis and lactate production increase. Associated with these early changes is local vasodilatation, which is part of a complex adjustment of regional cerebral blood flow to energy demand and nutrient supply, collectively called *autoregulation.* If the capacity of these metabolic and vascular mechanisms cannot compensate for a reduction in blood flow, mitochondrial respiration is further diminished, a larger proportion of the pyridine nucleotides becomes reduced, and the citric acid cycle reactants approach their equilibrium levels [15]. The level of ~P (ATP, phosphocreatine, and ADP) falls rapidly despite increased glycolysis. After 60 sec of complete ischemia, glycolytic substrates (glucose and glycogen) and ~P are reduced to low levels [19]. During this period, the lactate dehydrogenase reaction is the major source of NAD^+ required for continuing glycolysis, and lactate accumulates rapidly until the carbohydrate supply is exhausted. During ischemia the distribution of Na^+ and K^+ changes in the direction of their concentration gradients. The differential vulnerability of various brain regions to ischemia may be related to differences in blood supply as well as in total energy requirements [5, 13].

HYPOGLYCEMIA

With sudden decreases in blood glucose without interference in the availability of oxygen, the endogenous substrates of brain are sufficient to maintain energy metabolism for 20 to 25 min. As endogenous substrates are reduced below critical levels, the metabolic rate declines and the level of every metabolic intermediate from glucose to malate, including lactate, is diminished. Utilization of amino acids by transamination or deamination to keto acids results in a decrease of amino acids, particularly of glutamate and glutamine. Insulin hypoglycemia can induce a convulsion at normal brain concentrations of ~P, suggesting possible differences in accessibility for brain function between total ATP and that specifically generated by glycolytic citric cycle activity [15]. Diminished synthesis of neural transmitters (e.g., γ-aminobutyrate) may be the factor disrupting neuronal physiology during hypoglycemia, despite adequate energy stores [3, 5, 7].

HYPOXIA

In contrast to hypoglycemia, hypoxia causes metabolic changes almost identical (except for CO_2 accumulation) to ischemia. During brief periods of anoxia, when most lactate can be removed by the circulation, brain lactate levels are lower than during ischemia. CO_2 does not increase, so the pH falls less than during an equivalent brief period of ischemia. Because glucose is continually supplied by the intact circulation, the increase in glycolysis, which occurs as regulatory mechanisms respond to the failure of mitochondrial energy production, eventually results in greatly increased production and accumulation of lactate. Consequently, prolonged periods of hypoxia can result in a greater fall in pH than during ischemia when the circulation is entirely cut off [5].

SEIZURES

During a grand mal seizure, the combination of increased neural and muscular activity (producing increases in blood lactate levels and respiratory hypoxia), increased jugular venous pressure (resulting from an increase in intrathoracic pressure during the tonic phase), and decreased cardiac output results in a condition resembling ischemia. Usually, some energy reserves are maintained. If the extracerebral factors are controlled during the seizure (neuromuscular blockade and ventilation with oxygen), the metabolic effects are much less pronounced [11]. Although there may be as much as a fivefold increase in the metabolic demand of the brain associated with the neuronal discharge of a seizure, local vascular and metabolic regulatory mechanisms tend to maintain steady-state levels of ~P and substrates, mainly by increases in cerebral blood flow and oxygen and glucose uptake. Provided that oxygenation is maintained, there is no fall in ~P and glucose or rise in lactate, despite a marked increase in metabolic flux. During a clinical grand mal seizure, however, metabolic changes in brain reflect relative ischemia in the presence of increased metabolic demand. In either case, the course of the seizure is not affected. Thus, substrate depletion, metabolite accumulation, or anoxia cannot account for the termination of the seizure (see Chap. 15).

STARVATION

Although glucose is the major substrate of brain metabolism, this may not hold under conditions of starvation (Chap. 15). If no food is taken, the blood glucose level is maintained for about 24 hr by mobilization of liver glycogen and gluconeogenesis. Fasting beyond this period (starvation) leads to increasing fat mobilization and the appearance of acetoacetate and β-hydroxybutyrate (ketone bodies) in the blood. Under these circumstances, ketone bodies can be utilized by the brain. More than half of total cerebral oxygen consumption in starving man can be accounted for by the oxidation of ketone bodies. Their utilization depends on the oxidation of β-hydroxybutyrate to acetoacetate (mediated by β-hydroxybutyrate dehydrogenase) and the activation and subsequent cleavage of acetoacetate (see also Chap. 13). The capacity to metabolize ketone bodies exists in both immature and adult rat brain, and ketone bodies can be utilized for energy when blood levels rise. β-Hydroxybutyrate dehydrogenase levels, however, may vary in response to ketone levels. Rat milk is rich in fat, and ketone bodies are a major energy source to the young before they are weaned [16].

TEMPERATURE

A temperature rise of 10°C results in a 1.7 to 2-fold increase in the overall metabolic rate in brain. This corresponds to the temperature coefficients (Q_{10}) of glycolytic and Krebs cycle enzymes. Hypothermia thus enables the brain to sustain longer periods of diminished supply without damage, while higher temperatures make continued nutrient and oxygen supply more critical. Above 39°C, the activities of some enzymes begin to decline and metabolic rates are adversely affected. Diffuse functional changes, such as delirium, occur in the nervous system if the brain temperature rises above 40° to 41°C. In pathological states, a rise of 0.5°C or less may produce significant deterioration of function [5, 7, 19, 20].

ANESTHESIA AND SEDATIVE DRUGS

Agents that depress brain function also decrease oxygen and glucose consumption and frequently increase brain glycogen, as well. Several of these agents block specific metabolic steps (e.g., amytal affects the respiratory chain), so it might be thought that the agents produce their sedative effects by acting as metabolic inhibitors. At pharmacologically active concentrations little, if any, direct metabolic inhibition takes place. The evidence favors the concept that the decreased metabolic rates result from the reduction of energy demand brought about by interference with interneuronal interaction. In addition, brain glucose and glycogen increase beyond the levels expected from the decline in utilization. It might be expected that these agents would have a greater effect on gray matter than on white matter, and blood-flow studies appear to bear this out. There is, however, a greater inhibition of glycolytic flux in white matter than in gray. It may be that the high lipid solubility of these agents leads to higher local levels in white matter, but this is not clearly established [7].

HEPATIC DYSFUNCTION

It is currently believed that the major cause of the encephalopathy associated with hepatic disease is failure to detoxify ammonia, produced mainly from amino acid (protein) catabolism in the gut. The brain has a significant capacity for removing excess ammonia. When this regulation is exceeded, changes in brain metabolites (decreased glucose, glycogen, and ~P) can be correlated with functional disturbance [24]. It is not certain if the primary effect of ammonia involves energy metabolism, citric acid cycle metabolism, or some other mechanism. For instance, the metabolism of glutamate and glutamine may be affected because the synthesis of these amino acids is a major pathway for reducing local levels of ammonia [10].

RENAL DYSFUNCTION

Normal renal function is essential to the normal functions of the peripheral and central nervous systems. The brain metabolism of uremic humans or nephrectomized animals is slowed. Studies of metabolite changes in the brains of uremic rats suggest a decline in energy utilization rather than an interference with energy production. The biochemical changes thus resemble those of anesthesia.

ACKNOWLEDGMENTS

Supported by grant NS05368, the National Institute of Neurological Diseases and Stroke and grant MH17592, the National Institute of Mental Health, National Institutes of Health.

REFERENCES

BOOKS AND REVIEWS

1. Abood, L. G. Brain Mitochondria. In Lajtha, A. (Ed.), *Handbook of Neurochemistry.* Vol. 2. New York: Plenum, 1970.
2. Balazs, R. Carbohydrate Metabolism. In Lajtha, A. (Ed.), *Handbook of Neurochemistry.* Vol. 3. New York: Plenum, 1970.
3. Bradford, H. F. Carbohydrate and Energy Metabolism. In Davison, A. N., and Dobbing, J. (Eds.), *Applied Neurochemistry.* Philadelphia: Davis, 1968.
4. Lehninger, A. L. *Biochemistry.* New York: Worth, 1970, pp. 267–564.
5. Maker, H. S., and Lehrer, G. M. Effect of Ischemia. In Lajtha, A. (Ed.), *Handbook of Neurochemistry.* Vol. 6. New York: Plenum, 1971.
6. Matschinsky, F. M. Energy Metabolism of the Microscopic Structures of the Cochlea, the Retina, and the Cerebellum. In Costa, E., and Giacobini, E. (Eds.), *Advances in Biochemical Psychopharmacology.* Vol. 2. New York: Raven, 1970.
7. McIlwain, H. *Biochemistry and the Central Nervous System.* Boston: Little, Brown, 1966, pp. 1–126.
8. Greengard, P., and Costa, E. Role of Cyclic AMP in Cell Function. In Costa, E., and Giacobini, E. (Eds.), *Advances in Biochemical Psychopharmacology.* Vol. 3. New York: Raven, 1970.
9. Swaiman, K. F. Energy and Electrolyte Changes During Maturation. In Himwich, W. A. (Ed.), *Developmental Neurobiology.* Springfield, Ill.: Thomas, 1970, p. 311.
10. van den Berg, C. J. Glutamate and Glutamine. In Lajtha, A. (Ed.), *Handbook of Neurochemistry.* Vol. 3. New York: Plenum, 1970.

ORIGINAL SOURCES AND SPECIAL TOPICS

11. Collins, R. C., Posner, J. B., and Plum, F. Cerebral metabolic response to electroconvulsions in paralyzed ventilated mouse. *Trans. Am. Neurol. Assoc.* 94:242, 1969.
12. DeVellis, J., and Inglish, D. Hormonal control of glycerol phosphate dehydrogenase in the rat brain. *J. Neurochem.* 15:1061, 1968.
13. Gatfield, P. D., Lowry, O. H., Schulz, D. W., and Passonneau, J. V. Regional energy reserves in mouse brain and changes with ischaemia and anaesthesia. *J. Neurochem.* 13:185, 1966.

14. Goldberg, N. D., and O'Toole, A. G. The properties of glycogen synthetase and regulation of glycogen biosynthesis in rat brain. *J. Biol. Chem.* 244:3053, 1969.
15. Goldberg, N. D., Passonneau, J. V., and Lowry, O. H. Effects of changes in brain metabolism on the levels of citric acid cycle intermediates. *J. Biol. Chem.* 241:3997, 1966.
16. Hawkins, R. A., Williamson, D. H., and Krebs, H. A. Ketone-body utilization by adult and suckling rat brain *in vivo*. *Biochem. J.* 122:13, 1971.
17. Kauffmann, F. C., Brown, J. B., Passonneau, J. V., and Lowry, O. H. Effect of changes in brain metabolism on levels of pentose phosphate pathway intermediates. *J. Biol. Chem.* 244:3547, 1969.
18. Lehrer, G. M., Bornstein, M. B., Weiss, C., Furman, M., and Lichtman, C. Enzymes of carbohydrate metabolism in the rat cerebellum developing *in situ* and *in vitro*. *Exp. Neurol.* 27:410, 1970.
19. Lowry, O. H., and Passonneau, J. V. The relationships between substrates and enzymes of glycolysis in brain. *J. Biol. Chem.* 239:31, 1964.
20. Lowry, O. H., Passonneau, J. V., Hasselberger, F. X., and Schulz, D. W. Effect of ischemia on known substrates and cofactors of the glycolytic pathway in brain. *J. Biol. Chem.* 239:18, 1964.
21. O'Neill, J. J., Simon, S. H., and Shreeve, W. W. Alternate glycolytic pathways in brain. A comparison between the action of artificial electron acceptors and electrical stimulation. *J. Neurochem.* 12:797, 1965.
22. Passonneau, J. V., and Lowry, O. H. The Role of Phosphofructokinase in Metabolic Regulation. In Weber, G. (Ed.), *Advances in Enzyme Regulation.* Vol. 2. New York: Pergamon, 1964, pp. 265–274.
23. Prasannan, K. G., and Subrahmanyam, K. Enzymes of glycogen metabolism in cerebral cortex of normal and diabetic rats. *J. Neurochem.* 15:1239, 1968.
24. Schenker, S., McCandless, D. W., Brophy, E., and Lewis, M. S. Studies on the intracerebral toxicity of ammonia. *J. Clin. Invest.* 46:838, 1967.
25. Sloviter, H. A., and Kamimoto, T. The isolated, perfused rat brain preparation metabolism mannose but not maltose. *J. Neurochem.* 17:1109–11, 1970.
26. Sokoloff, L. The Action of Thyroid Hormones in Brain and Other Tissues. In McKhann, G. M., and Yaffe, J. J. (Eds.), *Drugs and Poisons in Relation to the Developing Nervous System.* Bethesda: Public Health Service Publication #1791, 1967.
27. Stewart, M. A., and Moonsammy, G. I. Substrate changes in peripheral nerve recovering from anoxia. *J. Neurochem.* 13:1433, 1966.

10

Transport and Metabolism of Amino Acids

Gordon Guroff

GENERALLY SPEAKING, the outlines of amino acid metabolism in brain are similar to those of amino acid metabolism in other tissues of the body. In many respects, however, the quantitative aspect of the amino acid economy is different. In a few cases the brain chemistry differs qualitatively. Put another way, most of the pathways found in brain are also found in liver and other organs. In many cases these pathways are working at substantially different rates in brain than elsewhere. In a few cases, amino acids or amino acid metabolizing enzymes found in brain do not appear elsewhere in the mammalian body.

AMINO ACID COMPOSITION

CONCENTRATION AND DISTRIBUTION

The amino acid composition of nervous tissue, the so-called free amino acid pool, has been the subject of a great deal of study [3, 4]. Clearly, the makeup of this pool of amino acids is dependent on a number of factors, including the rates of transport of the various amino acids into and out of brain, the speed of their metabolism, and their incorporation into protein. The concentration of each amino acid in the free amino acid pool is fairly constant, and the composition of the pool is characteristic of brain. This composition is not obviously related to the composition of brain protein nor to the composition of the blood or the cerebrospinal fluid (Table 10-1). Most amino acids are in higher concentration in brain than in blood.

Glutamate, glutamine, aspartate, *N*-acetylaspartate, and γ-aminobutyrate are the predominant amino acids in most, if not all, species. In fact, the high concentration and extensive metabolism of glutamate and its derivatives are among the most significant aspects of brain metabolism. In most species, these amino acids comprise more than two-thirds of the free α-amino nitrogen in the brain. Glutamate itself is in higher concentration in brain than in any other organ of the body.

TABLE 10-1. Free Amino Acids of Human Brain, Plasma, and Cerebrospinal Fluid

Amino Acid	Brain	Plasma	Cerebrospinal Fluid
	(μmoles per g or ml)		
Glutamic acid	10.6	0.05	0.225
N-Acetylaspartic acid	5.7	–	–
Glutamine	4.3	0.7	0.03
γ-Aminobutyric acid	2.3	–	–
Aspartic acid	2.2	0.01	0.007
Cystathionine	1.9	–	–
Taurine	1.9	0.1	–
Glycine	1.3	0.4	0.013
Alanine	0.9	0.4	0.017
Glutathione	0.9	–	–
Serine	0.7	0.1	0.01
Threonine	0.2	0.15	0.025
Valine	0.2	0.25	0.013
Lysine	0.1	0.12	0.014
Leucine	0.1	0.15	0.004
Proline	0.1	0.1	–
Asparagine	0.1	0.07	–
Methionine	0.1	0.02	0.003
Isoleucine	0.1	0.1	0.008
Arginine	0.1	0.1	0.006
Cysteine	0.1	0.1	0.002
Phenylalanine	0.1	0.1	0.010
Tyrosine	0.1	0.1	0.006
Histidine	0.1	0.1	0.003
Tryptophan	0.05	0.05	0.010

Sources: Sourkes, T. L., *Biochemistry of Mental Disease,* New York: Harper and Row, 1962, p. 21; Tallan, H. H., in *Amino Acid Pools* (J. T. Holden, Ed.). Amsterdam: Elsevier, 1962, pp. 472, 476, 478; Schreier, K., *ibid,* p. 227.

In spinal cord, as in brain, the concentration of glutamate and its related amino acids is high compared to that of other amino acids. Vertebrate peripheral nerve, on the other hand, is much lower in glutamate, glutamine, γ-aminobutyrate, and *N*-acetylaspartic acid content than is brain. In fact, γ-aminobutyrate (GABA) is nearly absent in vertebrate peripheral nerve, but is present in crustacean nerve. The content of other amino acids in peripheral nerve is not markedly different from that in brain or spinal cord.

As yet, no α-amino acid has been found to belong exclusively to the nervous system. Also, no amino acid found in other tissues can be shown to be completely missing from the nervous system. On the other hand, certain amino acid derivatives appear to be present in large quantities in nervous tissue and, as far as is known, only in nervous tissue.

Among the most abundant of these derivatives is *N*-acetylaspartic acid. Its concentration in most species is two to three times higher than that of aspartic acid itself. It occurs only in brain and generally its concentration is low at birth and rises to adult levels during development. The acetyl group appears to come from acetyl-CoA, but studies of its biosynthesis are conflicting and as yet incomplete. Its rate of turnover in brain at first appeared to be low, but more recent studies have revealed a more active metabolism. The function of this compound is unknown, but suggestions include: (1) a role as part of the intracellular "fixed" anion pool, (2) a reservoir of acetyl groups, and (3) a potential source of *N*-blocked end groups for certain brain proteins.

As mentioned above, GABA is a major component of the free amino acid pool of most species. It is the product of the α-decarboxylation of glutamic acid and is found almost exclusively in the nervous system. It is present in brains of various species at levels ranging from less than one to more than three μmoles per gram of wet tissue. There is substantial evidence that GABA functions as an inhibitory neural transmitter, i.e., a transmitter in inhibitory neurons. GABA appears in both neurons and glia and is found in high concentration in synaptosomes. Perhaps in some bound form, it is probably the major principle of the "factor I" preparations that are found to inhibit the crayfish stretch receptor. The overall metabolism of GABA is known as the GABA shunt, and is discussed later in this chapter.

γ-Guanidinobutyrate has been isolated from calf brain. It is present in rather small amounts (about 1/100th of the concentration of GABA), but its identification is unequivocal.

Putreanine, or *N*-(4-aminobutyl)-3-propionic acid, has been shown to be a component of the central nervous system of humans, rats, and birds. It is not found in other tissues, and it is found in greater concentration in white matter than in gray. In rats, the amino acid appears two weeks after birth and its concentration increases for several months. In humans, on the other hand, the concentration of putreanine in the parietal cortex is not substantially different between the ages of three days and 80 years.

Cystathionine is the product of the condensation of homocysteine and serine. It is an intermediate in the biosynthesis of cysteine and is found in brain. In the human, high concentrations are found in brain and much smaller amounts in other tissues. Human brain contains a much higher concentration than is found in the brains of other species.

Several other amino acid derivatives have been found in the brains of various species. Notable among these is taurine, which has been found in high concentrations in certain invertebrate nervous systems. Although taurine, like citrulline, hydroxylysine, cysteine sulfinic acid, hypotaurine, and a great number of others, is undoubtedly present in the free amino acid pool of brain, it and the other derivatives also appear as constituents of other tissue pools.

The amounts of most amino acids in brain change during development (Table 10-2). Glutamate, aspartate, GABA, and arginine increase as the brain matures. Most of the amino acids – particularly taurine – decrease to adult levels during

TABLE 10-2. Cerebral Amino Acid Content in Newborn and Adult Mice

Amino Acid	Newborn	Adult
	μmoles per g fresh brain	
Taurine	18.5	8.3
Glutamic acid	4.4	9.9
Aspartic acid	1.9	2.2
γ-Aminobutyric acid	1.2	1.8
Arginine	0.10	0.13
Glycine	2.3	0.91
Alanine	0.76	0.55
Lysine	0.23	0.19
Tyrosine	0.19	0.08
Histidine	0.14	0.08
Valine	0.11	0.08
Phenylalanine	0.11	0.08
Leucine	0.08	0.04
Ornithine	0.06	0.05
Methionine	0.05	0.04
Isoleucine	0.05	0.02

Source: Levi, G., Kandera, J., and Lajtha, A., Control of cerebral metabolite levels. I. Amino acid uptake and levels in various species. *Arch. Biochem.* 119:303, 1967.

brain development. The reason for these changes is not known, and their functional significance, if any, is a mystery. It is interesting, though, that animals which are relatively developed at birth, such as the guinea pig, show less change in the free amino acid pool in brain as the individual matures than do species that are not as developed neonatally.

Different parts of brain have free amino acid pools that differ somewhat in composition. The major, and possibly the only, generalization which can be drawn from such differences is that, although glutamate is the major amino acid of whole brain, GABA is the most abundant in midbrain and hindbrain. Other differences in the concentration of specific amino acids are known, but no correlation has yet been made between these differences and the functions of the various brain areas in which they occur.

ALTERATIONS IN AMINO ACID LEVELS

The free amino acid pool of brain is relatively insensitive to changes in the state of the rest of the body. Amino acid levels are not altered in any substantial way by severe starvation or dehydration. On the other hand, the administration of large amounts of one amino acid alters the concentrations of related amino acids. The basis of these changes is presumably the ability of the administered amino acid to inhibit the uptake of related amino acids into the brain by competing for the active transport systems, which are discussed in a later section of this chapter. Although few data are available, some information shows that the brain and cerebrospinal amino acid pools of individuals with amino acidurias, such as phenylketonuria, are different from those of normal individuals. The cause of these altered pools is presumably the same as in loading experiments, e.g., the high circulating content of phenylalanine alters the transportation into brain of related amino acids (see below). Whether these changes in the pool contribute to the behavioral problems found in these diseases is still a matter of speculation.

EFFECTS OF INSULIN ON AMINO ACID LEVELS

Several studies have attempted to observe changes in the brain pool of amino acids in response to the administration of various hormones or drugs. Administration of insulin to rats causes a reduction in the brain content of glutamate, glutamine, GABA, alanine, and glycine, and a rise in aspartate and ammonia. It is thought that the drop in glutamate indicates that it is used as a substitute energy source in hypoglycemia. The rise in aspartic acid seems to be due to a decreased availability of acetate and a decreased need for oxaloacetate as an acceptor in the Krebs cycle. The excess oxaloacetate could be transaminated to aspartate. The convulsions seen in insulin hypoglycemia could be the result of the lowering of GABA concentration. Alternatively, they could be caused by the increased ammonia levels in the brain. When the systemic level of ammonia rises, as in cases of severe liver damage or because of the administration of ammonium salts, coma results. Substantial increases in brain glutamine occur, along with some increase in histidine and some decrease in methionine. The mechanism of the coma is unknown, but it may be caused by a direct effect of ammonium ion on nerve transmission.

EFFECTS OF DRUGS ON AMINO ACID CONTENT

The effects of various drugs on the free amino acid pool of brain have been investigated. The results of the many studies with psychotropic drugs are inconsistent and confusing. For example, one study reported that chlorpromazine raises GABA concentrations in rat brain; another study reported that the concentrations were lowered. The data on reserpine are also contradictory.

The connection between brain amino acid levels and the effects of drugs that cause or alleviate convulsions also seems indirect. Certain convulsant drugs, such as the hydrazines and the semicarbazides, both of which are antimetabolites of vitamin B_6 (see Chap. 25), have opposite effects on the concentration of GABA. Semicarbazide lowers the GABA concentration in the brain and hydrazine raises it. Hydroxylamine is also a pyridoxal phosphate antagonist, but under certain conditions it is an anticonvulsant. Hydroxylamine raises the GABA concentration of whole brain, but it does not protect against the convulsant effect of thiosemicarbazide, even though GABA levels are higher than normal. In addition, the picture is even less clear with other compounds which have convulsive action because many of them have no effect on GABA concentrations at all. Indeed, the wide variety of materials that cause convulsions suggests that no common mechanism will be identified (see Chap. 8).

A number of other studies have provided data on the effects of various drugs, hormones, and environmental conditions on brain amino acids. The influences of such factors as anesthetics, poisons, environmental complexity, and hypoxia have been investigated. These studies have been uniformly disappointing in that no useful information has been obtained. Perhaps the only generalization that can be drawn is that the brain amino acid pool is remarkably stable, even in the face of drastic treatment of the whole animal.

METABOLIC EFFECTS OF PHENYLALANINE LOADING

When alterations in the pool of amino acids have been successful, the results on brain metabolism have been fairly predictable. Many of the studies have been concerned with the effects of raising the levels of phenylalanine in the brain and the effects of this treatment on serotonin levels. Interest in this experimental situation stems from the fact that the high levels of phenylalanine in blood of individuals afflicted with phenylketonuria are accompanied by alterations of the levels of indoles in blood and urine. The administration of phenylalanine to experimental animals also lowers levels of serotonin in brain. In this case it is not clear whether the cause is the inhibition of tryptophan uptake into brain by the increased phenylalanine in blood or a decreased serotonin synthesis because of some inhibitory effect of the increased level of brain phenylalanine on serotonin biosynthesis.

Changes in brain amino acid levels also affect the rate of synthesis of brain protein. Again, most of the data come from studies on phenylalanine administration. Large amounts of phenylalanine given in vivo lower the synthesis of protein in the brain measured either in vivo or in vitro. The reason for this is not completely clear, but substantial evidence is available to suggest that: (1) when phenylalanine is administered, levels of tryptophan are lowered in brain, (2) tryptophan in some way is limiting for brain protein synthesis, and (3) lowered tryptophan causes disaggregation of polysomes with a consequent reduction of normal rates of protein synthesis.

COMPARTMENTATION OF AMINO ACIDS

A large body of available data indicates that various different pools or compartments of amino acids exist in brain [1b]. These compartments have been described on a metabolic basis and, as yet, no anatomical or subcellular localization of such compartments has been successful. Basically, the observations leading to the concept of metabolic compartmentation are that, under certain conditions, the specific activity of a product exceeds the peak specific activity of its precursor, sometimes by several times. This observation seems to make inescapable the conclusion that the metabolism is carried on in some small active pool that is diluted by the larger, inactive amount of precursor in the tissue when the tissue is prepared for analysis.

Most of the experiments along these lines concern the glutamate-glutamine system. For example, when radioactive glutamate is injected into the bloodstream of a rat, the blood glutamine and the brain and liver glutamate and glutamine rapidly become labeled. After a short time, brain glutamine has a higher specific activity than brain glutamate. Clearly the glutamate from blood entered some small compartment of brain and was converted to glutamine. When the tissue was homogenized, the small pool of metabolically active glutamate was diluted by the larger amount of glutamate in the brain, lowering its specific activity below that of the formed glutamine.

This type of observation has been made repeatedly by a number of workers under different experimental conditions. The same results are obtained if the liver is excluded from the circulation or if the radioactive glutamate is introduced by intracerebral injection. Similar data are obtained upon infusion of isotopic NH_4^+ salts or HCO_3^-.

Metabolic compartmentation in brain can be shown when other substrates are administered. If radioactive glucose, lactate, or glycerol is administered, the specific activity of glutamate exceeds that of glutamine. If radioactive acetate, butyrate, propionate, or leucine is administered, the reverse is true; glutamine is higher than glutamate. These studies suggest that the glucogenic substrates are metabolized to amino acids in a different compartment than are the ketogenic substrates.

Metabolic compartmentation of this type is not present in the infant but develops within a few weeks after birth in most species tested. Such compartmentation can be demonstrated in properly prepared slices, but not in homogenates. The anatomical localization of the glutamate-glutamine metabolic pool has not been identified, but the endoplasmic reticulum would seem the likely site. Compartmentation probably is not due to the different cell types in the brain because HCO_3^- and NH_4^+ probably enter all cells and because both neurons and glia show such effects. The endoplasmic reticulum, on the other hand, contains a high level of glutamine synthetase. The results obtained with glucogenic and ketogenic energy sources suggest that the compartmentation of energy metabolism may reside in different fractions of a possibly heterogeneous mitochondrial population.

TRANSPORT

GENERAL PROPERTIES

Many of the amino acids required by the brain come from the bloodstream across the blood-brain barrier (see Chap. 16). The blood-brain barrier is virtually impenetrable to the diffusion of water-soluble charged molecules, such as amino acids. These molecules are obviously necessary for adequate brain functioning, so they must move into the brain by some other mechanism. It has been found that the brain has specific active transport systems, probably at the same site as the blood-brain barrier, to allow the capture of the required amino acids. These transport systems are similar to amino acid transport systems described in other cells, in that they are energy requiring, concentrative, structurally and sterically specific, and seem to be common to a family of amino acids rather than specific to a single amino acid [1c].

The transport of amino acids from blood into brain, then, is an active process and fulfills the criteria for a classic active transport system. Transport systems are as active in brain as in any other tissue and more active than in most. They do not, however, approach in quantitative terms the activity of the active transport systems for amino acids found in bacteria.

BIDIRECTIONAL TRANSPORT

In the whole animal, elevation of the concentrations of most amino acids in blood leads to an elevation of the level in brain. In other cases, notably that of glutamate in blood, this is not true. On the other hand, introduction of small amounts of labeled glutamate into the bloodstream leads to a rapid exchange and equilibration of the label across the blood-brain barrier, proving that glutamate is readily transported into brain. This apparent contradiction is resolved by the finding that active transport of amino acids also takes place in the opposite direction, from brain to blood. The extremely rapid transport of glutamate out of the brain prevents the brain concentration from rising in response to elevated blood levels. When brain levels of individual amino acids are raised by intracerebral injection, the amino acids are pumped out into the plasma against a concentration gradient. Similar experiments have shown that amino acids can be actively transported from the cerebrospinal fluid into the bloodstream. Thus, both brain and cerebrospinal fluid concentrations of amino acids are regulated by active transport mechanisms that apparently act in both directions.

DEVELOPMENTAL CHANGES AND REGIONAL DIFFERENCES

Changes in amino acid transport mechanisms have been investigated during development of the animal. In general, the brain of young animals takes up amino acids from the blood more actively than does that of adults. This may be due to more efficient amino acid transport systems in young animals, or it may be a reflection of the less pronounced blood-brain barrier generally found in immature individuals. Quantitative differences in the activity of amino acid transport mechanisms in various parts of the brain have been observed.

Little or no information is available about differences in transport of the various cell types found in brain. It has been shown, however, that synaptosomes, the pinched-off nerve ending particles, can transport many amino acids. This suggests that amino acids can enter neurons and may indicate that axonal flow is not a necessary mechanism for amino acid nutrition in those cells.

In general, since the detailed biochemical mechanisms of amino acid transport have not been dissected on a molecular level, these data comprise an interesting descriptive literature, but they have not yet provided a consistent theory of molecular events on a developmental, anatomical, functional, or phylogenetic level.

SPECIFIC TRANSPORT SYSTEMS

Use of various in vitro preparations, notably tissue slices, has led to a sharper understanding of the transport mechanism in brain. Much of this information has pertained to the specificity of the various transport mechanisms for amino acids, and much of it has been corroborated by work in vivo. Clearly, more than one transport mechanism is involved. On the other hand, there are no separate sites for each amino acid. At least five sites seem to exist for amino acid transport. Sites have been identified in brain, as in other mammalian cells, for (1) small, neutral amino acids; (2) large neutral amino acids, including aromatics; (3) small, basic amino acids; (4) large, basic amino acids; and (5) acidics. In addition, sites probably exist for GABA, imino acids, and so on. Although each group has high affinity for a given site, evidence is available to show that amino acids frequently have some affinity for the other sites as well. Several amino acids have affinity for a single site, so there may be competition among them for transport. For example, it is clear that all the aromatic amino acids share a common transport mechanism and that large amounts of one amino acid in the blood may interfere with the uptake of others in that group into the brain. This has been demonstrated experimentally by raising the level of a single amino acid in the blood of an experimental animal. It seems eminently possible that such competitive interactions play a role in the pathology of disease states in which blood levels of amino acids are elevated by lesions in the metabolic apparatus for amino acids.

REQUIREMENTS FOR TRANSPORT

Several other properties of amino acid transport have been investigated by using slices and intact animals. It is known that transport of amino acid in brain requires energy. Inhibitors of respiration or of phosphorylation inhibit transport, and at least a general correlation can be made between ATP levels in slices and transport activity under a variety of circumstances. Amino acid transport also requires Na^+ and is inhibited by ouabain. This shows that in brain, as in other organs, amino acid transport is linked in some way to the sodium pump. The D-isomers of most of the amino acids are taken up, and some are concentrated, by brain. Some evidence is available to suggest that they enter through the same mechanism as does the L-isomer. In most cases, the D-isomers have less affinity for these sites than do the L-isomers, but in a few cases they are accumulated to a greater degree. This apparent anomaly – lower affinity and higher concentration – may be caused by an even greater stereospecificity of the active efflux system. Finally, it is known that chemical alteration of either the amino or carboxyl groups of most of the transported amino acids significantly lowers their ability to be transported.

Observations of the differences between the characteristics of transport in vitro and in vivo are probably not too valuable because the slice system is artificial and can serve only as a guide to the physiological situation. Similarities between the uptake in slices and the uptake or the endogenous levels in vivo may be more fruitful. Thus, there is much evidence to show that a reasonable correlation exists between rates of transport of the various amino acids in slices and their uptake in vivo. Further, the competitive relationships and stereospecificity shown by slices are generally mirrored by the data collected in the whole animal. Probably, then, the slice is a valuable and accurate tool for the study of amino acid transport in brain.

REGULATING ACTION OF AMINO ACID TRANSPORT

Generally speaking, the most rapidly transported amino acids are those found in highest concentrations inside the cell. It seems, then, that transport rates are a major factor in determining the levels of amino acids in the cell. Some exceptions to this generalization are known, suggesting that other factors are involved. However, transport rates into and out of the brain cells certainly seem to be a major controlling influence on the concentrations of amino acids in the free amino acid pool. And, since the levels of amino acids in this pool have been shown to influence the amino acid metabolic rates and the rates of protein synthesis in the cell, the transport step must be considered in any evaluation of the amino acid economy of the brain.

METABOLISM

GLUTAMIC ACID DERIVATIVES

The metabolism of the glutamate family of amino acids is, quantitatively, the most significant aspect of amino acid metabolism in brain ([1a, 1d, 1g]; see also Chap. 8, this volume). In fact, the extensive metabolism of glutamate and its derivatives is probably the most characteristic part of the amino acid economy in brain. After the injection of labeled glucose, 70 percent of the isotope present in the soluble fraction of brain is in the amino acids, primarily as glutamate, glutamine, aspartate, and GABA. This conversion is much more pronounced in brain than in other tissues, and is partly due to the large pool of free glutamate in equilibrium with the α-ketoglutarate of the Krebs cycle.

The product of the α-decarboxylation of glutamic acid, γ-aminobutyric acid (GABA), is produced by a neural metabolic route known as the *GABA shunt* (see Fig. 8-1). The three enzymes of the GABA shunt are glutamic acid decarboxylase (GAD-I), γ-aminobutyric acid transaminase (GABA-T), and succinic semialdehyde dehydrogenase (SSA-D) [1a]. The first, GAD-I, is found almost exclusively in nervous tissue. Recently, a second glutamic acid decarboxylase (GAD-II) has been found in other tissues, but it seems to be a separate enzyme with wide distribution and quite different properties. GAD-I is a pyridoxal phosphate-requiring enzyme. When it is isolated, it readily loses its cofactor and becomes quite labile. The enzyme is found primarily in gray matter, as is GABA itself, and a good correlation exists between the regional distribution of the enzyme and that of its substrate. Like GABA, GAD-I is associated with synaptic structures, but it appears to be a soluble enzyme because it is released upon hypoosmotic shocking of the synaptosomes.

GABA-T also requires pyridoxal phosphate, but, whereas GAD-I has a loosely bound pyridoxal phosphate, the cofactor for GABA-T is tightly associated with the protein. GABA-T is a mitochondrial enzyme that is also present in tissues other than brain. GABA-T is specific for the amino group acceptor, α-ketoglutarate, but it is nonspecific for the amino donor, GABA, as β-alanine will also be transaminated. GABA-T has a distribution in brain similar to that of GAD-I, and follows a developmental course very much like that of the decarboxylase and, in fact, of GABA itself. The final enzyme, SSA-D, is regionally and subcellularly associated with the other two. It is specific for succinic semialdehyde and for NAD^+, and seems to be present in tissues other than brain, as well.

There are no good estimates of the quantitative importance of GABA-shunt metabolism in brain. Clearly, the function of the shunt is to provide GABA. It should be obvious, however, that the shunt metabolism also circumvents a phosphorylation at the substrate level in the Krebs cycle and obviates the need for a molecule of CoA. One reduced pyridine nucleotide is produced by the shunt, but one reduced pyridine nucleotide is also produced in the circumvented part of the Krebs cycle during the conversion of α-ketoglutarate to succinyl-CoA.

In most species the enzyme glutamine synthetase is in higher concentration in brain than in any other organ. It has been suggested that the formation of glutamine serves, among other things, as a mechanism for the detoxification of ammonia, to which the brain is sensitive. When the systemic level of ammonia rises, as during severe liver damage, coma results. The mechanism of this coma is unknown; but, as was mentioned earlier, ammonium ion may directly antagonize neural transmission.

ASPARTIC ACID

Acetylaspartic acid is formed from aspartate and acetyl-CoA. Little is known of the enzyme which hydrolyzes this derivative specifically, but acetylase enzymes have been described by several investigators. As mentioned before, the function of the compound is not known. The two brain enzymes involved in its metabolism do not seem to be unique.

HISTIDINE

Histidine is decarboxylated to histamine by two different enzymes in mammalian tissues. One enzyme, histidine decarboxylase, is specific for histidine, and one is general for aromatic amino acids. Both types are present in the nervous system, and it is not clear which is responsible for the formation of the histamine found in brain.

TYROSINE

The conversion of tyrosine to norepinephrine, although a minor pathway of tyrosine metabolism for the intact animal, seems to be the major route of tyrosine metabolism in brain and adrenal tissue ([3]; see also Chap. 6, this volume). The first step, catalyzed by tyrosine hydroxylase, has been shown to be rate-limiting for the whole process. The enzyme is a mixed-function oxidase that requires oxygen, reduced pteridine, and ferrous iron, and it has been shown to incorporate molecular oxygen into the 3 position of the tyrosine ring. The enzyme is inhibited by catechols, including norepinephrine, which suggests that it is under feedback control and that catecholamines control the rate of their own synthesis. The decarboxylation of DOPA to dopamine is carried out by a typical amino acid decarboxylase that requires pyridoxal phosphate. In brain, the enzyme is nonspecific, acting on a wide variety of aromatic amino acid substrates, including 5-hydroxytryptophan. Dopamine β-hydroxylase is present in brain, but it has not been thoroughly studied in this tissue because of difficulties in the assay and because of the presence of potent natural inhibitors. The corresponding enzyme from adrenal gland has been extensively purified and shown to require molecular oxygen and ascorbic acid. It contains copper ions and is markedly stimulated by fumaric acid. The enzyme is nonspecific and will catalyze the side-chain hydroxylation of a large number of β-phenylethylamine derivatives. The methylation of norepinephrine to epinephrine is catalyzed by an enzyme requiring *S*-adenosylmethionine as methyl donor.

TRYPTOPHAN

Conversion of tryptophan to serotonin in brain (see Fig. 6-5) is quantitatively relatively minor compared with the amount of tryptophan metabolized by the intact animal; it occupies about 1 percent of the total tryptophan metabolism, but is the major, if not the only, route of tryptophan metabolism in brain [1c]. The first step, hydroxylation in the 5 position, is catalyzed by tryptophan hydroxylase and appears to be rate-limiting. The enzyme requires molecular oxygen and reduced pteridine; it is stimulated by ferrous iron and high levels of mercaptoethanol. When assayed under optimal conditions, the enzyme can be shown to exist primarily in the soluble fraction of the cell. Tryptophan hydroxylases from brain and pineal gland have been studied and appear to differ in the details of their action. With both enzymes, however, catecholamines are strong inhibitors, indicating a close relationship between the catechol and the indole pathways of amine formation.

The second step is catalyzed by 5-hydroxytryptophan decarboxylase, a typical amino acid decarboxylase that may be identical to the DOPA decarboxylase in the biosynthesis of epinephrine (see Chap. 6). In the pineal gland (see Fig. 6-6), serotonin is acetylated by an acetyl transferase, using acetyl-CoA as cosubstrate, and then *O*-methylated by a methyl transferase, using *S*-adenosylmethionine. Both of the enzymes appear to be responsive to changes in the light-dark cycle of the animal and exhibit diurnal variation. These changes are reflected in a diurnal fluctuation in the concentration of melatonin, the pineal hormone, and may be important in the overall diurnal response exhibited by the whole animal.

OTHER PATHWAYS IN BRAIN

The pathways described above are the major routes of amino acid metabolism that are characteristic of brain. Of course, most of the amino acid metabolism found in neural tissue is also observed in liver and other tissues [1f]. Enzymes that are known to be present specifically in brain are L-amino acid oxidase, D-amino acid oxidase, and oxidases specific for glutamic acid and for proline. As mentioned above, hydroxylases for tyrosine, tryptophan, and dopamine have been demonstrated. *S*-Adenosylmethionine-dependent methylases for *N*-acetylserotonin, catechols, histamine, and phenylethanolamine have also been demonstrated, as has the enzyme responsible for *S*-adenosylmethionine formation. Enzymes are known that transfer the amidino group of arginine to glycine to form guanidinoacetate and to GABA to form γ-guanidinobutyrate. Most of the enzymes of the urea cycle are known to be present in brain. Transaminases for aspartic acid, alanine, glycine, tyrosine, leucine, ornithine, glutamine, phenylalanine, DOPA, tryptophan, 5-hydroxytryptophan, serine-*O*-phosphate, and GABA are also known to occur. The tyrosine transaminase from brain does not respond to steroids or to substrate, showing that its mode of regulation, if any, is different from that of the well-studied tyrosine transaminase in liver. Asparaginase and glutaminase activities have been demonstrated. Decarboxylases for glutamic acid, histidine, DOPA, 5-hydroxytryptophan, serine, GABA, leucine, and phenylalanine are found in brain. Several enzymes of sulfur metabolism, including cystathionine synthetase and cystathionase, are also present. As mentioned before, glutamine synthetase has an important brain function. The enzymes that activate amino acids for incorporation into protein by attaching them to transfer RNAs have been studied in brain. Finally, the enzymes that phosphorylate and dephosphorylate pyridoxal and pyridoxal phosphate, respectively, are known to occur in brain and provide a mechanism for regulating the levels of this important amino acid metabolic cofactor.

PEPTIDES

Several peptides, at least one of which is also generally distributed in the body and a number of which appear to be unique constituents of brain, occur in neural tissue [1e]. Glutathione, or γ-glutamylcysteinylglycine, is found to have a concentration of about 50 to 100 mg per 100 g of fresh brain. This tripeptide is present in all tissues and body fluids. In brain, it is found predominantly in the reduced form, but it is oxidized within minutes after the death of the animal. The function of this peptide in brain, as elsewhere, is unknown, although several enzymes utilize glutathione as cofactor.

A number of other γ-glutamyl peptides are found in brain. Among them are γ-glutamyl derivatives of glutamate, glutamine, glycine, alanine, β-aminoisobutyric acid, serine, and valine. These peptides are present in amounts ranging from 10 to 700 μg per gram of fresh tissue, and they may be formed by transpeptidation reactions from glutathione.

N-Acetyl-α-aspartylglutamic acid is also present in fairly large amounts in brain, and it is found only in brain. About 10 to 30 mg of this peptide are found per 100 g of fresh brain. Although a relation between this peptide and the large amount of free *N*-acetylaspartic acid in brain seems logical, such a relationship has not yet been clearly defined. No function has been attributed to this peptide, but it is known that it forms the N-terminal sequence of the actin molecule in muscle, and it may play a similar role in some brain protein.

Homocarnosine and homoanserine, two peptides of histidine, are found uniquely in brain. Homocarnosine is γ-aminobutyrylhistidine, the homolog of the long-known muscle constituent, β-alanylhistidine (carnosine). The γ-aminobutyryl derivative of anserine (β-alanyl-1-methylhistidine) is also present. These peptides are found only in brain and are present in much higher levels in this tissue than are their more widely distributed relatives, carnosine and anserine. Homocarnosine is highest in human brains (8 mg per 100 g of fresh brain tissue) and is also high in the brains of guinea pig, rabbit, and rat. It is low, but easily detectable, in the brains of cat, duck, and dog. It appears that these peptides are present as a consequence of the presence of their precursors, especially GABA, and not as a result of some unique brain enzyme. Carnosine synthetase from muscle will synthesize homocarnosine at a good rate if it is provided with GABA rather than with β-alanine.

"Substance P" is found in intestine as well as in brain. It is a polypeptide of some 20 to 30 amino acids and has been isolated in pure form. It was discovered because of its ability to stimulate the contraction of smooth muscle in vitro, and has been implicated in neural transmissions. Some investigators suggest that it functions as a centrally acting transmitter for inhibitory noncholinergic neurons.

PHARMACOLOGICAL AND PATHOLOGICAL ALTERATIONS

A number of compounds which, when administered, have severe neurological manifestations have been shown to affect amino acid levels or amino acid metabolism in brain. In no case, however, has it been absolutely proved that the altered amino acid economy is responsible for the neurological effects observed. On the other hand, in a few cases, substantial evidence is accumulating that links the altered amino acid picture with the neural effect.

Perhaps the most detailed picture, as indicated earlier, can be drawn around the actions of the various vitamin B_6 antagonists, but even here inconsistencies are found [1a]. Some drugs, such as semicarbazide, lower the level of GABA. Others, such as hydrazine, raise it. Both drugs cause convulsions. Hydroxylamine, which raises whole brain levels of GABA, has anticonvulsant properties. It may be that overall GABA levels are not the important factor and that changes in some specific small fraction of GABA more specifically affect cerebral excitability. This hypothesis lacks experimental proof.

A great number of other drugs exist that cause or prevent convulsions. In most cases these drugs do not have any effect on GABA concentrations. Methionine sulfoximine causes a characteristic behavioral disturbance in dogs, including grand mal seizures. It produces increases in blood Mg^{2+} and K^{+} concentrations, and its action can be reversed by methionine or glutamine. The compound may inhibit glutamine formation, but its exact mechanism of action is unclear. Even less is known about other convulsive drugs, such as metrazole or picrotoxin, or about the various anticonvulsants, but it seems fairly certain that they do not act primarily on the GABA shunt or, indeed, on other aspects of amino acid metabolism.

It is well known that most of the mood-affecting drugs are associated with changes in amine metabolism. This subject is discussed in detail in other sections of this book (see Chaps. 6 and 29). It should be mentioned that the biosynthesis of the amines can be selectively blocked by inhibiting the amino acid hydroxylases that catalyze the first steps in their biosynthetic pathways. Sedation results if the biosynthesis of norepinephrine is blocked selectively with α-methyltyrosine, a tyrosine hydroxylase inhibitor. When the biosynthesis of serotonin is blocked selectively with *p*-chlorophenylalanine, an inhibitor of tryptophan hydroxylase, more subtle changes in behavior are produced and must be observed over a long period of time. These changes may include alterations in aggressive and sexual behavior, but whether they are indeed due to the marked serotonin depletion or to other long-term alterations is still not known. In any case, the actions of these inhibitors of amino acid metabolism are reasonably consistent with the actions of other amine depleters, which work by entirely different mechanisms.

A number of genetic alterations in the amino acid metabolism of the whole animal also lead to severe alterations in brain function [2]. Phenylketonuria is the best characterized of these conditions. It is a genetically transmitted lesion which deletes phenylalanine hydroxylase, the liver enzyme. In the absence of this, the first enzyme in the only major route of phenylalanine catabolism, the increased blood phenylalanine floods some pathways of metabolism that are normally minor and enters some reactions that apparently do not use phenylalanine as substrate at all under normal conditions. The metabolites that are products of these pathways include phenylpyruvate, phenyllactate, phenylacetate, and phenylethylamine, all of which could arise through the action of enzymes known to be present in brain. Other products known to be formed from phenylalanine in phenylketonuria are *o*-hydroxyphenylacetic acid, phenylacetylglutamine, *N*-acetylphenylalanine, and hippuric acid. One of the most unusual products observed is the conjugate of phenylethylamine and pyridoxal, pyridoxylidene-β-phenylethylamine. These products are found in the blood, tissues, and urine of affected individuals. These subjects are discussed in more detail in Chapter 22.

REFERENCES

1. Lajtha, A. (Ed.), *Handbook of Neurochemistry,* New York: Plenum. See in particular:
   1a. Baxter, C. F. The Nature of γ-Aminobutyric Acid. Vol. 3, 1970, p. 289.

1b. Berl, S., and Clarke, D. D. Compartmentation of Amino Acid Metabolism. Vol. 2, 1969, p. 447.
1c. Guroff, G., and Lovenberg, W. Metabolism of Aromatic Amino Acids. Vol. 3, 1970, p. 209.
1d. Himwich, W. A., and Agrawal, H. G. Amino Acids. Vol. 1, 1969, p. 33.
1e. Pisano, J. J. Peptides. Vol. 1, 1969, p. 53.
1f. Seiler, N. Enzymes. Vol. 1, 1969, p. 325.
1g. Strecker, H. J. Biochemistry of Selected Amino Acids. Vol. 3, 1970, p. 173.

2. Paulson, G., and Allen, N. The Nervous System. In Goodwin, R. M. (Ed.), *Genetic Disorders of Man.* Boston: Little, Brown, 1970, p. 509.
3. Sourkes, T. L. *Biochemistry of Mental Disease.* New York: Harper and Row, 1962.
4. Tallan, H. H. A Survey of the Amino Acids and Related Compounds in Nervous Tissue. In Holden, J. T. (Ed.), *Amino Acid Pools.* Amsterdam: Elsevier, 1962, p. 471.

11

Chemistry and Metabolism of Brain Lipids

Kunihiko Suzuki

STUDIES OF LIPIDS in the nervous system form a significant part of neurochemical investigations. There are several obvious reasons for the importance of brain lipids, both as structural constituents and as participants in the functional activity of the brain. Among various body organs, the brain is one of the richest in lipids. It contains a unique structure, the myelin sheath, which has the highest concentration of lipid among any normal tissue or subcellular component except for adipose tissue, and which has been the subject of intensive and extensive studies in recent years. Another important aspect is the existence of a number of genetically determined metabolic disorders involving brain lipids. Identification of abnormally stored lipids and the search for underlying enzymatic defects have been giving strong impetus not only to the investigations of these pathological conditions but also to the study of chemistry and metabolism of brain lipids in general.

This chapter is designed to provide basic reference knowledge regarding brain lipids. The chapter, therefore, will cover only the most basic and elementary aspects of brain lipid, its chemistry, the lipid composition of normal brain, the peculiarities of brain lipids, and the major metabolic pathways. Many important aspects of biochemistry of brain lipids are covered elsewhere in this volume, such as the lipid and its metabolism in myelin (Chap. 18), developmental changes (Chap. 14), demyelinating diseases (Chap. 24), or inborn errors of sphingolipid metabolism (Chap. 23). There are also several excellent and reasonably up-to-date review articles on the general subject of brain lipids. They are given at the end of the reference list for this chapter and are recommended for further reading for more details of brain lipids and as the source of additional references.

CHEMISTRY OF MAJOR BRAIN LIPIDS

The lipid composition of the brain is unique not only in the high total lipid concentration but also in the types of lipids present. Three major categories include almost all of the lipids of normal brain: cholesterol, sphingolipids, and glycerophospholipids.

CHOLESTEROL

Cholesterol (Formula 11-A) is the only sterol present in normal adult brains in significant amounts. The alcohol group at position 3 may be esterified with a long-chain fatty acid. Cholesterol esters are not found in normal brain, however.

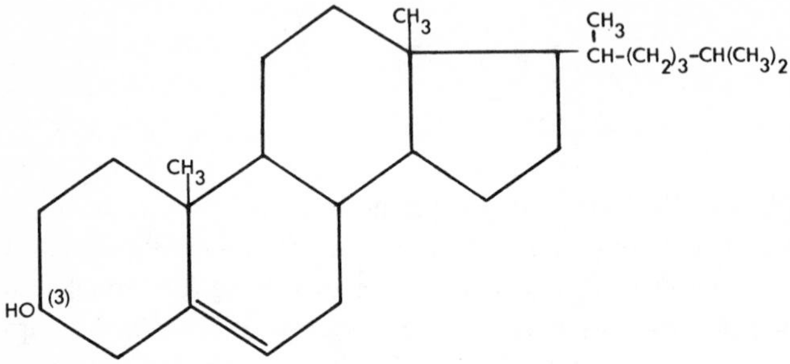

11-A Cholesterol

SPHINGOLIPIDS

Sphingosine

The basic building block of all sphingolipids is sphingosine, which is a long-chain amino diol with one unsaturated bond (Formula 11-B). The major sphingosine in the brain is C_{18}-sphingosine, but smaller amounts of C_{16}-, C_{20}-, and C_{22}-sphingosines are known to occur. Also, a small portion exists in the saturated form as dihydrosphingosine.

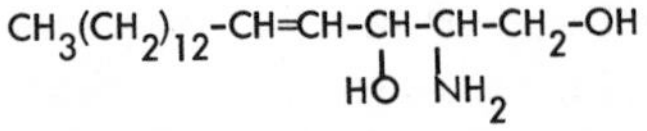

$$CH_3(CH_2)_{12}-CH{=}CH-\underset{\displaystyle HO}{CH}-\underset{\displaystyle NH_2}{CH}-CH_2-OH$$

11-B Sphingosine

Ceramide

The amino group of sphingosine is almost always acylated with a long-chain fatty acid, ranging from C_{14} to C_{26}. *N*-Acylsphingosine is generically called ceramide (Formula 11-C).

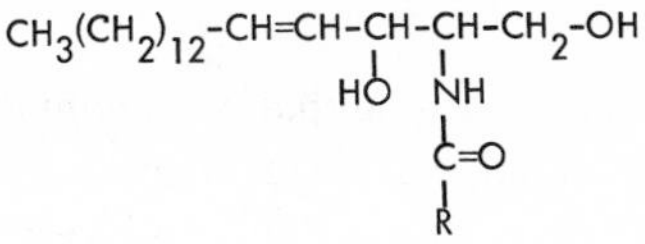

$$
\begin{array}{l}
CH_3(CH_2)_{12}\text{-}CH{=}CH\text{-}\underset{\substack{|\\ HO}}{CH}\text{-}\underset{\substack{|\\ NH\\ |\\ C=O\\ |\\ R}}{CH}\text{-}CH_2\text{-}OH
\end{array}
$$

$$R = -(CH_2)_nCH_3$$

11-C Ceramide (*N*-acylsphingosine)

The C-1 alcohol group of ceramide can be substituted with a variety of compounds to form different sphingolipids. Ceramide-A is any sphingolipid characterized by a substituent, A, at the terminal hydroxyl group of ceramide. We can now define individual sphingolipids in the brain by defining the substituent A.

Sphingomyelin

This is the only phospholipid in the brain that is also a sphingolipid; in this case, A is phosphorylcholine, so sphingomyelin may be defined as ceramide-phosphoryl-choline.

Cerebroside and Sulfatide

Cerebroside is a generic term for monohexosylceramide, i.e., the substituent A is a hexose. Ceramide-galactose is also known as galactocerebroside, and ceramide-glucose as glucocerebroside.

All the cerebroside in normal adult brain is galactocerebroside. Glucocerebroside occurs in the brain only in small amounts in certain pathological conditions, as well as in the immature brain. Sulfatide is galactocerebroside with an additional sulfate group at C-3 of the galactose moiety, i.e., sulfatide is ceramide-galactose-SO_4^-.

Ceramide Oligohexosides

A series of sugar-containing sphingolipids is known. Unlike cerebroside, all ceramide oligohexosides in the brain have a glucose moiety linked to ceramide. Although these compounds are practically absent in normal adult brains, they are present in measurable amounts in immature brains and in some pathological conditions. They are important in relation to the metabolism of brain ganglioside. Some of these substances are ceramide–Glc–Gal, or ceramide dihexoside, also called ceramide lactoside; ceramide–Glc–Gal–GalNAc, or ceramide trihexoside; and ceramide–Glc–Gal–GalNAc–Gal, or ceramide tetrahexoside (Glc = glucose; Gal = galactose; GalNAc = *N*-acetylgalactosamine).

Gangliosides

Gangliosides are defined as sphingoglycolipids that contain sialic acid. Sialic acid is a generic name for *N*-acylneuraminic acid (see below), and the acyl group of sialic acid in gangliosides of the brain is always the acetyl form. *N*-Acetylneuraminic acid is commonly abbreviated as NANA (Formula 11-D).

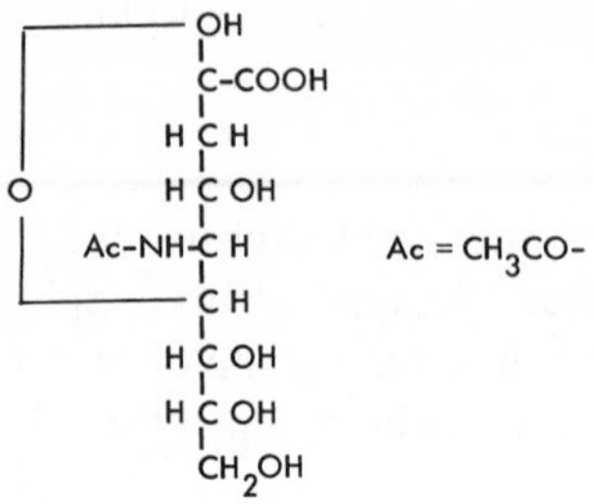

```
           ┌──────── OH
           │         |
           │         C-COOH
           │         |
           │       H C H
           │         |
           O       H C OH
           │         |
           │  Ac-NH-C H            Ac = CH3CO-
           │         |
           └──────── C H
                     |
                   H C OH
                     |
                   H C OH
                     |
                     CH2OH
```

11-D *N*-acetylneuraminic acid (NANA)

The series of gangliosides in the brain has the above ceramide oligohexosides as the backbone, with one or more NANA moieties attached. Major gangliosides of the brain are depicted in Figure 11-1.

GLYCEROPHOSPHOLIPIDS

Glycerophospholipids with two acyl ester linkages are termed phosphatidyl compounds (Formula 11-E). In Formula 11-E, R_1 and R_2 represent long-chain fatty acid moieties which form ester linkages to two alcohol groups of the parent glycerol molecule. The third alcohol group of the glycerol molecule is linked to a phosphate group. Like sphingolipids, one of the hydroxyl groups of the phosphate moiety is substituted with a substituent, A, to form various phosphatidyl compounds. When substituent A is hydrogen, the formula is that of phosphatidic acid. Other groups that may be substituted for A are listed under the basic formula.

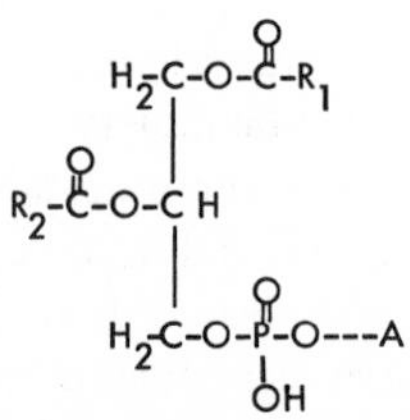

```
                    O
                    ‖
           H2-C-O-C-R1
              |
       O      |
       ‖      |
   R2-C-O-C H
              |
              |     O
              |     ‖
           H2-C-O-P-O---A
                    |
                    OH
```

11-E Phosphatidyl compounds
(Phosphatidic acid when A = H)

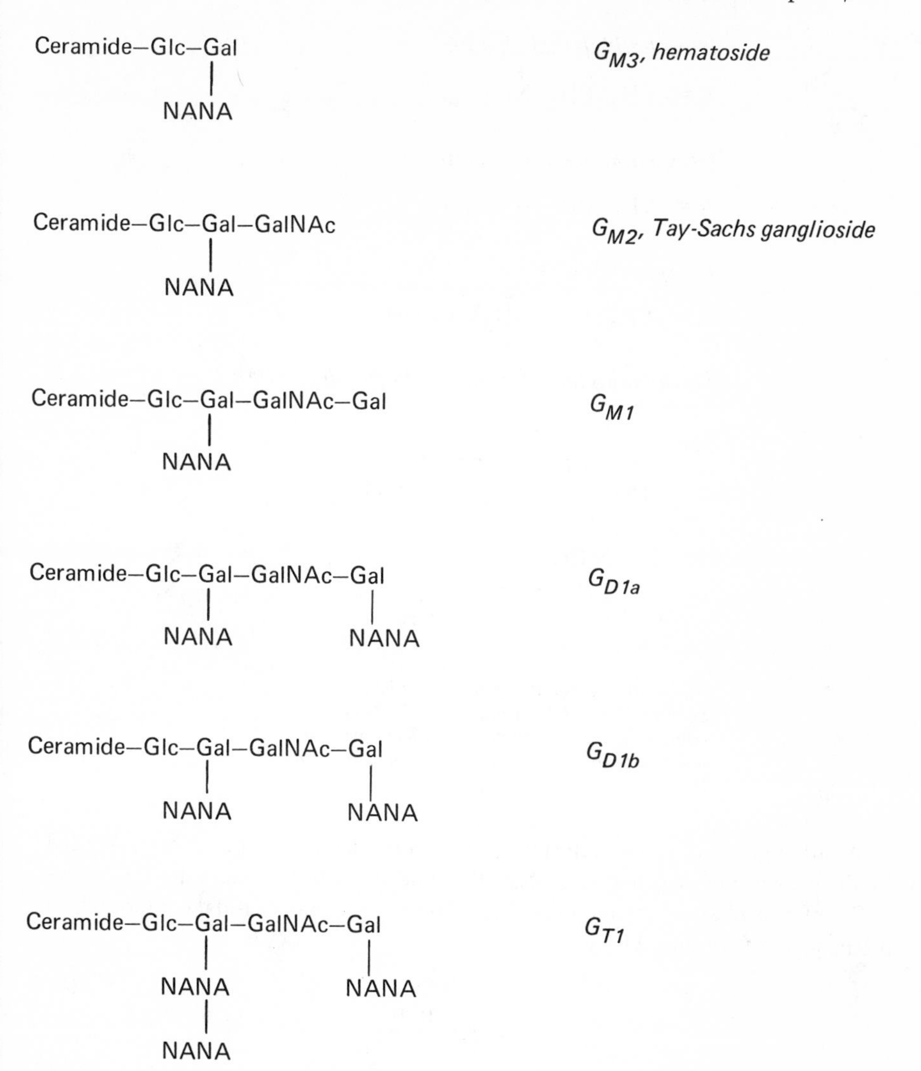

FIGURE 11-1. Major gangliosides of the brain. The nomenclature of gangliosides is that of Svennerholm [34], which is probably most widely used. For other nomenclatures, refer to Ledeen [18].

Phosphatidylethanolamine:

$$A = -CH_2-CH_2-\overset{+}{N}H_3$$

Phosphatidylcholine (lecithin):

$$A = -CH_2-CH_2-\overset{+}{N}(CH_3)_3$$

Phosphatidylserine:

$$A = -CH_2-CH-(\overset{+}{N}H_3)-COOH$$

Phosphatidylinositol (monophosphoinositide):

A =

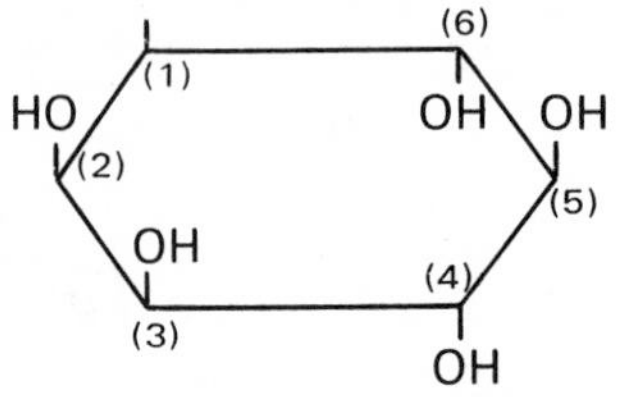

(6)
(1)
HO
OH OH
(2)
(5)
OH
(4)
(3)
OH

Additional phosphate groups may be present at position 4, or at positions 4 and 5. Then the compounds are called phosphatidylinositol-4-phosphate (diphosphoinositide) and phosphatidylinositol-4,5-diphosphate (triphosphoinositide).

Another group of glycerophosphatides of potential importance is characterized by the presence of an α,β-unsaturated ether linkage, which replaces the acyl ester at C-1 of phosphatidyl compounds. These are termed phosphatidal compounds or plasmalogens (Formula 11-F).

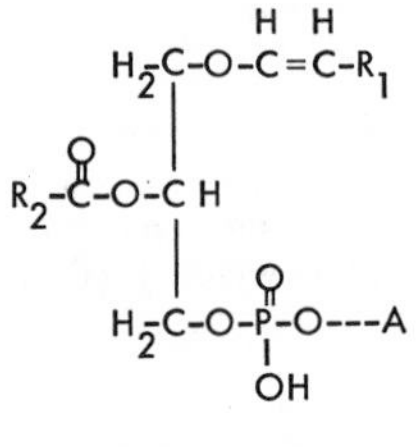

$$H_2C-O-CH=CH-R_1$$
$$R_2-\overset{O}{\overset{\|}{C}}-O-CH$$
$$H_2C-O-\overset{O}{\overset{\|}{P}}(OH)-O---A$$

11-F Phosphatidal compounds

Although substituent A may be any one of those described for phosphatidyl compounds, almost all of the phosphatidal lipid in the brain is phosphatidalethanolamine, with much smaller amounts of phosphatidalcholine and phosphatidalserine.

BASIC METHODOLOGY OF BRAIN LIPID ANALYSIS

TREATMENT OF BRAIN SPECIMENS

Freshly obtained specimens are always preferable, but this criterion cannot always be met with human specimens. Most brain lipids, with the exception of the polyphosphoinositides, appear to be sufficiently stable for reliable analysis, not only during the several-hour period of refrigeration which usually elapses between death and autopsy, but also during prolonged storage in the frozen state. A storage temperature of $-20^{\circ}C$ is often satisfactory but $-40^{\circ}C$ or lower is better. Specimens must be stored in air-tight containers or plastic bags to avoid loss of water, which occurs even at the lowest temperature in a deep freeze.

A large body of literature indicates that formalin preservation may be detrimental to lipid analysis. For instance, in addition to water loss, small molecules may be lost and the tissues may shrink, making analysis less reliable. Certain lipids are selectively destroyed in the fixative. For example, plasmalogen is exceedingly labile and disappears rapidly in formalin, and selective destruction of brain gangliosides by formalin is well documented [32]. Attempts have been made to correct the analytical results for the effect of formalin on pathological specimens by comparing them with normal control specimens that have been preserved for a length of time. In addition to the period of fixation, the effect of formalin depends on many factors, such as initial and subsequent pH, the ratio of fixative to tissue, or temperature changes, so the analysis of formalin-fixed material for lipids usually should be avoided.

GENERAL ANALYTICAL SCHEME

Although there are numerous minor modifications, the method of Folch, Lees, and Sloane-Stanley [11] is almost always the starting point for any investigation of lipids in the nervous system. The basic analytical scheme now being used in our laboratory for general analysis of brain lipid is depicted in Figure 11-2.

Polyphosphoinositides are not extracted by this procedure unless the insoluble residue is reextracted with an acidic mixture of chloroform and methanol. In order to extract ganglioside completely, it is necessary to reextract the chloroform-methanol insoluble residue with a mixture of chloroform-methanol of a reversed ratio (1:2, v/v) containing 5 percent water [32]. Otherwise the procedure is sufficiently quantitative for most purposes, so the final chloroform-methanol soluble fraction can be used for detailed lipid analysis and the retentate of the upper phase for ganglioside analysis.

The proteolipid protein-free lipid fraction and the ganglioside fraction are then ready for detailed analysis of individual lipids. It is not the purpose of this chapter to describe the analytical methods for individual lipids in any detail. Colorimetry and column and thin-layer chromatography are utilized extensively. Thin-layer chromatography has become established in the analytical investigations of lipids as one of the most convenient and rapid procedures. Figures 11-3 and 11-4 show

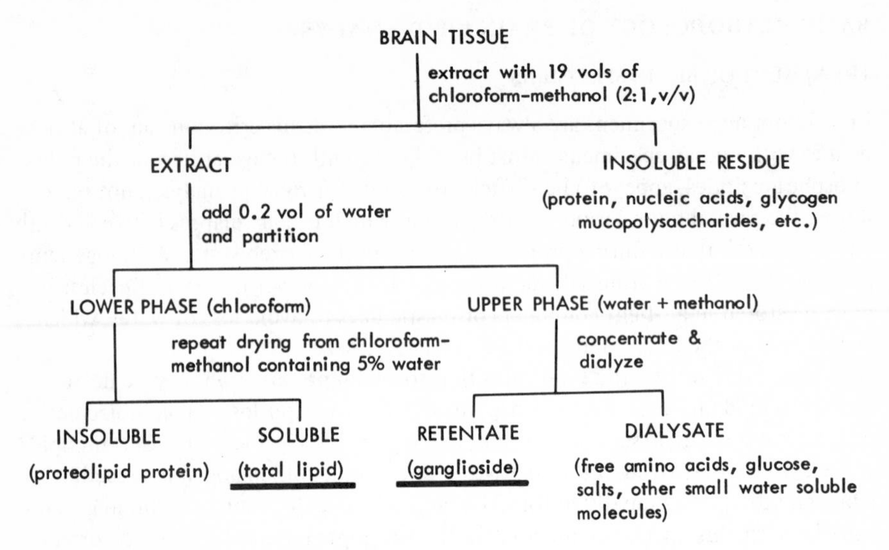

FIGURE 11-2. A typical scheme of brain lipid analysis.

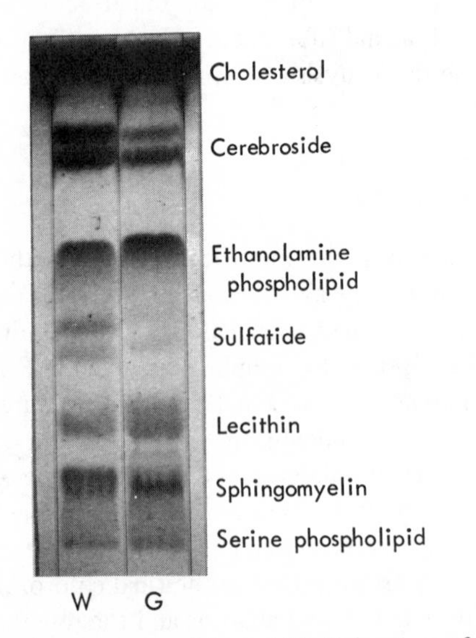

FIGURE 11-3. Thin-layer chromatogram of the total lipid fractions of normal gray *(G)* and white *(W)* matter. Approximately 250 μg of total lipid were spotted on a silica gel G plate 250 μ in thickness. Solvent system: chloroform-methanol-water (70:30:4, by volume). Spots were visualized by spraying with 50% sulfuric acid and heating. Note the greater amounts of cerebroside and sulfatide and lesser amounts of phospholipids in white matter. Serine phospholipid streaks from the origin to the area of sphingomyelin in this solvent.

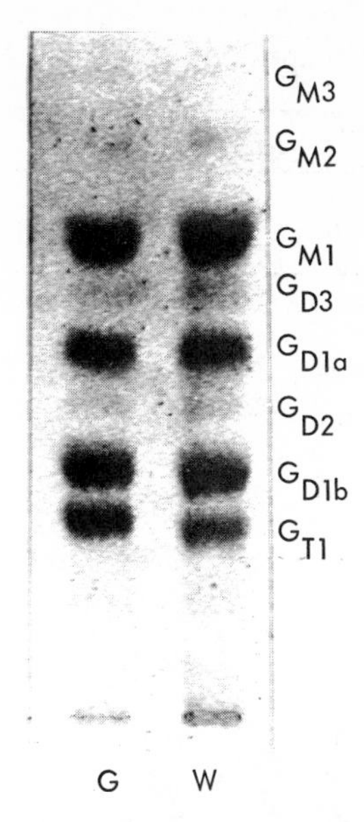

FIGURE 11-4. Thin-layer chromatogram of the ganglioside fraction of normal gray *(G)* and white *(W)* matter. The dialyzed upper phase containing approximately 30 μg of NANA was spotted on a silica gel G plate 250 μ in thickness. Chromatography was run in two solvent systems successively: the first solvent was chloroform-methanol-2.5 N ammonia (60:40:9, by volume), and the second solvent, *n*-propanol and water (7:3 by volume). Spots were visualized by 50% sulfuric acid spray and heating. The four major gangliosides are clearly visible, but other minor gangliosides also are present in very small amounts in normal brain.

separations of total lipid and gangliosides in gray and white matter. The compounds have been separated by thin-layer chromatography in some of the routine solvent systems.

Gas-liquid chromatography is now used extensively for the analysis of almost every constituent moiety of brain lipids. The combination of gas-liquid chromatography with mass spectrometry promises to be a useful tool for simultaneous definitive identification and quantitative determination of such compounds.

LIPID COMPOSITION OF NORMAL ADULT HUMAN BRAIN

LIPID COMPOSITION

Typical analytical results of normal adult human brain are given in Tables 11-1 and 11-2. Because the lipid compositions of gray and white matter differ both in total concentration and in the distribution of individual lipids, it is usually important to treat them separately. Depending on the purpose of investigation, selection of specific regions of the nervous system could become of fundamental importance. Equally important is the selection of the basis of reference, such as

TABLE 11-1. Typical Lipid Composition of Normal Adult Human Brain

	Gray Matter			White Matter		
Constituents	% fresh wt.	% dry wt.	% lipid	% fresh wt.	% dry wt.	% lipid
Water	81.9	–	–	71.6	–	–
Chloroform-methanol insoluble residue	9.5	52.6	–	8.7	30.6	–
Proteolipid protein	0.5	2.7	–	2.4	8.4	–
Total lipid	5.9	32.7	100	15.6	54.9	100
Upper phase solids	2.2	12.1	–	1.7	6.0	–
Cholesterol	1.3	7.2	22.0	4.3	15.1	27.5
Phospholipid, total	4.1	22.7	69.5	7.2	25.2	45.9
Ethanolamine phospholipid	1.3	7.2	22.7	2.3	8.2	14.9
Lecithin	1.6	8.7	26.7	2.0	7.0	12.8
Sphingomyelin	0.4	2.3	6.9	1.2	4.2	7.7
Monophosphoinositide	0.16	0.9	2.7	0.14	0.5	0.9
Serine phospholipid	0.5	2.8	8.7	1.2	4.3	7.9
Plasmalogen	0.7	4.1	8.8	1.8	6.4	11.2
Galactolipid, total	0.4	2.4	7.3	4.1	14.5	26.4
Cerebroside	0.3	1.8	5.4	3.1	10.9	19.8
Sulfatide	0.1	0.6	1.7	0.9	3.0	5.4
Ganglioside, total[a]	0.3	1.7	–	0.05	0.18	–

[a]The amounts of ganglioside were calculated from the total sialic acid, on the assumption that sialic acid constitutes 30% of ganglioside weight in a typical ganglioside mixture of normal brain.

whole brain, wet weight, dry weight, protein, DNA, or, in cases of lipid studies, total lipid. No single basis of reference is satisfactory or proper in all situations. In Table 11-1 values are given according to three bases of reference commonly used in analytical studies of brain lipids: fresh weight, dry weight, and lipid weight. Fresh weight is often the most proper reference in pathological conditions involving lipids because lipids are major constituents of the brain and often decrease drastically, thus substantially altering dry weight. When the water content of the tissue is in question, as is often the case in casually stored frozen specimens, dry weight may be the reference of choice. To compare relationships among various lipids, total lipid weight often provides the clearest picture.

White matter contains less water and much more lipid and proteolipid than does gray matter. On the basis of wet weight, there is almost three times as much lipid in white matter as in gray matter. There is almost twice as much on a dry weight basis. As a consequence, there is more of each of the individual lower phase lipids in white than in gray matter on the basis of wet weight, with the possible exception of monophosphoinositide. In contrast, gangliosides are characteristically gray matter lipids. Gray matter is ten times richer in gangliosides than is white matter, on dry weight basis.

TABLE 11-2. Composition of Major Gangliosides in Adult Human Brain[a]

Constituents	Gray Matter		White Matter	
	Average	Range	Average	Range
Total NANA (μg/g wet wt)	812	744–918	110	80–180
G_0	3.9	3.2–4.8	4.8	2.8–6.1
T_{T1}	19.7	15.8–25.7	19.1	14.1–21.2
G_{D1b}	16.7	14.3–19.9	14.8	12.2–18.1
G_{D2}	3.0	1.2–4.2	1.6	1.2–3.1
G_{D1a}	38.0	29.1–43.7	36.2	30.0–38.2
G_{D3}	2.0	1.0–2.8	3.2	1.2–5.0
G_{M1}	14.2	13.0–15.6	18.8	14.6–21.2
G_{M2}	1.7	1.5–2.0	1.0	0.6–2.0
G_{M3}	<1	–	<1	–

[a]Values are expressed as percent of total NANA in each ganglioside, except for total NANA. G_0 represents all sialic acid which has mobility slower than G_{T1}. G_{D2} is Tay-Sachs ganglioside (G_{M2}) with an additional NANA. G_{D3} is hematoside (G_{M3}) with an additional NANA.

When the lipid compositions of gray and white matter are compared, the most conspicuous difference is that white matter is relatively much richer in galactolipids and relatively poor in phospholipids. Galactolipids constitute 25 to 30 percent of the lipids in white matter, while they are only 5 to 10 percent of those in gray matter. Phospholipids account for two-thirds of the total lipids in gray matter, but less than half of those in white matter. Plasmalogen constitutes 75 to 80 percent of ethanolamine phospholipid in white matter, but less than half of this phospholipid in gray matter. There are no measurable amounts of sphingoglycolipids, other than galactocerebroside and sulfatide, in the lower-phase lipid of normal adult human brain. These glycolipids are primarily localized in white matter. In contrast, gangliosides are highly localized in neuronal membranes and consequently in gray matter, although the molecular distribution pattern is similar in gray and white matter (see Table 11-2).

Except for the enrichment of ganglioside in gray matter, all other differences in the lipid composition between gray and white matter appear to be due mostly to the presence of myelin in the latter. Myelin constitutes half or more of the total dry weight of white matter, and it is poor in water content and rich in proteolipid protein and galactolipids. Readers are referred to Chapter 18 for details of composition and metabolism of myelin lipids.

The lipid composition of the brain is essentially constant throughout adult life, and the above data may serve as a basis for judging conditions that are altered pathologically. It is essential to keep in mind, however, that substantial changes in lipid composition take place during early development, particularly during the period of active myelination. Before myelination, both gray and white matter have a similar lipid composition, which resembles adult gray matter composition.

During active myelination, the brain loses water, predominantly in white matter, the lipid content increases rapidly, and the differences between gray and white matter become more apparent. Characteristically, galactocerebroside is virtually nonexistent before myelination begins and increases concomitantly with the amount of myelin formed [7, 24, 35]. Ganglioside is also known to undergo substantial developmental changes [33]. For more details of developmental changes, the reader should refer to Chapters 14 and 18.

In addition to myelin, the lipid composition of specific subcellular fractions, such as nuclei, mitochondria, synaptic elements, or microsomes, have been investigated [10, 17, 29]. With the recent development of procedures to isolate relatively intact neurons and glial cells (Chap. 21), the lipid compositions of different cell types are also being investigated actively [12, 25].

CHARACTERISTICS OF BRAIN LIPIDS

The lipid of the brain possesses several unique characteristics. As mentioned earlier, the brain is one of the richest portions of the body in total lipid content; approximately half of the dry weight of the entire brain is lipid. Although by no means constant, the lipid composition of the brain tends to remain remarkably unaffected by various external factors, such as the nature of dietary intake, malnutrition, and other conditions that would alter drastically the lipid composition of systemic organs or plasma.

The brain is unique in its lack of certain lipids that are abundantly present elsewhere in the body. Triglycerides and free fatty acids constitute, at the most, only a few percent of the total lipid of the brain. Some of these are probably contributed by blood and blood vessels, rather than by neural tissues. Esterified cholesterol is always much less than 1 percent of the total cholesterol in normal brain, although it is present in slightly higher amounts just prior to active myelination. In pathological conditions in which massive myelin breakdown occurs, esterified cholesterol is often present in high concentrations. Because of this characteristic, the sum of cholesterol, phospholipid, and glycolipid comprises nearly the total lipid, an often useful criterion for judging analyses of brain lipid. If the sum of the three major classes of lipids is substantially less than total lipid weight (e.g., less than 85 percent), the analysis is technically faulty, unless it can be explained otherwise.

Glycolipids of the brain are unique in many ways. Normal adult brain contains only galactocerebroside, whereas almost all systemic organs contain glucocerebroside primarily and very little galactocerebroside. The kidney is the only systemic organ known to contain approximately equal amounts of galactocerebroside and glucocerebroside, but even in this organ the actual amount is almost negligible compared to that in the brain. Brain galactolipids contain high proportions of long-chain fatty acids, predominantly C_{24} to C_{26}, which are rarely found in systemic organs. Also, there are high proportions of α-hydroxy fatty acids in brain galactolipids; approximately two-thirds of cerebroside and one-third of sulfatide. α-Hydroxy fatty acids are very unusual in systemic organs.

Brain gangliosides are also unique in their high concentration and molecular distribution. The level of total ganglioside in the brain is rarely approached by any systemic organs; hematoside (G_{M3}) is usually the only ganglioside present in a significant amount in most systemic organs.

BIOSYNTHESIS AND CATABOLISM OF MAJOR LIPIDS

The biosynthetic and catabolic pathways of brain lipids are generally similar to those in systemic organs. Some of the general references listed at the end of the reference list contain excellent chapters on details of the metabolic pathways of individual lipids. Only the basic outline will be reviewed here.

FATTY ACIDS

Fatty acids are important constituents of all complex lipids of the brain, although free fatty acids are rare in brain. The mechanism of fatty acid biosynthesis in brain appears to be identical to that in such other tissues as liver [26]. De novo synthesis of palmitic acid ($C_{16:0}$) involves acetyl-CoA and malonyl-CoA, and the chain elongation takes place in mitochondria. Brain contains considerable amounts of polyunsaturated fatty acids. Desaturation of fatty acids appears to occur in the acyl-CoA derivative. In systemic organs, α-hydroxylation of fatty acid is an intermediate step in C-1 degradation of fatty acids through dehydrogenation and decarboxylation. In brain, α-hydroxy fatty acids are found as such in galactocerebroside and sulfatide.

CHOLESTEROL METABOLISM

Although not every step of the biosynthetic pathway of cholesterol known to occur in other tissues has been demonstrated in brain, it is safe to assume that cholesterol synthesis takes place in brain through the same pathway as in systemic organs. Acetate and its precursors are transformed through mevalonic acid to cholesterol (Fig. 11-5).

Desmosterol, the immediate precursor of cholesterol, is known to be present in brain in measurable amounts just prior to myelination [16, 27] and also in the myelin sheath itself in the early stage of myelination [30]. Biosynthesis of cholesterol in brain is most rapid during the period of active myelination, but adult brain retains the capacity to synthesize cholesterol when precursors such as acetate or mevalonate are available. Although most of the cholesterol in brain appears to be synthesized from endogenous precursors, experimental evidence indicates that a small amount of systemically injected cholesterol can be taken up intact and that the rate of uptake is greatest when the rate of cholesterol deposition in brain is most rapid, i.e., during active myelination [9]. Once deposited in brain, cholesterol, particularly that incorporated into myelin, is relatively inert metabolically [8, 15]. When radioactive cholesterol is injected into newly hatched chicks,

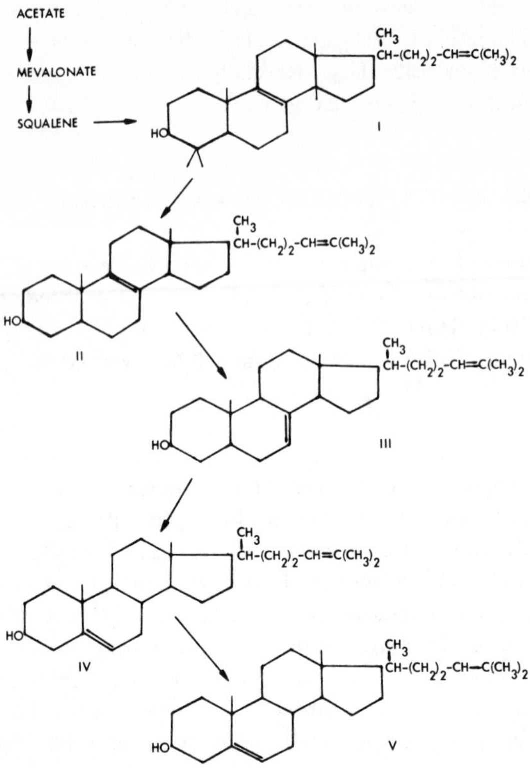

FIGURE 11-5. Outline of cholesterol biosynthesis. I = lanosterol; II = zymosterol; III = cholesta-7,24-dienol; IV = desmosterol (2,4-dehydrocholesterol); V = cholesterol.

considerable radioactivity remains in cholesterol of brain, whereas that of liver and plasma practically disappears within 3 to 8 weeks. This finding is consistent with the apparent lack in brain of an enzyme system for cholesterol degradation.

PHOSPHOLIPID METABOLISM

Metabolic pathways of phospholipid in brain are similar to those in systemic organs [28]. Figure 11-6 depicts the main pathways involving major brain diacylglycerophospholipids (phosphatidylphospholipid) and sphingomyelin.

There are two major synthetic pathways: one involves diglyceride and CDP-choline or CDP-ethanolamine for the formation of phosphatidylethanolamine, lecithin, and sphingomyelin; the other passes through CDP-diglyceride to form phosphoinositide and possibly phosphatidylserine (I and II in Fig. 11-6). The formation of phosphatidic acid is an important preliminary pathway common to both the major pathways (III in Fig. 11-6). Choline or ethanolamine is first

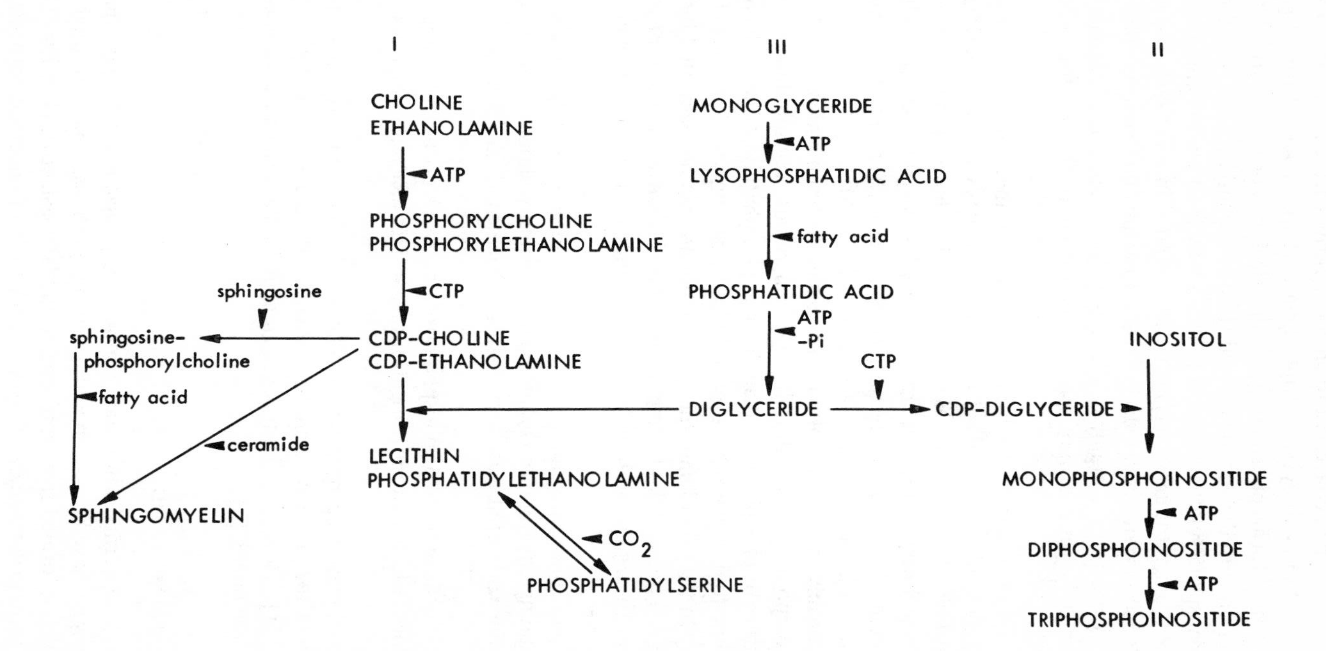

FIGURE 11-6. Major pathways of phospholipid biosynthesis.

phosphorylated and is then converted to an active form, CDP-choline or CDP-ethanolamine, by a reaction with cytidine triphosphate. This activated form of choline or ethanolamine then forms phosphatidylcholine (lecithin) or phosphatidylethanolamine by a reaction with diglyceride derived from phosphatidic acid. The direct conversion of phosphatidylethanolamine to lecithin by methylation apparently does not occur in the brain, although it has been demonstrated in some systemic organs. CDP-choline also reacts with ceramide to form sphingomyelin (ceramide-phosphorylcholine) [31]. At least in the brain, an alternate pathway appears to exist in which the first reaction is the formation of sphingosine phosphorylcholine, followed by its acylation [6].

Phosphoinositides are synthesized through a different mechanism (II in Fig. 11-6). Instead of forming CDP-compounds of inositol, CTP reacts with phosphatidic acid to form CDP-diglyceride, which, in turn, reacts with free inositol to form phosphatidylinositol. Polyphosphoinositides are synthesized by stepwise phosphorylation of monophosphoinositide. Interconversion of phosphatidylserine and phosphatidylethanolamine is known to take place in brain [1, 3, 20].

There are probably several hydrolytic enzymes in the brain which deacylate phosphatidyl compounds to lysophosphatidyl compounds. Phospholipase A deacylates the β position of phosphatidylethanolamine, phosphatidylserine, or lecithin. A specific enzyme to deacylate the α' position of these compounds has also been reported [13]. On the other hand, the lyso compounds can be reacylated by acyl-CoA to the original phosphatidyl compounds. This mechanism makes it possible for fatty acids of these phospholipids to turn over independently of the whole phospholipid molecule.

The first step of sphingomyelin degradation is catalized by sphingomyelinase, which cleaves sphingomyelin into ceramide and phosphorylcholine. The lack of this enzyme is the cause of at least some forms of Niemann-Pick disease (see Chap. 23).

The high metabolic activity of phosphoinositides is well documented and at present is the focus of intensive investigations regarding their physiological role in brain function.

Biosynthesis of plasmalogen in the brain has not been completely elucidated, but it appears that the pathway is similar to that for the diacyl form of glycerophospholipids [21]. However, plasmalogens might also be formed by reduction of phosphatidyl compounds [2].

CEREBROSIDE AND SULFATIDE

As stated earlier, the normal adult brain contains only galactocerebroside, and alternative pathways have been proposed for its biosynthesis. One is through psychosine, which is formed from sphingosine and UDP-galactose (I of Fig. 11-7). Psychosine, in turn, may be acylated by acyl-CoA to form galactocerebroside [4, 5]. More recent experimental data indicate, however, that galactocerebroside can be formed through ceramide and UDP-galactose, and the investigators were unable to confirm the acylation of psychosine [22, 23]. At present it appears more likely that the main biosynthetic pathway of brain galactocerebroside is pathway II of Figure 11-7. Biosynthesis of sulfatide occurs through cerebroside with the "active sulfate," 3'-phosphoadenosine 5'-phosphosulfate (PAPS), as the sulfate donor.

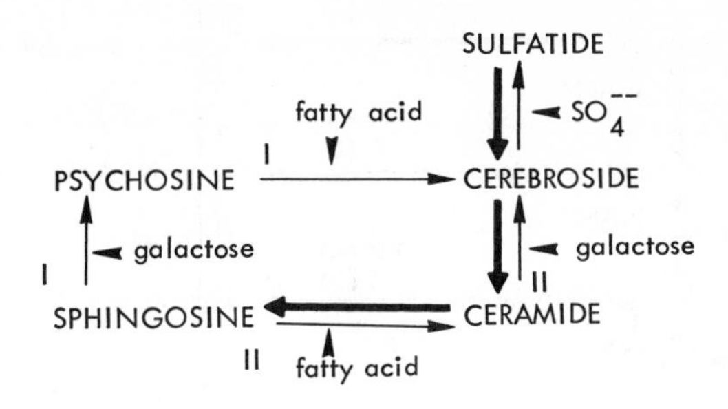

FIGURE 11-7. Outline of galactocerebroside biosynthesis.

The initial step of sulfatide degradation is removal of the sulfate group and its conversion back to galactocerebroside. The reaction is catalyzed by cerebroside sulfate sulfatase. The lack of this enzyme characterizes metachromatic leukodystrophy, in which excess sulfatide accumulation occurs (Chap. 23). Then galactocerebroside is degraded to ceramide and galactose by galactocerebroside β-galactosidase, the genetic lack of which causes Krabbe's globoid cell leukodystrophy (Chap. 23). Ceramide is further degraded to sphingosine and fatty acid by ceramidase.

As mentioned earlier, galactocerebroside is almost nonexistent in the brain before myelination. The rate of biosynthesis parallels that of myelin deposition and declines in adult brains. Once deposited, most of the brain cerebroside undergoes only slow turnover, as compared to many other phospholipids.

GANGLIOSIDE METABOLISM

Biosynthesis of gangliosides appears to take place by sequential addition of monosaccharides or NANA, catalyzed by glycosyl transferases specific for each step [14] (Fig. 11-8). The active forms of all monosaccharides are UDP compounds and that of NANA is CMP-NANA. The synthetic steps appear to occur in the microsomal fraction.

Degradation of brain gangliosides also proceeds by sequential removal of monosaccharides and NANA by glycosidases and neuraminidase [19]. These hydrolytic enzymes are localized in the lysosomal fraction. Although details of some of the pathways have not been completely established, the steps indicated by bold arrows in Figure 11-8 appear to be the main degradative pathway. Brain ganglioside

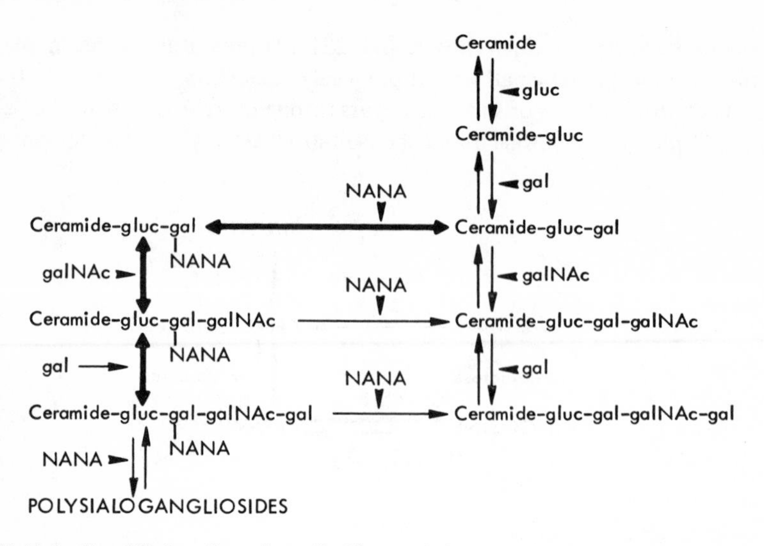

FIGURE 11-8. Simplified outline of ganglioside metabolism. In the direction of synthesis, hexose intermediates occur in the form of UDP esters. NANA is incorporated via CMP-NANA. Free sugars are released in the degradative pathways.

metabolism is of fundamental importance to the understanding of several of the most important sphingolipid storage disorders, including three enzymatically distinct G_{M2}-gangliosidoses, G_{M1}-gangliosidosis (possibly two forms), ceramide lactosidosis, and Gaucher's disease. The topic is covered elsewhere in this volume (Chap. 23).

Because of the high concentration in neuronal membranes, particularly in synaptic membranes, and because of the unusual chemical structure of both hydrophilic and lipophilic chains, the functional roles of brain gangliosides – such as their static role as a membrane constituent or their more dynamic roles in ion transport or nerve transmission – have often been implicated but have not been experimentally substantiated.

REFERENCES

ORIGINAL SOURCES AND SPECIAL TOPICS

1. Ansell, G. B., and Spanner, S. The incorporation of the radioactivity of [3–^{14}C] serine into brain glycerophospholipids. *Biochem J.* 84:12P, 1962.
2. Ansell, G. B., and Spanner, S. The metabolism of labelled ethanolamine in the brain of the rat *in vivo. J. Neurochem.* 14:873, 1966.
3. Borkenhagen, L. F., Kennedy, E. P., and Fielding, L. Enzymatic formation and decarboxylation of phosphatidylserine. *J. Biol. Chem.* 236:PC 28, 1961.
4. Brady, R. O. Studies on the total enzymatic synthesis of cerebrosides. *J. Biol. Chem.* 237:PC 2416, 1962.

5. Brady, R. O. Biosynthesis of Glycolipids. In Dawson, R. M. C., and Rhodes, D. N. (Eds.), *Metabolism and Physiological Significance of Lipids.* London: Wiley, 1964, p. 95.
6. Brady, R. O., Bradley, R. M., Young, O. M., and Kaller, H. An alternative pathway for the enzymatic synthesis of sphingomyelin. *J. Biol. Chem.* 240:PC 3693, 1965.
7. Cuzner, M. L., and Davison, A. N. The lipid composition of rat brain myelin and subcellular fractions during development. *Biochem. J.* 106:29, 1968.
8. Davison, A. N., Dobbing, J., Morgan, R. S., and Payling Wright, G. The deposition of [4–^{14}C] cholesterol in the brain of growing chickens. *J. Neurochem.* 3:89, 1958.
9. Dobbing, J. The entry of cholesterol into rat brain during development. *J. Neurochem.* 10:739, 1963.
10. Eichberg, J., Whittaker, V. P., and Dawson, R. M. C. Distribution of lipids in subcellular particles of guinea-pig brain. *Biochem. J.* 92:91, 1964.
11. Folch-Pi, J., Lees, M., and Sloane-Stanley, G. H. A simple method for the isolation and purification of total lipides from animal tissues. *J. Biol. Chem.* 226:497, 1957.
12. Freysz, L., Bieth, R., Judes, C., Sensenbrenner, M., Jacob, M., and Mandel, P. Quantitative distribution of phospholipids in neurons and glial cells isolated from rat cerebral cortex. *J. Neurochem.* 15:307, 1968.
13. Gatt, S. Purification and properties of phospholipase A_1 from rat and calf brain. *Biochim. Biophys. Acta* 159:304, 1968.
14. Kaufman, B., Basu, S., and Roseman, S. Studies on the Biosynthesis of Gangliosides. In Volk, B. W., and Aronson, S. M. (Eds.), *Inborn Disorders of Sphingolipid Metabolism.* Oxford: Pergamon, 1967, p. 193.
15. Khan, A. A., and Folch-Pi, J. Cholesterol turnover in brain subcellular particles. *J. Neurochem.* 14:1099, 1967.
16. Kritchevsky, D., Tepper, S. A., DiTullio, N. W., and Holms, W. L. Desmosterol in developing rat brain. *J. Am. Oil Chem. Soc.* 42:1024, 1965.
17. Lapetina, E. G., Soto, E. F., and DeRobertis, E. Lipids and proteolipids in isolated subcellular membranes of rat brain cortex. *J. Neurochem.* 15:437, 1968.
18. Ledeen, R. The chemistry of gangliosides: A review. *J. Am. Oil Chem. Soc.* 43:57, 1966.
19. Leibovitz, Z., and Gatt, S. Enzymatic hydrolysis of sphingolipids. VII. Hydrolysis of gangliosides by a neuraminidase from calf brain. *Biochim. Biophys. Acta* 152:136, 1968.
20. McMurray, W. C. Metabolism of phosphatides in developing rat brain. I. Incorporation of radioactive precursors. *J. Neurochem.* 11:287, 1964.
21. McMurray, W. C. Metabolism of phosphatides in developing rat brain. II. Labelling of plasmalogens and other alkali-stable lipids from radioactive cytosine nucleotides. *J. Neurochem.* 11:315, 1964.

22. Morell, P., Costantino-Ceccarini, E., and Radin, N. S. The biosynthesis by brain microsomes of cerebrosides containing nonhydroxy fatty acids. *Arch. Biochem. Biophys.* 147:738, 1970.
23. Morell, P., and Radin, N. S. Synthesis of cerebroside by brain from uridine diphosphate galactose and ceramide containing hydroxy fatty acid. *Biochemistry* 8:506, 1969.
24. Norton, W. T. The Myelin Sheath. In Shy, G. M., Goldensohn, E. S., and Appel, S. H. (Eds.), *The Cellular and Molecular Basis of Neurologic Disease.* Philadelphia: Lea & Febiger, in press.
25. Norton, W. T., and Poduslo, S. E. Neuronal perikarya and astroglia of rat brain: Chemical composition during myelination. *J. Lipid Res.* 12:84, 1971.
26. Olson, J. A. Lipid metabolism. *Annu. Rev. Biochem.* 35:559, 1966.
27. Paoletti, R., Fumagalli, R., and Grossi, E. Studies in brain sterols in normal and pathological conditions. *J. Am. Oil Chem. Soc.* 42:400, 1965.
28. Rossiter, R. J. Biosynthesis of Phospholipids and Sphingolipids in the Nervous System. In Rodahl, K., and Issekutz, B. (Eds.), *Nerve as a Tissue.* New York: Harper and Row, 1966, p. 175.
29. Seminario, L. M., Hren, H., and Gomez, C. J. Lipid distribution in subcellular fractions of the rat brain. *J. Neurochem.* 11:197, 1964.
30. Smith, M. E., Fumagalli, R., and Paoletti, R. The occurrence of desmosterol in myelin of developing rats. *Life Sci.* 6:1085, 1967.
31. Sribney, M., and Kennedy, E. P. The enzymatic synthesis of sphingomyelin. *J. Biol. Chem.* 233:1315, 1958.
32. Suzuki, K. The pattern of mammalian brain gangliosides. II. Evaluation of the extraction procedures, post-mortem changes and the effect of formalin preservation. *J. Neurochem.* 12:629, 1965.
33. Suzuki, K. The pattern of mammalian brain gangliosides. III. Regional and developmental differences. *J. Neurochem.* 12:969, 1965.
34. Svennerholm, L. Chromatographic separation of human brain gangliosides. *J. Neurochem.* 10:613, 1963.
35. Wells, M. A., and Dittmer, J. C. A comprehensive study of the postnatal changes in the concentration of the lipids of developing rat brain. *Biochemistry* 6:3169, 1967.

BOOKS AND REVIEWS

Ansell, G. B., and Hawthorne, J. N. *Phospholipids: Chemistry, Metabolism and Function.* Amsterdam: Elsevier, 1964. (A comprehensive monograph on phospholipids.)

Brady, R. O. Sphingolipid Metabolism in Neural Tissues. In Ehrenpreis, S., and Solnitzky, O. C. (Eds.), *Neurosciences Reserach.* Vol. 2. New York: Academic, 1969, p. 301.

Davison, A. N. Lipid Metabolism of Nervous Tissue. In Davison, A. N., and Dobbing, J. (Eds.), *Applied Neurochemistry.* Philadelphia: Davis, 1968.

Dawson, R. M. C. The Metabolism of Animal Phospholipids and Their Turnover in Cell Membranes. In Campbell, P. N., and Greville, G. D. (Eds.), *Essays in Biochemistry.* Vol. 2. New York: Academic, 1966, p. 69.

Eichberg, J., Hauser, G., and Karnovsky, M. L. Lipids of Nervous Tissue. In Bourne, G. H. (Ed.), *The Structure and Function of Nervous Tissue.* Vol. 3. New York: Academic, 1969, p. 185.

Lajtha, A. (Ed.). *Handbook of Neurochemistry.* New York: Plenum. See in particular:

D'Adamo, A. F., Jr. Fatty Acids. Vol. 3, 1970, p. 525.

Davison, A. N. Cholesterol Metabolism. Vol. 3, 1970, p. 547.

Hawthorne, J. N., and Kai, M. Metabolism of Phosphoinositides. Vol. 3, 1970, p. 491.

Paoletti, R., Grossi-Paoletti, E., and Fumagalli, R. Sterols. Vol. 1, 1969, p. 195.

Radin, N. S. Cerebrosides and Sulfatides. Vol. 3, 1970, p. 415.

Rapport, M. M. Lipid Haptens. Vol. 3, 1970, p. 509.

Rosenberg, A. Sphingomyelin: Enzymatic Reactions. Vol. 3, 1970, p. 453.

Rossiter, R. J., and Strickland, K. P. Metabolism of Phosphoglycerides. Vol. 3, 1970, p. 467.

Rouser, G., and Yamamoto, A. Lipids. Vol. 1, 1969, p. 121.

Svennerholm, L. Gangliosides. Vol. 3, 1970, p. 425.

Schettler, G. (Ed.). *Lipids and Lipidoses.* New York: Springer-Verlag, 1967. (The first seven chapters are devoted to chemistry, metabolism, and analytical methodology of mammalian lipids, with emphasis on brain lipids.)

Stanbury, J. B., Wyngaaden, J. B., and Frederickson, D. S. (Eds.). *The Metabolic Basis of Inherited Disease* (3rd ed.). New York: McGraw-Hill, 1972. (A standard reference book for biochemistry of inherited metabolic diseases, including those involving brain lipids.)

12

Protein Metabolism in the Nervous System

Samuel H. Barondes and
Gary R. Dutton

PROTEINS HAVE the widest diversity of functions of any class of biological compounds. This is made possible by the vast number of structures that can be generated by combining 20 different amino acids into long polymers. The resultant polypeptides assume unique conformations, largely determined by the interactions of their constituent amino acids. They then may aggregate with identical or different polypeptide chains to form a wide variety of functional units. The completed aggregated proteins can function as enzymes which catalyze biological reactions, as structural materials of cellular organelles, as other highly specialized substances mediating transport across cell membranes, or in determining the properties of the cell membrane and surface (e.g., receptor proteins).

Because of the enormous importance of maintaining precise levels of these specialized substances, an elaborate apparatus has evolved for the biosynthesis of proteins and for the regulation of their biosynthesis and functional state. The details of these processes have consumed the attention of molecular biologists for at least a decade and continue to be under intensive investigation. Studies of protein metabolism in the nervous system indicate that the overall mechanisms are shared with simpler systems, about which more detailed information is available. Because of this, a brief review of general aspects of protein metabolism is presented to provide a framework to assist in the interpretation of contemporary studies of brain protein metabolism. After these introductory considerations, attention is directed to an unusual aspect of brain protein metabolism – transport of protein from its site of synthesis in the cell body through the axon – and to several specific proteins that may play special roles in brain function. (Rather than document each point extensively, specific references are made primarily to review articles.) The goal in presenting this material is to clarify the aims and the problems presented by studies of protein metabolism in the nervous system. The relationship of brain protein metabolism to memory is considered in Chapter 31.

BIOSYNTHESIS AND REGULATION

The general scheme of protein synthesis has been well defined [1a, 1b]. The genetic material, DNA, directs the synthesis of three major classes of ribonucleic acids (RNA). These are (1) transfer RNAs (tRNA), each of which combines only with a single specific amino acid and carries it into the protein-synthesizing complex; (2) messenger RNAs (mRNA), whose polynucleotide sequence directs the insertion of individual amino acids into a growing polypeptide chain; and (3) ribosomal RNAs (rRNA), which, when combined with ribosomal proteins, form ribosomes that play a role in holding the protein-synthesizing complex together. A unit consisting of a group of ribosomes linked by a strand of mRNA is called a *polysome.* Transfer RNAs, each containing its specific amino acid, interact with the polysome and with a number of specific enzymes in the protein-synthesizing complex, and the appropriate amino acids are sequentially added to form a polypeptide chain. The polypeptide then twists and folds to take on its secondary and tertiary three-dimensional structure and interacts with other proteins to form active enzymes, active groups of enzymes, bits of membrane, or other cellular organelles.

This system is regulated in five major ways (Table 12-1). They have been well documented in the case of enzyme protein because enzymatic activity is easier to assess quantitatively than are other aspects of protein function. It is likely, however, that the same regulatory processes apply not only to enzymes but to proteins with nonenzymatic functions. Two forms of regulation are based on specific increases in the number of enzyme molecules. In one, the amount of mRNA is limiting, and synthesis of additional copies of specific mRNA results in increased biosynthesis of the protein for which it codes. This is the classic mechanism of enzyme induction, which has been shown both in bacteria and in mammalian cells, whereby an inducer – often the substrate of the enzyme in question – leads to the synthesis of more mRNA, which directs the synthesis of more of the enzyme. In mammalian cells, evidence for regulation beyond the biosynthesis of mRNA has also been found. In this case an existent mRNA is activated, thereby directing the synthesis of a greater number of the protein molecules for which it codes.

The role of such processes as enzyme induction in the nervous system is presently under investigation. Whereas it is clear that there are critical periods in the develop-

TABLE 12-1. Regulation of Functional Protein Levels in Cells

Mechanism	Effect
1. Synthesis of mRNA 2. Activation of existent mRNA	Altered synthesis of protein
3. Configurational ("allosteric") change 4. Addition of other groups (e.g., phosphate, sugars, lipids)	Altered protein structure
5. Altered degradation of protein	

ment of the nervous system, during which there are bursts of synthesis of specific enzymes and structural proteins (e.g., myelin proteins), the role of enzyme induction in regulating the adult nervous system is only now being investigated. Studies of the biosynthesis of norepinephrine by the adrenal medulla, which may be taken as a simple model of neuronal tissue, have shown that the amount of tyrosine hydroxylase, the limiting enzyme in norepinephrine biosynthesis, may be enhanced by prolonged nerve stimulation (see Chap. 6). There is evidence that this process is mediated by the biosynthesis of new mRNA, which in turn directs the synthesis of more of the enzyme in question [12]. The evidence suggests a similar regulation of tyrosine hydroxylase in the brain, and additional experimentation on the subject is being pursued actively.

Other major mechanisms of regulating protein function do not depend on protein synthesis but depend on changing the conformation of proteins or on the covalent addition of molecules to the protein. Conformational changes are the basis of "allosteric" regulation, whereby a small molecule alters the structure and function of a protein by combining with it. A well-studied example of this is feedback regulation. In this example the final product of a sequence of enzymatic reactions interacts with and inhibits an enzyme early in the sequence, thereby decreasing the activity of the total pathway. For example, norepinephrine can interact with tyrosine hydroxylase and inhibit its activity [12]. This is the rate-limiting enzyme in the sequence of reactions leading to the biosynthesis of norepinephrine, so norepinephrine biosynthesis is reduced.

Addition of molecules to proteins after they are released from polysomes is another mechanism for regulation. In brain, proteins may be altered structurally in many ways. Phosphates, sugars, or lipids may be linked to proteins to form phosphoproteins [11], glycoproteins [5b], or lipoproteins, respectively. Phosphorylation of proteins may alter their enzymatic properties. This is seen in the phosphorylation of the phosphorylase responsible for liver glycogenolysis by a protein kinase, which is activated by cyclic AMP [11]. Brain has been shown to be rich in protein kinase activity, but its function is not clear. Brain is also rich in enzymes that add sugars to proteins.

In addition to these mechanisms, functional protein levels can be controlled by modification of degradation rates of specific proteins. Proteins are actively degraded in brain and many have half-lives in the range of a few days. Although little is known about the mechanism of protein degradation in brain, there is good evidence that the amounts of several specific enzymes in liver can be altered by changing the rates at which they are degraded [9].

METHODS FOR STUDYING PROTEIN METABOLISM IN THE NERVOUS SYSTEM

The activation of certain organs – the pancreas for example – is accompanied by such an increase in the synthesis of certain proteins that a net increase in its total protein synthesis can be observed. General brain protein metabolism has been studied extensively to determine if it is altered in various functional states of the brain. Although these studies are of limited value because they give no information about types of protein or their localizations in the brain, the methodology used is presented here because it can also be applied to studies of specific purified proteins and specific brain regions.

The activity of a tissue in synthesizing protein can be determined by in vivo or in vitro methods. Both techniques can be applied conveniently to brain, as its protein metabolism is very active.

IN VIVO METHODS

In the in vivo method a precursor of protein – a radioactive amino acid – is injected into an animal either subcutaneously or in the cranium. There is an advantage to each method. In the former, the precursor is distributed relatively uniformly in the brain; the latter results in relatively high brain concentrations of the precursor. Measurements are then made of the degree to which the radioactive amino acid has become incorporated into protein. This is done by determining the amount of radioactivity incorporated into substances that are precipitable by trichloroacetic acid and are not extractable into lipid solvents. More detailed studies may be made by separating brain regions, by subcellular fractionation of a brain homogenate, or by purification of specific brain proteins after incorporation of the radioactive amino acid. Autoradiographic measurements can also be made of the incorporation of radioactive amino acid into the protein of specific brain regions, cell types, or cell parts after unincorporated amino acids have been washed out [6b]. With these methods, a measure of the activity of brain protein synthesis can be obtained.

A major assumption often made in these in vivo studies is that the rate of incorporation of the administered radioactive amino acid directly reflects the rate of protein synthesis. This would be true if the administered radioactive amino acid equilibrated with the endogenous amino acid precursor pool. It is never possible, however, to determine this with certainty. Awareness of this assumption is of critical importance for interpreting studies in which protein synthesis is compared under experimental and control conditions. Thus, one may administer a radioactive amino acid to animals and observe a change in the amount of labeled amino acid incorporated into brain protein as a result of some specific treatment. One must, in addition, determine whether the treatment influences the specific activity of the precursor pool. If a given treatment increases the amount of radioactive amino acid incorporated into brain protein, it could be the result of either an increase in ongoing brain protein synthesis or of increased access of the *labeled* amino acid to the protein synthesizing system.

Increased access might occur in any one of several ways. For example, the treatment being studied could so change blood flow to the brain that a larger proportion of the total amount of subcutaneously administered radioactive amino acid is taken up by the treated brain. This might produce a particularly high specific activity of the precursor pool of the administered amino acid; that is, there might be a higher ratio of labeled to unlabeled amino acid molecules. In that case, incorporation of more radioactivity into protein would reflect this higher specific activity of precursor and not necessarily an increased rate of protein synthesis. To evaluate the possible contribution of such access, one should measure both the amount of unlabeled amino acid and the amount of labeled amino acid present in the brains of both treated and untreated animals so that the specific activities of the precursor pools (radioactivity per mole of amino acid) in the brain are known. Ideally, this should be determined in different groups of animals at a number of times in the interval during which amino acid incorporation is being studied. If the specific activity differs in the treated and untreated conditions, appropriate corrections can then be made to improve the estimate of the influence of the treatment on protein synthesis per se.

Another important in vivo method for assessing the activity of protein metabolism is based on studies of the turnover of radioactive protein. In this method a "pulse" of radioactive amino acid is so administered that its incorporation into protein is completed within a short period of time, usually in minutes. The rate of decline of the specific activity of overall brain protein or specific proteins can then be studied by sacrificing groups of animals periodically after the pulse. If one assumes that none of the radioactive amino acid released by protein degradation is reutilized – an assumption that is never completely justified – an estimate may be made of the rate of protein synthesis required to replace this degraded protein. Proteins in the brain turn over at markedly different rates, so that the pulse-labeling technique preferentially labels the proteins that turn over most rapidly. Therefore, this technique overemphasizes the turnover of total brain protein. In studies of an individual protein assumed to have a uniform turnover rate, the technique yields fairly accurate results. As an example, the turnover of S-100 protein and of total soluble brain protein of the rat is shown in Figure 12-1.

IN VITRO METHODS

Brain protein synthesis also may be studied in vitro, using brain homogenates or subcellular fractions of brain [5e, 6a]. Usually, the brain is homogenized, the homogenate is centrifuged to remove nuclei, mitochondria, membrane fragments, and synaptosomes, and the resultant supernatant is centrifuged at high speed (e.g., 100,000 X *g* for 1 hr) to yield a microsomal pellet and the soluble fraction. By adjusting the pH of the soluble fraction to 5.0, enzymes required for protein synthesis and tRNA are precipitated to yield the "pH 5 fraction." By combining the pH 5 fraction with the microsomal fraction, the radioactive amino acid, and the cofactors, a system is produced which can incorporate radioactive amino acid into polypeptides. Mitochondrial [4] and synaptosomal [6a] fractions have also been used to study protein synthesis in vitro.

The in vitro protein-synthesizing system has been used to estimate the activity of brain protein synthesis in vivo. Its outstanding advantage is that the specific activity of the radioactive amino acid can be well controlled. Homogenization,

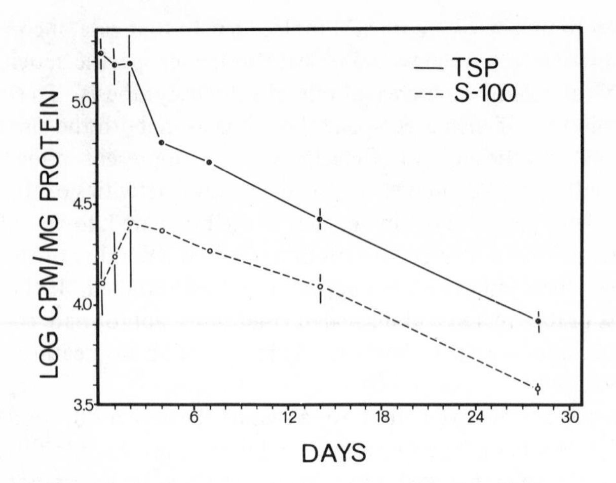

FIGURE 12-1. The log of the specific activity (count/min/mg of protein) of total soluble brain proteins (TSP) and the S-100 protein for each of the survival periods. Rats were perfused intraventricularly with ^{3}H-leucine. The specific activity of S-100 protein was determined directly at the times indicated by isolating a known amount of protein in a discrete band from polyacrylamide gels. Its quantity was concurrently determined in an identical sample by complement fixation, using a specific antibody. (From Cicero and Moore, *Science* 169:1333, 1970. Reprinted by permission; copyright 1970 by the American Association for the Advancement of Science.)

however, may dilute regulatory substances by disrupting the cells and so give misleading results. Usually it is assumed that the level of active mRNA is the limiting factor in protein synthesis. However, other components may be limiting in homogenates. Although in vitro studies may be particularly useful for studying the activity of specific components of the system (e.g., mRNA), great care must be taken in deducing the overall in vivo situation from in vitro studies.

STUDIES OF METABOLISM OF SPECIFIC PROTEINS

Because the amount of a specific protein may control a specific cellular function, regulation of the level of individual proteins has been of great general interest. There is presently a vast literature on the regulation of the synthesis of specific enzymes both in bacteria and in mammalian cells. Studies of the synthesis of a specific protein can, of course, be most conveniently carried out with an enzyme whose activity is usually the reflection of the number of enzyme molecules present in the cell. Under certain circumstances, however, changes in enzyme activity can be produced without the increased synthesis of enzyme molecules, e.g., by changing the levels of cofactors or by allosteric changes. Therefore, to establish that enzyme induction is occurring, the simplest approach is to demonstrate that the increase in enzyme activity is blocked when general protein synthesis is inhibited.

If enzyme induction is mediated through the synthesis of new mRNA which directs the synthesis of new protein, the entire system can be blocked by administering actinomycin D, which blocks DNA-dependent RNA synthesis. Whether or not new mRNA synthesis is involved, all enzyme induction based on synthesis of new proteins can be blocked by treatment with protein synthesis inhibitors, such as puromycin and cycloheximide. Thus, if one finds an increase in an enzyme level in response to a specific treatment and finds that this increase can be prevented by treatment with one of these protein synthesis inhibitors, the increase in enzyme activity probably is due to increased synthesis of specific protein molecules. Definitive proof is obtained only when an increase in isolated enzyme protein can be shown directly, but in many situations this is difficult to do.

Such inhibitor studies [3] may be made in the mouse brain because it has been shown that mice can tolerate cycloheximide-induced inhibition of 95 percent of cerebral protein synthesis over a period of many hours without exhibiting gross neurological changes. Cycloheximide may be administered intracerebrally or, in much larger doses, subcutaneously to produce this degree of inhibition. Inhibition of cerebral protein synthesis can also be produced with intracerebral puromycin, but it is accompanied by disturbances in cerebral electrical activity. Significant inhibition of cerebral RNA synthesis can be produced by intracerebral injection of actinomycin D. Mammals may survive a day or more after injection of large doses of this drug, but then neuronal degeneration and death occur. These drugs have been used extensively in an attempt to define the role of protein and RNA synthesis in a long-term memory storage, and are discussed in detail in Chapters 13 and 31.

Because their assays are based on purification, metabolism of proteins with no known enzymatic function is usually more difficult to study than is that of enzymes. The amount of an enzyme in the brain can often be determined from crude homogenates because enzymes can be assayed in the presence of other proteins as long as they have no effect on the reaction being studied. Study of other proteins in crude homogenates is possible if one can find an appropriate "marker." For example, microtubular protein can be assayed by its ability to bind colchicine [10]. For other proteins, however, purification is often necessary. This can be done by sequential application of a number of purification procedures. For example, some purification can usually be achieved by selective salt precipitation; by chromatography on columns that separate proteins on the basis of their charge (e.g., DEAE-cellulose) or on the basis of their molecular weight (e.g., Sephadex); by the use of sucrose density gradients for which molecular weight is the primary parameter; or by gel electrophoresis, which, under certain conditions, separates proteins on the basis of their charge and which, under other conditions, may be modified to separate proteins on the basis of molecular weight. Once a pure protein is obtained, it may prove possible to induce antibodies to it, which then provides an assay for it in crude extracts. Two important brain proteins – microtubular protein and S-100 protein – whose metabolism has been studied by physical or chemical (nonenzymatic) techniques are discussed in detail below (pp. 241–242).

SITES OF SYNTHESIS OF BRAIN PROTEIN

Proteins are synthesized on polysomes in all cells. Upon completion they are released and migrate within the cell, where they may remain free in the cytoplasm or may be organized into an organelle. In addition, a protein-synthesizing system is present in mitochondria and is directed by intramitochondrial DNA and mediated by mitochondrial RNA. These two processes, polysomal and mitochondrial protein synthesis, exist in the nervous system [4]. A unique characteristic of nerve cells is that they may have extremely long axonal processes, far removed from the DNA of the cell nucleus. The polysomes of neurons are clustered in the nerve cell body, and these clusters are visible by light microscopy as the so-called Nissl substance. Electron microscopic studies have demonstrated the high concentration of polysomes in the nerve cell body and have shown that there are no identifiable ribosomes in the axon except in the initial segment [5c]. Thus, the axon contains no morphologically identifiable protein-synthesizing system [8], except for the very limited protein-synthesizing system present in intraaxonal mitochondria. The nerve terminals at the ends of axons are also devoid of identifiable ribosomes, but they contain a high concentration of mitochondria.

Are the proteins of the axons and nerve endings metabolically active, and where do the newly synthesized proteins come from? This has been a major question in studies of protein metabolism in the nervous system. A large body of recently accumulated evidence indicates that proteins are transported from the nerve cell body to the axon and the nerve terminals. Some evidence has also been presented that supports the existence of local protein-synthesizing systems in axons and nerve terminals.

AXONAL PROTEINS

A substantial proportion of the protein synthesized by neurons is transported into axons. The best technique for demonstrating this is to apply a radioactive amino acid to a region rich in nerve cell bodies and to observe the appearance of the resultant labeled proteins in axons. The retinal ganglion cells [7] and the dorsal root ganglion cells are most commonly used for such studies [6c]. Radioactive amino acid is injected intraocularly or into a lumbar dorsal root ganglion, and the appearance of labeled protein in progressive segments of the optic nerve or the sciatic nerve is followed at intervals after injection of the radioactive precursor. The radioactive amino acid is generally incorporated rapidly into protein. In the ensuing hours and days, radioactive protein moves out of its site of synthesis in the nerve cell bodies and into the axon. This can be observed by preparing autoradiographs [6b] of segments of the peripheral nerve, taken at progressively greater distances from the nerve cell bodies, or by determining the amount of radioactive protein in such segments by precipitating with trichloroacetic acid, washing to remove any residual small molecules or precursors, and counting the resultant precipitated protein (Fig. 12-2).

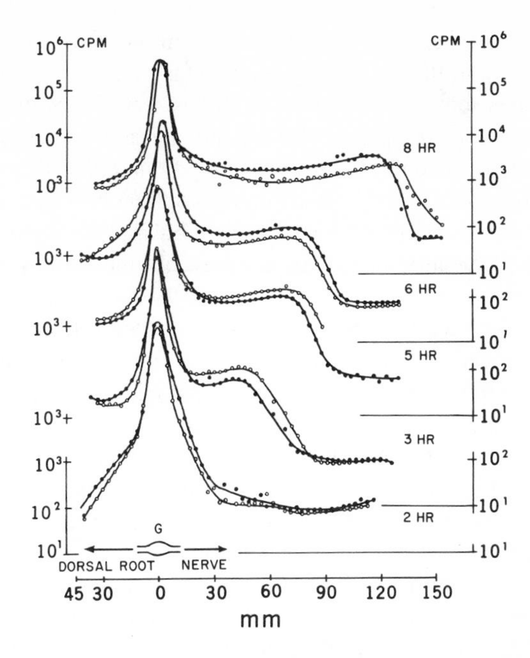

FIGURE 12-2. Distribution of ^{3}H-protein in the dorsal roots, ganglia, and sciatic nerves in five cats injected with ^{3}H-leucine in their lumbar 7th ganglia and studied at intervals from 2 to 8 hr afterward. The abscissa gives distance (in mm) from zero at center of the ganglia and the activities present in 3-mm segments are shown logarithmically on the ordinate. The ordinate scales are only partially represented for most pairs of nerves, except for the upper right set, taken 8 hr after injection. Fast transport is shown in the nerves from 3 hr onward by the crest of activity present distally and its more distal displacement with time after injection. The rate calculated from the displacement in the adult cat is 410 mm per day. (From Ochs et al., *Science* 163:686, 1969. Reprinted by permission; copyright 1969 by the American Association for the Advancement of Science.)

Both techniques demonstrate movement of labeled protein from the region of the nerve cell body through the axon in two major waves, which are referred to as "rapid flow" and "slow flow." The rapidly transported protein moves in the axon at rates in the range of several hundred millimeters per day (Fig. 12-2). When segments of the axon containing this material are homogenized and subjected to subcellular fractionation, about half of the rapidly transported protein is found to be particulate and the remainder soluble. The greatest part of the protein transported in the axon moves at a rate in the range of several millimeters per day and most of this protein is soluble.

The mechanism whereby proteins are transported in the axon is not presently known. Net flow is from the nerve cell body to the nerve terminals, but there is evidence for some retrograde flow. The proteins are not pushed passively into the axon as more proteins build up in the nerve cell body, because transport continues when protein synthesis in the nerve cell body is interrupted by the administration of inhibitors of protein synthesis. Furthermore, when segments of axon are isolated between a pair of ligatures, transport continues from the portion of the axon closest to the nerve cell body to the portion of the axon farthest from the nerve cell body, where it accumulates above the constriction. It has been shown that oxidative metabolism is required if transport is to continue. Because of the abundance of microtubules within the axon, it has been proposed that either they provide the motive force for transport or they organize the axoplasm into channels around which transport can occur. Disruption of microtubular structure in axons by local application of colchicine interferes with transport, but the reasons for this are not presently clear.

The techniques we have described are ideal for studies of the transport of proteins from the nerve cell body into the axon and to the nerve ending. They are not, however, applicable to studies of local protein synthesis in axons since the precursor of the radioactive amino acid is confined largely to the region of the nerve cell bodies. Axonal protein synthesis has been evaluated either by injecting radioactive amino acid in vivo or by incubating nerve segments in vitro [5c]. The in vivo studies have shown the rapid appearance of some labeled proteins in axons isolated from their nerve cell bodies, but it has been difficult to demonstrate with certainty that the amino acid is incorporated into protein in the axons rather than into contaminating glial cells or connective tissue. In vitro studies have also shown incorporation of radioactive amino acid into the protein of isolated peripheral nerves and the Mauthner axon [2c]. Because axons, even when stripped of myelin, are contaminated with glia and connective tissue cells, it has been difficult to establish conclusively that the incorporation is truly taking place in the axon. Several critical evaluations of these studies are listed in the bibliography [5c, 6a].

NERVE ENDING PROTEINS

The nerve ending is the site of accumulation of synaptic vesicles, mitochondria, and enzymes that synthesize neurotransmitters. As one component of the synapse, it is a site at which regulation of protein metabolism could regulate nervous system function directly. For this reason, the site of synthesis of nerve ending proteins and the rate of appearance of newly synthesized proteins at nerve endings in the brain have received particular attention. Accumulation of radioactive protein at nerve endings has been observed in studies of the transport of protein after intraocular injection of radioactive amino acid. However, most studies of protein metabolism in nerve endings have been made in the brain. The nerve endings can be isolated in the synaptosomal fraction after radioactive amino acid has been injected intracerebrally or after it has been incubated in vitro with brain tissue.

The site of synthesis of nerve ending proteins has been studied in vivo by determining the relative rate at which radioactive protein appears in components of the synaptosome fraction and in other subcellular fractions of brain at a series of times after the radioactive precursor has been administered [5a]. Particular attention is paid to the specific activity of the soluble protein obtained by water lysis of the synaptosomal fraction since this is less contaminated by protein from other parts of the brain than is the particulate synaptosomal protein. After intracerebral injection of radioactive leucine, maximal labeling of whole brain protein is observed in less than one hour. The specific activity decreases in the ensuing hours and days through degradation. In contrast, the specific activity of the soluble protein obtained by water lysis of an isolated synaptosome fraction is extremely low one hour after intracerebral injection of the precursor, but it increases progressively over several days. The marked lag between the completion of incorporation of labeled amino acid into whole brain and the appearance of labeled soluble protein in synaptosomes reflects the slow transport to nerve endings of soluble protein that was synthesized in the nerve cell body. These studies show that virtually all the soluble protein of nerve endings is transported from the nerve cell body.

In contrast to the soluble protein, the particulate component of the nerve ending fraction contains a significant amount of labeled protein within 15 min after injection of the precursor. In the ensuing hours and days, particulate protein associated with synaptic vesicles, mitochondria, and nerve ending membranes is transported to the nerve ending. Nevertheless, substantial radioactive particulate protein is found in synaptosomes 15 min after injection of the radioactive precursor. This was shown directly by electron microscope autoradiographs of the synaptosome fraction (Fig. 12-3). Shortly after precursor is administered, some of the radioactive protein in the synaptosomes is associated with mitochondria, but some is found in sections devoid of mitochondria. Therefore, radioactive protein, largely particulate, is present in nerve endings at the earliest time after injection of the precursor at which it is technically possible to make this observation.

The actual site of synthesis of the new protein, which rapidly becomes associated with the synaptosomes, cannot be determined from in vivo study. Being particulate, it may have been transported by "fast flow" through short axons after synthesis has taken place in the nerve cell body. However, in vitro studies have shown that synaptosomal fractions can incorporate amino acids into protein [6a]. Extensive experimentation supports the conclusion that some protein synthesis occurs in synaptosomes, rather than in microsomes which contaminate this fraction, and that protein synthesis is not confined to mitochondria in nerve endings. A summary of the evidence for protein synthesis by nerve endings is listed in the bibliography [6a], although this evidence remains controversial.

Active local metabolism of macromolecules in nerve endings is also suggested by in vivo studies which show that radioactive glucosamine and its carbohydrate derivatives are incorporated into polypeptides at the nerve endings after these polypeptides have been transported from their site of synthesis in the nerve cell body [2a]. In vitro studies also show incorporation of glucosamine into macromolecules at nerve endings. Therefore, by addition of carbohydrates, polypeptides can be structurally altered at a site distant from the polysome. This incorporation is mediated by glycosyl transferases, which are specific enzymes that add specific sugars to specialized sites on proteins or into carbohydrate chains, in the absence of an RNA template. The addition of carbohydrates to proteins is generally the prelude to the secretion of the glycoprotein onto the cell surface or into the extracellular space.

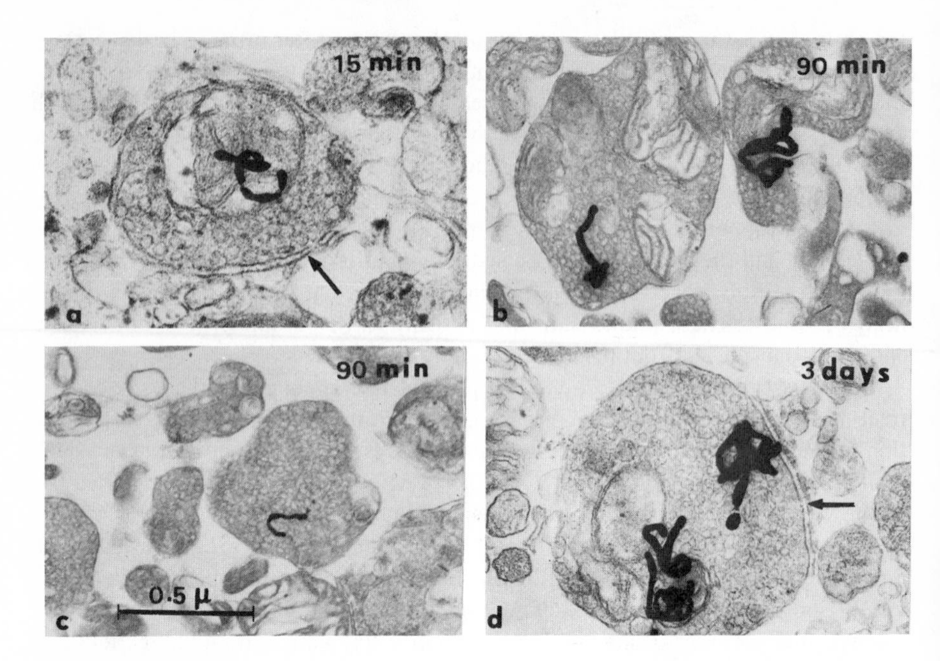

FIGURE 12-3. Electron microscope autoradiographs of nerve ending fractions prepared by sucrose-gradient centrifugation. Mice were sacrificed 15 min (a), 90 min (b) and (c), and 3 days (d) after intracerebral injection of a mixture of ^{3}H-lysine and ^{3}H-leucine. Their brains were homogenized and nerve ending (synaptosome) fractions were prepared. Unincorporated amino acids were removed by extensive washing. Contaminating membranes and mitochondria are intermingled with nerve endings. At 15 min (a) and 90 min (b and c) after injection, silver grains are located over and between mitochondrial profiles in nerve endings (a and b). Sections of labeled nerve endings which are free of mitochondria are also frequently observed (c). Three days later (d), nerve endings are often overlaid by two to four silver grains. The postsynaptic membrane, sometimes found attached to the presynaptic ending, is visible in (a) and (d) *(arrow).* (From Droz and Barondes, *Science* 165:1131, 1969. Reprinted by permission; copyright 1970 by the American Association for the Advancement of Science.)

All these studies show that protein metabolism is active at the nerve endings in brain and could play an important role in the regulation of nerve ending function. Regulation can occur via rapid transport of protein from the nerve cell body through the short axons of the brain, by local protein synthesis in mitochondria as well as in other undefined structures, and by modification of protein structure by addition of other groups, such as sugars. Local allosteric regulation of tyrosine hydroxylase in nerve endings has also been inferred [12].

SPECIAL BRAIN PROTEINS

Most brain proteins, such as the enzymes involved in glucose metabolism, are common to all the other cells of the organism. Of the proteins that are known to be specific to the brain most are enzymes involved in a specialized brain function. These include, for example, the enzymes involved in the biosynthesis of neurotransmitters and of certain complex lipids that are highly concentrated in the nervous system. The biosynthesis and metabolism of these enzymes are of great interest to those concerned with the general problem of protein metabolism in the brain, and these are discussed in appropriate sections of this book. Here we will consider studies of two brain proteins with no known enzymatic activity which are believed to play a special role in brain function. These are the S-100 protein, uniquely found in the nervous system of all species studied [5d], and microtubular protein [10], which is common to all cells but is most abundant in neurons. Studies of other special brain proteins, such as the myelin proteins, are discussed in Chapters 18 and 24.

S-100 PROTEIN

A classic method for separating soluble proteins from a homogenate is to precipitate classes of proteins progressively by adding increasing concentrations of ammonium sulfate. In the course of studies of soluble brain proteins, it has been found that some are soluble in 100 percent saturated ammonium sulfate solution. One is found only in brain. Because of its solubility in saturated ammonium sulfate solution, it is referred to as "S-100." S-100 is highly acidic, and because of its charge it can be separated on ion-exchange columns or by electrophoresis. Amino acid analysis of the purified protein shows that its charge is due to a large content of glutamic and aspartic acids. Its molecular weight is 24,000. The purified protein has been used to immunize rabbits, and the resultant antibody has been used subsequently to assay for the presence of the protein in various other tissues. It comprises approximately 1 percent of the total soluble protein in the brain. It is more highly concentrated in white matter than in gray matter and is synthesized abundantly by glioblastoma cultures. It is therefore apparent that the protein is concentrated in glia. Because of its specific interaction with calcium, it may play a role in binding calcium in the brain and regulating its availability.

Its abundance, its unusual physical properties, and the availability of specific antibody have made it possible to study the metabolism of this protein. These studies are shown in Figure 12-1.

MICROTUBULAR PROTEIN

In the course of an investigation of the mitotic apparatus, it was found that the protein of the microtubules of the mitotic spindle can be solubilized. It can then be identified by its binding of colchicine. A survey of the concentration of this protein in a variety of cell types showed that brain extracts, and specifically extracts of axoplasm, were the richest source of the colchicine-binding protein, which is presumed to be the subunit of microtubules [10]. The microtubular protein can be solubilized by homogenization in neutral phosphate buffer, and it can be purified readily from brain extracts. It is the major soluble protein in brain; it may also exist in a particulate form. It is fairly acidic and is made up of subunits with a molecular weight of approximately 60,000. It was shown that this protein can be quantitatively precipitated with vinblastine, although some other proteins are also precipitated with this reagent.

Because microtubules are a major component of axonal structure, it was of interest to determine whether their constituent proteins are metabolized actively. Mice were injected intracerebrally with radioactive amino acid and sacrificed at various times after injection. The brains were homogenized and the microtubular protein was extracted, partially purified by vinblastine precipitation, and identified by electrophoretic migration on polyacrylamide gel in a system that separates proteins on the basis of molecular weight. It was found that of all the soluble proteins being synthesized in the brain, at least 15 percent was microtubular protein, and even higher estimates were obtained from the developing brain. In both developing and adult brain, the half-life of this protein was shown to be about 4 days. Therefore microtubules are in flux rather than inert structural elements. Furthermore, by taking

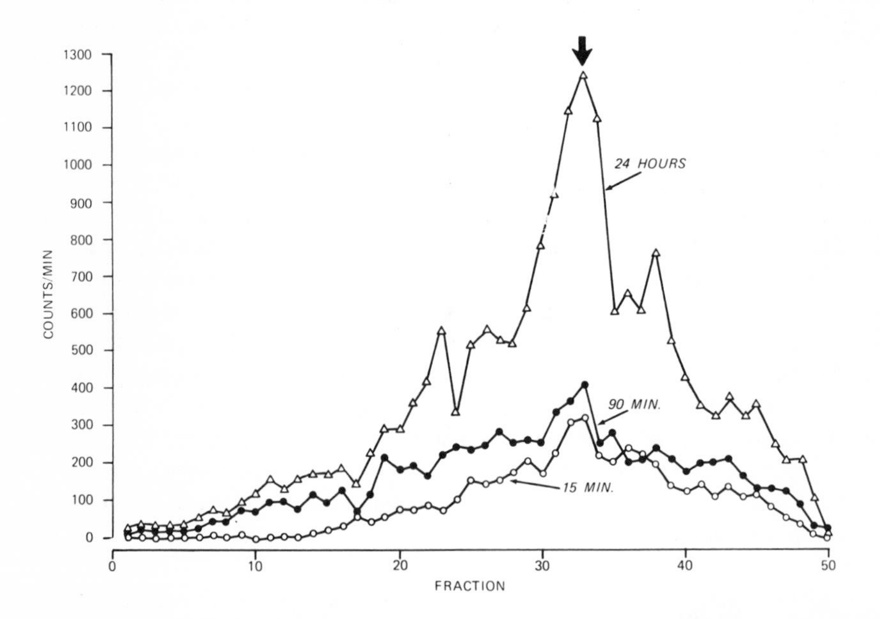

FIGURE 12-4. Transport of microtubular protein to the nerve ending in immature mouse brain. Five-day-old mice were injected intracerebrally with ^{3}H-leucine and were killed 15 min, 90 min, and 24 hr later. The soluble proteins of the nerve ending fraction were then separated according to molecular weight by polyacrylamide gel electrophoresis. The higher the molecular weight of the protein, the lower the fraction number in which it is found. Radioactive leucine incorporated into the protein of each fraction is shown in the ordinate. The heavy arrow indicates the peak of migration of ^{14}C-labeled microtubular protein, which was used as an internal standard to help calibrate each polyacrylamide gel. Note that radioactive microtubular protein increases strikingly between 90 min and 24 hr after injection of ^{3}H-leucine. This is due to its slow transport to the nerve ending. (For details see Feit, Dutton, Barondes and Shelanski, *J. Cell Biol.* 51:138, 1971.)

advantage of the protein's unique properties, it has been shown that it migrates with the slow component of axoplasmic transport in peripheral nerve and in mouse brain (Fig. 12-4). Although nerve endings do not characteristically contain any identifiable microtubules, a substantial proportion of the soluble component of isolated nerve ending fractions is microtubular protein. The reasons for the transport of this protein through the axon to the nerve terminal and for its active metabolism are not presently known.

ACKNOWLEDGMENT

Preparation of this paper was supported in part by grant MH-18282 from the National Institute of Mental Health.

REFERENCES

BOOKS AND REVIEWS

1. Anfinsen, C. B. (Ed.). *Aspects of Protein Biosynthesis.* New York: Academic, 1970. See in particular:
   1a. Nirenberg, M. The Flow of Information from Gene to Protein, p. 215.
   1b. Tomkins, G. M., and Kredich, N. M. Genetic Control of Protein Structure and the Regulation of Protein Synthesis, p. 1.
2. Barondes, S. H. (Ed.). *Cellular Dynamics of the Neuron.* New York: Academic, 1969. See in particular:
   2a. Barondes, S. H. Two Sites of Synthesis of Macromolecules in Neurons, p. 351.
   2b. Droz, B., and Koenig, H. L. The Turnover of Proteins in Axons and Nerve Endings, p. 35.
   2c. Edström, A. RNA and Protein Synthesis in Mauthner Nerve Fiber Components of Fish, p. 51.
3. Barondes, S. H. Cerebral protein synthesis inhibitors block long-term memory. *Int. Rev. Neurobiol.* 12:177, 1970.
4. Campbell, M. K., Mahler, H. R., Moore, W. J., and Tewari, S. Protein synthesis systems from rat brain. *Biochemistry* 5:1174, 1966.
5. Lajtha, A. (Ed.). *Handbook of Neurochemistry.* New York: Plenum, 1969. See in particular:
   5a. Barondes, S. H. Axoplasmic Transport. Vol. 2, 1969, p. 435.
   5b. Brunngraber, E. G. Glycoproteins. Vol. 1, 1969, p. 223.
   5c. Koenig, E. Nucleic Acid and Protein Metabolism of the Axon. Vol. 2, p. 423.
   5d. Moore, B. W. Acidic Proteins. Vol. 1, 1969, p. 93.
   5e. Roberts, S. Protein Synthesis in the Central Nervous System. Vol. 5, 1971, p. 1.

6. Lajtha, A. (Ed.). *Protein Metabolism of the Nervous System.* New York: Plenum, 1970. See in particular:
   - 6a. Austin, L., Morgan, I. G., and Bray, J. J. The Biosynthesis of Proteins Within Axons and Synaptosomes, p. 271.
   - 6b. Droz, B., and Koenig, H. L. Localization of Protein Metabolism in Neurons, p. 109.
   - 6c. Ochs, S. Fast Axoplasmic Flow of Proteins and Polypeptides in Mammalian Nerve Fibers, p. 291.
7. McEwen, B. S., and Grafstein, B. Fast and slow components in axonal transport of protein. *J. Cell Biol.* 38:494, 1968.
8. Palay, S. L., and Palade, G. E. The fine structure of neurons. *J. Biophys. Biochem. Cytol.* 1:69, 1955.
9. Schimke, R. J., and Doyle, D. Control of enzyme levels in animal tissues. *Annu. Rev. Biochem.* 39:777, 1970.
10. Shelanski, M. L., and Taylor, E. W. Biochemistry of neurofilaments and neurotubules. In Wolstenholme, G. E. W., and O'Connor, M. (Eds.), *Alzheimer's Disease and Related Conditions.* Ciba Foundation Symposium, London: Churchill, 1970, p. 249.
11. Walsh, D. A., Krebs, E. G., Reimann, E. M., Brostrom, M. A., Crobin, J. D., Hickenbottom, J. P., Soderling, T. R., and Perkins, J. P. The Receptor Protein for Cyclic AMP in the Control of Glycogenolysis. In Greenard, P., and Costa, E. (Eds.), *Advances in Biochemical Psychopharmacology.* Vol. 3. New York: Raven, 1970, p. 265.
12. Weiner, N. Regulation of norepinephrine biosynthesis. *Annu. Rev. Pharmacol.* 10:273, 1970.

13

Nucleic Acid Metabolism

Henry R. Mahler

NUCLEIC ACIDS AND BRAIN FUNCTION

Any attempt at formulating neurobiological phenomena in molecular terms must take cognizance of the nature, expression, and regulation of the genetic capabilities of the nervous system and its component parts. At one level there are systematic studies of how alterations in the genome can affect defined functions of the nervous system in neurological and behavioral mutants; such studies can be of the utmost importance in providing clues concerning the contributions made by single, defined, gene products to complex and integrated function and behavior (see Chap. 31). At another level we are concerned with emergent properties as a function of time as the independent variable. These will entail qualitative or quantitative changes in the pattern of the proteins present within a given cell, in a population of cells linked to one another structurally and functionally, or in even more complex systems. By their intervention as enzymes and membrane constituents, these proteins will influence all other constituents of the cells affected.

But such changes in protein patterns arise themselves as a consequence of shifts in gene expression and its regulation. This is the level we are dealing with here. Our concern is not just with those hypothetical or actual changes in response to electrical activity, sensory inputs, and behavioral challenges that produce or reflect specific products in and alter the properties of certain families of neurons and the synapses between them [17]. We must also be concerned with the events responsible for neuronal formation and localization in the course of development and differentiation. In biochemical terms, the question posed is equivalent to an inquiry into the parameters that govern the biosynthesis and degradation of DNA and RNA in brain and nerve, and that is the topic of this chapter.

DNA VS. RNA METABOLISM

DNA synthesis and turnover in all cells are usually a reflection of excision and repair of certain damages and appear to be directly correlated with the mitotic index of the cells, i.e., their ability to divide. Although there is now good evidence that certain brain cells (interneurons, astrocytes, glia) continue to divide into infancy or even adulthood (see below), the ordinary long-axoned neurons (macroneurons) of mammals cease dividing and proliferating shortly after birth or in early infancy (see Chap. 14). For instance, in rat cerebral cortex there is no net increase in DNA beyond the 18th day after birth. Furthermore, the amount of DNA per diploid nucleus and the genetic information residing therein is believed to be a constant for all the cells of any one individual. In consequence, most of the current literature has tended to neglect DNA and instead has been concerned with the synthesis, degradation, and functional involvement of various species of RNA found in the nervous system.

The coverage in this chapter reflects this state of the art. It finds its justification in the compelling evidence in favor of the continuity, stability, and identity of the genetic constitution of all cells of a metazoan organism in contrast to their structural and functional diversity, which, in turn, is not invariant as a function of time. This variability is itself caused by variability in the expression of the genetic information and its regulation, processes which in biochemical terms involve the transcription of DNA base sequences into RNA and the translation of some of the latter into the amino acid sequences of polypeptides. Although we might expect RNA and nucleic acid metabolism in brain to be similar to those in other tissues in its broad and general outline, quantitative differences should exist. In particular, we might anticipate that the existence of the blood-brain barrier would place serious and preferential constraints on the entry of certain precursors for macromolecular biosynthesis. Additional difficulties arise as a consequence of the inordinate cellular and regional complexity of brain tissue.

METABOLISM OF NUCLEIC ACID PRECURSORS

SYNTHESIS OF PURINE AND PYRIMIDINE NUCLEOTIDES

De Novo Synthesis

Qualitatively, the pathways established for the de novo biosynthesis of the purine and pyrimidine nucleotides in other tissues also appear to be operative in nerve and brain. This assertion rests on the demonstration of (1) the conversion of labeled precursors to the expected intermediates in the correct time course, (2) the presence of certain critical enzymes characteristic of and exclusive to the two pathways, and (3) the proper blocks by specific and characteristic inhibitors of key steps, such as the glutamine antagonists azaserine (*O*-diazoacetyl-L-serine) and DON (6-diazo-5-oxo-L-norleucine) for purine, and azauridine for pyrimidine biosynthesis.

Quantitatively, most of the nucleic acid purines probably arise de novo [20], although the operation of other routes (see below) appears to be of great regulatory significance. The situation may be different with the pyrimidines [3]: the rate of flux through the small pool of orotate actually present is probably insufficient to satisfy the demand for pyrimidines for nucleic acid and coenzyme synthesis, and the block by azauridine can be effectively circumvented by raising the level of available uridine. Furthermore, uridine, but not orotate, can serve as an adequate precursor of nucleic acid pyrimidines in vitro [19].

Salvage Pathways

These observations direct attention to the possibility that a substantial portion of the bases do not arise de novo in the brain, but instead originate in the liver and are provided by the circulation in the form of nucleosides, mainly uridine, in the case of pyrimidines. This, the so-called salvage pathway, appears to be of great potential significance both for normal function and in certain pathological states of the brain.

Geiger and Yamasaki [16] showed that administration of uridine and cytidine could protect a perfused cat brain from degradation and allowed it to exhibit normal carbohydrate metabolism and electrical activity for periods of up to 4 hours. Azauridine blocks the de novo formation of pyrimidine nucleotides on repeated injections [25]; a similar situation occurs in the autosomally inherited (X-chromosome linked) genetic lesion in humans that is called orotic aciduria. Azauridine, after conversion to the nucleotide, specifically inhibits orotidylate decarboxylase; in orotic aciduria both this enzyme and orotidylate pyrophosphorylase – the enzyme catalyzing the preceding step – drop to undetectable levels. Severe neurological and electrophysiological disturbances are observed in both instances. Both can be circumvented or reversed by the administration not only of uridine monophosphate (UMP), the absent direct product, but also of uridine, and this allows us to infer that the salvage pathway is operative.

Lesch-Nyhan Syndrome

Even more striking are the consequences of another genetic lesion [8] that affects hypoxanthine-guanine phosphoribosyl transferase, which is one of the enzymes of the salvage pathway for purines and is responsible for the formation of the nucleotides from hypoxanthine and guanine. Individuals homozygous for this recessive, autosomally inherited defect, called the Lesch-Nyhan syndrome, exhibit an elevated level of uric acid in blood and urine (hyperuricemia), spastic cerebral palsy (choreoathetosis), aggressive and destructive behavior leading to self-mutilation, as well as mental retardation (see Chap. 22). Although it is not yet certain to what extent these manifestations are consequences of the primary enzymatic defect rather than a response to the attendant deranged purine metabolism in general, the whole picture makes clear just how finely poised and closely interacting are the various pathways for purine biosynthesis and degradation, and points toward a crucial, perhaps regulatory, function of the salvage pathway.

Nucleotides and Behavior

Finally, it has been claimed that distinct behavioral effects can be elicited in animals and humans by feeding heterologous RNA, which is almost certainly degraded either prior to or in the course of administration to nucleotides and nucleosides. Because the most striking results were reported in correcting a memory defect in aged subjects [11], it is not impossible that upon stress or aging the demand on the salvage pathway becomes great enough that it can no longer be satisfied wholly by endogenous sources, and dietary supplementation might indeed prove beneficial.

NATURE OF PRECURSOR POOLS

For all practical purposes, the only kind of nucleic acid synthesis with which we need to concern ourselves takes place in the nucleus. The immediate precursor pools involved thus are constituted of the various nuclear (not the total cellular) nucleoside triphosphates, and the information that we require is their concentration both in the steady state and in its more-or-less transient response to a particular stimulus. Unfortunately, this information is not yet available. There are indications that the nuclear pools can be refilled from cytoplasmic nucleoside 5'-phosphates, but whether the diphosphates and triphosphates can also be transported across the nuclear membrane or whether intranuclear phospho transferases have to intervene are unknown factors, as are the levels and rates of the various enzymatic reactions involved. It is known [26] that the relative total concentrations in whole rat brain for the four triphosphates are ATP >> GTP ≈ UTP >> CTP at all ages, but whereas the concentrations for the first three remain relatively invariant during maturation (at about 200, 30, and 30 μmoles per 100 g wet weight, respectively), the last drops to very low levels (less than 1 μmole per 100 g).

FUNCTIONAL CHANGES

General Considerations in Tracer Experiments

The usual, and frequently the only practical, method for investigating the effect of perturbations on the rate and extent of synthesis of cellular, cytoplasmic, or nuclear RNA – either in its totality or in any of its subpopulations – consists in measuring the difference in uptake of some radioactive precursor, e.g., uridine (Urd). The following scheme, though simplified, depicts the flow of this precursor into various species of RNA.

$$\mathrm{RNA^3_{nucl}} \rightarrow \mathrm{RNA^3_{cyto}}$$

$$\mathrm{Urd_{circ}} \rightarrow \mathrm{Urd_{brain}} \rightarrow \mathrm{UMP_{cyto}} \rightarrow \mathrm{UMP_{nucl}} \rightarrow \mathrm{UDP_{nucl}} \rightarrow \mathrm{UTP_{nucl}} \rightarrow \mathrm{RNA^2_{nucl}} \rightarrow \mathrm{RNA^2_{cyto}}$$

$$\mathrm{RNA^1_{nucl}}$$

$$\mathrm{OMP_{cyto}}$$

Clearly, what needs to be done is (1) to establish the complete kinetics of the flux of label in the unperturbed state and its relevance to RNA synthesis, assuring that what is measured is indeed synthesis rather than transfer or the filling of one or more of the precursor pools, and (2) to extend this study to the much more difficult problem posed by the introduction of the perturbation. Very few of the studies that will be discussed meet these stringent criteria. What is commonly done is to take one or, at the most, a selected few time points subsequent to the administration of label and to compare the appearance of the label in some RNA (and in one or more of the intermediates in the better investigations) in appropriately selected experimental and control groups. The only permissible inference from studies of this type is the rather trivial one that any differences observed indicate differences in the net rate of appearance of label in the RNA, but they cannot be interpreted in terms of differences in the rate of synthesis of this species. In fact, more often than not the differences may be due to differences in the interconversion of precursors, in transport phenomena, or in rates of degradation of short-lived intermediates or products.

Electrical Stimulation

Electrical stimulation – either briefly at high intensity during electroshock in vivo, or at low intensity, but for a sustained period, on slices in vitro – elicits profound effects on the metabolism of the total nucleotide population in these systems. In the former case, the most striking results [32] are relatively persistent alterations in the concentrations of UTP and GTP, as well as in their rate of turnover. In the latter, investigations have been restricted so far to measurements of the conversion of uridine to its phosphorylated derivatives [29]. A diminution of some 40 percent coincident with maintenance of the total uridine-containing pool was found, and this was ascribed to effects on the cation pump that resulted in disturbances in the rate of influx of Na^+ into, and efflux of K^+ from, the cells.

To what extent such phenomena are relevant for a control of nuclear RNA synthesis in vivo is uncertain. In general, one might expect that the RNA polymerases would respond only if some ribonucleoside triphosphates either dropped to or rose from (1) extremely low levels (less than 10 percent of the K_m), in which case the regulation would be of the "on/off" variety, or (2) the region around its K_m, in which case the rate would be directly proportional to concentration. In view of results obtained by Mandel and Jacob [26] (see p. 248), it might prove fruitful to explore the response of nuclear GTP levels to stimulation in some well-defined system. In any event, any such regulation by availability of a critical substrate may be expected to be nonselective and to affect to the same extent the transcription of all active genes by any one polymerase.

SYNTHESIS AND TURNOVER OF DNA

LOCALIZATION AND SPECIES

The only well-authenticated sites at which DNA is found in the cells of all animals, including brain and nerves, are in the nucleus (this includes the nucleolus) and in the mitochondria. No evidence for tissue specificity – qualitative in the latter case, quantitative as well as qualitative in the former – has yet been adduced for either of these entities, in contrast to the compelling indications of their species specificity. In other words, the genetic constitution of all the somatic cells in a metazoan animal is invariant and the information encoded in their DNA is identical.

The DNA content of diploid mammalian cells, including neurons and other brain cells, is of the order of 6.5 picograms per cell (or about 1 μg per milligram wet brain weight). Virtually all of it is nuclear. This corresponds to a particle weight for the total chromosomal complement of such cells of 3.8×10^{12} daltons. The numbers are lower and subject to much greater variation for other animals. The base composition of all animal DNA is close to 60 percent guanine plus cytosine (G + C). This DNA in nondividing (interphase) nuclei is present as a complex or aggregate called chromatin, which consists of approximately 30 percent DNA, 10 percent RNA, and 60 percent protein by weight. The proteins are composed of both basic species (histones account for about 80 percent of the total) and acidic species. The DNA itself is not homogeneous, but varies with respect to size, base composition, and redundancy of its constituent or subunit species.

Nucleolar DNA is much more homogeneous with respect to all these criteria: its base composition is also different (about 70 percent G + C), all reflecting its principal function of serving as the depository for the information that encodes several hundred copies of the 45S (4×10^6 daltons) precursor of ribosomal RNA [5].

Mitochondrial DNA amounts to less than 0.1 percent of total cellular DNA. Its base composition is close to that of nuclear DNA, but mitochondrial DNA is much smaller and more uniform in size (5 μ in length or a particle weight of 10×10^6 daltons), and it is highly homogeneous with regard to base composition and lack of redundancy per genome [4]. Present in many copies per cell (corresponding, on the average, to one or two per mitochondrion), mitochondrial DNA is not tightly complexed with proteins. It can be isolated as a covalently closed, twisted circle (superhelix) and it may also exist this way intracellularly.

TURNOVER

As already mentioned, the nuclear DNA of mature macroneurons is not subject to metabolic turnover: replication is absent and repair synthesis, if it occurs at all, has so far escaped detection. Whether there is any further modification of bases in the intact polymer has not been determined. In contrast, mitochondrial DNA is subject to turnover [18]: its half-life in brain has been estimated at 31 days, as compared to 7.5 days in the liver of the same species.

DNA AND NEUROGENESIS

Autoradiographic Studies

Altman and his collaborators [1] have used autoradiography of cell nuclei labeled with ^{3}H-thymidine to determine the nature and extent of postnatal neurogenesis in rodents. By this technique, they were able to demonstrate that considerable DNA synthesis (or at least turnover) and cell proliferation, migration, and differentiation take place in the brain of infant rats, and that these events appear to be initiated by the cells of the outer granular layers of the cerebellum, hippocampus, olfactory bulb, and ventral cochlear nucleus (see Chap. 14). The formation and differentiation of cells, classified as microneurons, continues during infancy, although at declining rates in most structures. Thus, in the cerebellum, almost all the cells of the granular and molecular layers that form the precursors of granule, stellate, and basket cells are formed and completed during infancy, the last-named earlier than the first two types, but most of them completing their differentiation and migration by 21 days postnatally. In contrast, almost all Golgi cells have already been formed at birth. In the wall of the olfactory ventricle cell, multiplication continues at very high rates for 6 days and declines dramatically by 13 days, but it is still detectable even in the young adult. In the hippocampus, the bulk of the cell population is formed postnatally and the neurogenesis of granule and stellate cells in the dentate gyrus may continue over a prolonged span of time.

From these observations, Altman concluded that the long-axoned or macroneurons of the brain are formed prenatally, while many of the short-axoned or axonless microneurons are formed postnatally. He further proposes that this continued genesis of microneurons may be required for the modulation and feedback regulation of (1) the transmission of sensory information in primary afferent relay areas, (2) the execution of motor action in the cerebellar cortex, and (3) the motivational efficacy of need-catering (appetitive and consumatory activities) in the hippocampus and limbic system in general.

DNA Content and Ploidy

Perhaps related to some of these considerations is the fact that the chromosome number (ploidy), and hence the DNA content, of certain neuronal types in the higher integrative centers is about 13 pg per cell, i.e., twice that of the ordinary diploid brain cells [21]. The nuclei of these unusual cells also are larger than their standard counterparts: among them are the Purkinje cells of the cerebellum and the pyramidal cells of the hippocampus (but not its granule, interneuron, and glial cells).

Effects of Growth Hormone

The weight of the brain, its content of total DNA (i.e., its cell number), as well as the relative proportions of its neurons and glia in newborn or infant rats can be significantly increased if the mother is injected subcutaneously or intravenously with purified growth hormone (from bovine pituitary glands) during the period between the 7th and 20th days of pregnancy [43]. The significance of these findings and their generality as to strain and species remain to be explored (see Chap. 14).

Inhibitors

Arabinosylcytosine (cytosine arabinoside), an analog in which arabinose has taken the place of ribose or deoxyribose, is an effective and selective inhibitor of DNA synthesis in animals in vivo and in their cells in culture. Because it blocks the uptake of radioactive thymidine into various areas of goldfish brain, it appears to function in the expected manner in this system also. This block is, however, without any effect on certain behavioral parameters that are affected by blocks of protein synthesis [13]. Thus neurogenesis or other concomitants of DNA synthesis do not appear involved, at least in these particular behavioral tasks.

BIOSYNTHESIS OF RNA

ENZYMES IN BIOSYNTHESIS

Localization

In every cell not infected by an RNA virus, all RNA synthesis is absolutely dependent on the presence of a DNA template and is catalyzed by a family of enzymes called DNA-dependent RNA polymerases. From our considerations so far, we would expect RNA synthesis to be localized at only two intracellular sites in all cells of the nervous system: in the nucleus (including the nucleolus) and in the mitochondria, with the former providing for the bulk of this material both in variety and amounts. All the evidence so far available bears out these expectations.

Properties

Isolated nuclei of brain cells, particularly those of neurons, are capable of effective incorporation of radioactive precursors into RNA at rates at least equal to those observed in the nuclei of cells with a much higher mitotic index. These and other properties of the incorporation reaction, as observed with intact and disrupted nuclei as well as with different enzyme preparations at various stages of solubilization and purification, suggest a strong analogy (if not an identity) between the responsible enzymes in neuronal tissue and those in other tissues such as liver.

The latter are now known to consist of three species that differ in localization, requirements, inhibition pattern, type of product synthesized, and, presumably, function [9]. Polymerase I is localized in the nucleolus; polymerases II and III in the nucleoplasm. The former is active at low ionic strength and can utilize either Mg^{2+} or Mn^{2+} as a divalent cation cofactor; the latter exhibit optimal activity only at higher salt concentrations, 0.1 to 0.2 M $(NH_4)_2SO_4$ is usually employed, and

Mn^{2+} is the preferred cofactor. The reactions of all three enzymes are inhibited by actinomycin D (although perhaps to different extents), and they are resistant to the action of rifampicin and intercalating dyes (ethidium bromide, acridines) at low concentrations. The reaction of polymerase II alone is inhibited (with a $K_i \approx 10^{-8}$ M) by α-amanitin, a bicyclic octapeptide isolated from the poisonous mushroom *Amanita phalloides.* The product of polymerase I appears to be the large, guanine-rich, unmethylated precursor of ribosomal RNA (sedimentation coefficient equal to or less than 45S) in most animals. The product of polymerase II is much more heterogeneous, in agreement with its probable function as the principal enzyme responsible for the transcription of active genes in chromatin. The function of polymerase III is as yet unknown.

There is at present no information concerning mitochondrial RNA synthesis in brain. By analogy with the data obtained for other cells, one might expect it to be highly susceptible to inhibition by ethidium bromide and resistant to α-amanitin and, in intact particles, to actinomycin D; its stable products would then be the mitochondrial RNAs with sedimentation coefficients of about 12S and 18S.

TRANSPORT

Nucleus to Cytoplasm

Most of the RNA of neurons is localized in the perikaryon and there, as is true of all cells, the majority forms an integral constituent of ribonucleoprotein particles, which are themselves aggregated into polymeric structures (polyribosomes, polysomes). These structures exist either free in the cytosol or attached to membranes of the endoplasmic reticulum. They are held together by messenger RNA (mRNA), and, in conjunction with the soluble transfer RNA (tRNA) of the cytosol, constitute the protein-synthesizing machinery of the cell. After administration of a radioactive precursor, nuclear and nucleolar RNA are the first to become labeled, as might have been expected; mRNA enters the cytoplasm within a few minutes thereafter, probably in association with some proteins, and is followed within 30 min to 1 hr by the stable ribosomal RNAs already integrated into ribonucleoprotein particles.

Nucleus to Axons

In addition to their perikarya, many neurons probably contain some RNA in their axons, including the presynaptic thickening (nerve ending, bouton). In peripheral nerve, and particularly in the giant Mauthner neurons of fish [15] or the stretch receptor of the lobster, this axonal RNA may contribute some 0.5 mg per gram of wet weight, although in most other instances values are considerably lower. The origin of this axonal RNA is open to question. Because in addition to 4S RNA it consists of the usual ribosomal 18S and 28S species, and because its synthesis is blocked by actinomycin D, a mitochondrial origin of these stable species is not likely. In part, they may originate in the nucleus and then be transported down the axon by rapid somatoaxonal (axoplasmic) flow ([6, 12]; see also Chap. 12, this volume), probably as preformed ribonucleoprotein particles. In part, they may be of more local origin, arising from supporting cells (glia) or neighboring neurons. Evidence is accumulating for some protein synthesis in or on the axon as well as at the synaptic terminal, and the characteristics of this synthesis suggest an extramitochondrial origin for at least part of it [6]. Although it is generally believed that ribosomes and polysomes are absent in axons and nerve terminals, RNA-containing particles in low concentrations and integrated into the membranes might well have escaped detection.

ENVIRONMENTAL EFFECTS ON RNA SYNTHESIS

Electrical Stimulation in Vivo

The sea hare *Aplysia* provides a particularly useful system for correlating electrophysiological and biochemical findings: certain giant neurons are readily identifiable in its abdominal ganglia, which can be removed, stimulated, and monitored electrophysiologically. The giant neurons can, at the same time, be exposed to radioactive precursors of RNA. These precursors can be isolated subsequently, and the extent and the nature of the incorporation explored. Prolonged stimulation (about 90 min) of the ganglion, sufficient to elicit postsynaptic spikes in cell R2, results in a significant increase in the incorporation of labeled nucleosides into RNA as compared to unstimulated controls [7, 31]. The newly formed material is found in both the nucleus and the cytoplasm; both ribosomal and heterodisperse RNA (presumably mRNA appear to be affected [30].

Although changes in the specific activity of the precursor pool or in the rate of degradation might serve as alternate explanations for the results obtained, their interpretation in terms of an increase in the rate of synthesis is consistent with many reports on net increases of stable RNA as a result of stimulation in many species, including earthworm neurons, rat Purkinje cells, frog retinal ganglion, and cat sympathetic ganglion cells.

Electrical Stimulation in Vitro

Contrary to the results obtained in vivo, electrical stimulation of slices of rat or cat cerebral cortex in vitro [29, 33] decreases the rate of incorporation of labeled uridine into RNA by about 50 percent. This effect is blocked by 5 μM tetrodotoxin, a neurotoxin known to inhibit Na^+ influx during stimulation. An inhibition of the conversion of labeled UMP into UDP plus UTP can be demonstrated to accompany the inhibition of RNA labeling, so the most likely explanation, consistent also with a number of additional experiments, is that electrical activity affects the Na^+ pump, and a disturbance in the balance between Na^+ and K^+ is responsible for affecting the rate of phosphorylation and thereby the availability of labeled nucleoside triphosphates for RNA synthesis. Similar effects may also account for some of the reported decreases in RNA and its synthesis that result from gross stimulation of the cortex in vivo, particularly during electroshock (see below).

Visual Stimulation

When the incorporation of orotate into the RNA in the visual cortex of a totally blind strain of rats is compared with that of normal rats [14], both a delay and a decrease in the rate of labeling of ribosomal and of a heterogeneous 5S to 8S RNA are observed. There are no such differences in the labeling of the RNA in the motor cortex, while differences in the visual cortex are accentuated by using a flashing light rather than continuous illumination for the stimulation of the sighted rats. Thus, in this instance, there appears to be a positive correlation between neuronal activity and the kinetics of RNA labeling. Since the visual system of the blind animals might be expected to have suffered atrophy long before the start of the experiment, the ability of the system to synthesize certain enzymes required for the incorporation reaction also may have been impaired. In fact, it is known that blinded animals or those kept in the dark from birth show deficits in RNA in various cells of the visual system.

Visual imprinting of day-old chicks by flashing light is also reported to coincide with a relatively profound and rapid stimulation of RNA synthesis (as well as protein synthesis), particularly in the forebrain roof, as compared with slower increases elicited by continuous illumination, which does not result in imprinting [35].

Olfaction

Catfish brains in vivo or the experimental half of split-brain preparations in vitro have been shown to respond in a specific manner to odorants such as morpholine by affecting both the amount as well as the base ratios of nuclear RNA, but not of cytoplasmic RNA [34].

Drugs

1,3-Dicyanoaminopropene is reported [23] to increase the amount of RNA and its rate of synthesis in the nervous system. Because it also affects a variety of growth processes, can act as a general, mild stimulant of the CNS, and is completely inert insofar as affecting RNA synthesis by isolated brain nuclei, its mode of action is probably not specific for RNA but instead is more general, perhaps metabolic.

It has been claimed that magnesium pemoline, a combination of 2-amino-5-phenyl-4-oxazolidionone and magnesium hydroxide, affects incorporation of some precursors into nuclear RNA, but not into total brain RNA. It is not certain to what extent these effects, if real, are referable to which of the two components of the system. The parent mixture, as well as one prepared from the cyclopropyl derivative – a more effective CNS stimulant – is without effect on isolated brain nuclei.

Nerve Growth Factor (NGF)

When NGF is added to embryonic sensory or sympathetic nerve fibers in vivo or in vitro, their growth and proliferation are enhanced but remain largely restricted to axonal processes that are free of myelin sheaths (see Chap. 14). Under these conditions, autoradiography shows rapid incorporation of labeled uridine into the RNA of both the perikaryon and the axon, with a time course consistent with the uridine being the precursor of the RNA [2].

Training

Carefully controlled experiments seem to indicate that in mice the acquisition of a new behavioral pattern, at least in response to certain tasks, or of a specific emotional response elicited by or associated with this performance, results in an increased flux of labeled pyrimidines into nucleotide pools and thence into RNA [44]. Autoradiographic evidence suggests that these increases are restricted to the nuclei of the neurons of the limbic system, while decreases are observed in those of the outer layer of the neocortex [24]. The increases in radioactivity do not appear to be restricted to any one type of RNA and do not involve any changes in base ratio. Similar results have also been reported for rats, including actual increases in the amount (not just the rate of labeling) in the RNA of hippocampal nuclei [10]. (Additional discussion can be found in Chap. 31.)

MOLECULAR SPECIES OF RNA AND THEIR TURNOVER

If the internal or external milieu can bring about more-or-less selective effects on RNA synthesis, a qualitative or quantitative alteration of the profile made up of the various RNA species in the population will result. Some of the effects observed have already been mentioned in earlier sections: here we address ourselves more explicitly to this problem.

HYBRIDIZATION

With prokaryotic organisms (bacteria and viruses), it is possible, at least in principle, to determine by appropriate DNA-RNA hybridization techniques the nature and the length of a specific section of the total genome represented by any given RNA transcribed either in vivo or in vitro. The enormous complexity and the large number of fully or partially overlapping (reiterated or repetitive) long sequences of the nuclear DNA of metazoan animals, and especially of mammals, impose almost insurmountable obstacles to similar quantitative hybridization experiments, as well as formidable hazards in the interpretation of this type of experiment in general [27]. As ordinarily performed (i.e., at high RNA/DNA ratios), the restrictions mentioned virtually assure that what are measured are not locus-specific, unique sequences corresponding to what would be expected for mRNAs of structural genes. Instead, they are (a) the highly repetitive, redundant sequences with as yet unknown function, or (2) the extent of overlap between sequences that are partially homologous structurally but unrelated functionally. These reservations do not hold for properly performed determinations on ribosomal RNAs or their precursors, precisely because they are specified by clusters containing more than 100 genes for the various species. The values of these particular parameters obtained for brain agree reasonably well with those for other tissues. To conclude, although one would like to know the kind and number of structural genes (or even just the portion of the total genome) active in a neuron at any one time, this kind of information is not yet available for any somatic cell of any animal. In amphibian oocytes, where a beginning has been made in this direction, the fraction is of the order of 0.3 percent.

RAPIDLY LABELED RNA

As discussed earlier, immediately after administration of label to the brain of an animal or to even an isolated neuron, labeled (i.e., newly synthesized) RNA is restricted to the nucleus and nucleolus. As in other tissues, the RNA appears to be mainly of two kinds: ribosomal precursors that appear first in the nucleolus, with sedimentation coefficients of 45S, 35S, and 32S; and a polydisperse species, with sedimentation coefficients varying from about 50S to about 5S and a base composition close to that of DNA. The latter species is also capable of effective hybridization with DNA [39]. Most of this polydisperse species of RNA decays without even leaving the nucleus and, for this reason as well as because of its other properties, it cannot qualify as the direct precursor of the almost equally polydisperse mRNA found associated with the polysomes of the cytoplasm. Its function is so far totally unknown, but it may possibly be of great relevance in the regulation of the kind and amount of mRNA that does find its way into the cytoplasm [27].

Cytoplasmic mRNA makes its first appearance shortly (several minutes) after extensive labeling of nuclear RNA has already been accomplished. It is probably always associated with ribonucleoprotein particles, and it is destined to become associated with 80S ribosomes to form polysomes in order to discharge its function in translation. The labeling rate of the ribosomal RNA proper exhibits a lag and, although it is slower, it proceeds for a longer time, relative to mRNA, before reaching an isotopic steady state. This phenomenon presumably occurs because of gross differences in the kinetics of interconversion and the degradation of various macromolecular precursors, as well as of the final products themselves.

POLYSOME TURNOVER AND STATE OF AGGREGATION

Brain and neuronal polysomes exhibit properties quite similar to those of other tissues and cells with respect to both structure and function. Maintenance of their proper state of aggregation and configuration, and hence their activity, appears to be particularly finely poised and responsive to changes in environment: these parameters appear to be modulated in vitro by the level of Mg^{2+}, the presence and proper balance of K^+ and Na^+, and the concentrations of various amino acids and certain transmitters (e.g., γ-aminobutyrate).

Such effects may also be operative in vivo. In addition, we might anticipate from studies on other systems that changes in the functional state of a cell might affect the size and turnover profiles of polysomal populations of selected cell populations or brain regions if they influence (1) the rate of translation of any mRNA (as, for instance, in the presence of cycloheximide and its derivatives or other inhibitors of protein synthesis), (2) the rate or extent of its formation (by activation or inhibition of specific genes, as well as of transcription in general by agents such as actinomycin or amanitin), or (3) the rate or extent of its degradation (for example, by activation of specific nucleases or affecting accessibility). So far, studies have been restricted to whole brain or cerebral cortex. Studies that are extended to more localized areas and small groups of homogeneous neurons are badly needed.

In cortex we find polysomes in both glia and neuronal perikarya. In the latter, they appear both free and attached to endoplasmic reticulum; both types are capable of and responsible for effective protein synthesis. In adult rats the turnover rate of both the RNA and the proteins of the various constituent ribosomes is 12 days [40]. Thus, the particles are subject to turnover as a unit. It is known that protein synthesis is most extensive in the brain and in brain preparations of young rats and mice shortly after birth (it reaches a maximum at about 10 weeks for rats) and declines as the animals mature. This decline in activity is correlated with ATP levels in cortex slices; it is also correlated, at least in part, with a decrease in polysomal RNA, a change in the size distribution in favor of smaller aggregates, and decreased stability of isolated polysomes.

Coincident with these changes is a change in the base composition [42] as reflected in the ratio of guanine + cytosine to adenine + uracil [(G + C)/(A + U)]: this ratio has been reported to rise during maturation from 1.30 to 1.50 in whole rat brain and from 1.41 to 1.66 in single pyramidal cells of the hippocampus. In the latter, the amount of RNA per cell is reported to increase from 24 pg to 110 pg for mature rats and to drop to 53 pg per cell in very old rats. In the very old rats, the ratio (G + C)/(A + U) rises to 1.95.

The proportion of total ribosomal RNA found in polysomes has also been reported to increase as a function of environmental or behavioral stimulation. For instance, when adult rats were kept first in the dark for 3 days and then exposed to the lights and sounds of the laboratory for 15 min, the polysome to monomer ratio increased by 83 percent in cerebral cortex but was unaffected in the liver. On the other hand, electroshock had the opposite effect: polysomes were dissociated as a result.

BEHAVIOR AND CHANGES IN BASE RATIOS

The literature contains several reports concerning changes in base ratios resulting from or coincident with the acquisition of new behavioral patterns (see also Chap. 30). In general, the experiments have been of two types and both are subject to serious reservations. In the first, total RNA of small cell populations is analyzed by microtechniques and profound changes are observed [22], such as formation of RNA with (G + C)/(A + U) ratios in the range of 0.70 to 0.90. These, then, must have been the result of alterations in ribosomal RNA since it is the only species present in amounts large enough for its base composition to affect cellular base composition. (Messenger RNA accounts for about 5 percent of the total cellular RNA, and in its base composition, it must reflect the average of a large multiplicity of different base sequences. Even under hyperinduction in prokaryotes, a single species never accounts for more than 20 percent of the total population. Thus, total base composition could be altered by 1 percent in the most extreme case.) No such fluctuations in ribosomal RNA have ever been observed in any other system and are highly unlikely in view of the fact that these RNAs are, as we have mentioned, specified by a cluster of highly homogeneous genes.

In the second type of experiment [36], newly synthesized, rapidly labeled RNA is analyzed by determining the ratio of radioactivity in the uridylate and cytidylate fractions of RNA hydrolysates formed from labeled orotate. However, in the absence of data for nuclear UTP and CTP, the direct precursors, the results are not amenable to any ready interpretations. It is of interest to note in this context that the increased synthesis of RNA, which was discussed previously in regard to the effects of behavioral training on RNA synthesis (see p. 256), is not accompanied by any changes in base ratio.

EFFECTS OF ANTIBIOTICS AND OTHER DRUGS

ACTINOMYCIN D

As outlined earlier, this antibiotic, at least under the conditions usually employed in whole-animal experiments, is a general inhibitor for DNA-dependent RNA synthesis. As a result of this interference, it might be assumed that it will also block protein synthesis, but this is not necessarily true. (These differential effects are not observed in prokaryotic cells, in which transcription and translation are tightly coupled and mRNA turnover is much more uniform.) Synthesis supported by inherently unstable mRNAs or on polysomes in the perikaryon close to the site of the block will be affected much more immediately than will synthesis supported by relatively stable mRNAs or under conditions in which some of the nucleic acid may still be in transit to its site of utilization somewhere along the axon. This reasoning may explain some of the results on protein synthesis obtained with actinomycin D. Brief exposure to the drug, even at levels that inhibit RNA synthesis almost completely, elicits no effects on resting or action potentials, on acquisition of new tasks, or on short-term memory.

Interpretations of the retrograde amnesic effects (or their absence) obtained with the drug are complicated by its production of necroses and electrophysiological abnormalities as found, for instance, in the hippocampus of rodents. These reactions may occur either in addition to or, perhaps, sometimes in the absence of effects on RNA synthesis.

8-AZAGUANINE

Intracisternal injection of 8-azaguanine leads to the incorporation of this analog of guanine into the RNA of rat brain, but the qualitative and quantitative consequences of this intervention have not been explored. It is reported to exert behavioral effects, which, however, may be a secondary consequence of a generalized depression of motor activity.

6-AZAURACIL

The effects produced by repeated administration of this uracil analog have been described above under Salvage Pathways (see p. 247). The available evidence suggests that although 6-azauracil profoundly affects the levels of pyrimidine precursors available de novo, the operation and the quantitative importance of the salvage pathway insure their adequate supply, so the rate and extent of RNA formation becomes relatively insensitive to the inhibitor even when it is present in high concentration. Any neurophysiological and behavioral effects produced are therefore probably referrable to disturbances of metabolism rather than to the biosynthesis of RNA or protein.

5-FLUOROURACIL AND 5-FLUOROOROTATE

Unlike the substitution of nitrogen for carbon in 6-azauracil, the substitution of the isosteric fluoride for hydrogen in uracil results in a molecule that is sufficiently similar to its parent to permit its participation in all reactions, including the conversion to the nucleoside triphosphate and the incorporation of the resultant nucleotide into RNA at nearly normal rates and to the same extent [25]. This phenomenon, established in other than nervous tissue systems, also appears to occur in cat spinal cord, the only nervous tissue for which intensive studies are available. The most interesting result of exposure to 5-fluorouracil is its well authenticated ability to produce (or correct) miscoding, either at the transcriptional or the translational level, by virtue of "faulty" mRNAs or tRNAs containing 5-fluorouracil capable of forming base pairs with guanine rather than adenine. This in turn might result in the production of specific faulty proteins; however, this possible reaction has not yet been explored. The possible subtle behavioral consequences arising from such an alteration of protein synthesis have also not yet been investigated.

When injected into cats at a relatively high dose (about 10 mg), fluoroorotate or fluorouridylate (but not fluorouridine) produces severe neuroanatomical alterations in both the central and the peripheral nervous system.

URIC ACID

Because it is an end-product rather than a precursor of purine metabolism, uric acid probably cannot affect nucleic acid biosynthesis directly. Nevertheless it may serve as an indicator of disturbances in purine or nucleic acid catabolism (see Chap. 22).

ENZYME INDUCTION

The most precise and direct way of determining alterations in gene expression is to measure their effect on the formation of defined final products, i.e., specific proteins such as enzymes. Whether any changes in observed enzyme activity are due to specific de novo synthesis (induction proper), to activation of precursors, to relatively nonspecific stimulation of the synthesis of several proteins, or to the inhibition of their degradation is, however, a problem that must be solved in each particular instance. Three examples of enzymes of the nervous system are discussed here.

D(–)-BETA-HYDROXYBUTYRATE DEHYDROGENASE

This important enzyme of lipid catabolism, found tightly integrated into the mitochondrial membrane, is present at high levels in the brain of unweaned rats, but drops to low levels as the animal matures. When adults are starved for periods of up to 12 hr, their body weight drops concomitantly with a fivefold increase in the concentration of β-hydroxybutyrate in their blood because of the breakdown of their lipid stores. At the same time the activity of the dehydrogenase in brain increases by almost ten times [37]. Whether specific induction is actually involved has not been determined.

HYDROXYINDOLE-*O*-METHYL TRANSFERASE

The formation of the hormone melatonin (5-methoxy-*N*-acetyltryptamine) is localized specifically in the pineal glands of mammals due to the presence of an enzyme that catalyzes the *O*-methylation of *N*-acetylserotonin (see Chap. 6). Exposure of the pineal gland to light alters its morphology and reduces its mass and RNA content; darkness produces changes in the opposite direction. Levels of hydroxyindole-*O*-methyl transferase, but not of monoamine oxidase (the enzyme responsible for degradation of melatonin and also present in the pineal gland in high concentrations), respond to illumination in the same way as do mass and RNA, but the reductions are much more pronounced [41].

GLUTAMINE SYNTHETASE

The activity of this enzyme in the retina of embryonic chicks follows a characteristic developmental pattern typical for this tissue and is temporally and spatially correlated with other aspects of embryonic development. A characteristic rise in enzyme activity occurs which can be induced precociously with 11-β-hydroxycorticosteroids, either in the intact animal or in isolated retina in culture, without any generalized cell proliferation but within the context of other well-defined changes. The effects observed are therefore specific and characteristic [28]. They involve the de novo synthesis of the enzyme and its accumulation. Whether they are due to an increase in the rate of synthesis rather than to a decrease in the rate of degradation remains to be established; what has been ascertained by the use of actinomycin D is that the process also requires de novo synthesis of some species of RNA (but not protein) during the initial phases, although the process later becomes partially independent of it (neither high nor low concentrations of actinomycin D have an effect). These observations are consistent with analogous patterns obtained in the induction of a wide variety of enzymes by glucocorticoids and other hormones in many other systems.

ACKNOWLEDGMENTS

I would like to express my appreciation to my student, Mr. M. Maguire, for his invaluable help in surveying the voluminous recent literature on these topics and to Drs. E. Shooter and B. W. Agranoff for providing me with many stimulating suggestions. I would also like to express appreciation for being the recipient of a U.S. Public Health Service Career Award No. GM 05060 from the National Institutes of Health.

REFERENCES

ORIGINAL SOURCES

1. Altman, J., and Das, G. D. Autoradiographic and histological studies of postnatal neurogenesis. I. A longitudinal investigation of the kinetics, migration and transformation of cells incorporating tritiated thymidine in neonate rats, with special reference to postnatal neurogenesis in some brain regions. *J. Comp. Neurol.* 126:337, 1966. (See also Altman, loc. cit.)
2. Amaldi, P., and Rusca, G. Autoradiographic study of RNA in nerve fibers of embryonic sensory ganglia cultured *in vitro* under NGF stimulation. *J. Neurochem.* 17:767, 1970.
3. Appel, S. H., and Silberberg, P. H. Pyrimidine synthesis in tissue culture. *J. Neurochem.* 15:1437, 1968.
4. Ashwell, M., and Work, T. S. The biogenesis of mitochondria. *Annu. Rev. Biochem.* 39:251, 1970.
5. Attardi, G., and Amaldi, F. Structure and synthesis of ribosomal RNA. *Annu. Rev. Biochem.* 39:227, 1970.

6. Austin, L., Morgan, I. G., and Bray, J. J. The Biosynthesis of Proteins Within Axons and Synaptosomes. In Lajtha, A. (Ed.), *Protein Metabolism of the Nervous System.* New York: Plenum, 1970.
7. Berry, R. W. Ribonucleic acid metabolism of a single neuron: Correlation with electrical activity. *Science* 166:1021, 1970.
8. Bland, J. H. (Chairman). Proceedings of the seminars on the Lesch-Nyhan Syndrome. *Fed. Proc.* 27:1017, 1968.
9. Blatti, S. P., Ingles, C. J., Lindell, T. J., Morris, P. W., Weaver, R. F., Weinberg, F., and Rutter, W. J. Structure and regulatory properties of eukaryotic RNA polymerase. *Symp. Quant. Biol.* 35:649, 1970.
10. Bowman, R. E., and Strobel, D. A. Brain RNA metabolism in the rat during learning. *J. Comp. Physiol. Psychol.* 67:448, 1969.
11. Cameron, D. E., Sved, S., Solyom, L., Wainrib, B., and Barik, H. Effects of ribonucleic acid on memory defect in the aged. *Am. J. Psychiatry* 120:320, 1963.
12. Casola, L., Davis, G. A., and Davis, R. E. Evidence for RNA transport in rat optic nerve. *J. Neurochem.* 16:1037, 1969.
13. Casola, L., Lim, R., Davis, R. E., and Agranoff, B. W. Behavioral and biochemical effects of intracranial injection of cytosine arabinoside in goldfish. *Proc. Natl. Acad. Sci. U.S.A.* 60:1389, 1968.
14. Dewar, A. J., and Reading, H. W. Nervous activity and RNA metabolism in the visual cortex of rat brain. *Nature (Lond.)* 225:869, 1970.
15. Edström, A., Edström, J. E., and Hökfelt, T. Sedimentation analysis of ribonucleic acid extracted from isolated Mauthner nerve fiber components. *J. Neurochem.* 16:53, 1969.
16. Geiger, A., and Yamasaki, S. Cytidine and uridine requirement of the brain. *J. Neurochem.* 1:93, 1956.
17. Griffith, J. S. *A View of the Brain.* Oxford: Oxford University Press, 1967. (See also Griffith, J. S., and Mahler, H. R. DNA ticketing theory of memory. *Nature [Lond.]* 223:580, 1969.)
18. Gross, N. J., Getz, G. S., and Rabinowitz, M. Apparent turnover of mitochondrial deoxyribonucleic acid and mitochondrial phospholipids in the tissues of rat. *J. Biol. Chem.* 244:1552, 1969.
19. Guroff, G., Hogans, A. F., and Udenfriend, S. Biosynthesis of ribonucleic acid in rat brain slices. *J. Neurochem.* 15:489, 1968.
20. Held, I., and Wells, W. Observations on purine metabolism in rat brain. *J. Neurochem.* 16:529, 1969.
21. Herman, C. J., and Lapham, L. W. DNA content of neurons of the cat hippocampus. *Science* 160:537, 1968.
22. Hydén, H. *The Neuron.* Amsterdam: Elsevier, 1967.
23. Hydén, H. Behavior, neural function, and RNA. *Progr. Nucleic Acid Res. Mol. Biol.* 6:187, 1967.
24. Kahan, B. E., Krigman, M. R., Wilson, J. E., and Glassman, E. Brain function and macromolecules. VI. Autoradiographic analysis of the effect of a brief

training experience on the incorporation of uridine into mouse brain. *Proc. Natl. Acad. Sci. U.S.A.* 65:300, 1970.
25. Koenig, H. Neurobiological action of some pyrimidine analogs. *Int. Rev. Neurobiol.* 10:199, 1967.
26. Mandel, P., and Jacob, M. Regulation of transcription in nervous cells. In Lajtha, A. (Ed.), *Protein Metabolism of the Nervous System.* New York: Plenum, 1970.
27. McCarthy, B. J., and Church, R. B. The specificity of molecular hybridization reactions. *Annu. Rev. Biochem.* 39:131, 1970.
28. Moscona, A. A., Moscona, M. H., and Saenz, N. Enzyme induction in embryonic retina: The role of transcription and translation. *Proc. Natl. Acad. Sci. U.S.A.* 61:160, 1968.
29. Orrego, F. Synthesis of RNA in normal and electrically stimulated brain cortex slices *in vitro. J. Neurochem.* 14:851, 1967.
30. Peterson, R. P. RNA in single identified neurons of *Aplysia. J. Neurochem.* 17:325, 1970.
31. Peterson, R. P., and Kernell, D. Effects of nerve stimulation on the metabolism of ribonucleic acid in a molluscan giant neuron. *J. Neurochem.* 17:1075, 1970.
32. Piccoli, F., Camarda, R., and Bonaveta, U. Purine and pyrimidine nucleotides in the brain of normal and convulsant rats. *J. Neurochem.* 16:529, 1969.
33. Prives, C., and Quastel, J. H. Effects of cerebral stimulation in biosynthesis of nucleotides and RNA in brain slices *in vitro. Biochim. Biophys. Acta* 182:285, 1969.
34. Rappoport, D. A., and Daginauala, H. F. Changes in nuclear RNA of brain induced by olfaction in catfish. *J. Neurochem.* 15:991, 1968.
35. Rose, S. P. G., Bateson, P. P. G., Horn, A. L. D., and Horn, G. Effects of an imprinting procedure on regional incorporation of tritiated uracil into chick brain RNA. *Nature (Lond.)* 225:650, 1970.
36. Shashoua, V. E. RNA metabolism in goldfish brain during acquisition of new behavioral pattern. *Proc. Natl. Acad. Sci. U.S.A.* 65:160, 1970.
37. Smith, A. L., Satterthwaite, H. S., and Sokoloff, L. Induction of brain D(−)-β-hydroxybutyrate dehydrogenase activity by fasting. *Science* 163:79, 1969.
38. Stetten, D., Jr., and Hearon, J. Z. Intellectual level measured by Army classification battery and serum uric acid concentration. *Science* 129:1737, 1959.
39. Stévenin, J., Mandel, P., and Jacob, M. Relationship between giant-size dRNA and microsomal dRNA of rat brain. *Proc. Natl. Acad. Sci. U.S.A.* 62:490, 1969.
40. Von Hungen, K., Mahler, H. R., and Moore, W. J. Turnover of protein and ribonucleic acid in synaptic subcellular fractions from rat brain. *J. Biol. Chem.* 243:1415, 1968.
41. Wurtman, R. J., Axelrod, J., and Phillips, L. S. Melatonin synthesis in the pineal gland: Control by light. *Science* 142:1071, 1963.
42. Yamagami, S., and Mori, K. Changes in polysomes of developing rat brain. *J. Neurochem.* 17:721, 1970.

43. Zamenhof, S., Mosley, J., and Schuller, E. Stimulation of the proliferation of cortical neurons by prenatal treatment with growth hormone. *Science* 152: 1396, 1966.
44. Zemp, J. W., Wilson, J. E., Schlesinger, K., Boggan, W. O., and Glassman, E. Brain function and macromolecules. I. Incorporation of uridine into RNA of mouse brain during short-term training experience. *Proc. Natl. Acad. Sci. U.S.A.* 55:1423, 1966.

BOOKS AND REVIEWS

Altman, J. DNA Metabolism and Cell Proliferation. In Lajtha, A. (Ed.), *Handbook of Neurochemistry*. Vol. 2. New York: Plenum, 1969.

Glassman, E. Biochemistry of learning: An evaluation of the role of RNA and protein. *Annu. Rev. Biochem.* 38:605, 1969.

Koenig, E. Nucleic Acid and Protein Metabolism of the Axon. In Lajtha, A. (Ed.), *Handbook of Neurochemistry*. Vol. 2. New York: Plenum, 1969.

Rappoport, D. A., Fritz, R. R., and Myers, J. L. Nucleic Acids. In Lajtha, A. (Ed.), *Handbook of Neurochemistry*. Vol. 1. New York: Plenum, 1969.

SECTION IV

Physiological Integration

14

Neurochemistry of Development

Joyce A. Benjamins and
Guy M. McKhann

IN THIS CHAPTER, we present a summary of the metabolic processes that take place in the nervous system during its growth, and we will try to relate these events to morphological and functional changes, where possible. An outline of biochemical and metabolic changes is followed by a discussion of possible mechanisms that control development, emphasizing current approaches.

The events of nervous system development have often been divided into stages for ease of discussion and organization. These stages are artificial, at best, when applied to a dynamic system. In addition, the nervous system is heterogeneous in development from region to region in terms of time, types of cells, and complexity of interaction among these cell types. Davison and Dobbing [14] recently presented the following general scheme, which serves to indicate major changes applicable to all species:

1. Organogenesis and neuronal multiplication
2. The brain growth spurt
   A. Axonal and dendritic growth, glial multiplication, and myelination
   B. Growth in size
3. Mature, adult size
4. Aging

BIOCHEMICAL CHANGES IN DEVELOPMENT

The most obvious index of brain growth is weight. The most rapid growth may occur just before, after, or at the time of birth, depending on the species (Fig. 14-1). In general, brain weight reaches its maximum before body weight, as illustrated for rat brain in Figure 14-2. The period of maximal rate of weight increase takes place at various times in different regions of the nervous system. The cerebellum shows the sharpest rate of growth, the spinal cord the most

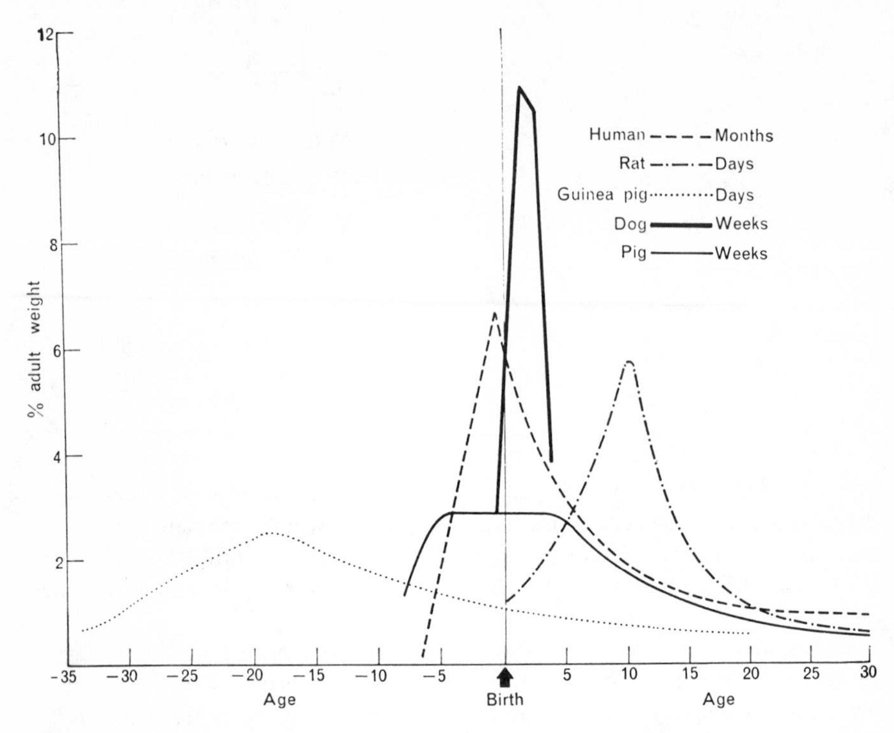

FIGURE 14-1. Rate curves of brain growth in relation to birth in different species. Values are calculated at different time intervals for each species. (From Davison and Dobbing [14]. Reprinted by permission; courtesy of Blackwell Scientific Publications Ltd.)

gradual. This growth spurt corresponds to the proliferation of neuronal processes and myelination. The increase in membranous structures is reflected by an increase in brain solids and a decrease in brain water. For example, in rat brain, water is 90 percent of the total brain weight at birth and 83 percent at maturity [29].

The developmental changes of four major constituents of brain – DNA, RNA, protein, and lipid – are summarized for the forebrain of rat in Figure 14-2. For ease of comparison, the values are expressed as percentages of adult levels. In Figure 14-3, changes in brain weight, DNA, and cholesterol are expressed in terms of change, emphasizing the time of most rapid increase. The sequential changes in these four components illustrate the events that take place in the developing nervous system. Early proliferation of cells is indicated by synthesis of DNA. As differentiation occurs, DNA replication is followed by increased transcription of DNA to RNA, then translation of RNA to protein. With the appearance of cell-specific enzymes and structural proteins, each cell type acquires its unique metabolism and morphology. Maturation of such elements as neuronal processes, synaptic endings, and myelin involves deposition of lipids into these membranous structures.

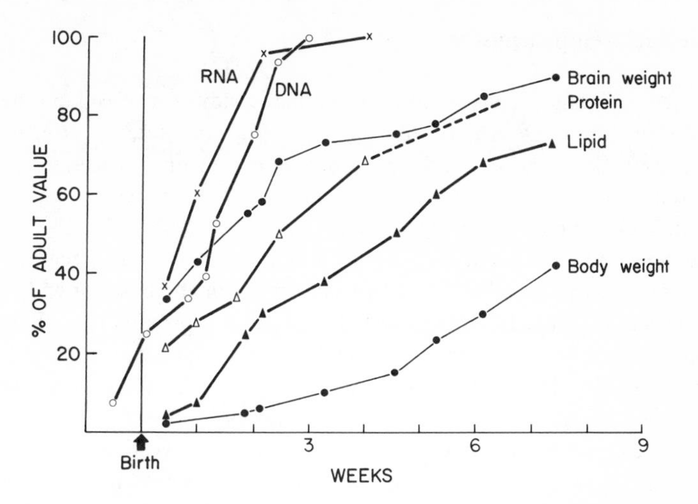

FIGURE 14-2. Increases in body weight and in several brain constituents during development in rat. Values are expressed as percentages of maximal (adult) values. Results for body weight, brain weight, and brain lipid have been calculated from the data of Kishimoto, Davies, and Radin [35], DNA values from the data of Fish and Winick [21], RNA values from the data of Mandel [42], and protein values from the data of Himwich [29, 30].

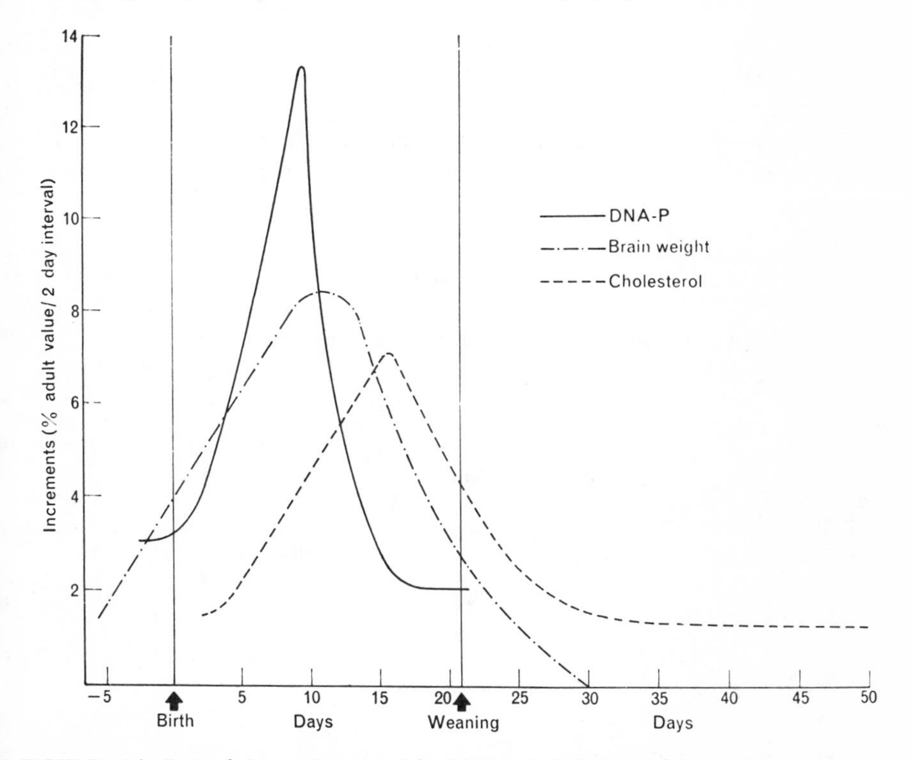

FIGURE 14-3. Rate of changes in wet weight, DNA, and cholesterol of developing rat brain. All values are calculated as increments (percentages of adult values) at 2-day intervals. (From Davison and Dobbing [14]. Reprinted by permission; courtesy of Blackwell Scientific Publications Ltd.)

DNA content of brain is considered to be a reliable indicator of cell number, and the ratio of protein to DNA indicates cell size. This ratio increases after neuronal division ends, reflecting, in part, the arborization of neuronal processes. Increasing lipid content indicates compartmentation and membrane formation, particularly of synaptic endings and myelin membranes.

These developmental events occur at different ages in various species (Table 14-1). It is obviously impossible to give exact times for the onset or end of these events; the ages in Table 14-1 are intended to serve as approximate guidelines. Cortex was chosen because more complete data are available for this brain region than for others. In some cases, only a few ages were studied; in others, subjective judgment plays a large role. For example, determination of the onset of myelination depends on the stain used and the particular area investigated. Thorough investigation is time-consuming; thus, examination of brains at many ages is prohibitive. In such a complex structure as cortex, generalizations may be misleading. We may generalize that outgrowth of axons and dendrites precedes myelination; however, axons and dendrites growing into a region from outside neurons may be myelinated before the neurons in that region send out processes.

NUCLEIC ACIDS

The first increases in brain DNA are due to proliferation of preneuronal cells. Later increases are due to the appearance of glial precursors [15]. An excellent description of cell proliferation in developing brain has been presented recently by

TABLE 14-1. Comparison of Developmental Milestones in Cortex of Several Species[a]

Species	Rapid Increase in Maturity and Number of Nerve Processes	Histological Appearance of Myelin	DNA at Maximal Level	Protein at Maximal Level
Rat	6–24 days [16]	10–40 days [34]	20 days [21]	99 days
Mouse	2–17 days [36]	9–30 days [68]	16 days [30]	80–90 day
Guinea pig	(41–45 days) [15]	(63 days) - 7 days [15]	7 days [15]	55 days [3(
Man	(6 months) - 3 months [30, 70]	(7 months) - 4 years [74]	6–8 months [73]	2 years [73

[a]Values in parentheses represent days or months of gestation, the other values the time after birth. Pertinent references are in brackets.

Altman [4]. He proposes that the primitive neuroepithelium gives rise to differentiating cells which become macroneurons (long-axoned, input-output neurons) and a second pool of proliferating cells. A portion of this secondary pool differentiates into microneurons (short-axoned, modulating interneurons) while another portion leaves the pool, becomes dispersed throughout the brain, and finally differentiates into glial cells. In most species, DNA in brain accumulates in a linear manner until the adult level is reached.

In human brain, two major periods of cell proliferation have been detected by measuring DNA levels [15]. The first period begins at 15 to 20 weeks of gestation, and corresponds to neuroblast proliferation. The second begins at 25 weeks and continues into the second year of postnatal life. This period corresponds to multiplication of glial cells, and it also may include a second wave of neurogenesis, restricted to microneurons in certain regions of the brain (in the rat, these are primarily the granular layer of the cerebellar cortex, the olfactory lobe, and hippocampus). Perhaps these two phases of DNA accumulation have not been detected in other species because of the faster development of the brain and the lack of sufficient data at critical time points [15].

The rate of increase in DNA content is most rapid in the cerebellum, less in the cerebrum, and least in the spinal cord [14]. DNA increases five times as rapidly in the cerebellum as in the cerebrum [21] of postnatal rat brain. In human brain, the most rapid accumulation of DNA in the cerebrum and cerebellum takes place prenatally. Cerebral DNA increases exponentially between 10 and 13 weeks of gestation and then becomes linear, whereas cerebellar DNA is still increasing exponentially at 30 weeks of gestation [31]. The postnatal rate of accumulation of DNA in human brain is similar in the cerebrum and cerebellum [73].

Obviously, histological and autoradiographic techniques are better suited than chemical techniques for detecting the exact timing and location of cell division in the various regions of brain. A general idea of regional differences may be obtained by sequential measurement of DNA in different areas of brain [21]. The cerebrum, brain stem, and hippocampus have nearly reached their adult levels of DNA at birth, whereas cerebellum undergoes rapid cell division during the first 2 weeks after birth. Data on the prenatal accumulation of DNA in the other three regions of brain are needed in order to compare the periods of rapid cell division in these regions with that in cerebellum.

One method of measuring DNA synthesis is by the incorporation of radioactive thymidine. By this method, labeling of DNA in rat cerebellum in vitro is maximal at day 6 and declines to 10 percent of the peak value by 18 days [64]. At day 6, labeling of DNA in cerebellum is 20 times more rapid than is that in cerebrum or brain stem, as would be expected from the rates of change in DNA content.

As DNA accumulation ends, the ratio of RNA to DNA increases [55]. On a cellular level, neuroblast division ends before the appearance of Nissl substance (ribosomal RNA) and the subsequent outgrowth of axons and dendrites [2, 76]. Similar observations have been made for differentiating glioblasts. RNA synthesis is limited to the nucleus until the nucleolus differentiates; then RNA increases rapidly in the cytoplasm [56].

Characterization of the RNA of rat brain is still incomplete. A larger proportion of RNA synthesis results in messenger RNA (mRNA) synthesis in brain as compared to liver [50]. In developing brain, RNA characteristics are compatible with the rapid rate of protein synthesis observed; compared with adult brain, (1) the mRNA turns over more rapidly, (2) more of the ribosomal RNA is membrane-bound (polysomal), and (3) these polysomes are more stable [75].

The individual ribosomal monomers of rat brain may acquire structural and functional differences by interaction with mRNA and the endoplasmic reticulum [50]. Any factor affecting the relative proportions of free and membrane-bound ribosomes may change the rate of protein synthesis. There may be qualitative differences as well; ribosomes attached to the membrane may synthesize protein for secretion (or axoplasmic flow, in the case of neurons), whereas free ribosomes may synthesize proteins for more local intracellular use [50].

Regional differences in protein synthesis may be related to the content and degree of aggregation of the ribosomes in various cell types [56]. In large neurons, ribosomes are concentrated near the nucleus, with others scattered throughout the cytoplasm. Some are seen in proximal dendrites, but not in the distal axons. Many of the ribosomes are in monomeric form or in small aggregates. In glial cells and microneurons, the ribosomes appear to be distributed more evenly through the cytoplasm, and a larger proportion is attached to the endoplasmic reticulum than in macroneurons.

PROTEINS

In developing rat brain, protein increases more rapidly than does wet weight. As previously mentioned, the ratio of protein to DNA indicates cell size, and this ratio is elevated during development. The time at which the protein content of brain reaches its maximal level in various species is summarized in Table 14-1. As with most other constituents, proteins increase most rapidly in rat cortex during the first 2 to 3 weeks of postnatal life. In this same period, there is a shift from water-soluble to membrane-bound proteins. This shift has been studied most extensively in mouse brain: about 60 percent of the protein is extractable by isotonic and hypotonic aqueous solutions at birth; the remainder is extractable with 1 percent sodium lauryl sulfate [27]. By 40 days, about 40 percent is water-soluble and the rest is detergent-soluble.

Changes in both membrane and soluble proteins during development are revealed by acrylamide-gel electrophoresis (Fig. 14-4). The functional significance of these changes in proteins remains to be determined. Continued systematic analyses, including studies of protein composition in cell types and subcellular fractions, promise to yield new insights into the role of specific proteins in development.

The subcellular fraction studied most extensively during development in brain is the myelin membrane (see Chap. 17). Changes in the protein composition of

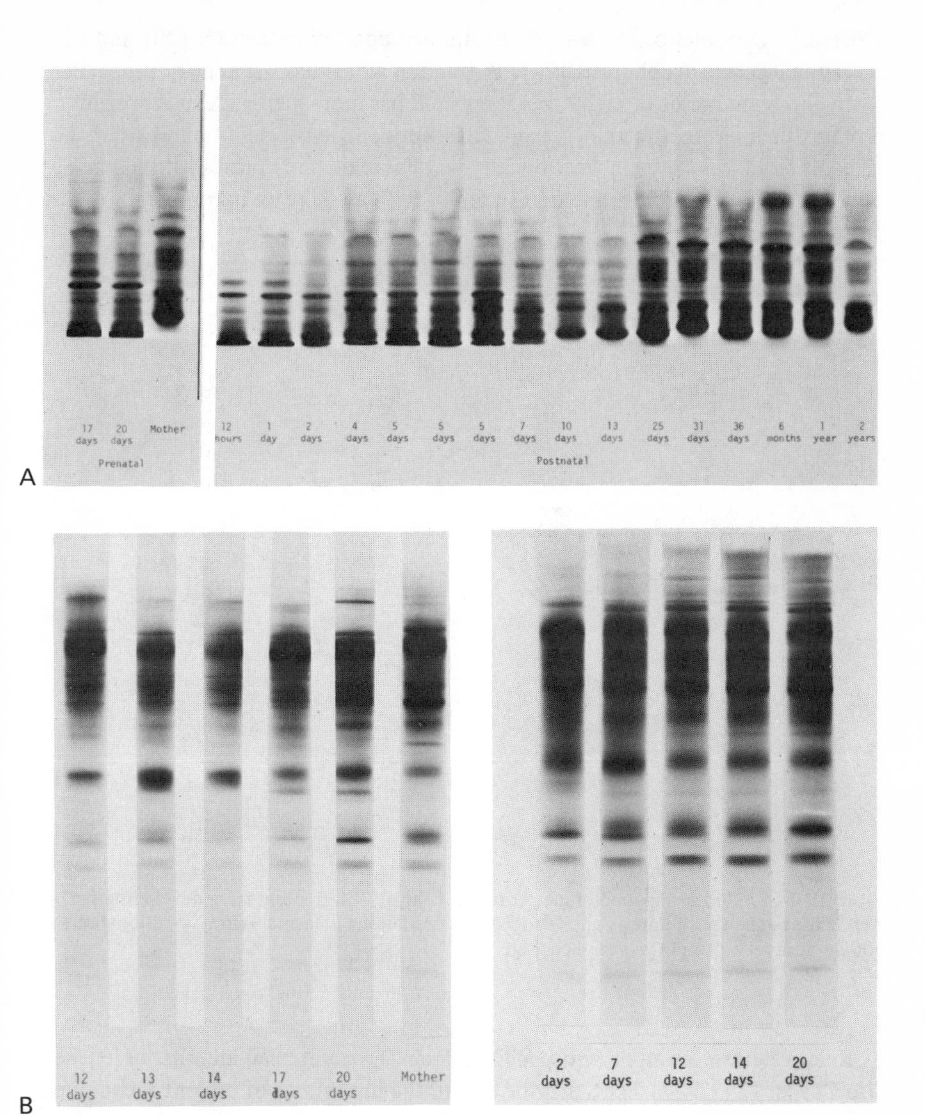

FIGURE 14-4. Developmental changes in (A) membrane proteins and (B) soluble proteins from mouse brain [27, 61]. Tissue from animals of different ages was treated with 1.0% Triton X-100 and then with 0.1% sodium dodecyl sulfate to extract membrane proteins. Patterns from prenatal brains are on the left of each figure and patterns from postnatal brains are on the right. (From Grossfeld and Shooter [27, 59]. Reprinted by permission; Figure 4-A courtesy of The Rockefeller University Press.)

isolated myelin fractions have been determined both by extraction [20] and by acrylamide-gel electrophoresis [26]. A fraction which has the density properties of myelin from brains of rats 9 to 15 days old has more high-molecular weight proteins than has that from older animals. Acrylamide-gel electrophoresis of the proteins from this "myelin fraction" shows almost no basic protein at 9 days; by 28 days, the pattern looks like that of adult, with an increased proportion of basic protein (Fig. 14-5).

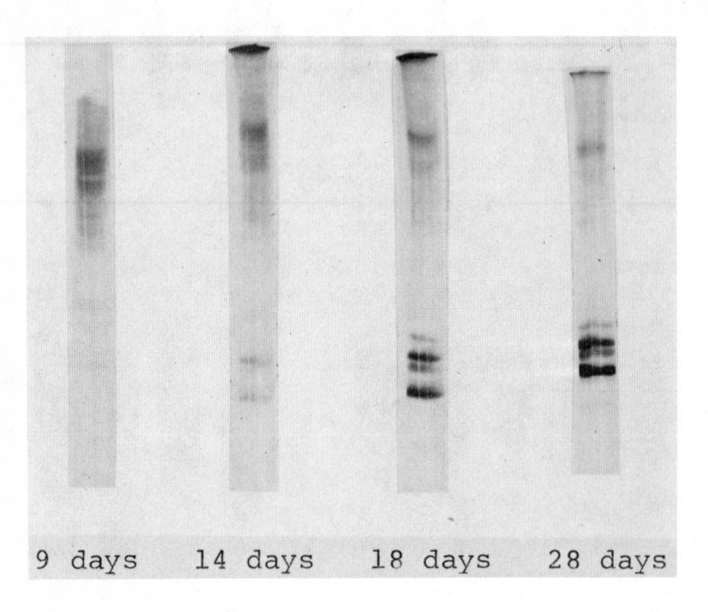

FIGURE 14-5. Changes in membrane proteins of the myelin fraction from developing rat brain. Acrylamide-gel electrophoresis on 5% to 7% gels containing urea and Triton X-100. About 300 μg of myelin protein was applied.

Relatively few brain- or cell-specific proteins have yet been identified ([9]; see also Chaps. 12 and 24). This may be due to the limitations of present techniques in detecting minor differences among proteins. Among the brain-specific proteins so far identified is one that possibly is related to dendritic arborization [9]. Two brain-specific proteins which appear about the time myelin formation begins are the acidic S-100 protein, which apparently is primarily glial, and the basic protein(s) of myelin. Other myelin proteins may be unique to nervous tissue, but they must be isolated and studied further before this can be determined.

Developmental changes in a number of enzymes have been observed, both histochemically [2, 23] and biochemically [2, 58]. The sequence of appearance of proteins with enzymatic activity determines the metabolic properties of a given cell (metabolic changes occurring in brain during development will be discussed later; see p. 281). Interaction of enzyme subunits may contribute to developmental changes. For example, fructose 1,6-diphosphate aldolase in adult human brain has five isozymes with two types of subunits, A and C. In fetal brain, only the isozyme with four A subunits is present. With maturation, isozymes with C subunits appear, so that adult brain has all five possible tetramers of A and C subunits [38]. Whether this has functional significance has yet to be determined. Similar changes occur with lactic dehydrogenase in developing brain [22].

Synthesis of structural proteins and their subsequent interaction with lipids to form membranes directs the developing morphology and the compartmentation of a given cell. Self-assembly of membranes has been demonstrated in protein and lipid units of mitochondria. Conformational changes take place in the proteins as they interact with lipids and with each other. There is increasing evidence that myelin proteins may direct the assembly and the lipid composition of myelin; similar mechanisms may be postulated for synaptic complexes and other highly specialized membranes in the nervous system.

In the rat, incorporation of amino acids into brain proteins is most rapid during the fetal and newborn periods, as has been found both in vivo and in vitro [48]. The rate of protein synthesis in brain slices decreases sharply at the end of the first week; the incorporation of radioactive valine into protein at day 16 is only one-sixth the rate found at day 7 [41]. Determination of the rate of protein synthesis in vivo with radioactive precursors is complicated because the available amino acid pool may change during development (see Chap. 10). If correction is made for the specific activity of the labeled amino acid in the brain, there appear to be two peaks of protein synthesis in the weanling rat, at 5 to 7 days and at 19 days [48]. In addition to decreasing rates of protein synthesis with development, the average half-lives of proteins become greater, indicating decreasing rates of breakdown [56].

LIPIDS

The most rapid increase in lipid content of brain begins after the periods of greatest increase of DNA and protein. Changes in specific lipids have been well documented [57, 72]. Changes in fatty acid chain length and degree of saturation have been determined for several groups of lipids [52, 57].

The rapid increase in lipid content is closely related to the onset of myelination. About 50 percent of all white matter lipid [6], or 30 percent of total brain lipid [63], has been estimated to belong to myelin sheaths in rat brain (see Chap. 11).

Comprehensive analysis of lipids in developing rat brain indicates that they may be divided into four groups on the basis of their period of most rapid change with respect to myelination: some lipids undergo marked shifts before myelination begins; others show marked, moderate, or small increases during myelination (Table 14-2) [72].

The two groups of lipids which change most in concentration during the first 6 postnatal days in rats are sterol esters and gangliosides. Sterol esters are the only lipids measured which decrease during development, falling from a concentration

TABLE 14-2. Changes in Lipids During Development in Rat Brain[a]

Lipid	Percent of Adult Level at 3 Days	Time of Maximal Increase	Localization or Possible Role
A. Sterol esters	?	Prenatal	Possible donor of fatty acids [47]
Gangliosides	27.0	3–12d	Neuronal processes
B. Cerebroside	0.2	6–42d	Myelin primarily
Sulfatide	3.3	6–24d	Myelin primarily
Sphingomyelin	5.6	6–24d	Myelin primarily
Triphosphoinositide	7.5	12–42d	Myelin primarily
Phosphatidic acid	10.0	12–42d	Intermediate in phospholipid metabolism
C. Choline plasmalogen	12.0	18–42d	Myelin primarily
Ethanolamine plasmalogen	17.0	6–24d	Myelin primarily
Cholesterol	26.0	Gradual	Myelin primarily
Phosphatidylserine	34.0	Gradual	All membranes
D. Phosphatidylethanolamine	50.0	Gradual	All membranes
Phosphatidylinositol	55.0	Gradual	All membranes
Phosphatidylcholine	59.0	Gradual	All membranes

[a]Data were recalculated from those of Wells and Dittmer [72]. The lipid content of rat brain was examined at 3, 6, 12, 18, 24, 42, 180, and 330 days after birth. The lipids are grouped according to their sequence of discussion in the text. The sterol esters decreased after birth; no significant data were obtained after 3 days.

of 2 μmoles per gram wet weight at 3 days to less than 5 percent of that value by 6 days. In the rat, this decrease occurs long before myelination begins, so it may be related to cell proliferation or very early differentiation of glial cells. The gangliosides are 27 percent of their adult value on day 3, and increase rapidly to 90 percent of their adult value by day 24. The pattern of gangliosides (see Chap. 11, p. 211, for nomenclature) in rat brain also changes during this period. G_{M1} is high at birth and then falls, while G_{D1a} increases from 30 mole percent to 50 mole percent of the total ganglioside during the first 3 weeks and then decreases to 30 mole percent by adulthood [14]. Similar changes occur in human cortex (Fig. 14-6): G_{M1} is the major ganglioside in the 10-week-old fetus and G_{D1a} rises to peak concentration at 30 weeks [70]. Increases in gangliosides appear to be related temporally to the outgrowth of axons and dendrites (Table 14-3); this association is also in keeping with the localization of much of the ganglioside in nerve processes [14].

A second category of lipids is characterized by their low concentration at 3 days (less than 10 percent of the adult level), followed by dramatic increases between 12 and 18 days. This category includes cerebroside, cerebroside sulfate, sphingo-

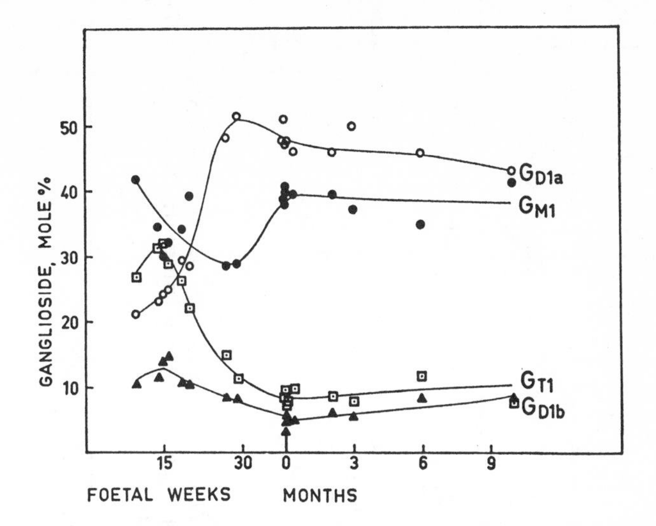

FIGURE 14-6. Ganglioside pattern in human frontal brain cortex during development. Gangliosides were separated by thin-layer chromatography, detected by resorcinol spray, then scraped from the plates and assayed quantitatively, using the resorcinol procedure. (From Vanier et al. [70]. Reprinted by permission; courtesy of Pergamon Press.)

myelin, triphosphoinositide, phosphatidic acid, galactosyldiglyceride, and inositol plasmalogen. The first five lipids are known to be highly enriched in the myelin membrane. The low concentration of these lipids at birth suggests that they are localized in specialized membrane structures which appear in brain at the time myelination starts. Triphosphoinositide and phosphatidic acid differ from the other lipids in this group because they continue to increase sharply after day 24; by that time, the others have reached nearly adult levels.

Lipids in a third category occur at 12 to 34 percent of their adult value at birth and increase during myelination, although not so much as those in the second category. Ethanolamine and choline plasmalogens, cholesterol, phosphatidylserine, ethanolamine phosphorylglycerol ether, and diphosphatidylglycerol are in this group. These lipids must be associated with nonmyelin membranes in the brain of newborn rat. Whether the increases seen during the period of myelination are due to elaboration of nonmyelin membranes or to myelination is not clear. The first three lipids, however, are localized to some extent in the myelin membrane, so at least part of their increase is related to myelination.

Three of the major phospholipids – phosphatidylethanolamine, phosphatidylinositol, and phosphatidylcholine – comprise the fourth category. These lipids are 50 to 59 percent of their adult value at 3 days and increase only moderately in concentration during development. They are known to be ubiquitous components of most membrane structures; their temporal pattern of development is not related to any specific morphological change, but in general reflects the increasingly membranous structure of cells in the brain.

TABLE 14-3. Ganglioside Patterns and Early Cerebral Development[a]

Period	Duration		Cell Growth Characteristics	Ganglioside Concentration	Ganglioside Pattern
	Rat	Human			
I	Until birth	Until 25th fetal week	Multiplication of neurons and glial cells	Moderate increase of gangliosides	G_{M1} and G_{T1} predominant fractions. Onset of G_{D1a} increase
II	0–10 days	25th fetal week to term	Multiplication of glial cells and microneurons. Outgrowth of dendrites and axons. Establishment of neuronal connections	First phase of rapid increase of ganglioside concentration	Maximal increase of G_{D1a}. Decrease of G_{T1} and G_{M1}
III	10–20 days	Birth to 8 months of age	Further extension of neuronal connections. Onset of myelination	Second phase of rapid increase of ganglioside concentration	Stabilized pattern with G_{D1a} as predominant fraction

[a]Ganglioside nomenclature according to Svennerholm, with growth stages as described by McIlwain [44].
Source: From [70]. (By permission of the authors and the *Journal of Neurochemistry*.)

In addition to changes in the types of lipids found in the brain during development, changes in fatty acid composition have been studied for several lipid classes [52]. In general, maturation is accompanied by an increase in long-chain saturated fatty acids. This is due in large part to myelin formation because lipids in the myelin membrane have a preponderance of saturated C_{19} to C_{26} fatty acids, compared with other membranes in brain. Hydroxy fatty acids are found predominantly in the cerebrosides, cerebroside sulfate, and sphingomyelin of myelin. The concentration of these fatty acids increases during myelination. Fatty acid aldehydes are found primarily in phosphatidalethanolamine and phosphatidalserine; these plasmalogens are also enriched in myelin, so fatty acid aldehydes of brain increase with the long-chain saturated fatty acids and the hydroxy fatty acids.

In summary, elevation of the glycolipid and plasmalogen content of brain correlates well with the morphological appearance of myelin, and increases in gangliosides appear to be related to the increasing arborization of neurons. In the rat, the maximal rate of change in ganglioside content takes place at about day 10, while the maximal rate of change in content of cerebroside plus sulfatide occurs about day 20 (Fig. 14-7). These times correspond, respectively, to the middle of the periods characterized by the outgrowth of neuronal processes and by the formation of myelin (see Table 14-1).

METABOLIC CHANGES IN DEVELOPMENT

Overall changes in oxygen consumption and uptake of substrates in developing brain were first measured by determining arteriovenous differences. Utilization of metabolites has most often been investigated in vivo by administration of radioactive precursors, followed by measurement of the appropriate products. Enzymatic activity in vitro has been measured in slices, homogenates, subcellular fractions, and single cells. Obviously, increased enzymatic activity in vitro does not prove that the enzyme is functionally increased in vivo, where its activity may be controlled by substrate availability, turnover, and other factors. In many cases, however, the observed increment in enzymatic activity is accompanied by a parallel increase in the product of the enzymatic reaction.

Many of the enzymatic changes that take place during development can be placed in one of three categories that emphasize their relation to morphological or functional events. These categories include (1) energy metabolism and substrate utilization, (2) metabolism of compounds related to transmission, and (3) myelin formation.

ENERGY METABOLISM

In fetal brain, glucose is metabolized primarily to lactate or pyruvate through the glycolytic pathway [14, 65]; predominance of this anaerobic pathway in the brain of newborn rat permits short-term survival under anoxic conditions. With maturation,

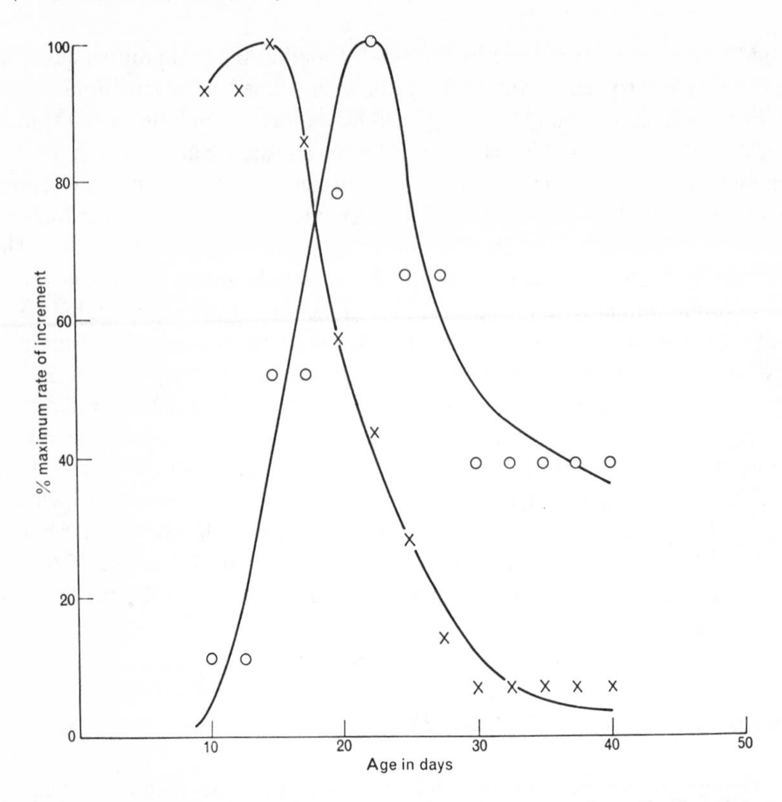

FIGURE 14-7. Rates of change in gangliosides and in cerebrosides plus sulfatides of the developing rat brain [14]. Results have been calculated from the data of Kishimoto, Davies, and Radin [35], and are expressed as (o) mg of galactolipid (cerebroside and sulfatide) or as (x) mg of ganglioside stearate per 2.5 days in whole rat brain. Stearate was found to be the major ganglioside fatty acid (83% to 91% of the total fatty acids) [35]. (From Davison and Dobbing [14]. Reprinted by permission; courtesy of Blackwell Scientific Publications Ltd.)

aerobic metabolism dominates, oxidizing glucose completely to CO_2 with increased energy production [14, 65]. In the rat, the transition occurs shortly after birth, with an accompanying increase in the enzymes of the tricarboxylic acid cycle [65]. For example, succinic dehydrogenase increases sharply in rat brain at about day 7, succinic semialdehyde dehydrogenase is increasing by day 6, and malic dehydrogenase increases a few days later. During the same period, enzymes of the glycolytic pathway, such as lactic dehydrogenase and glyceraldehyde phosphate dehydrogenase, show little change from their levels of activity at birth. One exception appears to be phosphofructoaldolase, which increases threefold in activity from birth to 40 days. Many of these enzymes, especially the dehydrogenases, are well suited for histochemical studies.

In general, a caudal to rostral increase in activity occurs; much detailed information has been gathered on the development of enzyme activities in various brain nuclei, as well as in the layers of cortex and cerebellum [23, 65]. Although glucose is the major source of energy for brain metabolism, ketone bodies (3-hydroxybutyrate and acetoacetate) also serve as major energy sources in rats before weaning [53]. Comparison between suckling and adult rats shows greater utilization of ketone bodies in the young: the arteriovenous differences in ketone bodies are three to four times higher in the young than in the adult. This greater utilization is determined by the higher concentration of ketone bodies in the blood (four to six times higher than in the adult), as well as by the higher levels of the enzymes that metabolize ketone bodies. Two of these enzymes, 3-hydroxybutyrate dehydrogenase and 3-oxoacid-CoA transferase, show a fivefold increase in activity during the first 3 weeks after birth. After rats are weaned (at 3 weeks), the activities fall to about half of their maximal level during the next few weeks and remain at that level into adulthood.

In addition to the tricarboxylic acid cycle, another aerobic pathway for the utilization of glucose is the hexose monophosphate shunt. The significance of this pathway in developing brain is open to question; data from histochemical, in vitro, and in vivo studies are often in disagreement, but immature brain appears to have a greater capacity to metabolize glucose by this pathway than does mature brain [65]. It could provide substrates for lipid synthesis as well as ribose precursors of nucleic acids in developing brain.

As expected from the increase in aerobic metabolism in the developing brain, mitochondria increase in both number and activity [65]. Oxygen consumption, oxidative phosphorylation, electron transport, and protein synthesis in mitochondria all increase during development.

NEURAL TRANSMISSION

Several groups of compounds in brain arise concomitantly with increasing electrical activity and the outgrowth of axons and dendrites. These compounds include acetylcholine, monoamines, certain amino acids, and gangliosides. Some of the compounds, such as acetylcholine, serve as synaptic transmitters. Others, primarily the gangliosides, have been related to transmission on the basis of their localization and time of appearance in the brain. Most of these compounds and the enzymes involved in their metabolism have been found in nerve endings either by histochemical staining or by isolation of nerve endings on density gradients.

Acetylcholine is known to be an excitatory transmitter [46]. Choline acetylase, the enzyme responsible for its synthesis, increases in activity in rat brain during the first few weeks after birth [43], along with the levels of acetylcholine and acetylcholinesterase [19, 43]. Acetylcholinesterase is found in both microsomal and synaptosomal fractions in the developing rat brain; some of this enzyme is localized in the synaptic membrane, degrading acetylcholine released into the synaptic cleft [46]. The amount of acetylcholine isolated in the synaptosomes increases gradually during the first 4 weeks after birth, while the specific activity of acetylcholinesterase increases fourfold between 5 and 15 days in parallel with a sharp increase in synaptosomal protein [1].

Norepinephrine, dopamine, and 5-hydroxytryptamine (serotonin) act as inhibitory transmitters in most cases, although some excitatory effects have been reported [46]. In rat brain, adult levels of 5-hydroxytryptamine are reached by day 25 and of norepinephrine by day 40, while dopamine is about 70 percent of the adult level by day 40 [3]. As expected, the enzymes which synthesize and degrade these amines also show increased activity during this period, as do specific transport systems believed to be responsible for inactivation of these compounds by reuptake into the synapse.

Recent studies provide an elegant correlation of the biochemical and morphological maturation of noradrenergic nerve endings [13]. In rat cortex, synaptic boutons proliferate rapidly just before birth. The number of mitochondria and synaptic vesicles in these endings then increase rapidly, approaching adult levels by day 16. Finally synaptic junctional complexes appear. On a biochemical level, β-dopamine hydroxylase, the last enzyme in the synthetic pathway for norepinephrine, first appears in rat brain at 15 days before birth and rises in activity until day 21. The amount of norepinephrine in rat brain increases fivefold between birth and 28 days, while the quantity isolated in synaptic endings increases 16-fold. Most of the norepinephrine in nerve endings is stored in synaptic vesicles, so the levels of norepinephrine in nerve endings are proportional to the number of storage vesicles. A specific, high-affinity transport system for norepinephrine is first detected at 17 days of gestation and reaches adult levels by day 28. The appearance of this transport system precedes the significant increase of norepinephrine localized in the synaptosomal fraction, indicating that the development of storage vesicles lags behind the development of the membrane mechanism for uptake of norepinephrine at the synapse. The parallel increase in transport and levels of norepinephrine (Fig. 14-8) suggests that uptake is a valid indicator of the outgrowth and maturation of noradrenergic terminals [13].

Of the amino acids, glutamate and aspartate act in an excitatory fashion at synapses, whereas γ-aminobutyrate (GABA) and glycine exhibit postsynaptic inhibition.

With maturation of the brain, increasing amounts of glucose entering the tricarboxylic acid (TCA) cycle are converted to glutamate from α-ketoglutarate by glutamate dehydrogenase [3]. Increased transamination leads to greater synthesis of glutamate from other amino acids, as well as greater conversion of glutamate to aspartate. Some of the glutamate is metabolized to glutamine by glutamine synthetase, some to GABA by decarboxylation. Many of the enzymes that metabolize these amino acids show similar developmental patterns, increasing fourfold from birth to adulthood in rats [69].

Immature brain has low concentrations of glutamate, aspartate, and GABA. These three metabolites increase considerably in rat brain during the first 3 weeks after birth; adult concentrations are reached by the end of this period [3]. For example, glutamate has a concentration of about 5 μmoles per gram of wet weight at birth and about 12 μmoles per gram of wet weight by day 21. This same pattern is repeated in mouse and rabbit brains, in which most of the functional and morphological maturation also takes place after birth. By way of contrast, guinea pig brain

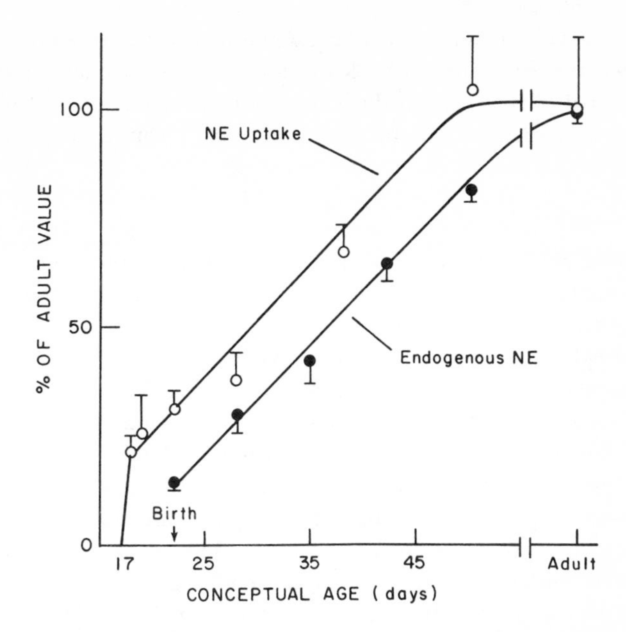

FIGURE 14-8. Correlation between the development of the high-affinity uptake of norepinephrine and the level of endogenous norepinephrine in rat brain at various stages of development [13]. Values are expressed as percentages of the adult value. Initial uptake is expressed in terms of the percentage of the adult V_{max} of transport. Adult norepinephrine: 0.47 ± 0.01 μg/g. Adult V_{max} of uptake: 2.216 ± 0.423 nmoles/g in 5 min. The upper curve represents uptake of norepinephrine by whole-brain homogenates (open circles); the lower curve the content of norepinephrine in whole brain (closed circles). (From Coyle and Axelrod [13]. Reprinted by permission; courtesy of Pergamon Press.)

at birth has adult concentrations of these three compounds, in keeping with its prenatal maturation. While glutamate and aspartate are increasing in concentration, other amino acids, such as glycine and alanine, are present in high concentrations at birth in rat brain but decrease significantly during subsequent development.

Evidence that any of the amino acids are transmitters is less clear than for acetylcholine and the monoamines [46]. Of these amino acids, GABA appears to be the most specifically related to maturation of synapses. Changes in glutamic acid decarboxylase are closely related to changes in its product, GABA. In the cortex of rabbit, increases in glutamic acid decarboxylase coincide with the fastest growth of the surface area of dendrites, while GABA levels correlate most closely with the proportional volume of the dendrites [7].

One enzyme that appears to play a fundamental role in synaptic transmission, and possibly in transport of transmitters, is (Na^+, K^+)-activated ATPase. Some of its activity is localized in nerve endings and increases sharply in rat brain during the first 2 weeks after birth [1, 14].

The localization and developmental changes occurring in gangliosides have been discussed in the section on lipids. Both synthetic and catabolic enzymes for the metabolism of gangliosides have been found in isolated synaptic endings, as well as in other subcellular fractions. The role of these compounds in determining the surface properties of cells or nerve endings is open to question, but available evidence suggests they are associated with receptor sites in the synaptic membranes [14].

MYELINATION

Many of the metabolic changes occurring with myelination probably reflect the increased capacity of oligodendroglial cells to synthesize proteins and lipids for the myelin membrane. Myelination is preceded by the proliferation and differentiation of oligodendroglial cells [2]. Just prior to myelination, these cells show histochemical increases for a number of enzymes, for example, oxidative enzymes, $NADH_2$-tetrazolium reductase, and 5'-nucleotidase [2]. At the same time, increasing numbers of mitochondria appear [2]. In human brain, lipid droplets in oligodendroglial cytoplasm become prominent, although they appear to be absent in other mammalian species [47]. In addition to a general increase in the metabolic activity of oligodendroglial cells, there are histochemical and biochemical elevations of the enzymes which synthesize and degrade myelin lipids. Cerebroside sulfotransferase activity is low until day 9 to 10 in the rat, increases rapidly until day 20, then decreases to 15 to 20 percent of its maximal activity by day 50 [45] (Fig. 14-9). Incorporation of galactose into cerebroside reaches a maximum at day 15 in rat brain in vivo [14]. Cholesterol synthesis and fatty acid elongation increase during this same period [14]. On the degradative side, an arylsulfatase, which probably degrades cerebroside sulfate, reaches its maximal activity about day 15, then decreases to about 70 percent of its maximal level by day 90 [54]. Cerebrosidase increases in activity more gradually; it reaches its maximal activity by day 90 and maintains this level until adulthood [54]. This agrees with histochemical obser-

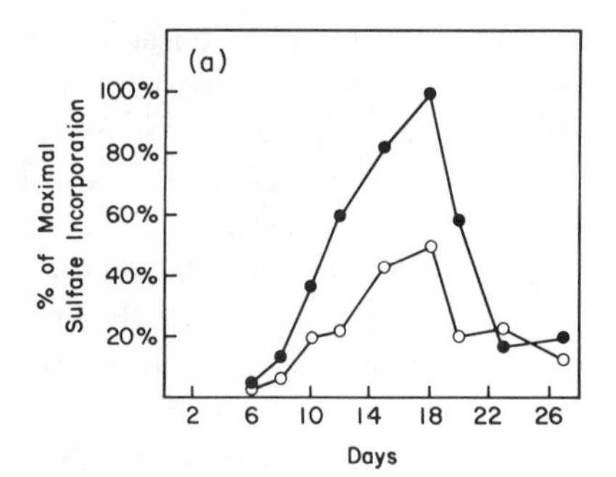

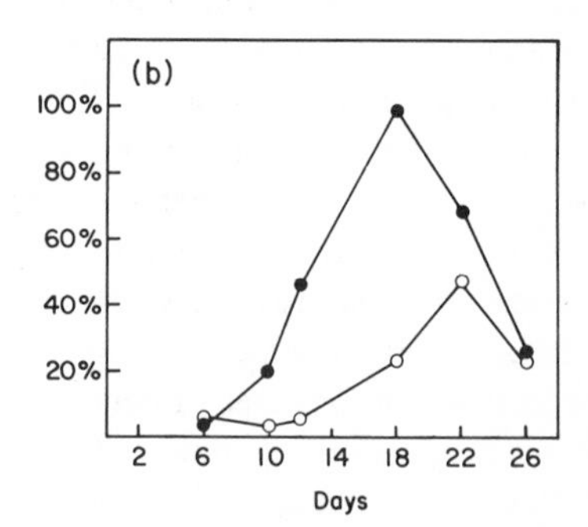

FIGURE 14-9. Incorporation of ^{35}S-sulfate into sulfatide of rat brain during development. (A) Comparison of malnourished animals (open circles) with their normal controls (closed circles) [12]. (B) Comparison of hypothyroid animals (open circles) with their normal controls (closed circles) [71]. Data have been recalculated using the value for control animals at day 18 as 100%.

vations in human brain that unmyelinated white matter does not stain for β-galactosidase (cerebrosidase?), while mature myelin sheaths are brilliantly stained, suggesting localization of this enzyme within the sheath itself or, more likely, in close proximity to it [47]. Esterase activity shows the same developmental changes as does β-galactosidase, in keeping with the disappearance of cholesterol esters from human brain during myelination [47]. (As previously discussed, disappearance of cholesterol esters does not seem to coincide with active myelination in the rat brain.)

Synthesis of individual myelin proteins in developing brain has not yet been studied in detail. However, several in vitro investigations with brain slices show that incorporation of amino acids into total chloroform-methanol soluble protein is maximal at 8 to 10 days. Much of this protein fraction is probably myelin protein [14].

When radioactive precursors of both proteins and lipids are injected into developing rat brain, the radioactivity remains in myelin membrane long after it has disappeared from other subcellular fractions. This indicates the general metabolic stability of the myelin sheath. However, individual lipids and proteins in myelin differ in their half-lives. Also, precursors are incorporated to a small extent into adult myelin, indicating that myelin components metabolize to some extent, even in adult brain. One interesting experiment indicates that membranes formed long before myelination begins are able to contribute lipid to the myelin membrane: radioactive sulfate injected into 4-day-old rats appeared in the sulfatide of microsomal and other membrane fractions [14]. At 10 days, this labeled sulfatide appeared to be transferred to the newly forming myelin membrane.

Another approach to studying the metabolism of the forming myelin membrane involves the use of inhibitors of protein synthesis [8, 61]. When total brain protein synthesis is greatly inhibited, appearance of new protein in myelin is depressed to the same extent. Lipid synthesis is not inhibited for several hours. Also, lipids appear to enter myelin at a normal rate, suggesting that the entrances of lipid and protein are not synchronous events, at least for this short period.

SUMMARY OF DEVELOPMENTAL EVENTS IN BRAIN

The two most prominent events in developing brain are neuronal maturation and myelination. Table 14-4 correlates some of the available morphological, histochemical, and biochemical findings related to these two events.

CONTROL OF GROWTH AND DIFFERENTIATION IN THE NERVOUS SYSTEM: EXPERIMENTAL APPROACHES

The data presented in the preceding sections summarize developmental changes in composition and metabolism of the whole brain, but tell very little about these

TABLE 14-4. Summary of Events in Neuronal and Glial Maturation

A. Neuronal Maturation

Morphology [2, 4]	Events [2, 4]	Histochemistry [2]	Composition Changes	Metabolism
Matrix cell	Mitosis	Nucleus Feulgen +[a]	DNA increases	DNA and RNA synthesis rapid
Neuroblast	Mitosis or migration	Nucleus Feulgen +	DNA increases	DNA and RNA synthesis rapid
Neuron	Differentiation	Nucleus Feulgen –; appearance of Nissl substance	RNA and protein per cell increase	RNA synthesis in nucleus and cytoplasm [56]; active endoplasmic reticulum [56]; protein synthesis rapid [50]
Neuronal processes	Axoplasmic flow	....	Gangliosides increase [70]; lipid per cell increases	....
Synapses	Transmission	Appearance of synaptic endings and vesicles	Gangliosides increase; dicarboxylic acids increase [30]; transmitters increase [43]	Activities of (Na^+, K^+)-ATPase, ACh esterase, other enzymes related to transmitter metabolism increase [14, 43]

[a]Feulgen$^+$ staining indicates presence of DNA in condensed or heterochromatin form.

B. Oligodendroglial Maturation

Morphology [2, 4]	Events [2, 4]	Histochemistry	Composition Changes	Metabolism
Matrix cell	Mitosis	Nucleus Feulgen + [2]	DNA increases	DNA and RNA synthesis rapid
Spongioblast	Mitosis or migration	....	....	DNA and RNA synthesis rapid
Oligodendrocyte	Differentiation	Large, pale nuclei; prominent nucleoli [47]	....	....
	Hyperplasia	Lipid droplets [47]; increase in mitochondria [2]; neutral lipids increase [47]	Increase in cholesterol esters	Carbonic anhydrase and pseudo-cholinesterase increase [2]
Myelin sheath	Myelination	Nucleus decreases in size and nucleolus disappears [47]; increase in 5′-nucleotidase and oxidative enzymes [2]	Decrease in cholesterol esters [47]; appearance of myelin proteins; increase in sulfatide, cerebroside, and plasmalogens [14]	Increase in cerebroside sulfotransferase activity [45]

changes at a cellular level or the control of these changes. During the period of early embryogenesis, the brain develops from a ridge of ectodermal cells into a complex organ with numerous cell types. The subcellular mechanisms of this development have not been clearly defined in any mammalian system, much less in one so complex as the nervous system.

Some approaches to dissecting the origin and development of various cell types have been made. Even less is known about the factors that control the sequence of appearance of cell types and the selective vulnerability of cell populations to internal enzymatic defects or external insults at critical stages of development. In most cells, acquisition of specific and unique characteristics occurs after rapid proliferation ceases. This suggests that DNA replication and differentiation cannot take place simultaneously, or that some event related to rapid proliferation triggers differentiation [17].

It has been suggested that transcriptional control, or selective gene activation, directs and maintains differentiation [67], leading to the development of cell-specific proteins, surface membranes, and morphology. After differentiation, translational control and subsequent interaction of proteins or their subunits would dominate in regulating the cell.

CELL PROLIFERATION AND DIFFERENTIATION

Control of DNA replication in mammalian systems has been studied primarily in early embryonic or regenerating systems. Until more is known about the molecular events at each stage of the mitotic cell cycle, the mechanism for programming the onset and duration of proliferation for a given population of cells must remain a matter of speculation. Rapidly dividing cells are extremely vulnerable to damage by such diverse factors as malnutrition, hypoxia, and viral attack [4].

At present the cerebellum offers the most advantages for studies of neurogenesis in the central nervous system because of its well-defined geometry and the detailed information available on its structure and function [18]. Neurogenesis in the peripheral nervous system and the transition of glial precursors into formed glial elements are less well explored [4].

The division and migration of cells in the developing cerebellum have been studied by injecting ^{3}H-thymidine and then following the labeled cells by autoradiography [4]. Those cells undergoing DNA replication at the time of injection will incorporate ^{3}H-thymidine; the label will be diluted in those cells which continue to divide. At some point, depending on species and brain region, neuroblasts are formed from primitive stem cells and cease mitotic division. Those neuroblasts undergoing "differentiating mitosis" at the time of injection of ^{3}H-thymidine will remain heavily labeled throughout the life of the animal [18].

It is not known why neuroblasts lose their ability to divide. The neuroblast may irreversibly lose some of its DNA either by deletion or by permanent repression [16]. On the other hand, two lines of study indicate that a cytoplasmic factor may be responsible for DNA stability after neurons have differentiated. Nuclei from cells of adult amphibian brain transplanted to oocytes begin to divide, although at a somewhat lower rate than do nuclei from less mature stages [28]. Fusion of adult neurons with an established fibroblast line results in reactivation of DNA synthesis in the neuronal nuclei [32].

Differentiation involves activation of a particular series of genes in a sequence unique to that developing cell type. Gene activation has been postulated to occur by a number of mechanisms, among them the unmasking of DNA by release of histones, the release of a repressor, the appearance of an activator, or an extracellular stimulus. It is impossible to dissect these events in a complex system of heterogeneous cell types at various stages of development. Organ and tissue culture of specific brain regions and cell types may permit finer biochemical dissection of events leading to differentiation. Gliomas and neuroblastomas have been established as clonal lines in culture; whether differentiation occurs remains to be determined [51]. Differentiation of neurons and glia, formation of synapses, and myelination have been observed in organ culture of spinal cord, cerebellum, and cerebral cortex [49]. Organ cultures from spinal cord may be held in a premyelinating state for long periods of time in the presence of low levels of antibody to myelin basic protein. When the antibody is washed out, oligodendrocytes appear and myelination occurs [10]. This system may be useful in determining the critical events controlling one aspect of differentiation in the nervous system.

Finally, many mutations in mice show modifications of neural development ranging from induction of the nervous system, to neural tube formation and differentiation, to differentiation of specific regions [24]. The genetic linkages of most of these mutants are known, but in many cases the pathological and biochemical abnormalities are not [60]. Comparison of the biochemistry of normal and mutant mice promises to answer many questions about the development of the nervous system.

CELLULAR INTERACTIONS

Cells interact continuously at every stage of development, and such interaction is being studied by a variety of experimental approaches.

Jacobson has studied retinal-brain connections in amphibians by mapping the relationship of retinal cells to tectal cells using electrophysiological techniques [33]. Specification of the gradients that determine the connections takes place after DNA synthesis has ceased and precedes fiber outgrowth in retinal cells. He concluded that "each axon in the optic nerve probably has a unique characteristic which is related to its function and to the position of its ganglion cell in the retina, and which specifies the connections which it makes in the optic tectum."

Another example of cellular interactions may be the triggering of myelin formation by contact between the axon and the oligodendroglial cell [11]. Antigenically similar basic proteins in the myelin sheath and in the region around the nucleus of neurons have been demonstrated in both peripheral and central nervous systems of guinea pig [37]. This may be one of the first clues to the mechanism of recognition and subsequent triggering of myelination, but developmental studies have not yet been done.

EXTRACELLULAR FACTORS

Perhaps the best example of a substance known to have a direct and specific effect on nervous system maturation is nerve growth factor (NGF). This protein may be required for survival and maturation of sympathetic and sensory ganglia, as demonstrated by the devastating effects of antibody to NGF on the development of the sympathetic nervous system [39]. NGF has been extensively purified and characterized [59]. It stimulates outgrowth of fibers from embryonic chick dorsal root ganglia, with an accompanying increase in the incorporation of precursors into nucleic acids, proteins, gangliosides, and other lipids [25, 39]. Inhibition of protein and RNA synthesis prevents these effects.

Although NGF stimulates the general metabolism of dorsal root ganglia, two observations point to effects on specific components of neurons. First, incorporation of radioactive precursors into gangliosides increases about 30 percent per milligram of protein in dorsal root ganglia incubated with NGF [25]. Also, in superior cervical ganglia of young rats, NGF produces a threefold to fivefold increase in the specific activities of two enzymes localized in noradrenergic nerve endings: tyrosine hydroxylase and dopamine β-hydroxylase [66].

NGF stimulates fiber outgrowth in dorsal root ganglia from chick embryos only during 8 to 11 days of incubation, which suggests a critical period for the factor's action [59]. Antisera to NGF injected into newborn rats destroys neurons in the sympathetic ganglia [39]. In adult animals, the neurons atrophy and neurofibrillar material disappears [5]. Norepinephrine uptake is inhibited, indicating localization of the damage at the nerve endings; these changes are reversible when the antisera is removed. These observations again indicate that there is a period during which the neuron is highly sensitive to NGF.

During specific critical periods of development, hormones may act directly on the central nervous system to produce permanent changes. The same hormone may have different actions at different times; the time at which it is most effective will depend on the sequence of developmental events in the species under study. Most hormones produce changes in the rat brain only when administered in the prenatal or immediate postnatal period [40]. Different regions or cell types may be affected preferentially. For example, testosterone administered to newborn rats apparently acts on the hypothalamus to cause male sexual behavior at maturity, while its absence leads to female sexual behavior [40].

A number of hormones – among them corticosterone, thyroid hormone, growth hormone, and insulin – have more widespread, less specific, effects on the developing nervous system. For example, thyroid hormone (thyroxin) affects the maturation and metabolism of many mammalian tissues [62]. Its presence is necessary for the maturation of brain [62]; however, once developed, the brain no longer responds to thyroid hormone. Administration of thyroid hormone to newborn rats results in accelerated physical and behavioral maturation, along with increased synthesis of myelin lipids. Deprivation of thyroid hormone can be accomplished by thyroidectomy or by the administration of ^{131}I or of thyroid depressants. When rats are thyroidectomized at birth, the density of axons and dendrites decreases and, concomitantly, the number of interneuronal connections decreases. Neuronal bodies are decreased in number and are smaller in size, and myelination is retarded. The accompanying electrophysiological and behavioral disturbances in development have also been characterized.

One objection to these experiments is that animals deprived of thyroid hormone show significant losses in both body and brain weights, so that any changes observed may be due to general weight loss rather than to a specific effect of the lack of hormone. For example, both malnutrition [12] and the administration of a thyroid depressant (methimazole) [71] result in decreased incorporation of radioactive sulfate into sulfatide, primarily a myelin lipid. With malnutrition, the enzyme which synthesizes sulfatide shows decreased activity when assayed in vitro. In hypothyroid animals, this enzyme has normal activity in vitro, but the formation of PAPS, the compound which donates its high energy sulfate to sulfatide, is considerably reduced. The developmental curves for sulfatide labeling in these two conditions are different (see Fig. 14-9). With malnutrition, incorporation is decreased throughout the first 3 weeks, but the peak occurs at day 18, as it does in control animals. With deprivation of thyroid hormone, the peak is delayed by several days, indicating possible retardation of development.

The mechanism of action of thyroid hormone in the developing rat brain has been studied extensively by Sokoloff [62]. Immature rat brain, both in vivo and in vitro, responds to thyroid hormone with an increase in protein synthesis, followed by a general increase in metabolic rate. The immediate stimulation of protein synthesis is directly related to the level of thyroid hormone in the tissue and precedes any changes in RNA. The stimulation in a cell-free system in vitro depends upon the presence of mitochondria from immature brain. This observation led to the suggestion that these mitochondria "contain a functional site which is the locus of the primary action of thyroid hormones" [62]. This site is then presumably lost in mitochondria from mature brain and with it, the ability of the tissue to respond to thyroid hormone.

Another factor which affects the development of the nervous system is the nutritional status of the animal. Winick [73] has detailed the effects of malnutrition on the developing brain. (This subject is discussed in Chap. 25.)

ACKNOWLEDGMENTS

This report represents activities supported by funds from The John A. Hartford Foundation, Inc., and from the National Institute of Neurological Diseases and Stroke.

REFERENCES

BOOKS

Davison, A. N., and Dobbing, J. (Eds.). *Applied Neurochemistry.* Philadelphia: Davis, 1968.

Davison, A. N., and Peters, A. *Myelination.* Springfield, Ill.: Thomas, 1970.

Himwich, W. A. (Ed.). *Developmental Neurobiology.* Springfield, Ill.: Thomas, 1970.

Himwich, H. E., and Himwich, W. A. (Eds.). *Developing Brain.* Springfield, Ill.: Thomas, 1968.

Jacobson, M. *Developmental Neurobiology.* New York: Holt, Rinehart and Winston, 1970.

Minkowski, A. (Ed.). *Regional Development of the Brain in Early Life.* Oxford: Blackwell, 1967.

Pease, D. C. (Ed.). *Cellular Aspects of Neural Growth and Differentiation.* Los Angeles: University of California Press, 1971.

Schmitt, F. O. (Ed.). *The Neurosciences: Second Study Program.* New York: Rockefeller University Press, 1970.

Waelsch, H. (Ed.). *Biochemistry of the Developing Nervous System.* New York: Academic, 1955.

ORIGINAL SOURCES AND REVIEWS

1. Abdel-Latif, A. A., Smith, J. P., and Ellington, E. P. Subcellular distribution of sodium-potassium adenosine triphosphatase, acetylcholine and acetylcholinesterase in developing rat brain. *Brain Res.* 18:441, 1970.
2. Adams, C. W. M. *Neurohistochemistry.* New York: Elsevier, 1965.
3. Agrawal, H. C., and Himwich, W. A. Amino Acids, Proteins and Monoamines of Developing Brain. In Himwich, W. A. (Ed.), *Developmental Neurobiology.* Springfield, Ill.: Thomas, 1970.
4. Altman, J. DNA metabolism and cell proliferation. In Lajtha, A. (Ed.), *Handbook of Neurochemistry.* Vol. 2. New York: Plenum, 1969.
5. Angeletti, P. V., Levi-Montalcini, R., and Caramia, F. Analysis of the effects of the antiserum to the nerve growth factor in adult mice. *Brain Res.* 27:343, 1971.
6. Autilio, L. A., Norton, W. T., and Terry, R. D. The preparation and properties of purified myelin from the central nervous system. *J. Neurochem.* 11:17, 1964.
7. Baxter, C. F. The Nature of γ-Aminobutyric Acid. In Lajtha, A. (Ed.), *Handbook of Neurochemistry.* Vol. 3. New York: Plenum, 1970.
8. Benjamins, J. A., Herschkowitz, N., Robinson, J., and McKhann, G. M. The effects of inhibitors of protein synthesis on incorporation of lipids into myelin. *J. Neurochem.* 18:729, 1971.
9. Bogoch, S. Proteins. In Lajtha, A. (Ed.), *Handbook of Neurochemistry.* Vol. 1. New York: Plenum, 1969.

10. Bornstein, M. B., and Raine, C. S. Experimental allergic encephalomyelitis. Antiserum inhibition of myelination *in vitro*. *Lab. Invest.* 23:536, 1970.
11. Bunge, M. B., Bunge, R. P., and Ris, H. Ultrastructural study of remyelination in an experimental lesion in adult cat. *J. Biophys. Biochem. Cytol.* 10:67, 1961.
12. Chase, H. P., Dorsey, J., and McKhann, G. M. The effects of malnutrition on the synthesis of a myelin lipid. *Pediatrics* 40:551, 1967.
13. Coyle, J. T., and Axelrod, J. Development of the uptake and storage of L-[^{3}H] norepinephrine in the rat brain. *J. Neurochem.* 18:2061, 1971.
14. Davison, A. N., and Dobbing, J. The Developing Brain. In Davison, A. N., and Dobbing, J. (Eds.), *Applied Neurochemistry*. Philadelphia: Davis, 1968.
15. Dobbing, J., and Sands, J. Timing of neuroblast multiplication in developing human brain. *Nature (Lond.)* 226:639, 1970.
16. Eayrs, J. T., and Goodhead, B. Postnatal development of the cerebral cortex in the rat. *J. Anat.* 93:385, 1959.
17. Ebert, J. D., and Kaighn, M. E. The Keys to Change: Factors Regulating Differentiation. In Locke, M. (Ed.), *Major Problems in Developmental Biology*. New York: Academic, 1966.
18. Eccles, J. C. Neurogenesis and morphogenesis in the cerebellar cortex. *Proc. Natl. Acad. Sci. U.S.A.* 66:294, 1970.
19. Elkes, J., and Todrick, A. On the Development of Cholinesterases in the Rat Brain. In Waelsch, H. (Ed.), *Biochemistry of the Developing Nervous System*. New York: Academic, 1955.
20. Eng, L. F., Chao, F.-C., Gerstl, B., Pratt, D., and Tavaststjerna, M. G. The maturation of human white matter myelin. *Biochemistry* 12:4455, 1968.
21. Fish, I., and Winick, M. Effect of malnutrition on regional growth of the developing rat brain. *Exp. Neurol.* 25:534, 1969.
22. Flexner, L. B. Lactic dehydrogenases of the developing cerebral cortex and liver of the mouse and guinea pig. *Develop. Biol.* 2:313, 1960.
23. Friede, R. L. *Topographic Brain Chemistry*. New York: Academic, 1966.
24. Gluecksohn-Waelsch, S. Genetic Factors and the Development of the Nervous System. In Waelsch, H. (Ed.), *Biochemistry of the Developing Nervous System*. New York: Academic, 1955.
25. Graves, M., Varon, S., and McKhann, G. M. The effect of nerve growth factor (NGF) on the synthesis of gangliosides. *J. Neurochem.* 16:1533, 1969.
26. Greenfield, S., Norton, W. T., and Morrell, P. Quaking mouse: Isolation and characterization of myelin protein. *J. Neurochem.* 18:2119, 1971.
27. Grossfeld, R. M. The Extraction and Fractionation of Mouse Brain Proteins. Ph.D. thesis, Stanford University, 1968.
28. Gurdon, J. B. Transplanted nuclei and cell differentiation. *Sci. Amer.* 161:24, 1968; Changes in somatic cell nuclei inserted into growing and maturing amphibian oocytes. *J. Embryol. Exptl. Morph.* 20:401, 1968.
29. Himwich, W. Appendix. In Lajtha, A. (Ed.), *Handbook of Neurochemistry*. Vol. I. New York: Plenum, 1969.

30. Himwich, W. A. Biochemistry and Neurophysiology of the Brain in the Neonatal Period. In Pfeiffer, C. C., and Smythies, J. R. (Eds.), *International Review of Neurobiology*. Vol. 4. New York: Academic, 1962.
31. Howard, E. DNA, RNA, and cholesterol increases in cerebrum and cerebellum during development of human fetus. *Brain Res.* 14:697, 1969.
32. Jacobson, C.-O. Reactivation of DNA synthesis in mammalian neuron nuclei after fusion with cells of an undifferentiated fibroblast line. *Exp. Cell Res.* 53:316, 1968.
33. Jacobson, M. Development of specific neuronal connections. *Science* 163:543, 1969.
34. Jacobson, S. Sequence of myelinization in the brain of the albino rat. A. Cerebral cortex, thalamus and related structures. *J. Comp. Neurol.* 121:5, 1963.
35. Kishimoto, Y., Davies, W. E., and Radin, N. S. Developing rat brain: Changes in cholesterol, galactolipids, and the individual fatty acids of gangliosides and glycerophosphatides. *J. Lipid Res.* 6:532, 1965.
36. Kobayashi, T., Inman, O. R., Buno, W., and Himwich, H. E. Neurohistological studies of developing mouse brain. *Progr. Brain Res.* 9:87, 1964.
37. Kornguth, S. E., Anderson, J. W., and Scott, G. Temporal relationship between myelinogenesis and the appearance of a basic protein in the spinal cord of the white rat. *J. Comp. Neurol.* 127:1, 1966.
38. Lebherz, G., and Rutter, W. J. Distribution of fructose diphosphate aldolase variants in biological systems. *Biochemistry* 8:109, 1969.
39. Levi-Montalcini, R., and Angelletti, P. U. Nerve growth factor. *Physiol. Rev.* 48:534, 1968.
40. Levine, S., and Mullins, R. F. Hormonal influences on brain organization in infant rats. *Science* 152:1585, 1966.
41. Lipmann, F. Effect of Electrical and Chemical Stimulation on Protein Synthesis in Brain Slices. In Lajtha, A. (Ed.), *Protein Metabolism in the Nervous System.* New York: Plenum, 1970.
42. Mandel, P., Rein, H., Harth-Edel, S., and Mardell, R. Distribution and Metabolism of Ribonucleic Acid in the Vertebrate Central Nervous System. In Richter, D. (Ed.), *Comparative Neurochemistry.* New York: Macmillan, 1964.
43. McCaman, R. E., and Aprison, M. H. The Synthetic and Catabolic Enzyme Systems for Acetylcholine and Serotonin in Several Discrete Areas of Developing Rat Brain. In Himwich, W. A., and Himwich, H. E. (Eds.), *The Developing Brain.* New York: Elsevier, 1964.
44. McIlwain, H. *Biochemistry and the Central Nervous System* (3rd ed.). Boston: Little, Brown, 1966, p. 270.
45. McKhann, G. M., and Ho, W. The *in vivo* and *in vitro* synthesis of sulfatides during development. *J. Neurochem.* 14:717, 1967.
46. McLennan, H. *Synaptic Transmission.* Philadelphia: Saunders, 1970, pp. 78 and 106.
47. Mickel, H. S., and Gilles, F. H. Changes in glial cells during human telencephalic myelinogenesis. *Brain* 93:337, 1970.

48. Miller, S. A. Protein Metabolism During Growth and Development. In Munro, H. N. (Ed.), *Mammalian Protein Synthesis.* Vol. 3. New York: Academic, 1969.
49. Murray, M. R. Nervous Tissues *in Vitro.* In Willmer, E. M. (Ed.), *Cells and Tissues in Culture.* New York: Academic, 1965.
50. Murthy, M. R. V. Membrane-bound and Free Ribosomes in the Developing Rat Brain. In Lajtha, A. (Ed.), *Protein Metabolism of the Nervous System.* New York: Plenum, 1970.
51. Nelson, P., Ruffner, W., and Nirenberg, M. Neuronal tumor cells with excitable membranes grown *in vitro. Proc. Natl. Acad. Sci. U.S.A.* 64:1004, 1969.
52. O'Brien, J. S. Lipids and Myelination. In Himwich, W. A. (Ed.), *Developmental Neurobiology.* Springfield, Ill.: Thomas, 1970.
53. Page, M. A., Krebs, H. A., and Williamson, D. H. Activities of enzymes of ketone-body utilization in brain and other tissues of suckling rats. *Biochem. J.* 121:49, 1971.
54. Radin, N. S. Brain Cerebroside Metabolism and Possible Implications for Clinical Problems. In Bernsohn, J., and Grossman, H. J. (Eds.), *Lipid Storage Diseases.* New York: Academic, 1971.
55. Rappoport, D. A., Fritz, R. R., and Myers, J. L. Nucleic Acids. In Lajtha, A. (Ed.), *Handbook of Neurochemistry.* Vol. 1. New York: Plenum, 1969.
56. Roberts, S., Zomzely, C. E., and Bondy, S. C. Protein Synthesis in the Nervous System. In Lajtha, A. (Ed.), *Protein Metabolism of the Nervous System.* New York: Plenum, 1970.
57. Rouser, G., and Yamamoto, A. Lipids. In Lajtha, A. (Ed.), *Handbook of Neurochemistry.* Vol. 1. New York: Plenum, 1969.
58. Seiler, N. Enzymes. In Lajtha, A. (Ed.), *Handbook of Neurochemistry.* Vol. 1. New York: Plenum, 1969.
59. Shooter, E. M. Some Aspects of Gene Expression in the Nervous System. In Schmitt, F. O. (Ed.), *The Neurosciences: Second Study Program.* New York: Rockefeller University Press, 1970.
60. Sidman, R. L., Green, M. C., and Appel, S. H. *Catalog of the Neurological Mutants of the Mouse.* Cambridge, Mass.: Harvard University Press, 1965.
61. Smith, M. E., and Hasinoff, C. M. Biosynthesis of myelin proteins *in vitro. J. Neurochem.* 18:739, 1971.
62. Sokoloff, L. The Mechanism of Action of Thyroid Hormones on Protein Synthesis and Its Relationship to the Differences in Sensitivities of Mature and Immature Brain. In Lajtha, A. (Ed.), *Protein Metabolism of the Nervous System.* New York: Plenum, 1970.
63. Sperry, W. M. Biochemistry of Brain During Development. In Elliott, K. A. C., Page, I. H., and Quastel, J. H. (Eds.), *Neurochemistry.* Springfield, Ill.: Thomas, 1962.
64. Sung, S. DNA synthesis in developing rat brain. *Can. J. Biochem.* 47:47, 1969.

65. Swaiman, K. F. Energy and Electrolyte Changes During Maturation. In Himwich, W. A. (Ed.), *Developmental Neurobiology.* Springfield, Ill.: Thomas, 1970.
66. Thoenen, H., Angeletti, P. U., Levi-Montalcini, R., and Kettler, R. Selective induction by NGF of tyrosine hydroxylase and dopamine-β-hydroxylase in the rat superior cervical ganglia. *Proc. Natl. Acad. Sci. U.S.A.* 65:1598, 1971.
67. Tompkins, G. Control of specific gene expression in higher organisms. *Science* 166:1474, 1969.
68. Uzman, L. L., and Rumley, M. K. Changes in the composition of developing mouse brain during early myelination. *J. Neurochem.* 3:171, 1958.
69. Van den Berg, C. J. Glutamate and glutamine. In Lajtha, A. (Ed.), *Handbook of Neurochemistry.* Vol. 3. New York: Plenum, 1970.
70. Vanier, M. T., Holm, M., Öhman, R., and Svennerholm, L. Developmental profiles of gangliosides in human and rat brain. *J. Neurochem.* 18:581, 1971.
71. Walravens, P., and Chase, H. P. Influence of thyroid on formation of myelin lipids. *J. Neurochem.* 16:1447, 1969.
72. Wells, M. A., and Dittmer, J. C. A comprehensive study of the postnatal changes in the concentration of the lipids of developing rat brain. *Biochemistry* 6:3169, 1967.
73. Winick, M., Rosso, P., and Waterlow, J. Cellular growth of cerebrum, cerebellum and brain stem in normal and marasmic children. *Exp. Neurol.* 26:393, 1970.
74. Yakovlev, P. I., and Lecours, A.-R. The Myelogenetic Cycles of Regional Maturation of the Brain. In Minkowski, A. (Ed.), *Regional Development of the Brain.* Oxford: Blackwell, 1967.
75. Yamagami, S., and Mori, K. Changes in polysomes of developing rat brain. *J. Neurochem.* 17:721, 1970.
76. Zamenhof, S., Van Martheus, E., and Margolis, F. L. DNA (cell number) and protein in neonatal brain: Alteration by maternal dietary protein restriction. *Science* 160:322, 1968.

15

Circulation and Energy Metabolism of the Brain

Louis Sokoloff

THE BIOCHEMICAL PATHWAYS of energy metabolism in the brain are, in most respects, similar to those of other tissues. Special conditions that are peculiar to the central nervous system in vivo limit quantitative expression of its biochemical potentialities. In no tissue are the discrepancies between in situ and in vitro results greater, or the extrapolations from in vitro data to conclusions about in vivo metabolic functions more hazardous. Valid identification of the normally used substrates and the products of cerebral energy metabolism, as well as reliable estimation of their rates of utilization and production, can be obtained only in the intact animal; in vitro studies serve to identify pathways of intermediary metabolism, mechanisms, and potential rather than actual performance.

In addition to the usual causes of differences between in vitro and in vivo studies that pertain to all tissues, there are two unique conditions that pertain only to the central nervous system.

First, in contrast to cells of other tissues, nerve cells do not function autonomously. They are generally so incorporated into a complex neural network that their functional activity is integrated with that of various other parts of the central nervous system and, indeed, with most somatic tissues as well. It is obvious, then, that any procedure which interrupts the structural and functional integrity of the network or isolates the tissue from its normal functional interrelationships would inevitably grossly alter – at least quantitatively and, perhaps, even qualitatively – its normal metabolic behavior.

The second is the phenomenon of the blood-brain barrier, which selectively limits the rates of transfer of soluble substances between blood and the brain. This barrier, which probably developed phylogenetically to protect the brain against noxious substances, serves also to discriminate among various potential substrates for cerebral metabolism. The substrate function is confined to those compounds in the blood which not only are suitable substrates for cerebral enzymes but which also can penetrate from blood to the brain at rates adequate to support the brain's considerable energy demands. Substances which can be readily oxidized by brain slices, minces, or homogenates in vitro and which are effectively utilized in vivo when formed endogenously within the brain are often incapable of supporting cerebral energy metabolism and function when present in the blood because of restricted passage through the blood-brain barrier. The in vitro techniques establish only the existence and potential capacity of the enzyme systems required for the utilization of a given substrate; they do not define the extent to which such a pathway is actually utilized in vivo. This can be done only by studies in the intact animal, and it is this aspect of cerebral metabolism with which this chapter is concerned.

STUDIES OF CEREBRAL METABOLISM IN VIVO

A variety of suitable methods have been used to study the metabolism of the brain in vivo; these vary in complexity and in the degree to which they yield quantitative results. Some require such minimal operative procedures on the experimental animal that no anesthesia is required, and there is no interference with the tissue except for the effects of the particular experimental condition being studied. Some of these techniques are applicable to normal, conscious human subjects, and consecutive and comparative studies can be made repeatedly in the same subject. Other methods are more traumatic and either require killing the animal or involve such extensive surgical intervention and tissue damage that the experiments approach an in vitro experiment carried out in situ. All, however, are capable of providing specific and relevant information.

BEHAVIORAL PHYSIOLOGICAL AND CHEMICAL CORRELATIONS

The simplest way to study the metabolism of the central nervous system in vivo is to correlate spontaneous or experimentally produced alterations in the chemical composition of the blood, spinal fluid, or both, with changes in cerebral physiological functions or gross CNS-mediated behavior. The level of consciousness, reflex behavior, or the electroencephalogram are generally used to monitor the effects of the chemical changes on the functional and metabolic activities of the brain. For example, such methods first demonstrated the need for glucose as a substrate for cerebral energy metabolism; hypoglycemia produced by insulin or other means altered various parameters of cerebral function that could not be restored to normal by the administration of substances other than glucose.

The chief virtue of these methods is their simplicity, but they are gross and nonspecific and do not distinguish between direct effects of the agent on cerebral metabolism and those secondary to changes produced initially in somatic tissues. Also, negative results are often inconclusive, for there always remain questions of insufficient dosage, inadequate cerebral circulation and delivery to the tissues, or impermeability of the blood-brain barrier.

BIOCHEMICAL ANALYSES OF TISSUES

The availability of analytical chemical techniques makes it possible to measure specific metabolites and enzyme activities in brain tissue at selected times during or after exposure of the animal to an experimental condition. (These techniques are described in Chap. 20.) This approach has been very useful in studies of the intermediary metabolism of the brain. It has permitted the estimation of the rates of flux through the various steps of established metabolic pathways and the identification of control points in the pathways where regulation may be exerted. Such studies have helped to define more precisely the changes in energy metabolism associated with altered cerebral functions produced, for example, by anesthesia, convulsions, or hypoglycemia. Although these methods require killing the animal and analyzing tissue samples, they are in vivo methods in effect since they attempt to describe the state of the tissue while it is still in the animal at the moment of sacrifice. These methods have encountered their most serious problems regarding this point. Postmortem changes in brain are extremely rapid and are not always completely retarded by the most rapid freezing techniques available. Nevertheless, these methods have proved to be very valuable, particularly in the area of energy metabolism. (A detailed discussion of results obtained with these methods is presented in Chap. 9.)

RADIOISOTOPE INCORPORATION

The technique of administering radioactive precursors followed by the chemical separation and assay of products in the tissue has added greatly to the armamentarium for studying cerebral metabolism in vivo. Labeled precursors are administered by any one of a variety of routes; at selected later times the brain is removed and the precursor and the various products of interest are isolated. The radioactivity and quantity of the compounds in question are assayed. Such techniques facilitate the identification of metabolic routes and the rates of flux through various steps of the pathway. In some cases the comparison of the specific activities of the products and precursors has led to the surprising finding of higher specific activities in the products than in the precursors. This is conclusive evidence of the presence of compartmentation [20]. These methods have been used effectively in studies of amine and neurotransmitter synthesis and metabolism, lipid metabolism, protein synthesis, amino acid metabolism, and the distribution of glucose carbon through the various biochemical pathways present in the brain.

The methods are particularly valuable for studies of intermediary metabolism that generally are not feasible by most other in vivo techniques. They are without equal for the qualitative identification of the pathways and routes of metabolism. They suffer, however, from a disadvantage; only one set of measurements per animal is possible because the animal must be killed. Quantitative interpretations are often confounded by the problems of compartmentation. Also, they are all too frequently misused; unfortunately, quantitative conclusions are often drawn on the basis of radioactivity data without appropriate consideration of the specific activities of the precursor pools.

POLAROGRAPHIC TECHNIQUES

The oxygen electrode has been employed for measuring the amount of oxygen consumed locally in the cerebral cortex in vivo [22]. The electrode is applied to the surface of the exposed cortex, and the local pO_2 is measured continuously before and during occlusion of the blood flow to the local area. During occlusion the pO_2 falls linearly as the oxygen is consumed by the tissue metabolism, and the rate of fall is a measure of the rate of oxygen consumption locally in the cortex. Repeated measurements can be made successively in the animal, and the technique has been used to demonstrate the increased oxygen consumption of the cerebral cortex and the relation between the changes in the EEG and the metabolic rate during convulsions [22]. The technique is limited to measurements in the cortex and, of course, to oxygen utilization.

ARTERIOVENOUS DIFFERENCES

The primary function of circulation is to replenish the nutrients consumed by the tissues and to remove the products of their metabolism. This function is reflected in the composition of the blood traversing the tissue. Substances taken up by the tissue from the blood are higher in concentration in the arterial inflow than in the venous outflow, and the converse is true for substances released by the tissues. The convention is to subtract the venous concentration from the arterial concentration so that a positive arteriovenous difference represents net uptake and a negative difference means net release. In nonsteady states there may be transient, but significant, arteriovenous differences which reflect reequilibration of the tissue with the blood. In steady states, in which it is presumed that the tissue concentration remains constant, positive and negative arteriovenous differences mean net consumption or production of the substance by the tissue. Zero arteriovenous differences indicate neither. This method is useful for all substances in blood that can be assayed with enough accuracy, precision, and sensitivity to enable the detection of arteriovenous differences. The method is useful only for the tissues from which mixed representative venous blood can be sampled. Arterial blood has essentially the same composition throughout and can be sampled from any artery. On the other hand, venous blood is specific for each tissue, and to establish valid arteriovenous differences the venous blood must represent the total outflow or the flow-weighted average of all the venous outflows from the tissue under study, uncontaminated by blood from any other tissue. It is not possible to fulfill this condition for many tissues.

The method is fully applicable to the brain, particularly in man, in whom the anatomy of venous drainage is favorable for such studies. Representative cerebral venous blood, with no more than about 3 percent contamination with extracerebral blood, is readily obtained from the superior bulb of the internal jugular vein. The venipuncture can be made percutaneously under local anesthesia [33, 43] so the measurements can be made during the conscious state, undistorted by the effects of general anesthesia. The monkey is similar, although the vein must be surgically exposed before puncture. Other common laboratory animals are less suitable because extensive communications between cerebral and extracerebral venous beds are present, and uncontaminated, representative venous blood is difficult to obtain from the cerebrum without major surgical intervention. In these cases, one can obtain relatively uncontaminated venous blood from the torcular Herophili even though it does not contain representative blood from the brain stem and some of the lower portions of the brain.

The chief advantages of these methods are their simplicity and applicability to unanesthetized man. They permit the qualitative identification of the ultimate substrates and products of cerebral metabolism. They have no applicability, however, to those intermediates which are formed and consumed entirely in the brain without being exchanged with blood, nor to those substances which are exchanged between brain and blood with no net flux in either direction. Furthermore, they provide no quantitation of the rates of utilization or production because arteriovenous differences depend not only on the rates of consumption or production by the tissue but also on the blood flow (see below). Blood flow affects all the arteriovenous differences proportionately, however, and comparison of the arteriovenous differences of various substances obtained from the same samples of blood reflects their relative rates of utilization or production.

COMBINATION OF CEREBRAL BLOOD FLOW AND ARTERIOVENOUS DIFFERENCES

In a steady state, the tissue concentration of any substance utilized or produced by the brain is presumed to remain constant. When a substance is exchanged between brain and blood, the difference in its rates of delivery to the brain in the arterial blood and removal in the venous blood must be equal to the net rate of its utilization or production by the brain. This relation can be expressed as follows:

$$CMR = CBF\ (A - V)$$

where $(A - V)$ is the difference in concentration in arterial and cerebral venous blood, CBF is the rate of cerebral blood flow in volume of blood per unit time, and CMR (cerebral metabolic rate) is the steady-state rate of utilization or production of the substance by the brain.

Therefore, if both the rate of cerebral blood flow and the arteriovenous difference are known, the net rate of utilization or production of the substance by the brain can be calculated. This has been the basis of most quantitative studies of the cerebral metabolism in vivo [3, 7, 9, 10].

There are a number of methods for determining the cerebral blood flow, many of them of questionable value or validity. One type of technique involves isolating the cerebral circulation by surgical means; the brain is then perfused at a fixed rate with blood or perfusate, or the rate of flow in the cerebral vascular bed is measured by any one of a variety of flowmeter devices [9]. Such methods are, of course, limited to animals. They can be carried out usually only under general anesthesia and require extensive surgical procedures that make truly normal preparations impossible.

A thorough discussion of the methods for measuring cerebral blood flow is beyond the scope of this chapter, but the subject has been comprehensively reviewed in recent years [2, 6, 7, 9, 11].

The most reliable method for determining cerebral blood flow is the inert gas method of Kety and Schmidt [2, 33]. It was originally designed for use in studies of conscious, unanesthetized man, and it has been most widely employed for this purpose, but it also has been adapted for use in animals [45]. The method is based on the Fick principle, or the law of conservation of matter, and it utilizes low concentrations of a freely diffusible, chemically inert gas as a tracer substance. The original gas was nitrous oxide, but subsequent modifications have substituted other gases, such as ^{85}Kr, ^{79}Kr, or hydrogen [6, 11, 26, 38, 40], which can be measured more conveniently in blood. During inhalation of 15 percent N_2O in air, for example, arterial and cerebral venous blood samples are withdrawn at intervals and analyzed for N_2O content. The cerebral blood flow (in milliliters per 100 g of brain tissue per minute) can be calculated from the equation:

$$\mathrm{CBF} = 100\lambda V_{10} \Big/ \int_0^{10} (A-V)\,dt$$

where A and V = arterial and cerebral venous blood concentrations of N_2O
V_{10} = concentration of N_2O in venous blood at 10 min
λ = partition coefficient for N_2O between brain tissue and blood
t = time of inhalation in minutes
$\int_0^{10} (A-V)dt$ = integrated arteriovenous difference in N_2O concentration over 10 min of inhalation.

The partition coefficient for N_2O is approximately one when equilibrium has been achieved between blood and brain tissue. It has been found that at least 10 minutes of inhalation are required to establish equilibrium. At the end of this interval, the N_2O concentration in brain tissue is about equal to the cerebral venous blood concentration.

Because the method requires sampling of both arterial and cerebral venous blood, it lends itself readily to the simultaneous measurement of arteriovenous differences of substances involved in cerebral metabolism. This method and its modifications have provided most of our knowledge of the rates of substrate utilization or product formation by the brain in vivo.

REGULATION OF CEREBRAL METABOLIC RATE

NORMAL CONSUMPTION OF OXYGEN BY THE BRAIN

The brain is metabolically one of the most active of all the organs of the body. This is reflected in its relatively enormous rate of oxygen consumption, which provides the energy required for its intense physicochemical activity. The most reliable data on cerebral metabolic rate have been obtained in man, although the rates in lower mammals appear to be comparable on a unit-weight basis. Cerebral oxygen consumption in normal, conscious, young men is about 3.5 ml/100 g brain/minute (Table 15-1); the rate is similar in young women. The rate of oxygen consumption by an entire brain of average weight (1400 g) is then about 49 ml O_2 per minute.

TABLE 15-1. Cerebral Blood Flow and Metabolic Rate in Normal Young Adult Man

Function	Rate	
	Per 100 g Brain Tissue	Per Whole Brain[a]
Cerebral blood flow (ml/min)	57	798
Cerebral O_2 consumption (ml/min)	3.5	49
Cerebral glucose utilization (mg/min)	5.5	77

[a]Brain weight assumed to be 1400 g.
Source: Based on data derived from literature [10].

The magnitude of this rate can be more fully appreciated when it is compared with the metabolic rate of the body as a whole. The average man weighs 70 kg and consumes about 250 ml O_2 per minute in the basal state. Therefore, the brain alone, which represents only about 2 percent of total body weight, accounts for 20 percent of the resting total body oxygen consumption. In children the brain takes an even larger fraction [31].

Oxygen is utilized in the brain almost entirely for the oxidation of carbohydrate [5, 10]. The energy equivalent of the total cerebral metabolic rate is, therefore, approximately 20 watts, or 0.25 kcal per minute. If it is assumed that this energy is utilized mainly for the synthesis of high-energy phosphate bonds, that the efficiency of the energy conservation is about 20 percent, and that the free energy of hydrolysis of the terminal phosphate of ATP is about 7 kcal per mole, this energy expenditure then can be estimated to support the steady turnover of close to 7 mmoles, or about 4×10^{21} molecules, of ATP per minute in the entire brain.

The brain normally has no respite from this enormous energy demand. The cerebral oxygen consumption continues unabated day and night. Even during sleep there is no reduction in cerebral metabolic rate; indeed, it may even be markedly increased in rapid eye movement (REM) sleep (see below).

There has been considerable speculation about the nature of the functions responsible for the brain's energy demands. The brain does not do mechanical work, like that of cardiac and skeletal muscle. It does not do osmotic work, as does the kidney in concentrating urine. It does not have the complex, energy-consuming metabolic functions of liver, nor, despite the synthesis of a few hormones and neurotransmitters, is it noted for its biosynthetic activities. Recently, considerable emphasis has been placed on the extent of macromolecular synthesis in the central nervous system, an interest stimulated by the recognition that there are some proteins with short half-lives in the brain. These represent, however, relatively small numbers of molecules, and, in fact, the average protein turnover and the rate of protein synthesis in the mature brain are slower than in most other tissues, except, perhaps, muscle [18]. Clearly, the functions of nervous tissues are mainly excitation and conduction, and these are reflected in the unceasing electrical activity of the brain. The electrical energy is ultimately derived from chemical processes, and it is likely that most of the brain's energy consumption is utilized for active transport of ions to sustain and restore the membrane potentials discharged during the processes of excitation and conduction.

ROLE OF CEREBRAL CIRCULATION

Not only does the brain utilize oxygen at a very rapid rate; it is absolutely dependent on continuously uninterrupted oxidative metabolism for maintenance of its functional and structural integrity. There is a large Pasteur effect in brain tissue [42], but even at its maximum rate anaerobic glycolysis is unable to provide sufficient energy to meet the brain's demands. Since the oxygen stores of the brain are extremely small compared with its rate of utilization, the brain requires the continuous replenishment of its oxygen by the circulation. If the cerebral blood flow is completely interrupted, consciousness is lost within less than 10 sec, or the amount of time required to consume the oxygen contained within the brain and its blood content [49]. Loss of consciousness as a result of anoxemia, caused by anoxia or asphyxia, takes only a little longer because of the additional oxygen present in the lungs and still-circulating blood. There is evidence that the critical level of oxygen tension in the cerebral tissues, below which consciousness and the normal EEG pattern are invariably lost, lies between 15 and 20 mm Hg [9]. This appears to be so whether the tissue anoxia is achieved by lowering the cerebral blood flow or the arterial oxygen content.

Cessation of cerebral blood flow is followed within a few minutes by irreversible pathological changes within the brain, readily demonstrated by microscopic anatomical techniques. It is well known, of course, that in medical crises, such as cardiac arrest, damage to the brain occurs earliest and is most decisive in determining the degree of recovery.

The cerebral blood flow must be able to maintain the brain's avaricious appetite for oxygen. The average rate of blood flow in the brain as a whole is about 57 ml/100 g tissue/min (see Table 15-1). For the whole brain this amounts to almost 800 ml per minute or approximately 15 percent of the total basal cardiac output. This level must be maintained within relatively narrow limits, for the brain cannot tolerate any major drop in its perfusion. A fall in cerebral blood flow to half of its normal rate is sufficient to cause loss of consciousness in normal, healthy young men [7, 9, 24]. There are, fortunately, numerous reflexes and other physiological mechanisms to sustain adequate levels of arterial blood pressure at the head level and to maintain the cerebral blood flow, even when arterial pressure falls in times of stress [7, 9]. There are also mechanisms to adjust the cerebral blood flow to changes in cerebral metabolic demand.

Regulation of the cerebral blood flow is achieved mainly by control of the tone or the degree of constriction or dilatation of the cerebral vessels. This, in turn, is controlled mainly by local chemical factors, such as pCO_2, pO_2, and pH. High pCO_2, low pO_2, and low pH – products of metabolic activity – tend to dilate the blood vessels and increase cerebral blood flow; changes in the opposite direction constrict the vessels and decrease blood flow [7, 9, 17]. The cerebral blood flow is regulated through such mechanisms to maintain homeostasis of these chemical factors in the local tissue. Their rates of production depend on the rates of metabolism, so cerebral blood flow is, therefore, also adjusted to the cerebral metabolic rate [7, 9, 17].

LOCAL BLOOD FLOW AND METABOLISM

The rates of blood flow and metabolism presented in Table 15-1 and discussed above represent the averages in the brain as a whole. The brain, however, is not a homogeneous organ, but is composed of a variety of tissues and discrete structures which often function independently or even inversely in regard to one another. There is little reason to expect that their metabolic rates would be similar. The evidence clearly indicates that they are not. Measurement of the individual metabolic rates in the component structures of the brain have not yet been fully accomplished in situ. A method for measuring local glucose utilization, although still under development, has provided preliminary results that suggest a wide range of rates of glucose utilization throughout the brain. The rates in gray matter are generally several times more rapid than are those in white matter. These results are consistent with the distribution of the values for blood flow observed in the individual structures of the brain. Such measurements reveal that blood flow rates in gray matter vary over a wide range but are at least four to five times more rapid in gray matter than in white matter [25, 36, 53].

Changes in functional activity produced by physiological stimulation or anesthesia produce corresponding changes in the blood flow to the structures involved in the functional change [53]. Because normally blood flow is adjusted to the local cerebral metabolic demand, it is likely that the wide range in local perfusion rates reflects comparable variation in the metabolic rates of the individual component structures of the brain.

SUBSTRATES OF CEREBRAL METABOLISM

NORMAL SUBSTRATES AND PRODUCTS

In contrast to most other tissues, which exhibit considerable flexibility in regard to the nature of the foodstuffs absorbed and consumed from the blood, the normal brain is restricted almost exclusively to glucose as the substrate for its energy metabolism. Despite long and intensive efforts, the only incontrovertible and consistently positive arteriovenous differences demonstrated for the human brain under normal conditions have been for glucose and oxygen [5, 10]. One report [19] of glutamate uptake and equivalent release of glutamine has never been substantiated and must at present be considered doubtful. Negative arteriovenous differences, significantly different from zero, have been found consistently only for carbon dioxide, although it is likely that water, which has never been measured, is also produced [10]. Pyruvate and lactate production have been observed occasionally, certainly in aged subjects with cerebral vascular insufficiency [42], but irregularly in subjects with normal oxygenation of the brain [10].

It appears, then, that in the normal in vivo state, glucose is the only significant substrate for the brain's energy metabolism. Under normal circumstances no other potential energy-yielding substance has been found to be extracted from the blood in more than trivial amounts. The stoichiometry of glucose utilization and oxygen consumption is summarized in Table 15-2. The normal, conscious human brain consumes oxygen at a rate of 156 μmoles/100 g tissue/minute. Carbon dioxide production is the same, leading to a respiratory quotient of 1.0, further evidence that carbohydrate is the ultimate substrate for oxidative metabolism. The O_2 consumption and CO_2 production are equivalent to a rate of glucose utilization of 26 μmoles glucose/100 g tissue/minute, assuming 6 μmoles of O_2 consumed and CO_2 produced for each μmole of glucose completely oxidized to CO_2 and H_2O. The glucose utilization actually measured is, however, 31 μmoles/100 g/minute, which indicates that glucose consumption is not only sufficient to account for the total O_2 consumption but is in excess by 5 μmoles/100 g/minute. For the complete oxidation of glucose, the theoretical ratio of oxygen-to-glucose utilization is 6.0; the excess glucose utilization is responsible for a measured ratio of only 5.5 μmoles of O_2 per μmole of glucose. The fate of the excess glucose is unknown, but it is probably distributed in part in lactate, pyruvate, and other intermediates of carbohydrate metabolism that are released from the brain into the blood, each in insufficient amount to be detectable in significant arteriovenous differences. Probably some of the glucose is also utilized not for the production of energy but for the synthesis of other chemical constituents of the brain.

Some oxygen is known to be utilized for the oxidation of substances not derived from glucose, as, for example, in oxygenase reactions involved in the synthesis of

TABLE 15-2. Relationship Between Cerebral Oxygen Consumption and Glucose Utilization in Normal Young Adult Man

Function	Value
O_2 consumption (μmoles/100 g brain tissue/min)	156
Glucose utilization (μmoles/100 g brain tissue/min)	31
O_2/glucose ratio (mole/mole)	5.5
Glucose equivalent of O_2 consumption (μmoles glucose/100 g brain tissue/min)	26[a]
CO_2 production (μmoles/100 g brain tissue/min)	156
Cerebral respiratory quotient	0.97

[a]Calculated on the basis of 6 moles of O_2 required for complete oxidation of 1 mole of glucose.

Source: Values are the median of the values reported in the literature [10].

the neurotransmitters, norepinephrine and 5-hydroxytryptamine. The amount of oxygen utilized for these processes is, however, extremely small and is undetectable in the presence of the enormous oxygen consumption used for carbohydrate oxidation. Also, the amount of amino acids taken up for these purposes is so small as to be undetected in the arteriovenous differences.

The combination of a cerebral respiratory quotient of unity, an almost stoichiometric relationship between oxygen uptake and glucose consumption, and the absence of any significant arteriovenous difference for any other energy-rich substrate is strong evidence that the brain normally derives its energy from the oxidation of glucose. In this respect, cerebral metabolism is unique because no other tissue, with the possible exception of the testis [28], has been found to rely only on carbohydrate for energy. This does not imply that the pathways of glucose metabolism in the brain lead, like combustion, directly and exclusively to oxidation. Various chemical and energy transformations occur between the uptake of the primary substrates, glucose and oxygen, and the liberation of the end-products, carbon dioxide and water. Various compounds derived from glucose or produced through the energy made available from glucose catabolism are intermediates in the process. Glucose carbon is incorporated, for example, into amino acids, protein, lipids, glycogen, and so on. These are turned over and act as intermediates in the overall pathway from glucose to carbon dioxide and water. There is clear evidence from studies with ^{14}C-glucose that the glucose is not entirely oxidized directly, and that at any given moment some of the carbon dioxide being produced is derived from sources other than the glucose that enters the brain at the same moment or just prior to that moment [1, 8]. That oxygen and glucose consumption and carbon dioxide production are essentially in stoichiometric balance and no other energy-laden substrate is taken from the blood means, however, that the net energy made available to the brain must ultimately be derived from the oxidation of glucose. It should be noted that this is the situation in the normal state; as will be discussed later, other substrates may be used in special circumstances or in abnormal states.

OBLIGATORY UTILIZATION OF GLUCOSE

The brain normally derives almost all its energy from the aerobic oxidation of glucose, but this does not distinguish between preferential and obligatory utilization of glucose. Most tissues are largely facultative in their choice of substrates and can use them interchangeably more or less in proportion to their availability. This does not appear to be so in brain. The present evidence indicates that, except in some unusual and very special circumstances, only the aerobic utilization of glucose is capable of providing the brain with enough energy to maintain normal function and structure. The brain appears to have almost no flexibility in its choice of substrates in vivo. This conclusion is derived from the following evidence.

Effects of Glucose Deprivation

It is well known clinically that a drop in blood glucose content, if of sufficient degree, is rapidly followed by aberrations of cerebral function. Hypoglycemia, produced by excessive insulin or occurring spontaneously in hepatic insufficiency, is associated with changes in mental state ranging from mild, subjective sensory disturbances to coma, the severity depending on both the degree and the duration of the hypoglycemia. The behavioral effects are paralleled by abnormalities in EEG patterns and cerebral metabolic rate. The EEG pattern exhibits increased prominence of slow, high-voltage delta rhythms, and the rate of cerebral oxygen consumption falls. In studies of the effects of insulin hypoglycemia in man [34], it was observed that when the arterial glucose concentration fell from a normal level of 70 to 100 mg per 100 ml to an average level of 19 mg per 100 ml, the subjects became confused and their cerebral oxygen consumption fell to 2.6 ml/100 g/minute or 79 percent of the normal level. When the arterial glucose level fell to 8 mg per 100 ml, a deep coma ensued, and the cerebral oxygen consumption decreased even farther to 1.9 ml/100 g/minute.

These changes are not caused by insufficient cerebral blood flow, which actually increases slightly during the coma. In the depths of the coma, when the blood glucose content is very low, there is almost no measurable cerebral uptake of glucose from the blood. Cerebral oxygen consumption, although reduced, is still far from negligible, and there is no longer any stoichiometric relationship between glucose and oxygen uptakes by the brain – evidence that the oxygen is utilized for the oxidation of other substances. The cerebral respiratory quotient remains approximately one, however, indicating that these other substrates are still carbohydrate, presumably derived from the brain's endogenous carbohydrate stores. The effects are clearly the result of hypoglycemia and not some other direct effect of insulin in the brain. In all cases, the behavioral, functional, and cerebral metabolic abnormalities associated with insulin hypoglycemia are rapidly and completely reversed by the administration of glucose. The severity of the effects is correlated with the degree of hypoglycemia and not the insulin dosage, and the effects of the insulin can be completely prevented by the simultaneous administration of glucose with the insulin.

Similar effects are observed in hypoglycemia produced by other means, such as hepatectomy. The inhibition of glucose utilization at the phosphoglucoisomerase step with 2-deoxyglucose also produces all the cerebral effects of hypoglycemia despite an associated elevation in blood glucose content [10]. It appears, then, that when the brain is deprived of its glucose supply in an otherwise normal individual, no other substance present in the blood can satisfactorily substitute for it as the substrate for the brain's energy metabolism.

Other Substrates in Hypoglycemia

The hypoglycemic state provides convenient test conditions to determine whether a substance is capable of substituting for glucose as a substrate of cerebral energy metabolism. If it can, its administration during hypoglycemic shock should restore consciousness and normal cerebral electrical activity without raising the blood glucose level. Numerous potential substrates have been tested in man and animals. Very few can restore normal cerebral functions in hypoglycemia, and of these all but one appear to operate through a variety of mechanisms to raise the blood glucose level rather than by serving as a substrate directly (Table 15-3).

Mannose appears to be the only substance that can be utilized by the brain directly and rapidly enough to restore or maintain normal function in the absence of glucose [51]. It traverses the blood-brain barrier and is converted to mannose 6-phosphate. This reaction is catalyzed by hexokinase as effectively as the phosphorylation of glucose. The mannose 6-phosphate is then converted to fructose 6-phosphate by phosphomannose isomerase, which is active in brain tissue [51]. Through these reactions mannose can enter directly into the glycolytic pathway and replace glucose.

Maltose has also been found to be effective occasionally in restoring normal behavior and EEG activity in hypoglycemia, but only by raising the blood glucose level through its conversion to glucose by maltase activity in blood and other tissues [10, 51].

Epinephrine is effective in producing arousal from insulin coma, but this is achieved through its well-known stimulation of glycogenolysis and the elevation of blood glucose concentration. Glutamate, arginine, glycine, *p*-aminobenzoate, and succinate also are effective occasionally, but they probably act through adrenergic effects which raise the epinephrine level and consequently the glucose concentrations of the blood [10].

It is clear, then, that no substance normally present in the blood can replace glucose as the substrate for the brain's energy metabolism. Thus far, the one substance found to do so – mannose – is not normally present in blood in significant amounts and is, therefore, of no physiological significance. It should be noted,

TABLE 15-3. Effectiveness of Various Substances in Preventing or Reversing the Effe[cts] of Hypoglycemia or Glucose Deprivation on Cerebral Function and Metabolism

Effectiveness	Substance	Comments
Effective	Epinephrine	Raises blood glucose concentration
	Maltose	Converted to glucose and raises blood glucose level
	Mannose	Directly metabolized and enters glycolytic pathway
Partially or occasionally effective	Glutamate Arginine Glycine *p*-Aminobenzoate Succinate	Occasionally effective by raising blood glucose level
Ineffective	Glycerol Ethanol Lactate Pyruvate Glyceraldehyde Hexosediphosphates Fumarate Acetate β-Hydroxybutyrate Galactose Lactose Inulin	Some of these substances can be metabolized to varying extent by brain tissue and could conceivably be effective if it were not for the blood-brain barrier

Source: Summarized from literature [10].

however, that failure to restore normal cerebral function in hypoglycemia is not synonymous with an inability of the brain to utilize the substance. Many of the substances that have been tested and found ineffective are compounds normally formed and utilized within the brain and are normal intermediates in its intermediary metabolism. Lactate, pyruvate, fructose 1,6-diphosphate, acetate, β-hydroxybutyrate, and acetoacetate are such examples. These can all be utilized by brain slices, homogenates, or cell-free fractions, and the enzymes for their metabolism are present in the brain. Enzymes for the metabolism of glycerol [52] or ethanol [47], for example, may not be present in sufficient amounts. For other substrates, for example, β-hydroxybutyrate and acetoacetate [29, 46, 54], the enzymes are adequate, but the substrate is not available to the brain because of inadequate blood level or restricted transport through the blood-brain barrier.

Nevertheless, the functioning of the nervous system in the intact animal depends on substrates supplied by the blood, and no satisfactory, normal, endogenous substitute for glucose has been found. Glucose must, therefore, be considered essential for normal physiological behavior of the central nervous system.

STARVATION

Recent developments indicate that the brain may be somewhat more flexible in its choice of nutrients than was once believed and that, in special circumstances, it may fulfill its nutritional needs partly, although not completely, with substrates other than glucose. Normally there are no significant cerebral arteriovenous differences for D-β-hydroxybutyrate and acetoacetate [5, 10], which are ketone bodies formed in the course of the catabolism of fat. Owen et al. [44] observed, however, that when human patients were treated for severe obesity by complete fasting for several weeks, there was considerable uptake of both substances by the brain. Indeed, if one assumed that the substances were completely oxidized, their rates of utilization would have accounted for more than 50 percent of the total cerebral oxygen consumption – more than that accounted for by the glucose uptake. The uptake of D-β-hydroxybutyrate was several times greater than that of acetoacetate, a reflection of its higher concentration in the blood. The enzymes responsible for their metabolism, D-β-hydroxybutyric dehydrogenase, acetoacetate-succinyl-CoA transferase, and acetoacetyl thiolase, have been demonstrated to be present in brain tissue in sufficient amounts to convert them into acetyl-CoA and to feed them into the tricarboxylic cycle at a sufficient rate to satisfy the brain's metabolic demands [29, 46, 54].

Under normal circumstances, when there is ample glucose and the levels of the ketone bodies in the blood are very low, the brain apparently does not resort to their use in any significant amounts. In prolonged starvation, the carbohydrate stores of the body are exhausted, and the rate of gluconeogenesis is insufficient to provide glucose fast enough to meet the requirements of the brain; blood ketone levels rise as a result of the rapid fat catabolism. The brain then apparently turns to the ketone bodies as the source of its energy supply.

Cerebral utilization of the ketone bodies appears to follow passively their levels in the blood. In diabetic acidosis, however, ketosis is just as severe as in starvation, and yet there is no evidence of ketone body utilization in that condition [5]. The reason for the discrepancy is unclear, but it may be that the combination of the high glucose level in the blood with the low cerebral metabolic rate in diabetic acidosis or coma is in some way responsible for the brain's failure to utilize the ketone bodies.

It should be noted that D-β-hydroxybutyrate is incapable of maintaining or restoring normal cerebral function in the absence of glucose in the blood [50]. This suggests that, although it can partially replace glucose, it cannot fully satisfy the cerebral energy needs in the absence of some glucose consumption. One possible explanation may be that the first product of D-β-hydroxybutyrate oxidation, acetoacetate, is further metabolized by its displacement of the succinyl moiety of succinyl-CoA to form acetoacetyl-CoA. A certain level of glucose utilization may be essential to drive the tricarboxylic cycle and provide enough succinyl-CoA to permit the further oxidation of acetoacetate and hence pull along the oxidation of β-hydroxybutyrate.

INFLUENCE OF AGE AND DEVELOPMENT ON CEREBRAL ENERGY METABOLISM

The energy metabolism of the brain and the blood flow which sustains it vary considerably from birth to old age. Data on the cerebral metabolic rate obtained directly in vivo are lacking for the early postnatal period, but the results of in vitro measurements in animal brain preparations [27] and inferences drawn from cerebral blood flow measurements in intact animals [30] suggest that the cerebral oxygen consumption is low at birth, rises rapidly during the period of cerebral growth and development, and reaches a maximal level at about the time maturation is completed (Chap. 14). This rise is consistent with the progressive increase in the levels of a number of enzymes of oxidative metabolism in the brain (Chap. 9) [35]. The rates of blood flow in different structures of the brain reach peak levels at different times, depending on the maturation rate of the particular structure [30]. In the structures that consist predominantly of white matter, the peaks coincide roughly with the times of maximal rates of myelination [30]. From these peaks, blood flow and, probably, cerebral metabolic rate decline to the levels characteristic of adulthood.

Reliable quantitative data on the changes in cerebral circulation and metabolism in man from the middle of the first decade of life to old age are summarized in Table 15-4. By 6 years of age, the cerebral blood flow and oxygen consumption have already attained their high levels, and they decline thereafter to the levels of normal young adulthood [31]. A cerebral oxygen consumption of 5.2 ml/100 g

TABLE 15-4. Cerebral Blood Flow and Oxygen Consumption in Man from Childhood to Old Age and Senility

Life Period and Condition	Age (years)	Cerebral Blood Flow (ml/100 g/min)	Cerebral O_2 Consumption (ml/100 g/min)	Cerebral Venous O_2 Tension (mm Hg)
Childhood	6[a]	106[a]	5.2	–
Normal young adulthood	21	62	3.5	38
Aged:				
Normal elderly	71[a]	58	3.3	36
Elderly with minimal arteriosclerosis	73[a]	48[a]	3.2	33[a, b]
Elderly with senile psychosis	72[a]	48[a, b]	2.7[a, b]	33[a, b]

[a]Statistically significant difference from normal young adults ($P < 0.05$).
[b]Statistically significant difference from normal elderly subjects ($P < 0.05$).
Source: From [13] and [31].

brain tissue/minute in a 5- to 6-year-old child corresponds to a total oxygen consumption by the brain of approximately 60 ml per minute, or more than 50 percent of the total body basal oxygen consumption, a proportion markedly greater than that occurring in adulthood. The reasons for the extraordinarily high cerebral metabolic rates in children are unknown, but presumably they reflect the extra energy requirements for biosynthetic processes associated with growth and development.

Despite reports to the contrary [4], cerebral blood flow and oxygen consumption normally remain essentially unchanged between young adulthood and old age. In a population of normal elderly men in their eighth decade of life – who were carefully selected for good health and freedom from all disease, including vascular – both blood flow and oxygen consumption were not significantly different from those of normal young men 50 years younger (see Table 15-4) [13]. In a comparable group of elderly subjects, who differed only by the presence of objective evidence of minimal arteriosclerosis, cerebral blood flow was significantly lower. It had reached a point at which the oxygen tension of the cerebral venous blood declined, which is an indication of cerebral hypoxia. Cerebral oxygen consumption, however, was still maintained at normal levels through removal of larger-than-normal proportions of the arterial blood oxygen. In senile psychotic patients with arteriosclerosis, cerebral blood flow was no lower, but cerebral oxygen consumption also had declined. These data suggest that aging per se need not lower the cerebral oxygen consumption and blood flow, but that when blood flow is reduced, it probably is secondary to arteriosclerosis, which produces cerebral vascular insufficiency and chronic hypoxia in the brain. Because arteriosclerosis is so prevalent in the aged population, most individuals probably follow the latter pattern.

CEREBRAL METABOLIC RATE IN VARIOUS PHYSIOLOGICAL STATES

CEREBRAL METABOLIC RATE AND FUNCTIONAL ACTIVITY

In organs such as the heart or skeletal muscles, which perform mechanical work, increased functional activity clearly is associated with increased metabolic rate. In nervous tissues outside the central nervous system, the electrical activity is an almost quantitative indicator of the degree of functional activity, and in structures such as sympathetic ganglia and postganglionic axons [10, 37] increased electrical activity produced by electrical stimulation is definitely associated with increased utilization of oxygen. On the other hand, the electrical activity of heterogeneous units in the central nervous system is integrated into a composite record; EEG data cannot always be interpreted readily in terms of total functional activity, and the relationship between functional activity and metabolic rate may be obscured.

Convulsive activity, induced or spontaneous, has often been employed as the indicator of increased functional and electrical activity of the brain. Davies and Rémond [22] used the oxygen electrode technique in the cerebral cortex of cat and found increases in oxygen consumption during electrically induced or drug-induced convulsions. Because the increased oxygen consumption either coincided with or followed the onset of convulsions, it was concluded that the elevation in metabolic rate was the consequence of the increased functional activity produced by the convulsive state. Similar results have been obtained in the perfused cat brain in which cerebral oxygen consumption was determined from the combined measurements of blood flow and arteriovenous oxygen differences [1].

In the lightly anesthetized monkey, Dumke and Schmidt [23] found an excellent correlation between cerebral oxygen consumption and cerebral functional activity; the latter was judged by muscular movements, ocular reflexes, character of respiration, and level of arterial pressure. Changes in functional activity either occurred spontaneously or were caused (1) by altering cerebral blood flow by means of hemorrhage, transfusions, or epinephrine infusion, or (2) by producing convulsions with analeptic drugs. Cerebral oxygen consumption during the convulsions was double that of the preconvulsive state and fell to half the resting level during the postconvulsive state of depression. Similar increases in cerebral oxygen consumption have been observed in human epileptic patients during seizures [14].

Convincing correlations between cerebral metabolic rate and mental activity have been obtained in man in a variety of pathological states of altered consciousness [3, 10, 16]. Regardless of the cause of the disorder, graded reductions in cerebral oxygen consumption were accompanied by parallel graded reductions in the degree of mental alertness, all the way to profound coma (Table 15-5).

TABLE 15-5. Relationship Between Level of Consciousness and Cerebral Metabolic Rate

Level of Consciousness	Cerebral Blood Flow (ml/100 g/min)	Cerebral O_2 Consumption (ml/100 g/min)
Mentally alert:	54	3.3
Normal young men		
Mentally confused:	48	2.8
Brain tumor		
Diabetic acidosis		
Insulin hypoglycemia		
Cerebral arteriosclerosis		
Comatose:	57	2.0
Brain tumor		
Diabetic coma		
Insulin coma		
Anesthesia		

Source: From [3].

Mental Work

It is difficult to define or even to conceive of the physical equivalent of mental work. A common view equates concentrated mental effort with mental work, and it is also fashionable to attribute a high demand for mental effort to the process of problem-solving in mathematics. Nevertheless, there appears to be no increased energy utilization by the brain during such processes. Cerebral blood flow and oxygen consumption remain unchanged from the resting levels during the exertion of the mental effort required to solve complex arithmetical problems (Table 15-6) [15]. It may be that the assumptions which relate mathematical reasoning to mental work are erroneous, but it seems more likely that the areas which participate in the processes of such reasoning represent too small a fraction of the brain for changes in their functional and metabolic activities to be reflected in the energy metabolism of the brain as a whole.

TABLE 15-6. Cerebral Blood Flow and Metabolic Rate in Normal Young Men During Sleep and Mental Work

	Cerebral Blood Flow (ml/100 g/min)		Cerebral O_2 Consumption (ml/100 g/min)	
Condition	Control	Exptl.	Control	Exptl.
Deep (slow-wave) sleep	59	65[a]	3.5	3.4
Mental arithmetic	69	67	3.9	4.0

[a]Statistically significant difference from control level ($P < 0.05$).
Source: From [15] and [41].

Sleep

Sleep is a naturally occurring, periodic, reversible state of unconsciousness, and the EEG pattern in deep sleep is characterized by high-voltage, slow rhythms very similar to those often seen in pathological comatose states. Nevertheless, in contrast to the depressed cerebral energy metabolism seen in pathological coma, there is surprisingly no change in the oxygen consumption of the brain as a whole in deep, slow-wave sleep (see Table 15-6) [15, 41]. There are no comparable data in man for the state of "paradoxical" sleep (REM sleep), the stage characterized by rapid, low-voltage frequencies in the EEG and believed to be associated with the dream state. However, measurements of local blood flow in most areas of the brain indicate an enormous increase in blood flow throughout the brain in REM sleep [48]. The basis for this increase is unknown. It cannot be accounted for by any of the known extracerebral factors which could change cerebral blood flow to this degree, and it must be tentatively attributed to a comparable increase in cerebral metabolic rate.

CEREBRAL ENERGY METABOLISM IN PATHOLOGICAL STATES

The cerebral metabolic rate is normally relatively stable and varies little under physiological conditions. There are, however, a number of pathological states of the nervous system and other organs that affect the functions of the brain either directly or indirectly, and some of these have profound effects on the cerebral metabolism.

PSYCHIATRIC DISORDERS

In general, disorders which alter the quality of mentation but not the level of consciousness – for example, the functional neuroses, psychoses, and psychotomimetic states – have no apparent effects on the blood flow and oxygen consumption of the brain. Thus, no changes in either function are observed in schizophrenia [15, 34] or LSD intoxication (Table 15-7) [15]. There is still uncertainty about the effects of anxiety, mainly because of the difficulties in evaluating quantitatively the intensity of anxiety. It is generally believed that ordinary degrees of anxiety or "nervousness" do not affect the cerebral metabolic rate, but severe anxiety or "panic"

TABLE 15-7. Cerebral Blood Flow and Metabolic Rate in Schizophrenia and in Normal Young Men During LSD-induced Psychotomimetic State

Condition	Cerebral Blood Flow (ml/100 g/min)	Cerebral O_2 Consumption (ml/100 g/min)
Normal	67	3.9
LSD intoxication	68	3.9
Schizophrenia	72	4.0

Source: From [15].

may increase the cerebral oxygen consumption [3, 15]. This may be related to the level of epinephrine circulating in the blood. Small doses of epinephrine that raise heart rate and cause some anxiety do not alter cerebral blood flow and metabolism, but large doses that are sufficient to raise the arterial blood pressure cause significant increases in the levels of both [15].

Psychosis related to parenchymatous damage in the brain is usually associated with depression of the cerebral metabolic rate. Thus, although normal aging per se has no effect on cerebral blood flow and oxygen consumption, both are reduced in the psychoses of senility (see Table 15-4) [13, 15, 39]. Senile dementia is always the result of degenerative changes in the brain tissue. In most cases these are secondary to cerebral vascular disease and the ensuing chronic tissue hypoxia. In some cases there is primary parenchymatous degeneration which may even occur prematurely, as in Alzheimer's disease. In either case, cerebral blood flow and metabolic rate are reduced.

CONVULSIVE DISORDERS

Convulsions have been discussed previously in regard to the relation of cerebral metabolic rate to functional activity. In most types of convulsions studied, the seizures were induced by electroshock or analeptic drugs, or they occurred spontaneously in epileptic disease. In all these, cerebral blood flow and oxygen consumption are significantly increased in the course of the seizure [14]. Increased energy utilization is not, however, an essential component of the convulsive process. Convulsions may also be induced by impairment of the energy-generating system. Oxygen or glucose deficiency and fluoroacetate poisoning, all associated with a deficient availability or utilization of essential substrates for the brain's metabolism, cause seizures without increasing cerebral metabolic rate. Hypoglycemic convulsions are, in fact, associated with lowered rates of oxygen and glucose utilization [34] with no apparent depletion of high-energy phosphate stores [14].

Pyridoxine deficiency (see Chap. 25) produces a substrate-deficiency type of convulsion with reductions in cerebral blood flow and oxygen consumption [14]. The reason is unknown, but since pyridoxine deficiency leads to lowered γ-aminobutyric acid levels in the brain, it has been presumed that the lowered cerebral metabolic rate reflects impairment of the α-ketoglutarate-to-glutamate-to-γ-aminobutyric acid-to-succinic semialdehyde-to-succinate shunt pathway around the α-ketoglutarate-to-succinate step in the tricarboxylic acid cycle in the brain (see Chaps. 8 and 10).

COMA

Coma is correlated with depression of cerebral oxygen consumption; progressive reductions in the level of consciousness are paralleled by corresponding graded decreases in cerebral metabolic rate (see Table 15-5). There are almost innumerable derangements that can lead to depression of consciousness. Table 15-8 includes only a few typical examples that have been studied by the same methods and by the same or related groups of investigators.

Inadequate cerebral nutrient supply leads to decreases in the level of consciousness, ranging from confusional states to coma. The nutrition of the brain can be limited by lowering the oxygen or glucose levels of the arterial blood, as in anoxia or hypoglycemia, or by impairment of their distribution to the brain through lowering of the cerebral blood flow, as in brain tumors. Consciousness is then depressed,

TABLE 15-8. Cerebral Blood Flow and Metabolic Rate in Humans with Various Disorders Affecting Mental State

Condition	Mental State	Cerebral Blood Flow (ml/100 g/min)	Cerebral O_2 Consumption (ml/100 g/min)
Normal	Alert	54	3.3
Increased intracranial pressure (brain tumor)	Coma	34[a]	2.5[a]
Insulin hypoglycemia:			
Arterial glucose level:			
74 mg/100 ml	Alert	58	3.4
19 mg/100 ml	Confused	61	2.6[a]
8 mg/100 ml	Coma	63	1.9[a]
Thiopental anesthesia	Coma	60[a]	2.1[a]
Postconvulsive state:			
Before convulsion	Alert	58	3.7
After convulsion	Confused	37[a]	3.1[a]
Diabetes:			
Acidosis	Confused	45[a]	2.7[a]
Coma	Coma	65[a]	1.7[a]
Hepatic insufficiency	Coma	33[a]	1.7[a]

[a]Denotes statistical significant difference from normal level ($P < 0.05$).
Source: All studies listed were carried out by Kety and/or his associates, employing the same methods. For references see [10].

presumably because of inadequate supplies of substrate to support the energy metabolism necessary to sustain the appropriate functional activities of the brain.

In a number of conditions, the causes of depression of both consciousness and the cerebral metabolic rate are unknown and must, by exclusion, be attributed to intracellular defects in the brain. Anesthesia is one example. Cerebral oxygen consumption is always reduced in the anesthetized state regardless of the anesthetic agent used, whereas blood flow may or may not be decreased and may even be increased [12]. This reduction is the result of decreased energy demand and not an insufficient nutrient supply or a block of intracellular energy metabolism [16]. There is evidence that general anesthetics interfere with synaptic transmission, thus reducing neuronal interaction and functional activity and, consequently, metabolic demands [16, 37].

Several metabolic diseases with broad systemic manifestations are also associated with disturbances of cerebral functions. Diabetes mellitus, when allowed to progress to states of acidosis and ketosis, leads to mental confusion and, ultimately, to deep coma, with parallel proportionate decreases in cerebral oxygen consumption (see Table 15-8) [32]. The abnormalities are usually completely reversed by adequate insulin therapy. The cause of the coma or depressed cerebral metabolic rate is unknown. Deficiency of cerebral nutrition cannot be implicated because the blood glucose level is elevated and cerebral blood flow and oxygen supply are more than adequate. Neither is insulin deficiency, which is presumably the basis of the systemic manifestations of the disease, a likely cause of the cerebral abnormalities since no absolute requirement of insulin for cerebral glucose utilization or metabolism has been demonstrated [10]. Ketosis may be severe in this disease, and there is disputed evidence that a rise in the blood level of at least one of the ketone bodies, acetoacetate, can cause coma in animals [16]. In studies of human diabetic acidosis and coma, a significant correlation between the depression of cerebral metabolic rate and the degree of ketosis has been observed, but there is an equally good correlation with the degree of acidosis [32]. It is possible that ketosis, acidosis, or the combination of both may be responsible for the disturbances in cerebral function and metabolism.

Coma is occasionally associated with severe impairment of liver function, or hepatic insufficiency. In human patients in hepatic coma, cerebral metabolic rate is markedly depressed (see Table 15-8) [16]. Cerebral blood flow is also moderately depressed, but not sufficiently to lead to limiting supplies of glucose and oxygen. The blood ammonia level is usually elevated in hepatic coma, and significant cerebral uptake of ammonia from the blood is observed. Ammonia toxicity has, therefore, been suspected as the basis for cerebral dysfunction in hepatic coma. Because ammonia can, through glutamic dehydrogenase activity, convert α-ketoglutarate to glutamate by reductive amination, it has been suggested that ammonia might thereby deplete α-ketoglutarate and thus slow the Krebs cycle [21]. The correlation between the degree of coma and the blood ammonia level is far from convincing, however, and coma has, in fact, been observed in the absence of an increase in blood ammonia concentration [16]. Although ammonia may be involved in the mechanism of hepatic coma, the mechanism remains unclear, and other causal factors are probably involved [16].

Depression of mental functions and cerebral metabolic rate has been observed in association with kidney failure, uremic coma [16]. The chemical basis of the functional and metabolic disturbances in the brain in this condition also remains undetermined.

In the comatose states associated with these systemic metabolic diseases, there is depression of both the level of conscious mental activity and cerebral energy metabolism. From the available evidence, it is impossible to distinguish which, if either, is the primary change. It is more likely that the depressions of both functions, although well correlated with each other, are independent reflections of a more general impairment of neuronal processes by some unknown factors incident to the disease.

REFERENCES

BOOKS AND REVIEWS

1. Geiger, A. Correlation of brain metabolism and function by use of a brain perfusion method *in situ*. *Physiol. Rev.* 38:1, 1958.
2. Kety, S. S. Quantitative Estimation of Cerebral Blood Flow in Man. In Potter, V. R. (Ed.), *Methods in Medical Research.* Vol. 1. Chicago: Year Book, 1948.
3. Kety, S. S. Circulation and metabolism of the human brain in health and disease. *Am. J. Med.* 8:205, 1950.
4. Kety, S. S. Human cerebral blood flow and oxygen consumption as related to aging. *Res. Publ. Assoc. Res. Nerv. Ment. Dis.* 35:31, 1956.
5. Kety, S. S. The General Metabolism of the Brain *in Vivo.* In Richter, D. (Ed.), *The Metabolism of the Nervous System.* London: Pergamon, 1957.
6. Kety, S. S. The Cerebral Circulation. In Field, J. (Ed.), *Handbook of Physiology – Neurophysiology.* Vol. 3. Washington, D.C.: American Physiological Society 1960.
7. Lassen, N. A. Cerebral blood flow and oxygen consumption in man. *Physiol. Rev.* 39:183, 1959.
8. Sacks, W. Cerebral Metabolism *in Vivo.* In Lajtha, A. (Ed.), *Handbook of Neurochemistry.* Vol. 1. New York: Plenum, 1969.
9. Sokoloff, L. The action of drugs on the cerebral circulation. *Pharmacol. Rev.* 11:1, 1959.
10. Sokoloff, L. The Metabolism of the Central Nervous System *in Vivo.* In Field, J., Magoun, H. W., and Hall, V. E. (Eds.), *Handbook of Physiology–Neurophysiology.* Vol. 3. Washington, D.C.: American Physiological Society, 1960.
11. Sokoloff, L. Quantitative Measurements of Cerebral Blood Flow in Man. In Bruner, H. D. (Ed.), *Methods in Medical Research.* Vol. 8. Chicago: Year Book, 1960.
12. Sokoloff, L. Control of Cerebral Blood Flow: The Effects of Anesthetic Agents. In Papper, E. M., and Kitz, R. J. (Eds.), *Uptake and Distribution of Anesthetic Agents.* New York: McGraw-Hill, 1963.
13. Sokoloff, L. Cerebral circulatory and metabolic changes associated with aging. *Res. Publ. Assoc. Res. Nerv. Ment. Dis.* 41:237, 1966.
14. Sokoloff, L. Cerebral Blood Flow and Energy Metabolism in Convulsive Disorders. In Jasper, H. H., Ward, A. A., and Pope, A. (Eds.), *Basic Mechanisms of the Epilepsies.* Boston: Little, Brown, 1969.
15. Sokoloff, L. Cerebral Circulation and Behavior in Man: Strategy and Findings. In Mandell, A. J., and Mandell, M. P. (Eds.), *Psychochemical Research in Man.* New York: Academic, 1969.
16. Sokoloff, L. Neurophysiology and Neurochemistry of Coma. In Polli, E. (Ed.), *Neurochemistry of Hepatic Coma.* New York/Basel: S. Karger. [*Exp. Biol. Med.* 4:15, 1971.]
17. Sokoloff, L., and Kety, S. S. Regulation of cerebral circulation. *Physiol. Rev.* *40 (Suppl. 4)* :38, 1960.

18. Waelsch, H., and Lajtha, A. Protein metabolism in the nervous system. *Physiol. Rev.* 41:709, 1961.

ORIGINAL SOURCES AND SPECIAL TOPICS

19. Adams, J. E., Harper, H. A., Gordon, G. S., Hutchin, M., and Bentinck, R. C. Cerebral metabolism of glutamic acid in multiple sclerosis. *Neurology* 5:100, 1955.
20. Berl, S., and Clarke, D. D. Compartmentation of Amino Acid Metabolism. In Lajtha, A. (Ed.), *Handbook of Neurochemistry.* Vol. 2. New York: Plenum, 1969.
21. Bessman, S. P., and Bessman, A. N. The cerebral and peripheral uptake of ammonia in liver disease with an hypothesis for the mechanism of hepatic coma. *J. Clin. Invest.* 34:622, 1955.
22. Davies, P. W., and Rémond, A. Oxygen consumption of the cerebral cortex of the cat during metrazole convulsions. *Res. Publ. Assoc. Res. Nerv. Ment. Dis.* 26:205, 1946.
23. Dumke, P. R., and Schmidt, C. F. Quantitative measurements of cerebral blood flow in the Macacque monkey. *Am. J. Physiol.* 138:421, 1943.
24. Finnerty, F. A., Jr., Witkin, L., and Fazekas, J. F. Cerebral hemodynamics during cerebral ischemia induced by acute hypotension. *J. Clin. Invest.* 33:1227, 1954.
25. Freygang, W. H., and Sokoloff, L. Quantitative measurements of regional circulation in the central nervous system by the use of radioactive inert gas. *Adv. Biol. Med. Phys.* 6:263, 1958.
26. Gotoh, F., Meyer, J. S., and Tomita, M. Hydrogen method for determining cerebral blood flow in man. *Arch. Neurol.* 15:549, 1966.
27. Himwich, H. E., and Fazekas, J. F. Comparative studies of the metabolism of the brain of infant and adult dogs. *Am. J. Physiol.* 132:454, 1941.
28. Himwich, H. E., and Nahum, L. H. The respiratory quotient of testicle. *Am. J. Physiol.* 88:680, 1929.
29. Itoh, T., and Quastel, J. H. Acetoacetate metabolism in infant and adult rat brain *in vitro. Biochem. J.* 116:641, 1970.
30. Kennedy, C., Grave, G. D., Jehle, J. W., and Sokoloff, L. Blood flow to white matter during maturation of the brain. *Neurology* 20:613, 1970.
31. Kennedy, C., and Sokoloff, L. An adaptation of the nitrous oxide method to the study of the cerebral circulation in children; normal values for cerebral blood flow and cerebral metabolic rate in childhood. *J. Clin. Invest.* 36:1130, 1957.
32. Kety, S. S., Polis, B. D., Nadler, C. S., and Schmidt, C. F. The blood flow and oxygen consumption of the human brain in diabetic acidosis and coma. *J. Clin. Invest.* 27:500, 1948.
33. Kety, S. S., and Schmidt, C. F. The nitrous oxide method for the quantitative determination of cerebral blood flow in man: Theory, procedure, and normal values. *J. Clin. Invest.* 27:476, 1948.

34. Kety, S. S., Woodford, R. B., Harmel, M. H., Freyhan, F. A., Appel, K. E., and Schmidt, C. F. Cerebral blood flow and metabolism in schizophrenia. The effects of barbiturate semi-narcosis, insulin coma, and electroshock. *Am. J. Psychiatry* 104:765, 1948.
35. Klee, C. B., and Sokoloff, L. Changes in D(–)-β-hydroxybutyric dehydrogenase activity during brain maturation in the rat. *J. Biol. Chem.* 242:3880, 1967.
36. Landan, W. H., Freygang, W. H., Rowland, L. P., Sokoloff, L., and Kety, S. S. The local circulation of the living brain; values in the unanesthetized and anesthetized cat. *Trans. Am. Neurol. Assoc.* 80:125, 1955.
37. Larrabee, M. G., Ramos, J. F., and Bülbring, E. Effects of anesthetics on oxygen consumption and synaptic transmission in sympathetic ganglia. *J. Cell. Comp. Physiol.* 40:461, 1952.
38. Lassen, N. A., and Munck, O. The cerebral blood flow in man determined by the use of radioactive krypton. *Acta Physiol. Scand.* 33:30, 1955.
39. Lassen, N. A., Munck, O., and Tottey, E. R. Mental function and cerebral oxygen consumption in organic dementia. *Arch. Neurol. Psychiatry* 77:126, 1957.
40. Lewis, B. M., Sokoloff, L., Wechsler, R. L., Wentz, W. B., and Kety, S. S. A method for the continuous measurement of cerebral blood flow in man by means of radioactive krypton (Kr^{79}). *J. Clin. Invest.* 39:707, 1960.
41. Mangold, R., Sokoloff, L., Conner, E. L., Kleinerman, J., Therman, P. G., and Kety, S. S. The effects of sleep and lack of sleep on the cerebral circulation and metabolism of normal young men. *J. Clin. Invest.* 34:1092, 1955.
42. Meyer, J. S., Ryu, T., Toyoda, M., Shinohara, Y., Wiederholt, I., and Guiraud, B. Evidence for a Pasteur effect regulating cerebral oxygen and carbohydrate metabolism in man. *Neurology* 19:954, 1969.
43. Myerson, A., Halloran, R. D., and Hirsch, H. L. Technique for obtaining blood from the internal jugular vein and internal carotid artery. *Arch. Neurol. Psychiatry* 17:807, 1927.
44. Owen, O. E., Morgan, A. P., Kemp, H. G., Sullivan, J. M., Herrera, M. G., and Cahill, G. F., Jr. Brain metabolism during fasting. *J. Clin. Invest.* 46:1589, 1967.
45. Page, W. F., German, W. J., and Nims, L. F. The nitrous oxide method for measurement of cerebral blood flow and cerebral gaseous metabolism in dogs. *Yale J. Biol. Med.* 23:462, 1951.
46. Pull, I., and McIlwain, H. 3-Hydroxybutyrate dehydrogenase of rat brain on dietary change and during maturation. *J. Neurochem.* 18:1163, 1971.
47. Raskin, N. H., and Sokoloff, L. Alcohol dehydrogenase activity in rat brain and liver. *J. Neurochem.* 17:1677, 1970.
48. Reivich, M., Isaacs, G., Evarts, E. V., and Kety, S. S. The effect of slow wave sleep and REM sleep on regional cerebral blood flow in cats. *J. Neurochem.* 15:301, 1968.
49. Rossen, R., Kabat, H., and Anderson, J. P. Acute arrest of cerebral circulation in man. *Arch. Neurol. Psychiatry* 50:510, 1943.

50. Sloviter, H. A. Personal communication, 1971.
51. Sloviter, H. A., and Kamimoto, T. The isolated, perfused rat brain preparation metabolizes mannose but not maltose. *J. Neurochem.* 17:1109, 1970.
52. Sloviter, H. A., and Suhara, K. A brain-infusion method for demonstrating utilization of glycerol by rabbit brain *in vivo. J. Appl. Physiol.* 23:792, 1967.
53. Sokoloff, L. Local Cerebral Circulation at Rest and During Altered Cerebral Activity Induced by Anesthesia or Visual Stimulation. In Kety, S. S., and Elkes, J. (Eds.), *The Regional Chemistry, Physiology and Pharmacology of the Nervous System.* Oxford: Pergamon, 1961.
54. Williamson, D. H., Bates, M. W., Page, M. A., and Krebs, H. A. Activities of enzymes involved in acetoacetate utilization in adult mammalian tissues. *Biochem J.* 121:41, 1971.

16

Blood-Brain-CSF Barriers

Robert Katzman

ALTHOUGH THE BRAIN obtains its required constant supply of oxygen and glucose from the bloodstream, other substances are not readily absorbed from the blood. For example, fructose cannot be substituted for glucose in the treatment of hypoglycemic coma, although brain slices will metabolize it as well as glucose. The central nervous system of mammals appears to require an ultrastable internal environment in order to function effectively, and there are special controls involved in the transport of many materials in the CNS. The sum total of these special transport mechanisms is called the *blood-brain barrier.* Similar special transport mechanisms between blood and cerebrospinal fluid (CSF) constitute the *blood-CSF barrier.*

Many substances either do not cross brain capillaries (e.g., acidic dyes) or they cross at a very slow rate (e.g., ions). The classic experiments of Paul Ehrlich in 1882 are said to be the first to demonstrate that animals given vital dyes become intensely stained by the dye in all parts of the body except the brain. In 1909 Goldman studied the vital staining by trypan blue and again noted that, after parenteral administration of a dye, only the brain remains unstained. He found that even after injection of such large amounts of trypan blue the animal tissues became intensely stained, the brain remained "snow-white." However, when he instilled a small amount of trypan blue into the subarachnoid space, the brain became intensely stained. The processes that hindered the movement of trypan blue from the bloodstream to the CNS but did not hinder the movement of the dye from the spinal fluid into the brain became known as the *blood-brain barrier.* It is now known that trypan blue quickly binds to albumin in the bloodstream. Hence, the impermeant molecule studied by Goldman was the albumin-dye complex. The morphological basis of this blood-brain barrier to proteins will be discussed shortly.

The concentration of many substances in brain and CSF is independent of their concentration in the blood. Homeostatic mechanisms produce an ultrastable internal environment in the brain as other mechanisms maintain the stable internal environment of the body. The stability of the internal environment may be due either to special transport processes (utilizing mediated transport, active clearance, and other mechanisms) or to the dynamic state of cerebral metabolites. It may be worthwhile to review briefly some of the mechanisms involved in transport. (See also the Davson monograph listed under Books and Reviews.)

1. *Bulk Flow.* Bulk flow means that solutes of various sizes move together with the solvent as a bulk liquid. This concept is important in discussing circulation and absorption in the CSF. Significant bulk flow through the extracellular fluid of the brain has not yet been shown by any adequate experimental evidence.

2. *Diffusion.* Diffusion consists in the movement of a solute from a region of higher to one of lower concentration as a result of the random motion of the solute molecule. Diffusion occurs both in the CSF and in the extracellular spaces of the brain. In addition, diffusion across plasma membranes is important, especially diffusion across the membranes of the capillary endothelial cells. This is the major route of movement of water, urea, and gases into the brain.

3. *Pinocytosis.* In this process, fluid that includes large molecules and even particles is engulfed by invaginating cell membranes, forming a vesicle which then separates from the membrane. This vesicle can now transport its contents across the cell. Under ordinary conditions, pinocytosis is a slow and uncertain process in brain capillaries and probably moves few molecules.

4. *Mediated Transport.* Mediated transport requires a carrier molecule with specific sites for the substrate involved; this permits the substrate molecule to move readily across either the plasma membrane of a single cell or a cellular membrane composed of a sheet of cells in continuity. When the limited number of sites on the carrier molecule are filled, as, for example, when the concentration of substrate molecules on one side of the membrane is sufficient, the number of molecules transported can no longer increase. With further increases of the substrate molecule, the carrier is then saturated and transport is independent of concentration. The kinetics of carrier-substrate transport are identical to those of enzyme-substrate complexing. Moreover, mediated transport implies that another molecule of sufficiently close chemical and steric similarity may occupy the site intended for a given substrate molecule. In the mammalian blood-brain or blood-CSF barrier, such carriers are hypothesized to explain the stereospecificity and concentration independence of transport of hexoses and amino acids, but as yet carrier molecules per se have not been isolated.

5. *Active Transport.* Active transport implies that energy is utilized in the transport of a molecule. To prove the existence of active transport, either the energy-utilizing process must be specifically identified or it must be shown that the molecule is transported against an electrochemical gradient. Although it is possible that the transport of many molecules by carriers does require energy, this must be demonstrated for each molecular species studied. The best examples of active transport are the halides and the small organic molecules which can be actively cleared from CSF, even though the blood level of such molecules is much higher than that in CSF.

6. *Stability due to Transport Processes.* The question arises of how the various transport processes can be combined to provide a high degree of stability for the

constituents of the CSF and the brain extracellular fluid. Bradbury [2] has recently defined stability of the blood-CSF systems as follows. If a substance is present in CSF at concentration C_{CSF} and in plasma at concentration C_{pl}, stability occurs when as a result of a change in plasma concentration a new steady state is reached, so that

$$\Delta C_{CSF} < \Delta C_{pl} \tag{1}$$

At steady state the flux of this substance from plasma to CSF, J_{in}, must equal its flux out, J_{out}, so that for any change in plasma concentration, ΔC_{pl}, stability of CSF will occur when

$$\Delta J_{in}/\Delta C_{pl} < \Delta J_{out}/\Delta C_{CSF} \tag{2}$$

J_{in} and J_{out} represent transport processes that need not be identical. For instance, one might be passive and one active. If the carrier involved in J_{in} is saturated at usual plasma concentrations, then the ratio $\Delta J_{in}/\Delta C_{pl}$ will approach zero. Such carrier mediated transport is probably the most common mechanism controlling the flow of nonlipid soluble substances from the capillary lumen to the brain, but carrier systems have also been found to operate for outward flux. Here, the greatest stability is achieved when the carrier system operates well below saturation, so that the ratio $\Delta J_{out}/\Delta C_{CSF}$ is a positive number. Such asymmetric carrier mechanisms have been implicated in the maintenance of the control of the stability of K^+ in CSF and may also exist for other molecules.

7. *Stability due to Dynamic State of Cerebral Metabolites.* The processes involved in maintaining constant brain levels of molecules that are metabolized within the brain system are obviously complex and depend upon the interplay of dynamic processes. For example, the level of dopamine in brain tissue is relatively constant under normal conditions. This level is the result of an equilibrium between the synthesis of dopamine via tyrosine hydroxylase and DOPA decarboxylase; its degradation by dopamine β-hydroxylase, monoamine oxidase, and catechol-*O*-methyl transferase; and its storage in granules, which isolates it from these degradative enzymes. Synthesis is usually rate-limited by the activity of tyrosine hydroxylase. If this is bypassed by administration of large amounts of the amino acid L-DOPA, the brain dopamine will increase; a similar increase can be obtained by inhibiting the degradative enzymes. If dopamine in storage granules is released by the administration of reserpine, it will be destroyed quickly by the degradative enzymes, and the total brain level will fall dramatically. Finally, it should be noted that the maintenance of specific levels of metabolites in the brain may result, in part, from the action of one of the metabolic products in controlling the rate of an enzyme. For example, the level in the brain tissue of DOPA, a normal product of tyrosine hydroxylase, will control the rate of action of tyrosine hydroxylase. Hence, the interplay of factors in maintaining levels of metabolizable compounds is exceedingly complex and relatively difficult to analyze.

CHARACTERISTICS OF THE BLOOD-BRAIN BARRIER

MORPHOLOGY

For many years there was confusion about the nature of the blood-brain barrier since it did not seem conceivable that the capillaries within the brain could be different from capillaries elsewhere in the body. However, ultrastructural studies using markers such as peroxidase, the reaction product of which can be stained in situ with osmium, have shown that the capillary endothelium is a continuous layer, with tight junctions between contiguous cells that do not permit the passage of protein markers from the capillary lumen to the basement membrane that surrounds the capillary endothelium [14]. Recently, studies using microperoxidase have shown that molecules with molecular weights as low as 2000 are excluded by these tight junctions [13]. Whether the tight junctions also exclude small hydrophilic molecules of the molecular weight of sugars and amino acids has not been established morphologically. In contrast, if the protein marker enters the extracellular space of the brain (for example, if it is placed in the subarachnoid or ventricular space), it will diffuse through this extracellular space until it reaches the capillary endothelium. Under these circumstances, the tight junctions will prevent the molecules from diffusing into the capillary lumen.

DIFFUSION

Water is the most important substance entering the brain by diffusion. When D_2O is administered intravenously as a tracer, the half-time of exchange of brain water varies between 12 and 25 sec, depending upon the vascularity of the region studied. The calculated permeability constant of the capillary wall to the diffusion of D_2O is about the same as that estimated for the diffusion of water across the lipid membranes. Water also moves freely into or out of the brain as the osmolality of the plasma changes. This phenomenon is clinically useful, since intracranial pressure can be reduced by the dehydration of the brain after plasma tonicity has been elevated by such substances as mannitol. It has been calculated that when plasma osmolality is raised from 310 mOsm to 344 mOsm, for example, a 10 percent shrinkage of the brain will result, with half of the shrinkage taking place in 12 min. The permeability constant calculated for this osmotic flow of water is slightly larger than that for diffusional flow. However, the movement of water under osmotic load in the example given is slower because the concentration gradient for the osmotic load is only 34 mOsm but for D_2O, the solvent, it is 55 Osm.

Gases such as CO_2, O_2, N_2O, Xe, Kr, and volatile anesthetics diffuse rapidly into brain. As a consequence, the rate of concentration at which the brain comes into equilibrium with plasma concentration is limited primarily by the cerebral blood flow. Hence, the inert gases – N_2O, Xe, and Kr – have been widely used to measure cerebral blood flow, both by the Kety technique for total cerebral blood flow and the Lassen-Ingvar technique for regional blood flow (Chap. 15).

Lipid-soluble substances also move freely, diffusing across the plasma membranes in proportion to their lipid solubility. The permeability of ethanol is greatest of all solutes measured. The movement of antipyrine into the brain is sufficiently rapid to be used to monitor cerebral blood flow. The permeability constant of such substances as thiopental and aminopyrine is similar to that of antipyrine; barbital is somewhat slower, and urea and salicylic acid are still slower. In a compound such as salicylic acid, the unionized form diffuses across the membranes, and hence the dissociation constant and the pH of the blood are of importance [3].

MEDIATED TRANSPORT

There is evidence that glucose, amino acids, and ions are transported by carrier systems. Due to the steric specificity of the glucose transport system, *d*-glucose but not *l*-glucose will enter the brain. Such hexoses as mannose and maltose will also be transported rapidly into the brain; the uptake of galactose is intermediate, whereas fructose is taken up very slowly [9]. 2-Deoxyglucose, on the other hand, is taken up quickly and will competitively inhibit the transport of glucose. There is great variability in the rate of movement of amino acids into the brain [9]. Phenylalanine, leucine, tyrosine, isoleucine, tryptophan, methionine, histidine, and DOPA may enter as rapidly as glucose, whereas alanine, proline, glutamic acid, aspartic acid, γ-aminobutyric acid, and glycine are virtually excluded. Lysine, arginine, threonine, and serine are intermediate in their rate of uptake. Competitive inhibition is easy to demonstrate. Although there may be five or six major carriers, several amino acids probably have affinity for more than one carrier. It should be noted that the amino acids that are taken up rapidly are primarily essential amino acids not synthesized in the brain, whereas amino acids readily synthesized from glucose metabolites are virtually excluded. Simple charged ions will exchange with brain ions, although noticeably slower than with other tissues. For example, intravenously administered $^{42}K^+$ exchanges with muscle in 1 hr, but K^+ exchange in brain is only half completed in 24 to 36 hr. The rate of K^+ flux shows little change as plasma K^+ is varied. Ca^{2+} and Mg^{2+} exchange as slowly as does K^+. Na^+ is somewhat faster, with half exchange occurring in 3 to 8 hr, but Na^+ exchange is much slower between blood and brain than between blood and the Na^+ of other tissues.

A good contrast is found between the effects of CO_2, a gas which diffuses freely, and H^+, which moves very slowly into brain. Consequently, the brain pH will reflect blood pCO_2 rather than blood pH. In a patient with a metabolic acidosis and a secondary respiratory alkalosis, the brain then will tend to become alkalotic.

It is not known which of these carrier-mediated transport processes are immediately linked to metabolism. In many tissues, the transport of organic molecules such as amino acids requires the presence of Na^+; whether this is true for brain tissue has not been established.

In addition to carrier-mediated transport into brain tissue, carrier-mediated transport out of such tissue has been established for some amino acids. Mediated transport of organic acids and certain halogen ions from CSF into the blood has been well established. There is some suggestive, but not entirely conclusive, evidence that this is also true for movement of these molecules from brain into blood.

ENZYMATIC BARRIERS IN CAPILLARY ENDOTHELIUM

For at least two molecules, γ-aminobutyrate and L-DOPA, transport between blood and brain tissue is retarded by enzymatic degradation of these substances within the capillary endothelium. In both instances, the rate of degradation is such that with administration of very large amounts of γ-aminobutyrate or L-DOPA, there will be an increase in the amount of these substances reaching brain tissue [10]. Such enzymatic degradation can be demonstrated histochemically.

PERMEABILITY OF CAPILLARY ENDOTHELIUM

The existence of the tight endothelial junctions impedes water-soluble molecules that are much larger than urea. The evidence from ultrastructural studies shows that molecules with molecular weights as low as 2000 do not enter. Physiological studies in which the osmotic effect of different solutes upon the brain has been monitored have shown that there is little diffusion of hexoses [5]; the uptake of hexoses noted previously is presumably via mediated transport.

Because of the failure of proteins to move into the brain, molecules that are bound to protein are, therefore, impermeant, even if intrinsically lipid-soluble. Examples of these are dyes such as trypan blue and the important molecule, bilirubin. If sulfa drugs are administered to an infant or young animal with elevated bilirubin, the bilirubin will be displaced from the albumin by the sulfa drug. This can lead to an increase in kernicterus in jaundiced infants. Some movement of impermeant molecules can occur by pinocytosis, but this is slight in the cerebral capillary endothelium as compared with that in other capillaries.

INFLUENCES ON BLOOD-BRAIN BARRIER

DEVELOPMENT

Many of the features of the blood-brain barrier, including the presence of tight junctions in the capillary endothelium and the exclusion of protein molecules, are present in newborn animals. Transport of certain substances – for example, simple ions – is moderately faster in the newborn animal. There is a very great increase in the rate of uptake of certain actively metabolized substances in the rapidly growing brain, which may result either from transport per se or from the high rate of turnover of the metabolites.

CHEMICALS AND DRUGS

As already discussed, competitive inhibitors may interfere with carrier-mediated transport of molecules into the brain. Drugs such as dilute mercuric chloride given intravascularly may increase the permeability of the blood-brain barrier. There is some possibility that greatly elevated levels of pCO_2 may also increase blood-brain barrier permeability [8].

PATHOLOGY

A characteristic of almost any focal injury to the brain – whether it is produced by excessive convulsions, knife wounds, freezing, heating, electrical currents, tumors, inflammatory agents, toxins, or whatever – is the breakdown of the blood-brain barrier with subsequent diffusion of protein molecules into brain tissue. This phenomenon has been widely used to trace such areas of injury, since administration of dyes such as trypan blue bound to protein or fluorescein bound to protein enables histological identification of those areas in which the blood-brain barrier has broken down. The mechanism of traumatic injury to the capillary endothelium is still under investigation. Apparently, the tight junctions between cells are sometimes ruptured; in other instances, the capillary endothelial cells are simply torn, so that plasma proteins may pour into the area of injury. But in some instances, at least in certain brain tumors, the capillary endothelium thins out, forming fenestrations across which proteins may move.

REGIONAL DIFFERENCES IN THE BLOOD-BRAIN BARRIER

There are, of course, significant differences in the distribution of metabolites throughout the brain. Hence, the blood-brain barrier, in a broad sense of the term, must be different in different regions. Movement of substances into given regions will depend upon such factors as capillary density. Most important, however, are various regions that are "excluded" from the blood-brain barrier. These are regions in which the capillary endothelium contains fenestrations across which proteins and small organic molecules may move from the blood into the adjacent tissue. Examples of such areas are the area postrema, the median eminence of the hypothalamus, the line of attachment of the choroid plexus, and the pineal gland. These areas may have special functional significance in that they may be where the brain samples the contents of the blood. The area postrema is close to what has been called the "vomiting" center of the brain, and the hypothalamus is involved in the regulation of the body's metabolic activity. Therefore, these areas that lack the blood-brain barrier may provide sites at which neuronal receptors may sample plasma directly.

CHARACTERISTICS OF BLOOD-CSF BARRIER

COMPOSITION OF CSF

Cerebrospinal fluid has been characterized as an "ideal" physiological solution. It differs from plasma in that it is almost free of protein, and it differs from an ultrafiltrate of plasma by maintaining the concentrations of various ions at different levels (Table 16-1).

TABLE 16-1. Typical Plasma and CSF Levels of Various Substances[a]

Substance	Plasma	CSF
Na^+	145.0	150.0
K^+	4.8	2.9
Ca^{2+}	5.2	2.3
Mg^{2+}	1.7	2.3
Cl^-	108.0	130.0
HCO_3^-	27.4	21.0
Lactate	7.9	2.6
PO_4^{3-}	1.8	0.5
Protein	7000.0	20.0
Glucose	95.0	60.0

[a]Protein and glucose concentrations in mg/100 ml; all others in meq/liter.

CSF FORMATION

CSF is constantly being formed and removed. The chief site of formation is in the choroid plexuses of the ventricles, where the rate of secretion of CSF has been estimated to be 0.3 to 0.4 ml per minute. In normal subjects, this is about one-third the rate of formation of urine. The fluid elaborated in the ventricles contains only 5 to 10 mg of protein per 100 ml.

There is considerable evidence that the choroid plexus acts to secrete the CSF. Histochemical and electron microscopic investigations indicate that the cells have morphological features similar to other secretory cells. Formation of CSF within the ventricles has been deduced from the fact that if the foramen is obstructed, fluid rapidly accumulates and the ventricle enlarges. From time to time, neurosurgeons have reported seeing drops of fluid form on the surfaces of the choroid plexus.

The formation of CSF by the choroid plexus has been studied more directly. Welch [16] has been able to cannulate the artery and vein of a single choroid plexus in the rabbit. By measuring the average differences of radioisotopes and of the hematocrit, he demonstrated that the formation of CSF and the movement of $^{24}Na^+$ are stopped when a carbonic anhydrase inhibitor (diamox) is applied to the plexus. Thus, CSF is not simply a serum ultrafiltrate, but is controlled by enzymatic processes. Ames et al. [1] were able to collect and analyze the fluid formed at the choroid plexus after filling the ventricle with oil. In the cat, they found that the concentration of electrolytes, particularly K^+, Ca^{2+}, and Mg^{2+}, differs from that of a plasma ultrafiltrate. However, the choroid plexus fluid was again found to change very slightly in its electrolyte concentration by the time it reaches the cisterna magna.

The rate of formation of CSF has been measured by various means. One simple measurement, carried out more than 30 years ago by Masserman, was to determine the time for replacement of CSF after a known amount had been drained. The rate in man was 0.35 ml per minute. This measurement was criticized on the basis that the drainage altered the intracranial pressure relationships and, therefore, probably altered (increased) the rate of CSF formation. Recently, a sophisticated method of measuring CSF formation was introduced by Pappenheimer and co-workers [6, 12]. In this method, simulated spinal fluid is perfused between the ventricle and the cisterna magna, and inulin is added to the perfusion fluid. Because inulin diffuses very slowly into tissue during such a perfusion, the elution of inulin is taken as a measure of the rate of formation of new spinal fluid. Therefore, if the perfusion at a rate of V_i ml per minute is carried out until a steady state is reached, and if the initial concentration of inulin is C_i and the outflow concentration of inulin is C_o, the rate of formation of spinal fluid, V_f, is given by the equation:

$$V_f = V_i (C_i - C_o)/C_o \tag{3}$$

Such perfusions have been carried out in a wide variety of species. Typical rates of spinal fluid formation are: rabbit, 0.001 ml per minute; cat, 0.02 ml per minute; rhesus monkey, 0.08 ml per minute; and goat, 0.19 ml per minute. Recently, both Cutler et al. [4] and Rubin et al. [15] carried out measurements in patients undergoing ventriculocisternal perfusion with antitumor drugs and found an average value of 0.37 ml per minute. This corresponds rather closely to the value previously determined from drainage experiments.

By using the ventriculocisternal perfusion method, it has been possible to demonstrate that CSF is formed at about the same rate, despite moderate changes in intracranial pressure [7]. Moreover, CSF is formed at a normal rate and osmolality even when the fluid perfused through the ventricle is moderately hypertonic or hypotonic. However, when the fluid is very hypotonic, CSF formation may cease.

The total volume of CSF is not precisely measurable. It is estimated to be 100 to 150 ml in normal adults.

The circulation of CSF begins with the elaboration of fluid in the ventricles. It is aided by arterial pulsations of the choroid plexuses. These pulsations are transmitted throughout the CSF and can be seen on the manometer at the time of lumbar spinal tap. The usual manometer, however, is filled with spinal fluid, and this displacement of fluid tends to dampen the pulsations. More prominent pulsations are recorded if a strain gauge is used.

The fluid circulation is from the lateral ventricles into the third ventricle and then into the fourth ventricle. If obstructions are placed at the foramen between these ventricles, the ventricle upstream from the obstruction will enlarge significantly, producing obstructive hydrocephalus. Thus, if a foramen of Monro is obstructed, the lateral ventricle will enlarge. If the aqueduct is obstructed, both lateral ventricles and the third ventricle will enlarge. The fluid passes from the fourth ventricle to the cisterna magna and then circulates into the cerebral and spinal subarachnoid spaces.

The CSF is removed at multiple sites. Among the important sites are nerve root sheaths in the spinal column and the villi and granulations over the large venous sinuses in the skull. There is evidence that the villi act as valves, permitting the one-way flow of CSF from the subarachnoid spaces into sinuses. Occasionally, disease processes will affect these removal sites. For example, abnormal proteinaceous material may clog the villi, or thrombosis of the sinuses will prevent the clearance of fluid. When this occurs, CSF pressure will increase, and hydrocephalus may develop without any obstruction in the ventricular foramina. This is called communicating hydrocephalus.

Physiologically, absorption appears to be via a valvular mechanism. CSF absorption does not occur until CSF pressure exceeds the pressure within the sinuses. Once this threshold pressure has been reached, the rate of absorption is roughly proportional to the difference between CSF and sinus pressures. A normal human can absorb CSF at a rate up to four to six times the normal rate of CSF formation with only a moderate increase in intracranial pressure. This phenomenon has been used in the development of a constant infusion manometric test for diagnosis of occult communicating hydrocephalus. The CSF spaces also show elasticity or distensibility: that is, the volume of the space changes in a nonlinear fashion with changes in CSF pressure.

Throughout the circulation of CSF, exchange of substances between it and blood occurs. This exchange appears to involve metabolic processes that serve to maintain the concentration of substances within the CSF at relatively fixed values.

SIMILARITY TO THE BLOOD-BRAIN BARRIER

The movement of substances from the blood into CSF is, in many ways, analogous to that from blood into the brain, even though there are differences in the development of the choroid plexus and of the capillary system in the brain. There is free movement of molecules such as water, gases, and lipid-soluble substances from the blood into the CSF. Substances important for metabolism and maintenance of CSF electrolytes, such as glucose, amino acids, and cations (including K^+, Ca^{2+}, and Mg^{2+}), are transported by saturable carrier-mediated processes. The transport of Na^+ from blood into CSF is significantly reduced when carbonic anhydrase is inhibited. With Na^+, however, it has not been possible to demonstrate the phenomenon of saturability because blood Na^+ cannot safely be elevated much above its usual concentrations. Finally, macromolecules such as proteins and most hexoses other than glucose are impermeable and do not enter the CSF.

ACTIVE TRANSPORT FROM CSF

Active transport from CSF has been especially well studied because it is possible to manipulate the concentration gradient of substances between the CSF and the bloodstream. It has been shown that iodide, thiocyanate, and organic molecules, including diodrast and penicillin, are transported from the CSF by saturable carrier mechanisms, can be competitively inhibited (for example, perchlorate will inhibit iodide transfer), involve the choroid plexus, and must be active since the transport can be carried out against unfavorable electrochemical gradients. The combination of the bulk absorption of CSF and the active transport of molecules from it has been termed the "sink" function of the CSF. This implies that molecules reaching the extracellular fluid of the brain may diffuse into CSF and then be removed either by bulk absorption, by active transport, or by both mechanisms. It has been postulated that this will explain such phenomena as the low concentration of iodide in both brain and CSF, as compared with plasma concentrations. The possibility that brain capillaries may transport the same substances out of the brain as the choroid plexus does out of the CSF has not been ruled out.

CSF-BRAIN INTERFACE

The absence of tight junctions between some ependymal and pial cells permits diffusion of proteins and other hydrophilic molecules from the CSF into the brain, and vice versa. Frequently, the diffusional gradients near the interface are very steep, and many substances penetrate only superficially. Some recent evidence shows that the diffusion across the ependyma is slightly faster than that across the pial-glial surface; however, molecules do move across both surfaces by diffusion.

Quantitative studies of the movement of substances between CSF and brain have led some investigators to postulate that the concentration of ions in extracellular fluid in brain is similar to that of CSF. Because the concentrations of K^+, Ca^{2+}, and Mg^{2+} in CSF are quite different from those in plasma, the verification of this postulate would be of considerable physiological significance. It might be argued then that one of the functions of CSF, in addition to mechanical protection of the brain and its "sink" function, may be to provide a reservoir of extracellular fluid.

The CSF-brain interface may play a special role in the regulation of respiration. There is suggestive evidence that some respiratory neurons or their processes are located close enough to the ependymal surface to be responsive both to the HCO_3^- level in the CSF and to capillary pCO_2 [11].

REFERENCES

ORIGINAL SOURCES

1. Ames, A., Sakanoue, M., and Endo, S. Na, K, Ca, Mg, and Cl concentrations in choroid plexus of fluid and cisternal fluid compared with plasma ultrafiltrate. *J. Neurophysiol.* 27:672, 1964.

2. Bradbury, M. W. B., and Stulcova, B. Efflux mechanism contributing to the stability of the potassium concentration in cerebrospinal fluid. *J. Physiol. (Lond.)* 208:415, 1970.
3. Crone, C. The permeability of brain capillaries to non-electrolytes. *Acta Physiol. Scand.* 64:407, 1965.
4. Cutler, R. W. P., Page, L., Galicich, J., and Watters, G. V. Formation and absorption of cerebrospinal fluid in man. *Brain* 91:707, 1968.
5. Fenstermacher, J. D., and Johnson, J. A. Filtration and reflection coefficients of the rabbit blood-brain barrier. *Am. J. Physiol.* 211:341, 1966.
6. Heisey, S. R., Held, D., and Pappenheimer, J. R. Bulk flow and diffusion in the cerebrospinal fluid system of the goat. *Am. J. Physiol.* 203:775, 1962.
7. Katzman, R., and Hussey, F. A simple constant-infusion manometric test for measurement of CSF absorption. 1. Rationale and method. *Neurology* 20:534, 1970.
8. Mayer, S., Maickel, R. P., and Brodie, B. B. Kinetics of penetration of drugs and other foreign compounds into cerebrospinal fluid and brain. *J. Pharmacol.* 127:205, 1959.
9. Oldendorf, W. H. Brain uptake of radiolabeled amino acids, amines, and hexoses after arterial injection. *Am. J. Physiol.* 221:1629, 1971.
10. Owman, C., and Rosengren, E. Dopamine formation in brain capillaries – an enzymic blood-brain barrier mechanism. *J. Neurochem.* 14:547, 1967.
11. Pappenheimer, J. R. The ionic composition of cerebral extracellular fluid and its relation to control breathing. *Harvey Lect.* 61:71, 1967.
12. Pappenheimer, J. R., Heisey, S. R., and Jordan, E. F. Active transport of diodrast and phenolsulfonphthalein from cerebrospinal fluid to blood. *Am. J. Physiol.* 200:1, 1961.
13. Reese, T. S., Feder, N., and Brightman, M. W. Electron microscopic study of the blood-brain and blood-cerebrospinal fluid barriers with microperoxidase. *J. Neuropathol. Exp. Neurol.* 30:137, 1971.
14. Reese, T. S., and Karnovsky, M. J. Fine structural localization of a blood-brain barrier to exogenous peroxidase. *J. Cell Biol.* 34:207, 1967.
15. Rubin, R. C., Henderson, E. S., Ommaya, A. K., Walker, M. D., and Rall, D. P. The production of cerebrospinal fluid in man and its modification by acetazolamide. *J. Neurosurg.* 25:430, 1966.
16. Welch, K. Secretion of cerebrospinal fluid by choroid plexus of the rabbit. *Am. J. Physiol.* 205:617, 1963.

BOOKS AND REVIEWS

Davson, H. *Physiology of the Cerebrospinal Fluid.* Boston: Little, Brown, 1967.

Lajtha, A. (Ed.). *Handbook of Neurochemistry.* Vol. 2. *Structural Neurochemistry.* New York: Plenum, 1969. (In particular see Chapter 1, Pappius, H. M., Water Spaces, and Chapter 2, Katzman, R., and Schimmel, H., Water Movement.)

Lajtha, A. (Ed.). *Handbook of Neurochemistry*. Vol. 4. *Control Mechanisms in the Nervous System.* New York: Plenum, 1970. (In particular see Chapter 14, Katzman, R., Ion Movement.)

Lajtha, A., and Ford, D. H. (Eds.). *Brain Barrier Systems.* Amsterdam: Elsevier, 1968.

Tschirgi, R. D. Chemical Environment of the Central Nervous System. In Field, J., Magoun, H. W., and Hall, V. E. (Eds.), *Handbook of Physiology, Section 1: Neurophysiology*. Vol. 3. Baltimore: Williams & Wilkins, 1960.

17

Hypothalamic-Pituitary Regulation

George J. Siegel and
Joseph S. Eisenman

THE NEUROENDOCRINE SYSTEM is broadly defined as those structures in the central nervous system that are concerned with the regulation of endocrine function. With the known exceptions of the direct innervations of the adrenal medulla and the pineal gland, CNS influences on endocrine functions are mediated by hypothalamo-hypophysial connections. The main paths in mammals are: (1) the neural tracts from the supraoptic and paraventricular nuclei to the posterior pituitary lobe and (2) the portal circulatory link between the median eminence and the anterior pituitary lobe or adenohypophysis. At these interfaces between the CNS and the endocrine system, coding of information is transformed from neural action potentials to the secretion of specific chemicals into the bloodstream. In the first case, oxytocin and vasopressin are liberated from nerve terminals in the neurohypophysis (posterior pituitary lobe) into the systemic circulation. In the second instance, hypothalamic nerve terminals in the median eminence secrete several different hypophysiotropic hormones into the pituitary portal system which act on the adenohypophysis.

NEUROHYPOPHYSIS

The hypothalamo-neurohypophysial complex consists of the supraoptic and paraventricular nuclei, their axons coursing in the supraoptico-neurohypophysial tract through the infundibular stalk to the neurohypophysis (posterior pituitary lobe) and the posterior lobe itself. Within the posterior lobe, the axons end adjacent to perivascular spaces. Glial cells, or pituicytes, are found intermingled among the axons and nerve terminals.

Neurohypophysial extracts from different vertebrate species have yielded several nonapeptide hormones that are classified into two groups according to their main biological activity: the oxytocic–milk-ejecting and vasopressor–antidiuretic groups. Mammalian pituitaries contain oxytocin, which represents the first group, and arginine-vasopressin or lysine-vasopressin, which represent the second group. Lysine-vasopressin is found only in porcine tissue [12]. The chemistry and synthesis of these peptides and their analogs originated from the pioneering work by du Vigneaud and his school and have been fully described [16]. Detailed reviews of the chemistry, biology, and pharmacology of these peptides have been compiled recently [6].

NEUROSECRETION

Early workers who extracted vasopressin and oxytocin from the posterior pituitary assumed that the pituicytes were the source of these factors and that these cells were controlled by hypothalamic neurons. Largely through the work of Scharrer, Scharrer, and Bargmann it was recognized that this assumption is incorrect. The pituicytes are not glandular but glial cells, and the terminals of the supraoptico-hypophysial tract are not in apposition to the pituicytes but rather to vascular channels [14]. These authors developed the concept that the hypothalamic neurons themselves are the sites of hormone synthesis and that their terminals represent hormone storage sites.

Strong support for this concept was obtained by using Gomori histological stains, which identify aggregates of proteinaceous material; such aggregates, or granules, are typical of products destined for secretion within glandular cells. In these studies, Bargmann traced secretory material from the perikarya within the hypothalamic nuclei along axons to their terminals in the posterior lobe. Subsequent experiments showed that the quantity of secretory material that was demonstrated as Gomori-positive granules correlated with the amount of antidiuretic activity in the hypothalamo-neurohypophysial complex. Furthermore, when the pituitary stalk was severed, secretory material accumulated rostral to the severance. Bargmann and Scharrer postulated that "the pars nervosa of the vertebrate hypophysis stores, but does not produce, the stainable material which it contains. This material originates in the neurosecretory cells of the nuclei supraopticus and paraventricularis in the higher vertebrates...; it passes to the pars nervosa by way of the hypothalamo-hypophysial tracts. There is evidence that this stainable material carries the antidiuretic, oxytocic, and vasopressor principles from the site of origin in the hypothalamic nuclei to the place of storage in the pars nervosa" [20].

The production and release of hormones by neurons that have the cytological features of secretory cells is called *neurosecretion;* it has been well characterized in the neurohypophysial system. Gomori-positive secretory granules have been isolated from hypothalamic-median eminence tissue as well as from the posterior pituitary. These membrane-bound granules, 1000 to 3000 Å in diameter, are rich in hormone activity. Within the granules, the peptide hormones are noncovalently bonded to a carrier protein, neurophysin, which actually constitutes the bulk of the Gomori-positive material.

BIOSYNTHESIS

Biochemical studies show that the hypothalamic-median eminence tissue is the primary site for vasopressin and neurophysin biosynthesis. For example, the specific radioactivity of vasopressin and neurophysin isolated from dog hypothalamic tissue after infusion of ^{35}S-cysteine into the third ventricle is several times greater than in the posterior lobe. In addition, guinea pig hypothalamic-median eminence tissue, but not neurohypophysial tissue, has been found capable of vasopressin biosynthesis in vitro. However, it is considered likely that some modifications of the hormone or its complex with neurophysin may take place during transport within the axon or in the axon terminal. Information on the initial steps in vasopressin biosynthesis suggests there is a precursor molecule that is biologically inactive [12].

NEUROPHYSIN

The carrier protein, neurophysin, appears to be synthesized simultaneously with the peptide hormones, and is also released with them. Neurophysin has the property of binding vasopressin and oxytocin specifically, and it presumably functions in the storage of these hormones [12]. Neurophysin isolated from bovine neurosecretory granules consists of two major proteins of similar amino acid composition that are referred to as neurophysin-I and neurophysin-II [26]. Under acid conditions (pH 3 to 4), which favor catheptic enzyme activity, several other proteins, which are believed to be proteolytic products, are obtained. Although both neurophysins are capable of binding vasopressin and oxytocin, it is thought that they might occur in separate cells, one neurophysin for each hormone. There is evidence that, in vivo, neurophysin-II may be associated with vasopressin. Neurophysin-II, which has been most extensively studied, has a molecular weight of 10,000, and the monomer binds vasopressin in the ratio of 1:1 [30].

The complete amino acid sequence of bovine neurophysin-II, which consists of 97 residues with a molecular weight of 10,043, has been achieved recently [48]. The structure of a porcine neurophysin that contains 92 residues has been obtained; this structure is a fairly close homolog of bovine neurophysin-II [49].

NEURAL REGULATION OF HORMONE RELEASE

The hormones of the posterior lobe are released in response to a variety of stimuli. Osmotic stimuli, such as infusion of hyperosmotic solutions into the carotid artery, or changes in vascular volume by graded hemorrhages, lead mainly to release of vasopressin. Pain and certain emotional states may also affect vasopressin release. The antidiuretic and pressor activities of vasopressin help maintain constant blood volume and osmotic content, as well as blood pressure. Oxytocin, on the other hand, is mainly released in response to suckling, coitus, and parturition, with the attendant effects of milk ejection and uterine contractions. Under some conditions, however, oxytocic and vasopressor activities in the plasma are increased simultaneously. Hence, the pathways for regulation of these two principles, although independent, show some convergence.

It was deduced from the classic experiments of Verney that osmoreceptors exist in the anterior hypothalamus, which is the location of the supraoptic and paraventricular nuclei. In addition, there is evidence that several groups of receptors within the cardiovascular system participate in regulating vasopressin release in response to changes in volume, pressure, and oxygenation of the blood. These receptors apparently are important in integrating the regulation of blood pressure and blood volume. The physiology and pharmacology of such receptors have been reviewed in detail [4, 6, 15].

Little is known about the precise neuronal pathways that mediate the various physiological responses, but the final path is obviously through the hypothalamus. It has been amply demonstrated that acute secretory stimuli and electrical stimulation of the supraoptic and paraventricular nuclei and pituitary stalk result in rapid release of vasopressin and oxytocin. The supraoptic nucleus is more concerned with vasopressin and the paraventricular nucleus more with oxytocin, although this demarcation is not complete. Stimuli which provoke hormone release also increase the firing frequency of neurons in these hypothalamic nuclei. The stimuli excite hypothalamic units in which potentials also can be evoked antidromically by electrical stimulation of the pituitary stalk. This latter procedure identifies hypothalamic neurons afferent to the pituitary [28]. Iontophoretic application of acetylcholine increases and norepinephrine decreases single-unit firing rates of antidromically identified neurosecretory cells in the cat supraoptic nucleus [21].

EXCITATION-SECRETION COUPLING

The last link in the regulation of hormone release is mediated by the propagated action potentials in the supraoptico-neurohypophysial tract. The mechanism for the coupling of neuronal conduction to secretion appears to depend upon depolarization that leads to Ca^{2+} uptake into the neuron. It has been found that, in the presence of Ca^{2+}, the isolated pars nervosa fails to respond to osmotic stimuli but does release hormone upon electrical stimulation or increased K^+ concentration. The hormone release is strictly dependent on Ca^{2+} and is associated with Ca^{2+} uptake. The precise role of Ca^{2+} in the release process has not been defined. Although Ca^{2+} inhibits the binding of hormone to neurophysin, the physiological role of this action is unknown, and there is no evidence that this effect constitutes the Ca^{2+} function in the physiological release mechanism. On the contrary, there is evidence that neurophysin is released together with hormone, and one hypothesis holds that the entire granule content is discharged in a process of exocytosis [12]. The role of acetylcholine in hormone release appears to be mainly through its neurotransmitter action on the hypothalamic neurons because the isolated neurohypophysis fails to respond to acetylcholine. The physiological role of acetylcholine found in the neural lobe is not known [37].

The neurosecretory function of the hypothalamic-neurohypophysial tract provides an important model for the study of neuronal control of biochemical events; the output of vasopressin, for example, is specific for this precisely localized nerve bundle and is sensitive to fairly well-defined stimuli, such as dehydration. Hypothalamic-median eminence slices from guinea pig deprived of water for 4 days incorporate two to five times more radioactivity into vasopressin than do similar slices from nondehydrated animals. Neurons of rat supraoptic nuclei show increases in RNA content and in numbers of ribosomes when the animals are subjected to chronic osmotic stimuli or dehydration. Thus, prolonged neural receptor activity is capable of inducing major synthesis, not just of an end-product, but of the entire system of participating biosynthetic machinery [12]. Attempting to uncover the intervening biochemical events directly induced by membrane potential changes is one of the most exciting problems in modern neurobiology.

ADENOHYPOPHYSIS

HYPOTHALAMIC-ADENOHYPOPHYSIAL COMMUNICATION

The anterior pituitary secretes the tropic hormones that control the function of the gonads, adrenal, thyroid, and mammary glands, as well as secreting growth hormone (somatotropin), which has widespread actions in the body (Table 17-1). The reader is referred to standard texts on endocrinology for full accounts of their actions on target organs.

Although many of the pathways have not been well defined anatomically, there is abundant physiological evidence for neural input to the hypothalamus from other CNS areas [8b]. Stimulation of peripheral nerves, midbrain reticular formation, or limbic structures evokes changes in the activity of hypothalamic neurons. Examples of neural inputs influencing endocrine function include cold-induced thyroxine

TABLE 17-1. Hypothalamic and Pituitary Hormones

Hypothalamic Hormones	Pituitary Hormones	Target Organs
	Adenohypophysis	
LH-RF and FSH-RF	LH and FSH	Gonads
PIF	Prolactin	Mammary glands
CRF	ACTH	Adrenal cortex
TRF	Thyrotropin	Thyroid gland
GH-RF	Growth hormone	Many tissues
	Pars Intermedia	
MSH-IF and MSH-RF	MSH	Melanocytes
	Neurohypophysis	
Oxytocin	(Site for storage and secretion)	Mammary glands, uterus
Vasopressin		Kidneys, vascular smooth muscle

secretion, stress-induced adrenocorticoid secretion, and ovulation induced by mechanical stimulation of the vagina. Many neural paths converge upon the hypothalamus, which is then the final link in controlling the adenohypophysis.

Hypothalamic neurons send their axons to the median eminence or infundibulum, where they end in close proximity to the capillary loops of the portal system [5]. The median eminence is not hypothalamic tissue, but rather is a part of the neurohypophysis. It contains mainly the longitudinally coursing axons of hypothalamic neurons, glial cells, and the capillary loops of the portal system. It does not appear to have neuronal cell bodies. The blood-brain barrier is absent here, as it is in the neural lobe of the pituitary gland. This feature is most important because it allows substances released from the axon terminals in these areas to diffuse into the circulation: into the pituitary portal system from the median eminence and into the general circulation from the neurohypophysis. The absence of a blood-brain barrier also means that the axon terminals will be exposed to blood-borne compounds. Whether this has any functional significance is not known.

Isolation of the pituitary gland from the hypothalamus by stalk section or by transplantation to another site in the body depresses the secretion of the adenohypophysial tropic hormones except prolactin secretion, which is increased. Electrical stimulation in the median eminence induces release of the tropic hormones except prolactin, the secretion of which is inhibited. Lesions in this region have the reverse effect. Thus, it is clear that the neural terminals in the median eminence contain substances – the releasing factors (RF) or hypothalamic hypophysiotropic hormones (HHH) – which cause secretion of all the anterior pituitary tropic hormones except prolactin, whose release is inhibited by the hypothalamic hormone (prolactin inhibitory factor, PIF).

Because the median eminence forms the final common pathway for control of the anterior pituitary lobe, most attempts to isolate and purify the releasing factors begin with extracts of the stalk-median eminence (SME) tissue. The disadvantage in using this material is that all of the releasing factors are found in it and the purification procedures must separate the individual factors from each other as well as from inactive components. An advantage to using SME extracts is that the specific activity of the releasing factors may be much higher than in the diffusely organized hypothalamic nuclei.

Injection of SME extracts in animals previously prepared by section of the pituitary stalk causes inhibition of prolactin secretion and release of all the other tropic hormones. Addition of these extracts to pituitary glands in vitro produces similar effects. These results again demonstrate the presence of prolactin inhibitory factor and releasing factors in the median eminence [8c].

GENERAL FEATURES OF THE HYPOTHALAMIC HORMONES

It is now generally accepted that the hypothalamic control of anterior pituitary lobe activity is mediated entirely through the releasing factors carried by the hypophysial portal circulation to the pituitary gland from hypothalamic nerve terminals in the median eminence.

Although most evidence indicates that a different releasing factor exists for each of the hormones produced by the anterior lobe, there is recent evidence that at least in the pig the factors for follicle-stimulating hormone (FSH) and luteinizing hormone (LH) release may be the same [41].

The hypothalamic areas – which, when stimulated or destroyed, most clearly influence adenohypophysial function – have been collectively termed the *hypophysiotropic area* by Halász. This area includes the ventromedial tuberal hypothalamus (median eminence, arcuate nucleus, and parts of the ventromedial nucleus) as well as the ventromedial anterior hypothalamic and suprachiasmatic portions [4a]. The somata of neurons that synthesize the releasing factors appear to be located in various parts of the hypophysiotropic area. For some releasing factors, the cells seem to be localized in a fairly discrete area, while for others they are more widely dispersed.

One method for determining the locations of these neurons is to place a lesion in a part of the hypothalamus and, after allowing a few days for degeneration to progress, to assay the concentration of releasing factor in the median eminence. Using this technique, it was found that follicle-stimulating hormone releasing factor (FSH-RF) disappeared from the median eminence after lesions were made in the area of the paraventricular nuclei. On the other hand, some reduction in thyrotropin releasing factor (TRF) in the median eminence took place after lesions were made anywhere in the ventral tuberal area. Luteinizing hormone releasing factor (LH-RF) can be decreased by lesions in two distinct regions: the suprachiasmatic-preoptic portion or the arcuate nucleus-ventromedial portion. Whether the factors are also synthesized or modified within the axons of these neurons is not known. The exact localization of the sites of synthesis and storage of the individual releasing factors awaits the development of histochemical techniques for identification of these substances.

The secretory response of the anterior lobe to the hypophysiotropic hormones has a short latency (5 to 10 min). The need for Ca^{2+} for release of TSH, LH, and FSH from pituitary glands in vitro has been demonstrated. A role for 3',5'-cyclic AMP has been proposed based on the observations that this substance induces TSH, FSH, and LH release from pituitary glands in vitro and that theophylline (a drug that prevents inactivation of 3',5'-cyclic AMP by inhibiting phosphodiesterase) potentiates the action of FSH-RF. Theophylline, when injected into the cerebral ventricles, induces somatotropin (GH) release [9c, 17].

Most evidence supports the view that the hypophysiotropic hormones primarily cause release of stored tropins, but they may also act to increase the rate of synthesis of the adenohypophysial hormones. This effect, however, may be secondary to the discharge of the hormone store.

SYNAPTIC TRANSMITTERS IN NEUROENDOCRINE CONTROL

In common with other CNS neurons, hypothalamic cells are acted upon by presumed synaptic transmitter compounds. Histofluorescence techniques have demonstrated the presence of norepinephrine, dopamine, and serotonin (5-HT) in the axon terminals that end in most parts of the hypothalamus. The cell bodies of these neurons are situated in various structures of the limbic and reticular activating systems. Amine-containing somata have also been observed in some hypothalamic areas, particularly in the arcuate nucleus, from which arises a dopaminergic pathway – the tubero-infundibular tract – that traverses the median eminence and infundibulum to terminate at the portal plexus [9a]. Acetylcholinesterase has been demonstrated histochemically in several hypothalamic areas, indicating the presence of cholinergic pathways, as well. The hypothalamus also contains histamine in high concentrations. Thus, many of the substances commonly implicated in synaptic transmission are found in the hypothalamus; some or all of these undoubtedly act as synaptic transmitters in the hypothalamic neural pathways.

One special feature that deserves consideration here is the role of the dopaminergic terminations of the tubero-infundibular tract within the median eminence. Much of the dopamine in the median eminence is contained in terminals that end on other synaptic boutons. This is of interest because studies on excised hypothalamic fragments have shown that dopamine causes the secretion of LH-RF and FSH-RF [1b]. Activity in this pathway might affect the secretion of releasing factors via axo-axonal junctions; dopamine, acting on the terminals of the neurosecretory cells, could modify the amount of releasing factor secreted when the neurosecretory cell discharges.

It was originally suggested that amines in the median eminence may, in fact, act as releasing factors themselves since they could enter the portal circulation and be carried to the anterior pituitary gland. Except for prolactin secretion (see below), there is good evidence that the amines do not act directly on the pituitary gland as releasing factors. A possibility remaining to be explored is that amines could modify the response of the anterior lobe to releasing factors.

In addition to the suggested action within the median eminence, dopamine may also act elsewhere in the hypothalamus. Infusion of dopamine into the third ventricle of rats produces elevated FSH-RF, LH-RF, and PIF activity in hypophysial portal blood. Intraventricular dopamine also causes elevated plasma FSH and LH and decreased plasma prolactin, as expected. Regarding gonadotropin release, norepinephrine was found to be very much less effective than dopamine, and serotonin was inhibitory. Injections of dopamine into the pituitary gland had none of these effects [33–35]. These observations support the hypothesis that dopamine stimulates the release of FSH-RF, LH-RF, and PIF, but it is difficult to localize its site or sites of action precisely.

FEEDBACK CONTROL OF NEUROENDOCRINE FUNCTION

A general principle that applies to a greater or lesser degree to all endocrine glands is that the control of secretory activity involves a feedback loop. That is, some change in the periphery related to secretory activity is detected by the control mechanism, and this information is used to adjust the output of the control mechanism in an appropriate way. Negative or inhibitory feedback is the usual means for maintaining a particular level of output in the face of uncontrolled or unpredictable disturbances. For example, an increase in the blood level of adrenocorticosteroids will cause a decrease in the release of adrenocorticotropic hormone (ACTH) from the anterior pituitary lobe. This, in turn, will decrease the secretion level of adreno-corticosteroids toward its original value. Negative feedback by target-organ hormones is also exhibited by the gonadal steroids and thyroxine.

Although negative feedback of this type is most common, there is evidence for positive or stimulatory feedback, especially in the control of gonadotropic hormones. Thus, implantation of estrogen in the rat pituitary during a critical period of the estrus cycle causes advancement of ovulation, presumably due to an increase in the plasma level of LH which, by its action on the developing ovarian follicle, leads to a further increase in the estrogen level [2]. Unlike the negative feedback system, positive feedback controls are inherently unstable in that, once started, they will shift rapidly to a high level of activity. The major advantage of positive feedback is that it forces the system to attain this level of activity very rapidly. Although it is far from established, estrogen stimulation of LH secretion may be responsible for the LH output peak that triggers ovulation at the end of follicular maturation [2]. An inhibitory mechanism is required for turning off a positive feedback control loop. LH can inhibit its own secretion, so this effect may represent such a mechanism [9e].

The common means of monitoring endocrine output uses receptors sensitive to the levels of the various hormones in the general circulation. For many of the target-organ secretions, sensitivity to hormonal levels has been found in the anterior pituitary lobe itself. In addition, the hypothalamus and other areas in the CNS contain cells sensitive to these hormones. The hypothalamus has also been shown to be sensitive to the tropic hormones secreted by the anterior pituitary lobe. Feedback loops involving these structures are described below. (Detailed reviews and bibliographies are found in [4b], [9e], and [17].)

Long Feedback Loops

Basal secretion of tropic hormones does not depend on intact pathways from the hypothalamus to the hypophysis. After transection of the pituitary stalk, the levels of target-gland hormones drop significantly, but low levels of secretion are maintained. Only after complete hypophysectomy does total atrophy of the peripheral organs take place. Thus, even in the absence of stimulation by hypothalamic releasing factors, the pituitary gland will secrete tropic hormones. Furthermore, a measure of feedback control exists after stalk secretion. For example, repeated injection of exogenous thyroxine into an animal with an isolated pituitary will cause atrophy of its thyroid gland, just as in the intact animal. This is taken to indicate that the isolated pituitary gland, in addition to maintaining a low rate of TSH secretion, is sensitive to the level of thyroxine in the general circulation. Increasing the thyroxine level inhibits the TSH release and produces atrophy of the target organ. After stalk section, however, the system is unresponsive to many of the stimuli that would have increased thyroid activity in the intact animal, such as environmental stress. The input to the pituitary-thyroid axis from these stimuli is neural and is mediated by the hypophysial portal system. Similar observations have been made for the adrenocortical and gonadal tropic hormones. Interestingly, the thyroid response to physical trauma (laparotomy) is maintained in animals with transected pituitary stalks. In this instance, the tissue damage may produce a change in some humoral or hormonal agent which can act directly on the isolated pituitary gland.

It appears that one level of feedback control of the anterior pituitary lobe involves a reciprocal interaction between the pituitary gland and its target organs, this interaction taking place in the anterior lobe itself. However, the level of activity and the dynamic range of control are greatly enhanced by the addition of hypothalamic input. All neurally mediated stimuli to pituitary endocrine activity are channeled through the hypothalamus and the hypothalamo-hypophysial interface.

Another level of long feedback input is to the hypothalamus itself. Hypothalamic neurons "sense" circulating hormone levels and modify the synthesis or release of the hypophysiotropic hormones. It is not known if the detector cells are identical to those producing the releasing factors or if specialized detector cells are present. Although it is experimentally difficult to demonstrate hormone specificity at the cellular level, it may be presumed that independent control of the tropic hormones requires independent detector mechanisms for each control loop.

The existence of detector cells has been determined by implanting target-organ hormones into various parts of the hypothalamus and noting the effect on the release of tropic hormones. Areas of the anterior hypothalamus are sensitive to ovarian steroids, while portions of the medial hypothalamus appear sensitive to adrenocortical hormones. There is still some doubt as to the existence of somatotropin and thyroxine sensitivity in the hypothalamus. Precise localization of hormonal sensitivity is difficult to achieve by this method because of diffusion of the implanted materials.

Short Feedback Loops

Increasing evidence shows that the hypothalamus contains elements sensitive to the adenohypophysial tropic hormones themselves. This feedback loop has been termed "internal," or "short," as opposed to the external, longer feedback loop from target organ to the hypothalamus or the anterior lobe. There is evidence that each of the tropic hormones except prolactin may act on the hypothalamus in such an inhibitory short loop. Prolactin, on the other hand, stimulates PIF. There is some evidence that FSH-RF can inhibit its own secretion within the hypothalamus. This would constitute an ultrashort feedback loop. Another form of ultrashort feedback loop that has been proposed is an inhibitory action of the adenohypophysial tropins themselves on the adenohypophysis.

NEURAL "OVERRIDE" OR "RESET" OF THE CONTROL SYSTEM

The negative feedback controls discussed above provide a mechanism for maintaining endocrine activity at some fixed level in the face of various uncontrollable disturbances that might change that level. This is the essence of a feedback control mechanism. Obviously, situations arise in which the body economy requires a change in endocrine gland output. In such situations, it becomes necessary to "override" the basic control loop and to drive the hormonal concentration to a new level. This may be accomplished by a simple override: a parallel input to the hypothalamo-hypophysial pathways which would activate tropic hormone release independently of (or in spite of) feedback inhibition. Another mechanism would involve changing the operating level, or set-point, of the controller so that activity is controlled at a new level. This reset mechanism is comparable to resetting the thermostat of a home heating system to give a different room temperature; the override technique is comparable to building a fire and ignoring the thermostat.

The adjustments of endocrine activity in response to neurally mediated inputs probably represent control of both these types. Examples include the thyroid response to cold, the adrenal response to stress, and the gonadotropin response to coitus in nonspontaneously ovulating animals.

In most instances, studies have not been performed in order to distinguish override from reset mechanisms. Simple override can probably account for most transient responses to environmental stimuli. There are two cases in which a reset mechanism has been postulated: the adrenocortical response to chronic stress and the increase in gonadotropin secretion at puberty. In the immature animal, the gonadotropin feedback control seems to operate to keep ovarian steroid levels low; the hypothalamic steroid-sensitive cells respond to low concentrations of hormone by inhibiting the secretion of gonadotropin releasing factors. At puberty, some mechanism, presumably neural, decreases the sensitivity of the detector cells, thus allowing circulating sex hormone levels to rise.

EXTRAHYPOTHALAMIC AREAS IN NEUROENDOCRINE CONTROL

Various CNS areas other than the hypothalamus play important roles in regulating the activity of the pituitary gland. CNS structures that form part of the afferent pathways to the hypothalamus will, of course, be involved in the neurally mediated response, as discussed above. In addition, it has been shown that some areas contain hormone-sensitive cells whose activity influences tropic hormone release. For instance, implants of cortisone into the midbrain reticular formation, amygdala, and hippocampus modify adrenal steroid secretion. Furthermore, ablation and stimulation of limbic system structures have been found to alter adrenocortical and gonadal activity. Because the limbic system is also important for emotional reactions and sexual behavior, it is presumed to be a portion of the neural structures involved in the interactions between hormones and emotional states.

GONADOTROPIN RELEASING FACTORS (FSH-RF AND LH-RF)

Numerous investigators have demonstrated both FSH-RF and LH-RF in hypothalamic and SME extracts from a number of species, including rats, sheep, pigs, cattle, and humans. Extracts obtained from some species have been shown to have physiological activity in man, thus indicating a lack of species specificity [10]. The direct measurement of LH-RF elevation in hypophysial portal blood subsequent to stimulation of the median eminence has been accomplished recently in rats by means of bioassay of plasma collected from the cut end of the pituitary stalk [31].

There has been some controversy over the chemical nature of FSH-RF and LH-RF and whether they are actually separate molecules. In highly purified porcine extracts, the two activities have not been separable. Recently, Schally and his co-workers have achieved a two-millionfold purification of LH-RF from porcine hypothalami, and the apparently homogeneous product contained FSH-RF activity as well. Their amino acid analysis shows this hypothalamic hormone to be a decapeptide of the following structure: (pyro) Glu–His–Trp–Ser–Tyr–Gly–Leu–Arg–Pro–Gly–NH_2. The synthetic decapeptide also has FSH and LH releasing activities [41].

It is well known that ovulation, even in cyclically ovulatory animals, depends on hypothalamic control of the release of gonadotropin from the pituitary. Ovulation may be induced by local vaginal stimulation with a glass rod or by electrical stimulation of the preoptic area in the brain. In both cases, ovulation is accompanied by increased multiunit neuronal activity in the arcuate nucleus-median eminence area, which is recorded by stereotactically implaced electrodes [13]. In addition, the plasma levels of both LH and FSH, as measured by radioimmunoassay, are elevated after the hypothalamus has been stimulated electrically in a zone extending from the septal region through the preoptic nucleus, anterior hypothalamic area, and median eminence-arcuate complex [25]. It is thought that the axons mediating both gonadotropin releasing factors arise rostrally near the preoptic nucleus and extend caudally to traverse the median eminence to the portal plexus, where their factors are secreted. However, there may be separate fiber tracts for FSH-RF and LH-RF. Earlier studies showed that anterior deafferentation of the medial basal hypothalamus blocks ovulation, despite the maintenance of basal gonadotropin secretion [4a]. According to the current view, the neural signals involved in producing ovulation (which requires a surge of gonadotropins at a critical period of estrus) activate neuronal somata in the preoptic area to increase the synthesis or release of the releasing factors.

Gonadotropin secretion is subject to both stimulatory and inhibitory feedback regulation by the ovarian steroids, estrogen and progesterone. Feedback control is discussed above, and some specific examples are cited here.

Elevated plasma levels of the ovarian steroids, probably acting together, are important in suppressing FSH and LH secretion in the postovulatory phase of estrus and during pregnancy. This represents a long, or external, feedback loop. In ovariectomized rats, subcutaneous administration of ovulation-inhibiting steroids lowers the plasma concentrations of FSH and LH. In such animals, these substances do not block the stimulatory effects of FSH-RF and LH-RF on the pituitary, indicating that the steroids have an action on the hypothalamus or perhaps on some proximal brain center [42]. When the ovaries are not previously removed, progesterone suppresses the action of administered releasing factors, showing that under certain conditions the pituitary itself may also be sensitive to target-organ feedback [19]. It is thought by some investigators that the pituitary becomes more sensitive to releasing factors after ovariectomy, probably as a consequence of the lack of suppressive action by the ovarian steroids.

Under certain conditions, estrogen may be stimulatory. For example, estrogen implants in the hypothalamus can precipitate precocious puberty in immature female rats. In adult rats, estrogen implanted into the pituitary during a particular 4-hr period on the second day of the estrus cycle (diestrus-2 of a 5-day cycle) produces an advancement of ovulation [2]. It would appear that both the brain and the pituitary may be responsive to ovarian hormones and may be either stimulated or inhibited. These various responses depend upon the background of neuronal and endocrine events in a manner that has not yet been elucidated.

An example of the short-loop feedback mechanism is provided by the experiments in which LH, when implanted into the median eminence of castrated female rats, produces decreases in pituitary and plasma levels of LH [4b, 9e]. This action of LH might be involved in the rapid return of LH secretion to basal levels after ovulation. Recently, the ultrashort feedback loop was postulated by Motta and colleagues [9e], who demonstrated that subcutaneous injection of rat SME extract into male rats that previously had been castrated and hypophysectomized reduced the hypothalamic stores of FSH-RF.

PROLACTIN INHIBITORY FACTOR (PIF)

Maintenance of lactation in the postpartum state depends on pituitary secretion of prolactin. This hormone mainly acts on the mammary glands (in promoting pubertal development and lactation) and on the corpus luteum (in sustaining the formed luteum and its secretion of the ovarian steroids). There is ample evidence that the suckling stimulus required to maintain prolactin secretion and lactation in the postpartum period is mediated through neural pathways that ascend in the spinal cord. Electrical stimulation of the lateral mesencephalic tegmentum, the rostral-central gray matter, and the posterior hypothalamus has induced lactogenesis in pseudopregnant rabbits [46].

In mammals, it is well substantiated that the predominant or net effect of the hypothalamus on prolactin secretion is one of inhibition. When the pituitary is removed from hypothalamic control by section of the hypophysial stalk, transplantation, or incubation of the pituitary in vitro, prolactin secretion is greatly increased. This is in contrast to the secretions of the other pituitary hormones, which are decreased under these circumstances. In avian species, however, the hypothalamus appears to exert stimulatory control over prolactin secretion. It has been suggested that there may be a prolactin stimulatory factor in mammals as well.

A factor that inhibits prolactin secretion (PIF) has been found in hypothalamic extracts from several mammalian species. PIF has been greatly purified, but in ovine tissue it has not been substantially separated from LH-RF by column chromatography. On the other hand, purified LH-RF without PIF activity has been obtained from porcine hypothalami. Little is known regarding the chemical structure of PIF [9f].

Under physiological circumstances, suckling maintains a high serum prolactin level. When the litter of a lactating rat is removed for a 12-hr interval, her serum prolactin falls. After 30 min of suckling, the serum prolactin again rises while, at the same time, the pituitary is depleted of prolactin. However, after prolonged suckling (more than 3 hr) prolactin contents in both the serum and the pituitary remain high [18]. It would appear from these data that an acute suckling stimulus primarily increases the release of prolactin, and prolonged stimuli increase synthesis as well.

If PIF is injected intraperitoneally 1 to 2 min before the 30-min suckling period, the prolactin depletion in the pituitary and the prolactin increase in serum are completely prevented [27]. This effect is used as a bioassay of PIF activity. Another bioassay technique takes advantage of the fact that when the cervix in rats, for example, is stimulated with a vibrating glass rod for 2 min, pituitary depletion of prolactin takes place within 30 min. This effect is also blocked by prior administration of PIF. SME extracts can also block the pituitary depletion of prolactin that occurs subsequent to bleeding or severe stress. However, as noted above, pituitary prolactin content is not always inversely related to serum prolactin levels. A lack of species specificity among mammalian sources is indicated by the fact that ovine, bovine, and porcine hypothalamic extracts have exhibited PIF activity in rats.

Recent radioimmunological measurements of levels of prolactin in the serum of rats during their various reproductive phases have shown that the levels are highest during estrus, tend to be slightly lower during proestrus and metestrus, and are about 60 percent lower during diestrus [18]. Thus, the serum prolactin tends to be highest when the estrogen levels are high and low when the estrogen levels are low. Estrogen administration results in elevated serum and pituitary prolactin. In vivo, estrogen or progesterone implants in the anterior hypothalamus or estrogen implants in the hypophysis deplete PIF from the rat hypothalamus and stimulate prolactin secretion. Estrogen, but not progesterone, also stimulates release of prolactin from anterior pituitary tissue incubated in vitro. These observations show that estrogen acts both on the hypothalamus to reduce PIF and directly on the pituitary to increase prolactin release. Progesterone, on the other hand, appears to act only on the hypothalamus. High concentrations of progesterone inhibit prolactin secretion. These data are useful in explaining the high levels of prolactin in serum during estrus (high estrogen) and the low levels during most of pregnancy (low estrogen coupled with very high progesterone) in the rat.

Prolactin secretion may also be regulated in a short feedback loop through stimulation of PIF. Prolactin implants in the median eminence lead to depressed prolactin levels in serum, decreased prolactin content in the pituitary, elevated FSH and LH in serum, and the termination of pseudopregnancy [47].

As mentioned earlier, at least a part of the neural path involved in prolactin regulation appears to involve dopamine [35]. Reserpine or phenothiazines implanted in the median eminence stimulate prolactin release [9f]. This may be attributed to the action these drugs have in depleting monamines from nerve terminals assuming that a monoamine is required for PIF release (monamine depletion is discussed in Chap. 29). However, the precise role of the amines is not resolved. One hypothesis holds that PIF is, in fact, norepinephrine (see [17]).

THYROTROPIN RELEASING FACTOR (TRF)

There are several hypothalamic structures that have been implicated in thyrotropin regulation, including the suprachiasmatic, paraventricular, and arcuate-ventromedial hypothalamic areas. Destruction of these regions is associated with decreased thyroid function and decreased hypothalamic TRF [39]. Electrical stimulation of the medial-basal hypothalamus in rats produces an increase in serum TSH as measured directly by radioimmunoassay. This takes place within 5 min, and peak responses are seen in 10 to 25 min [38].

Thyroxine inhibits the release of thyrotropin from pituitary tissue incubated in vitro and blocks the stimulating action of TRF on the pituitary in vitro and in vivo. The thyroxine inhibition of TRF action appears to depend on synthesis of a specific protein because its inhibition can be blocked by actinomycin D and puromycin [9d]. Whether thyroxine has an additional action on the hypothalamus to reduce TRF activity is still an open question.

TRF is the first of the releasing factors to be identified chemically and synthesized in vitro. Schally and his colleagues [43] discovered that porcine TRF consists of a tripeptide, Glu–His–Pro. Burgus and co-workers found the same composition for ovine TRF [23]. However, the synthetic peptide was devoid of physiological activity until it was acetylated; the active product contained a pyroglutamyl residue in which the glutamate was cyclized rather than open. Further investigations [22] with a series of synthetic peptides showed that preparations of L-(pyro)Glu–L-His–L-Pro (NH_2) exhibit chemical and biological characteristics indistinguishable from those of natural TRF, and that porcine and ovine TRF are identical. The synthetic hormone is active orally and parenterally in mice, rats, and humans. It stimulates TSH release in vivo and from anterior pituitary tissue incubated in vitro; both actions are blocked by triiodothyronine. When administered to euthyroid humans, it elicits increases of serum thyrotropin within 5 min. Preliminary results indicate that synthetic TRF will provide a useful clinical test of human pituitary function [32].

The demonstration of TRF biosynthesis in vitro by rat hypothalamic fragments has also been accomplished recently [40]. The sole precursors of a TRF that corresponds chromatographically to the structure described above were found to be glutamate, histidine, and proline, indicating that the glutamate was cyclized and the proline amidated biologically. The TRF-synthesizing tissue was found to be widely distributed in the hypothalamus, including the ventral and dorsal portions, as well as the median eminence.

GROWTH HORMONE RELEASING FACTOR (GH-RF)

The normal maintenance of growth hormone secretion by the pituitary depends upon the integrity of the hypothalamus and the hypothalamic-pituitary portal circulation. Destruction of the pituitary stalk or median eminence blocks growth hormone secretion in response to (1) insulin-induced hypoglycemia, (2) blocking of glucose metabolism with 2-deoxyglucose, and (3) arginine infusion. Bilateral lesions limited to the ventromedial nuclei of the hypothalamus are sufficient to cause growth impairment and decreased plasma and pituitary growth hormone, as measured by radioimmunoassay in weanling rats. Electrical stimulation in the ventromedial nuclei, but not in the lateral hypothalamus, of rats evokes increases in growth hormone in plasma within 5 min [29]. Although some specific cell group or fiber bundle subserving growth hormone regulation may be situated in these nuclei, possible roles of other hypothalamic structures are not yet known.

Injections of SME extracts cause depletion of growth hormone levels in pituitary and elevation of the levels in plasma, as measured in bioassay techniques. Attempts to confirm bioassay results with radioimmunoassay in the rat have not always been successful, although there are reports of elevated immunoassayable growth hormone in other species after administration of crude SME extracts [11a].

Despite major controversy regarding the specificity and reliability of bioassay methods, there is wide agreement that SME extracts contain a growth hormone releasing factor. Bioassayable GH-RF purified from porcine hypothalami appears to be an acidic polypeptide. Although it has not been firmly established, some hypothalamic fractions have been reported to exhibit inhibition of hormone release from incubated pituitary glands [11b].

There is evidence for a feedback loop for growth hormone regulation [17]. In monkeys, growth hormone release provoked by the administration of insulin or vasopressin is impaired by prior infusions of human growth hormone. In addition, implantation of growth hormone in various regions of the brain causes decreases in pituitary weight and growth hormone content. Hormone secretion in response to electrical stimulation of the rat hypothalamus is greater in those animals with initially lower growth hormone levels in serum [29]. This inhibitory feedback effect may be on either the pituitary or the hypothalamus.

It should be noted that other factors enter into the regulation of growth hormone secretion. In men, for example, estrogen potentiates the growth hormone release stimulated by arginine infusion; thyroid hormone and glucocorticoids appear to be important in the response of the pituitary to GH-RF.

Little is known of the neural circuits that participate in the activation of GH-RF, although there is evidence that catecholamines are involved, possibly as neurotransmitters [1a]. It has been found that epinephrine, norepinephrine, and dopamine, but not serotonin, cause depletion of pituitary growth hormone when injected in rat lateral ventricles. These agents also lead to the disappearance of GH-RF activity from the hypothalamus and, in hypophysectomized rats, its appearance in the plasma. Serotonin has no effect and norepinephrine is the most potent. It is known that agents such as reserpine and phenothiazines, which deplete monoamines, suppress the release of hormone that normally follows insulin-induced hypoglycemia [44]. In addition, α-adrenergic blocking agents decrease, while β-adrenergic blocking agents increase, growth hormone releasing responses to provocative tests [17].

CORTICOTROPIN RELEASING FACTOR (CRF)

Corticosteroid secretion is increased in response to a variety of stress stimuli including specific physical stress, such as limb fracture, pain, or bleeding and emotional stress, such as fear or anxiety. The adrenal hormones are intimately related to emotional reactions, which is discussed in Chapter 30 (see also [3]).

A number of nervous system structures have been implicated in ACTH regulation. Steroid responses to a specific stress, such as limb fracture, can be abolished by contralateral spinal cord hemisection. Stimulation in the brain stem reticular formation can evoke steroidal output.

Although the net influence of the hypothalamic efferent path is stimulatory to the adenohypophysis, the hypothalamus itself appears subject to a net tonic inhibitory input. Complete isolation of a hypothalamic island with preserved median eminence and stalk leads to increased ACTH secretion. Inhibitory zones may be located in portions of the hippocampus and brain stem.

There is evidence that the neuronal path which sets the diurnal rhythm of steroid secretion is different from that mediating stress-induced responses. For example, destruction of periventricular and arcuate nuclei of the anterior hypothalamus of animals blocks the afternoon rise in corticosteroids but not the rise subsequent to acute stress. Destruction of the posterior tuber cinereum, on the other hand, blocks the stress response but not the circadian variation. The afternoon rise in corticosteroids is also abolished by Halász knife cuts through the anterior hypothalamus, presumably because of transection of the anterior input to the hypothalamus [4a].

There is still considerable controversy over the identity and the production site of corticotropin releasing factor (CRF). Posterior lobe extracts contain at least four factors with CRF activity: vasopressin, β-CRF (related to vasopressin), and α_1-CRF and α_2-CRF (which are related to alpha melanocyte stimulating hormone, α-MSH). In addition, CRF activity can be obtained from hypothalamic extracts free of vasopressin. The hypothalamic CRF can be separated from other releasing factors and the neurohypophysial hormones by gel filtration; in contrast to vasopressin, it is resistant to inactivation by thioglycollate, which reduces disulfide bonds. It appears to be a peptide since it is inactivated by proteolytic digestion. It is believed that the hypothalamic CRF, rather than those in posterior lobe extracts, is more important physiologically [9b]. The site of CRF production within the hypothalamus has not been localized. Upon subcellular fractionation, it is associated with synaptosome and granule fractions, which suggests that it is stored in granules within nerve endings [3b].

Little is known of the physiological functions of the other factors that show CRF activity. Of these, vasopressin is the most studied. Vasopressin activates adrenal responses when given to animals with a sectioned median eminence, and it causes release of ACTH in vitro and from pituitary tumor after injection into the anterior lobe. It thus has a direct action on the adenohypophysis. There is evidence, however, that at least a portion of the vasopressin effects may result from potentiation of hypothalamic CRF [17].

It is well known that corticosteroids exert an inhibitory feedback control of ACTH secretion, which is an example of an external, or long, feedback loop. Whether the physiological sites of this cortisone action are within the brain, the pituitary, or both, is still conjectural [9b, 17]. Many experiments have shown that implants of corticosteroids in the midbrain, medial basal hypothalamus, septum, amygdala, or hippocampus inhibit ACTH secretion. However, it has been demonstrated that implanted steroids in areas which are close to the ventricular system can diffuse rapidly into the pituitary. Nevertheless, the possibility of physiologically active cortisone receptor sites within the brain still exists.

On the other hand, it is clear that corticosteroids are capable of acting directly on the pituitary to block the action of administered CRF. This has been shown in vivo in rats and dogs with intravenous and intrapituitary injections of a synthetic corticosteroid, dexamethasone, and in vitro with rat pituitary tissue. The dexamethasone suppression of the CRF effect on rat tissue may be abolished by prior treatment with actinomycin D, either in vivo or in vitro, thus implicating DNA-dependent RNA synthesis in the dexamethasone block (see [17]).

In view of the multiplicity of neural pathways that regulate ACTH secretion through both inhibitory and stimulatory influences, it is not surprising that several classes of putative neurotransmitters are capable of modifying pituitary-adrenal function. Stimulation of cortisone secretion has been produced by implants of carbachol, norepinephrine, serotonin, and γ-aminobutyrate in various loci within the cat hypothalamus and limbic system. Although the various zones show some selectivity in their responses to the different agents, the median eminence is responsive to all four [36]. Intraventricular norepinephrine, dopamine, and carbachol also produced ACTH secretion [3a].

On the other hand, there is some indirect evidence suggesting that amines may also inhibit ACTH secretion. For instance, reserpine administration causes first a strong activation of ACTH release, followed by the blockade of response to stress and a reduction in hypothalamic CRF content. These data have been interpreted to mean that amine depletion induced by reserpine releases the CRF secretion from some inhibitory influence and that the CRF stores are finally exhausted. Presently, it is difficult to assign specific physiological roles to the various transmitters implicated in these observations [3a].

PARS INTERMEDIA

CONTROL OF MELANOCYTE-STIMULATING HORMONE (MSH) RELEASE

In amphibians, reptiles, and fish, MSH is important in adaptive color changes of skin. In these animals, MSH produces dispersion of melanin within melanophores, which results in darkening of the skin. Comparatively little attention has been given to studying the relationship of melanocyte-stimulating hormone to hypothalamic function, principally because the biological role of MSH in mammals is not known. Nevertheless, MSH is present in mammalian species, and recent observations regarding its regulation may prove to represent a new principle in pituitary hormone control that has broad applicability.

MSH is produced in the pars intermedia of the pituitary. Based largely on studies of amphibian tissue, the anatomical communication of the hypothalamus with this lobe differs from that with the anterior pituitary in two notable features: (1) generally, the pars intermedia is poorly vascularized and no significant blood flow from the median eminence has been found to enter the intermedia and (2) the intermedia, in contrast to the pars anterior, is invaded by many nonmyelinated nerve fibers derived from the hypothalamus, most of which contain one of two kinds of granules – a large, peptidergic type (1000 to 3000 Å) or smaller ones (700 Å) – and appear to form synapses with intermedia cells and presumably control their hormone release [8a].

The release of MSH from mammalian and amphibian pituitary glands is subject to a net inhibitory control by the hypothalamus that resembles that of prolactin release. Thus, in rats and frogs, destruction of the hypothalamus or transplantations of the pituitary to sites remote from the hypothalamus result in decreased pituitary MSH content and, in the frog, darkening of the skin. However, both releasing factors (MSH-RF) and release-inhibiting factors (MSH-IF) are found in hypothalamic extracts. It would appear that the regulation of MSH release in amphibian and mammalian species depends on the balance between these inhibitory and stimulatory factors [7].

Some insight into a possible enzymatic control of this balance and the possible structure of MSH-IF has been gained from studies recently reported by Celis, Taleisnik, and Walter [24]. These workers have found that incubation of oxytocin with a rat SME microsome preparation blocks the usual MSH-RF activity exhibited by these preparations. Employing a number of acyclic oxytocin intermediates and other neurohypophysial hormones and analogs, they concluded that the incubation product responsible for the inhibition of MSH release should be L-prolyl-L-leucylglycinamide, the C-terminal portion of oxytocin.

All the above peptides tested by direct administration are incapable of inhibiting MSH release, with the exception of the C-terminal tripeptide. This peptide is effective in nanogram quantities in both in vivo and in vitro assays. The microsomal enzymic activity, presumed to represent an exopeptidase, was found in male and female rats and was constant during the estrus cycle. In addition, a mitochondrial-bound enzyme was found in female SME extracts, which, upon incubation with microsomes and oxytocin, prevents the appearance of MSH-IF activity. This factor varies with the estrus phase, being lower during proestrus and diestrus and highest during estrus. Celis et al. proposed that oxytocin functions as a "prohormone" which yields MSH-IF to an extent depending on the balance between the activities of two different enzymes.

The variations of the enzyme activity described above are in keeping with the known changes in pituitary MSH content, which is highest during proestrus and lowest on the day of estrus. In female rats, pituitary MSH content has been found to be increased by ovariectomy and decreased by administration of estrogen. Estrogen, although acting directly on the pituitary, also depletes MSH-IF activity in SME extracts. Finally, although SME extracts from ovariectomized rats contain MSH-IF activity, extracts of SME tissue taken from intact animals in estrus produce decreases in pituitary MSH content [45]. These observations are consistent with those of the mitochondrial-bound factor described above and suggest that estrogen inhibits the net activity or synthesis of MSH-IF through activation of such a factor.

REFERENCES

BOOKS AND REVIEWS

1. Bargmann, W., and Scharrer, B. (Eds.), *Aspects of Neuroendocrinology.* Berlin: Springer, 1970. See in particular:
   1a. Muller, E. E. Brain Catecholamines and Growth Hormone Release, p. 206.
   1b. Schneider, H. P. G., and McCann, S. M. Dopaminergic Pathways and Gonadotropin Releasing Factors, p. 177.
2. Davidson, J. M., Weick, R. F., Smith, E. R., and Dominguez, R. Feedback mechanisms in relation to ovulation. *Fed. Proc.* 29:1900, 1970.
3. De Wied, D., and Weignen, J. A. W. M. (Eds.), *Pituitary, Adrenal and the Brain.* Amsterdam: Elsevier, 1970. See in particular:
   3a. Marks, B. H., Hall, M. M., and Bhattacharya, A. N. Psychopharmacological Effects and Pituitary-Adrenal Activity, p. 57.
   3b. Mulder, A. H. On the Subcellular Localization of Corticotropin Releasing Factor (CRF) in the Rat Median Eminence, p. 33.
4. Ganong, W. F., and Martini, L. (Eds.), *Frontiers in Neuroendocrinology, 1969.* New York: Oxford University Press, 1969. See in particular:
   4a. Halász, B. The Endocrine Effects of Isolation of the Hypothalamus from the Rest of the Brain, p. 307.
   4b. Motta, M., Fraschini, F., and Martini, L. "Short" Feedback Mechanisms in the Control of Anterior Pituitary Function, p. 211.
5. Haymaker, W., Anderson, E., and Nauta, W. J. H. *The Hypothalamus.* Springfield, Ill.: Thomas, 1969.
6. Heller, H., and Pickering, B. T. (Eds.), *Pharmacology of the Endocrine System and Related Drugs: The Neurohypophysis.* Vol. 1. Oxford: Pergamon, 1970.

7. Kastin, A. J., and Schally, A. V. MSH Release in Mammals. In Riley, V. (Ed.), *Pigmentation: Its Genesis and Biologic Control.* New York: Appleton-Century-Crofts, 1971.
8. Martini, L., and Ganong, W. F. (Eds.), *Neuroendocrinology.* New York: Academic. See in particular:
   8a. Etkin, W. Relation of the Pars Intermedia to the Hypothalamus. Vol. 2, 1967, p. 261.
   8b. Ganong, W. F. Neuroendocrine Integrating Mechanisms. Vol. 1, 1966, p. 1.
   8c. McCann, S. M., and Dhariwal, A. P. S. Hypothalamic Releasing Factors and the Neurovascular Link Between the Brain and the Anterior Pituitary. Vol. 1, 1966, p. 261.
9. Martini, L., Motta, M., and Fraschini, F. (Eds.), *The Hypothalamus.* New York: Academic, 1970. See in particular:
   9a. Fuxe, K., and Hökfelt, T. Central Monoaminergic Systems and Hypothalamic Function, p. 123.
   9b. Ganong, W. F. Control of Adrenocorticotropin and Melanocyte-Stimulating Hormone Secretion, p. 313.
   9c. Jutisz, M., de la Llosa, M. P., Bérault, A., and Kerdelhué, B. Concerning the Mechanisms of Action of Hypothalamic Releasing Factors on the Adenohypophysis, p. 293.
   9d. McKenzie, J. M., Adiga, P. R., and Solomon, S. H. Hypothalamic Control of Thyrotropin, p. 335.
   9e. Motta, M., Piva, F., and Martini, L. The Hypothalamus as the Center of Endocrine Feedback Mechanisms, p. 463.
   9f. Pasteels, J. L. Control of Prolactin Secretion, p. 385.
10. McCann, S. M. Neurohormonal correlates of ovulation. *Fed. Proc.* 29:1888, 1970.
11. Meites, J. (Ed.), *Hypophysiotropic Hormones of the Hypothalamus: Assay and Chemistry.* Baltimore: Williams & Wilkins, 1970. See in particular:
   11a. Daughaday, W. H., Peake, G. T., and Machlin, L. J. Assay of the Growth Hormone Releasing Factor, p. 151.
   11b. Schally, A. V., Arimura, A., Wakabayashi, I., Sawano, S., Barrett, J. F., Bowers, C. Y., Redding, T. W., Mittler, J. C., and Saito, M. Chemistry of Hypothalamic Growth Hormone-Releasing Hormone (GRH), p. 208.
12. Sachs, H. Neurosecretion. In Lajtha, A. (Ed.), *Handbook of Neurochemistry.* Vol. 4. New York: Plenum, 1970, p. 373.
13. Sawyer, C. H. Electrophysiological correlates of release of pituitary ovulating hormones. *Fed. Proc.* 29:1895, 1970.
14. Scharrer, E., and Scharrer, B. Hormones produced by neurosecretory cells. *Recent Progr. Horm. Res.* 10:183, 1954.
15. Share, L. Vasopressin, its bioassay and the physiological control of its release. *Am. J. Med.* 42:701, 1967.
16. Walter, R., Rudinger, J., and Schwartz, I. L. Chemistry and structure-activity relations of the antidiuretic hormones. *Am. J. Med.* 42:653, 1967.

17. Yates, F. E., Russell, S. M., and Maran, J. W. Brain-adenohypophysial communication in mammals. *Annu. Rev. Physiol.* 33:393, 1971.

ORIGINAL SOURCES

18. Amenomori, Y., Chen, C. L., and Meites, J. Serum prolactin levels in rats during different reproductive states. *Endocrinology* 86:506, 1970.
19. Arimura, A., and Schally, A. V. Progesterone suppression of LH-releasing hormone-induced stimulation of LH release in rats. *Endocrinology* 87:653, 1970.
20. Bargmann, W., and Scharrer, E. The site of origin of the hormones of the posterior pituitary. *Am. Sci.* 39:255, 1951.
21. Barker, J. L., Crayton, J. W., and Nicoll, R. A. Supraoptic neurosecretory cells: Adrenergic and cholinergic sensitivity. *Science* 171:208, 1971.
22. Bowers, C. Y., Schally, A. V., Enzmann, F., Bøler, J., and Folkers, K. Porcine thyrotropin releasing hormone is (pyro) glu-his-pro (NH_2). *Endocrinology* 86:1143, 1970.
23. Burgus, R., Dunn, T., Desiderio, D., Ward, D., Vale, W., and Guillemin, R. Characterization of ovine hypothalamic hypophysiotropic TSH-releasing factor. *Nature (Lond.)* 226:321, 1970.
24. Celis, M. E., Taleisnik, S., and Walter, R. Regulation of formation and possible structure of the factor inhibiting the release of melanocyte-stimulating hormone. *Proc. Natl. Acad. Sci. U.S.A.* 68:1428, 1971.
25. Clemens, J. A., Shaar, C. J., Kleber, J. W., and Tandy, W. A. Areas of the brain stimulatory to LH and FSH secretion. *Endocrinology* 88:180, 1971.
26. Dean, C. R., Hollenberg, M. D., and Hope, D. B. The relationship between neurophysin and the soluble proteins of pituitary neurosecretory granules. *Biochem. J.* 104:8C, 1967.
27. Dhariwal, A. P. S., Grosvenor, C. E., Antunes-Rodrigues, J., and McCann, S. M. Studies on the purification of ovine prolactin-inhibiting factor. *Endocrinology* 82:1236, 1968.
28. Dyball, R. E. J., and Koizumi, K. Electrical activity in the supraoptic and paraventricular nuclei associated with neurohypophysial hormone release. *J. Physiol. (Lond.)* 201:711, 1969.
29. Frohman, L. A., Bernardis, L. L., and Kant, K. J. Hypothalamic stimulation of growth hormone secretion. *Science* 162:580, 1968.
30. Furth, A. J., and Hope, D. B. Studies on the chemical modification of the tyrosine residue in bovine neurophysin-II. *Biochem. J.* 116:545, 1970.
31. Harris, G. W., and Ruf, K. B. Luteinizing hormone releasing factor in rat hypophysial portal blood collected during electrical stimulation of the hypothalamus. *J. Physiol. (Lond.)* 208:243, 1970.
32. Hershman, J. M., and Pittman, Jr., J. A. Response to synthetic thyrotropin-releasing hormone in man. *J. Clin. Endocrinol. Metab.* 31:457, 1970.
33. Kamberi, I. A., Mical, R. S., and Porter, J. C. Effect of anterior pituitary perfusion and intraventricular injection of catecholamines on prolactin release. *Endocrinology* 88:1012, 1971.

34. Kamberi, I. A., Mical, R. S., and Porter, J. C. Effect of anterior pituitary perfusion and intraventricular injection of catecholamines on FSH release. *Endocrinology* 88:1003, 1971.
35. Kamberi, I. A., Mical, R. S., and Porter, J. C. *In vivo* demonstration of FRF, LRF and PIF in hypophysial portal blood after intraventricular injection of dopamine. *Endocrinology* 88:A-72, 1971.
36. Krieger, H. P., and Krieger, D. T. Chemical stimulation of the brain: Effect on adrenal corticoid release. *Am. J. Physiol.* 218:1632, 1970.
37. Lederis, K., and Livingston, A. Neuronal and subcellular localization of acetylcholine in the posterior pituitary of the rabbit. *J. Physiol. (Lond.)* 210:187, 1970.
38. Martin, J. B., and Reichlin, S. Thyrotropin secretion in rats after hypothalamic electrical stimulation or injection of synthetic TSH-releasing factor. *Science* 168:1366, 1970.
39. Mess, B. Intrahypothalamic localization and onset of production of thyrotrophin releasing factor (TRF) in the albino rat. *Hormones* 1:332, 1970.
40. Mitnick, M., and Reichlin, S. Thyrotropin-releasing hormone: Biosynthesis by rat hypothalamic fragments *in vitro*. *Science* 172:1241, 1971.
41. Schally, A. V., Arimura, A., Kastin, A. J., Matsuo, H., Baba, Y., Redding, T., Nair, R. M. G., Debeljuk, L., and White, W. Gonadotropin-releasing hormone: One polypeptide regulates secretion of luteinizing and follicle-stimulating hormones. *Science* 173:1036, 1971.
42. Schally, A. V., Parlow, A. F., Carter, W. H., Saito, M., Bowers, C. Y., and Arimura, A. Studies on the site of action of oral contraceptive steroids. II. Plasma LH and FSH levels after administration of antifertility steroids and LH-releasing hormone (LH-RH). *Endocrinology* 86:530, 1970.
43. Schally, A. V., Redding, T. W., Bowers, C. Y., and Barrett, J. F. Isolation and properties of porcine thyrotropin-releasing hormone. *J. Biol. Chem.* 244:4077, 1969.
44. Sherman, L., Sooseng, K., Benjamin, F., and Kolodny, H. Effect of chlorpromazine on serum growth hormone concentration in man. *N. Engl. J. Med.* 284:72, 1971.
45. Taleisnik, S., and Tomatis, M. E. Effect of estrogen on pituitary melanocyte – stimulating-hormone content. *Neuroendocrinology* 5:24, 1969.
46. Tindal, J. S., and Knaggs, G. S. An ascending pathway for release of prolactin in the brain of the rabbit. *J. Endocrinol.* 45:111, 1969.
47. Voogt, J. L., and Meites, J. Effects of an implant of prolactin in median eminence of pseudopregnant rats on serum and pituitary LH, FSH and prolactin. *Endocrinology* 88:286, 1971.
48. Walter, R., Schlesinger, D. H., Schwartz, I. L., and Capra, J. D. Complete amino acid sequence of bovine neurophysin II. *Biochem. Biophys. Res. Commun.* 44:293, 1971.
49. Wuu, T. C., Crumm, S., and Saffran, M. Structure of a hormone-binding neurohypophysial polypeptide. *Fed. Proc.* 30:1241, 1971.

18

Myelin

William T. Norton

THE MORPHOLOGICAL DISTINCTION between white matter and gray matter is one that is also useful for the neurochemist. White matter is composed of myelinated axons, glial cells, and capillaries. Gray matter contains, in addition, the nerve cell bodies with their extensive dendritic arborizations, and quite different ratios of the other elements. The predominant element of white matter is the myelin sheath, which comprises about 50 percent of the total dry weight. Myelin is mainly responsible for the gross chemical differences between white and gray matter. It accounts for the glistening white appearance, high lipid content, and relatively low water content of white matter.

The myelin sheath is a greatly extended and presumably modified plasma membrane which is wrapped around the nerve axon in a spiral fashion. The myelin membranes originate from, and are part of, the Schwann cell in the peripheral nervous system (PNS), and the oligodendroglial cell in the central nervous system (CNS). In the mature myelin sheath, these membranes have condensed into a compact, paracrystalline structure in which each unit membrane is closely apposed to the adjacent one; the protein layers of each unit membrane fuse with the proteins of the unit in apposition. The sheath is not continuous for the whole length of an axon because each myelin-generating cell furnishes myelin for only a segment of the axon. Between these segments, short portions of the axon are left uncovered. These periodic interruptions, called nodes of Ranvier, are critical for the function of myelin.

PATTERNS OF MYELINATION

Myelination does not proceed in all parts of the nervous system at the same time, but follows the order of phylogenetic development. Portions of the PNS myelinate first (although many of its fibers never myelinate), then the spinal cord, and the brain last (excluding fibers in areas of the cerebellum that also do not myelinate).

Even within the brain, different areas myelinate at different rates, the intracortical association areas being the last to do so. It is generally true that pathways in the nervous system become myelinated before they become completely functional. This relationship is apparently reciprocal, however, and function also appears to stimulate myelination. For example, optic nerve myelination can be retarded by preventing the use of an eye. The relationship of myelination to function is somewhat clouded because the period of maximum myelination may also coincide with many other, less-known changes in the nervous system. It is not seriously doubted, although somewhat difficult to prove, that much of the loss of function in demyelinating diseases is a result of loss of myelin. (The reader will find diseases of myelin discussed in Chap. 24.)

The importance of myelin to proper performance of the nervous system can easily be seen by comparing the different capabilities of newborns of various species. The period at which myelination takes place varies considerably among different species. For instance, the CNS of such nest-building animals as rats myelinates largely postnatally and the animals are quite helpless at birth. Myelin is still being deposited in the rat brain up to 425 days of age and possibly longer; this is partially related to the continued increase in brain weight, which continues for many months. Grazing animals, such as horses, cows, and sheep, have considerable myelin in the CNS at birth and a correspondingly much higher level of complex activity immediately postnatally.

The maximal rate of myelination in humans takes place during the perinatal period. The motor roots begin to myelinate in the fifth fetal month. The brain is almost completely myelinated by the end of the second year of life, although apparently myelination continues in the human brain through the end of the second decade.

FUNCTION

The conventional view that myelin acts as an electrical insulator that surrounds axons much as insulation surrounds a wire is probably true, at least in part. However, the main function of this insulating sheath appears to be to facilitate conduction in axons rather than to prevent "short circuiting" between adjacent fibers. The nature of this facilitation has no exact analogy in electrical circuitry. In unmyelinated nerves, impulse conduction is propagated by local circuits that flow between the resting and active nerve in and out of the axon through the axonal membrane. The electrical resistance of the myelin sheath is 10 to 20 times higher than extracellular salt solutions, and this prevents local current flow. Conduction in myelinated fibers depends upon the sheath being interrupted periodically at the nodes of Ranvier. The bioelectrical current generated at the low-resistance region of the node acts through the medium external to the sheath to activate the axonal membrane at the next node, rather than causing continuous sequential depolarization of the membrane through the high-resistance sheath. This results in "saltatory" conduction; that is, the impulse jumps from node to node. Current flow by saltatory conduction is approximately six times faster than current flow in a comparably sized unmyelinated fiber. Saltatory conduction requires only 1/300th of the sodium ion flux required for conduction in an unmyelinated nerve of the same diameter and an equivalent or greater reduction in energy.

Myelin also lowers the capacitance per unit length of axon, which results in increased speed of local circuit-spreading. The calculated ratio of axon diameter to myelinated fiber diameter for optimal current flow between nodes is 0.6 to 0.7. This ratio is close to that observed in myelinated peripheral nerves. Conduction velocity can also be increased by increasing the diameter of the unmyelinated axon. In unmyelinated fibers, however, conduction velocity is proportional to the square root of the diameter, whereas in myelinated fibers the velocity is proportional to the diameter of the fiber (including the myelin sheath). Thus, the main function of myelin appears to be to facilitate conduction, and, at the same time, to conserve space and energy.

There is an exception to the rule that myelinated fibers conduct impulses more rapidly than do unmyelinated ones. If myelinated fibers are smaller than 1 μ in diameter (according to theoretical calculations), their rate of impulse transmission falls below that of unmyelinated fibers of the same diameter. This is because the axons of these myelinated fibers are smaller than those of unmyelinated fibers of equal diameter. This critical 1-μ dimension is close to the observed minimum limit of myelinated fibers in peripheral nerves. In the CNS, the minimum diameter of myelinated fibers is about 0.3 μ – considerably less than this critical size.

ULTRASTRUCTURE

The existence of a sheath surrounding nerve fibers has been known since the early days of light microscopy (Fig. 18-1). Myelin, as well as many of its structural features, such as nodes of Ranvier and Schmidt-Lantermann clefts, can be seen readily in the light microscope. Before the 1930s, sufficient chemical and histological work had been done to indicate that myelin was primarily lipoidal in nature but having a protein component as well. Our current view of myelin as a system of condensed plasma membranes with alternating protein-lipid-protein lamellae is derived mainly from studies by three physical techniques: polarized light, x-ray diffraction, and electron microscopy. The earliest physical studies in the latter half of the nineteenth century showed that myelin was birefringent when examined by polarized light. This property indicates a considerable degree of long-range order. As early as 1913, Göthlin showed that there was both a lipid-dependent and a protein-dependent birefringence, and that the lipid-dependent type predominated (see [20a] for review). Further work with polarized light by Schmidt and Schmitt and co-workers in the 1930s established that myelin was built up of layers. They also found that the lipid components of these layers were oriented radially to the axis of the nerve fiber while the protein component was

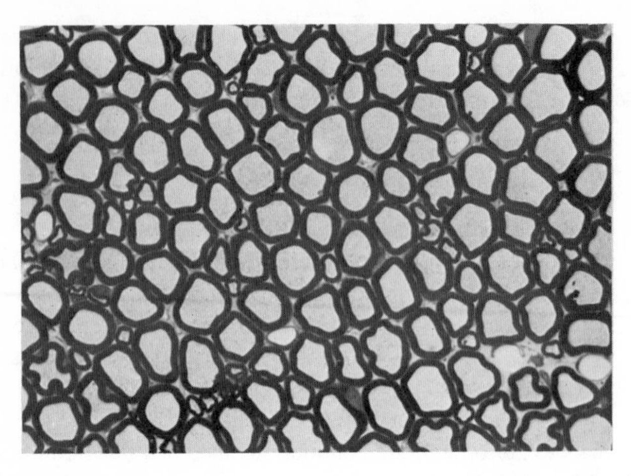

FIGURE 18-1. Light micrograph of a one-micron epon section of rabbit peripheral nerve (anterior root), stained with toluidine blue. The myelin sheath appears as a thick black ring around the pale axon. × 600, before 30% reduction. (Courtesy of Dr. Cedric Raine.)

oriented tangentially to the nerve (see [20a] for review). Danielli and Davson had already formulated their concept of the cell membrane as a bimolecular lipid leaflet coated on both sides with protein, so the results of the latter studies were interpreted with the awareness of the possible membrane nature of myelin.

During the same period, in pioneering studies with x-ray diffraction, Schmitt and co-workers [20] found that peripheral nerve myelin had a radial repeating unit of 170 to 180 Å, a distance sufficient to accommodate two bimolecular leaflets of lipid together with the associated protein. In 1939, Schmitt and Bear [20] concluded that the configuration of the lipid and protein in the myelin sheath was as follows: "The proteins occur as thin sheets wrapped concentrically about the axon, with two bimolecular layers of lipoids interspersed between adjacent protein layers," a description nearly consistent with our current view.

The x-ray diffraction studies were extended and elaborated by a series of investigations by Finean and co-workers [9]. Low-angle diffraction studies of peripheral nerve myelin provided an electron density plot of the repeating unit showing three peaks and two troughs, with a repeat distance of 180 Å. The peaks were attributed to protein plus lipid polar groups and the troughs to lipid hydrocarbon chains. The dimensions and appearance of this repeating unit were consistent with a protein-lipid-protein-lipid-protein structure in which the lipid portion is a bimolecular leaflet and adjacent protein layers are different in some way.

Similar electron density plots of mammalian optic nerve showed a repeat distance of 80 Å (Fig. 18-2); i.e., adjacent protein layers reacted identically to the x-ray beam. Because 80 Å can accommodate one bimolecular layer of lipid (about 50 Å) and two protein layers (about 15 Å each), this represents the width of one unit membrane, and the main repeating unit of two fused unit membranes is twice this figure, or 160 Å.

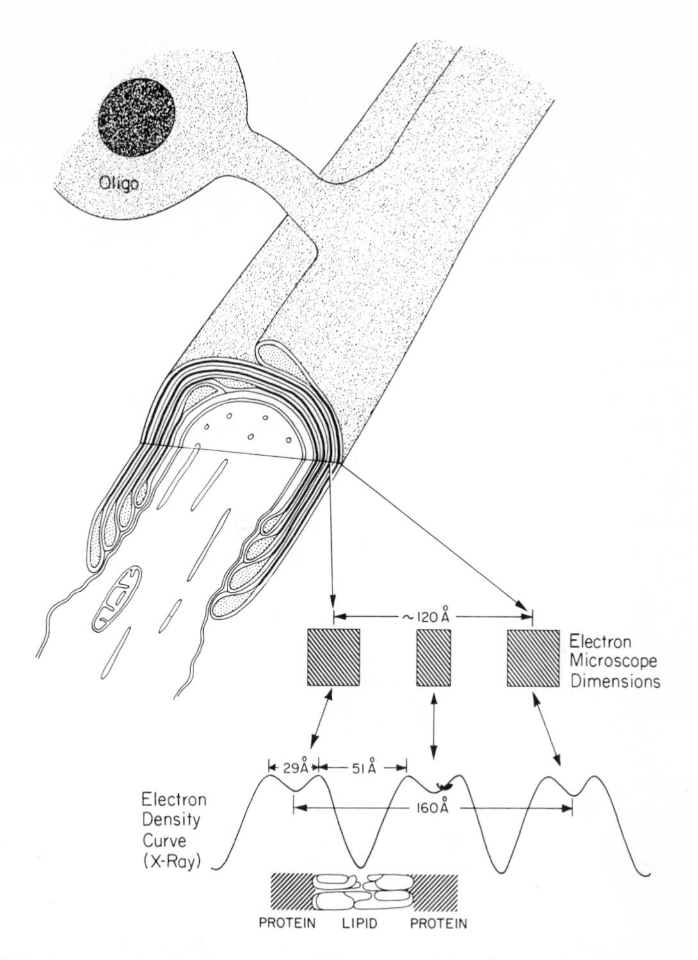

FIGURE 18-2. A composite diagram summarizing some of the ultrastructural data on CNS myelin. At the top an oligodendroglial cell is shown connected to the sheath by a process. The cutaway view of the myelin and axon illustrates the relationship of these two structures at the nodal and paranodal regions. (Only a few myelin layers have been drawn for the sake of clarity.) At the internodal region, the cross section reveals the inner and outer mesaxons and their relationship to the inner cytoplasmic wedges and the outer loop of cytoplasm. Note that in contrast to PNS myelin, there is no full ring of cytoplasm surrounding the outside of the sheath. The lower part of the figure shows roughly the dimensions and appearance of one myelin repeating unit as seen with fixed and embedded preparations in the electron microscope. This is contrasted with the dimensions of the electron density curve of CNS myelin obtained by x-ray diffraction studies in fresh nerve. The components responsible for the peaks and troughs of the curve are sketched below. (Reprinted courtesy of Lea & Febiger, publishers.)

The conclusions regarding myelin ultrastructure derived from these two techniques are fully supported by electron microscope studies. Myelin is now routinely seen in electron micrographs as a series of alternating dark and less dark lines separated by unstained zones (Figs. 18-3, 18-4, and 18-5). The stained or osmiophilic lines are thought to represent the protein layers and the unstained zones the lipid hydrocarbon chains (see Fig. 18-2). The asymmetry in the staining of the protein layers results from the way the myelin sheath is generated from the cell plasma membrane (see following section and Fig. 18-6). The less dark, or intermediate period line represents the fused, outer protein coats of the original cell membrane; the dark, or major period line is the fused, inner protein coats of the cell membrane.

The x-ray diffraction data and the electron microscope data correlate very well. Myelin in the PNS, when swollen in hypotonic solutions, is found to split only at the interperiod line, and the electron density plots show that the broadening occurs at the wider of the three peaks in the repeating unit. This combined approach shows the continuity of the fusion of the minor period with the extracellular space and proves that the wide electron density peak in peripheral nerve plots corresponds

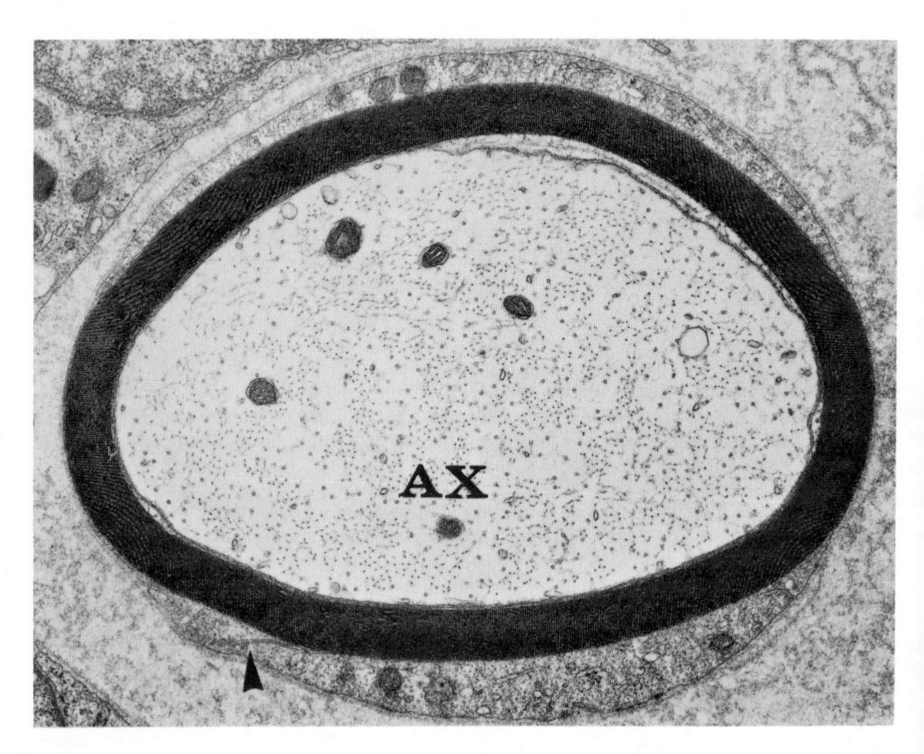

FIGURE 18-3. Electron micrograph of a single peripheral nerve fiber from rabbit. Note that the myelin sheath has a lamellated structure and is surrounded by Schwann cell cytoplasm. The outer mesaxon (arrow) can be seen in lower left. X 18,000. (Courtesy of Dr. Cedric Raine.)

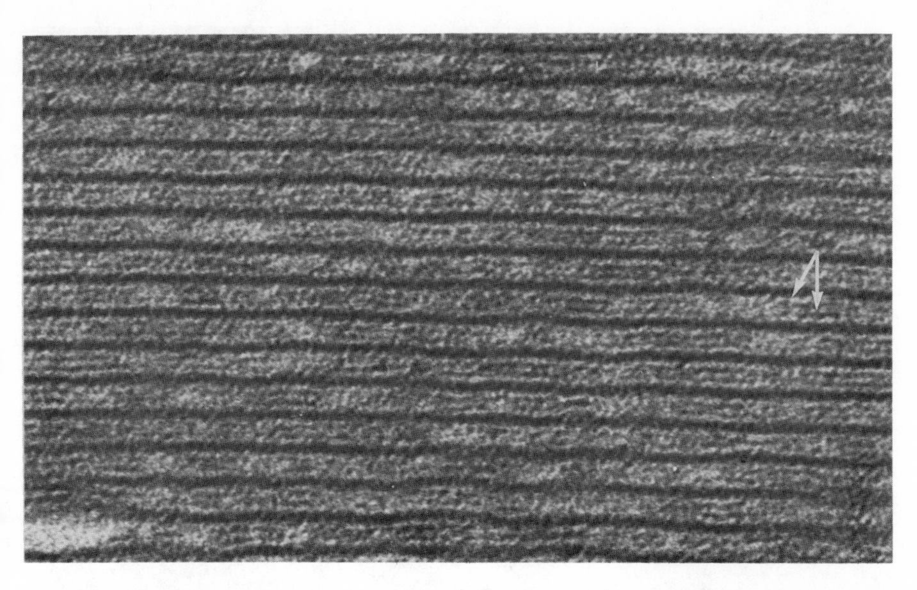

FIGURE 18-4. Magnification of the myelin sheath of Figure 3. Note that the interperiod line (arrows) at this high resolution is a double structure. X 350,000. (Courtesy of Dr. Cedric Raine.)

to the intermediate period lines seen in electron micrographs. It also indicates that the troughs correspond to the light zones between the two dark lines in electron micrographs (see Fig. 18-2). From both x-ray and electron microscope data, it can be seen that the smallest radial subunit that can be called myelin is a five-layered structure of protein-lipid-protein-lipid-protein. This unit, comprising the center-to-center distance between two major period lines, has a spacing in fixed and imbedded preparations of about 120 Å. This repeat distance is lower than the 160 to 180-Å period given by x-ray data because of the considerable shrinkage that takes place after fixation and dehydration.

Although a significant difference was found between central and peripheral myelin in the low-angle x-ray diffraction data, the electron micrographs were much the same for both; each showed a major repeat period of about 120 Å with the intermediate, minor period line of low electron density (compare Figs. 18-3 and 18-5). Recently it has been shown that peripheral myelin has an average repeat distance of 119 Å and the central myelin of 107 Å, confirming the x-ray data for the differences in the size of the period. However, the detailed appearance of electron micrographs is highly dependent on the processing procedures, and some fixation procedures favor retention of the relatively large period seen in x-ray diffraction. Also, in many cases the interperiod line is seen to be two lines rather than one (e.g., see Fig. 18-4).

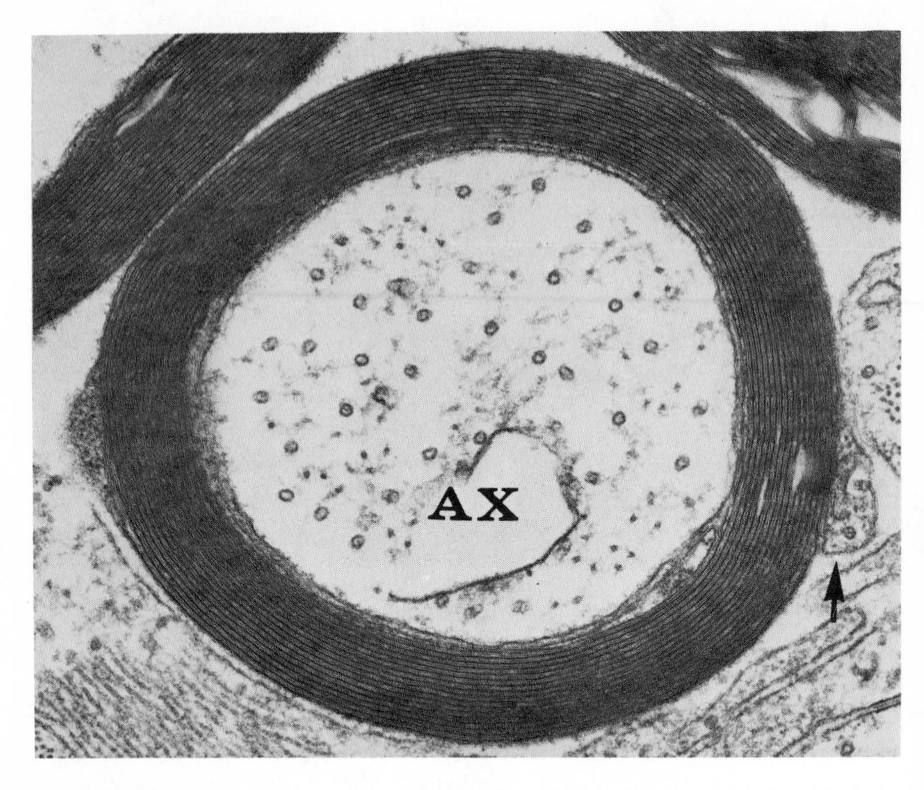

FIGURE 18-5. Electron micrograph of a single CNS myelinated fiber from adult dog spinal cord. The arrow indicates the outer tongue of glial cytoplasm. The inner loop is situated just opposite, between the sheath and the axon. Adjacent myelin sheaths touch because they are not covered with cytoplasm. × 72,000. (Courtesy of Dr. Cedric Raine.)

So far we have only discussed the ultrastructure of the major portion of the sheath in the internodal region. As we have said, two adjacent segments of myelin on one axon are separated by a node of Ranvier, and in this region the axon is uncovered. These nodes are present in both central and peripheral myelin and apparently have a similar structure. At the perinodal region, the membrane sheaths open up at the major period line and loop back upon themselves, enclosing Schwann or glial cell cytoplasm within the loop. The myelin loops ending at the node form membrane complexes with the axolemma, whereas myelin in the internodal region is separated from the axon by a gap of extracellular space. These membrane complexes are believed to be helical structures that seal the myelin to the axolemma but provide, by the intervening spaces, a path from the extracellular space to the periaxonal space. The morphology of this region is complex and can be best understood by referring to Figure 18-2. Peripheral nerves also have Schmidt-Lantermann clefts, which are cone-shaped separations of myelin layers that occur at major period lines. The best way to visualize the structure of both the nodes and the Schmidt-Lantermann clefts is to refer to the diagram of Hirano and Dembitzer, which shows myelin unrolled from its axon as a flat sheet [11].

CELLULAR FORMATION OF MYELIN

Myelination in the PNS is preceded by Schwann cells invading the nerve bundle, rapid multiplication of these cells, and segregation of the individual axons by Schwann cell processes. Smaller axons (less than 1 μ), which will remain unmyelinated, are segregated and several may be enclosed in one cell, each within its own pocket, similar to the structure shown in Figure 18-6A. Large axons (greater than 1 μ) destined for myelination are enclosed singly; one cell per axon per internode.

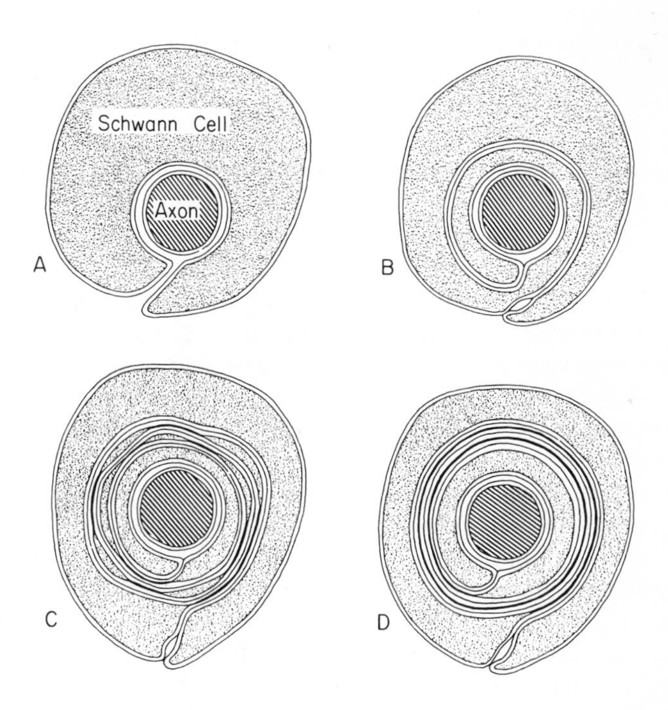

FIGURE 18-6. Myelin formation in the peripheral nervous system. (A) The Schwann cell has surrounded the axon but the external surfaces of the plasma membrane have not yet fused in the mesaxon. (B) The mesaxon has fused into a five-layered structure and spiralled once around the axon. (C) A few layers of myelin have formed but are not completely compacted. Note the cytoplasm trapped in zones where the cytoplasmic membrane surfaces have not yet fused. (D) Compact myelin showing only a few layers for the sake of clarity. Note that Schwann cell cytoplasm forms a ring both inside and outside of the sheath. (Reprinted courtesy of Lea & Febiger, publishers.)

These cells line up along the axons with intervals between them; these intervals become the nodes of Ranvier.

Geren [10] showed that before myelination the axon lies in an invagination of the Schwann cell (see Fig. 18-6). The plasmalemma of the cell then surrounds the axon and joins to form a double membrane structure that communicates with the cell surface. This structure, previously noted in unmyelinated fibers and called the "mesaxon," then elongates around the axon in a spiral fashion (see Fig. 18-6). Geren postulated that mature myelin is formed in this jelly-roll fashion; the mesaxon winds about the axon, and the cytoplasmic surfaces condense into a compact myelin sheath.

In early electron micrographs, cell membranes appeared as single dense lines, the mesaxon as two lines, and myelin as a series of repeating dense lines 120 Å apart. Robertson [18] was later able to show that the Schwann cell membrane is composed of two dense lines, as illustrated in Figure 18-6. When the two portions of this membrane come together to form the mesaxon, the two external surfaces fuse to form a single line that eventually becomes the myelin interperiod line. As a mesaxon spirals into the compact myelin layers, the cytoplasmic surfaces of the mesaxon fuse to form the major dense line. It was thus shown beyond doubt that peripheral myelin is morphologically an extension of the Schwann cell membrane. The mesaxon is thus the smallest myelin subunit and has the five-layered structure previously described.

It was reasonable to assume that myelin in the CNS was formed in a similar fashion by the oligodendroglial cell. However, these nerve fibers are not separated by connective tissue nor are they surrounded by cell cytoplasm, and specific glial nuclei are not obviously associated with particular myelinated fibers. In 1960 it was shown independently by Maturana and Peters that central myelin is a spiral structure similar to peripheral myelin; it has an inner mesaxon and an outer mesaxon that ends in a loop or tongue of glial cytoplasm (Fig. 18-2; see also [6] for a review of this work).

Unlike peripheral nerve, where the sheath is surrounded by Schwann cell cytoplasm, the cytoplasmic tongue in the CNS is restricted to a small portion of the sheath. It was assumed that this glial tongue was eventually connected in some way to the glial cell, but confirmation was difficult. Finally Bunge [2] and colleagues showed that the central myelin sheath is continuous with the plasma membrane of the oligodendroglial cell through slender processes. They also showed that one glial cell apparently can myelinate more than one axon. Peters has recently calculated that in the rat optic nerve, one oligodendroglial cell myelinates, on the average, 42 separate axons (see Chap. 24 and [6]).

The actual mechanism of myelin formation is still obscure. In the PNS, a single axon may have up to 100 myelin layers, and it is therefore improbable that myelin is laid down by a simple rotation of the Schwann cell nucleus around the axon. In the CNS, such a postulate is precluded by the fact that one glial cell can myelinate several axons. During myelination, there are increases in the length of the internode, the diameter of the axon, and the number of myelin layers. Myelin is therefore expanding in all planes at once, and any mechanism to account for this growth must assume the membrane system is flexible. It must be able to expand and contract, and layers probably slip over each other.

ISOLATION

Much biochemical information has been inferred from studies of organized tissue. For more definitive studies, myelin can readily be isolated in high yield and high purity by conventional methods of subcellular fractionation. It is the only homogeneous, well-defined, pure membrane fraction of any tissue that can be obtained easily in large amounts. Methods of isolation and fractionation are described in Chap. 21.

Criteria of purity have been difficult to set for myelin preparations because there is no a priori way of knowing what the intrinsic myelin constituents are. The types of criteria for myelin should be the same as for any other subcellular fraction: typical ultrastructure, the absence or minimization of markers characteristic of other particles, and the maximization of markers characteristic of myelin. Most investigators have used the obvious criterion of electron microscopic appearance. Isolated myelin retains the typical five-layered structure and repeat period of about 120 Å seen in situ. The difficulty of identifying small membrane vesicles of microsomes in a field of myelin membranes and the well-known sampling problems inherent in electron microscopy make this characterization unreliable after a certain purity level has been reached. Useful markers for contamination are succinic dehydrogenase (for mitochondria), Na^+,K^+-activated ATPase, glucose 6-phosphatase, 5'-nucleotidase, and nucleic acid (for cell membranes, microsomes, ribosomes, and nuclei), all of which are very low in purified myelin.

Markers characteristic of myelin are fewer, and some used previously are now to be regarded with suspicion. Purification by a final density-gradient centrifugation step establishes, within limits, the density and therefore the lipid-protein ratio. Many people have used the solubility, or near solubility, of the final product in 2:1 chloroform-methanol as an indicator of purity. It now appears that one class of myelin proteins is insoluble in this solvent; even so, most whole myelin preparations are about 95 percent soluble. Therefore, complete or nearly complete solubility is a good sign, but may be unreliable if new species or very young animals are being examined or if there is any reason to suspect that the protein composition may be altered. In recent years the enzyme 2',3'-cyclic nucleotide 3'-phosphohydrolase has been shown to be myelin-specific. This enzyme should prove very useful in devising new myelin isolation methods and in assaying myelin contamination in other fractions. With increasing knowledge of myelin proteins and with the development of new methods for their separation and measurement, determination of the myelin protein pattern by, for example, acrylamide-gel electrophoresis should be an excellent indicator of myelin purity.

COMPOSITION

In contrast to other subcellular fractions, myelin is characterized by a low amount of water, low protein, and a high lipid content. The solids of myelin are 70 to 80 percent lipid and 20 to 30 percent protein; the lipids of mammalian CNS myelin are composed of 25 to 28 percent cholesterol, 27 to 30 percent galactosphingolipid, and 40 to 45 percent phospholipid. Early inferences about myelin composition were made from three types of indirect measurements: comparative analyses of gray and white matter, the measurement of brain constituents during the period of rapid myelination, and studies of brain and nerve composition during experimental demyelination. From such studies it became generally accepted that proteolipid protein, cerebrosides, and sulfatides were exclusively myelin constituents; that sphingomyelin and the plasmalogens were predominantly myelin constituents; that cholesterol and phosphatidylserine were major, but not exclusively, myelin lipids; and that lecithin was probably not a myelin lipid. These suppositions have now been shown to be only partially correct. Even so, in 1949 Brante [1] calculated that myelin sheath lipids were 25 percent cholesterol, 29 percent galactolipids, and 46 percent phospholipids, figures very close to those obtained by direct analysis of isolated myelin.

No direct determination of water can be made on myelin, although obviously myelin is a relatively dehydrated structure. The low water content of white matter (72 percent) as opposed to gray matter (82 percent) is largely due to the high myelin content of white matter. From x-ray diffraction studies on nerve tissue during drying, Finean [9] determined the water content to be about 40 percent. This is probably a fairly accurate calculation, and all the data on yields of myelin and the composition of myelin and white matter are consistent with myelin having about 40 percent water and the nonmyelin portions of white matter having about 80 percent water, similar to the rest of the nervous system.

CNS MYELIN LIPIDS

Table 18-1 lists the composition of bovine, rat, and human myelin compared to bovine and human white matter, human gray matter, and rat whole brain. (The classification and metabolism of brain lipids are discussed in Chap. 15.) It can be seen that all the lipids found in whole brain are also present in myelin; that is, there are no lipids localized exclusively in some nonmyelin compartment except, possibly, cardiolipin. We also know that the reverse is true; that is, there are no myelin lipids which are not also found in other subcellular fractions of the brain. Even though there are no "myelin-specific" lipids, cerebroside is the most typical of myelin. During development, the concentration of cerebroside in brain is directly proportional to the amount of myelin present. There are only minor differences between the lipid composition of myelin and the corresponding white matter, although myelin lipids tend to have somewhat more cholesterol, cerebrosides, and ethanolamine phosphatides than white matter lipids, and somewhat less sulfatides and lecithin

TABLE 18-1. Composition of CNS Myelin and Brain[a]

Substance[b]	Myelin			White Matter		Gray Matter	Whole Brain
	Human	Bovine	Rat	Human	Bovine	(Human)	(Rat)
Protein	30.0	24.7	29.5	39.0	39.5	55.3	56.9
Lipid	70.0	75.3	70.5	54.9	55.0	32.7	37.0
Cholesterol	27.7	28.1	27.3	27.5	23.6	22.0	23.0
Cerebroside	22.7	24.0	23.7	19.8	22.5	5.4	14.6
Sulfatide	3.8	3.6	7.1	5.4	5.0	1.7	4.8
Total galactolipid	27.5	29.3	31.5	26.4	28.6	7.3	21.3
Ethanolamine phosphatides	15.6	17.4	16.7	14.9	13.6	22.7	19.8
Lecithin	11.2	10.9	11.3	12.8	12.9	26.7	22.0
Sphingomyelin	7.9	7.1	3.2	7.7	6.7	6.9	3.8
Phosphatidylserine	4.8	6.5	7.0	7.9	11.4	8.7	7.2
Phosphatidylinositol	0.6	0.8	1.2	0.9	0.9	2.7	2.4
Plasmalogens[c]	12.3	14.1	14.1	11.2	12.2	8.8	11.6
Total phospholipid	43.1	43.0	44.0	45.9	46.3	69.5	57.6

[a]All average figures obtained on adults in the author's laboratory.
[b]Protein and lipid figures in % dry weight; all others in % total lipid weight.
[c]Plasmalogens are primarily ethanolamine phosphatides.

Figures expressed in this way give no information about lipid concentrations in either the dry or wet tissue. Thus, while human myelin and white matter lipids have similar galactolipid contents, the total galactolipid is 19.3 percent of the myelin dry weight but 14.5 percent of white matter dry weight. These differences are much greater if a wet-weight reference is used. Then galactolipid is 11.6 percent of myelin but only 4.1 percent of fresh white matter. As another example, total phospholipids are a larger percentage of gray matter lipid than of either myelin or white matter lipids. However, on a total dry-weight basis, phospholipids are 30.2 percent of myelin, 25.2 percent of white matter, and 22.7 percent of gray matter.

Figures in Table 18-1 show that many of the suppositions of the earlier deductive work are true. The major lipids of myelin are cholesterol, cerebrosides, and ethanolamine phosphatides in the plasmalogen form. However, lecithin is seen to be a major myelin constituent and sphingomyelin a relatively minor constituent. If the data for lipid composition are expressed in mole percent, most of the preparations analyzed so far contain cholesterol, phospholipid, and galactolipid in a molar ratio of 4:3:2. Thus, cholesterol constitutes the largest proportion of lipid molecules in myelin, although the galactolipids are usually a greater proportion of the lipid weight.

The composition of brain myelin from all mammalian species studied is very much the same. However, there are some obvious species differences. For example, rat myelin has less sphingomyelin than ox or human (Table 18-1). As one goes down the phylogenetic scale, the differences in myelin become much more apparent. Myelin from amphibians and fish apparently have less sphingolipid and more glycerylphospholipid than does myelin from mammals.

Besides the lipids listed in the table, there are several others of importance. If myelin is not extracted with acid organic solvents, the polyphosphoinositides remain tightly bound to the myelin protein, and therefore are not included in the lipid analysis. There is good evidence that the triphosphoinositide of brain is mainly localized in myelin and may therefore have some status as a myelin marker. These lipids are of considerable interest because triphosphoinositide has the highest turnover rate of any phospholipid of the brain, yet one of the main characteristics of myelin is its low metabolic activity. Triphosphoinositide accounts for between 4 and 6 percent of the total myelin phosphorus, and diphosphoinositide for 1 to 1.5 percent of the myelin phosphorus. Although low ganglioside levels have been used as an indicator of myelin purity, it is now apparent that there is an irreducible amount of ganglioside associated with myelin; this is in the order of 40 to 50 μg of *N*-acetylneuraminic acid per 100 mg of myelin (about 0.15 percent ganglioside). In the mature animal, the myelin gangliosides have a pattern completely unlike the pattern of gangliosides extracted from whole brain; the major monosialoganglioside, G_{M1}, accounts for 80 to 90 mole percent of the total myelin ganglioside.

The long-chain fatty residues of myelin are characterized by a very high proportion of fatty aldehydes. These fatty aldehydes, which are derived primarily from phosphatidylethanolamine and, to a lesser extent, from phosphatidylserine, constitute one-sixth of the total glycerylphosphatide fatty residues and 12 mole percent of the total hydrolyzable fatty chains of the myelin lipids. The phospholipid fatty acids differ considerably from one phospholipid to another, but are generally characterized by a high oleic acid content and a low level of polyunsaturated fatty acids. The glycosphingolipids, cerebrosides, and sulfatides have two classes of fatty acids – unsubstituted and α-hydroxy – both of which can be saturated or monounsaturated, whereas sphingomyelin has only unsubstituted fatty acids. The sphingolipid acids are primarily long-chain (22 to 26 carbon atoms) with varying amounts of 18:0. For example, human myelin glycosphingolipids have very little α-hydroxystearic acid but significant amounts of stearic acid, whereas bovine glycosphingolipids have both. Cerebrosides with unsubstituted acids and hydroxy fatty acids correspond to cerasine and phrenosine respectively, of the the older literature.

The previous discussion on composition refers primarily to myelin isolated from the brain. There is some evidence that myelin isolated from the spinal cord has a considerably higher lipid-protein ratio than that isolated from the brain of the same species. The differences between myelins isolated from different parts of the CNS are not well documented and deserve much further study. We do know, however, that myelin from the PNS has a different composition from myelin of the CNS. Peripheral nerve myelin has not received the same extensive documentation primarily because of the technical difficulty of homogenizing peripheral nerve. The few analyses that have been made show that PNS myelin has less cerebroside and sulfatide and considerably more sphingomyelin than CNS myelin.

NONLIPID CONSTITUENTS

Myelin cannot contain any substantial amount of polysaccharide, although it is possible that it does contain glycoproteins. Little enzymatic activity has been detected in myelin, with two exceptions. Some workers have reported reasonably high levels of various peptidases in isolated myelin. Other workers, however, claim they can dissociate the peptidase activity from myelin. The enzyme 2',3'-cyclic nucleotide 3'-phosphohydrolase appears to be definitely established as a myelin enzyme. Sixty percent of the enzyme activity of whole rat brain can be recovered in the purified myelin fraction. The increase of this enzyme in brain and spinal cord during development parallels the course of myelination, and very low levels have been found in the two mouse mutants, quaking and jumpy, that are characterized by deficient myelination. It would appear that this enzyme could have considerable utility as a myelin marker in future studies.

With the availability of purified myelin preparations, the once rather confused picture of myelin proteins began to be clarified somewhat. The studies of myelin proteins were slow in getting started because, with the exception of the "basic protein," the experimental allergic encephalomyelitis (EAE) antigen, the myelin proteins are insoluble in aqueous media and thus are not amenable to study by conventional protein techniques. Solutions of phenol, acetic acid, and water, or phenol, formic acid, and water, as well as aqueous buffers containing sodium dodecyl sulfate will completely dissolve myelin and myelin proteins, and the proteins can be separated by acrylamide-gel electrophoresis in such solvents. Myelin proteins can also be separated by solvent fractionation and by differential solubility in salt solutions or salt solutions containing detergents. These various techniques give reasonably consistent results, and it is now generally agreed that there are three main proteins or protein classes in CNS myelin: 30 to 50 percent proteolipid protein, 30 to 35 percent basic protein, and the remainder high molecular weight proteins soluble in acidified $CHCL_3$:CH_3OH.

The major protein, the classic Folch-Lees proteolipid protein, can be extracted from whole brain with chloroform-methanol (2:1) and can be solubilized in either chloroform or chloroform-methanol even though its lipid content is reduced to less than 10 percent. Proteolipid protein may be a single protein and has a molecular weight of 20,000 to 30,000. It contains about 40 percent polar amino acids and 60 percent nonpolar amino acids.

The basic protein of myelin has been the most extensively studied because, when it is injected into an animal, it elicits a cellular antibody response that produces an autoimmune disease of the brain called experimental allergic encephalomyelitis (EAE). This disease involves focal areas of inflammation and demyelination that resemble multiple sclerosis in some respects. The basic protein cannot be extracted from whole brain with organic solvents, but it dissolves when isolated myelin is treated with chloroform-methanol (2:1). It can readily be extracted from myelin or from whole brain by dilute acid or salt solutions, and, when so extracted, it is soluble in water.

The bovine protein is a highly basic protein (isoelectric point greater than pH 12), and it is highly unfolded, with essentially no tertiary structure. It has a molecular weight of around 18,000 and contains approximately 54 percent polar amino acids and 46 percent nonpolar amino acids. It has no cysteine and has one mole of tryptophan per mole of protein. This is in contrast to proteolipid protein, which is high in cysteine and methionine as well as being rich in tryptophan. Evidently only a very small portion of this molecule is necessary for encephalitogenic activity. Degradation studies have shown that a nine-residue peptide centering around the single tryptophan residue contains nearly all of the encephalitogenic activity of the parent protein. Modification of the single tryptophan residue in either the isolated peptide or in the parent molecule abolishes encephalitogenic activity of the whole protein.*

The complete amino acid sequence of both bovine and human basic proteins has now been determined [8]. The basic proteins of CNS myelin from all mammals except the rat and the mouse appear to be very similar and appear to consist of a single protein. CNS myelin of both rat and mouse has two basic proteins: the second is of lower molecular weight than the major basic protein.

The third class of myelin proteins are acid-soluble proteolipid proteins that have been called "Wolfgram proteins" after their discoverer. This protein fraction is water insoluble and is also apparently insoluble in neutral chloroform-methanol. It contains about 53 percent polar amino acids and 47 percent nonpolar amino acids. On disc gel electrophoresis, this fraction appears to consist of two or three major bands and possibly several minor ones, all of considerably higher molecular weight than the other two myelin protein fractions.

PNS myelin proteins are different from those of CNS myelin. It has been known for many years that peripheral nerve had little proteolipid protein. Recent electrophoretic studies show that PNS myelin has one major nonpolar protein, two lower molecular weight basic proteins, and 3 or 4 other prominent bands. With the exception of the larger basic protein, these bands may not correspond to any CNS myelin proteins.

COMPOSITIONAL CHANGES

DEVELOPING BRAIN

The developing nervous system is marked by several overlapping periods, each defined by one major event in brain growth and structural maturation. These periods can be determined by following the concentration of a specific marker. For example, the period of cellular proliferation can be followed by measuring the amount of DNA per whole brain, and the period of myelination by following a myelin marker such as cerebroside. In the rat, whose CNS undergoes considerable

*Note added in proof. These comments on encephalitogenic activity refer primarily to EAE in the guinea pig. Recent studies show that there are other encephalitogenic regions in the basic protein which are active in other species. See Chapter 24 for a more complete discussion.

development postnatally, the maximal rate of cellular proliferation occurs at 10 days. The period of rapid myelination overlaps with this period of cellular proliferation and is one of the most dramatic in nervous system development. The rat brain begins to form myelin at about 10 to 12 days postnatally. At 15 days of age, about 4 mg of myelin can be isolated from one brain. This amount increases sixfold during the next 15 days, and at 6 months of age, 60 mg of myelin can be isolated from one brain. This represents an increase of about 1500 percent over 15-day-old animals. During the same 5 1/2 month period, the brain weight increases by 50 to 60 percent.

It has been proved convincingly by several groups that the myelin that is first deposited has a very different composition from that of the adult. As the rat matures, the myelin galactolipids increase by about 50 percent and lecithin decreases by a similar amount. The very small amount of desmosterol declines, but the other lipids remain relatively constant. In addition, the polysialogangliosides decrease and the monosialoganglioside, G_{M1}, increases to 90 percent of the total gangliosides. These changes are not complete until the rat is about 2 months old. There is, however, a change in the composition of the protein portion as well as in the lipid portion. Earlier work suggested that the ratio of basic protein to proteolipid protein in human myelin increases during development. More recent studies using disc gel electrophoresis suggest that these protein changes, at least in rodents, may be quite complex, but that both protein and proteolipid protein increase in the myelin sheath during development while the amount of higher molecular weight protein decreases.

These studies of myelin during development are complicated by the presence of the fraction called the "myelinlike fraction," which is found in crude myelin from young animals. The myelinlike fraction can be separated from true myelin by a density-gradient centrifugation technique, and, when separated, it has physical properties similar to those of microsomes. It differs from myelin in lipid, protein, and enzyme composition, and its possible role in myelination is still obscure. The main reasons for the interest in this fraction are that it is closely associated with myelin from young animals but it is not present in crude myelin fractions from mature animals, and its enzyme profile is quite different from that of microsomes. It has been suggested that the myelinlike fraction may represent the transition form from an oligodendroglial cell plasma membrane to myelin, or perhaps even represents the first few layers that the glial cell forms during myelination. Brain development is described further in Chapter 17.

DISEASE

Besides the normal changes seen during development, myelin composition is also altered in certain neurological diseases (see Chap. 24). The myelin abnormalities in human diseases have been found to include both nonspecific and specific changes.

Abnormal myelin was first reported in a case of subacute sclerosing panencephalitis, a disease caused by an atypical measles virus infection of the brain. Analysis of the brain white matter showed a picture typical of many demyelinating diseases: a loss of major myelin constituents and a very high concentration of cholesterol esters, which are myelin degradation products. Grossly, the isolated myelin had a normal ultrastructural appearance. However, it was quite abnormal chemically, with very high cholesterol, low cerebroside, and low ethanolamine phosphatide (plasmalogen) contents.

Subsequently, similar abnormalities have been found in myelin from human brains with spongy degeneration, G_{M1} gangliosidosis, Tay-Sachs disease, and infantile Niemann-Pick disease. These diseases, although they all involve some form of demyelination (usually secondary) are quite different both in etiology and in pathology. Therefore, it is probable that the abnormal myelin is produced by a nonspecific degradative process.

Specific myelin changes have been found in metachromatic leukodystrophy and Refsum's disease. The former is caused by a lack of a sulfatase that degrades sulfatide to cerebroside, leading to abnormal accumulations of sulfatide in the brain and other organs. It is accompanied by a striking deficiency of myelin. Myelin isolated from several cases of this disease reflects the chemical defect seen in the brain as a whole, having a very high sulfatide content and low cerebroside content. In Refsum's disease, the victims lack the ability to degrade phytanic acid. This tetraisoprenoid acid, a normal dietary product, has been found to accumulate in myelin phospholipids as it does in other tissues of the body.

BIOCHEMISTRY

The principal biochemical features of myelin are its high rate of synthesis and turnover during the early stages of myelination and its relative metabolic stability in the adult. Although myelin is one of the most stable structures of the body when it is once formed, it is not by any means a completely inert tissue. At least one of its components, triphosphoinositide, has a very high rate of turnover in the adult, and there is now evidence that the other lipid constituents have varying metabolic activities. Before myelination begins, the immature brain has a relatively high concentration of cholesterol and phospholipids, but the amount of cerebroside, the lipid typical of myelin, is extremely low, as is the activity of the enzyme which synthesizes cerebrosides from UDP-galactose and ceramide. In the rat, the activity of the cerebroside-synthesizing system reaches a peak at 10 to 20 days, coinciding very well with the maximal rate of myelination. Most other lipid synthesizing systems also become most active at this time.

The concept of the metabolic stability of myelin in the adult has been largely developed by the studies of Davison and co-workers [6]. This concept originated in classic studies of Waelsch, Sperry, and Stoyanoff in the 1940s (for a review, see [5]). They found that heavy water is incorporated slowly into adult brain cholesterol and fatty acids, but it is incorporated much more rapidly in young myelinating rats. Davison and colleagues confirmed these observations in much greater detail, using various isotopic labels and several precursors of myelin constituents. Because these constituents were isolated from whole brain, the biochemical behavior of myelin itself had to be determined by inference. Davison and his group then pioneered in the biochemical study of isolated myelin using similar tracer techniques, and the earlier results were found to be valid for the most part. The conclusions from this work were that myelin shows considerably more long-term metabolic stability than do other structures in the nervous system, and that all of the myelin lipids turn over at about the same rate; therefore, the myelin membrane may be metabolized as a unit.

A more recent series of experiments by Smith and colleagues [21] confirm the relative metabolic stability of myelin but cast some doubt on the idea that myelin is metabolized as a unit. They showed that lipid precursors are incorporated into myelin and mitochondria at similar rates in young animals but that they are lost much more rapidly from the mitochondrial lipids. This finding is in accord with Davison's earlier work. However, long-term experiments show that the radioactivity is lost from individual myelin lipids at different rates. Three myelin lipids – phosphatidylinositol, lecithin, and phosphatidylserine – have half-lives of 5 weeks, 2 months, and 4 months, respectively, whereas the ethanolamine phospholipids – cholesterol, sphingomyelin, cerebrosides, and sulfatides – have half-lives of 7 months to more than 1 year. By contrast, the half-lives of the lipids in mitochondria ranged from 11 days (phosphatidylinositol) to 59 days (cerebroside).

The rates of synthesis of these lipids, as determined in short-term experiments, are in agreement with the rates of turnover, phosphatidylinositol and lecithin being labeled more than the others. These observations have been extended to long-term studies of adult rats, as well as to in vitro studies of the myelin of rat spinal cord slices, using ^{14}C-glucose as a precursor in both instances. These experiments also confirm the earlier data: phosphatidylinositol and lecithin have the most active metabolism, and sulfatide, cerebroside, sphingomyelin, and cholesterol the least. The incorporation of precursors into myelin in vitro is, as might be expected, age-dependent: higher activities are present in actively myelinating animals, but considerable uptake can still be measured in slices from 120-day-old animals.

It is possible that the relative inertness of myelin is an evolutionary adaptation that serves a useful purpose in the adult nervous system. It may contribute to stability of function and help buffer the effects of minor or short-lived traumas. The actual reasons for the greater metabolic stability of myelin compared with other brain constituents are unknown. However, we may need to look no farther than the peculiar geometry of the Schwann and oligodendroglial cells. Consider a typical large PNS axon of 5 μ diameter, with a myelin sheath 1.0 μ thick and an internode length of 1000 μ. Such an axon will have about 50 layers of myelin. A simple calculation shows that the total volume of myelin generated by one Schwann cell is 18,840 cu μ. The total area of a myelin double-unit membrane (assuming a thickness of 180 Å or 0.018 μ), if it were unrolled from the axon, is about 1×10^6 sq μ or 1 sq mm. This enormous amount of membrane is maintained by the remainder of the cell, which may be two orders of magnitude smaller. The situation with respect to the oligodendroglial cell is similar. Although CNS sheaths are thinner and internodes shorter, one cell having a perikaryon volume of about 500 cu μ myelinates an average of about 40 such internodes. It can easily be seen that the combination of the physical isolation of most of the sheath from enzymatic systems and the enormous ratio of membrane to cytoplasm may be sufficient to explain the average low metabolic activity.

MOLECULAR MODELS

Although myelin is an extension of a cell plasma membrane, it is probably quite different from it. The low protein content, low enzymatic activity, and the absence of polysaccharides indicate that in the process of differentiation to become myelin, the original membrane must have lost much of its specialized components, producing a sort of skeletal or minimal membrane. Nevertheless, its ultrastructural appearance suggests it is very much like most membranes, and evidence obtained from myelin studies has been used to construct detailed molecular models that may resemble other membranes as well.

Given the dimensions of myelin from x-ray data and the complete chemical composition, there are two main problems in constructing a model: the fatty acid chains are too long to fit in the bilayer space and there is not enough protein to cover the bilayers. The most detailed of such models [22] solves the first problem by allowing the long sphingolipid fatty acid chains to interdigitate across the bilayer. It solves the second by proposing an unorthodox extended β-keratin structure for the protein. The large amount of cholesterol in the structure is complexed with both sphingolipids and phosphatides.

It has been suggested that lipids with long-chain fatty acids, which produce physical stability, are also metabolically more stable, and it is true that the sphingolipids turn over more slowly than do the phosphatides, which have shorter fatty acids

In the last few years much dissatisfaction has arisen about the universality of the unit-membrane model, and a plethora of new membrane theories and models have been proposed. In the midst of this, myelin has stood out as being the one unshakeable example of the protein-lipid-protein sandwich type proposed by Danielli and Davson so many years ago.

ACKNOWLEDGMENT

I wish to thank Dr. Cedric Raine for supplying the elegant microphotogrphs that illustrate this chapter.

Myelin is usually isolated by centrifugation of tissue homogenates in sucrose solutions. During homogenization, myelin peels off the axons and re-forms in spherical vesicles of the size range of nuclei and mitochondria. (It is important to homogenize the tissue in media of low ionic strength, otherwise much of the myelin remains bound to the axon and the intact myelinated axon fragments present in the homogenate will contaminate the crude myelin preparations.) Because of their high lipid content, these myelin vesicles have the lowest intrinsic density of any membrane fraction of the nervous system. Most isolation methods utilize both of these properties – large vesicle size and low density.

One general class of methods involves the isolation, by differential centrifugation, of a crude mitochondrial fraction that contains mitochondria, synaptosomes, and myelin. This fraction is re-suspended in isotonic (0.3 M) sucrose and layered over 0.8 M sucrose. Myelin collects at the interface during centrifugation, whereas mitochondria and synaptosomes sediment through the dense layer. Unless the nuclear fraction is processed similarly, the myelin fragments in that fraction are lost. The other general class of methods bypasses the initial differential centrifugation step. An homogenate of nervous tissue in isotonic sucrose is layered directly onto 0.85 M sucrose and centrifuged at high speed. A crude myelin layer collects at the interface.

The crude myelin layer obtained by either of these methods is a varying purity, depending on the tissue from which the myelin is isolated. White matter from adult brain yields reasonably pure myelin; myelin from the whole brain of a very young animal might be quite impure. The major impurities are microsomes and axoplasm trapped in the vesicles during the homogenization procedure. Further purification is generally achieved by subjecting the myelin to osmotic shock in distilled water. This opens up the myelin vesicles, releasing trapped material. The larger myelin particles can then be separated from the smaller membranous material by low-speed centrifugation or by repeating the density gradient centrifugation on continuous or discontinuous gradients of sucrose or CsCl. On sucrose gradients myelin forms a band centering at approximately 0.65 M sucrose, equivalent to a density of 1.08 g per milliliter. In CsCl gradients, however, the myelin layers out at a higher density and in sucrose-Ficoll gradients (see Chap. 21) at a lower density. For comparison, mitochondria have a density of 1.2 (equivalent to about 1.55 M sucrose).

REFERENCES

[References 2, 4-7, 9, 12-19, 21, and 23 are either comprehensive reviews of a particular topic or summaries of an investigator's own work in the field. For more complete reviews of the chemistry and biochemistry of myelin, see references 4, 6, 7, 13, 14, 16, 17, and 21. The remaining references (1, 3, 8, 10, 11, 20, and 22) are either classic articles or contain important material not included in the reviews.]

1. Brante, G. Studies on lipids in the nervous system with special reference to quantitative chemical determinations and topical distribution. *Acta Physiol. Scand.* 18:(Suppl. 63), 1949.
2. Bunge, R. P. Glial cells and the central myelin sheath. *Physiol. Rev.* 48:197, 1968.
3. Danielli, J. F., and Davson, H. Contribution to the theory of permeability of thin films. *J. Cell. Comp. Physiol.* 5:495, 1935.
4. Davison, A. N. Myelination. In Linneweh, F. (Ed.), *Fortschritte der Padologie,* Band II. Berlin: Springer, 1968.
5. Davison, A. N., and Dobbing, J. *Applied Neurochemistry.* Philadelphia: Davis, 1968.
6. Davison, A. N., and Peters, A. *Myelination.* Springfield, Ill.: Thomas, 1970.

7. Eichberg, J., Hauser, G., and Karnovsky, M. L. Lipids of Nervous Tissue. In Bourne, G. H. (Ed.), *The Structure and Function of Nervous Tissue.* Vol. 3. New York: Academic, 1969.
8. Eylar, E. H. Amino acid sequence of the basic protein of the myelin membrane. *Proc. Natl. Acad. Sci. U.S.A.* 67:1425, 1970.
9. Finean, J. B. The Nature and Stability of the Plasma Membrane. In Fishman, A. P. (Ed.), *The Plasma Membrane.* New York: New York Heart Assn., 1962.
10. Geren, B. H. The formation from the Schwann cell surface of myelin in the peripheral nerves of chick embryos. *Exp. Cell Res.* 7:558, 1954.
11. Hirano, A., and Dembitzer, H. M. A structural analysis of the myelin sheath in the central nervous system. *J. Cell Biol.* 34:555, 1967.
12. Hodgkin, A. L. *The Conduction of the Nervous Impulse.* Springfield, Ill.: Thomas, 1964.
13. Mokrasch, L. Myelin. In Lajtha, A. (Ed.), *Handbook of Neurochemistry.* Vol. 1. New York: Plenum, 1969.
14. Norton, W. T. The Myelin Sheath. In Shy, G. M., Goldensohn, E. S., and Appel, S. H. (Eds.), *The Cellular and Molecular Basis of Neurologic Disease.* Philadelphia: Lea & Febiger, 1972.
15. Norton, W. T. Recent Developments in the Investigation of Purified Myelin. In Paoletti, R., and Davison, A. N. (Eds.), *Chemistry and Brain Development.* New York: Plenum, 1971.
16. O'Brien, J. S. Lipids and Myelination. In Himwich, H. E. (Ed.), *Developing Brain.* Springfield, Ill.: Thomas, 1970.
17. O'Brien, J. S. Chemical Composition of Myelinated Nervous Tissue. In Vinken, B., and Bruyn, G. (Eds.), *Handbook of Neurology*. Vol. 7. Amsterdam: N. Holland Publ. Co., 1970.
18. Robertson, J. D. Design Principles of the Unit Membrane. In Wolstenholme, G. E. W., and O'Connor, M. (Eds.), *Principles of Biomolecular Organization* (A Ciba Foundation Symposium). London: Churchill, 1966.
19. Rorke, L. B., and Riggs, H. E. *Myelination of the Brain in the Newborn.* Philadelphia: Lippincott, 1969.
20. Schmitt, F. O., and Bear, R. S. The ultrastructure of the nerve axon sheath. *Biol. Rev.* 14:27, 1939.
20a. Schmitt, F. O. Ultrastructure of Nerve Myelin and Its Bearing on Fundamental Concepts of the Structure and Function of Nerve Fibers. In Korey, S. R. (Ed.), *The Biology of Myelin.* New York: Hoeber-Harper, 1959, p. 1.
21. Smith, M. E. The metabolism of myelin lipids. *Adv. Lipid Res.* 5:241, 1967.
22. Vandenheuvel, F. A. The structure of myelin in relation to other membrane systems. *J. Am. Oil Chem. Soc.* 42:481, 1965.
23. Yakovlev, P., and Lecours, A. R. The Myelogenetic Cycles of Regional Maturation of the Brain. In Minkowski, A. (Ed.), *Regional Development of the Brain in Early Life.* Oxford: Blackwell, 1967.

19

Muscle

John Gergely

CONSIDERATION OF THE BIOCHEMICAL ASPECTS of muscle is virtually impossible without taking a close look at those structures which distinguish muscle cells from other cells, viz., those structures that constitute the contractile machinery. The existence of myofibrils has long been known to light microscopists, but it has been only during the last 15 years or so that more sophisticated optical and electron microscopic tools have revealed both their intimate structure and some of the changes that occur during contraction.

At one time, muscle contraction was considered in terms of long filaments that undergo changes similar to those underlying the contraction of a rubber band, viz., the transition of long macromolecules from an ordered into a disordered state. Due to the work of H. E. Huxley and J. Hanson [44] and A. F. Huxley and R. Niedergerke [40], the typical striation pattern of voluntary muscle can now be attributed to a regular arrangement of two sets of filaments, and contraction to the relative sliding motion of these filaments (Fig. 19-1). One set, referred to as the thin filaments (having a diameter of about 80 Å), appears to be attached to the Z bands and is found in the I band. The second set of so-called thick filaments (diameter about 150 Å) occupies the A band. The thick filaments seem to be connected crosswise by some material in the M zone. In cross section, the thick filaments constitute a hexagonal lattice and, in vertebrate muscle, the thin filaments occupy the centers of the triangles formed by the thick filaments [7].

The length of the muscle depends on the length of the sarcomeres and, in turn, the variation in sarcomere length is based upon the variation in the overlap between the two sets of filaments [27]. Although there is no permanent connection between the thick and the thin filaments, high-resolution micrographs have shown that cross-bridges emanate from the thick filaments, and it is thought that in active muscle these cross-bridges make temporary links with thin filaments [9].

Table 19-1 shows the protein constituents of the various structures of the contractile apparatus. There is very good evidence for myosin being the chief constituent of thick filaments and actin being the chief constituent of the thin filaments [30].

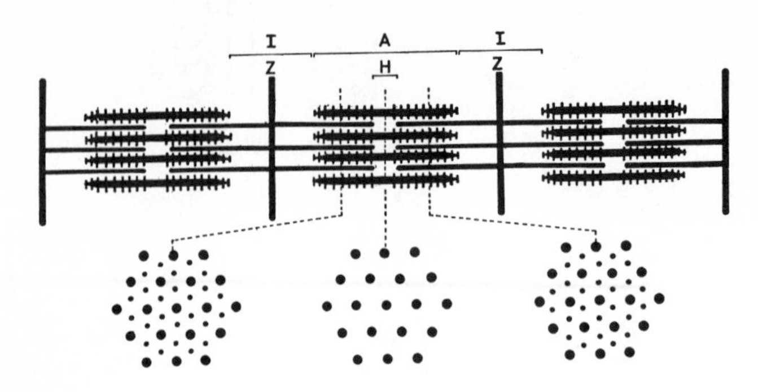

FIGURE 19-1. Schematic representation of structure of striated muscle. Actin-containing thin filaments originate at Z lines. Note thick myosin-containing filaments that bear cross-bridges. (The M disc lies in the center of the H band. See text.) (From [9]. Copyright 1969 by the American Association for the Advancement of Science.)

Evidence is steadily mounting to support the presence of tropomyosin and a complex of other proteins [28], subsumed under the term troponin [3], that play an important role in the regulation of muscle contraction, in the thin filaments. The proteins constituting the M and the Z bands have not been fully characterized [17].

ACTIN

Electron micrographs show that thin filaments of muscle contain globular subunits about 55 Å in diameter, arranged in a double helical structure [31]. These are identified with the protein actin which can be extracted from muscle after treatment with acetone. It is soluble in water and has a molecular weight of 47,000. On addition of salts, it undergoes a drastic change in viscosity, and negatively stained

TABLE 19-1. Myofibrillar Proteins

Protein	Localization	Function
Myosin	A band Thick filaments	Contraction; ATPase
Actin	I band (partial overlap with thick filaments)	Contraction
Tropomyosin	Thin filaments	Regulation of actin-myosin interaction; confer Ca^{2+} requirement
Troponin	Thin filaments	
M protein	Center of thick filaments	Structural?
α-Actinin	Z band	Structural?

electron micrographs reveal the presence of double helical filaments that are essentially identical with the thin filaments. The two states characterized by low and high viscosity are referred to respectively as G- and F-actin (G = globular; F = fibrous). Although actin is a simple molecule in terms of its shape, it is a complex entity in that it contains both tightly bound metal and a nucleotide. The nucleotide in G-actin is ATP [59], while that in F-actin is ADP. The transformation of ATP to ADP takes place during polymerization [51], a process that can be characterized by the following equation:

$$n(\text{G-actin-ATP}) \rightleftharpoons (\text{F-actin-ADP})_n + nP_i$$

As is discussed below in detail, muscle contraction depends on the hydrolysis of ATP as an energy source. It should be emphasized that, according to our current knowledge, the transformation of ATP to ADP that accompanies the polymerization of actin is not involved in muscle contraction. This reaction presumably takes place when actin filaments are laid down in the course of development, growth, or regeneration. From in vitro studies, it appears that both divalent metal and ATP are needed to stabilize the actin molecule, particularly in the globular form. Both the metal and the nucleotide are readily exchangeable in the globular form, indicating greater openness of the molecule, while in the polymerized form both are less accessible.

MYOSIN

Myosin, the chief constituent of the thick filaments, contrasts with actin in almost every respect. Myosin is a highly asymmetrical molecule having an overall length of about 1500 Å and a molecular weight of about 500,000 (e.g., see [46]). Its lateral dimension varies between about 20 Å and 100 Å along its length. In contrast to actin, which is made up of a single polypeptide chain, myosin consists of several chains. There are two so-called heavy chains, each with a molecular weight about 200,000, that run from one end of the molecule to the other. Along most of the length of the molecule, the two chains are intertwined to form a double α-helix; at one end of the molecule they separate, each forming an essentially globular portion. The two globular portions contain the sites responsible for the biological activity of myosin, viz., the ability to hydrolyze ATP and to combine with actin. In addition to the two main heavy chains, there are three or perhaps four light chains in each myosin molecule. These light chains have molecular weights of the order of 17,000 to 27,000 [55]. There are indications that the light chains are involved in the ATPase activity of myosin, and it has been reported that the chains can be reversibly dissociated with loss of ATPase activity and reassociated to form active molecules [17, 55].

The peptide chains of myosin can be separated from each other by reagents that are known to destroy hydrophobic interactions among proteins, such as guanidine, urea, and sodium dodecyl sulfate. Another tool useful in the study of the structure of myosin has been the limited digestion by proteolytic enzymes. The first reported limited proteolysis of myosin in which ATPase was left intact employed trypsin [26]. By using various enzymes, it has been possible to isolate various segments of the helical rod as well as the globular active portions of the molecule [46, 61]. The structure of the myosin molecule and the currently used terminology for the various parts are illustrated in Figure 19-2.

AGGREGATION OF MYOSIN

Myosin forms aggregates at low ionic strength. These are revealed by an increased turbidity, and the examination of negatively stained preparations shows that the aggregates formed are regular and in many ways resemble the structure of the thick filaments.

Myosin molecules have a tendency to form end-to-end aggregates involving the light meromyosin (LMM) rods, which then grow into larger structures. The polarity of the myosin molecules is reversed on either side of the center. The globular ends of the myosin molecules form projections on these aggregates similar to those seen on the thick filaments [44]. The use of fluorescent antibodies in combination with electron microscopic studies has greatly elucidated the structure of the myosin filaments, and the data suggest that the headpieces are attached to the filaments by means of flexible hinges [53]. This has important implications for the possible molecular mechanisms of contraction discussed below. In this context, it is also of interest that according to x-ray data, the cross-bridges are arranged in a helical fashion [39], there being six bridges – two at each level – for a repeat period of about 430 Å. This period differs from the half-pitch of the actin helix, which is about 380 Å (Fig. 19-3).

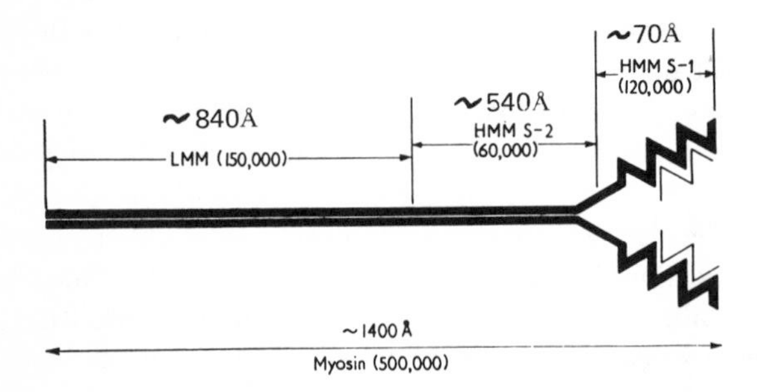

FIGURE 19-2. Schematic representation of myosin. Note the presence of two polypeptide chains running from one end of the molecule to the other. The vertical lines indicate sites at which limited proteolytic action produces the LMM, HMM S-2, and two HMM S-1 fragments. The segment LMM + HMM S-2 is almost 100% α-helical. Each of the HMM fragments possesses one ATPase site. The light zigzag lines in the HMM S-1 region indicate additional light subunits whose precise number per molecule is still under investigation. The numbers in parentheses are molecular weights for the double-stranded portions except for HMM S-1 for which the single-strand weight is given. (From [46].)

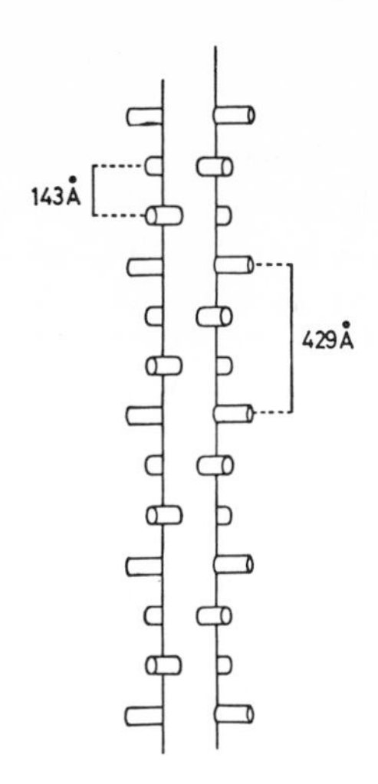

FIGURE 19-3. Helical arrangement of cross-bridges on myosin filaments. (From [40].)

MYOSIN-ACTIN INTERACTION AND ATPase ACTIVITY

The discovery of the ATPase activity of myosin by Engelhardt and Ljubimova [25] led to the recognition of the important interrelations between the structural and functional aspects of this protein and its role in muscle contraction. It should be noted that the protein originally termed "myosin" was, in the light of our current knowledge, a complex of actin and myosin [60]. The ATPase activity of myosin itself is stimulated by Ca^{2+} and is rather low in Mg^{2+}-containing media. If purified actin is added to myosin at low ionic strength in the presence of Mg^{2+}, considerable activation of ATPase activity takes place. The activation of the ATPase activity of myosin by actin is also accompanied by a remarkable change in the physical state of the system. There is an increase in the turbidity and, depending on the concentration, so-called superprecipitation. The latter refers to the appearance of a flocculent precipitate, which often shrinks into a contracted plug. Glycerol-extracted muscle fibers have also been found useful for the study of the interaction of myosin and actin without destroying the spatial relation existing in intact muscle. These fibers lack the energy supply system and the excitation-contraction coupling mechanism of intact muscle. Addition of ATP, however, elicits contraction accompanied by the hydrolysis of the ATP.

The interaction of actin and myosin is also revealed by the contraction, on addition of ATP, of threads consisting of actin and myosin (actomyosin) that are extruded into solutions of low ionic strength.

The combination of actin and myosin can also be observed in solutions of high ionic strength, as indicated by an increase of viscosity. Addition of ATP to this system results in a lowering of viscosity and a decrease of light scattering by the actomyosin solution, both of which are attributable to the dissociation of actomyosin into actin and myosin.

The dissociating effect of ATP, which can also be observed under some conditions at low ionic strength, and the stimulation of the myosin ATPase activity by actin are important links in our understanding of the mechanism of muscle contraction. Recent kinetic measurements by Taylor and his colleagues [47, 62] have suggested that, whether actin is present or not, the actual splitting of the phosphate bond in ATP is carried out by myosin without combining with actin. If myosin alone carries out the hydrolysis in the presence of Mg^{2+}, the rate of release of ADP, and perhaps of phosphate, is slow. Combination with actin accelerates the release of ADP. The complex is dissociated by ATP and the cycle starts again. It should be noted that, in this view, the direct effect of ATP is always dissociation of actomyosin, while the combination of the two proteins results from an inherent affinity between the two. How these processes lead to contraction is discussed later in this chapter.

TROPOMYOSIN AND TROPONIN

The picture presented so far is somewhat oversimplified. Within recent years it has become apparent that the presence of additional proteins exerts a great influence on the interaction of actin and myosin. Of these additional proteins, tropomyosin has been known for a long time, but its role in muscle contraction or its localization had not been elucidated. It is now reasonably clear that tropomyosin, which is an essentially purely α-helical protein consisting of two polypeptide chains, is associated with the thin filaments. Interestingly, a protein very similar to vertebrate tropomyosin, the so-called tropomyosin A or paramyosin, is present in molluscan muscles in close association with myosin, rather than with actin. The function of tropomyosin A in molluscs may be related to the so-called catch – the ability of molluscan muscles to maintain tension over long periods of time without expenditures of energy.

In typical striated muscle of vertebrates, important roles in the regulation of the interaction of actin and myosin are played by tropomyosin as well as by troponin [2, 3, 23], a protein complex whose detailed features are currently being unraveled [28] (Fig. 19-4). If the tropomyosin-troponin complex is present, actin cannot stimulate the ATPase activity of myosin unless the concentration of free Ca^{2+} exceeds about 10^{-6} M. The system consisting solely of purified actin and myosin does not show the dependence on Ca^{2+}. Thus, the actin-myosin interaction becomes controlled by Ca^{2+} in the presence of the regulatory troponin-tropomyosin complex. This can be demonstrated readily in vitro; the Ca^{2+} concentration can be varied by varying the ratio of total Ca^{2+} added and of chelators such as EGTA. There are strong indications that, in vivo, the interaction of actin and myosin is regulated by the intracellular concentration of Ca^{2+}. As is discussed below, the structure involved in the regulation of the intracellular Ca^{2+} level is the sarcoplasmic reticulum and it plays an important role in the mechanism of excitation-contraction coupling.

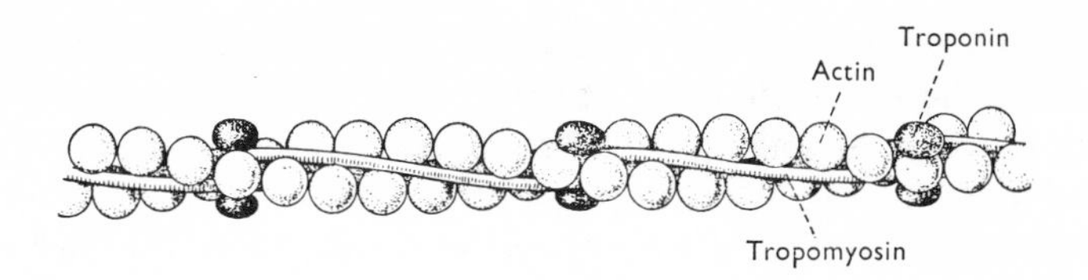

FIGURE 19-4. Model of arrangement of actin, tropomyosin, and troponin in the thin filament. Note that troponin itself is a complex of three or four proteins, and that its localization in relation to tropomyosin is somewhat hypothetical; its place may be in the middle of the tropomyosin molecule. (From [5].)

EXCITATION-CONTRACTION COUPLING

Motor-nerve impulses cause the release of acetylcholine at the neuromuscular junction. Acetylcholine initiates depolarization of the muscle cell membrane. Depolarization of the membrane penetrates into the interior of the cell via the transverse tubules that are continuous with the outer membrane [41]. Ultrastructural studies reveal the presence of a closed compartment within the muscle cell, the sarcoplasmic reticulum, which is in close contact with, but distinct from, the transverse tubules [12]. The sarcoplasmic reticulum and the transverse tubular system are the elements that, in appropriately directed sections, form the so-called triads noted in electron micrographs. It is currently supposed [16] that the sarcoplasmic reticulum stores Ca^{2+} in relaxed muscle and releases it into the sarcoplasm upon depolarization of the cell membrane and the transverse tubular system. The sarcoplasmic reticulum maintains the low intracellular concentration of resting muscle by means of an ATP-dependent Ca^{2+} pump [24, 33]. This has been extensively studied in vitro on fragmented sarcoplasmic reticulum preparations, which contain vesicles limited by lipoprotein membranes [14]. It seems that the formation of phosphorylated intermediates derived from ATP precedes the transport of Ca^{2+} into the interior of the vesicles.

MUSCLE CONTRACTION

ENERGETICS

The classic studies on the energetics of muscle contraction have shown that when muscle shortens under a load, extra energy is liberated in the form of work in addition to a certain amount of heat that is inevitably evolved. This extra energy liberation is known as the *Fenn effect.* Originally, A.V. Hill sought to describe the total energy liberated by a shortening muscle as a sum of three terms: (1) work, (2) activation heat, whose magnitude is independent of both the degree of shortening and the amount of work done, and (3) shortening heat, which is proportional only to the length changes and independent of the load, and hence, of work. Recent studies by Hill himself have shown that this analysis of the energy balance may involve some oversimplification [35–38].

There is now general agreement that ATP hydrolysis accompanies muscle contraction and is the immediate source of energy for it [21]. Although there is good correlation between the amount of ATP broken down and the amount of work performed at various lengths and speeds of contraction, no unequivocal chemical equivalent has been found of shortening heat in a single twitch [9]. In a series of contractions, the total energy liberated by a muscle, i.e., heat and work, agrees well with the calculated energy release (based on in vitro data) from creatine phosphate breakdown. The latter continually rephosphorylates the ADP resulting from the hydrolysis of ATP. Discrepancies, however, still exist in the early stages of contraction between actual measured energy production and that calculated from the heat content of compounds known to change during that phase. Current work is directed at the resolution of this puzzling problem.

MOLECULAR EVENTS UNDERLYING MUSCLE CONTRACTION

The sliding-filament theory and the role of the cross-bridges in tension production are supported by the agreement between the experimentally determined tension of single muscle fibers as a function of length and the tension that would follow from the sliding theory on the assumption that tension is proportional to the number of links formed between the thick and thin filaments [22, 27, 41].

X-ray work has shown that the cross-bridges of the thick filaments undergo a transient movement when muscle contracts [39]. Differences in the orientation of the cross-bridges can be seen clearly when electron micrographs of relaxed and rigor insect muscle were compared [54].

That the interaction of the cross-bridges with the actin filaments results in an undirectional movement, viz., contraction, seems to be based on two things: first, the myosin filaments change their polarity in the middle of the sarcomere owing to the end-to-end aggregation of the constituent molecules; second, there is a built-in polarity in the actin filaments on each side of the Z band, as shown in electron micrographs by the "arrowheads" formed when heavy meromyosin (HMM), or subfragment-1 (S-1), complexes with actin [8].

X-ray diffraction studies indicate that the distances among actin and myosin filaments increase as the sarcomere shortens. The flexible attachments of the myosin heads to the rod portions, discussed earlier, make it possible for the cross-bridges to interact with actin across a varying distance. The driving force of the contraction is most likely the interaction between actin and the S-1 portion of the myosin molecule. ATP apparently plays a role in the dissociation of the links between actin and myosin. When these links dissociate, ATP is then hydrolyzed by myosin (see above) and ADP remains bound. A new interaction at a different actin site can then take place, resulting in the displacement of ADP and a small relative movement between the two filaments. This movement would correspond to a changed angle of attachment between the myosin head and actin (Fig. 19-5). It was mentioned above that the pitch of the actin helix is slightly out of register with that of the cross-bridge helix. As a result, a slight movement at an attached

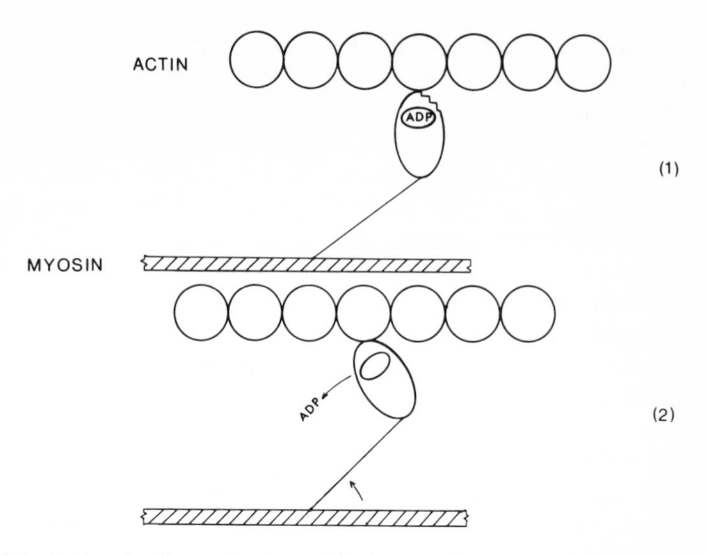

FIGURE 19-5. Model of interaction of the myosin head with actin, resulting in contraction, or at fixed length, in tension development. Myosin that bears ADP attaches to actin (1), resulting in the release of ADP (2). This is possibly accompanied by conformational changes in myosin which would permit closer apposition to actin. This movement is transmitted through cross-bridges and flexible joints to thick filaments (hatched in figure). Part 2 of the diagram shows the changed position of the cross-bridges after movement while remaining attached to the same actin monomer. Not shown in the figure is the subsequent combination of myosin with ATP that results in dissociation from actin, hydrolysis of ATP to ADP (see [47]), and change in myosin conformation, permitting restarting of the cycle.

bridge would create a favorable situation for attachment of another bridge several actin units away on the same filament.

MUSCLE METABOLISM

Muscle utilizes energy made available in the form of ATP, which is hydrolyzed to ADP and inorganic phosphate. The task of muscle metabolism, apart from the necessity of producing the specific constituents required for the construction of structural components, is to produce ATP. The metabolic pathways involved in the production of ATP in muscle are not essentially different from those present in other tissues, including nerve tissue. There are, however, some features of muscle metabolism that are closely related to the mechanism by which contraction is initiated, and others that are subject to control by metabolites arising in the course of contraction.

It has long been known that active muscle increases its metabolism by a factor greater than 100. This is particularly true in muscles that contain "white" fibers that are involved in rapid bursts of activity. In these muscles, the chief source of energy is the anaerobic breakdown of glycogen, and the enzymatic control of glycogen metabolism furnishes a good illustration of the complex way in which regulation takes place. Apart from the specific mechanisms discussed below, the increase in ADP and inorganic phosphate itself serves to increase the rate of ATP synthesis by providing more substrate.

MUSCLE PHOSPHORYLASE

The breakdown of glycogen is catalyzed by the enzyme phosphorylase, whose properties and structure have been clarified recently in considerable detail [4, 32, 34, 49]. Phosphorylase exists in two forms: phosphorylase *b,* the inactive form, which is transformed into the active form, phosphorylase *a,* by means of phosphorylation. ATP is the phosphate donor and the reaction is catalyzed by another enzyme, phosphorylase kinase. The latter enzyme also exists in an inactive and an active form, and the activation again is produced by phosphorylation catalyzed by a protein kinase. This enzyme, in turn, is activated by cyclic AMP, which is formed from ATP under the influence of adenyl cyclase. The cyclase responds to hormones and neurohumoral agents, such as epinephrine. It has been demonstrated in vivo that stimulation of muscle produces an increase in the active form of phosphorylase *a.* Although epinephrine stimulates muscles, no actual increase in the phosphorylated form of phosphorylase kinase has been demonstrated [48].

As has been discussed, the onset of activity in muscle is mediated by the release of Ca^{2+} from the sarcoplasmic reticulum. It now appears that Ca^{2+}, in roughly the same concentration as that required for the activation of the actomyosin system, also increases the activity of the activated phosphorylase kinase [52] by increasing the affinity to phosphorylase *b* [34]. Thus, as the actomyosin system becomes active, phosphorylase kinase produces phosphorylase at a higher rate. As muscle activity proceeds, more phosphorylated glucose, which in turn is broken down through the glycolytic pathway, is made available from glycogen. It has also been shown that phosphorylase phosphatase, the enzyme that reverses the activation of phosphorylase, is inactivated by Ca^{2+}. Thus, the economy of the operation of the system is insured by shutting off the activation mechanism of phosphorylase while there is a need for higher activity. Recent studies have shown that these control features are particularly well established in the complex formed by glycogen and the enzymes involved in glycogen metabolism. It is likely that these enzymes are present in vivo in the form of such a complex.

GLYCOGEN SYNTHETASE

A different enzyme system is involved in the biosynthesis of glycogen. This enzyme, glycogen synthetase, exists in two forms, one of which is active under physiological conditions while the other is inactive. The same protein kinase that activates phosphorylase kinase also phosphorylates glycogen synthetase I. In this case, however, the I form is the active form, and phosphorylation by the kinase converts it into an inactive D form. This insures that when increased muscle activity requires increased release of phosphorylated glucose units from glycogen, the reverse process, which is catalyzed by the synthetase, is shut off [57].

FRUCTOSE PHOSPHOKINASE

The activity of another enzyme of the glycolytic pathway, fructose phosphokinase, which introduces a second phosphate into the hexose molecule prior to its breakdown into two trioses, is under the control of the concentration of various metabolites, e.g., inorganic phosphate and AMP. This kind of control, whereby a small molecule binds to a site that is different from that which binds the substrate, has been referred to as allosteric regulation. Similar allosteric regulation applies to the normally inactive phosphorylase *b* that can be activated by high levels of AMP. The extent to which these regulations play a role under physiological conditions is not quite clear. Several muscle diseases are attributable to abnormalities in their metabolic processes, including the absence of certain specific enzymes involved in glycogen metabolism (see below).

MYOKINASE AND CREATINE KINASE

Mention should be made of two enzymes that play an important role in the nucleotide metabolism of muscle and hence in muscle energetics. One is myokinase, which catalyzes the reversible reaction

$$2\ \text{ADP} \rightleftharpoons \text{ATP} + \text{AMP}$$

This reaction assures better utilization of the energy stored in ATP by permitting the hydrolysis of both high-energy phosphate bonds. The other enzyme, creatine kinase, catalyzes the reaction

$$\text{creatine phosphate} + \text{ADP} \rightleftharpoons \text{ATP} + \text{creatine}$$

Since creatine phosphate is the chief store of high-energy phosphates in muscle, this reaction permits the rephosphorylation of ADP to ATP, which is the immediate source of energy in contraction. During rest, the metabolic processes, in turn, replenish the creatine phosphate stores. Interestingly, this reaction created considerable difficulty in an attempt to demonstrate the breakdown of ATP during single twitch because ATP broken down to ADP was immediately rephosphorylated to ATP and the small change in creatine phosphate could not be detected. Fortunately, the creatine kinase can be blocked selectively by small amounts of dinitrofluorobenzene, and under that condition a decrease in ATP and formation of ADP could be measured [21].

OXIDATIVE METABOLISM

During sustained activity, particularly in those muscles that are rich in mitochondria (see below) and in heart muscle, oxidative processes dominate. These involve the oxidation, via the Krebs cycle, of pyruvate formed in the anaerobic breakdown of glucose or glycogen and the oxidation of fatty acids which, as has now been generally recognized, serves as the chief source of energy in oxidative metabolism. As in other tissues, oxidative metabolism is highly efficient in producing ATP, with about three moles of ATP formed per atom of oxygen utilized.

COMPARISON OF MUSCLE TYPES

It has been known for about a hundred years that striated muscles differ in their velocity of contraction and that correlations exist among the color of the muscle, its velocity of contraction, and the frequency required to produce a tetanus. Red muscles in general are slower and require higher frequencies to produce tetanus than white muscles. Restricting our discussion for the moment to mammalian muscles, Table 19-2 summarizes the chief characteristics of each type. It should be emphasized that as a rule no muscle as a whole can be considered red or white and that the properties of each individual muscle are determined by the distribution of various fiber types to which the criteria listed in the table apply. There are wide differences when one compares various species. For instance, the soleus muscle in rabbit consists almost entirely of typically slow fibers, whereas in rat and man the

TABLE 19-2. Comparison of Properties of Muscle Fiber Types

Property	Type[a]	
	I	II
Velocity	Slow	Fast
Color	Red	White
Z band	Wider	Narrower
Fiber size	Small	Large
Metabolism	Aerobic	Anaerobic
Mitochondrial content	High	Low
Myoglobin content	High	Low
Phosphorylase	Low	High
Succinic dehydrogenase	High	Low
Myosin ATPase	Low	High
Myosin ATPase alkali-stable	No	Yes
Loss of ATPase at high pH	Less	More
Lactic dehydrogenase	More H type	Less H type
Lipid stain	Strong	Weak

[a]Fiber types have been found that fall into one group with respect to some characteristics but belong to the other with respect to other characteristics (see text).

soleus contains a mixture of fiber types. In addition to this simple classification, a number of other systems have been proposed recently, based on gradations between the appearance of the various fibers in histochemical tests for evaluating the enzyme activity of phosphorylase, succinic dehydrogenase, and NADH dehydrogenase. Typically, type I fibers show high dehydrogenase and low phosphorylase activity; in type II the whole staining pattern is reversed. This correlation does not consistently hold for intermediate fiber types [19, 29].

It appears that the difference between types of muscle depends not only on the presence of certain enzymes in large or small amounts; different types of muscle may contain the same enzyme in different forms. The different forms of an enzyme catalyzing the same reaction have been called *isoenzymes* or *isozymes.* The first molecule that was thus identified was lactic dehydrogenase. Lactic dehydrogenase consists of four subunits. Each subunit can be either the so-called H type, which is predominant in heart muscle, or the M type, predominant in white skeletal muscle. In embryonic muscle, a higher content of the H type has been found, although differences exist with respect to species and the muscles examined. Various muscles of the adult contain the two kinds of subunits in various proportions – again with species variations – and the complexes, of which there can be five, differ in their electrophoretic mobility.

During recent years, myosin has been discovered to exist in different forms [18, 58]. The ATPase activity of myosin from a typical red muscle is lower than that from white muscle, and a striking difference exists with respect to its stability at alkaline pH. White muscle myosin can be incubated for 10 min at pH 9 without any loss of activity, whereas myosin from red muscle loses most of its activity when so treated. More subtle differences exist with respect to stability at low pH. Both types of myosin lose activity at acid pH, but the red muscle type of myosin loses less than the white type. These in vitro differences in myosin stability at different pHs have been used recently for the histochemical classification of various fiber types. The degree of correlation between staining for oxidative enzymes and the pH effect on myosin ATPase varies depending on the species.

As discussed above, myosin contains two heavy chains and three or four subunits It appears that in white muscle there are three different kinds of small subunits, whereas in red muscle there are only two. Furthermore, the light-chain subunits found in red muscle differ from their counterparts in molecular weight, although their electrophoretic mobilities in polyacrylamide gels in the presence of sodium dodecyl sulfate are very similar to two of the small subunits found in white muscle [55]. It is interesting that the similarity between red muscle and cardiac muscle, first noted with respect to the lactic dehydrogenase isozyme pattern, appears to apply to the light chains of myosin as well. Cardiac myosin has a light-chain pattern similar to that of typical red muscles.

A striking difference between myosin from white and red muscles can be demonstrated by looking at the electron micrographs of negatively stained aggregates of the rodlike portion of the myosin molecule, LMM [51]. Whereas LMM aggregates from white muscle myosin show a striation pattern consisting of strongly stained wider segments alternating with lightly stained narrower segments with a main period of 430 Å, the situation is reversed in LMM aggregates from red muscle myosin. The lighter segments are wider than the electron-dense segments, and within the lightly stained zones there are four distinct, darkly stained lines. The precise molecular basis of this difference is not known, but it seems that it can serve as a useful "molecular" fingerprint of the two types of myosin. Again, cardiac LMM aggregates exhibit the same pattern as a red muscle LMM aggregate. It has been recently shown that myosin contains some unusual amino acids: 3-methylhistidine, monomethyllysine, and trimethyllysine (see Perry's chapter in [1]; also [46]). Interestingly, 3-methylhistidine occurs only in white muscle myosin and is absent from cardiac and red muscle myosin; however, the methylated lysines are present in all types of myosins investigated. Because these methylated amino acids are found in the S-1 portion that contains the active center(s) of the myosin molecule, functional differences in the different types of myosins may be related to differences in the methylation of histidine.

Considerable attention has been given to the possible role of neural effects in the differentiation of the different types of muscle fibers [6]. If the different types reflect the expression of a different set of genes, the question arises of which factors are responsible for activating one set of genes and repressing another. The role of neural influence has been demonstrated in recent experiments in which both the physiological properties and several biochemical characteristics of the muscles changed after the nerve supplies to a slow muscle and to a fast muscle were cut and the slow muscle was reinnervated with the central stump of the fast-muscle nerve, and vice versa [20]. Changing the biochemical characteristics from a fast muscle toward those of a slow muscle was more successful than in the reverse direction.

A final word of clarification may be in order. Very often the literature refers to slow muscles as "tonic" and fast muscles as "phasic." This identification is not correct for most vertebrate muscles. Most slow and fast muscles of vertebrates are so-called phasic muscles that exhibit a spread of action potential and exhibit innervation by a single end plate, the so-called *en plaque* innervation. On the other hand, true tonic muscles, such as the latissimus dorsi anterior of the chicken and some mammalian eye muscles, have multiple innervation, called *en grappe,* and there is no spreading action potential.

BIOCHEMICAL ASPECTS OF MUSCLE DISEASE

In only a few diseases of muscle does a close correlation exist between the clinical pathological picture and an identifiable biochemical disorder. In many instances in which biochemical changes are found in the muscle, the changes merely reflect the result of destruction brought about by the disease. Also, those diseases in which there are changes in the levels of enzymes present in blood serum may reflect destruction of muscle cells. In the latter type of disease, however, the underlying cause may be an abnormal permeability of the muscle membrane. Various enzymes involved in oxidative phosphorylation – lactic dehydrogenase, glutamic oxalic transaminase, glutamic pyruvate amino transferase and, particularly, creatine kinase – are to be noted.

Muscular dystrophy represents a number of genetically determined diseases that result in destruction of the muscle tissue. The primary biochemical defect is not known, but changes in membrane structure are possibly involved. In vitro studies have shown that abnormalities occur in the sarcoplasmic reticulum in human dystrophy and in the hereditary form of mouse dystrophy. Although more work will be needed in this field, it seems that the rate of protein synthesis increases in dystrophy, and thus it appears that the disappearance of the contractile machinery is due to an increased destruction of the proteins. There are reports concerning increased proteolytic activity in dystrophic muscle, perhaps due to the breakdown of lysosomes [10, 15].

The appearance of elevated serum enzymes, particularly creatine kinase, is a typical finding in muscular dystrophy; this enzyme is also found in elevated amounts in the serum of female carriers who do not actually have the disease. Increased urinary excretion of creatine, which is also found in muscular dystrophy, is due to the inability of the muscle to retain creatine. Normally, creatine is transformed into creatinine in the muscle for excretion.

Metabolic diseases of muscle take many forms. Several types of muscle diseases are associated with a genetic absence or deficiency in any one of several specific enzymes involved in glycogen metabolism. Another group of muscle diseases is characterized by myoglobinuria, either as a result of muscle destruction due to ischemia, injury, or intoxication, or as a result of an abnormal membrane permeability of unknown origin. Finally, there is a group of hereditary diseases characterized by periodic muscle paralysis and either normal, lower, or higher blood K^+ level (for a detailed discussion, see McArdle's chapter in [16]).

An interesting type of muscle disease has been found in nemaline myopathy [56]. Investigators have described rodlike bodies that appear to be connected to a Z band and whose electron microscopic structure bears similarities to that of tropomyosin crystals. When this disease was discovered, it was generally assumed that tropomyosin is a constituent of the Z band, hence it was assumed that tropomyosin is the protein responsible for nemaline bodies.

We have discussed only some of the aspects of the connection between muscle diseases and biochemical disorders. The reader is referred to [15] for a detailed review.

REFERENCES

1. Briskey, E. J., Cassens, R. G., and Marsh, B. B. *Muscle as a Food* (2nd ed.). Madison: University of Wisconsin Press, 1970. (Covers a wide range of topics. Written by specialists, but aimed at those interested in obtaining a good general review. Chapters cover muscle-cell development, comparative aspects, ultrastructure, proteins, relaxation, and adaptation.)

2. Ebashi, S., and Endo, M. Calcium ion and muscle contraction. *Progr. Biophys. Mol. Biol.* 18:123, 1968.
3. Ebashi, S., Endo, M., and Ohtsuki, I. Control of muscle contraction. *Rev. Biophys.* 2:351, 1969.
4. Fischer, E. H., Pocker, A., and Saari, J. C. The Structure, Function and Control of Glycogen Phosphorylase. In Campbell, P. N., and Dickens, F. (Eds.), *Essays in Biochemistry.* Vol. 6. London: Academic, 1970.
5. Gergely, J. Contractile proteins. *Annu. Rev. Biochem.* 35:691, 1966.
6. Guth, L. Trophic influence of nerve on muscle. *Physiol. Rev.* 48:645, 1968.
7. Hanson, J. Recent x-ray diffraction studies of muscle. *Rev. Biophys.* 1:177, 1968.
8. Huxley, H. E. Muscle Cells. In Brachet, J., and Mirsky, A. E. (Eds.), *The Cell.* New York: Academic, 1960.
9. Huxley, H. E. The mechanism of muscle contraction. *Science* 164:1356, 1969.
10. Milhorat, A. (Ed.). Exploratory Concepts of Muscular Dystrophy. *Internatl. Congr. Ser. 147.* Amsterdam: Excerpta Medica Foundation, 1966. (Proceedings of a conference devoted to muscular dystrophy. Contains a good deal of basic material on muscle biochemistry.)
11. Mommaerts, W. F. H. M. Energetics of muscle contraction. *Physiol. Rev.* 49:427, 1969.
12. Perry, S. V. The structure and interactions of myosin. *Progr. Biophys. Mol. Biol.* 17:327, 1967.
13. Sandow, A. Skeletal muscle. *Annu. Rev. Physiol.* 32:87, 1970.
14. Smith, D. S. The organization and function of the sarcoplasmic reticulum and T-system of muscle cells. *Progr. Biophys. Mol. Biol.* 16:107, 1966.
15. Walton, J. (Ed.). *Diseases of Voluntary Muscle* (2nd ed.). London: Churchill, 1969. (Aimed primarily at the clinician. Chapters written by "acknowledged experts in the field." The first part of the book is devoted to modern views of the structure and function of muscle; the second to changes, both structural and biochemical, which may occur in disease; the third and fourth with clinical, genetic, and methodological aspects of disease.)
16. Weber, A. Energized Calcium Transport and Relaxing Factors. In Sanadi, D. R. (Ed.), *Current Topics in Bioenergetics.* Vol. 1. New York: Academic, 1966.
17. Young, M. Molecular basis of muscle contraction. *Annu. Rev. Biochem.* 38:913, 1969.

ORIGINAL SOURCES

18. Bárány, M. Activity of myosin correlated with speed of muscle shortening. *J. Gen. Physiol.* 50(No. 6, Pt. 2):197, 1967. (Demonstration of ATP breakdown in single muscle.)
19. Brooke, M. H., and Kaiser, K. K. Muscle fiber types: How many and what kind? *Arch. Neurol.* 23:369, 1970. (Good discussion of histochemical classification of fiber types.)

20. Buller, A. J., Mommaerts, W. F. H. M., and Seraydarian, K. Enzymic properties of myosin in fast and slow twitch muscles of the cat following cross innervation. *J. Physiol. (Lond.)* 205:501, 1969. (Demonstration of changes in enzymatic properties of myosin corresponding to changes in physiological properties of cross-innervated muscle.)
21. Cain, D. F., and Davies, R. E. Breakdown of adenosine triphosphate during a single contraction of working muscle. *Biochem. Biophys. Res. Commun.* 8:361, 1962.
22. Davies, R. E. A molecular theory of muscle contraction: Calcium-dependent contractions with hydrogen bond formation plus ATP-dependent extensions of part of the myosin-actin cross-bridges. *Nature (Lond.)* 199:1068, 1963. (Detailed theory of molecular changes underlying contraction.)
23. Ebashi, S., and Ebashi, F. A new protein component participating in the super-precipitation of myosin B. *Biochemistry* 55:604, 1964. (Fundamental paper in the field of troponin-tropomyosin; Ca^{2+} regulation of actin-myosin interaction.)
24. Ebashi, S., and Lipmann, F. Adenosine triphosphate linked concentration of calcium ions in a particulate fraction of rabbit muscle. *J. Cell Biol.* 14:389, 1962. (Ca uptake by fragmented sarcoplasmic reticulum.)
*25. Engelhardt, W. A., and Ljubimova, M. N. Myosin and adenosine triphosphatase. *Nature (Lond.)* 144:668, 1939. (First report on enzymatic activity of myosin. The material used was actomyosin in terms of our current nomenclature.)
26. Gergely, J. Studies on myosin-adenosinetriphosphatase. *J. Biol. Chem.* 200: 543, 1953. (Tryptic digestion of myosin leaves ATPase intact.)
27. Gordon, A. M., Huxley, A. F., and Julian, F. J. The variation in isometric tension with sarcomere lengths in vertebrate muscle fibers. *J. Physiol. (Lond.)* 184:170, 1966. (Changes of tension with sarcomere length in single fibers agree well with the sliding filament theory.)
28. Greaser, M., and Gergely, J. Reconstitution of troponin activity from three protein components. *J. Biol. Chem.* 246:4226, 1971.
29. Guth, L., and Samaha, F. J. Qualitative differences between actomyosin ATPase of slow and fast mammalian muscle. *Exp. Neurol.* 25:138, 1969. (Correlation is sought between actomyosin ATPase, histochemistry, velocity of contraction, and fiber types.)
30. Hanson, J., and Huxley, H. E. Quantitative studies on the structure of cross-striated myofibrils. II. Investigations by biochemical techniques. *Biochim. Biophys. Acta* 23:250, 1957. (Important information on correlation of proteins in filaments.)
31. Hanson, J., and Lowy, J. The structure of F-actin and of actin filaments isolated from muscle. *J. Mol. Biol.* 6:46, 1963. (Electron-microscopic demonstration of actin in a double helical arrangement of globular subunits.)

*A classic.

32. Haschke, R., Heilmeyer, M., Meyer, F., and Fischer, E. Control of phosphorylase activity in a muscle glycogen particle. III. Regulation of phosphorylase activity in a muscle glycogen particle. (See also refs. [34] and [49]. These three papers strongly suggest the existence of glycogen – enzyme complexes that play an important role in vivo.)
33. Hasselbach, W., and Makinose, M. Uber den mechanismus des calcium transportes durch die membranen des sarkoplasmiatischen reticulums. *Biochem. Zeit.* 339:94, 1963. (Suggests that Ca^{2+} transport by sarcoplasmic reticulum is coupled to ATPase activity.)
34. Heilmeyer, M., Meyer, F., Haschke, R., and Fischer, E. Control of phosphorylase activity in a muscle glycogen particle. II. Activation by calcium. *J. Biol. Chem.* 245:6649, 1970.
35. Hill, A. V. The effect of load on the heat of shortening of muscle. *Proc. R. Soc. Ser. B.* 159:297, 1964.
36. Hill, A. V. The efficiency of mechanical power development during muscular shortening and its relation to load. *Proc. R. Soc. Ser. B.* 159:319, 1964.
37. Hill, A. V. The effect of tension on prolonging the active state in a twitch. *Proc. R. Soc. Ser. B.* 159:589, 1964.
38. Hill, A. V. The variation of total heat production in a twitch with velocity of shortening. *Proc. R. Soc. Ser. B.* 159:596, 1964. (Important papers reviewing some of the concepts put forward by Hill himself about 25 years before.)
39. Huxley, A. F. Muscle structure and theories of contraction. *Progr. Biophys.* 7:255, 1957. (Mathematical analysis of cross-bridge interaction.)
*40. Huxley, A. F., and Niedergerke, R. Structural changes in muscle during contraction. *Nature (Lond.)* 173:971, 1954. (Experiments that form the basis of the sliding-filament theory.)
41. Huxley, A. F., and Taylor, R. E. Local activation of striated muscle fibers. *J. Physiol. (Lond.)* 144:426, 1958. (Demonstration of physiological importance of T tubules in excitation-contraction coupling.)
42. Huxley, H. E. Electron microscopic studies on the structure of natural and synthetic protein filaments from striated muscle. *J. Mol. Biol.* 7:281, 1963. (Electron-microscopic demonstration of arrowheads – heavy myosin-actin complexes and thick models formed by myosin aggregation.)
43. Huxley, H. E., and Brown, W. The low-angle x-ray diagram of vertebrate striated muscle and its behavior during contraction and rigor. *J. Mol. Biol.* 30:383, 1967. (These studies show important changes in the x-ray diffraction pattern of live muscle, depending on whether it is resting, contracting, or in rigor, which can be attributed to changes in the cross-bridges.)
*44. Huxley, H. E., and Hanson, J. Changes in the cross-striations of muscle during contraction and stretch and their structural interpretation. *Nature (Lond.)* 173:973, 1954. (See comments on ref. [40].)

*A classic.

45. Kuehl, W. M., and Adelstein, R. S. The absence of 3-methylhistidine in red, cardiac and fetal myosin. *Biochem. Biophys. Res. Commun.* 39:956, 1970. (Comparative studies on chemistry of myosin.)
46. Lowey, S., Slayter, H. S., Weeds, A. G., and Baker, H. Substructure of the myosin molecule. I. Subfragments of myosin by enzymic degradation. *J. Mol. Biol.* 42:1, 1969. (Contains a great deal of current information on the structure of myosin, based on limited digestion with proteolytic enzymes.)
47. Lymn, R. W., and Taylor, E. W. Transient-state phosphate production in the hydrolysis of nucleoside triphosphates by myosin. *Biochemistry* 9:2975, 1970.
48. Mayer, S. E., and Krebs, E. G. Studies on the phosphorylation and activation of skeletal muscle phosphorylase kinase in vivo. *J. Biol. Chem.* 245:3153, 1970.
49. Meyer, F., Heilmeyer, M., Haschke, R., and Fischer, E. Control of phosphorylase activity in a muscle glycogen particle. I. Isolation and characterization of the protein-glycogen complex. *J. Biol. Chem.* 245:6642, 1970.
50. Mommaerts, W. F. H. M. The molecular transformation of actin. III. The participation of nucleotides. *J. Biol. Chem.* 198:469, 1952.
51. Nakamura, A., Sréter, F., and Gergely, J. Comparative studies of light meromyosin paracrystals derived from red, white and cardiac muscle myosins. *J. Cell Biol.* 49:883, 1971.
52. Ozawa, E., Hosoi, K., and Ebashi, S. Reversible stimulation of muscle phosphorylase b kinase by low concentrations of calcium ions. *J. Biochem.* 61:431, 1967.
53. Pepe, F. A. The myosin filament. II. Interaction between myosin and actin filaments observed using antibody staining in fluorescent and electron microscopy. *J. Mol. Biol.* 27:227, 1967. (The combination of the use of electron microscopy and fluorescent antibodies has yielded important information regarding the molecular architecture of muscle in relation to function.)
54. Reedy, M. K., Holmes, K. C., and Tregear, R. T. Induced changes in orientation of the cross bridges of glycerinated insect flight muscle. *Nature (Lond.)* 207:1276, 1965.
55. Sarkar, S., Sreter, F. A., and Gergely, J. Light chains of myosins from fast, slow and cardiac muscles. *Proc. Natl. Acad. Sci. U.S.A.* 68:946, 1971.
56. Shy, G. M., Engel, W. K., Somers, J. E., and Wanko, T. Nemaline myopathy: New congenital myopathy. *Brain* 86:793, 1963. (Muscle diseases with rodlike bodies apparently connected with Z bands. Presence of tropomyosin in these structures has been suggested.)
57. Soderling, T. R., Hickenbottom, J. P., Reimann, E. M., Hunkeler, F. L., Walsh, D., and Krebs, E. G. Inactivation of glycogen synthetase and activation of phosphorylase kinase by muscle adenosine 3-,5-monophosphate-dependent protein kinase. *J. Biol. Chem.* 245:6317, 1970. (An interesting aspect of the reciprocal control of glycogen synthesis and breakdown.)
58. Sreter, F. A., Seidel, J., and Gergely, J. Studies on myosin from red and white skeletal muscle of the rabbit. I. Adenosine triphosphate activity. *J. Biol. Chem.* 241:5772, 1966. (Molecular differences between myosin from slow and fast muscles.)

59. Straub, F. B., and Feuer, G. Adenosine triphosphate: The functional group of actin. *Biochim. Biophys. Acta* 4:455, 1950.

*60. Szent-Györgyi, A. Studies on muscle. *Acta Physiol. Scand.* 9(Suppl. 25), 1945. (Covers pioneering work on the isolation of myosin and actin and their properties.)

61. Szent-Györgyi, A. Meromyosins. The subunits of myosin. *Arch. Biochem.* 42:305, 1953. (Description of meromyosins – tryptic fragments of myosin.)

62. Taylor, E. W., Lymn, R. W., and Moll, G. Myosin-product complex and its effect on the steady state rate of nucleoside triphosphate hydrolysis. *Biochemistry* 9:2984, 1970. (This paper introduces important new concepts in the analysis of the mechanism of myosin ATPase.)

---

*A classic.

SECTION V

Investigative Techniques

20

Analytical Methods

David B. McDougal, Jr.

PROGRESS IN NEUROCHEMISTRY must be based on hard facts that are obtained by accurate, specific methods. Data collected with inadequate methods may prove to be worse than inadequate. Although investigators have occasionally reached correct conclusions on the basis of invalid data, it is far more common for them to be led astray. Progress in methodology may lead to refinements in experimental results, or it may open up whole new areas of research.

Neurochemistry is not unique in regard to its experimental methods. Many, perhaps most, of the biochemical methods used in the study of the brain are the same as those used in the study of any other organ. However, the nervous system has certain peculiarities that make the problem of obtaining samples representative of the in vivo state difficult. Perhaps the most important of these peculiarities of the brain are its high metabolic rate, sensitivity to external stimuli, and residence within a bony casing. In addition, the nervous system is anatomically complex, so analyzing samples small enough to be reasonably homogeneous becomes a problem in microchemistry. This chapter, therefore, is primarily addressed to problems of sampling and to methods of microchemical analysis.

SAMPLING THE BRAIN IN SITU

METHODS COMPATIBLE WITH SURVIVAL OF THE ANIMAL

Metabolism of the brain as a whole can be studied by measurements of cerebral blood flow coupled with measurements of arterial and internal jugular venous blood levels of substances such as O_2, CO_2, glucose, and lactate. These methods disturb the brain least of all, and can be carried out in conscious animals, including man.

Cerebral Blood Flow

The following descriptions include only a few of the many methods by which cerebral blood flow has been measured (see also Chap. 15).

1. In man, cerebral blood flow is determined by administering an inert gas and simultaneously measuring it as a function of time in arterial and internal jugular venous blood. The gas may be radioactive, which facilitates its measurement. The calculations are based on the *Fick principle,* a special statement of the law of conservation of mass, which may be stated thus:

> The amount of inert substance taken up by a tissue in a given time is equal to the amount of the substance brought to the tissue in the arterial blood minus that taken away in the venous blood in the given time.

The derivation of the equations used and the assumptions employed are discussed by Kety [6].

2. A technically simpler method involves measuring the radiation emitted from the head after administering a radioactive gas; the gas may be added to the respired air or dissolved in saline and injected into the carotid artery. The disappearance of radiation from the head can be monitored continuously, and the slope of the curve, plotted semilogarithmically, can be used to calculate the cerebral flow rate when the partition coefficient of the gas between blood and brain is known [5]. If a large radiation detector is placed correctly, essentially all the brain is monitored. If a small detector is used, studies of blood flow can be made over different regions of the brain surface. Both regional and whole-brain studies performed in this manner give evidence for two compartments in brain with different flow rates: the curves obtained are not monoexponential, but can be resolved into two components with different slopes.

In experimental animals, the same sorts of results have been obtained with a small counter positioned over the exposed brain. When the cortical gray matter is dissected away carefully, the component with the faster rate disappears from the record, suggesting the identification of the compartment with the faster flow rate as gray matter and that with the slower as white matter. If this analysis is correct, the blood flow in gray matter is, on the average, four times as fast as that in white (65 to 80 ml/100 g/min vs. 16 to 21 ml/100 g/min in normal man [5]).

3. Neither method 1 nor method 2 can be used to provide a continuous record of cerebral flow. This can be obtained with the use of a thermoelectric blood-flow recorder, which consists of two needles with thermocouples. One needle is heated at the tip, and the temperature difference between the two needle tips is measured by the thermocouples. This difference is recorded continuously. The cooling effect of the flow of blood past the heated needle tip tends to reduce the temperature difference between the two tips; this cooling effect is greater when the blood flow increases [5]. The thermoelectric blood-flow recorder has two advantages: the probe can be inserted deep within the brain and it provides a continuous record. Its disadvantage is that it does not give flow rates in absolute terms.

When measurements of cerebral blood flow are coupled with measurements of O_2, CO_2, glucose, and lactate made on samples of arterial and jugular vein blood collected at the same time (or under similar conditions), several important parameters of cerebral metabolism can be defined (Table 20-1). Measurements such as these demonstrate that glucose serves as the major energy source in brain. They also give evidence for the very high metabolic rate of brain, which, as is discussed later, can give rise to serious problems when one tries to estimate the in vivo levels of certain substances in brain.

Other Cerebral Measurements

Advances in the development of selective electrodes have enabled the measurement of regional tissue levels of pO_2, pCO_2, and pH [4]. Miniaturization of electrodes has made it possible to record changes in circumscribed regions.

Surface fluorometry is another kind of measurement that can be made upon intact brain. By the use of appropriate filters, the method can be made specific for reduced pyridine nucleotides. Thus, transient changes in the levels of these substances can be observed and an approximate quantitation can be achieved [12].

TABLE 20-1. Cerebral Blood Flow and Arteriovenous Differences in Key Metabolites in Man

Parameter	Rate (ml/100 g brain/min)
1. Cerebral blood flow[a]	54 ± 12
2. Cerebral oxygen consumption[a]	3.3 ± 0.4
	Ratio of O_2 consumed to CO_2 produced
3. Cerebral respiratory quotient[a]	0.99 ± 0.09
	A-V differences (mg/100 ml blood)
4. Glucose[b]	−9.8 ± 1.7
5. Lactic acid[b]	+1.6 ± 0.9
	Rate (mmole/kg brain/min)
6. Glucose consumption (calculated from 1 and 4)	0.29
7. Lactic acid produced (calculated from 1 and 5)	0.096
8. Glucose oxidized (calculated from 2, 6, and 7)	0.24
9. Oxygen consumed (calculated from 2)	1.47
10. Oxygen required to oxidize the amount of glucose in 8	1.45

[a]Kety, S. S., and Schmidt, C. F., *J. Clin. Invest.* 27:476, 1948; Method 1 (see text), using nitrous oxide as the inert gas; mean ± S.D. for 34 observations on 14 young men.

[b]Gibbs, E. L. et al., *J. Biol. Chem.* 144:325, 1942; mean ± S.D. for 50 men aged 18–29 years.

METHODS FOR ANALYSIS OF INTERMEDIARY METABOLISM

The intermediary metabolism of any tissue may be studied by measuring the levels of relevant metabolites and the changes in those levels under appropriate experimental conditions. In brain, the purpose of such experiments is to establish the normal cerebral levels of substances in a particular biochemical pathway, such as glucose metabolism, acetylcholine metabolism, or protein metabolism, and to observe their responses to disturbing influences, such as changes in neuronal activity, drugs, poisons, or disease processes. Such investigations may give useful information both about the metabolic process under study and about the disturbing influence, whether it be disease, drug, or dream.

However, two general obstacles complicate the determination of the amount of a substance in brain: biological lability and chemical lability.

Problems of Biological Lability

Biological lability is a function of the role of the molecule in the life of the cell. Consider, for example, brain glucose. When a sample of brain is removed from the body, the blood supply – and hence the supply of glucose and oxygen – to the sample is interrupted. Because the cerebral metabolic rate is high, anoxia develops rapidly (in man, the brain accounts for 20 to 25 percent of the total oxygen consumed, although the brain represents only 2 percent of the total body weight). In the mouse, the oxygen in the brain at the time of decapitation can be expected to be consumed in less than 1 sec after interruption of the blood supply. In the frog, which has a more leisurely life style, a minute might elapse before all the oxygen is consumed. In the course of becoming anaerobic, the brain undergoes striking biochemical changes [16] and shifts rapidly to a glycolytic economy. Glycolysis plus oxidation of glucose produces 38 moles of high-energy phosphate per mole of glucose, but glycolysis alone produces only 2. In an attempt to compensate, glucose utilization speeds up enormously in ischemic tissue (switching to a rate of 6 to 7 mmoles/kg/min in mouse brain from an aerobic rate of 0.8 mmoles/kg/min). Since the initial level of glucose is 1.5 mmoles per kilogram, a delay of only 10 sec during sampling introduces an error of 50 percent in the measured brain glucose level. Notice that, in spite of its biological lability, *chemically* glucose is a stable molecule.

To minimize the biological lability of glucose, the brain must be frozen as rapidly as possible. This may be done by immediately freezing either the whole animal, the head immediately upon decapitation, or, in the case of a larger animal, a sample removed from the surface of the exposed brain. Freezing is usually done by immersing the animal, head, or sample in liquid nitrogen (–196°C) or in CCl_2F_2 kept at its freezing point (–150°C). This method of tissue fixation is also satisfactory for other biologically labile metabolites (α-ketoglutarate and phosphocreatine, for example) when the object frozen is small. Naturally, not all substances are so labile, but it is not always possible to predict which substances will be stable and which will not.

Problems of Chemical Lability

Chemical lability becomes a problem during the extraction and assay of certain substances in the brain. In the assay of metabolites and other small molecules, advantage is taken of the lability or insolubility of proteins at the extremes of the pH scale to obtain extracts that are essentially free of contaminating enzymes. However, some small molecules are also unstable at very high or low pH. For example, NADH is extremely unstable in acid, even in the cold, and must be extracted with an alkaline medium. Acetylcholine, on the other hand, is unstable at an alkaline pH, and extracts for this material are best made, and kept, acid. Many metabolic intermediates are quite stable in cold acid; even adenosine triphosphate (ATP), which can be assayed as "acid-labile phosphate," will survive extraction with cold perchloric acid. After the tissue has been extracted and the proteins removed, the pH of portions of the extract can be adjusted to those most favorable for preservation of the substances under study.

In general, enzymes from brain are surprisingly stable after death of the animal [20], and, in most cases, the brain may be dissected from the head without preliminary freezing. Tissue preparations for enzyme assay are usually made in buffer or sucrose solutions at a pH near 7, although special conditions may be needed for particular enzymes (it should go without saying that each new enzyme is a new problem in this regard). It is evident from the discussion of biologically labile metabolites that if the properties of an enzyme are radically changed under conditions such as those which accompany decapitation or death from cervical dislocation, rapid freezing techniques may again be needed. An example is glycogen phosphorylase, which rapidly converts from the *b* to the *a* form in brain in response to anoxia (see Chap. 9). In this case, not only must freezing be very rapid, but the tissue must be homogenized and assayed under conditions that minimize the activities of the two enzymes that catalyze the interconversion of the *a* and *b* forms [10].

Biological Lability and the Measurement of Metabolic Rates

Decapitation not only makes the brain anoxic, but also converts it into a closed system. Therefore, it is possible to measure the cerebral metabolic rate by measuring the changes in the levels of ATP and phosphocreatine plus either the disappearance of glucose and glycogen or the appearance of lactate at brief intervals (5 to 10 sec in mouse, 2 to 5 min in frog) [16]. It is assumed (and seems to be true) that during the first few seconds (or, in frog, minutes) after decapitation there is relatively little change in metabolic activity, and that the metabolic rate measured at this time is a good approximation of the in vivo rate. The appropriate equation is as follows:

$$\text{mmoles} \sim\text{P/kg/min} = (1/t)\,(2\Delta_t[\text{ATP}] + \Delta_t[\text{P-creatine}] + 2.9\Delta_t[\text{glycogen}] + 2\Delta_t[\text{glucose}])$$

where ~P/kg/min is the rate of utilization of high-energy phosphate bonds, t is the time interval from the time of decapitation, Δ_t[ATP] is the difference in ATP concentration from the time of decapitation to the subsequent time, t, and so on. The accuracy may be improved by adding Δ_t[ADP] to the right-hand side of the equation. If glycogen and glucose are not measured directly, their concentration differences may be replaced in the equation with minus $c\Delta_t$[lactate]; the factor c is 1 when all the lactate comes from glucose and 1.45 when it all comes from glycogen. The rate of ~P utilization can be converted to oxygen consumption by dividing by 6.

It has thus been possible to examine such effects as those of drugs, stimulation, development, and aging upon the cerebral metabolic rate of small animals and to compare metabolic rates in various regions of brain, even with samples the size of single large neurons (see below). An extension of the method has provided an alternative to oxygen consumption methods for measuring metabolic rates in isolated peripheral nerve and sympathetic ganglion (see below).

SAMPLING THE BRAIN IN VITRO

ISOLATED, SURVIVING TISSUE

The details of any metabolic pathway can be investigated in samples of tissue that contain cells or cell processes if the integrity of the tissue is intact, or partially so, and the tissue is incubated in media of known composition. From such experiments, data may be obtained that are obtainable in no other way.

The supply of oxygen to the tissue must be adequate. With surviving nerve, particularly from poikilotherms such as frog, this presents no problem because the tissue has a small diameter and a very low metabolic rate (0.025 mmole O_2/kg wet weight/min at 25°C for frog sciatic nerve). In this instance, depriving the nerve of oxygen is much more difficult than keeping it adequately supplied. With mammalian cerebral cortex slices, however, the metabolic rate is at least 40 times higher. The problem of providing an adequate supply of oxygen is obviated by making the slices sufficiently thin (about 0.3 mm) so that the diffusion of oxygen into the tissue is not rate-limiting and by bubbling gas containing 95 to 100 percent O_2 through the medium [7].

Interpretation of data from this sort of experiment is made more difficult because many cells are inevitably damaged during preparation of the slice; for example, in cerebral cortex the cytoplasmic extensions of many neurons reach from the pial surface to the white matter and many more go more than half way from one surface to the other. In addition, the slice is deprived of its normal supply of afferent impulses, and thus the resting metabolic rate of the slice in vitro is much slower than the resting cerebral metabolic rate in vivo. Electrical stimulation of the slice helps remedy the latter situation, and the metabolic rate of stimulated slices approaches that seen in normal "resting" brain [7]. The in vitro rate cannot be made to exceed that observed in vivo at rest, however, although the metabolic rate can increase by several times in stimulated brain in vivo.

There may be other problems, as well. For example, slices of cerebral cortex respiring in a bath invariably convert more of the glucose they metabolize to lactate and less to CO_2 and water than does brain in situ and in vivo, although the oxygen supply to such slices appears to be adequate (e.g., no increase in tissue lactate, and levels of ATP and phosphocreatine are well maintained). The bath seems to act as a lactate trap. The effects of this trap, if any, on other aspects of metabolism, such as the pentose phosphate shunt, do not seem to have been studied.

Examples of surviving tissue preparations that have been used include isolated invertebrate giant axons, invertebrate and vertebrate peripheral nerve, sympathetic ganglia, nerve-muscle preparations, brain slices, and choroid plexus. Electrical stimulation has been used with all but the last of these. Biochemically, the studies have concerned such topics as energy metabolism, the pentose phosphate shunt, and glutamate metabolism, and more recently studies of the metabolism of biogenic amines in cerebral slices have appeared.

The biochemical study of the nervous system in tissue and organ culture has only just begun. It would appear to have a promising future.

SEPARATION OF BRAIN INTO COMPONENT PARTS

Cellular Fractionation

The technique of mass separation of the cells of the brain into more-or-less homogeneous groups is of quite recent origin [7]. Fractions that are enriched with glial cells and those that contain mostly neurons have been obtained. So far, the yield of cells has been low, and it may be that the cellular processes have all been more-or-less shortened during isolation. Much can be expected of such a method, however, particularly in those areas of neurochemistry in which single-cell methods are most difficult to apply.

Subcellular Fractionation

The methods of subcellular fractionation depend upon cell fragmentation followed by repeated centrifugations at different speeds and in varying media ([8]; see also Chap. 21, this volume). For brain, in particular, combinations of velocity sedimentation with isopycnic sedimentation are employed. Usually, homogenates of the tissue are subjected to preliminary purification by velocity sedimentation. Then the crude fraction that is richest in the desired organelle is layered over a continuous or discontinuous density gradient, usually made with sucrose solutions or with Ficoll (a polysaccharide of high molecular weight), and is then sedimented at high speed until the various components of the sample attain density equilibrium with the medium.

By combining sedimentation methods with light and electron microscope techniques, investigators have been able to isolate nuclei, mitochondria, lysosomes, and synaptic endings from brain [7]. The synaptic endings, in turn, can be lysed in water and refractionated to give synaptic vesicles, the synaptic apparatus, and mitochondria. The synaptic endings also can be fractionated on another gradient in order to separate endings that contain different transmitters.

In addition to light and electron microscopic monitoring of subcellular fractions, assays of various chemical constituents are also performed to determine the purity of the fractions. Lactate dehydrogenase is distributed largely in the soluble fraction of the tissues, whereas succinate dehydrogenase is an almost purely mitochondrial enzyme. Synaptic endings contain enzymes characteristic of both the soluble and mitochondrial fractions, as well as increased concentrations of certain special components, such as transmitter substances and the enzymes involved in their synthesis and degradation.

MICROCHEMICAL METHODS

The nervous system is enormously complex and heterogeneous: the complexity is evident from anatomical and physiological studies; the heterogeneity is perhaps best shown in studies of the development and the pathology of the nervous system. The heterogeneity, perhaps more than the complexity, of the nervous system decreases the probability that techniques of cell and particle separation will soon produce sizeable, truly homogeneous samples.

Therefore, in order to study even reasonably homogeneous samples, microchemical techniques have had to be devised and employed. The problem may be illustrated by considering the measurement of a metabolite and an enzyme in very small amounts of tissue: glucose, the main energy source in brain, and hexokinase, the first enzyme in the path of glucose metabolism. Under optimal analytical conditions, the activity of hexokinase in brain is equivalent to about 1 mole of glucose phosphorylated per kilogram of brain (wet weight) per hour. The perikaryon of a large neuron might weigh as much as 0.1 μg. It will contain, on the average, enough enzyme to phosphorylate 1×10^{-10} mole of glucose in an hour. If the enzyme product, glucose-6-P, were measured in a volume of 1 ml, its concentration at the time of the final measurement would be 1×10^{-7} M. The glucose concentration in mouse brain, for example, is 1 mmole per kilogram. One large neuron contains perhaps 1×10^{-13} moles. In the course of measurement, glucose is also converted to glucose-6-P. If the amount of glucose-6-P derived from the glucose content of one neuron were measured in a volume of 1 ml, its concentration would be 1×10^{-10} M. The analytical sensitivity required to measure this metabolite, then, is one thousand times greater than that needed to assay the enzyme.

Different enzymes vary widely in their measurable activities in brain. Many that are involved in energy metabolism have activities of the same order of magnitude as hexokinase; a few are much more active. Acetylcholinesterase is about as active as hexokinase, but choline acetylase is much less active, with an activity of 3 to 4 mmoles/kg/hr in whole brain. Low activities tend to be characteristic of enzymes that may be considered to have synthetic roles. Similarly, many metabolites and other small molecules are present at average concentrations of from 0.1 to 10 mmoles per kilogram. Few are as high as 20 mmoles per kilogram, but many are quite low indeed: acetylcholine, 20 to 30 μmoles per kilogram; acetyl-CoA, 5 μmoles per kilogram; and glyceraldehyde-3-P, 1 μmole per kilogram or less. RNA and DNA are also present at low levels (0.05 to 0.2 percent of the wet weight), but lipids represent 5 to 10 percent of the wet weight of brain. Thus, cholesterol may be present at levels of 50 mmoles per kilogram wet weight in average brain. Proteins also make up about 10 percent of the wet weight of brain, but there are so many varieties that no single protein represents more than a small fraction of the total. Even the protein S-100, which is relatively abundant in brain, represents only 2.5 percent of the total protein.

Total lipid in brain tissue samples may be measured as the weight lost during solvent extraction of the sample. Fractional solvent extraction can also be used, and measurements of lipid phosphorus can then be employed to quantitate the amounts of each type of lipid present. Specific lipids in fairly small samples can be measured by gas chromatography.

Protein is adequately represented in even the smallest samples by the weight remaining after the dry sample has been extracted with fat solvents. On samples of 1 μg or larger, protein can also be measured photometrically.

Ingenious microspectrophotometric assays have been developed for the measurement of RNA in amounts of 10^{-10} to 10^{-11} g. RNA also can be resolved into its constituent bases, which can then be determined separately [3].

Some enzymes produce or consume a gas as a consequence of their activity; for example, succinic oxidase consumes O_2 and decarboxylases produce CO_2. Certain other enzymes can also be determined indirectly by gas reactions. For example, cholinesterase produces acetic acid as one product of its activity, and this in turn will release CO_2 from a bicarbonate solution. The Cartesian diver is a most sensitive instrument for enzyme analyses using such a principle, and it has been developed to a high state of refinement by Giacobini [7]. It is now possible to measure acetylcholinesterase or succinic oxidase, for example, in anatomically homogeneous fragments of retina and even in single neurons and segments of single axons.

Micro flame photometric methods have also been developed to measure Na^+ and K^+ single large neurons [15].

QUANTITATIVE HISTOCHEMISTRY

The principles of quantitative histochemistry are simply those of analytical biochemistry [2a, 2b, 2c]. The constituent of interest is measured in a tissue sample of known origin and anatomical composition, and the measurements are referred to some measure of sample size – most often wet weight, dry weight, or protein content, and less often DNA content or some other measure of cell number. The differences between quantitative histochemistry and general analytical biochemistry are generated almost entirely by differences in sample size. Special techniques for sample handling and weighing had to be devised and biochemical assays of high sensitivity are required just to handle samples as small as 1 μg dry weight. Such samples still contain hundreds of cells or parts of cells, but, even so, they begin to enable meaningful anatomical separations and chemical comparisons. With improvements in the techniques of dissection and tissue handling, in microbalances, and in sensitivity of analytical methods, it has been possible to reduce the sample size to that of individual single neurons and even, for some analyses, to parts of single neurons.

Tissue may be obtained by fresh dissection or dissection of the frozen sample, depending upon the properties of the substances to be measured (see above). Frozen sections are made of the region under study and then are dried overnight at −35° to −40°C under vacuum. The dried sections can be stored in the cold in vacuo. To obtain samples for assay, sections are warmed to room temperature. The structure desired can be dissected, usually freehand, under a dissecting microscope and weighed on a quartz-fiber ("fishpole") balance of appropriate sensitivity. After a little practice, and, perhaps, preliminary histological staining of a few sections for orientation, the various structures of the frozen, dried sections can usually be dissected without staining.

Assays for enzymes or metabolites are carried out in small volumes of suitable reagents delivered from Lang-Levy constriction pipets. These can be constructed of glass or quartz to deliver volumes from several hundred microliters to 0.001 μl, as needed. The large pipets (more than 100 μl) can be made to deliver reproducibly to 0.1 percent, the smaller pipets to 1 percent.

If incubations are to be carried out in volumes of 1 μl or less, it is much more convenient to use "oil wells" instead of small test tubes. These are flat-bottomed cavities drilled almost through a block of Teflon of suitable thickness (3 to 4 mm) and filled with a mixture of hexadecane and light mineral oil. When small aqueous droplets are placed in these wells, they fall to the bottom and can be incubated for considerable periods of time, even at elevated temperatures, without appreciable volume loss by evaporation.

Great analytical versatility and sensitivity have been achieved in these analyses with the use of nicotinamide adenine dinucleotides (NAD and NADP). The advantages stem from three properties of these substances, two chemical, the third biological. The chemical advantages are given in Table 20-2. When sufficient NADH or NADPH has been produced in a reaction, it may be measured directly in the presence of a 1000-fold excess of the NAD^+ or $NADP^+$ from which it came. Furthermore, if necessary, either the oxidized or the reduced form of a nicotinamide adenine dinucleotide can be destroyed without damaging the other form. Either form can then be made to fluoresce strongly, and thus fairly dilute solutions are easily measured.

The biological properties of the nicotinamide adenine dinucleotides that are of analytical importance include the wide range of enzymatic reactions in which they play a role. Thus, many enzymes can be measured because they oxidize or reduce a nicotinamide adenine dinucleotide. These same enzymes, after purification (many are commercially available), can also be used to measure the substances in

TABLE 20-2. Chemical Properties of Nicotinamide Adenine Dinucleotides

Form of Dinucleotide	Acidic Solution (0.2 to 0.4 N)	Neutral (pH 7)	Basic Solution	
			0.02 to 0.04 N	6 N
NAD^+ and $NADP^+$	Stable	Not fluorescent	Unstable; destroyed in 15 min at 60°C	Strongly fluorescent product after 15 min at 60°C, measurable from 5×10^{-8} to 1×10^{-6} M
NADH and NADPH	Unstable even at 0°C	Fluorescent; measurable at 5×10^{-7} to 1×10^{-5} M	Stable even at 60°C	Forms same fluorescent product in 0.03% H_2O_2 as NAD^+ and $NADP^+$

tissue that are their substrates. By using the enzymes as auxiliaries, enzymatic reactions which do not involve nicotinamide adenine dinucleotides can be measured by coupling them to those that do. Furthermore, by the use of appropriate combinations of enzymes, a metabolic intermediate which is not involved in a nicotinamide adenine dinucleotide-dependent reaction can be made to oxidize or reduce a nicotinamide adenine dinucleotide one or two steps away from the enzyme reaction in which it is substrate. Scores of enzymes are measurable in tissue samples as small as 0.1 μg with these methods.

The methods of freeze-drying, dissecting, and weighing can be combined with other analytical methods of adequate sensitivity: for example, both choline acetyl transferase and glutamic decarboxylase can be measured in small samples by radiometric techniques [9, 17], and the unique nervous system protein, S-100, has been measured in small samples by using a micro complement fixation assay [19].

To measure metabolic intermediates in small samples or enzymes in very small samples, a further step is required to produce adequate amplification. A widely used method is called "cycling." In this method, a product of the assay reaction, usually a nicotinamide adenine dinucleotide, is used to catalyze two or more reactions in such a way that it is not used up but repeatedly regenerated. After an appropriate interval, the reaction is stopped and a product, which may have accumulated to a concentration many times (100 to 20,000 times) that of the cycled product of the assay reaction, is now measured.

An outline for the determination of glucose in a 1 μg sample of brain is given in Table 20-3. It is evident that the amplification achieved during the cycling step was 11,000-fold (compare volumes and concentrations in steps 5 and 7, Table 20-3) and that 80 percent of the sample was not used (of the 10 μl available, only 2 μl were

TABLE 20-3. Assay of Glucose in Brain Tissue by Enzymatic Cycling

	Volume (μl)		
Step[a]	Incu-bation	Carried to Next Step	Conc. of Substances[a]
1. Dissect tissue containing **glucose** (1 μg freeze-dried tissue)	–	–	6.5×10^{-3} moles/kg dry wt.
2. Heat in 0.01 N HCl (60°C, 10 min) to destroy tissue enzymes and NADPH	1	all	6.5×10^{-6}
3. Assay reaction (25°C, 30 min)[b] **Glucose** + *ATP* —*hexokinase*→ Glucose-6-P + ADP **Glc-6-P** + *$NADP^+$* —*Glc-6-P dehydrogenase*→ Gluconate-6-P + NADPH	6	all	1.1×10^{-6}
4. Heat in alkali (60°C, 15 min) to destroy $NADP^+$, enzymes	10	2	0.5×10^{-6}
5. Cycling step (38°C, 60 min)[c] *α-Keto-Glu* + *NH_3* + **NADPH** —*glutamate dehydrogenase*→ Glutamate + $NADP^+$ Gluconate-6-P + NADPH ←*Glc-6-P dehydrogenase*— $NADP^+$ + *Glc-6-P*	100	all	1.0×10^{-8}
6. Heat to destroy cycling enzymes (100°C, 2 min)	–	–	–
7. Reaction of product of cycling[d] **Gluconate-6-P** + *$NADP^+$* —*gluconate-6-P dehydrogenase*→ Ribulose-5-P + NADPH	1100	–	1×10^{-5}

[a]Substances in *italic* type are reagents in each reaction; those in **boldface** type are products of reaction of the preceding step, and their concentrations are given in the right-hand column.

[b]Step 3 is stoichiometric; the amount of NADPH produced is exactly equal to the amount of glucose in the sample.

The total amount of glucose-6-P formed would give the concentration shown in the right-hand column; however, since the two reactions are run together, the concentration of glucose-6-P never actually gets this high and at the end of the reaction is essentially zero.

[c]Step 5 is a rate step. The amount of gluconate-6-P produced is (1) proportional to the amoun of NADPH introduced and the length of time the cycling process is allowed to run, and (2) a func of the concentrations of two cycling enzymes.

The figure in the right-hand column is the initial concentration of NADPH. When the reaction starts, NADPH will fall and $NADP^+$ will rise to equilibrium levels determined by the relative amou of the two cycling enzymes that are added and their affinities for NADPH and $NADP^+$. The equil rium levels will be maintained either for the duration of the incubation period or until one of the substrates becomes limiting.

[d]Step 7 is another stoichiometric reaction; the amount of NADPH produced is equal to the am of gluconate-6-P introduced. The NADPH is measured fluorometrically.

taken from step 4 to step 5, Table 20-3). Therefore, a sample weighing 0.2 μg could have been used and would have given the same final concentration of NADPH. The only change necessary would have been to reduce the volumes used in the first four steps by 80 percent and to take the whole sample from step 4 to step 5.

To assay samples much smaller than that used in the example, e.g., 0.01 μg, double cycling would be necessary. Volumes would be greatly reduced, and the reaction mixture present at the completion of step 7 would be treated with heat and alkali to destroy gluconate-6-P dehydrogenase and $NADP^+$. The NADPH would then be recycled, and the resulting gluconate-6-P assayed (i.e., steps 5 through 7 would be repeated). This could give an overall amplification of 10^8 times.

SLIDE HISTOCHEMISTRY

Interpretation of the role of a substance in the life of the cell is greatly facilitated if one knows its precise location. Slide histochemistry makes its greatest contribution in the realm of localization, especially when coupled with electron microscopy.

The method encompasses four general approaches to making specific tissue constituents visible: staining, immunofluorescence, enzyme histochemistry, and autoradiography [1]. The subject is obviously an exceedingly broad one, and each approach has its own special merits and pitfalls. When a specific method is successful, a clear picture of the distribution of the compound is seen, as, for example, in the conversion of catecholamines to insoluble fluorescent compounds [13] or in many enzyme histochemical techniques [1]. It is usually difficult to give a quantitative assessment to the amounts of color or fluorescence seen, and one is restricted to phrases like "more than" and "less than," or a series of plus signs in which the quantitative relationships between grades are unknown.

Probably the single most troublesome problem, however, is that of localization itself. Many of the substances one might want to demonstrate are at least partially soluble at one or more stages in the procedure employed to make them visible. This engenders two possible difficulties.

1. Some of the material may be lost during the preparation. If the loss is regionally specific, a misleading distribution pattern will result. For example, several enzyme activities take place both inside and outside mitochondria. The extramitochondrial portion is much more likely to escape from the sample than is the particulate-bound portion.

2. Another problem of partial solubility is that a substance may migrate during sample preparation from its in vivo locus to a neighboring cellular structure for which it has more affinity. The product of an enzyme reaction may, similarly, be adsorbed to a structure for which its affinity is high, rather than coming to rest where the enzyme itself is located. Considerable effort and ingenuity must be expended to discover whether the final picture and the in vivo situation correspond. A survey and critique of methods applied to nervous tissue is provided by Adams [1].

Probably the greatest possibilities for specificity and precision of localization are offered by enzyme reactions. At the same time, because of the indirect nature of the indicator reaction, the possibilities of introducing artifacts are also great. A good example of the difficulties attending the establishment of a new enzyme histochemical method is provided in a recent study of choline acetyl transferase [11].

When the experimental situation is particularly propitious – for example, if the anatomy of the cell is very regular or if there is appreciable anatomical separation of cell functions – autoradiography may be very useful. Effective use of this method is demonstrated in a recent study of the assembly of the outer segment of the rod cell in the frog retina [14]. Autoradiography is also valuable in studies of nervous system development since dividing cells that take up tritiated thymidine into DNA can be clearly identified using this technique [17, 18].

ACKNOWLEDGMENT

The author is a Career Development awardee, Grant 5K3NB18005, from the National Institute of Neurological Diseases and Stroke.

REFERENCES

BOOKS AND REVIEWS

1. Adams, C. W. M. (Ed.). *Neurochemistry.* Amsterdam: Elsevier, 1965.
2. Dubach, U. C. (Ed.). *Proceedings of the International Conference on Quantitative Histochemistry.* Bern: Hans Huber, 1970. See in particular:
   2a. Lowry, O. H., and Passonneau, J. V. Some Recent Refinements of Quantitative Histochemical Analysis.
   2b. Matschinsky, F. M. Quantitative Histochemistry of Glucose Metabolism in the Islets of Langerhans.
   2c. Passonneau, J. V., and Lowry, O. H. Metabolic Flux in Single Neurons During Ischemia and Anesthesia.
3. Edström, J. E. Microextraction and Microelectrophoresis for Determination and Analysis of Nucleic Acids in Isolated Cellular Units. In Prescott, D. (Ed.), *Methods in Cell Physiology.* Vol. 1. New York: Academic, 1964, p. 417.
4. Feder, W. (Ed.). Bioelectrodes. *Ann. N.Y. Acad. Sci.* 148:1, 1968.
5. Ingvar, D. H., and Lassen, N. A. (Eds.). Regional cerebral blood flow – an international symposium. *Acta Neurol. Scand.* 41:(Suppl. 14), 1965.
6. Kety, S. S. The theory and applications of the exchange of inert gas at the lungs and tissues. *Pharmacol. Rev.* 3:1, 1951.
7. Lajtha, A. (Ed.). *Handbook of Neurochemistry.* Vol. 2. *Structural Neurochemistry.* New York: Plenum, 1969.
8. Mahler, H. R., and Cordes, E. H. *Biological Chemistry.* New York: Harper & Row, 1966. (An adequate treatment of the principles of centrifugation may be found in Chapter 9 of that volume.)

ORIGINAL SOURCES

9. Albers, R. W., and Brady, R. O. The distribution of glutamic decarboxylase in the nervous system of the rhesus monkey. *J. Biol. Chem.* 234:926, 1959.
10. Breckenridge, B. M., and Norman, J. H. The conversion of phosphorylase b to phosphorylase a in brain. *J. Neurochem.* 12:51, 1965.
11. Burt, A. M. A histochemical procedure for the localization of choline acetyltransferase activity. *J. Histochem. Cytochem.* 18:408, 1970.
12. Chance, B., and Legallais, V. A spectrofluorometer for recording of intracellular oxidation-reduction states. *IEEE Trans. Biomed. Eng.* 10:40, 1963.
13. Corrodi, H., and Jonsson, G. The formaldehyde fluorescence method for the histochemical demonstration of biogenic amines. A review on the methodology. *J. Histochem. Cytochem.* 15:65, 1967.
14. Hall, M. O., Bok, D., and Bacharach, A. D. E. Biosynthesis and assembly of the rod outer segment membrane system. Formation and fate of visual pigment in the frog retina. *J. Mol. Biol.* 45:397, 1969.
15. Katzman, R., Lehrer, G. M., and Wilson, C. E. Sodium and potassium distribution in puffer fish supramedullary nerve cell bodies. *J. Gen. Physiol.* 54:232, 1969.
16. Lowry, O. H., Passonneau, J. V., Hasselberger, F. X., and Schulz, D. W. The effect of ischemia on known substrates and cofactors of the glycolytic pathway in the brain. *J. Biol. Chem.* 239:18, 1964.
17. McCamen, R. E., and Hunt, J. M. Microdetermination of choline acetylase in nervous tissue. *J. Neurochem.* 12:253, 1965.
18. Miale, I. L., and Sidman, R. L. An autoradiographic analysis of histogenesis in the mouse cerebellum. *Exp. Neurol.* 4:277, 1961.
19. Moore, B. W., Perez, V. J., and Gehring, M. Assay and regional distribution of a soluble protein characteristic of the nervous system. *J. Neurochem.* 15:265, 1968.
20. Smith, D. E., Robins, E., Eydt, K. M., and Daesch, G. E. The validation of the quantitative histochemical method for use on postmortem material. 1. The effect of time and temperature. *Lab. Invest.* 6:447, 1957.

21

Cellular and Subcellular Fractionation

Stanley H. Appel, Eugene D. Day, and Don D. Mickey

THE AIM OF NEUROCHEMISTRY is to describe neuronal and glial cell behavior at the molecular level. Because of the cellular heterogeneity of the brain, most neurochemical research problems must begin with the separation of subcellular organelles into relatively pure fractions that retain metabolic activity and structural integrity. We shall assume that the reader has access to one or more of the general presentations [1, 4, 9b, 10, 12, 33, 34, 39–41, 58, 60–62, 67, 69] and only briefly review the basic principles before going on to specific cellular and subcellular fractionation schemes.

To separate cell constituents, one must disrupt cells by such various means as shearing, shocking, swelling, and sonicating. Depending upon the method chosen, one must sacrifice the yield of the more sensitive particles in order to dissociate the more structurally sound units. No single disruption technique will yield all particles intact. It is remarkable that the simple method of tissue homogenization in buffered isoosmotic sucrose or saline works as well as it does. The tolerance between the Teflon pestle and glass tube may be important for one particle (e.g., a loose fit for synaptosomes); the number of up-and-down strokes may be critical for some tissues (several for lung, few for brain); the viscosity of the medium can be significant for others (high for tumor cells); and the adjustment of bivalent cation concentrations and pH has an optimal and sometimes critical effect upon certain particles (e.g., whether they disperse or aggregate, swell or shrink, absorb or discharge components).

The object of the initial disruption and separation step is not to accomplish dissociation and purification all at once, but to treat tissues mildly and to separate their parts into a few major categories (see Fig. 21-1, right ordinate). Before separation in the centrifuge, the isolation and purification of any particular component may also require more definitive steps, such as mild osmotic shock in hypotonic media to release nerve ending mitochondria from their parent synaptosomes.

The two properties of subcellular particles most important to the design of biophysical separations are size and density. In fact, there are enough differences among cell constituents with respect to size and density to make it theoretically possible to

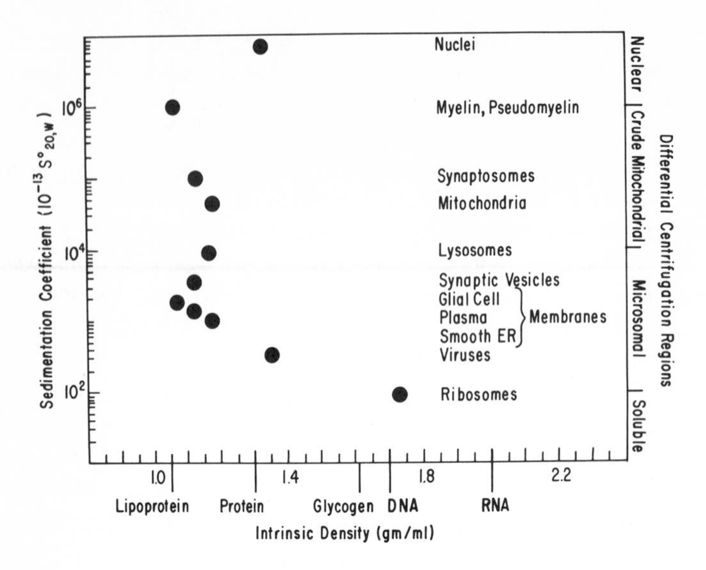

FIGURE 21-1. Size and density of subcellular particles in a typical brain homogenate as ordered by sedimentation coefficient and intrinsic density.

define each type uniquely. One may order the placement of subcellular constituents of neurons and glia within a typical suspension of disrupted cells according to sedimentation coefficient (one parameter of size) and intrinsic density, as shown in Figure 21-1. Because of the limitations of our current techniques, however, organelles with closely related sizes and densities are difficult to separate. This difficulty results from our present crude methods of cell disruption and subsequent particle-particle interaction, rather than an expression of overlap in physical properties of particle types in their actual native states. There is, of course, recognizable second-order heterogeneity within a given type. Improvement in particle preservation during separation will enable us to prepare subfractions of any one type (e.g., small, medium, and large mitochondria) and to obtain greater homogeneity of size as well as density within a given fraction.

The process of homogenization and release of subcellular particles into a suspending medium will change particle properties through solvation, osmotic action, and ion effects.

1. Solid particles (e.g., membranes) that carry a solvation mantle have a density intermediate between the medium and the particles themselves until the isopycnic point is reached, i.e., the point at which the medium and the particles have the same density. At that point the "anhydrous," or intrinsic, density of the particles can be measured.

2. Vesicular particles that exhibit osmotic behavior may eventually become equilibrated with the medium in which they move as they approach an isopycnic density. They, too, will then exhibit an intrinsic density value independent of fluid

content or mantle. Pure osmotic behavior, however, is rarely observed among subcellular components, perhaps because we cannot often wait long enough to achieve equilibrium and because other effects are also involved; nevertheless, mitochondria and synaptosomes both approach the ideal as osmometers.

3. Vesicular particles, such as cell nuclei, that respond less readily to changes in osmotic pressure may exhibit volume changes that are contrary to expectation (e.g., swelling when taken from hypotonic 0.01 M $CaCl_2$ into hypertonic 0.88 M ion-free sucrose [16]). Such a response may be related to the ionic environment and the ion-exchange capacity of the particles. In such cases, there may be considerable difficulty in obtaining intrinsic densities or even a stabilized extrinsic one that is independent of the operational procedure.

In none of these situations will the sedimentation coefficient be descriptive of a particle devoid of environmental factors, and the coefficient must be interpreted as being more operational than definitive. However, the purpose of subcellular fractionation procedures is not so much to duplicate the intracellular milieu as it is to reach sufficient artifactual stability so that experimentation can be conducted and meaningful questions asked.

The classic method of subcellular fractionation, which was developed for liver cells and carried over into brain cell fractionation with only slight modification, involves only the sedimentation velocity parameter. In differential centrifugation, particles are both dissociated and separated in a medium of relatively low and fixed density, usually isotonic salt or sugar solutions. Four fractions are obtained by sequential centrifugation steps of increasing rotational velocity and time: a low-speed, short-term nuclear pellet; a medium-speed, medium-time, crude mitochondrial pellet; and, by a high-speed, prolonged spin, a microsomal pellet and a supernatant fluid of soluble components.

In spite of many drawbacks, this technique is probably more widely used as an initial step than any other. One serious defect that is inherent in the differential centrifugation process is that there is inevitable contamination of low-speed pellets with high-speed components since, as Anderson has pointed out [1], "a few particles of every size, including the smallest ones, are present at the bottom before centrifugation starts." In general, the degree of contamination of a pelleted dense fraction by a light one is proportional to the ratio of their sedimentation rates. Extensive washing of pellets by repeated centrifugation is necessary to bring about adequate purification. Sufficient dilution of suspensions of both original homogenates and subsequent pellet washes is required to limit the number of contaminants already at the bottom before centrifugation begins.

Isopycnic or buoyant density centrifugation exploits only the density parameter and orders the separation of constituents solely on the basis of initial extrinsic particle density and final intrinsic particle density. Usually this is accomplished by equilibrium density-gradient centrifugation, in which each constituent will either sediment down to or float up to the density of the medium at which it is isopycnic and at which it is weightless.

The relative amount of four main constituents – RNA, DNA, protein, and lipid constituents – determines the intrinsic density of any one particle. The intrinsic density of a particle is a fixed property and is not subject to much change. The actual extrinsic density of a given particle at any given time is subject to fluctuation and regulation since it depends upon the volume and concentration of inner fluids as well as upon the nature of surface-adsorbed components. To obtain a true intrinsic density value experimentally (e.g., in CsCl) one must sacrifice metabolic activity; it follows that mild separation techniques depend more on the knowledge of extrinsic or operationally defined densities and their possible regulation than upon the fixed parameters.

Exploitation of the density parameter in the design of processes of particle separation has been reserved mainly for the vertical procedures subsequent to the initial horizontal differential separation. Thus, crude mitochondrial pellets become the source of lysosomes and synaptosomes as well as of mitochondria in a procedure that involves "isopycnic" banding of the particles in sucrose density gradients [1, 4, 13, 17, 31, 33, 39, 40, 58, 67, 69]. Sucrose was found to yield synaptosomes of less than optimal purity and metabolic activity. Isoosmotic sucrose-Ficoll* has been substituted for sucrose, but the horizontal differential separation and the vertical isopycnic banding remained the same [13, 17, 32] (Fig. 21-2).

Rate-zonal centrifugation makes use of the velocity-density differential among components. It is based on the fact that the sedimentation coefficient decreases as the density difference between particle and medium narrows when the particle moves into denser regions of a gradient. Thus, although both dense and light particles are slowed down, the ratio between their respective sedimentation velocities continuously widens and makes the clear separation of one from the other much easier to obtain than by differential procedures.

Some of the practical advantages of density-gradient centrifugation procedures are as follows:

1. When the sample layer is placed on top of the medium, the desired purified component is sedimented away from the initial sample. (By contrast, see differential centrifugation, above.)
2. Multiple washings are not required.
3. Several components are separated in a single centrifugation step.
4. Improvement of separation can be made according to three parameters – sedimentation velocity, density, and time.

The best way to design specific density-gradient procedures for both the horizontal and vertical steps involves exploration with the zonal ultracentrifuge [1]. In fact, in this laboratory, once we have established the correct procedural design with the zonal centrifuge, we find it much more convenient to adapt our findings to a conventional centrifuge for routine preparative runs (Fig. 21-3). The zonal gradient volume is large (1670 ml in the B-XV rotor) compared with the weight of

*Pharmacia Co., Inc., Piscataway, New Jersey.

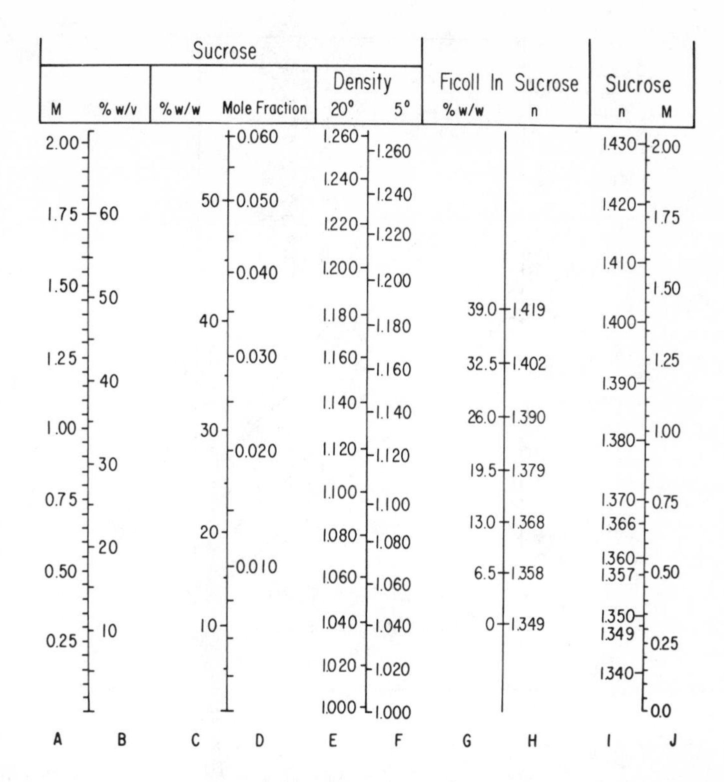

FIGURE 21-2. Sucrose solutions. Comparison of methods of expression: (A) and (J) molarity, and (B) weight-volume % at 25°C; (C) weight-weight %; (D) mole fraction; (E) density at 20°C; (F) density at 5°C. Comparison with isoosmotic Ficoll-sucrose (of 0.19 g/liter osmosity or 0.32 M sucrose): (G) weight-weight %; (H) refractive index measured at 20°C. (I) Refractive index of sucrose alone. Line up straight edge for the same values in (A) and (J). (From [1, 2, 5–7, 67].)

original tissue (0.1 to 5 g), the digital recording of $\omega^2 t$ insures a precise reproduction of angular velocity-time components from one run to the next, and the light-scatter profile of components in the gradient is easily obtained after a given run since the contents of the spinning rotor are pumped through a recording spectrophotometer and into collection vessels (Fig. 21-4).

CRITERIA OF PURITY

To attempt to set up criteria of purity for isolated subcellular particles of brain at this point would be more of a hindrance than an aid. Unlike the liver, which may now be dissected and described in subcellular terms by both qualitative and quantitative means, the brain displays too much heterogeneity of cell types to make such

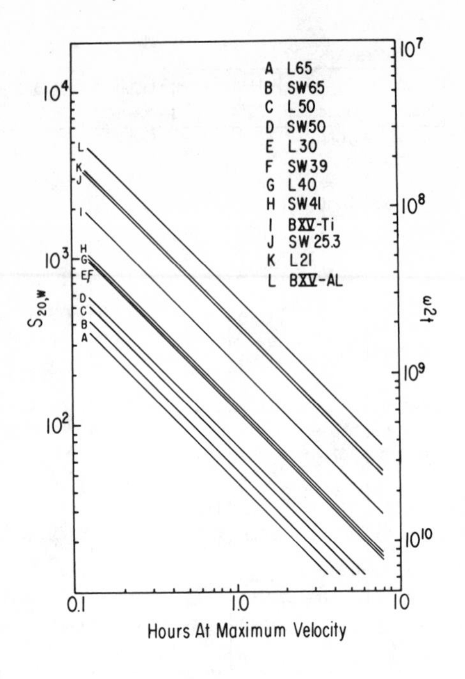

Rotor Designation	Rotor Radius (min.–max.)	Max. Velocity, 10^3 rpm	RCF at Max. Vel., 10^3 *g* (min.–max.)	Volume (no. tubes X ml/tube)
A L65	3.7–7.8	65	175–368	8 X 13.5
B SW65	3.8–8.9	65	180–420	3 X 5
C L50	3.7–7.1	50	103–198	10 X 10
D SW50	4.7–9.8	50	131–274	3 X 5
E L30	5.0–10.5	30	50–106	12 X 38.5
F SW39	4.7–9.8	40	84–175	3 X 5
G L40	3.8–8.1	40	68–145	12 X 13.5
H SW41	6.3–15.2	41	118–286	6 X 13.2
I BXV-Ti	2.0–8.2	40	37–150	1 X 1670
J SW25.3	6.3–16.2	25	44–113	6 X 17
K L21	6.0–12.0	21	30–59	10 X 94
L BXV-Al	2.0–8.2	25	14–59	1 X 1670

FIGURE 21-3. Time to sediment particles of varying sedimentation coefficients from top to bottom of tubes in different rotors. The values for $\omega^2 t$ are given for particles that move to twice their starting radius in a rotor. Under such conditions, S = $\omega^2 t$ X ln 2. L = angular; SW = swinging bucket; B = zonal.

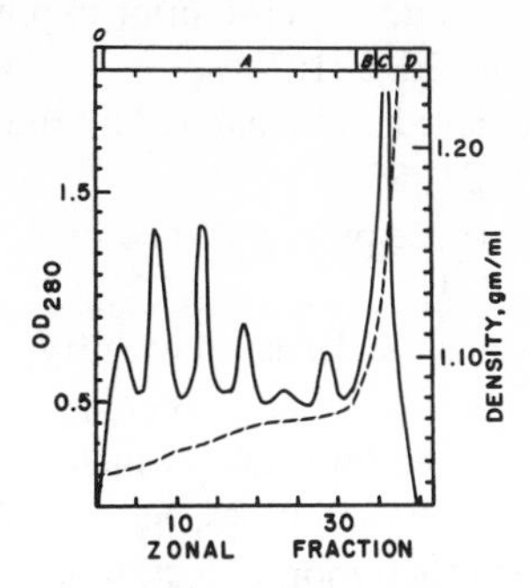

FIGURE 21-4. Light-scatter profile of whole rat brain homogenate after zonal and centrifugation in Ficoll-sucrose gradient. The $\omega^2 t$ value at the time of gradient removal from the BXV-Al zonal rotor was 40 X 10^{10} radians^2/sec. Density line is dashed.

a description meaningful. A case in point is the well-known, universal marker of mitochondria: cytochrome oxidase activity. It is difficult to say how one can establish the purity of brain mitochondria by the specific activity of such an enzyme system when its activity per unit of protein varied widely from brain region to brain region, e.g., high in cerebral cortex, medium in the thalamus, and low in the corpus callosum [65]. Another case concerns the use of proteolipid protein as a marker for myelin in spite of its presence in other subcellular fractions of the brain and in subcellular fractions from other tissues ([66]; see also Chap. 18).

In the ensuing discussions, therefore, the descriptions of subcellular particles must be understood in terms of quantitative enrichment rather than qualitative separation and mutual exclusion of contaminating organelles. Even where extensive enrichment is achieved, the problems of heterogeneity – white vs. gray matter, region vs. region – make such enrichment apparent only as identified by ultrastructural uniformity; in the study of ultrastructure only a small fraction of the particles present is sampled.

SEPARATION OF NEURONS FROM GLIA

Brain tissue consists principally of neurons and glia, the relative proportions of which change significantly with the age of the brain. In immature rats, glia represent a small percentage of cortical cells, whereas in adult animals, glia outnumber neurons [24]. Attempts have been made to isolate various fractions by hand dissection, micromanipulation, and macroscale preparations. As reviewed by Johnston and Roots [8], neuronal perikarya were first isolated by Deiters in 1865. Subsequently, the studies of Lowry in 1953, Chu in 1954, and Hydén in 1959 demonstrated the usefulness of microdissection techniques to sample neuronal and glial populations. At least four types of procedures have been described for macroscale preparations:

1. Brain is fixed in acetone and glycerol, prior to preparation of cellular suspension and centrifugation [36, 57];
2. Brain suspensions are sieved and separated on sucrose or high molecular weight dextran gradients [11, 21, 55];
3. Brain is disrupted in a tissue press and then the suspensions are separated by zonal centrifugation [35];
4. Brain is disrupted by enzyme digestion with trypsin as well as by sieving [51].

Such techniques can be divided into stages which consist in preparing suspensions of neurons and glia, separating the cell types, and assessing the enrichment of the fractions with respect to both their morphological purity and their functional integrity. Fixation of the brain in acetone and glycerol prior to preparing the suspension may provide preparations of higher purity, but it has not been widely employed because it precludes the isolation of metabolically active cellular fractions. In the technique described by Norton and Poduslo [51], minced rat brain is softened by treatment with trypsin and is then disrupted through nylon and steel meshes to produce a suspension of free-floating cells and debris; their method is as follows:

> Brains from 10- to 30-day-old rats are chopped fine in cold medium containing 5% glucose, 5% fructose, and 1% bovine serum albumin (BSA) in 100 mM KH_2PO_4-NaOH buffer, pH 6.0. Trypsin (1%) is added to the suspended tissue and incubation is carried out under O_2 for 90 min at 37°C. The softened, disassociated tissue is diluted with 0.4 volume of chilled 90% calf serum and 10% phosphate buffer (pH 6.0) and is centrifuged at 140 X *g* for 5 min. The tissue is washed twice with medium by suspension and centrifugation through 150-mesh nylon bolt cloth and is then passed through a 200-mesh stainless steel screen (74 μ) to complete the disruption.

The isolation of various cell types from suspension has been accomplished by different investigators either in sucrose or in Ficoll-sucrose solutions. For example, the experiments of Flangas and Bowman [35] provide the following method:

> Centrifugation was performed in the B-XV zonal rotor at $5.384 \times 10^{10}\ \omega^2 t$. Following the technique of Norton and Poduslo (above), separation is achieved via sucrose gradients buffered with hexose-albumin-phosphate, as described above. The gradients represent fractions of 0.9, 1.35, and 1.55 M sucrose, and a cushion of 2.0 M sucrose. After centrifugation a very small pellet remains. Floating upon the 2.0 M sucrose is the neuronal fraction, which is used without further purification. Floating upon the 1.55 M layer is a mixture of glia, neurons, and capillaries. Banding just above the 1.35 M is a crude glial fraction. Above the 0.9 M layer are myelin and small particulate debris. The crude glial fraction can be further purified on a second gradient consisting of 0.9 M sucrose laid over a 1.4 M sucrose solution. Purified glial cells are collected at the interface between 0.9 M and 1.4 M sucrose.

Other methods of separating cell suspensions are reviewed by Johnston and Roots [8].

Recently, Norton has modified the separation technique to produce oligodendroglia in high yield and purity [53].

Calf corpus callosum and centrum semiovale are used as the starting materials. Tissue is trypsinized, washed, and disrupted in 0.9 M sucrose by filtration through 150-mesh nylon and 200-mesh steel screens. The cell suspension is centrifuged at 330 X *g* for 15 min in a three-step gradient, and the oligodendroglia are collected at a 1.40–1.55 M sucrose interface. The preparation consists of greater than 90% round cells of about 8 μ diameter and with an average weight of 25 pg. The weight recovery is 0.2 to 0.3 mg dry weight per gram tissue.

The identification, purity of cells, and the integrity of cellular constituents are monitored by electron microscopy. Such preparations offer an important approach to the understanding of myelination.

One of the greatest difficulties is how to assess the purity of fractions reported by various investigators. As pointed out by Johnston and Roots [8], a major difficulty lies in the adoption of uniform criteria for the description of fractions and the relative importance of the terms *pure* and *rich* to different investigators. Through light microscopy, it has been relatively easy to identify the neuronal fraction and determine its intactness and relative enrichment, although minor contaminants are difficult to detect. Glia, however, are more difficult to identify and assess with respect to degree of purity. The use of enzymes as markers presents some difficulty, since even slight glial contamination of the neuronally enriched fraction may lead to erroneous evaluation of enzyme content [15]. In the isolation described by Norton [51], the neuronal fraction was claimed to be more than 90 percent pure by particle count. The cells retained only stumps of axons and dendrites and were readily identified by their size, by large nuclei with prominent nucleoli, and by abundant cytoplasm with characteristic morphology. The glial fraction was probably contaminated by small neuronal processes.

Norton and Poduslo [52] found a surprisingly similar lipid composition for both separated neuronal perikarya and astroglia. The most unexpected finding was the considerable ganglioside content of astroglia, possibly attributable to their high ratio of surface membrane to mass. Tiplady and Rose [63] demonstrated a higher in vivo incorporation of radioactive lysine into separated neurons than into isolated neuropil fractions. However, in such studies it is extremely difficult to know how extensively contaminating particles contributed to the incorporation.

Another approach to cellular integrity is a morphological one. Johnston and Roots [8] demonstrated that the cytoplasmic membrane may lack a surface-membrane image under conditions of the separations cited above. However, others have interpreted such results as an artifact of the techniques employed for electron microscopy. The technique of Norton and Poduslo presumably yields cell bodies with intact cellular membranes [52].

General, definitive statements about relative functional integrity must await confirmation of the neuronal and glial separation techniques, the extent of consistent "enrichment" of various fractions, further evaluation of fraction contamination, and the possible alteration of cell structure introduced by tryptic digestion and cell separation. With the satisfactory resolution of these problems, metabolic studies will assume greater significance. An excellent start has been made in this field, and the recent separations reported by Norton's laboratory appear promising.

ISOLATION OF CELL NUCLEI

Attempts to isolate nuclei from the nervous system [9d] have been complicated by the heterogeneous population of cells present in the brain and the varied array of particles produced by homogenization. Although neurons may differ considerably in size, their morphological characteristics are sufficiently similar to permit light and electron microscopy to be employed as criteria of purity and integrity. The hallmark of the neuron is the single, dense, spherical nucleolus and the diffuse chromatin network. Astrocytic and oligodendroglial nuclei differ from each other considerably when examined in vivo, but morphological differentiation is rather difficult in isolated fractions and we must be content at present with separation of neuronal from glial nuclei [44].

The isolation of nuclei is facilitated by their high DNA and RNA content with resulting high intrinsic density, as indicated in Figure 21-1. Other structures, such as myelin, synaptosomes, synaptic vesicles, and membranes, are less dense and thus should be easily separable. The major problems are (1) to assure homogenization that is adequate to disrupt plasma membranes and free nuclei of attached cytoplasmic contamination, (2) to prevent aggregation or disruption by the ionic constituents and pH of the homogenization and isolation media, and (3) to minimize osmotic changes that would alter the extrinsic density of the nuclei in unpredictable ways. It has been claimed that homogenization in isotonic sucrose preserves the integrity of nuclei if Ca^{2+} or Mg^{2+} ions in concentrations of approximately 1mM and pHs between 6.2 and 6.7 are employed [46]. Other investigators have found that the presence of K^{+} helped preserve nuclear structure [60]. Rappoport, Fritz, and Moraczewski [54] employed Triton X-100 to solubilize cytoplasmic membranes, but other investigators have commented that such a procedure also solubilizes the outer membrane of the nuclei [41].

After dispersion of nuclei in homogenates, most techniques take advantage of the high intrinsic density of nuclei by recentrifuging the pellet from a low-speed spin (1000 × *g* for 10 min) in 1.5 to 2.5 M sucrose and obtaining the nuclear fraction in the pellet. Other laboratories homogenize the brain directly in 2.2 M sucrose containing 1 mM $MgCl_2$ and 10 mM potassium succinate, and then centrifuge at 78,000 × *g* for 50 min [48]. The nuclei remain as a pellet and all other cellular debris floats. These crude nuclear fractions were further fractionated on discontinuous sucrose gradients of 2.2, 2.4, 2.6, and 2.8 M sucrose to separate neuronal and glial nuclei [46]. With most of these techniques, the density of the nuclei as well as of other particles actually increases in the presence of the hypertonic sucrose solutions, and the crude nuclear fractions could possibly be contaminated by other particles with altered extrinsic density.

MYELIN

Myelin may be defined morphologically as the lamellar structure resulting from the concentric layers of "glial cell membranes" that are deposited around axons. This definition, then, would exclude any material from the rest of the glial cell and any material from the axon itself. However, because of the strong physical bonding of the glial cell membranes to the axonal membranes, many of the isolation procedures for "pure" myelin yield myelin that includes axonal membrane as a contaminant.

Myelin, which comprises about 50 percent of the white matter of the brain [50], has been extensively studied because it is one of the most stable membranes known and can be prepared in good yield. It is a lipid-rich membrane in which the lipid and fatty acid compositions are known (see Chap. 18).

The methods of preparing myelin from the central nervous system usually begin with a brain or spinal cord homogenate. Myelin is separated from the other subcellular constituents either by differential centrifugation or by gradient centrifugation in either swinging bucket rotors or zonal rotors. The separation is enhanced because myelin membranes are the most buoyant of the brain membrane fractions. Pure myelin is difficult to prepare, however, because of the tendency of the concentric myelin membranes to wrap themselves around small amounts of other subcellular materials during the separation procedures.

The earliest and simplest separation techniques consisted of differential centrifugation of brain tissue homogenized in 0.32 M sucrose into fractions of crude nuclei, crude mitochondria, microsomes, and supernatants [40]. The "microsomal" fraction was then pelleted, resuspended, and spun through a discontinuous sucrose density gradient consisting of 0.8, 1.0, 1.2, and 1.4 M sucrose solutions. After sedimentation at 75,000 X *g* for 2 hr on this gradient, the material that banded on top of the 0.8 M sucrose solution was collected and designated as myelin fragments. Starting with the "crude mitochondrial" fraction, deRobertis et al. [34] prepared myelin in a very similar fashion. Slight modifications of this method are still being used to isolate myelin fractions prior to morphological and biochemical studies [56]. Other workers [47] modified the centrifugation techniques slightly and isolated myelin from the "crude nuclear" fraction by flotation in 1.5 M sucrose. However, the myelin prepared by this method was somewhat contaminated with axonal fragments and cytoplasm.

Autilio et al. [18] isolated somewhat purer myelin by dissecting white matter from ox brains prior to homogenization and passing the homogenate through cheesecloth. After two sedimentations on a discontinuous sucrose gradient, the crude myelin that banded at the density of 0.656 M sucrose was collected and spun for 10 min to remove microsomal material. After two additional centrifugations on a discontinuous sucrose gradient, the final sedimentation gave two distinct, pure myelin bands, which these workers termed "light" and "heavy" myelin.

The introduction of the zonal rotor and the use of continuous density gradient centrifugation for myelin isolation enabled the examination of a continuum of relevant zones rather than just a selected few. Cotman et al. [28] made use of this technique to study the distribution of brain subfractions in the zonal rotor. However, their fractions were isolated under two restrictive influences: First, their starting material consisted of a crude mitochondrial fraction and they may have lost components that were present in a whole-brain homogenate; second, the sucrose gradient created high osmotic pressures on the membranes and may have caused impure separations.

Day et al. [32] compared sucrose and isoosmotic Ficoll-sucrose gradients in a zonal rotor with respect to the distribution of myelin and other subcellular particles within the gradient after various periods of centrifugation. They showed that myelin from a whole rat-brain homogenate did not band at a single density in Ficoll-sucrose, even if centrifuged in the B-XV zonal rotor at $4.64 \times 10^{10}\ \omega^2 t$. Rather, the myelin was distributed into three distinct and pure fractions at separate densities, none of which was even close to the density of 0.656 M sucrose. They concluded that myelin membranes are unstable in hyperosmotic solutions of sucrose, and stable in isoosmotic Ficoll-sucrose. A modification of this Ficoll-sucrose gradient zonal separation has been made so that myelin may be prepared from whole-brain homogenates in conventional centrifuges.

In the method of Day et al., crude myelin was obtained from the isoosmotic Ficoll-sucrose step gradient used to separate mitochondria (see p. 442) by removing the material floating as a band on top of a 1.067 density layer. The myelin was then enriched by diluting with 0.32 M sucrose, layering on another isoosmotic Ficoll-sucrose step gradient (densities of 1.049, 1.058, 1.063, 1.069, and 1.080), centrifuging at $68{,}108 \times g$ for 20 min, and permitting deceleration without braking. The three fractions floating upon the 1.058, 1.063, and 1.069 density layers were collected and termed myelins I, II, and III, respectively. Each of these myelin fractions was analyzed for lipid and fatty acid content and found to conform to the published data on myelin. Each was distinct, however, due to minor but definite quantitative differences in phospholipid content. The three myelin fractions were also tested immunologically and were shown, by adsorption tests, to react specifically with antimyelin antibodies. When thin sections of each myelin fraction were examined for ultrastructural features by electron microscopy, all three preparations were found to consist of the classic concentric-ring configurations typical of myelin preparations made by other workers. The criteria for purity of myelin fractions are discussed in Chap. 18.

NERVE ENDING PARTICLES (SYNAPTOSOMES)

Homogenization of brain tissue results in a shearing of nerve terminals from their axonal connections as well as from surrounding glial elements. The synaptic terminals are relatively resistant to such mechanical stresses and can reseal to form "synaptosomes" [37]. Often the intersynaptic cleft and a postsynaptic process can be isolated as a portion of the same particle. The presynaptic terminals may then be used to study transmitter metabolism, macromolecular synthesis, presynaptic terminal mitochondria, vesicles, and plasma membrane structure and function.

The synaptosomal fraction has presented a unique opportunity to study neuronal properties in isolation without glial contamination. Studies on the isolation of synaptosomes have been performed primarily with CNS tissue. The yield of such particles from peripheral tissues is low, and isolation of presynaptic terminals from outside the CNS is not presently a practical technique. Studies by deRobertis [34] and Whittaker [9e, 69] have demonstrated the usefulness of synaptosomes for the study of synaptic enzyme localization. These investigators employed both morphological data and enzyme-specific activity to demonstrate intactness and relative purity.

Homogenization of brain tissue has been performed with a Teflon and glass homogenizer having a clearance of 0.025 cm, using a 10 percent (w/v) suspension in 0.32 M sucrose. A variety of techniques have been described in the literature for isolating the synaptosomal fraction from crude brain homogenates. These techniques take advantage of the fact that synaptosomes are predominantly in the crude mitochondrial fraction.

Attempts to sacrifice speed for increased yield of fractions may increase the disruption of synaptosomes. For example, if the crude nuclear pellet that is usually sedimented at 1000 X *g* for 10 min is washed, the mitochondrial and synaptosomal yield will be increased, but the functional intactness of the synaptosomes may be impaired by the prolonged sitting at 4°C, even in sucrose. The myelin contamination of the mitochondrial fraction may be minimized either by using young animals (10 to 20 days old), or by attempting to separate gray from white matter before homogenization. After a spin at 1000 X *g* for 10 min, the supernatant is sedimented at 20,000 X *g* for 15 min. At these low speeds, the yield of synaptosomes will be decreased by loss of smaller size particles into the "microsomal fraction." On the other hand, the microsomal contamination of this fraction would thereby be minimized. Other sedimentations of the crude mitochondrial fraction (as high as 17,000 X *g* for 55 min) have been used to maximize the yield, but these also increase the yield of microsomes in the crude mitochondrial fraction.

Fractionation of the crude mitochondrial pellet has been performed in two ways. One is the standard technique devised by Whittaker [39, 40] and by deRobertis [33]:

> Sedimentation is carried out in a discontinuous sucrose gradient made up in equal volumes of 0.8 and 1.2 M sucrose. The crude mitochondrial pellet is suspended in 0.32 M sucrose and centrifuged in a gradient at 53,000 X *g* for 2 hr. With this gradient, small myelin fragments have been found above the 0.8 M fraction. Synaptosomes are isolated at the 0.8 and 1.2 M sucrose fraction interface, and mitochondria are recovered in the pellet below 1.2 M sucrose.

In experiments employing these procedures, we find that there are synaptosomal contaminants in the small myelin fragment, as well as in the 1.2 M mitochondrial pellet. In addition, there is mild contamination of the 0.8–1.2 M interface by mitochondria. This contamination may well be related to the interaction of myelin with membrane fragments and synaptosomes in the presence of hyperosmotic sucrose [32]. The efficiency of the density-gradient separation will depend on the differential osmotic effects on synaptosomes and mitochondria. For example, mitochondria may be more rapidly dehydrated than synaptosomes and initial differences in density would be enhanced.

Another method is that employed by our laboratory and introduced initially by Abdel-Latif [13]. It involves the use of Ficoll made isoosmotic in 0.32 M sucrose, which minimizes the lability of membrane interaction and permits an isopycnic centrifugation [17, 32]. We isolated synaptosomes with a density of 1.072, as defined by zonal-centrifugation profile analysis. Another fraction of synaptosomes was isolated at 1.152 gm per milliliter, but this density is so close to that of mitochondria (1.176 gm per ml) that it has not been practical to employ this fraction for functional studies. Our isolation procedure is as follows:

The crude mitochondrial pellet is resuspended in 0.32 M sucrose using 3 ml per gram of original tissue, and then layered on Ficoll gradients. The gradients are prepared by layering 9.0 ml each of 5%, 7.5%, 10%, 13%, and 20% Ficoll in 0.32 M sucrose. The gradients are allowed to equilibrate at room temperature for 1 hr, and then cooled at 4°C for 30 min prior to use. The gradients are centrifuged in an SW-25.2 Spinco rotor at 64,145 X *g* for 45 min. The two synaptosomal fractions that form layers between 7.5% and 13% are combined for subsequent study. They are diluted with 0.32 M sucrose and centrifuged at 40,000 X *g* for 30 min. The pellet is washed once in 0.32 M sucrose to remove the Ficoll, and the final pellet is suspended in a loose Dounce homogenizer in 0.32 M sucrose at 0.4 ml per gram of original tissue.

To assess the purity of the synaptosome fractions, we have employed electron microscopy to determine both enzyme-specific activity and morphology. The synaptosomes appear to be enriched in acetylcholinesterase and (Na^+, K^+)-ATPase. In addition, they show lowered specific activities in fractions commonly associated with soluble cytoplasmic constituents, such as lactic dehydrogenase, as well as in those commonly associated with mitochondrial fractions, such as succinic dehydrogenase. We have determined by electron microscopy that fractions isolated in Ficoll-sucrose are enriched with presynaptic terminals. A small percentage of the population (10 percent) consists of both presynaptic and postsynaptic particles, each bound by a membrane with an intersynaptic cleft. In this fraction, we found minimal contamination by mitochondria or myelin. However, there is at least 20 percent contamination by membranes of unknown origin. These membrane structures increase in concentration when attempts are made to increase the yield by prolonging the preparation time for synaptosomes or to purify the synaptosome fraction further by recentrifugation through Ficoll-sucrose gradients. We have been able to fractionate synaptosomes more rapidly in 7.5 percent Ficoll and 0.32 M sucrose overlying a 13 percent Ficoll, 0.32 M sucrose base, and have recovered synaptosomes at the interface of the two Ficoll fractions.

The difficulty in using Ficoll is its variability from batch to batch, depending upon the ion concentration and the particle size. To study ion flux, it was necessary to remove the high salt content by dissolving the Ficoll in a small amount of water and reprecipitating it from 95 percent ethanol. The ethanol was removed by lyophilization for prolonged periods of time.

SUBFRACTIONATION OF NERVE ENDING PARTICLES

Nerve ending particles can easily be disrupted by changing the osmotic pressure of the medium. The most popular technique for separating subsynaptosomal constituents is that of Whittaker et al. [69]:

A washed, crude, mitochondrial pellet is subjected to osmotic shock in water; approximately 2 ml water is used per amount of pellet derived from 1 g of tissue. The resuspended pellet is then sucked up and down in a pipet and layered on a discontinuous sucrose density gradient consisting of equal volumes of 0.4, 0.6, 0.8, 1.0, and 1.2 M sucrose fractions. The density gradient is centrifuged at 53,000 X *g* for 2 hr to yield the following fractions: a pellet containing mitochondria, a layer at the 1.0–1.2 M interface containing partially disrupted synaptosomes, and a particulate layer consisting of synaptic plasma membranes at the 0.8–1.0 M interface. In the 0.4 and 0.6 M fractions, synaptic vesicles and occasional microsomal contaminants predominate. In the clear top layer are soluble cytoplasmic constituents.

Some laboratories, including our own, have used the synaptosomal fraction, rather than the crude mitochondrial fraction, as the starting point for the osmotic shock. When the synaptosomal fraction is isolated and subjected to osmotic shock, a longer period of exposure to hypoosmotic treatment is necessary.

The synaptosomal fraction is homogenized in water (2 ml per gram of starting tissue) in a loose homogenizer. The suspension is then kept at room temperature for approximately 2 hr. Thereafter, it is placed on a 0.4, 0.6, 0.8, 1.0, 1.1, and 1.2 M discontinuous sucrose gradient, and centrifuged for 1 hr at 53,000 X *g*. The introduction of a 1.1 M fraction, as well as 10^{-5} M $CaCl_2$ and 10^{-3} M Tris-HCl (pH 7.4), has assured more reliable separation of the synaptic membrane fraction at the 0.8–1.0 M interface.

Assessment of the relative purity of the various fractions has been obtained by enzyme-specific activities. The synaptic plasma membrane at the 0.8–1.0 M interface was found to have the highest specific activity of acetylcholinesterase and (Na^+, K^+)-ATPase. Furthermore, it was determined by electron microscopy that the 1.2 M pellet consists of mitochondria and the 0.4 and 0.6 M fractions consist of small vesicular-shaped structures.

Unfortunately, unless special precautions are taken, the 1.2 M pellet will be contaminated significantly with synaptosomal particles. As a result, metabolic studies of heterogeneous 1.2 M sucrose fractions will share some of the characteristics of both synaptosomal and mitochondrial fractions. For the greatest enrichment of mitochondria, we employed the techniques listed on pp. 441–443.

VESICLES

As mentioned above, the gradients devised by Whittaker [9e, 69] enabled the separation of synaptic vesicles in his "fraction D" at a particle density equivalent to 0.4 M sucrose. Because of their relative resistance to changes in osmotic pressure, vesicles can be further purified by relayering them on a continuous gradient of 0.4 to 0.6 M sucrose above a layer of 1.6 M sucrose [9e].

Small vesicles of approximately 500 Å diameter isolated at a sucrose density of 0.4 to 0.5 M contain approximately 50 percent of the acetylcholine found in the synaptosome fraction. The remainder of the acetylcholine is found in the supernatant fraction as the so-called "labile pool." That the acetylcholine is within the synaptic vesicles is evident by the fact that it can be liberated by osmotic shock. During the initial osmotic shock to the synaptosomes, the synaptic vesicles apparently are protected by the presence of the synaptic membrane. However, if the synaptic vesicles are directly subjected to osmotic shock in the presence of distilled water, they will release their constituents into the medium.

There is similar evidence that nerve endings from tissues enriched in catecholamines possess vesicles with high catecholamine content. According to the data of Whittaker, the yield of synaptic vesicles is approximately 3.8×10^{12} vesicles per gram of original tissue (guinea pig cortex) [9e]. The fraction was claimed to represent a 15 to 16 percent yield of the vesicles present in the initial homogenate. The membranes of the synaptic vesicles have different protein constituents from other membranes, as indicated by the studies of Cotman, Mahler, and Hugli [27], as well as by those conducted in our own laboratory [31]. In addition, they are thought to have a Mg^{2+}-activated ATPase. Markers of the various fractions include gangliosides, which are relatively rich in the synaptic plasma membrane and poor in the other fractions, and glycoproteins, which are also relatively rich in the synaptic plasma membrane.

LYSOSOMES

Brain lysosomes, like those of other tissues, are membrane-limited cytoplasmic organelles with acid hydrolase activity [4]. They comprise a heterogeneous population of granules that vary widely in size, structure, physical properties, and chemical composition. The reason for presenting a brief discussion of lysosomes in this chapter is to point out that such organelles contaminate almost all fractions and may impair the integrity of isolated particles.

Lysosomes contain enzymes, usually glycoproteins, with optimal pHs in the acid range. They sediment in most of the particulate fractions, and to release optimal enzyme activity, they usually require disruption by various techniques. Using acid hydrolases as markers, Koenig [9c] demonstrated that an enriched lysosomal fraction could be isolated from a crude mitochondrial fraction as a 1.4 M sucrose pellet after centrifugation through a discontinuous gradient of 0.8, 1.0, 1.2, and 1.4 M sucrose. Some less dense lysosomal particles were isolated at the 1.2–1.4 M sucrose interface, especially when the microsomal fraction was used as starting material. Histochemical and electron microscopic studies have confirmed this isolation procedure.

BRAIN MITOCHONDRIA

Mitochondria isolated from neural tissue possess both qualitative and quantitative differences from mitochondrial preparations of both heart and liver. The methods used for the isolation of neural mitochondria were not capable of distinguishing between the mitochondria of glial and those of neuronal origins, and it is very possible that "neural" mitochondria were not homogeneous in all of their functional and biochemical characteristics [20, 43]. Neural tissue is a rich source of mitochondria since the brain utilizes 25 percent of the total body oxygen and glucose and the function of mitochondria is oxidative metabolism and the production of ATP. In fact, 15 percent of the total brain protein content is mitochondrial protein [9a]. There is a greater proportion of mitochondria in neurons than in glia. Therefore, the gray matter of the brain contains more mitochondria than does the white matter. Mitochondria may be found throughout the cytoplasm, axons, and dendrites of neurons, but are particularly abundant near the synaptic regions. Among the cells derived from glia, most of the mitochondria are found in the oligodendrocytes, with relatively few in the astrocytes and microglia. Schwann cells are also particularly rich in mitochondria, with mitochondrial clusters around the nuclei. (For more details concerning the ultrastructure, chemical composition, enzyme complexes, and metabolism of mitochondria, see the reviews in [9a, 10, and 12].)

Procedures for the isolation of neural mitochondria invariably start with a brain homogenate, a heterogeneous complex of cells with different sizes, shapes, and densities. A requisite of any subcellular isolation technique is the morphological and biological preservation of the involved plasma membranes. In this respect, the isolation and purification procedures for mitochondria are facilitated because mitochondria possess tough outer membranes that are not easily destroyed by osmotic and other physical forces, whereas the cellular plasma membranes are fragile and easily disrupted. Most of the isolation procedures for neural mitochondria have utilized rat, mouse, rabbit, or guinea pig brains, but bovine, frog, and chicken brains have also been used.

The isolation procedures have included the use of differential centrifugation techniques and continuous and step gradients, both in swinging bucket rotors and in zonal rotors. The earliest isolation procedures utilized simple differential centrifugation of the brain homogenate in isotonic (0.25 M) sucrose solutions [9a, 25]. These separation methods were simply a modification of the liver mitochondrial isolation procedures of Schneider and Hogeboom [58]. Although this method produced large yields of brain mitochondria, they were always contaminated with substantial amounts of myelin and neuronal fragments. In spite of this cross-contamination, recent biochemical studies have been made of "pure" mitochondria, utilizing isotonic sucrose in differential centrifugation separation techniques [22, 42, 64]. One recent paper examined analytical differential centrifugation as a guide to the proper centrifugation procedures for the determination of the sedimentation properties of mitochondria as well as of synaptosomes and lysosomes [29].

The next refinement in neural mitochondrial separation procedures was introduced by Gray and Whittaker [39] and deRobertis et al. [33]. Their technique involved the use of sucrose step gradients ranging from 5 to 30 percent sucrose. The mitochondrial preparations resulting from these procedures were still contaminated with myelin and synaptic material. Also, the increased osmotic properties of the higher concentrations of sucrose disrupted many of the mitochondrial membranes, thus increasing contamination and decreasing the yield. Combinations of differential centrifugation and sucrose step gradients are still being used by some investigators as a technique for intraneural mitochondrial preparations [23].

Tanaka and Abood [62] introduced the use of Ficoll in conjunction with a continuous sucrose gradient as an improved method for isolating mitochondria. Ficoll can impart increasing densities to the sucrose while maintaining isoosmotic or near isoosmotic conditions. Abdel-Latif [13, 14] used a Ficoll-sucrose step gradient, instead of a continuous gradient, in a simple method for isolating nerve ending particles. His technique was adapted by Autilio et al. [17] to isolate other subcellular particles, including mitochondria. Ficoll was also included at a final concentration of 8 percent (w/v) in a differential centrifugation scheme by Stahl et al. [61] to isolate bovine brain mitochondria. Ficoll-sucrose step gradients are currently being used in the isolation of brain mitochondria prior to biochemical analysis [26].

The adaptation of the zonal centrifuge to isolation and separation of brain homogenates was pioneered by Barber et al. [3] and Anderson et al. [2]. They first utilized sucrose gradients in zonal rotors. The increased volumes of the gradients, the precise control of the angular velocity achieved, and the ability to remove the "bands" while the rotor was spinning enabled cleaner separations of the subcellular particles. Day [30, 31] and Shapira et al. [59] have utilized sucrose gradients in the zonal rotor for separating and isolating subcellular components of brain homogenates, including mitochondria. Day et al. [32] have now introduced the use of isoosmotic Ficoll-sucrose gradients in the zonal rotor to stabilize the bands of separated subcellular material.

The excellent resolution of a brain homogenate into subcellular particles by zonal rotor separation can also be easily obtained in a swinging bucket rotor separation, using isoosmotic Ficoll-sucrose step gradients. The adaptation permits separations in laboratories that are not equipped with zonal centrifuge facilities. The following techniques have been worked out by two of us (D. M. and E. D.) for mitochondrial isolation.

Individual rat brains are hand homogenized in a 0.32 M sucrose solution (containing 5×10^{-5} M $CaCl_2$ and 2.5×10^{-3} M Tris, adjusted to pH 7.4) in the proportion of 1 volume of brain to 9 volumes of sucrose solution. Isoosmotic Ficoll-sucrose [32] is used to prepare step gradients with densities of 1.049, 1.067, 1.080, and 1.120. Samples of the homogenate are centrifuged on this gradient for 20 min at 68,108 $\times g$ and allowed to decelerate without braking. The fraction of the homogenate that collects as a band on top of 1.120 Ficoll-sucrose is removed, diluted with 0.32 M sucrose to a density of 1.062, and layered on another Ficoll-sucrose step gradient with densities of 1.080, 1.120, and 1.156. After centrifugation for 20 min at 68,108 X *g*, the material that bands at density 1.156 is removed; this is the mitochondrial fraction.

Because the intrinsic isopycnic point of mitochondria is 1.176, the mitochondria banding at 1.156 may represent either another population of mitochondria or a rate-zonal centrifugation. The purity of these brain mitochondria has been tested by biochemical, immunological, and structural methods. They are apparently not contaminated with synaptic membranes because cycloheximide, an inhibitor of ribosomal and microsomal protein synthesis, did not inhibit uptake of labeled amino acids into protein. On the other hand, puromycin, a potent inhibitor of mitochondrial protein synthesis, greatly inhibited the uptake of labeled amino acid in this mitochondrial preparation. As an immunological test of purity, antibodies were produced in rabbits against the preparation and, by adsorption experiments, the antibody was shown to react specifically with brain mitochondria. As a structural check for purity, samples were taken for electron microscopy, and their sections were shown to be composed primarily of brain mitochondria.

PURIFICATION OF INTACT MICROTUBULES FROM BRAIN

The isolation of microtubules has been approached from two different starting points. Kirkpatrick [45] has attempted to purify intact microtubules and then analyze the constituents of the structure. He employed a morphological assay in which the typical ultrastructure could be assessed as the purification proceeded. An alternate procedure for isolating microtubules has been to purify brain "tubulin," which is considered by many to be the structural subunit of microtubules. The circumstantial evidence that such a subunit probably belongs to microtubules is that colchicine, which disrupts microtubules, binds to a component called tubulin. Furthermore, tubulin can be extracted from various tissues in amounts roughly proportional to the prominence of microtubules in the tissue after such microtubules have been dissolved. Although tubulin can be identified as a subunit of microtubules, it has not yet been successfully aggregated into microtubules.

Kirkpatrick's technique depends upon maintaining the labile microtubular structure intact for sufficient periods of time to purify the samples by differential centrifugation. He employed hexylene glycol, together with low temperature, to accomplish this purification:

> A 16.7% homogenate of brain tissue in 1 M hexylene glycol buffered with 20 mM KH_2PO_4, pH 6.4, at 1°C is centrifuged at 48,000 X *g* for 30 min. The supernatant is then sedimented in a discontinuous sucrose gradient consisting of sucrose dissolved in buffered hexylene glycol to obtain solutions with density of 1.16 layered over a solution of density 1.19. The most highly purified microtubules were obtained after centrifugation at 150,000 X *g* for 1 hr from the interface of the 1.16 and 1.19 fractions.

According to Kirkpatrick's data, this represented a purification of greater than 100-fold. Interestingly, this fraction showed similar bands on sodium dodecyl sulfate and acrylamide-gel electrophoresis as does tubulin prepared according to the techniques of Weisenberg et al. [68].

More recent studies have demonstrated that vinblastine can induce the formation of "microtubule crystals" in vivo [19] and can precipitate tubulin in vitro [49]. Vinblastine is believed to induce a reversible aggregation of tubulin. Colchicine, which disassociates microtubules in vivo, is believed to increase the aggregation of tubulin protein induced by divalent ion. Further work will be necessary to determine if constituents other than tubulin are necessary for the aggregation of protein into intact microtubules. It is of interest that the extent of tubulin aggregation depends upon the divalent-ion concentration (Mg^{2+} or Ca^{2+}), that tubulin can bind guanosine nucleotides, and that it apparently participates in a phosphorylation reaction that is stimulated by 3',5'-cyclic AMP [38].

Extension of these cellular and subcellular fractionations are in progress in many laboratories. Their success will have far-reaching consequences upon our understanding of brain structure and function in health and disease. Often it will be necessary to combine several procedures in order to isolate a particular structure. For example, the separation of neurofilaments will be simplified by the extension of the techniques for cell separation to those for axons. The developments in this area should be extremely fruitful in the next decade.

ACKNOWLEDGMENTS

This work was supported in part by Grant No. NB-07872 from the National Institute of Neurological Diseases and Stroke, the Robert McManus Memorial Grant No. 558-B-3 from the National Multiple Sclerosis Society, Contract AT-(40-1)-3195 with the Atomic Energy Commission, and by Grant No. MH-08394, Postdoctoral Training Grant from U.S. Public Health Service.

REFERENCES

1. Anderson, N. G. An introduction to particle separations in zonal centrifuges. *Natl. Cancer Inst. Monogr.* 21:9, 1966.
2. Anderson, N. G., Harris, W. W., Barber, A. A., Rankin, C. T., Jr., and Chandler, E. L. Separation of subcellular components and viruses by combined rate- and isopycnic-zonal centrifugation. *Natl. Cancer Inst. Monogr.* 21:253, 1966.
3. Barber, A. A., Rankin, C. T., Jr., and Anderson, N. G. Lipid peroxidation in rat tissue particulates separated by zonal centrifugation. *Natl. Cancer Inst. Monogr.* 21:333, 1966.
4. deDuve, C. The separation and characterization of subcellular particles. *Harvey Lect.* 59:49, 1965.

5. *Handbook of Biochemistry*. Selected Data for Molecular Biology. Cleveland: Chemical Rubber Co., 1968, pp. J248-J251. (Viscosity and density tables of sucrose in water.)
6. *Handbook of Chemistry and Physics* (47th ed.). Concentrative Properties of Sucrose. Cleveland: Chemical Rubber Co., 1942, pp. D175-D176.
7. *Handbook of Chemistry and Physics.* Index of Refraction of Aqueous Solutions of Sucrose. Cleveland: Chemical Rubber Co., 1942, pp. 2121-2122.
8. Johnston, P. V., and Roots, B. I. Neuronal and glial perikarya preparation: An appraisal of present methods. *Int. Rev. Cytol.* 29:265, 1970.
9. Lajtha, A. *Handbook of Neurochemistry*. Vol. 2. New York: Plenum, 1969. See in particular:
   9a. Abood, L. G. Brain Mitochondria, p. 303.
   9b. deRobertis, E., and Rodriguez de Lores Arnaiz, G. Structural Components of the Synaptic Region, p. 365.
   9c. Koenig, H. Lysosomes, p. 255.
   9d. Rappoport, D. A., Maxcy, P., Jr., and Daginawala, H. F. Nuclei, p. 241.
   9e. Whittaker, V. P. The Synaptosome, p. 327.
10. Lehninger, A. *The Mitochondrion.* New York: Benjamin, 1965.
11. Rose, S. P. R. The Biochemistry of Neurons and Glia. In Davison, A. N., and Dobbing, J. (Eds.), *Applied Neurochemistry*. Philadelphia: Davis, 1968.
12. Slater, E. C., Kaniuga, Z., and Wojtczak, L. (Eds.). *Biochemistry of Mitochondria.* New York: Academic, 1967.

ORIGINAL SOURCES

13. Abdel-Latif, A. A. A simple method for isolation of nerve ending particles from rat brain. *Biochim. Biophys. Acta* 121:403, 1966.
14. Abdel-Latif, A. A., and Abood, L. G. Biochemical studies on mitochondria and other cytoplasmic fractions of developing rat brain. *J. Neurochem.* 11:9, 1964.
15. Abood, L. G., Gerard, R. W., Banks, J., and Tschirgi, R. D. Substrate and enzyme distribution in cells and cell fractions of the nervous system. *Am. J. Physiol.* 168:728, 1952.
16. Anderson, N. G., and Wilbur, K. M. Studies on isolated cell components. IV. The effect of various solutions in the isolated rat liver nucleus. *J. Gen. Physiol.* 35:781, 1952.
17. Autilio, L. A., Appel, S. H., Pettis, P., and Gambetti, P. Biochemical studies of synapses *in vitro.* I. Protein synthesis. *Biochemistry* 7:2615, 1968.
18. Autilio, L. A., Norton, W. T., and Terry, R. D. Preparation and some properties of purified myelin from the CNS. *J. Neurochem.* 11:17, 1964.
19. Bensch, K. G., and Malawista, S. Microtubular crystals in mammalian cells. *J. Cell Biol.* 40:95, 1969.
20. Blokhuis, G. G., and Veldstra, H. Heterogeneity of mitochondria in rat brain. *FEBS Letters* 11:197, 1970.

21. Blomstrand, C., and Hamberger, A. Protein turnover in cell-enriched fractions from rabbit brain. *J. Neurochem.* 16:1401, 1969.
22. Bosmann, H. B. Asparaginase action: Inhibition of protein synthesis in rat liver mitochondria and inhibition of glycoprotein synthesis in liver brain mitochondria by asparaginase. *Life Sci.* 9:851, 1970.
23. Bosmann, H. B., and Hemsworth, B. A. Intraneural mitochondria: Incorporation of amino acids and monosaccharides into macromolecules by isolated synaptosomes and synaptosomal mitochondria. *J. Biol. Chem.* 245:363, 1970.
24. Brizzee, K. R., Vogt, J., and Kharetchko, X. Postnatal changes in glia/neuron index with a comparison of methods of cell enumeration in the white rat. *Progr. Brain Res.* 4:136, 1964.
25. Brody, T. M., and Bain, J. A. A mitochondrial preparation from mammalian brain. *J. Biol. Chem.* 195:685, 1952.
26. Clark, J. B., and Nicklas, W. J. The metabolism of rat brain mitochondria, preparation and characterization. *J. Biol. Chem.* 245:4724, 1970.
27. Cotman, C., Brown, D. H., Harrel, B. W., and Anderson, N. G. Analytical differential centrifugation: An analysis of the sedimentation properties of synaptosomes, mitochondria, and lysosomes from rat brain homogenates. *Arch. Biochem. Biophys.* 136:436, 1970.
28. Cotman, C., Mahler, H. R., and Anderson, N. G. Isolation of a membrane fraction enriched in nerve-end membranes from rat brain by zonal centrifugation. *Biochim. Biophys. Acta* 163:272, 1968.
29. Cotman, C., Mahler, H. R., and Hugli, T. E. Isolation and characterization of insoluble proteins of the synaptic plasma membrane. *Arch. Biochem. Biophys.* 126:821, 1968.
30. Day, E. D. Myelin as a locus for radioantibody absorption *in vivo* in brain and brain tumors. *Cancer Res.* 28:1335, 1968.
31. Day, E. D., and Appel, S. H. The biologic half-life of brain-localized antisynapse radioantibodies. *J. Immunol.* 3:710, 1970.
32. Day, E. D., McMillan, P. N., Mickey, D. D., and Appel, S. H. Zonal centrifuge profiles of rat brain homogenates, instability in sucrose, stability in iso-osmotic Ficoll-sucrose. *Anal. Biochem.* 39:29, 1971.
33. deRobertis, E., delraldi, A. P., Rodriguez de Lores Arnaiz, G., and Gomez, C. On the isolation of nerve endings and synaptic vesicles. *J. Biophys. Biochem. Cytol.* 9:229, 1961.
34. deRobertis, E., delraldi, A. P., Rodriguez de Lores Arnaiz, G., and Salganicoff, L Cholinergic and non-cholinergic nerve endings in rat brain. I. Isolation and subcellular distribution of acetylcholine and acetyl cholinesterase. *J. Neurochem.* 9:23, 1962.
35. Falngas, A. L., and Bowman, R. E. Neuronal perikarya of rat brain isolated by zonal centrifugation. *Science* 161:1025, 1968.
36. Freysz, L., Bieth, R., Judes, C., Sensenbrenner, M., Jacob, M., and Mandel, P. Quantitative distribution of phospholipids in neurons and glial cells isolated from rat cerebral cortex. *J. Neurochem.* 15:307, 1968.

37. Gfeller, E., Kuhar, M. J., and Snyder, S. H. Neurotransmitter-specific synaptosomes in rat corpus striation morphological variations. *Proc. Natl. Acad. Sci. U.S.A.* 68:155, 1971.
38. Goodman, D. B. P., Rasmussan, H., diBella, F., and Guthrow, C. E., Jr. Cyclic adenosine 3'-5' monophosphate-stimulated phosphorylation of isolated neurotubule subunits. *Proc. Natl. Acad. Sci. U.S.A.* 67:652, 1970.
39. Gray, E. G., and Whittaker, V. P. The isolation of synaptic vesicles from the central nervous system. *J. Physiol. (Lond.)* 153:35P, 1960.
40. Gray, E. G., and Whittaker, V. P. The isolation of nerve endings from brain: An electron microscopic study of the cell fragments of homogenation and centrifugation. *J. Anat.* 96:79, 1962.
41. Hadjiolov, A. A., Tencheva, Z. S., and Bojadjieva-Mikhailova, A. G. Isolation and some characteristics of cell nuclei from brain cortex of adult cat. *J. Cell Biol.* 26:383, 1965.
42. Haldar, D. Protein synthesis in mammalian brain mitochondria. *Biochem. Biophys. Res. Commun.* 40:129, 1970.
43. Hamberger, A., Blomstrand, C., and Lehninger, A. L. Comparative studies on mitochondria isolated from neuron-enriched and glia-enriched fractions of rabbit and beet brain. *J. Cell Biol.* 45:221, 1970.
44. Kato, T., and Kurokawa, M. Isolation of cell nuclei from the mammalian cerebral cortex and their assortment on a morphological basis. *J. Cell Biol.* 32:649, 1967.
45. Kirkpatrick, J. B., Haynes, L., Thomas, V. L., and Howley, P. M. Purification of intact microtubules from brain. *J. Cell Biol.* 47:384, 1970.
46. Lovtrup-Rein, H., and McEwen, B. S. Isolation and fractionation of rat brain nuclei. *J. Cell Biol.* 30:405, 1966.
47. Mandel, P., Borkowski, T., Harth, S., and Mardell, R. Incorporation of 32_p in ribonucleic acid of subcellular fractions of various regions of the rat central nervous system. *J. Neurochem.* 8:126, 1961.
48. Mandel, P., Dravid, A. R., and Pete, N. Poly C synthetase activity in the particulate fraction of rat brain nuclei. *J. Neurochem.* 14:301, 1967.
49. Marantz, R., Ventilla, M., and Shelanski, M. Vinblastine induced precipitation of microtubule protein. *Science* 165:498, 1969.
50. Norton, W. T., and Autilio, L. A. The lipid composition of bovine brain myelin. *J. Neurochem.* 13:213, 1966.
51. Norton, W. T., and Poduslo, S. E. Neuronal soma and whole neuroglia of rat brain: A new isolation technique. *Science* 167:1144, 1970.
52. Norton, W. T., and Poduslo, S. E. Neuronal perikarya and astroglia of rat brain: Chemical composition during myelination. *J. Lipid Res.* 12:84, 1971.
53. Norton, W. T., Poduslo, S. E., and Raine, C. S. The isolation, composition and ultrastructure of bovine oligodendroglia. *Trans. Am. Soc. Neurochem.* 2:98, 1971.
54. Rappoport, D. A., Fritz, R. R., and Moraczewski, A. Biochemistry of the developing rat brain. I. Soluble enzymes in isolated neonatal brain nuclei. *Biochim. Biophys. Acta* 74:42, 1963.

55. Rose, S. P. R., and Sinha, A. K. Some properties of isolated neuronal cell fractions. *J. Neurochem.* 16:1319, 1969.
56. Ross, L. L., Andreoli, V. M., and Marchbanks, R. M. A morphological and biochemical study of subcellular fractions of the guinea pig spinal cord. *Brain Res.* 25:103, 1971.
57. Satake, M., Hasegawa, S., Abe, S., and Tanaka, R. Preparation and characterization of nerve cell perikaryon from pig brain stem. *Brain Res.* 11:246, 1968.
58. Schneider, W. C., and Hogeboom, G. H. Cytochemical studies of mammalian tissues: The isolation of cell components by differential centrifugation. *Cancer Res.* 11:1, 1951.
59. Shapira, R., Brinkley, F., Kibler, R. F., and Wundram, I. J. Preparation of purified myelin of rabbit brain by sedimentation in a continuous sucrose gradient. *Proc. Soc. Exp. Biol. Med.* 133:238, 1970.
60. Sporn, M., Wauko, T., and Dingman, W. The isolation of cell nuclei from rat brain. *J. Cell Biol.* 15:109, 1962.
61. Stahl, W. L., Smith, J. C., Napolitano, L. M., and Basford, R. E. Brain mitochondria. I. Isolation of bovine brain mitochondria. *J. Cell Biol.* 19:293, 1963.
62. Tanaka, R., and Abood, L. G. Studies on adenosine triphosphatase of relatively pure mitochondria and other cytoplasmic constituents of rat brain. *Arch. Biochem. Biophys.* 105:554, 1964.
63. Tiplady, B., and Rose, S. P. R. Amino acid incorporation into protein in neuronal cell body and neuropil fractions *in vitro*. *J. Neurochem.* 18:549, 1971.
64. Tjioe, S., Bianchi, C. P., and Hangaard, N. The function of ATP in Ca^{2+} uptake by rat brain mitochondria. *Biochem. Biophys. Acta* 216:270, 1970.
65. Tolani, A. J., and Mokrasch, L. C. Incorporation of ^{14}C amino acids into proteolipid protein of subcellular fractions from rat brain, heart, and liver. *Life Sci.* 6:1771, 1967.
66. Tolani, A. J., and Talwar, G. P. Differential metabolism of various brain regions. Biochemical heterogeneity of mitochondria. *Biochem. J.* 88:357, 1963.
67. Trautman, R., and Cowan, K. M. Preparative and analytical ultracentrifugation. *Meth. Immunol. Immunochem.* 2:81, 1968.
68. Weisenberg, R. C., Borisy, G. C., and Taylor, E. W. The colchicine binding protein of mammalian brain and its relation to microtubules. *Biochemistry* 7:4466, 1968.
69. Whittaker, V. P., Michaelson, I. A., and Kirkland, R. J. A. The separation of synaptic vesicles from nerve ending particles. *Biochem. J.* 90:293, 1964.

TWO

MEDICAL NEUROCHEMISTRY

22

Inherited Disorders of Amino Acid, Carbohydrate, and Nucleic Acid Metabolism

Y. Edward Hsia

METABOLIC DISORDERS can cause profound disturbances of mental and neurological function. The nervous system is so dependent on the constancy of the internal milieu and on a continuing supply of nutrients and oxygen that its delicately tuned state can be radically upset by systemic metabolic illnesses. These may produce acute neurological symptoms, such as convulsions, and eventually may produce chronic brain damage. In infants, the rapidly growing nervous system is uniquely vulnerable to metabolic trauma, with delayed and disrupted brain growth leading to psychomotor and intellectual retardation.

At the functional level, the neurotransmitter system is vulnerable to both impeded energy metabolism and abnormal electrolyte distribution, and it is sensitive to altered concentrations of neuroactive amino acids, such as glycine and glutamate (see Chap. 8), and of biogenic amine derivatives of the amino acids, such as γ-amino butyric acid, histamine, serotonin, and the catecholamines. These alterations may affect not only neurotransmission, but also psychosocial behavior [1]. In addition, in metabolic disorders, there may be accumulation of metabolites toxic to certain biochemical reactions and thus neurological function may be seriously disturbed.

At the structural level, lesions may be produced by the prolonged effects of metabolic disorders that prevent normal cell growth and differentiation or that cause cell damage and cell death, with resulting demyelination and brain degeneration. Lesions may also arise from toxic inhibition of protein and neurolipid synthesis, which can cause faulty myelination and defective development of the brain [22, 51, 56, 58].

At the clinical level, neurological reactions to metabolic disorders generally present nonspecific symptoms, such as irritability, lethargy, hypotonicity, convulsions, and eventual mental retardation. Some metabolic disorders, however, produce highly specific symptoms, such as cerebellar ataxia or choreoathetosis, which suggests that the neurological lesions are localized and are caused by focal toxic processes. Abnormal accumulation of certain macromolecules in the brain occurs in defects of degradative enzymes or of lysosomal function. Clinically, these conditions result in degenerative brain diseases with progressive loss of neurological functions (see Chap. 23).

Many inborn errors of metabolism are innocuous to the nervous system. The association of other errors with neurological lesions may be a coincidental consequence of intensive screening for metabolic abnormalities in patients with neurological deficits. Those metabolic errors which cause specific damage to the nervous system are of great significance because these rare experiments of nature have great potential for furthering the understanding of normal neurochemical function in man. Effective dietary or biochemical treatment for these metabolic disorders may provide, in addition, bases for rational treatment of other neurological illnesses.

It is possible that some gross structural malformations of the nervous system are caused by metabolic disturbances during fetal neural differentiation. For instance, the increased incidences of fetal deformities in children of diabetic mothers and of microcephaly and retardation in children of phenylketonuric mothers could be secondary to abnormal intrauterine environments. The occasional findings of brain malformations in infants with inborn metabolic errors support the concept that biochemical disorders can disturb normal embryogenesis of the nervous system [13, 17a].

DISORDERS OF AMINO ACID AND ORGANIC ACID METABOLISM

The primary disorders of amino acid metabolism that cause neurological lesions are enzyme deficiencies or membrane abnormalities that are genetically determined. The defects may block the transport or catabolism either of an amino acid itself or of one of its metabolites. The aminoacidopathies that are believed to be associated with neurological lesions are grouped into those causing metabolic acidosis (Table 22-1), those causing hyperammonemia (Table 22-2), and other aminoacidopathies (Table 22-3).

Excess dietary protein precipitates or exacerbates many of these metabolic and neurological disorders. In such cases, dietary restriction of protein or of the specific amino acid whose catabolism is blocked has often proved beneficial.

METABOLIC ACIDOSIS*

Disorders of Branched-Chain Amino Acids

Leucine, isoleucine, and valine are individually deaminated and the corresponding keto acids are decarboxylated by a common enzyme system. Leucine thus produces isovalerate, which is oxidized to β-methylcrotonate. This is carboxylated to β-methylglutaconate, which is oxidized and then split to form acetoacetate and acetate. Isoleucine produces α-methylbutyrate; this is progressively oxidized to

*For discussion of inherited abnormalities of carbohydrate metabolism or of lysosomal enzymes that cause acidosis, see pp. 473, 478 and Table 22-6.

TABLE 22-1. Disorders of Amino Acids and Organic Acids That Cause Severe Metabolic Acidosis

Condition	Biochemical Defect	Diagnostic Features	Metabolic Disturbances	Neurological Lesions	Treatment
Defects of Branched-Chain Amino Acid Catabolism (see also Table 22-3):					
Maple-syrup urine disease [3, 4] and variants (see text)	Branched-chain α-keto acid decarboxylase	Characteristic odor, leucine, isoleucine, valine, and their α-keto acids in urine	Severe ketoacidosis, hypoglycemia	Vomiting, anorexia, hypotonia, rigidity, seizures, retardation	Dietary restriction of branched-chain amino acids
Isovalericacidemia [3, 4]	Isovaleryl-CoA dehydrogenase	Odor of "sweaty feet," isovaleric acid in blood and urine	Severe metabolic acidosis	Drowsiness, cerebellar ataxia, hyperreflexia, psychomotor retardation	Restricted leucine diet
β-Methylcrotonylglycinuria and β-hydroxyisovaleric-aciduria [16][a]	β-Methylcrotonyl-CoA carboxylase (?)	Odor of "cat's urine"	(?)	Spinal muscular atrophy	Dietary restriction of leucine (?)
Biotin-responsive variant [20a]	Variant of above (?)	Same as above	Metabolic keto-acidosis	Irritability	Biotin
Hypervalinemia [4][a]	Valine "trans-aminase" (?)	Valinemia	(?)	Retardation	Dietary restriction of valine (?)
α-Methylacetoacetic-aciduria [15][a]	α-Methylaceto-acetic thiolase (?)	α-Methylaceto-acetate, α-methyl-β-hydroxyacetate, and butanone in urine	Ketoacidosis	No retardation	Dietary restriction of isoleucine (?)

TABLE 22-1 (continued)

Condition	Biochemical Defect	Diagnostic Features	Metabolic Disturbances	Neurological Lesions	Treatment
Defects of the Propionate Pathway:					
Propionicacidemia [26] (ketotic hyperglycinemia)	Propionyl-CoA carboxylase	Propionicacidemia, hyperglycinemia, blocked oxidation of propionate by leukocytes	Ketoacidosis, lacticacidosis, hypoglycemia, hyperammonemia	Vomiting, lethargy, prostration, seizures, retardation	Restricted isoleucine, valine, methionine, and threonine in diet
Nonketotic hyperglycinemia variant [35][a]	Propionyl-CoA carboxylase	As for propionicacidemia	Milder than propionicacidemia	Retardation	Low-protein diet
Biotin-dependent variant [7][a]	(?)	Isoleucine loading test with biotin	Ketoacidosis, hyperglycinemia	Lethargy, hypotonia, athetosis, retardation	Biotin, diet
Methylmalonicacidemia [3, 4]	Methylmalonyl-CoA mutase	Methylmalonic acid in blood and urine; otherwise as for propionicacidemia	As for propionicacidemia	As for propionicacidemia	Low-protein diet
Vitamin B_{12}-dependent variant [25]	(?)	Methylmalonicaciduria decreased by vitamin B_{12}	As for propionicacidemia	As for propionicacidemia	Vitamin B_{12}
Methylmalonicaciduria with cystathioninemia; homocystinemia [35a]	Vitamin B_{12} reductase (?)	Homocystine and cystathionine in blood and urine; low tissue methionine; methylmalonate in urine	As for propionicacidemia	May have cerebellar lesions; one patient was clinically normal	Vitamin B_{12} (?)

Defects of Lactate and Pyruvate Metabolism:					
Familial lacticacidosis [22]	(?)	Lactate and pyruvate raised in blood and in urine, with α-ketoglutarate	Lacticacidosis	Hypotonia, tachypnea, ataxia, convulsions, retardation, cataracts	(?)
Subacute necrotizing encephalomyelopathy [24, 46]	Thiamine triphosphate formation (?); pyruvate carboxylase (?) [10]	Lactate raised in blood; neuropathological lesions of Wernicke's encephalopathy	Lacticacidosis	Nystagmus, cranial nerve lesions, ataxia, dementia	Lipoic acid (?)
Pyruvicacidemia with hyperalaninemia [34]	(?)	Pyruvate, lactate, and alanine raised in blood and urine	Lacticacidosis	Optic atrophy, ataxia, confusion	Thiamine
Other Causes of Metabolic Acidosis:					
Succinyl-CoA:3-ketoacid CoA transferase deficiency [62][a]	Specific transferase deficiency	Persistent ketonemia	Profound ketoacidosis	(?)	(?)
Cystinosis [3, 4]	Intracellular cystine transport	Intracellular cystine accumulation; generalized renal tubular dysfunction	Uremic acidosis, renal rickets	Photophobia, failure to thrive	(?)
Lowe's syndrome [3]	(?)	Cataracts, aminoaciduria, proteinuria, hypotonia, retardation	Metabolic acidosis, renal rickets	Hypotonia, seizures, retardation	(?)

[a]Disorder described in only one patient or one family.

TABLE 22-2. Disorders of Amino Acid and Intermediary Metabolism That Cause Hyperammonemia

Condition	Biochemical Defect	Diagnostic Features	Metabolic Disturbances	Neurological Lesions	Treatment
Hyperammonemia [18][a]	Carbamyl phosphate synthetase (?)	Hyperammonemia; demonstration of enzyme defect in liver	High ammonia levels in blood and cerebrospinal fluid; hyper-glycinemia	Vomiting, lethargy, retardation	Low nitrogen diet
Hyperammonemia [11]	Ornithine trans-carbamylase	Hyperammonemia; demonstration of enzyme defect in liver	High ammonia levels in blood and cerebrospinal fluid	Vomiting, anorexia, lethargy, irritability, retardation	Low nitrogen diet
Citrullinemia [4, 52]	Argininosuccinic acid synthetase	Huge excess of citrulline in blood and urine	Postprandial hyperammonemia	Vomiting, irritability, seizures, retardation	Low nitrogen diet
Argininosuccinic-aciduria [58]	Agininosuccinase	Argininosuccinic acid accumulation in blood and urine and particularly in cerebrospinal fluid; trichorrhexis nodosa of hair	Postprandial hyperammonemia	Seizures, ataxia, severe mental retardation	(?)
Argininemia [61][a]	Arginase	Elevated arginine in blood and cerebro-spinal fluid	Hyperammonemia	Seizures, spastic diplegia, severe retardation	(?)

Hyperornithinemia [4][a]	(?)	Hyperammonemia, hyperornithinemia, homocitrullinuria	Hyperammonemia	Vomiting, anorexia, irritability, myoclonic spasms	Low nitrogen diet
Familial protein intolerance [36]	Glutaminase (?)	Hyperammonemia	Hyperammonia	Retardation	(?)
Lysine intolerance [3, 4]	Lysine dehydrogenase	Hyperammonemia, elevated lysine and arginine in blood on high-protein diet	Hyperammonemia	Vomiting, spasticity, seizures, coma, retardation	(?)
Disorders of the propionate pathway (see Table 22-1)					

[a]Disorders described in only one patient or one family.

TABLE 22-3. Other Aminoacidopathies with Neurological Lesions

Condition	Biochemical Defect	Neurological Lesion
Aromatic Amino Acids		
Phenylketonuria [4]	Phenylalanine hydroxylase	Retardation, oligophrenia, seizures
Methylmandelicaciduria [48][a]	(?)	Ataxia, seizures, retardation
Tyrosinosis [3, 4, 28]	*p*-Hydroxyphenylpyruvate oxidase	Irritability, anorexia, lethargy, mild retardation
Goitrous cretinism (see Table 22-4)		
Indole Amino Acids		
Tryptophanuria [3, 4][a]	Transport defect (?)	Dwarfism, ataxia, retardation
Tryptophanemia [3, 4][a]	(?)	Retardation, ataxia
Blue diaper syndrome [3, 4]	Transport defect	Irritability, retardation
Hartnup disease [3, 4]	Transport defect	Cerebellar ataxia, tremors, mild retardation
Xanthurenicaciduria [3, 50]	Kynureninase (?)	Infantile spasms, retardation (?)
Hyperserotoninemia [3, 12]	(?)	Instability, ataxia, seizures, retardation
Indolylacrylylglycinuria [3][a]	(?)	Moderate retardation
Imidazole Amino Acids and β-Alanine		
Histidinemia [4, 13]	Histidase	Retardation, speech disorders, hypotonia, seizures
Hyper-β-alaninemia [4, 32]	(?)	Somnolence, lethargy, hypotonia, seizures, retardation
Carnosinemia [4, 63]	Carnosinase	Myoclonic seizures, retardation
Imino Acids		
Hyperprolinemia (type I) [3, 4]	Proline oxidase	Possibly deafness, retardation
Hyperprolinemia (type II) [55]	Pyrroline-5-carboxylate dehydrogenase	Convulsions, retardation

Sulfur Amino Acids and Sulfur Metabolism		
Homocystinuria [4, 54]	Cystathionine synthetase	Retardation, dislocated eye lens
Hypermethioninemia [3, 4]	Probably none	Irritability, drowsiness; secondary to liver disease (?)
Oasthouse urine disease [3, 4]	Transport defect for methionine	Hyperpnea, hypertonia, convulsions, retardation
Cystathioninuria [3, 4]	Cystathioninase	Probably benign
β-Mercaptolactatecystinuria [14][a]	(?)	Retardation
Glutathionemia [21][a]	γ-Glutamyltranspeptidase	Retardation
Sulfite oxidase deficiency [4][a]	Sulfite oxidase	Psychomotor retardation, ectopia lentis
Dibasic Amino Acids (see also Table 22-2)		
Hyperlysinuria [3, 43]	Transport defect	Retardation
Hyperlysinemia [3, 14a]	(?)	Hypotonia, convulsions, retardation
Hydroxylysinuria [3, 23]	(?)	Retardation, seizures, hyperactivity, tremors
Saccharopinuria [3, 4][a]	(?)	Low I.Q., abnormal electroencephalograph
Pipecolatemia [20][a]	(?)	Hypotonia, tremors, retardation
Hyperglycinemias and Miscellaneous Disorders (see also Table 22-1)		
Nonketotic hyperglycinemia [3, 8]	(?)	Seizures, retardation
Sarcosinemia [4][a]	Sarcosine dehydrogenase	Retardation
Hyperglutamatemia [3][a]	(?)	Mental retardation
Aspartylglycosaminuria [47][a]	Cleaving enzyme	Severe mental defect
Trimethylaminuria [26a] (fish-odor syndrome)[a]	(?)	(?)

[a]Disorders described in only one patient or one family.

α-methylacetoacetate, which is cleaved to form propionate and acetate. Valine produces isobutyrate; the isobutyrate is oxidized to β-hydroxyisobutyrate, which in turn is oxidized either to methylmalonate or to propionate. More than a dozen enzyme defects and their variants have been described that affect the catabolism of these branched-chain amino acids and the propionate pathway (see below). Although hypervalinemia and β-hydroxyisovalericaciduria are not associated with metabolic acidosis, they are reviewed here with the other disorders of branched-chain amino acids.

Maple-Syrup Urine Disease. This disease results in grave metabolic disturbances that are lethal in infancy unless they are reversed effectively by prompt treatment. The neurological damage might be mainly secondary to ketoacidosis, hypoglycemia, and precarious nutrition, but the disease also exhibits low platelet serotonin and depletion of glutamic acid and γ-aminobutyric acid in the brain. Furthermore, the neuropathological findings of decreased myelin, status spongiosus, and focal astrocytosis with decreased cerebroside content of the brain suggest specific toxic lesions, perhaps related to ammonia intoxication or to depletion of the biogenic amines. The lesions are much less prominent in infants started on diets in which leucine, isoleucine, and valine are restricted than in untreated patients [56]. In these patients, dietary leucine produces ataxia and convulsions most readily; although the enzyme defect blocks oxidative decarboxylation of the keto acids of all three branched-chain amino acids.

Direct evidence for a toxic metabolite has been provided by Silberberg [56], who showed that α-ketoisocaproic acid is the only metabolite which accumulates in this disease that is toxic to myelinating cultures of newborn rat cerebellum. In his animal experiments, this keto acid derivative of leucine, in concentrations comparable to those found in the blood and brain of patients with maple-syrup urine disease, caused cellular degeneration and inhibited myelin formation, but it did not destroy previously formed myelin. The α-keto acids might interfere with neurological function in another way: by inhibiting decarboxylation of pyruvate and decelerating the Krebs cycle, α-ketoglutarate is no longer available to maintain brain concentrations of glutamate and γ-aminobutyrate.

One variant of maple-syrup urine disease is of intermediate severity; another variant exhibits episodic or intermittent ketoacidosis, which is associated with an intermittent motor disorder [4]. A vitamin-responsive variant has recently been described [53] in which the metabolic disturbance was corrected by large doses of thiamine.

Isovalericacidemia. This condition, which is associated with occasional ketotic attacks, may also be neurotoxic because of α-ketoisocaproate accumulation [4], since this is the immediate precursor of isovalerate.

Several other rare disorders are listed in Table 22-1.

Defects of the Propionate Pathway

Propionate is derived from methionine, threonine, odd-chain and branched-chain fatty acids, part of the pyrimidine nucleus, and the cholesterol molecule. Propionate is carboxylated to methylmalonate by a biotin-carboxylase. After racemization, methylmalonate is isomerized to succinate by a mutase that requires the deoxyadenosyl coenzyme form of vitamin B_{12} as a cofactor. In all defects of the propionate pathway, there is an intolerance to dietary protein, with accompanying ketoacidosis, lacticacidosis, hyperammonemia, and hypoglycemia, and the appearance of long-chain ketone by-products of isoleucine breakdown in the urine. The neurological disorders, which range from mild retardation with seizures to acute neonatal collapse and death, could be attributable to (1) general metabolic disturbances, (2) the accumulation of ammonia, (3) possible α-ketoisocaproic acid toxicity, or (4) other toxic metabolites that may appear in defects of this important pathway. The intermittent hyperglycinemia associated with these disorders has not been shown to have any neurological consequences, despite the fact that glycine is an important neuroinhibitory substance (see Chap. 8).

Propionicacidemia. The condition formerly known as ketotic hyperglycinemia has been identified as propionicacidemia [26]. Spongy degeneration of cerebral white matter was found in five patients from two families who appeared to have this metabolic disorder [51]. A patient has been described with a milder, nonketotic variant, who nonetheless was retarded [35]. Another patient has been reported to have a biotin-responsive variant [7].

Methylmalonicacidemia. Methylmalonicaciduria was first noted to occur in vitamin B_{12} deficiency. Inborn defects of metabolism are now known to occur in the mutase enzyme and in a vitamin B_{12}-responsive variant. This variant is not due to faulty apoenzyme-coenzyme interaction, but to defective formation of the B_{12} coenzyme. Another variant, occurring with homocystinemia and abnormalities of transmethylation as well as of methylmalonate isomerization, has been shown to be caused by an earlier block which affects the biosynthesis of both the mutase coenzyme and the transmethylation coenzyme, methyl B_{12} [35a].

Vitamin B_{12} deficiency, with its resultant blocking of propionate and methylmalonate catabolism (see Chap. 25), is associated with long-tract degenerations of the spinal column, peripheral neuropathies, and cerebral derangements, as well as megaloblastic anemia. These lesions must be caused by effects of vitamin B_{12} deficiency remote from the propionate pathway since they do not occur in methylmalonicacidemia. In vitamin B_{12}-responsive methylmalonicacidemia, therapy with restricted diet and B_{12} injections has resulted in dramatic reversal of developmental retardation [25].

Defects of Lactate and Pyruvate Metabolism

Decarboxylation of pyruvate to acetyl-CoA is by a multienzyme complex with many cofactor requirements, including thiamine pyrophosphate, NAD, lipoic acid, and CoA. This is the final step between the glycolytic breakdown of carbohydrate and the Krebs cycle. A block here would lead to pyruvicacidemia and lacticacidemia since pyruvate is readily converted to lactate by lactic acid dehydrogenase. A block would also lead to increased levels of serum alanine, which is formed from pyruvate transamination. A block beyond acetyl-CoA would lead to ketosis, because entry of fatty acids into the Krebs cycle would also be blocked. Carboxylation of pyruvate to oxaloacetate is a key reaction in priming the Krebs cycle and in gluconeogenesis. A block of this reaction would lead to lacticacidosis and ketosis too.

Besides the acute increased production of lactic acid in strenuous exercise, hypoxia, and some malignant conditions such as leukemia [42], lacticacidosis is a feature of several groups of metabolic disorders, including some of the aminoacidopathies, glycogen storage diseases, and a group of conditions in which no other primary biochemical lesions have been identified. There are probably several different related conditions in this last group, and some may represent primary disorders of lactate and pyruvate metabolism.

Familial Lacticacidosis. This clinical entity, in which no enzyme defect has been identified as yet, is considered to be caused by a partial defect of pyruvate decarboxylation. The lacticacidosis is episodic and associated with normal pyruvate-lactate ratios. The clinical features, although varying in severity, consist of tachypnea, marked muscular hypotonia, tetany, convulsions, and mental retardation. The neuropathological findings have included dilated lateral ventricles, necrotizing encephalopathy, and spongy degeneration of the brain [22].

Subacute Necrotizing Encephalopathy. The neuropathological lesions in this condition are reminiscent of those found in Wernicke's encephalopathy, which are attributed to thiamine deficiency. Lacticacidosis and pyruvicacidosis, which occur in thiamine deficiency, have both been found in this disease. It is therefore thought to be caused by defective pyruvate or thiamine metabolism. One patient was reported to have had a defect in the carboxylation of pyruvate to oxaloacetate [24]. A curious block in the conversion of thiamine pyrophosphate to thiamine triphosphate has been found in the brain tissue of another patient who died with this disease [46]. The function of thiamine triphosphate in mammalian brain is unknown, but it may be the neurophysiologically active form of thiamine, and it may play a role in ion movement in nervous tissue (see Chap. 25). These metabolic disorders characteristically cause necrosis and macrophage response in the regions of the mamillary bodies, hypothalamus, and brain stem; they cause destruction of myelin but, remarkably, isolated neurons and axons are spared. These lesions are not compatible with ischemia or edema and are strongly indicative of focal neurotoxic damage [17]. Spongy degeneration and demyelination of the white matter, although it often occurs in these conditions, may be a nonspecific agonal result of carbon dioxide formation from glycogen breakdown.

Attempted therapy of chronic lacticacidosis and of necrotizing encephalopathy has included use of lipoic acid, pantothenic acid, thiamine, and nicotinamide, none of which has resulted in convincing benefit, except for some chemical lowering of blood pyruvate with lipoic acid and abolition of a serum inhibitor of thiamine triphosphate formation by large doses of thiamine.

Pyruvicacidemia with Alaninemia. A metabolic defect in pyruvate decarboxylation, with elevated blood and urine pyruvate, lactate, and alanine, has been described [34]. This condition is associated with optic atrophy, intermittent episodes of cerebellar ataxia, and mental confusion [10]. It has been reported that one patient responded chemically to large doses of thiamine [34].

Another condition has been reported in two sisters, who were microcephalic and severely retarded and who had lacticacidosis with pyruvicacidemia and alaninemia. In infancy, they developed frank diabetes mellitus, which responded to insulin [60]. It remains to be determined whether these conditions are related to the other metabolic errors of pyruvate and lactate metabolism and how variants of these disorders are related to each other.

Other Inherited Causes of Metabolic Acidosis

Succinyl-CoA:3-Ketoacid-CoA Transferase Deficiency. One infant with severe ketoacidosis has been described whose condition was temporarily improved by a protein-free diet. After the infant succumbed, his brain and other tissues were found to lack CoA transferase [62].

Renal. Any inherited disorder that causes chronic renal failure, such as the polycystic kidney diseases, will lead to renal acidosis and uremic poisoning, with blunting of the intellect. Some inherited causes of renal disease are associated with peculiar neurological features.

Cystinosis. This aminoacidopathy causes renal failure in infancy with failure to thrive and renal dwarfism. Although the major damage in cystinosis is to renal tubular function, it is a generalized defect in which cystine accumulates in subcellular lysosomal particles. It is associated with photophobia and a retinopathy. Possibly, cystine accumulation in the eye produces the retinopathy responsible for the photophobia [4].

Oculo-Cerebro-Renal Syndrome, or Lowe's Syndrome. This is an ill-defined entity, associated with cataracts, hypotonia, seizures, mental retardation, and renal lesions that lead to renal acidosis and renal rickets. Its biochemical basis is obscure, as is the reason for association of the renal disease with cataracts and brain abnormalities [3].

HYPERAMMONEMIA

In the liver, urea is synthesized from amino nitrogen via the arginine cycle. Carbamyl phosphate is synthesized from bicarbonate, an amino radical, and high-energy phosphate. It is then transcarbamylated with ornithine to form citrulline, which is condensed with aspartate to form argininosuccinate. Fumarate is then split off to form arginine, which liberates urea to regenerate ornithine. Ammonia can be formed from urea, notably in the kidney, or by deamination of amino acids, and it may accumulate if the urea cycle is defective. How it accumulates in other inborn errors, such as the propionicacidemias (see above), is not yet clear.

Blood ammonia levels are elevated in acute liver failure or when there is marked portosystemic shunting, due either to chronic portal hypertension or to surgical shunting operations. The portal encephalopathy that ensues, with irritability, disorientation, a flapping tremor, and coma, is probably caused by the toxic effects of ammonia on the brain since this substance has been shown experimentally to be epileptogenic and to produce coma rapidly (see Chap. 15). Some inherited disorders, such as tyrosinosis or galactosemia, may result in secondary hyperammonemia and neurotoxicity because they lead to severe liver damage. The key histopathological feature of ammonia toxicity appears to be nuclear swelling of the astrocytes [58]. Similar neuropathological effects along with arrest of mental development occur in infants with inherited disorders of urea cycle enzymes. It is odd that of these enzymes, only argininosuccinate synthetase, argininosuccinase, and arginase are present in the brain, where their function is obscure.

Carbamyl Phosphate Synthetase. This enzyme has been reported to be lower than normal in one patient who had vomiting, ketosis, hyperglycinemia, and hyperammonemia, which were exacerbated by a high nitrogen intake. The possibility that this patient actually had propionicacidemia has not been convincingly excluded [18].

Ornithine Transcarbamylase. Low levels of this enzyme have been found in the livers of infants with a syndrome of massive hyperammonemia and protein intolerance. The blood ammonia levels, which may be over 1,000 μg per 100 ml, appear to be solely responsible for the neurotoxicity of this disease. It can be lethal in the neonatal period, and its incidence is likely to be higher than published data would indicate. Generally, this active enzyme is present in affected female patients at only 10 percent of the normal amount, which should not be rate-limiting for the urea cycle, but infant boys with almost complete absence of this enzyme have been reported [11]. Also, urea formation and turnover are not totally blocked in this or any of the other defects of the urea cycle. Treatment of affected girls with a low-protein diet has led to biochemical improvement and some recovery from the retardation caused by this disease, even when the ammonia levels in blood are not fully reduced to normal.

Citrullinemia. Argininosuccinic acid synthetase is abnormally low in this disorder. The mutant enzyme has a much lower affinity for the substrate than the normal one. The clinical presentation includes vomiting, irritability, seizures, and mental deterioration associated with a transient, severe, postprandial hyperammonemia. Low blood urea has been found in one patient with citrullinemia. There is no evidence that citrulline itself is toxic. Attempts to induce increased activity of this enzyme with thyroid extract appeared to benefit one patient [4]. Recently, another patient with citrullinemia – a retarded adult – was described. He had normal blood urea and ammonia, and excreted large amounts of homocitrulline and homoarginine [52].

Argininosuccinicaciduria. This is the most distinctive disorder of the urea cycle. All patients are severely retarded, most have ataxia and seizures, and the hair has a characteristic nodular friability, trichorrhexis nodosa, with abnormal composition. There are gross increases of argininosuccinic acid in blood, cerebrospinal fluid, and urine, with elevated postprandial blood ammonia. Increased citrulline was found in the blood of patients from one family. The cerebrospinal fluid concentration of argininosuccinic acid normally exceeds that in the blood, and argininosuccinase has been shown to be defective in all neural tissue in this disease. The brain shows extensive failure of myelination, which might be due to a toxic effect of argininosuccinate accumulation [58].

Argininemia. One inbred family has been reported in which two siblings had higher than normal blood and cerebrospinal fluid arginine concentrations, hyperammonemia, and a pattern of increased lysine and cystine excretion in the urine. They were retarded, had spastic diplegia, and suffered from seizures. Unfortunately, no information was given about urea levels in blood, but arginase, which cleaves arginine to urea and ornithine, was defective in the patients' blood cells [61]. An attempt has been made to infect the affected patients with Shope papilloma virus in order to introduce viral arginase, in the hope that this would reverse their metabolic defect.

Hyperornithinemia. This has been found in one boy who showed irritability, anorexia, screaming attacks, myoclonic seizures, and mental deterioration. Blood ornithine and ammonia were raised, and homocitrullinuria was present. The precise metabolic defect has not been located in this patient, but he improved significantly on a low-protein diet. Perhaps the myoclonic spasms were due to a toxic effect of ornithine; the other neurological features are indistinguishable from those of ammonia intoxication [4].

Familial Protein Intolerance. Several Scandinavian families have been reported in whom there is no defect of a urea cycle enzyme, but in whom hyperammonemia followed high protein meals. Lysinuria and argininuria, low blood lysine, arginine, and ornithine, and intolerance to alanine loading were found. The defect may be of glutaminase [36] because glutamine is a major supplier of amino nitrogen for urea biosynthesis.

Lysine Intolerance. Other families with protein intolerance show marked elevation of blood lysine, arginine, and ammonia. Vomiting, seizures, episodic coma, and retardation are also features of this disorder, which has been shown to be due to a defect of lysine dehydrogenase [4]. Although lysine inhibits arginase, in other conditions with elevated blood lysine there is no elevation of arginine and ammonia (see Table 22-3 and p. 471). Toxicity in this condition might be due to the hyperammonemia.

OTHER AMINOACIDOPATHIES ASSOCIATED WITH NEUROLOGICAL DISORDERS

Aromatic Amino Acids

Phenylketonuria. Formerly known as oligophrenic phenylpyruvic idiocy, this is the most common of all the neuropathological aminoacidopathies. The disease should become only a clinical curiosity if neonatal screening programs and preventive dietary treatment are successfully instituted. The clinical picture, in untreated patients, is of eczematous, severely retarded, microcephalic, hyperactive, hypertonic individuals with pigment dilution, who are permeated with a pungent, mousy odor.

The central nervous system shows generalized failure of myelination. The basic biochemical defect is of a component of the hepatic enzyme complex that hydroxylates phenylalanine to tyrosine. The untreated disorder is associated with high phenylalanine in blood and urine, depressed concentrations of the other amino acids, low blood levels of serotonin, and increased production of by-products of phenylalanine, such as phenylpyruvic acid and phenylacetic acid (see also Chap. 10).

The neurological lesions in phenylketonuria all appear to be secondary to accumulation of phenylalanine and its derivatives, although the specific neurotoxic agent and its mode of action have yet to be identified. Many possibilities have been proposed. Excess phenylalanine and phenylpyruvate (1) are supposed to inhibit brain uptake of tryptophan [55a] or to suppress 5-hydroxytryptophan decarboxylation to serotonin, (2) depress tyrosine levels in blood and nervous tissue and thus may decrease formation of catecholamines, (3) have been reported to disaggregate brain polyribosomes and hence may impair protein synthesis, (4) suppress energy metabolism in brain by inhibiting hexokinase and pyruvate kinase [18a, 64], and (5) decrease cerebroside synthesis. In addition, glutamine, which was low in the blood of one retarded sibling with phenylketonuria but normal in his affected but unretarded brother [45], and homocarnosine accumulation have both been postulated to be the cause of brain damage in this disorder.

Pregnancies of phenylketonuric mothers have a high incidence of abortions and fetal malformations. Infants who survive are all severely retarded, although none of them has had the homozygous phenylketonuric genotype. Hence, whatever the biochemical mechanism may be that leads to retardation in phenylketonuria, it can harm even the unborn child of a phenylketonuric mother by altering his environmental milieu in utero.

The neuropathological consequences of phenylketonuria, although not reversible in later life, are totally preventable by the careful administration, starting with the newborn period, of a diet limited in phenylalanine. Furthermore, evidence is accumulating which indicates that optimal dietary control of a pregnant mother with phenylketonuria may protect her unborn child from damage. One result of the mass screening programs for phenylketonuria is the discovery of variants of this disorder which are biochemically milder and which are not associated with neurological damage [4]. These variants probably do not require dietary control, but it is difficult diagnostically to differentiate them safely from true phenylketonuria in the newborn infant.

Methylmandelicaciduria. Two brothers with ataxia, seizures, and mental retardation were found to secrete methylmandelic acid. Their symptoms were alleviated by a low-protein diet. Return to a normal protein diet induced nystagmus and ataxia within a few days, but increasing phenylalanine intake caused convulsions within 12 hr. Two older brothers may have died of the same disorder [48]. The mechanism of neurotoxicity is obscure. Mandelic acid has been postulated to be derived from phenylalanine via phenylethanolamine.

Tyrosinosis. This condition is not associated with any general metabolic disturbance except hypoglycemia and abnormalities secondary to liver damage. Tyrosinosis has been mimicked by fructosemia [33] (see p. 475). It is usually not associated with neurological lesions, but one patient has been reported who developed the biochemical and clinical features of porphyria, with abnormal movements, convulsions, loss of consciousness, and possibly barbiturate hypersensitivity. She had elevated δ-aminolevulinate synthetase activity as well as *p*-hydroxyphenylpyruvate oxidase deficiency. Increased δ-aminolevulinate synthetase is believed to have been induced by an abnormal metabolite of tyrosine and to have caused the porphyria [28].

Goitrous Cretinism. Defects of iodine and tyrosine metabolism that prevent biosynthesis of the thyroid hormones present as goitrous cretinism, with the neurological sequelae of hypothyroidism in an infant.

Thyroid hormone is essential for normal energy-yielding reactions, its role apparently being to control oxidative phosphorylation. Deficiency of the hormone decelerates metabolic activities and may cause hypoglycemia. The nervous system appears to be highly dependent upon normal supplies of this hormone for its activity. For instance, an index of the hypothyroid state is a delay in muscle relaxation after stimulation of the tendon stretch reflex, which indicates that thyroid is required for normal neuromuscular reactions. When there is thyroid deficiency in utero or in an infant, severe retardation of all growth processes, delayed neural maturation, and brain hypoplasia result. These lesions can be reversed in their early stages by supplying exogenous thyroid.

Endemic iodine deficiency prevents synthesis of thyroxine and triiodothyronine, leading to the goitrous enlargement of the thyroid gland. When there is no iodine deficiency, the formation of hormonally active iodinated thyronines depends on (1) normal transport and concentration of iodides in the thyroid gland, (2) oxidation and organification of iodine, and formation of monoiodotyrosine and diiodotyrosine, and (3) the coupling of iodotyrosines to form the iodothyronines. Genetically determined enzyme defects that cause goitrous cretinism have been described at each of these stages. Also, failure of deiodination and salvage of iodine from the iodotyrosines and an abnormal thyroxine-binding protein in the blood have been recognized as causes of familial goiters [5] (Table 22-4).

Indole Amino Acids and Related Compounds

Tryptophanuria, Tryptophanemia, and the "Blue-Diaper Syndrome." These are errors in tryptophan transport or metabolism. Whether there is a causal association between these conditions and neurological lesions is doubtful [3, 4]. Tryptophan is the precursor of serotonin (5-hydroxytryptamine) via 5-hydroxylation and decarboxylation. It usually provides about half the daily requirement for nicotinic acid, one of its degradation products.

Hartnup Disease. This condition is due to a defect in intestinal and renal transport of tryptophan and the other neutral amino acids: alanine, serine, threonine, valine, leucine, isoleucine, phenylalanine, tyrosine, and histidine. Bacterial breakdown of intestinal tryptophan gives rise to indicanuria and indoluria. Either because

TABLE 22-4. Enzymatic Defects of Thyroxine Metabolism

Condition	Cause
Iodide-transport defect; mild goitrous cretinism	Inability to concentrate inorganic iodide in thyroid, salivary glands, or gastric mucosa
Organification defect; goitrous cretinism with dwarfism	Readily displaceable organic iodide stores in the thyroid gland
Pendred syndrome; nerve deafness, euthyroid state	Partially displaceable iodide stores; iodine transferase defect (?)
Coupling defect; goitrous cretinism	Inability to couple monoiodo- and diiodotyrosine to form iodothyronines
Deiodinase defect; goitrous cretinism with dwarfism	Inability to salvage iodine from monoiodo- and diiodotyrosine with depletion of body iodine stores
Abnormal serum thyroxine-binding protein; goitrous cretinism	Inability to release protein-bound thyroxine normally

Source: From [5].

of toxic inhibition by these bacterial breakdown products or because of persistent inadequate absorption of tryptophan leading to nicotinamide deficiency, some patients with this disorder develop late cerebellar ataxia, tremors, and mild retardation. The symptoms may improve either spontaneously or upon treatment with nicotinamide [4]. The presumption is that the mild neurological lesions, like the pellagra rash that appears in Hartnup disease, is due to a secondary nicotinamide deficiency that interferes with generalized energy metabolism, as this vitamin is a constituent of the NAD coenzymes.

Kynurenicacidurias. In pyridoxine deficiency, there is an abnormality of tryptophan metabolism, and the kynurenines – 3-hydroxykynurenin and xanthurenic acid, which are breakdown products of tryptophan – are increased in the urine. In xanthurenicaciduria there is a defect of kynureninase, which converts kynurenine to anthranilic acid, and 3-hydroxykynurenine to 3-hydroxyanthranilic acid. Clinically, this condition may be associated with seizures and early retardation. In at least two brothers with xanthurenicaciduria, in which a defect of kynureninase was associated with abnormal tryptophan metabolism and who had severe retardation, large doses of pyridoxine temporarily corrected the metabolic abnormalities. Defective hepatic kynureninase from one of these brothers was found to be activated by large amounts of its cofactor, pyridoxal phosphate [50]. How these defects of tryptophan catabolism cause retardation is not known.

Hyperserotoninemia. 5-Hydroxytryptamine (serotonin) has been reported to be abnormally elevated in one retarded girl who had hypertonia and hyperreflexia [12]. Other inborn errors of metabolism might have their effects on the central nervous system by altering the activity and turnover of this substance.

Imidazole Amino Acids and Beta-Alanine

Histidinemia. This condition, due to a defect of histidase, has been believed to be associated with a peculiar speech disorder. Since patients who have no retardation or speech disorder have been demonstrated to have this condition, the association may be spurious. Platelet serotonin, however, is reduced in histidinemia, and it is restored by dietary restriction of histidine. The retardation, therefore, might be due to depletion of serotonin in the brain, and thus have a common basis with phenylketonuria, maple-syrup urine disease, and some of the disorders of indole metabolism [13].

Beta-Alaninemia. Somnolence, lethargy, and seizures in one patient, deafness in another, as well as hypotonia and retardation have been reported as occurring in conjunction with persistent elevation of β-alanine in the blood, brain, and urine. In one patient, there was an excess of taurine and β-aminoisobutyric acid in urine, γ-aminobutyric acid was increased in brain and urine, and carnosine (β-alanylhistidine) was increased in all tissues. The accumulation of carnosine was presumably secondary to the block in β-alanine metabolism. The neurotoxic effect of β-alanine may have been due to inhibition of the metabolism of γ-aminobutyric acid, or perhaps a common transaminase, acting on both substrates, was defective. Another patient with only β-alaninuria may have had a different defect [32].

Carnosinemia. Carnosine is synthesized from β-alanine and histidine, and homocarnosine is synthesized from γ-aminobutyric acid and histidine. Both of these dipeptides are present in human brain, but their biological function is not known. Patients have been described with twitching, progressive mental deterioration, and myoclonic seizures who have had carnosinuria or carnosinemia resulting from deficiency of carnosinase [63]. The similarities between β-alaninemia and carnosinemia and their close association with homocarnosine strongly suggest that the neurotoxic effects of these disorders are all via carnosine accumulation or disturbed γ-aminobutyric acid metabolism. Homocarnosine content also was very low in the brain of a patient with pipecolatemia [20] (Table 22-3).

Imino Acids

Hyperprolinemia Type I. Hereditary renal disease, deafness, and mental retardation have been associated with defective proline oxidase, which converts proline to pyrroline-5-carboxylic acid. Hyperprolinemia then results in urinary excretion of proline, hydroxyproline, and glycine. Some siblings of affected patients show hyperprolinemia without any neurological or renal abnormalities, so these associations with hyperprolinemia may be coincidental [3, 4].

Hyperprolinemia Type II. This variant is characterized by excretion of pyrroline-5-carboxylic acid and by greater elevations of blood proline than in type I. The postulated defect is the dehydrogenation of pyrroline-5-carboxylic acid to glutamic acid. The condition has been associated with convulsions and mild retardation. The fact that proline is degraded to glutamic acid suggests that glutamate equilibrium also might be disturbed in this condition. Normally pyrroline-5-carboxylic acid and ornithine are reversibly interconverted, so in hyperprolinemia the urea cycle may also be disturbed [55].

Sulfur Amino Acids and Sulfur Metabolism

Homocystinuria. This is associated with marfanoid habitus and ectopia lentis, and it is often associated with mild mental retardation, as well. It is probably the second most common aminoacidopathy after phenylketonuria. Homocystine and methionine are elevated in the blood and cystine is lowered. Cystathionine, normally abundant, is virtually absent from the brain tissue of these patients. Cystathionine synthetase, which catalyzes the condensation of homocysteine and serine, is defective in this condition. Sometimes this enzyme can be reactivated in vitro by large supplements of its cofactor, pyridoxal phosphate, and in vivo by pyridoxine. Some patients have been reported in whom pyridoxine supplements have caused chemical improvements without directly reactivating cystathionine synthetase. However, these patients showed adverse neuropsychiatric reactions, raising the possibility that abnormal alternate pathways of sulfur amino acid metabolism had been stimulated [54]. The role of cystathionine in the central nervous system is not known, so whether its depletion in this condition is neurotoxic remains an open question.

Oasthouse Urine Disease. This syndrome, resulting from defective methionine absorption, is associated with hyperpnea, convulsions, and retardation. The patients are reported to have white hair, edema, and an unpleasant odor described as resembling that of dried celery. Dietary methionine causes diarrhea in this syndrome, with excess methionine and α-hydroxybutyric acid appearing in the stool and urine, as well as urinary excesses of phenylalanine, tyrosine, the branched-chain amino acids, and their keto acids. The basic defect appears to be in intestinal methionine transport, analogous to the defect of neutral amino transport in Hartnup disease [3, 4]. The neurotoxic consequences of this disease may be due to intracellular deficiency or depletion of this essential amino acid. Methionine is the chief donor of methyl groups in the biosynthesis of choline and epinephrine. It is also the methyl donor for catecholamine inactivation through orthomethylation (see Chap. 6).

Several other rare disorders are listed in Table 22-3.

Dibasic Amino Acids

Hyperlysinuria. This has been reported to be part of a defective transport of dibasic amino acids in two different disorders. One family had dominant inheritance of lysinuria with no mental abnormality; the other variant had a recessive pattern of inheritance, and some of the affected children were retarded both mentally and physically. In this second type, not only was there an increased excretion rate of cystine, lysine, arginine, and ornithine, but homocitrulline, probably a derivative of lysine, was also excreted in excess [4]. The retardation in these patients might have been caused by a cellular deficiency of the dibasic amino acids. One infant who has been treated since birth with a low lysine diet has not developed any psychomotor complications [31].

Hyperlysinemia. Other than in the syndrome of lysine intolerance with hyperammonemia (see Table 22-2 and above), lysinemia has been described in additional patients with inconstant clinical features. Some patients have been severely retarded, with hypotonia and elevated levels of lysine in blood, cerebrospinal fluid, and urine, as well as evidence of an undefined block in lysine catabolism [43]. Whether elevated lysine in the brains of these patients is toxic is unknown. It is the only amino acid that does not contribute to the common amino pool; its degradative pathway in mammals is still incompletely defined.

Hydroxylysinuria. Five patients in three families have been reported to excrete hydroxylysine. The clinical syndrome is of seizures in childhood and profound retardation, with great restlessness. Of four similarly retarded siblings in one family with excitability, tremors, myoclonus, and seizures, only two had hydroxylysinuria. In another family, one child had similar neurological features. In a third, two brothers had seizures and retardation with hyperactivity [23]. Hydroxylysine is a constituent of connective tissue and it is excreted normally in the urine in small amounts as an oligopeptide.

Glycine

Hyperglycinemia. Elevation of blood glycine in ketotic hyperglycinemia is clearly secondary to propionicacidemia or methylmalonicacidemia (see above). Some patients with nonketotic hyperglycinemia who have seizure disorders or retardation may also have propionicacidemia [35]. Others with seizures and severe retardation associated with nonketotic hyperglycinemia, have no defect of propionate metabolism, but they have been shown to have defective oxidation of glycine. It has been postulated that such patients have a defect in methyl transfer from the C-2 carbon of glycine [8]. The original suggestion of a block in oxalate formation has been refuted.

One family has been reported in which three brothers had progressive combined upper and lower motor neuron disease with isolated hyperglycinemia. No cause for the hyperglycinemia was found, and its association with the neuromuscular disorder has not yet been determined [37].

Other rare disorders – such as saccharopinuria, pipecolatemia, and sarcosinemia – have been described [3, 4, 20, 47] and are listed in Table 22-3.

INBORN DISORDERS OF VITAMIN METABOLISM

The B vitamins are either cofactors or precursors of cofactors for enzyme reactions (see Chap. 25). Many enzymes of intermediary metabolism utilize these cofactors; for instance, pyridoxal phosphate in transamination reactions and biotin in carboxylation reactions. Of the disorders of amino acid and carbohydrate metabolism that involve enzymes with cofactors, a few are known in which apoenzyme-cofactor interaction or cofactor biosynthesis is abnormal and in which excess of the cofactor or its precursor has improved the metabolic block [50]. These include pyridoxine-responsive cystathioninuria, homocystinuria, and xanthurenicacidemia; thiamine-responsive maple-syrup urine disease; biotin-responsive propionicacidemia; biotin-responsive β-methylcrotonylglycinuria [20a]; and vitamin B_{12}-responsive methylmalonicacidemia, each of which has been reviewed in its appropriate section.

PYRIDOXINE DEFICIENCY AND DEPENDENCY

Dietary deficiency of pyridoxine in infants causes seizures that are cured by the restoration of pyridoxine in their diets. One clinical syndrome of neonatal seizures without pyridoxine deficiency responds dramatically to pyridoxine, and retardation is prevented by prolonged pyridoxine treatment. This might be due to an abnormality of brain glutamic acid decarboxylase (see Chap. 25), which is the enzyme that forms γ-aminobutyric acid from glutamate (depletion of γ-aminobutyrate has been reported to lead to seizures). There is evidence that the excitatory action of glutamate and the inhibitory action of γ-aminobutyrate on nerve membranes are both dependent on the direct mediation of pyridoxal phosphate [59], so the action of pyridoxal phosphate might be faulty in pyridoxine-dependent seizures. This syndrome is therefore of great interest (1) clinically, because it is a preventable cause of seizures and retardation; (2) biochemically, because it is an example of vitamin dependency in which only one of the many enzymes for which pyridoxal phosphate is a cofactor is involved, and in which the defect can be overcome by increasing the concentration of the cofactor; and (3) neurologically, because it might be related to the vital importance of γ-aminobutyric acid in neural transmission (see also Chap. 8).

FOLATE METABOLISM

Folic acid is of fundamental importance in protein and nucleotide biosynthesis (see Chap. 25). It is thought to be absorbed as a polyglutamate conjugate, and it is present at higher concentrations in cerebrospinal fluid than in plasma. Folic acid deficiency can be caused by poor intake, by malabsorption, or by use of antifolate drugs, including some anticonvulsants. The deficiency state in adults is associated with megaloblastic anemia.

Inborn errors of folate absorption have been reported in two families: in one, there were two siblings with megaloblastic anemia, ataxia, physical and mental retardation, and convulsions; in another, there was a young adult, who, since infancy, had suffered from megaloblastic anemia, athetotic movements of the upper limbs, severe retardation, seizures, and calcification of the basal ganglia [30]. In both conditions, there was a deficient transport of folate across both intestinal and cerebrospinal membranes. In the young adult, correction of anemia with large doses of folate aggravated the seizures, suggesting that folate lowered the seizure threshold in susceptible patients and that some anticonvulsants may act primarily because of their antifolate properties.

In Japan, five infants have been found with high serum folate levels, physical and mental retardation, and cerebral atrophy [6]. Four of these are thought to have formimino transferase deficiency and one to have methyltetrahydrofolate transferase deficiency. One patient with anemia, but no neurological abnormalities, probably had deficiency of folic acid reductase. Experimentally, folate deficiency in infancy appears to delay maturation, as indicated electroencephalographically, and folate antagonists in pregnant animals have produced microcephaly and dilated cerebral ventricles in fetuses. This supports the hypothesis that folate is essential for purine biosynthesis and brain growth in utero.

DISORDERS OF CARBOHYDRATE METABOLISM

In contrast to the numerous aminoacidopathies that are known to be associated with neurological disorders, fewer genetic defects of carbohydrate metabolism have been discovered so far that result in specific neurological damage. Disturbances of carbohydrate metabolism which lead to hypoglycemia (see Table 22-5 and Chap. 9) may all interrupt neuronal function and cause seizures, coma, permanent brain damage, or death, since neural tissue requires a constant uninterrupted source of energy, usually glucose [2]. Chronic hypoglycemia, however, may permit metabolic adjustments to increase blood concentrations of lactate, ketones, or fatty acids, all of which can be utilized by brain cells as substituents for glucose [57].

DIABETES MELLITUS

Diabetes mellitus is by far the most common of all the disorders reviewed in this chapter. Yet, despite extensive investigations into its biochemical pathology, its pathogenesis remains a matter for dispute [5], and the causes of its many associated neurological syndromes are not clear. Acute irrational behavior, loss of consciousness, and even permanent brain damage or death can result from the metabolic disturbances which produce ketoacidosis and hyperosmotic hyperglycemia in this disease (or from hypoglycemia due to insulin overdosage). In addition, peripheral neuropathy, retinopathy, and autonomic nervous system and certain central nervous system complications of diabetes mellitus are likely to be secondary to focal vascular degeneration, which is a characteristic of this disease.

TABLE 22-5. Causes of Hypoglycemia

Dietary:
Prolonged starvation; amino acid imbalances; vitamin B deficiencies; high carbohydrate diets

Drugs and Toxic Agents:
Insulin; sulfonylureas; diguanides; salicylates; ganglion blocking agents; ethyl alcohol; methyl alcohol; akee nuts

Endocrine:
Hypopituitarism, hypoadrenalism; congenital adrenal hyperplasia; hypothyroidism
Hyperinsulinism; reactive (postprandial) hypoglycemia; insulin hypersensitivity; islet-cell hyperplasia or tumors; insulin-producing malignancies

Inborn Errors of Metabolism:
Disorders of carbohydrate metabolism (see Table 22-6); inborn errors of branched-chain amino acids and of propionate pathway (see Table 22-1); tyrosinosis

Hepatic and Gastrointestinal:
Acute liver neurosis; viral hepatitis; Reye's encephalopathy; postgastrectomy dumping syndrome; malabsorption syndromes

Other Causes:
Transient hypoglycemia of infancy; idiopathic hypoglycemia; ketotic hypoglycemia; leucine sensitivity; Beckwith's macroglossia and organomegaly; acute encephalitis; focal lesions of brain stem or hypothalamus; bacteremias; viremias; malignancies

Source: From [2].

GALACTOSE METABOLISM

Galactosemia

This is the most important cause of infant mortality or severe mental retardation among the inherited disorders of carbohydrate metabolism (Table 22-6). The block is in the conversion of galactose 1-phosphate to glucose 1-phosphate via their uridine diphosphate derivatives. Upon feeding milk, galactose accumulates in body fluids, and galactose 1-phosphate accumulates intracellularly. Possibly maternal blood galactose can be toxic in utero, but the major damage is in postnatal life. Acute symptoms in galactosemia include vomiting, enteritis, seizures, hypotonia, and coma. There is liver damage, kidney damage, and cataract formation. If the child does not succumb, severe brain damage will follow unless a galactose-free diet is started promptly. Part of the metabolic effect of galactosemia is due to dehydration, acute liver failure, and hypoglycemia. The retardation and neuropathology in galactosemia are consistent with the combined effects of hypoglycemia and acute cerebral edema followed by secondary degenerative changes. Cataract formation and cerebral edema apparently arise because of intracellular accumulation of galactose 1-phosphate or of galactitol, an alcoholic derivative of galactose, with subsequent osmotic disruption of cellular membranes [27]. The neurological consequences of this metabolic error are completely avoidable with continual administration of a galactose-free diet from early infancy.

Galactokinase Deficiency

A rarer disorder, this defect prevents the phosphorylation of galactose. It is also associated with cataract formation, confirming experimental data that the eye lesion, at least, can be due to galactitol and not galactose 1-phosphate accumulation. Brain damage has not been proved to be a feature of galactokinase deficiency, implying that galactose 1-phosphate, and not galactose or galactitol, may be the toxic cause of brain damage.

FRUCTOSE INTOLERANCE

Hereditary fructosemia is due to a defect of the aldolase that splits fructose 1-phosphate to glycerol and a triose phosphate. There is severe gastrointestinal upset upon ingestion of fructose, followed by hypoglycemia and metabolic acidosis. If continued, dietary fructose will lead to liver and kidney damage. Patients may develop hypermethioninemia secondary to liver failure and even an aminoacidopathy that mimics tyrosinosis, with depression of *p*-hydroxyphenylpyruvate oxidase activity, presumably caused by the biochemical derangement of fructose metabolism. One child had apneic spells and died after generalized convulsions. His brain showed retarded myelination and lipid accumulation in glial cells [33].

GLYCOGEN STORAGE DISEASES

The hepatic glycogenoses [4a] (Table 22-6), of which glucose 6-phosphatase deficiency is the most severe, are characterized by a hypoglycemia that may be profound and life-threatening. Compensatory mechanisms, however, can lead to an asymptomatic metabolic state with blood glucose levels as low as 15 mg per 100 ml, but with no permanent detriment to brain development. In these conditions, the neurological symptoms again appear to be due entirely to metabolic blocks that prevent normal maintenance of blood sugar. The ability to mobilize adequate compensatory mechanisms demonstrates that the brain can adapt to other energy fuels satisfactorily [57], that brain damage is not inevitable with chronic hypoglycemia, and that lacticacidosis and ketoacidosis are not necessarily toxic to the nervous system.

Type II glycogenosis, in which there is deficiency of lysosomal acid maltase, is associated with severe systemic disease and with amyotonia, constipation, and occasionally mental retardation. Glycogen accumulates in intracellular vacuoles

TABLE 22-6. Inherited Disorders of Carbohydrate Metabolism Associated with Neurological Dysfunction

Condition	Biochemical Defect	Diagnostic Features	Metabolic Disturbances	Neurological Lesions
Galactosemia	Galactose 1-phosphate uridylyl transferase	Cataracts; hepatomegaly and jaundice; galactosuria. Confirmed by enzyme assay in red cells	Dehydration; hypoglycemia	Vomiting, lethargy, seizures, coma, retardation
Fructose intolerance [33]	Fructose 1-phosphate aldolase	Violent gastrointestinal reaction to fructose ingestion; symptoms of hypoglycemia; fructosuria	Hypoglycemia; hypophosphatemia; hyperuricemia	Vomiting, drowsiness, coma, retardation
Glycogenosis O [4a]	Glycogen synthetase	Hepatomegaly, hypoglycemia	Hypoglycemia	Symptoms of hypoglycemia; retardation
Glycogenosis I [4a] Von Gierke's	Glucose 6-phosphatase	Hepatorenomegaly, severe hypoglycemia. Diagnosis by liver or kidney biopsy	Hypoglycemia; lacticacidosis; ketoacidosis; hyperuricemia; lipidemia	Symptoms of hypoglycemia; hypotonia; retardation
Glycogenosis II (Pompe's)	Lysosomal acid maltase	Cardiomegaly; generalized organomegaly. Diagnosis by leukocyte or fibroblast assay	Usually none	Vomiting, amyotonia, various neurological defects may be present
Glycogenosis III (Cori's)	Debranching enzyme system	Hepatomegaly. Diagnosis by analysis of glycogen and assay of enzyme in leukocytes	Milder than type I	Milder than type I
Glycogenosis IV (Andersen's)	Branching enzyme	Cirrhosis. Diagnosis by analysis of glycogen in blood cells	Liver disease	Retardation

Glycogenosis V (McArdle-Schmid-Pearson)	Muscle phosphorylase	Muscle cramps and failure of serum lactate to rise in ischemic exercise	Glycogen deposition in skeletal muscle	Myopathy
Glycogenosis VI (Hers')	Liver phosphorylase	Hepatomegaly. Diagnosis by enzyme assay in leukocytes	Milder than type I	Milder than type I
Glycogenosis VII	Muscle phosphofructokinase	Muscle cramps, myoglobinuria. Diagnosis by enzyme assay in erythrocytes	Glycogen deposition in skeletal muscle	Myopathy
Glycogenosis VIII[a]	Muscle phosphohexosisomerase (?)	Muscle fatigue, myoglobinuria. Diagnosis by muscle biopsy	As for type VII	Myopathy
Glycogenosis IX	Liver phosphorylase kinase	As for type VI	Milder than type I	Milder than type I
Fructose 1,6-diphosphatase deficiency [6a]	Fructose 1,6-diphosphatase	Severe fasting and postprandial hypoglycemia	Hypoglycemia; metabolic ketoacidosis; lacticacidosis	Vomiting, hypotonia, seizures, possible retardation
Lysosomal acid phosphatase deficiency [39][a]	Lysosomal acid phosphatase	Enzyme assay in cultivated fibroblast	Metabolic acidosis; hypoglycemia	Vomiting, lethargy, hypotonic convulsions; retardation
Total acid phosphatase deficiency [38][a]	All cellular acid phosphatases	Same as above	Acidosis	Vomiting, lethargy, opisthotonos
β-Xylosidase deficiency [44][a]	Lysosomal β-xylosidase	Enzyme assay in leukocytes	(?)	Seizures, choreoathetosis, retardation

[a]Disorder described in only one patient or one family.

in all tissues, including the neurons of the brain, spinal cord, and sympathetic ganglia [23a]. This disruption of neuronal structure must be related to the neurological abnormalities found in this disease [40].

LYSOSOMAL DEFECTS

Intracellular lysosomes, crucial for the degradation of many cytoplasmic structures, may have genetically determined enzyme defects that are responsible for many storage diseases, including those affecting cystine, glycogen, and gangliosides (the last of which are described in Chap. 23). In addition, some case reports have described defects in lysosomal acid phosphatase [39], total acid phosphatase [38], and β-xylosidase [44] (see Table 22-6).

DISORDERS OF NUCLEIC ACID METABOLISM

In humans, the pathways for the biosynthesis of nucleic acids from simple molecules have been well defined. The degradative pathways for the purines is complicated by salvage pathways which permit their reutilization. The pyrimidine nucleus can be fully broken down to yield β-alanine and other products, but the purine nucleus can only be broken down as far as the relatively insoluble oxypurine, uric acid. (Nucleic acid metabolism is discussed in Chap. 13.)

PURINES

Uric acid levels in the blood and urine are determined by an intricate interplay of (1) dietary intake, (2) de novo synthesis, (3) nucleoprotein catabolism, (4) degree of purine reutilization, and (5) integrity of renal excretory mechanisms. Normally, uric acid turnover is in regulated equilibrium, but uric acid concentrations can be increased by alteration of any of the above factors. Hypouricemia occurs but rarely, for instance in xanthine oxidase deficiency, and has no known neurological associations. Hyperuricemia, however, occurs in many metabolic disorders and is associated with various neurological findings.

Gout

The classic irascibility of the gouty patriarch, formerly ascribed to his excruciating arthritis, has now been linked by some investigators to intelligence and aggressive personality traits, which are associated with gout in some unexplained manner. So-called primary familial hyperuricemia, the common cause of gout, is an expression of many biochemical lesions. In one group of these patients, there is no demonstrable acceleration of urate production; in another group, urate biosynthesis is distinctly increased. Within this second group, a few families have been found to have decreased activity of hypoxanthine-guanine phosphoribosyl transferase, the enzyme in the salvage pathway that re-forms guanylic acid and inosinic acid from their respective purines. Deficient production of these two ribonucleotides decreases feedback inhibition of phosphoribosyl-1-amine formation, which is the first and rate-limiting step in de novo synthesis of the purines. This is an example of an enzyme defect that can lead to overproduction of a metabolite. A few of these patients have had nystagmus, tremors, spasticity, or retardation, but the association is inconstant, and there is biochemical evidence of different molecular species of enzyme abnormality among these families [29].

X-Linked Automutilation

The Lesch-Nyhan syndrome of hyperuricemia, choreoathetosis, spasticity, and compulsive automutilation with retardation is caused by genetically determined deficiency of hypoxanthine-guanine phosphoribosyl transferase activity [5]. This enzyme is much more active in nervous tissue, especially the basal ganglia, than elsewhere. The reasons for this are obscure, as is the neurological basis for the automutilation, which is not a feature of other spinocerebellar diseases. No neuropathological lesions are associated with this unique syndrome. Biochemically, deficiency of the enzyme results in markedly increased de novo synthesis of purines and an associated increase in adenine phosphoribosyl transferase activity. Uric acid itself is excluded from the central nervous system, but its precursors, hypoxanthine and xanthine, accumulate more in the cerebrospinal fluid than elsewhere. They are not known to be toxic, but the probability of a neurotoxic purine derivative being produced in this disease is suggested by the observation that caffeine induces automutilation in rats. Caffeine is a methylated xanthine that also inhibits cyclic phosphodiesterase and thereby potentiates the action of 3',5'-cyclic AMP, so perhaps abnormal activity of this cyclic nucleotide mediates some of the toxic effects of the disease. Attempted suppression of the overproduction of purines in this disease has failed to modify its neurological aspects. Milder spinocerebellar lesions, or even complete absence of neurological symptoms, have been found in a few patients who have partial deficiency of hypoxanthine-guanine phosphoribosyl transferase [29].

Other syndromes of hyperuricemia with odd behavioral and neurological features have been reported. One boy had inappropriate volatile mood changes, lack of speech, retardation, and hypotonia. His hypoxanthine-guanine phosphoribosyl transferase activity was unimpaired, but his adenine phosphoribosyl transferase activity was increased [41]. In another family, hyperuricemia, sensorineural deafness, and spinocerebellar degeneration were present but both phosphoribosyl transferase enzymes were normal [49].

PYRIMIDINES

Oroticaciduria

In this disorder, orotic acid is excreted in excess due to a genetically determined block of de novo biosynthesis of the pyrimidines. The block results in pyrimidine deficiency, affecting synthesis of both nucleic acids and pyrimidine cofactors. The clinical presentations are megaloblastic anemia, severe retardation, and oroticacid-crystalluria. In most cases, there have been deficiencies in both orotidylic pyrophosphorylase and orotidylic decarboxylase, suggesting a regulator gene defect, but one patient had a deficiency of only the decarboxylase. He had increased activity of the pyrophosphorylase, which returned to normal after treatment with uridine [17a]. Since oroticaciduria is considered an example of a control-gene defect because of the concomitant absence of two related enzymes, this one boy must have a mutation involving the structural gene for the decarboxylase alone.

Uridine, which is formed from orotic acid by the sequential action of these two enzymes, will produce a dramatic reversal of megaloblastic anemia, and may reverse or prevent the retardation, as well. This provides an excellent example of overcoming an enzyme block by replacement of the missing end-product. However, uridine as such does not penetrate the blood-brain barrier. In one patient, therapy with the pyrimidine base, uracil, was attempted [9], but within 3 weeks, anemia reappeared. This demonstrates the inadequacy of the salvage pathway from uracil to uridine and the complete reliance, at least of the hemopoietic system, upon a constant supply of uridine.

ACKNOWLEDGMENT

Supported in part by Training Grant HD 00198 of the National Institutes of Child Health and Human Development.

REFERENCES

BOOKS AND REVIEWS

1. Bryson, G. Biogenic amines in normal and abnormal behavioral states. *Clin. Chem.* 17:5, 1971.
2. Greenberg, R. E., and Christiansen, R. O. The critically ill child: Hypoglycemia. *Pediatrics* 46:915, 1970.
3. O'Brien, D. (Ed.). Rare Inborn Errors of Metabolism in Children with Mental Retardation. U.S. Dept. of Health, Education, and Welfare PHS Publication No. 2049, 1970.
4. Rosenberg, L. E., and Scriver, C. R. Disorders of Amino Acid Metabolism. In Bondy, P. K., and Rosenberg, L. E. (Eds.), *Duncan's Diseases of Metabolism* (6th ed.). Philadelphia: Saunders, 1969.

4a. Sidbury, J. B. The Glycogenoses. In Gardner, L. I. (Ed.), *Endocrine and Genetic Diseases of Childhood.* Philadelphia: Saunders, 1969.

5. Stanbury, J. B., Wyngaarden, J. B., and Fredrickson, D. S. (Eds.), *The Metabolic Basis of Inherited Disease* (3rd ed.). New York: McGraw-Hill, 1972.

ORIGINAL SOURCES

6. Arakawa, T. Congenital defects in folate utilization. *Am. J. Med.* 48:594, 1970.

6a. Baker, L., and Winegrad, A. I. Fasting hypoglycemia and metabolic acidosis associated with deficiency of hepatic fructose 1,6-diphosphatase activity. *Lancet* 2:13, 1970.

7. Barnes, N. D., Hull, D., Balgobin, L., and Gompertz, D. Biotin-responsive propionicacidemia. *Lancet* 2:244, 1970.

8. Baumgartner, R., Ando, T., and Nyhan, W. L. Nonketotic hyperglycinemia. *J. Pediatr.* 75:1022, 1969.

9. Becroft, D. M. O., Phillips, L. I., and Simmonds, A. Hereditary orotic aciduria: Long-term therapy with uridine and a trial of uracil. *J. Pediatr.* 75:885, 1969.

10. Blass, J. P., Avigan, J., and Uhlendorf, B. W. A defect in pyruvate decarboxylase in a child with an intermittent movement disorder. *J. Clin. Invest.* 49:423, 1970.

11. Campbell, A. G. M., Rosenberg, L. E., Snodgrass, P. J., and Nuzum, C. T. Lethal neonatal hyperammonaemia due to complete ornithine-transcarbamylase deficiency. *Lancet* 2:217, 1971.

12. Coleman, M., Barnet, A., and Boullin, D. J. Parachlorophenylalanine administration to a retarded patient with high blood serotonin levels. *Trans. Am. Neurol. Assoc.* 95:224, 1970.

13. Corner, B. D., Holton, J. B., Norman, R. M., and Williams, P. M. A case of histidinemia controlled with a low histidine diet. *Pediatrics* 41:1074, 1968.

14. Crawhall, J. C., Parker, R., Sneddon, W., and Young, E. P. β-Mercaptolactate-cysteine disulfide in the urine of a mentally retarded patient. *Am. J. Dis. Child.* 117:71, 1969.

14a. Dancis, J., Hutzler, J., Cox, R. P., and Woody, N. C. Familial hyperlysinemia with lysine-ketoglutarate reductase insufficiency. *J. Clin. Invest.* 48:1447, 1969.

15. Daum, R., Delvin, E., Goldman, H., Lamm, P., Mamer, O., and Scriver, C. R. A new inherited defect of isoleucine catabolism. *Pediatr. Res.* 5:392, 1971 (Abstract).

16. Eldjarn, L., Jellum, E., Stokke, O., Pande, H., and Waaler, P. E. β-Hydroxyisovaleric aciduria and β-methylcrotonylglycinuria: A new inborn error of metabolism. *Lancet* 2:521, 1970.

17. Feigin, I., Pena, C. E., and Budzilovich, G. The infantile spongy degenerations. *Neurology* 18:153, 1968.

17a. Fox, R. M., O'Sullivan, W. J., and Firkin, B. G. Orotic aciduria: Differing enzyme patterns. *Am. J. Med.* 47:332, 1969.

18. Freeman, J. M., Nicholson, J. F., Schimke, R. T., Rowland, L. P., and Carter, S. Congenital hyperammonemia: Association with hyperglycinemia and decreased levels of carbamyl phosphate synthetase. *Arch. Neurol.* 23:430, 1970.

18a. Gallagher, B. B. The effect of phenylpyruvate on oxidative-phosphorylation in brain mitochondria. *J. Neurochem.* 16:1071, 1969.

19. Gambetti, P., DiMauro, S., and Baker, L. Nervous system in Pompe's disease. *J. Neuropath. Exp. Neurol.* 30:412, 1971.
20. Gatfield, P. D., Taller, E., Hinton, G. G., Wallace, A. C., Abdelnour, G. M., and Haust, M. D. Hyperpipecolatemia: A new metabolic disorder associated with neuropathy and hepatomegaly. *Can. Med. Assoc. J.* 99:1215, 1968.
20a. Gompertz, D., Draffin, G. H., Watts, J. L., and Hull, D. Biotin-responsive β-methylcrotonylglycinuria. *Lancet* 2:22, 1971.
21. Goodman, S. I., Mace, J. W., and Pollak, S. Serum gamma-glutamyl transpeptidase deficiency. *Lancet* 1:234, 1971.
22. Greene, H. L., Schubert, W. K., and Hug, G. Chronic lactic acidosis of infancy. *J. Pediatr.* 76:853, 1970.
23. Hoefnagel, D., and Pomeroy, J. Hydroxylysinuria. *Lancet* 1:1342, 1970.
23a. Hogan, G. R., Gutmann, L., Schmidt, R., and Gilbert, E. Pompe's disease. *Neurology* 19:894, 1969.
24. Hommes, F. A., Polman, H. A., and Reerink, J. D. Leigh's encephalomyelopathy: An inborn error of gluconeogenesis. *Arch. Dis. Child.* 43:423, 1968.
25. Hsia, Y. E., Lilljeqvist, A-Ch., and Rosenberg, L. E. Vitamin B_{12} dependent methylmalonicaciduria. *Pediatrics* 46:497, 1970.
26. Hsia, Y. E., Scully, K. J., and Rosenberg, L. E. Inherited propionyl-CoA carboxylase deficiency in "ketotic hyperglycinemia." *J. Clin. Invest.* 50:127, 1971.
26a. Humbert, J. R., Hammond, K. B., Hathaway, W. E., Marcoux, J. G., and O'Brien, D. Trimethylaminuria: The fish-odour syndrome. *Lancet* 2:770, 1970.
27. Huttenlocher, P. R., Hillman, R. E., and Hsia, Y. E. Pseudotumour cerebri in galactosemia. *J. Pediatr.* 76:902, 1970.
28. Kang, E. S., and Gerald, P. S. Hereditary tyrosinemia and abnormal pyrrole metabolism. *J. Pediatr.* 77:397, 1970.
29. Kelley, W. N., Greene, M. L., Rosenbloom, F. M., Henderson, J. F., and Seegmiller, J. E. Hypoxanthine-guanine phosphoribosyltransferase deficiency in gout. *Ann. Intern. Med.* 70:155, 1969.
30. Lanzkowsky, P. Congenital malabsorption of folate. *Am. J. Med.* 48:580, 1970.
31. Levy, H. L., Feingold, M., and Letsoie, A. Early dietary therapy for hyperlysinemia. *Soc. Pediatr. Res.* p. 179, 1970 (Abstract).
32. Lieberman, L., Shaw, K. N. F., Donnell, G. N., Gutenstein, M., and Jacobs, E. E. β-Alaninuria associated with chronic renal disease. *Soc. Pediatr. Res.* p. 163, 1970 (Abstract).
33. Lindemann, R., Gjessing, L. R., Merton, B., Löken, A. C., and Halvorsen, S. Amino acid metabolism in hereditary fructosemia. *Acta. Paediatr. Scand.* 59:141, 1970.
34. Lonsdale, D., Faulkner, W. R., Price, J. W., and Smeby, R. R. Intermittent cerebellar ataxia associated with hyperpyruvic acidemia, hyperalaninemia, and hyperalaninuria. *Pediatrics* 43:1025, 1969.
35. Mahoney, M. J., Hsia, Y. E., and Rosenberg, L. E. Propionyl-CoA carboxylase deficiency (propionicacidemia): A cause of non-ketotic hyperglycinemia. *Pediatr. Res.* 5:395, 1971 (Abstract).

35a. Mahoney, M. J., and Rosenberg, L. E. Inherited defects of B_{12} metabolism. *Am. J. Med.* 48:584, 1970.
36. Malmquist, J., Jagenburg, R., and Lindstedt, G. Familial protein intolerance: Possible nature of enzyme defect. *N. Engl. J. Med.* 284:997, 1971.
37. Morrow, G., Aksu, A., Bank, W., Rowland, L. P., and Barness, L. A. Familial neuromuscular disease and non-ketotic hyperglycinemia. *Pediatr. Res.* 4:480, 1970 (Abstract).
38. Nadler, H. L. Genetic heterogeneity in acid phosphate deficiency. *Pediatr. Res.* 5:421, 1971 (Abstract).
39. Nadler, H. L., and Egan, T. J. Deficiency of lysosomal acid phosphatase: A new familial metabolic disorder. *N. Engl. J. Med.* 282:302, 1970.
40. Nihill, M. R., Wilson, D. S., and Hugh-Jones, K. Generalized glycogenosis type II (Pompe's disease). *Arch. Dis. Child.* 45:122, 1970.
41. Nyhan, W. L., James, J. A., Teberg, A. J., Sweetman, L., and Nelson, L. G. A new disorder of purine metabolism with behavioral manifestations. *J. Pediatr.* 74:20, 1969.
42. Oliva, P. B. Lactic acidosis. *Am. J. Med.* 48:209, 1970.
43. Oyanagi, K., Miura, R., and Yamanouchi, T. Congenital lysinuria: A new inherited transport disorder of dibasic amino acids. *J. Pediatr.* 77:259, 1970.
44. Payling-Wright, C. R., and Evans, P. R. A case of β-xylosidase deficiency. *Lancet* 2:43, 1970.
45. Perry, T. L., Hansen, S., Tischler, B., Bunting, R., and Diamond, S. Glutamine depletion in phenylketonuria: A possible cause of the mental defect. *N. Engl. J. Med.* 282:761, 1970.
46. Pincus, J. H., Itokawa, Y., and Cooper, J. R. Enzyme-inhibiting factor in subacute necrotizing encephalomyelopathy. *Neurology* 19:841, 1969.
47. Pollitt, R. J., Jenner, F. A., and Merskey, H. Aspartylglycosaminuria: An inborn error of metabolism associated with mental defect. *Lancet* 2:253, 1968.
48. Rennert, O., Julius, R., Aylsworth, A., Williams, C., and Greer, M. A new disorder of phenylalanine metabolism associated with ataxia, convulsions and retardation: Methylmandelic aciduria. *Pediatr. Res.* 5:652, 1971 (Abstract).
49. Rosenberg, A. L., Bergstrom, L.-V., Troost, B. T., and Bartholomew, B. A. Hyperuricemia and neurologic deficits: A family study. *N. Engl. J. Med.* 282:992, 1970.
50. Rosenberg, L. E. Inherited aminoacidopathies demonstrating vitamin dependency. *N. Engl. J. Med.* 281:145, 1969.
51. Rushton, D. I. Spongy degeneration of the white matter of the central nervous system associated with hyperglycinemia. *J. Clin. Path.* 21:456, 1968.
52. Scott-Emuakpor, A., Higgins, J. V., and Kohrman, A. F. Citrullinemia: A new case, with implications concerning adaptation to defective urea synthesis. *Soc. Pediatr. Res.* p. 201, 1971 (Abstract).
53. Scriver, C. R., Mackenzie, S., Clow, C. L., and Delvin, E. Thiamine responsive maple-syrup-urine disease. *Lancet* 1:310, 1971.

54. Seashore, M. R., Durant, J. L., and Rosenberg, L. E. Studies of the mechanism of pyridoxine-responsive homocystinuria. *Pediatr. Res.* 6:187, 1972.
55. Selkoe, D. J. Familial hyperprolinemia and mental retardation: A second metabolic type. *Neurology* 19:494, 1969.
55a. Siegel, F. L., Aoki, K., and Colwell, R. E. Polyribosome disaggregation and cell-free protein synthesis in preparations from cerebral cortex of hyperphenylalaninemic rats. *J. Neurochem.* 18:537, 1971.
56. Silberberg, D. H. Maple syrup urine disease metabolites studied in cerebellum cultures. *J. Neurochem.* 16:1141, 1969.
57. Smith, A. L., Satterthwaite, H. S., and Sokoloff, L. Induction of brain d (–) β-hydroxybutyrate dehydrogenase activity by fasting. *Science* 163:79, 1969.
58. Solitare, G. B., Shih, V. E., Nelligan, D. J., and Dolan, T. F. Arginino succinic aciduria: Clinical, biochemical, anatomical and neuropathological observations. *J. Ment. Defic. Res.* 13:153, 1969.
59. Steiner, F. A. L-Glutamic acid, gamma-aminobutyric acid and pyridoxal-5'-phosphate at the level of the single unit in rat brain. *Ann. N.Y. Acad. Sci.* 166:199, 1969.
60. Stimmler, L., Jensen, N., and Toseland, P. Alaninuria, associated with microcephaly, dwarfism, enamel hypoplasia, and diabetes mellitus in two sisters. *Arch. Dis. Child.* 45:682, 1970.
61. Terheggen, H. G., Schwenk, A., Lowenthal, A., van Sande, M., and Columbo, J. P. Argininaemia with arginase deficiency. *Lancet* 2:748, 1969.
62. Tildon, J. T., and Cornblath, M. Succinyl-CoA:3-ketoacid CoA transferase deficiency: A cause of ketoacidosis in infancy. *J. Clin. Invest.* 51:493, 1972.
63. van Heeswijk, P. J., Trijbels, J. M. F., Schretlen, E. D. A. M., van Munster, P. J. J., and Monnens, L. A. H. A patient with a deficiency of serum-carnosinase activity. *Acta Paediatr. Scand.* 58:584, 1969.
64. Weber, G. Inhibition of human brain pyruvate kinase and hexokinase by phenylalanine and phenylpyruvate: Possible relevance to phenylketonuric brain damage. *Proc. Natl. Acad. Sci. U.S.A.* 63:1365, 1969.

23

Sphingolipidoses

Roscoe O. Brady

THE STUDY of the lipid storage diseases began with the recognition by clinicians of the disparate disease entities which are now known collectively as the sphingolipidoses. In 1881, Warren Tay, a British ophthalmologist, described changes in the macular region of the eyes of very young children. Later, Bernard Sachs (1887), an American neurologist, found that the condition was associated with a generalized arrest of cerebral development. In time, this condition was named in honor of these clinicians and is now called Tay-Sachs disease (Table 23-1). Very soon after Tay's paper, the French physician Phillipe C. E. Gaucher described what he believed to be an epithelioma of the spleen in patients who presented a consistent series of clinical findings. Only a few years later, Fabry described another clinical entity, which he, a dermatologist, considered to be a primary disease of the skin. Since then, numerous identifications of various lipodystrophies have been made, the most recent being the description of a patient in whom lactosylceramide (Fig. 23-1) is the primary accumulating material. The major clinical findings in patients with various sphingolipid storage diseases are summarized in Table 23-2.

Another aspect of this study has been the identification of the substances that are stored in tissues of patients with various sphingolipidoses. In 1934, Aghion identified the material that accumulates in patients with Gaucher's disease as glucocerebroside [1]. This finding was substantiated by Halliday and co-workers a few years later. As early as 1935, Klenk had clearly shown that the lipid which accumulates in patients with Niemann-Pick disease is sphingomyelin. In time, the lipids that are known to accumulate in all of the other sphingolipidoses have been identified (Table 23-3). An exception to this statement appears to occur in the condition known as globoid leukodystrophy. Austin and co-workers [3] recognized that there was an increase in the ratio of galactocerebroside to sulfatide in patients with this condition, although, probably because of attendant demyelination, there was an overall net decrease in the amount of galactocerebroside in the central nervous system of these patients when compared with suitable control specimens. The details of the recent resolution of this dilemma are as follows.

TABLE 23-1. Chronology of the Clinical Descriptions of Lipid Storage Diseases

Disease	Clinicians[a]	Date
Tay-Sachs disease	W. Tay	1881
	B. Sachs	1887
Gaucher's disease	P. C. E. Gaucher	1882
Fabry's disease	J. Fabry	1898
Niemann-Pick disease	A. Niemann	1914
	L. Pick	1927
Globoid leukodystrophy	K. Krabbe	1916
Metachromatic leukodystrophy	W. Scholz	1925
Generalized gangliosidosis	R. M. Norman et al.	1959
	B. H. Landing et al.	1964
Variant form of Tay-Sachs disease	H. Pilz et al.	1968
Ceramide lactoside lipidosis	G. Dawson	1970

[a]References to the respective individual contributions are documented in a previous review [8].

A major advance in the study of lipidoses occurred in 1965 with the conclusive demonstration of the nature of the metabolic defect in Gaucher's disease [11, 13]. It had been known for some time that overproduction of glucocerebroside probably was not a factor in the pathogenesis of this disease [36]. The demonstration of a normal rate of synthesis, combined with the identification of an impaired catabolic reaction, provided the keystone for our present understanding of the underlying enzymatic lesions in all of the sphingolipidoses; there is a deficiency of a specific hydrolytic enzyme that normally catalyzes the immediately succeeding step in the biodegradation of the accumulating lipid (Table 23-4).

But even this generalization has apparently a single, perhaps subtle, exception. It has recently been shown that the metabolic defect in Tay-Sachs disease is a deficiency of a hexosaminidase that normally catalyzes the hydrolysis of the terminal molecule of *N*-acetylgalactosamine of Tay-Sachs ganglioside [24]. However, the enzyme that catalyzes the hydrolytic cleavage of the *N*-acetylneuraminic acid moiety of Tay-Sachs ganglioside is apparently normally active in tissues of patients with this disease. We are left, therefore, with the necessity of providing an explanation of why the *entire* molecule of Tay-Sachs ganglioside with its sialic acid component intact, rather than the corresponding asialorceramidetrihexoside, accumulates in patients with this disease. Recent experimental data have provided helpful

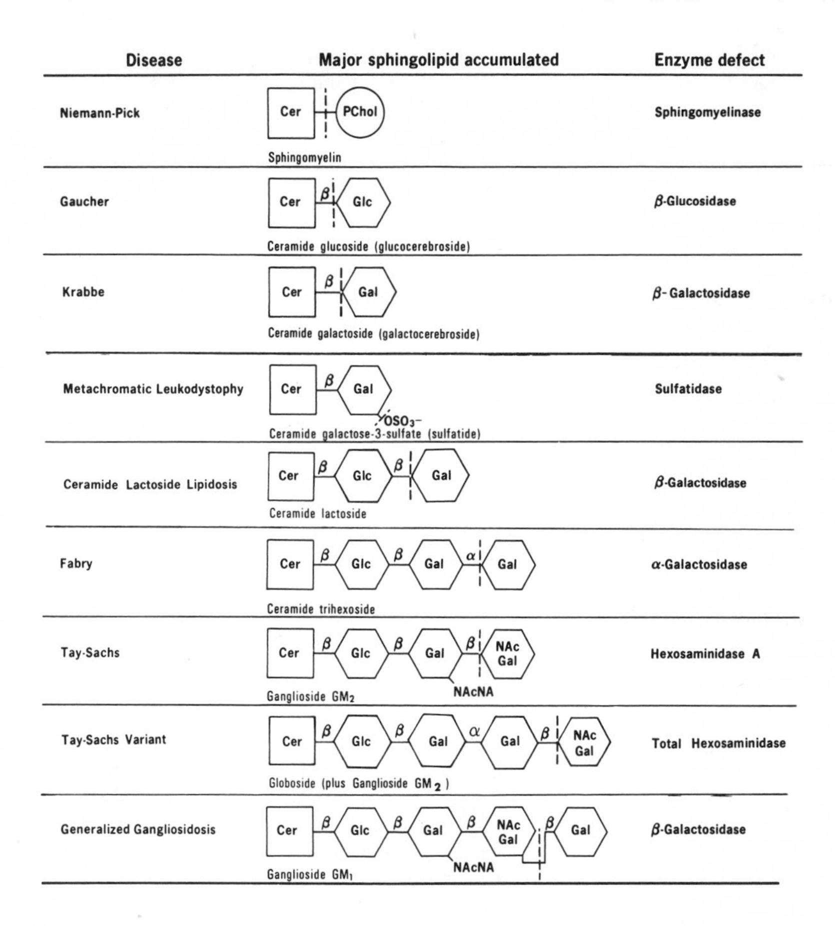

FIGURE 23-1. Metabolic diseases characterized by inabilities to degrade sphingolipids. (Cer = *N*-acylsphingosine, or ceramide; NAcNA = *N*-acetylneuraminic acid; Gal = galactose; Glc = glucose; PChol = phosphorylcholine; NAcGal = *N*-acetylgalactosamine.)

insights toward the solution of this problem. The hydrolysis of the *N*-acetylneuraminyl moiety of Tay-Sachs ganglioside appears to be one of two alternative pathways for ganglioside catabolism in brain. This enzymatic reaction is inhibited by other gangliosides. Therefore, since hexosaminidase is deficient and the reaction via sialidase is inhibited and limiting, Tay-Sachs ganglioside will accumulate in tissues of patients with Tay-Sachs disease [14].

In general, the identification of the metabolic lesions in the sphingolipidoses hinged upon the preparation of the accumulating substance with a radioactive

TABLE 23-2. Clinical Aspects of Lipodystrophies

Tay-Sachs disease (classic and variant forms)	Mental retardation; amaurosis; cherry-red spot in macula; neuronal cells distended with "membranous cytoplasmic bodies"
Gaucher's disease	Mental retardation (infantile form only); hepatosplenomegaly; hip and long-bone involvement; oil red and PAS-positive lipid-laden (Gaucher) cells in bone marrow
Niemann-Pick disease	Generally similar to Gaucher's disease; 30% with cherry-red spot in macula; marrow cells (foam cells) stain for both lipid and phosphorus
Fabry's disease	Reddish-purple maculopapular rash in umbilical, inguinal, and scrotal areas; renal impairment; corneal opacities; peripheral neuralgias and abnormalities of ECG
Globoid leukodystrophy	Mental retardation; "globoid bodies" in brain tissue sections
Metachromatic leukodystrophy	Mental retardation; psychological disturbances (adult form); decreased nerve-conduction time; nerve biopsy shows yellow-brown droplets when stained with cresyl violet (metachromasia)
Generalized gangliosidosis	Mental retardation, cherry-red spot in macula; hepatomegaly; bone marrow involvement
Ceramide lactoside lipidosis	Slowly progressing CNS impairment; organomegaly; macrocytic anemia, leukopenia, and thrombocytopenia due to involvement of bone marrow and spleen

tracer in a critical portion of the molecule, e.g., the leaving group in the hydrolytic reaction. An important ancillary benefit that arose as a result of the availability of these labeled complex lipids was the development of convenient diagnostic tests for the identification of patients with the various lipid storage diseases. This aspect came about through the demonstration that enzyme assays for glucocerebrosidase and sphingomyelinase performed on circulating leukocytes obtained from a small sample of venous blood showed exactly the same degree of attenuation as that seen in solid-tissue biopsy specimens from patients with Gaucher's disease and Niemann-Pick disease, respectively [22]. This technique has been extended for accurate diagnosis and, indeed, even for estimating the prognosis in other sphingolipidoses (Table 23-5). Knowledge of the degree of enzymatic deficiency enables the clinician to estimate the rapidity of progression of the clinical course in such patients.

TABLE 23-3. Identification of Materials Accumulating in Lipid Storage Diseases

Disease	Lipid	Discoverer[a]
Gaucher's disease	Glucocerebroside	A. Aghion (1934) N. Halliday et al. (1940)
Niemann-Pick disease	Sphingomyelin	E. Klenk (1935)
Tay-Sachs disease	Monosialoceramidetrihexoside (ganglioside G_{M2})	E. Klenk (1942) L. Svennerholm (1962)
Metachromatic leukodystrophy	Sulfatide	H. Jatzkewitz (1958)
Globoid leukodystrophy (Krabbe's disease)	Increased ratio of galactocerebroside to sulfatide	J. Austin et al. (1961)
Fabry's disease	Ceramidetrihexoside	C. Sweeley and B. Klionsky (1963)
Generalized gangliosidosis	Monosialoceramidetetrahexoside (ganglioside G_{M1})	R. Ledeen et al. (1965) J. O'Brien et al. (1965)
Variant form of Tay-Sachs disease	Ceramidetetrahexoside (globoside) and G_{M2}	K. Sandhoff et al. (1968)
Ceramide lactoside lipidoses	Lactosylceramide	G. Dawson (1970)

[a]See footnote to Table 23-1.

Individuals with a moderate amount of residual sphingolipid hydrolytic activity for the particular substance involved develop their symptomatic manifestations more slowly than do those with lesser residual enzyme activity.

Although diagnostic studies can be performed with greatest certainty with labeled authentic lipids as substrates, such artificial substrates as nitrocatechol sulfate and nitrophenyl-β-D-*N*-acetylgalactosamine are also reliable in some instances and are becoming increasingly important diagnostic reagents for identifying conditions such as metachromatic leukodystrophy [32] and Tay-Sachs disease [28], respectively. These substances may be determined by automated analytical techniques, and thereby may be extremely useful in screening large populations for such diseases.

Two additional benefits have been derived from the development of accurate diagnostic methods. The first is the ability to detect heterozygous carriers in many instances. This possibility was first shown by jejunal biopsy of the mothers of patients afflicted with Fabry's disease [9], and now has been extended to metachromatic leukodystrophy [4] and very recently to Tay-Sachs disease [28] and

TABLE 23-4. Metabolic Defects in Sphingolipidoses

Disease	Enzyme Deficiency	References
Gaucher's disease	Glucocerebroside glucosidase	[11, 13]
Metachromatic leukodystrophy	Sulfatide sulfatase	[2, 27]
Niemann-Pick disease	Sphingomyelin phosphorylcholine phosphohydrolase	[12]
Fabry's disease	Ceramidetrihexoside galactosidase	[15]
Generalized gangliosidosis	G_{M1} ganglioside galactosidase	[30]
Variant form of Tay-Sachs disease	Total hexosaminidase activity	[33]
Tay-Sachs disease	Hexosaminidase A G_{M2} ganglioside *N*-acetylgalactosaminidase	[31] [24]
Globoid leukodystrophy	Galactocerebroside galactosidase	[35]
Ceramide lactoside lipidosis	Lactosylceramide galactosidase	[17]

Gaucher's disease [5]. The second, and extremely important, development is the observation that enzyme assays on cultured fetal cells are feasible. This procedure should enable attending physicians to have an antenatal determination in cells obtained from fetuses at risk to learn if they are homozygous for the defect in question. The cells can be obtained by amniocentesis in the fourth month of pregnancy, and in a few weeks a sufficient quantity of cells can be grown in tissue culture for diagnostic enzyme assay [7, 15, 18, 29]. In general, the sphingolipid hydrolases are not present in sufficient concentration in amniotic fluid per se for diagnostic testing. Thus, the fetal cells must be cultured. However, the sensitivity of these assays can be improved by concentrating the enzyme in the amniotic fluid through the use of suitable dehydrating substances, such as Sephadex, or by conventional enzyme fractionation procedures, such as with ammonium sulfate.

Thus, a useful framework has been developed for the identification, diagnosis, antenatal recognition, and detection of carriers of most of the lipid storage diseases. However, rational procedures for ameliorating these diseases have not yet been established. Various therapeutic possibilities are summarized in Table 23-6. The most obvious approach is that of attempting enzyme replacement, provided suitable quantities of highly purified enzymes become available. At the present time, this is a formidable task. Probably the main reason for the difficulty is that in tissues these degradative enzymes seem to be primarily of lysosomal origin [37],

TABLE 23-5. Diagnostic Tests

Disease	Tissue Source	Identification of Heterozygotes	References
Gaucher's disease	Circulating leukocytes	Yes	[5, 22]
	Cultured skin fibroblasts	Yes	[6]
	Cultured amniotic cells	Yes	[16]
Niemann-Pick disease	Circulating leukocytes	Yes	[10, 22]
	Cultured skin fibroblasts	Yes[a]	[34]
	Cultured amniotic cells	...	[18]
Fabry's disease	Jejunal biopsy	Yes	[9]
	Circulating leukocytes	Yes	[23]
	Cultured amniotic cells	Yes	[15]
Metachromatic leukodystrophy	Circulating leukocytes	Yes	[4, 32]
	Cultured amniotic cells	...	[8]
Tay-Sachs disease	Muscle biopsy	...	[24]
	Fibroblast cultures	Yes	[31]
	Serum	Yes	[28]
	Cultured amniotic cells	Yes	[25, 28]
Generalized gangliosidosis	Circulating leukocytes	Yes	[38]
	Cultured amniotic cells	...	[25]
Globoid leukodystrophy	Circulating leukocytes	...	[26]
	Solid tissue biopsy	Yes	[35]
Variant form of Tay-Sachs disease	Serum and cultured skin fibroblasts	Yes	[25, 28]
Ceramide lactoside lipidosis	Skin fibroblasts	Yes	[17]

[a]Brady, R. O. Biochemical and Metabolic Basis of Familial Sphingolipidoses. *Seminars in Hematology,* Vol. 9, July, 1972.

and they must be separated from macromolecular lysosomal membrane components. We have attacked this problem by attempting dissociation by chemical agents, as well as by physical means. However, progress has been regrettably slow, and we are currently investigating other sources of lysosomal enzymes (e.g., urine).

When the requisite enzymes have been obtained in a sufficiently purified state, there is the problem of how to administer them. It seems likely that exogenously administered enzymes will be taken up by the reticuloendothelial cells, and as far as the peripheral organs are concerned, proper intracellular localization can be obtained thereby [21]. However, it is possible that exogenously administered enzymes will not reach the central nervous system in sufficient quantities to be

TABLE 23-6. Therapeutic Possibilities

A. Enzyme replacement
   1. Parenteral administration of purified enzyme from human or animal sources
   2. Encapsulated enzyme in biodegradable microspherules
      a. Native enzyme
      b. Polymer-stabilized enzyme
   3. Organ transplantation
B. Extracorporeal therapy: Percolation of plasma over columns of bound enzyme
C. Administration of materials that convey metabolic cooperativity
D. Genetic engineering
   1. Administration of messenger RNA
   2. Administration of DNA
   3. Hybridization and replacement of patient's cells

effective; one report of this has appeared [20]. Accordingly, an alternative procedure may be the passage of the patient's plasma over columns or sheets of stably bound enzyme. It is hoped that such procedures may reduce the level of circulating sphingolipid and promote egress of the intracellularly accumulated material. An additional advantage in this form of therapy is that foreign proteins are not administered to the recipient and thereby the possibility of sensitivity reactions is avoided.

Recently, another potential form of therapy has been proposed: the administration of material(s) that convey metabolic cooperativity in tissue culture. It has been shown that cells from patients with genetic deficiencies maintain these insufficiencies in culture and, in some cases, even accumulate the involved lipid to some extent. When such cells are grown in the presence of normal cells, or even in the presence of cells with a different genetic abnormality, the defect is corrected in those cells that are in proximity with each other. The substance involved in this correction is probably an enzyme [19], and it seems to be a good material with which to try to effect a cure. Attempts to cure genetic diseases by such substances will probably be forthcoming in the near future.

Finally, consideration has been given, and more still will be devoted, to the possibility of genetic engineering. Most investigators are presently reluctant to undertake this form of correction, even if rational procedures could be devised, and they are probably correct in maintaining such an attitude. We should prefer to make available other less tenuous forms of therapy, and it is to this goal that efforts are currently being directed.

REFERENCES

1. Aghion, A. La maladie de Gaucher dans l'enfance. Thése, Paris, 1934.
2. Austin, J., Armstrong, D., and Shearer, L. Metachromatic form of diffuse cerebral sclerosis. V. The nature and significance of low sulfatase activity:

A controlled study of brain, liver, and kidney in four patients with metachromatic leukodystrophy (MLD). *Arch. Neurol. Psychiat.* 13:593, 1965.
3. Austin, J., Lehfeldt, D., and Maxwell, W. Experimental globoid bodies in white matter and chemical analyses in Krabbe's disease. *J. Neuropathol. Exp. Neurol.* 20:284, 1961.
4. Bass, N. H., Witmer, E. Z., and Dreifuss, F. E. A pedigree study of metachromatic leukodystrophy. Biochemical identification of the carrier state. *Neurology* 20:52, 1970.
5. Beutler, E., and Kuhl, W. Detection of the defect of Gaucher's disease and its carrier state in peripheral blood leukocytes. *Lancet* 1:612, 1970.
6. Brady, R. O. Cerebral lipidoses. *Annu. Rev. Med.* 21:317, 1970.
7. Brady, R. O. The prenatal diagnosis of lipid storage diseases. *Clin. Chem.* 16:811, 1970.
8. Brady, R. O. The Ups and Downs of Complex Lipid Metabolism. In Bernsohn, J. (Ed.), *Lipid Storage Diseases: Enzymatic Defects and Clinical Implications.* New York: Academic, 1971.
9. Brady, R. O., Gal, A. E., Bradley, R. M., Martensson, E., Warshaw, A. L., and Laster, L. Enzymatic defect in Fabry's disease. Ceramide-trihexosidase deficiency. *N. Engl. J. Med.* 276:1163, 1967.
10. Brady, R. O., Johnson, W. G., and Uhlendorf, B. W. Identification of heterozygous carriers of lipid storage diseases. Current status and clinical application. *Am. J. Med.* 51:423, 1971.
11. Brady, R. O., Kanfer, J. N., Bradley, R. M., and Shapiro, D. Demonstration of a deficiency of glucocerebroside-cleaving enzyme in Gaucher's disease. *J. Clin. Invest.* 45:1112, 1966.
12. Brady, R. O., Kanfer, J. N., Mock, M. B., and Fredrickson, D. S. The metabolism of sphingomyelin. II. Evidence of an enzymatic deficiency in Niemann-Pick disease. *Proc. Natl. Acad. Sci. U.S.A.* 55:366, 1966.
13. Brady, R. O., Kanfer, J. N., and Shapiro, D. Metabolism of glucocerebrosides. II. Evidence of an enzymatic deficiency in Gaucher's disease. *Biochem. Biophys. Res. Commun.* 18:221, 1965.
14. Brady, R. O., and Kolodny, E. H. Disorders of Ganglioside Metabolism. In Steinberg, A. G., and Bearn, A. G. (Eds.), *Progress in Medical Genetics,* Vol. 8. New York: Grune & Stratton, 1972, pp. 225–241.
15. Brady, R. O., Uhlendorf, B. W., and Jacobson, C. B. Fabry's disease: Antenatal detection. *Science* 172:174, 1971.
16. Brady, R. O., Uhlendorf, B. W., and Johnson, W. G. Unpublished data, 1970.
17. Dawson, G., and Stein, A. O. Lactosyl ceramidosis: Catabolic enzyme defect in glycosphingolipid metabolism. *Science* 170:556, 1970.
18. Epstein, C. J., Brady, R. O., Schneider, E. L., Bradley, R. M., and Shapiro, D. *In utero* diagnosis of Niemann-Pick disease. *Am. J. Med. Genet.* 23:533, 1971.
19. Fratantoni, J. C., Hall, C. W., and Neufeld, E. F. The defect in Hurler and Hunter syndromes. II. Deficiency of specific factors involved in mucopolysaccharide degradation. *Proc. Natl. Acad. Sci. U.S.A.* 64:360, 1969.

20. Greene, H. L., Hug, G., and Schubert, W. K. Metachromatic leukodystrophy. Treatment with arylsulfatase-A. *Arch. Neurol.* 20:147, 1969.
21. Hug, G., and Schubert, W. K. Lysosomes in type II glycogenosis. Changes during administration of extract from *Aspergillus niger. J. Cell Biol.* 35:C1, 1967.
22. Kampine, J. P., Brady, R. O., Kanfer, J. N., Feld, M., and Shapiro, D. Diagnosis of Gaucher's disease and Niemann-Pick disease with small samples of venous blood. *Science* 155:86, 1967.
23. Kint, J. A. Fabry's disease: Alpha-galactosidase deficiency. *Science* 167:1268, 1970.
24. Kolodny, E. H., Brady, R. O., and Volk, B. W. Demonstration of an alteration of ganglioside metabolism in Tay-Sachs disease. *Biochem. Biophys. Res. Commun.* 37:526, 1969.
25. Kolodny, E. H., Uhlendorf, B. W., Quirk, J. M., and Brady, R. O. Unpublished observations, 1970.
26. Malone, M. J. Deficiency in a degradative enzyme system in globoid leukodystrophy. *Trans. Am. Soc. Neurochem.* 1:56, 1970.
27. Mehl, E., and Jatzkewitz, H. Evidence for a genetic block in metachromatic leukodystrophy (ML). *Biochem. Biophys. Res. Commun.* 19:407, 1965.
28. O'Brien, J. S., Okada, S., Chen, A., and Fillerup, D. L. Tay-Sachs disease. Detection of heterozygotes and homozygotes by serum hexosaminidase assay. *N. Engl. J. Med.* 283:15, 1970.
29. O'Brien, J. S., Okada, S., Fillerup, D. L., Veath, M. L., Adornato, B., Brenner, P. H., and Leroy, J. G. Tay-Sachs disease: Prenatal diagnosis. *Science* 172:61, 1971.
30. Okada, S., and O'Brien, J. S. Generalized gangliosidosis: Betagalactosidase deficiency. *Science* 160:1002, 1968.
31. Okada, S., and O'Brien, J. S. Tay-Sachs disease: Generalized absence of a beta-D-*N*-acetylhexosaminidase component. *Science* 165:698, 1969.
32. Percy, A. K., and Brady, R. O. Metachromatic leukodystrophy: Diagnosis with samples of venous blood. *Science* 161:594, 1968.
33. Sandhoff, K., Andreae, U., and Jatzkewitz, H. Deficiency hexosaminidase activity in an exceptional case of Tay-Sachs disease with additional storage of kidney globoside in visceral organs. *Pathol. Eur.* 3:278, 1968.
34. Sloan, H. R., Uhlendorf, B. W., Kanfer, J. N., Brady, R. O., and Fredrickson, D. S. Deficiency of sphingomyelin-cleaving enzyme activity in tissue cultures derived from patients with Niemann-Pick disease. *Biochem. Biophys. Res. Commun.* 34:582, 1969.
35. Suzuki, K., and Suzuki, Y. Globoid cell leukodystrophy (Krabbe's disease): Deficiency of galactocerebroside-β-galactosidase. *Proc. Natl. Acad. Sci. U.S.A.* 66:302, 1970.
36. Trams, E. G., and Brady, R. O. Cerebroside synthesis in Gaucher's disease. *J. Clin. Invest.* 39:1546, 1960.

37. Weinreb, N. J., Brady, R. O., and Tappel, A. L. The lysosomal localization of sphingolipid hydrolases. *Biochim. Biophys. Acta* 159:141, 1968.
38. Wolfe, L. S., Callahan, J., Fawcett, J. S., Andermann, F., and Scriver, C. R.

24

Diseases of Myelin

Pierre Morell, Murray B. Bornstein, and William T. Norton

MYELIN DEPENDS on the normal functioning of the Schwann cell or oligodendrocyte of which it is a part for its maintenance. It also depends on the integrity of the axon it ensheaths. Many diseases of the nervous system which affect general cell functioning and all of those which result in the death of neurons involve some loss of myelin. Because myelin accounts for approximately 50 percent of the weight of white matter, a reduction in the amount of myelin or the presence of an active demyelinating process is morphologically easy to detect by special staining methods. It is not always clear whether the loss of myelin should be considered a significant factor in the disease process or whether it is incidental to some other pathological process. There are, however, diseases in which demyelination or a failure to myelinate normally are major factors.

CLASSIFICATION

There have been many attempts over the years to classify diseases involving myelin. Such schemes include categories described on the basis of pathology, etiology, and biochemistry.

The usual classification includes demyelination as a major category. For purposes of convenience, demyelinating diseases are subdivided, primarily on the basis of histological criteria, into categories of primary and secondary demyelination. *Primary demyelination* involves the destruction of myelin with relative sparing of the axons. Such disorders (referred to as myelinoclastic) attack a normal nervous system and destroy normally constituted myelin. Examples include multiple sclerosis (MS), Guillain-Barré syndrome, experimental allergic encephalomyelitis (EAE), and diphtheritic neuritis.

Secondary demyelination is any loss of normal myelin after destruction of the axon. Examples would be Wallerian degeneration after nerve section and amyotrophic lateral sclerosis.

Certain demyelinating diseases, such as some cases of subacute sclerosing panencephalitis and canine distemper, are difficult to classify as primary or secondary because both axons and myelin show pathology. In such cases, a picture typical of either primary or secondary demyelination may predominate in different microscopic sections. Variables include the region of the brain studied, case-to-case variability with respect to given clinical disease entity, the staining methods used to study the specimen, and so on. No clear, overall picture emerges that makes possible an unambiguous classification as demyelination with sparing of axons, or vice versa.

The concept of dysmyelination was put forth by Poser [7] as "an expression of a genetically determined enzymatic disorder in myelinogenesis." It seems from this definition that any hereditary disease or "inborn error of metabolism" that results in loss of myelin would be included in this category. However, Poser also stated that his theory of dysmyelination implies that the "myelin initially formed is abnormally constituted, thus inherently unstable, vulnerable and liable to degeneration."

As our understanding of the biochemistry of nervous system diseases progressed, and especially with the advent of reliable procedures for isolating and characterizing myelin, it became apparent that many of the hereditary diseases do not fit a composite definition of dysmyelinating diseases, i.e., hereditary diseases in which abnormal myelin is formed and subsequently destroyed. At the present time, only two such diseases are known: metachromatic leukodystrophy (MLD) and Refsum's syndrome.

We propose a third major class of myelin diseases: the hypomyelinating conditions (this term was introduced by Samorajski, Friede, and Reimer [50]). These would be inherited diseases in which, for various reasons, the myelination process is arrested or impeded and insufficient myelin is produced. If there is an almost complete arrest of myelinogenesis early in development, only small amounts of an immature, uncompacted myelin, chemically and ultrastructurally different from mature myelin, are formed. Examples include the CNS of the "quaking" mouse and possibly Pelizaeus-Merzbacher disease. In other cases, the myelin sheath may be similar in chemistry and ultrastructure to normal controls, except that there is less total myelin (some inborn errors of amino acid metabolism probably fall in this category).

It is obvious that there are many unknown factors regarding the etiology, pathological mechanism, and biochemistry of many of the diseases of myelin, and therefore no completely satisfactory classification of myelin disorders is possible at present.

Among the many disorders involving myelin, EAE plays a unique role. Because of the manner in which this disease can be induced, it is clearly immunological in nature and there is general agreement that myelin is a primary target in this disorder. Therefore, the immunological basis of EAE and its possible relationship to multiple sclerosis are discussed separately in detail.

DEMYELINATING DISEASES

CHEMICAL FINDINGS

The analytical findings in whole brain (or in white matter, which usually shows more pronounced symptoms) are similar in many types of demyelination, whether primary or secondary. These are an increase of water, a decrease of myelin constituents – especially proteolipid, cerebroside, phosphatidalethanolamine, and cholesterol – and the presence of cholesterol esters. All of these changes can be explained by the breakdown and gradual loss of myelin and its replacement by tissue (e.g., extracellular fluid, astrocytes, inflammatory cells) that is much more hydrated, relatively lipid-poor, and, of course, free of myelin-specific constituents. The magnitude of these changes varies considerably among different specimens of the same disease and even from one area of brain to another. These variations undoubtedly reflect the severity, duration, and activity of the disease process at the time of biopsy or necropsy.

In addition to these changes in the composition of whole brain, myelin composition itself is altered in many diseases in which it is involved. These abnormalities have been found to include nonspecific changes in demyelination, as well as specific changes in two cases of dysmyelination to be discussed later. Abnormal myelin was first reported in a case of subacute sclerosing panencephalitis (SSPE), a disease caused by an atypical measles virus infection (so-called slow-virus, or slow-acting virus, infection) of the brain [46]. It is probable that this disease involves destruction of both neurons and oligodendroglia. The brain white matter showed the typical changes mentioned previously for severe demyelination. The isolated myelin has a grossly normal ultrastructural appearance and a normal lipid-protein ratio. However, it was found to have much more cholesterol, less cerebroside, and less phosphatidalethanolamine than normal human myelin (Table 24-1). No cholesterol esters were found in the myelin, although cholesterol esters were abundant in the white matter.

It has been postulated that the abnormal myelin represents a partially degraded form and that therefore such nonspecific changes might be seen in other demyelinating conditions. Subsequently, similar abnormalities were found in myelin from human brains with spongy degeneration (Canavan's disease) [33], G_{M1} gangliosidosis and Tay-Sachs disease [55], Niemann-Pick disease [32], and Schilder's disease [56]. These diseases, although they all involve some form of demyelination (probably secondary in the case of the lipidoses, possibly primary in Canavan's and Schilder's disease), differ in etiology, pathology, and biochemistry. These data are support for the postulate that the abnormality is produced by a nonspecific degradative process. When there is an accumulation of considerable cholesterol ester, such as in SSPE or Schilder's disease, these compounds are not found in the myelin itself. The purification of myelin involves discontinuous sucrose gradients, and a light fraction, containing cholesterol esters, is separated from the denser myelin.

Interestingly, abnormally constituted myelin has not been found in some demyelinating diseases. Several groups have examined isolated myelin in cases of

TABLE 24-1. Human Myelin Composition in Two Diseases Compared with Controls

Lipids[a]	Control	SSPE	MLD
Total lipid, % dry weight	70.0	73.7	63.2
Cholesterol	27.7	43.7	21.2
Cerebrosides	22.7	18.8	9.0
Sulfatides	3.8	2.8	28.4
Total phospholipids	43.1	36.6	36.1
Ethanolamine phosphatides	15.6	9.7	8.1
Lecithin	11.2	10.4	10.7
Sphingomyelin	7.9	8.8	7.1
Serine phosphatides	4.8	4.6	3.8
Phosphatidylinositol	0.6	1.4	3.1
Plasmalogen	12.3	9.1	5.3

[a]Individual lipids expressed as weight % of total lipid.

MS, the prototypical myelinoclastic disease, and have not, as yet, found any significant changes. Apparently, in this disease the islands of demyelination are separated by areas of normal white matter. Presumably, the partially degraded myelin, which must be present at the edges of plaques, is discarded either during dissection or during isolation of myelin, or else it occurs to such a small extent in the final preparation that its effect on the analytical figures is negligible. These findings are evidence against the long-standing speculation that myelin in MS is abnormally formed in childhood and therefore is susceptible to degradation by environmental agents to which normal myelin is resistant.

Another disease in which the myelin has been found to be grossly normal is globoid cell leukodystrophy (Krabbe's disease) [25]. This genetic disorder involves dramatic pathological alteration of the white matter with a severe loss of myelin and of oligodendroglia. The genetic block recently has been shown to be expressed as a lack of galactosylceramide β-galactosidase, the enzyme which degrades cerebroside to ceramide. Curiously, excess amounts of cerebroside are not found upon whole-brain analysis, but they do accumulate in the special storage cell, the globoid cell. Poser [7] considered this disease to be a dysmyelinating disease and one would expect that the myelin would contain abnormal accumulations of cerebroside (analogous with the findings in MLD, see below). The reason it does not is unclear, but it is believed that the myelin that is present is material formed early in myelination, before the enzyme deficiency has fully manifested itself. It has been postulated that the accumulation of cerebroside then elicits the globoid body response, followed by death of oligodendroglia and little further production of myelin, normal or otherwise. Thus, this disease might be better classified as a hypomyelinating disease.

MECHANISMS OF DEMYELINATION

Very little is known about the actual mechanism initiating the demyelinating process in most diseases. Degradation of myelin might be accelerated either by action of the cells and enzymes normally involved in metabolic turnover of myelin, or by an immunological or toxin-mediated process. In autoimmune or allergic types of demyelination (e.g., EAE, experimental allergic neuritis or EAN, Guillain-Barré syndrome, and possibly MS), demyelination takes place in the presence of inflammatory cells (lymphocytes, plasma cells, and mononuclear cells) that arise from the circulation. The mononuclear cells appear to attack ultrastructurally normal myelin lamellae directly, splitting them away from the axon [62].

Tissue culture studies (see below) indicate a possible role for circulating antibodies in this process. In other types of primary demyelination (e.g., carbon monoxide intoxication or polyneuropathy caused by diphtheria toxin), myelin damage apparently precedes the appearance of macrophages in a lesion, and demyelination occurs in the absence of lymphocytes or plasma cells. In either case, however, large segments of the myelin sheath eventually are engulfed by macrophages. Here the myelin is degraded by lysosomal enzymes to small molecules, which are absorbed in the circulation or reused in metabolic pathways. Cholesterol, however, is apparently the one myelin constituent that cannot be degraded to smaller units. This compound is esterified and remains in phagocytes for some time at the site of the lesion. Cholesterol esters are essentially absent in mature brain, so their presence is considered indicative of active demyelination. These compounds are also responsible for the neutral fat-staining, or sudanophilia, demonstrated histochemically in many demyelinating diseases. It is possible to have considerable demyelination with little cholesterol ester present if the disease process has been inactive for some time before analysis. Presumably, the macrophages have had time to leave the site of the lesion. For example, in demyelinated areas (or plaques) in MS, the center of the plaque will not contain any macrophages or cholesterol esters, while the margins of the plaque will have both.

Numerous hydrolytic enzymes have been studied in preparations from MS and control brains. Both qualitative and quantitative changes in nonspecific esterases are observed. In homogenates from demyelinating white matter, there are losses of soluble enzymes that hydrolyze certain naphthol esters, as well as of 5′-nucleotidase activity [12]. The meaning of these changes is obscure. More relevant is the observation that there is a four-fold increase in acid proteinase activity in MS plaque regions [24]. This proteinase activity might be involved in releasing encephalitogenic polypeptides into the circulation and promoting an immunological reaction.

The enzymological changes that take place during Wallerian degradation have been studied in a number of laboratories. In degenerating nerves of both the peripheral and central nervous systems, there is an increase in a number of lysosomal enzymes [43]. Some of these changes are considerable: β-galactosidase activity in the tibial nerve of the rabbit increases 15-fold after nerve section, for example. In homogenates of rat sciatic nerve prepared 14 days after onset of Wallerian degeneration, there is a significant increase in incorporation of radioactive choline into choline-containing phospholipids [13]. This is an unexpected result since it is known that the concentration of these lipids progressively declines during the course of Wallerian degeneration. This seeming anomaly has been explained by the observation that the choline-containing precursor, cytidine diphosphate-choline, is both synthesized and degraded more rapidly in homogenates from the degenerating nerve. Thus, although the absolute level of this precursor is lower, its specific activity and that of its phospholipid products are much higher. Interpretation of these data with regard to demyelination is difficult because it is not clear in what cellular type – neuron, Schwann cell, or phagocytic cells – the enzymatic activities are located. Hydrolysis of UDP-glucose, presumably a precursor of UDP-galactose and therefore of cerebroside, is also increased.

During Wallerian degeneration, there is a progressive decrease of cholesterol ester hydrolase [45]. This observation might offer a partial explanation for the observed accumulation of cholesterol esters. It is also possible that loss of cholesterol esterase activity reflects only the disappearance of myelin since a considerable portion of this enzyme may be localized in myelin. $NADP^+$-dependent isocitrate dehydrogenase activity also increases during the same period. This enzyme has been implicated in the process of myelinogenesis by its developmental pattern. Activity of the enzyme reaches a peak during the period of maximal myelination, as shown by Klee and Sokoloff [35], and its increase under these conditions may represent the Schwann cell proliferation that takes place after nerve section.

DYSMYELINATING DISEASES

The category of dysmyelinating diseases was initially thought to include most inborn errors of metabolism involving the nervous system primarily. However, only two disorders clearly fit into this category, each characterized by virtually complete absence of a degradative enzyme. In metachromatic leukodystrophy (MLD), there is an almost total absence of a sulfatase that degrades cerebroside sulfate to cerebroside [44], with the consequent accumulation of sulfatides in the brain and other organs. This disease, more properly considered a lipidosis or sulfatide storage disease, is accompanied by an almost total lack of myelin, at least in its final stages. The small amount of myelin isolated from several cases of MLD shows a very high sulfatide content and a low cerebroside content [5, 46a] (see Table 24-1). There are other variations in lipid composition, but the lipid-protein content is relatively normal and the myelin has a grossly normal appearance.

In Refsum's disease, patients lack the ability to degrade branched-chain fatty acids by α-oxidation [54]. Although this enzyme is not essential for degradation of normal fatty acids, it is necessary for the degradation of phytanic acid (3,7,11,14-tetramethyl hexadecanoic acid). This tetraisoprenoid fatty acid, a normal dietary product that comes from the degradation of pigments from green plants, has been found to accumulate in myelin phospholipids [41] as it does in other tissues of the body.

The possibility cannot be eliminated that some diseases which are really dysmyelinating disorders have been classified as demyelinating diseases because current analytical methods are not sensitive enough to detect the abnormality in myelin.

HYPOMYELINATING DISEASES

Certain disorders of amino acid metabolism, notably phenylketonuria, might cause a degree of hypomyelination. Upon histological examination, the amount of myelin in brains of patients suffering from phenylketonuria appears abnormally low, and the white matter may be up to 40 percent deficient (as compared to controls) in such myelin-specific components as proteolipid protein and cerebroside [48]. However, in certain cases of phenylketonuria the deficit in myelin-specific components may be considerably less [31]. The related model system of experimental hyperphenylalaninemia (induced by injecting animals with large doses of phenylalanine) is characterized by a decreased in vitro and in vivo ability to incorporate radioactive precursors into cerebral lipids and protein [42, 53]. This inhibition may not be the result of the phenylalanine itself but of the abnormal, deaminated metabolites that accumulate.

The adult "quaking" mouse mutant has a severe CNS myelin deficit that is demonstrable histologically and ultrastructurally. Although the oligodendroglial cells appear relatively normal, axons of adult quaking mice have only a few lamellae of myelin instead of the thick sheath of twenty or more lamellae present in similar axons of litter-mate controls [61]. The ultrastructural picture (as well as the lipid and protein analyses of isolated myelin) of adult quaking mouse is characteristic of the developing 7- to 10-day-old CNS of normal mice. Enzymatic activity for cerebroside formation (UDP-galactose-ceramide galactosyl transferase) is more than 60 percent depressed in the mutants as compared with control litter-mates at various age points, but the regulatory enzymes that are involved in biosynthesis of sphingolipids found in gray matter are completely normal. Two other murine mutants, characterized histologically as having pathologically low levels of myelin, are defective in enzymatic activity for the biosynthesis of myelin-specific lipids. Although these observations are certainly relevant to an understanding of the pathology involved, there is no reason to believe that the enzymatic deficits are primary.

Pelizaeus-Merzbacher disease may be analogous to a hypomyelination disorder in one of the murine mutants, but no in vitro studies of biosynthesis of myelin lipids have been carried out with material from human disorders.

IMMUNOLOGICAL BASIS OF EXPERIMENTAL ALLERGIC ENCEPHALOMYELITIS AND ITS RELATIONSHIP TO MULTIPLE SCLEROSIS

The first suggestion of a possible autoallergic cause for multiple sclerosis arose from observations of patients who had been treated with the Pasteur antirabies vaccine which contained both attenuated virus and the nerve tissue in which it had been prepared. A small minority of patients developed an acute, severe, sometimes fatal, postinoculation encephalomyelitis, characterized histologically by areas of demyelination. These lesions somewhat resembled those in MS brain, and they were unrelated to the pathological changes observed in an untreated case of rabies. It was therefore suggested that the demyelination was due to the injected brain tissue rather than to the virus. The laboratory counterpart to the clinical observations began in the 1920s, when a number of investigators noted that animals produce organ-specific, complement-fixing, antibrain antibodies in response to inoculations of CNS tissue, and that a series of such inoculations can lead to neurological symptoms of paralysis, ataxis, and sphincter disturbances, accompanied by a characteristic lesion consisting of perivascular and diffuse lymphocytic infiltrations, as well as demyelination with sparing of axons. If the nervous tissue is first suspended in Freund's adjuvant – a combination of mineral oil, killed mycobacteria, and a binding agent – the allergenic potency of tissue is increased and the onset of the disease appears within weeks, rather than months, after inoculation. The significance of the relation of EAE to MS is based largely on the similarity of the histological lesions in EAE cases to those seen in some acute forms of MS. Much of this information can be found in the reports of two symposia, one edited by Kies and Alvord in 1959 [4] and another by Scheinberg, Kies, and Alvord in 1965 [51]. Paterson's chapter [6] is also particularly informative. There are three factors involved in the demyelination disorders of MS and EAE: the antigen, the immunological reaction, and the pathophysiological response in the nervous system.

THE ANTIGEN

The earliest investigators induced EAE by injecting whole brain tissue or white matter into experimental animals, a method that is still widely used in conjunction with Freund's adjuvant. Since then, a considerable effort has been directed toward the isolation and identification of a specific antigen as the causative agent in EAE. The search for such an encephalitogenic substance required quantitative techniques which could evaluate the relative potency of various fractions. Kies and Alvord [4] succeeded in establishing an assay system which utilized an index based on the appearance and severity of both clinical symptoms and histopathological lesions in guinea pigs. With the aid of quantitative evaluation, the water-soluble protein fractions of myelin became recognized as an antigen more potent than whole myelin. Eventually, the major encephalitogenic component in the CNS was identified as a basic protein with a molecular weight of about 18,000.

Intensive research on the amino acid sequence of the basic protein of myelin, particularly with regard to the antigenic sequences, has taken place in the last few years. The complete sequences of human and bovine basic protein have been elucidated by Eylar [26], and the sequence of human basic protein was independently determined by Carnegie [21]. The sequence of amino acids in bovine and human myelin differs by only 11 residues. The complete sequences of bovine and human basic protein are shown in Figure 24-1. (There is some disagreement about certain parts of this sequence, but the discrepancies are relatively minor.)

An unusual amino acid, a methylated arginine (107 in the sequence of Fig. 24-1), has been described by Baldwin and Carnegie [11] and by Brostoff and Eylar [18]. It is known that certain peptides derived from the basic protein can retain encephalitogenic activity [39]. A peptide of nine amino acids (114 to 122 in Fig. 24-1) has been isolated from a pepsin digest of the basic protein of bovine spinal cord, its amino acid sequence has been determined, and it has been shown to produce EAE in guinea pigs [28]. A similar peptide from human basic protein has been isolated by Carnegie [21]. A study using chemically synthesized peptides [59] demonstrated that the minimal sequence requirement for this peptide to induce EAE was Trp–...–...–...–...–Gln–Lys (Arg); various amino acid substitutions for the four amino acids between tryptophan and glutamine did not eliminate encephalitogenic activity. Peptides containing this sequence induce EAE in rabbits and guinea pigs, but are inactive in monkeys.

Another peptide, which does not contain the tryptophan region and which is active in rabbits but inactive in guinea pigs, was first studied by Kibler et al. [34]. Eylar, Westall, and Brostoff [30] demonstrated that this peptide corresponds in sequence to amino acids 44 to 89 in Figure 24-1. Still a third EAE-inducing peptide, active in monkeys but not in rabbits or guinea pigs, has been isolated [27]. The sequence in this peptide corresponds to amino acids 134 to 170 in Figure 24-1. The clinical implications of these studies will be mentioned later.

THE IMMUNOLOGICAL REACTION

Evidence for the participation of an immunological mechanism was obtained by the demonstration that EAE appeared in a normal rat that had been joined in parabiotic union to an inoculated rat [38]. Taking this as a clue, Paterson [6] and his co-workers began a series of experiments which established beyond doubt that EAE can be transferred from a treated donor animal to a normal recipient by means of sensitized lymph node cells. Later, these experiments were confirmed and extended to different species by other investigators. All of the experiments employed viable lymph node cells as the transferring agent. Finally, Wenk, Levine, and Warren [58] supplied data proving that circulating cellular elements, obtained from the peripheral blood of rats during the development of EAE, possess the ability to produce characteristic symptoms and histological lesions when introduced parenterally in sufficient quantity into histocompatible recipients.

An in vitro model system, the antigen-induced inhibition of migration of sensitized cells, may be of considerable value in examining the cellular aspects of EAE as a model of delayed hypersensitivity. In the experiments reported by David and Paterson [23], the peritoneal exudate cells of guinea pigs sensitized to spinal cord from adult guinea pigs were inhibited from migrating out of a capillary tube in the presence of encephalitogen-containing brain extracts. They were not inhibited by kidney or by neonatal brain that did not contain encephalitogen. The reverse

```
                                   His–Gly                           Thr
NAc–Ala–Ser–Ala–Gln–Lys–Arg–Pro–Ser–Gln–Arg–Ser–Lys–Tyr–Leu–Ala–Ser–Ala–Ser–Thr–Met–Asp–
                                    10                                          20
                                                                  Ile
His–Ala–Arg–His–Gly–Phe–Leu–Pro–Arg–His–Arg–Asp–Thr–Gly–Ile–Leu–Asp–Ser–Leu–Gly–Arg–
                               30                                  40
           Gly                                                    Ser       Pro
Phe–Phe–Gly–Ser–Asp–Arg–Gly–Ala–Pro–Lys–Arg–Gly–Ser–Gly–Lys–Asp–Gly–His–His–Ala–Ala–Arg–
                             50                                    60
  Ala                                                       Thr
Thr–Thr–His–Tyr–Gly–Ser–Leu–Pro–Gln–Lys–Ala–Gln–Gly–His–Arg–Pro–Gln–Asp–Glu–Asn–Pro–Val–
                     70                                     80
                                                                                 Methyl
                                                                                   |
Val–His–Phe–Phe–Lys–Asn–Ile–Val–Thr–Pro–Arg–Thr–Pro–Pro–Pro–Ser–Gln–Gly–Lys–Gly–Arg–Gly–
           90                                     100
                                                   Arg
Leu–Ser–Leu–Ser–Arg–Phe–Ser–Trp–Gly–Ala–Glu–Gly–Gln–Lys–Pro–Gly–Phe–Gly–Tyr–Gly–Gly–Arg–
  110                                   120                                          130
                                  Phe          Val
Ala–Ser–Asp–Tyr–Lys–Ser–Ala–His–Lys–Gly–Leu–Lys–Gly–His–Asp–Ala–Gln–Gly–Thr–Leu–Ser–Lys–
                                 140                                  150

Ile–Phe–Lys–Leu–Gly–Gly–Arg–Asp–Ser–Arg–Ser–Gly–Ser–Pro–Met–Ala–Arg–Arg–COOH
                             160                              170
```

FIGURE 24-1. Amino acid sequence of basic protein of bovine myelin. Substitutions in human basic protein are shown above the appropriate position of the bovine sequence. Between positions 10 and 11, a His–Gly dipeptide occurs in the human protein. (After Eylar [26].)

experiment was also reported; cells sensitized to newborn (nonencephalitogenic) rat cord also were not inhibited by adult guinea pig brain. Since Paterson's demonstration of the passive transfer of EAE, data from other fields of investigation [8, 57] have implicated the cell-associated antibody as the sole immunological agent and have excluded circulating antibodies from playing a significant role either in EAE or in other reactions of this type. Until recently, there has been no evidence that any of the clinical symptoms or histological signs of EAE could be transferred by means of serum from an affected to a normal animal. This failure to obtain a passive transfer of EAE in animals can be explained by assuming either that the mechanism involved is a cell-mediated, delayed type of hypersensitivity or that, if circulating factors are the causative agent, the blood-brain barrier protects the CNS.

Now a more sensitive assay method for studies of circulating factors involved in EAE and MS has been made available with the development of a tissue culture system [17]. Explanted fragments of rat cerebellum organize in vitro in such a way that the neurological and neuronal elements interact to form myelin. This offers a specific test object for the evaluation of animal and human serum for demyelinating activity. No barrier exists between the culture and its nutrient medium. In the first experiments [14], samples of serum were obtained from rabbits before and after they had been inoculated with whole spinal cord in Freund's adjuvant. The preinoculation sera, as well as those obtained postinoculation from unaffected rabbits and samples from rabbits inoculated with rat kidney in adjuvant, produced no observable change in cultured rat cerebellum. On the other hand, all samples of serum from EAE-affected animals, whether they were obtained before or after the onset of clinical signs, produced a specific and characteristic pattern of demyelination. The same response has also been frequently observed in cultures exposed to the sera of patients experiencing an active phase of MS, and rarely in other patients or normal controls [1].

It should be noted that consistent demyelination in tissue cultures was observed only when serum was obtained from animals sensitized against homogenates of whole CNS tissue. Serum from animals sensitized against purified basic protein, although containing antibodies against basic protein, did not cause demyelination [52]. The tissue culture system can also be used to demonstrate the concurrent role of sensitized lymph node cells and circulating antibodies in demyelination [16].

The organ specificity of EAE has since been confirmed using the tissue culture system. The same pattern of demyelination has been obtained with serum from EAE-affected guinea pigs, rats, and mice, and the test object has been varied to include rat spinal cord and mouse spinal cord, cerebellum, and cerebrum. In no instance has demyelination been produced in peripheral nerve cultures exposed to similar concentrations of EAE serum for similar periods of time. This observation is parallel to its counterparts in experimental allergic neuritis (EAN), in which peripheral, but not central, myelin is affected in culture by serum or lymph node cells obtained from affected animals [60, 63].

The culture system has also been employed to determine the immunological characteristics of the active factors in EAE serum [10]. It was demonstrated that complement was necessary for the reaction to take place because its removal by exposure to heat, NH_4OH, or heparin destroyed the demyelinating potency of the nutrient solutions. The demyelinating potency could then be restored by the addition of complement in the form of fresh guinea pig serum.

Analysis of the serum revealed that the complement-dependent activity resides in the γ_2-globulin (7S) fraction of rabbit serum. The albumin and α-globulin (19S) fractions were inactive. The myelinotoxic activity could also be removed by exposing the whole serum to its homologous or heterologous CNS tissue, whereas red blood cells, liver, lung, or kidney had no such effect. These studies, therefore, provide evidence that the myelinotoxic factors in EAE serum are antibodies [10]. Lumsden [39] has recently obtained evidence for the localization of circulating immunoglobulins and complement at the actively demyelinating border of MS plaques.

The relative roles of circulating antibodies and of cell-mediated response in the pathogenesis of EAE and MS is a topic of intensive research and considerable controversy. These factors are not mutually exclusive nor necessarily independent. It is possible that immunocompetent cells migrate to the site of demyelination and release soluble antibodies which rapidly combine with antigen (presumably a component of myelin). Thus, the antibodies, although potentially circulating factors, might not actually enter the bloodstream in any appreciable amount.

THE PATHOPHYSIOLOGICAL RESPONSE

The existence of perivascular infiltrations as a characteristic "lesion" in EAE already presupposes that the protective blood-brain barrier has been affected. Direct experimental evidence of an increased vascular permeability has been revealed by trypan blue, radioactive iodinated bovine albumin, and ^{125}I-labeled γ-globulin and leukocytes. Permeability to γ-globulin was occasionally the earliest evidence of vascular pathology on the first day of clinical signs. In 1960 Levine [36] showed the importance of increased permeability by producing a cyanide lesion, which simultaneously acted to localize the EAE lesions about the permeable vessels.

In a crucial experiment by Oldstone and Dixon [47], Lewis strain rats were sacrificed serially after having been inoculated with encephalitogen. Immunohistochemical methods for fibrinogen, γ-globulin, and β_{1c}-globulin were used to examine many organs and have revealed increased vascular permeability specifically in the CNS. Abnormal perivascular and intramural deposits of fibrinogen were already apparent by the fifth to sixth day after inoculation, and by the seventh day exudations of fibrinogen and γ-globulin were readily seen. In 10 to 15 percent of the sections, the globulins were also deposited in myelin tracts, without particular relations to blood vessels. However, histological lesions, e.g., perivascular cellular infiltrates, were not found until the tenth day. This increased vascular permeability of the CNS in EAE prior to the appearance of detectable inflammatory cells suggests that a humoral factor may initiate the sequence of pathological events.

In the previous sections, the in vivo demyelinating response of the CNS to sensitized lymph node cells (as well as the response of CNS tissue cultures to EAE serum) was described. For many years it was thought that remyelination did not take place in the mammalian CNS. However, the tissue culture system has been used to demonstrate that remyelination can follow total demyelination if the tissue is washed free of the EAE or MS serum and returned to its normal nutrient medium [14]. Bunge, Bunge, and Ris [19] also reported a degree of remyelination in the spinal cord of adult cats previously exposed to demyelination induced by repeated withdrawal and reinjection of cerebrospinal fluid.

Since then, many reports have appeared of remyelination in both animals and humans subjected to a demyelinating experience. At a certain point during a demyelinating experience, the ability to remyelinate may be lost. In the analogous situation in tissue culture, the exposure to demyelinating serum can be prolonged to the point at which the capacity for remyelination is lost [49]. Concomitant ultrastructural investigation has demonstrated a state of sclerosis characterized by the sequestration of axons by encircling astrocytic processes and a marked increase in gliofibrillar content of the astrocytes. The clinical implications of these studies are clear with respect to the ability of the CNS to repair damage to myelin if a pathological process can be halted in time.

As has been discussed previously, the fact that myelin is destroyed in a particular disease does not imply that this is the initial lesion or the primary cause for malfunction of the nervous system. The cultured tissues offer a model system for correlating demyelination with other pathological events. The neurons in the cultured fragments form synaptic interrelationships and develop complex patterns of bioelectrical activity that signal the propagation of the nervous impulse along axons and the establishment of intricate interneuronal networks [2, 22]. Cultures of cerebrum and spinal cord were exposed to sera obtained from animals with EAE or from patients during an acute exacerbation of MS [15]. The presence of either serum led to an extensive alteration in the bioelectrical properties of the tissue long before any morphological changes had been detected. The complex responses characteristic of synaptic transmission often disappeared within a few minutes after the application of the serum, but simple axon spikes still could usually be evoked. The complex polysynaptic functions returned within minutes to hours after removal of the offending serum. Complement was needed for the blocking action since its removal abolished and its replacement restored the serum's potency. Carels and Cerf [20] have reported a detailed confirmation of the tissue culture studies of MS serum. They employed a preparation of isolated spinal cord of frog as the test object. This was exposed to MS serum and to normal serum. The MS serum produced a rapid, reversible, complement-dependent depression of the polysynaptic responses of the spinal cord.

These data suggest that the pathophysiology of EAE and MS may involve at least two sets of circulating antibodies: antiglial antibodies, which demyelinate and eventually produce an irreversible state of sclerosis, and antineuronal antibodies, which produce no visible structural or ultrastructural change but instead a rapid and reversible block of neuronal function.

TREATMENT AND SUPPRESSION OF EAE

Various techniques have been employed to suppress or modify EAE. These experiments have been directed primarily along immunological lines and have, consequently, increased our knowledge of the mechanisms involved. They have also supplied clues of potential therapeutic value in the treatment of patients with MS.

Many immunosuppressive agents have been tried in attempts to affect the course of EAE. Gabrielson and Good [3] have recently prepared an exhaustive review of the substances used to suppress adaptive immunity, many of which have appeared in the literature on EAE. They include salicylates, adrenal steroid hormones, alkylating agents, folic acid antagonists, purine base antimetabolites, antibiotics, plant alkaloids, antilymphocyte serum, and others. The action of a number of these, particularly the purine and pyrimidine analogs, the folic acid antagonists, and the alkylating agents, have been reviewed recently [6, 9]. Usually, the action of these compounds is limited to the period of treatment and for a short time thereafter.

A countertheme that has run throughout the study of EAE is that the encephalitogen itself might provide the means of altering or preventing EAE. The encephalitogen-induced inhibitions of EAE have usually been caused by whole tissue. Levine, Hoenig, and Kies [37] have demonstrated recently that the intravenous injection of basic protein into recipient rats will prevent the passive transfer of EAE by means of sensitized lymph node cells obtained from donor rats inoculated with the same basic protein. Einstein, Csejty, Davis, and Rauch (personal communication) have also recently examined the protective action of various basic proteins isolated from human and bovine neural tissue. They used both encephalitogenic basic protein and chemically modified, nonencephalitogenic basic protein to protect guinea pigs challenged with basic protein in a complete adjuvant. Both proteins were partially successful in preventing clinical signs and histological lesions. Eylar et al. [29] have succeeded in suppressing EAE induced in monkeys by injection of bovine or human myelin basic protein in complete Freund's adjuvant. Even after onset of clinical symptoms, clinical recovery could be achieved by a course of injections of basic protein in an incomplete Freund's adjuvant.

This line of investigation has obvious therapeutic implications, and it is hoped that knowledge of the amino acid sequence and the location of antigenic sites will enable investigators to design an antigenic but nonencephalitogenic preparation that is able to suppress EAE. The possibility of suppression of EAE is the result of years of research in many laboratories and has important implications for both basic and clinical sciences. A major clinical question is what results this type of therapy might have on patients suffering from MS (who constitute possibly 0.05 percent of the population of the United States alone). It should, however, be emphasized that the correlation between EAE and MS (besides the morphological similarity of the lesions) is based largely on evidence that serum of patients with MS contains circulating factors directed against various components of the CNS (e.g., the neuroglial cell membrane, as evidenced by demyelination in tissue culture, and the neuronal cell membrane, as demonstrated by alteration in electrical properties). In EAE, neurological symptoms and lesions appear and basic protein is clearly implicated as the causative factor for the immune response. The evidence that immunological response to basic protein is primarily responsible for MS is far from being firmly established. It is possible that antigens other than basic protein may also be responsible for the lesions and clinical symptoms of MS.

The characterization and investigation of interrelationships between various antibodies specific to EAE or MS, the sensitized cells, the antigenic components of the CNS, and the various states of response or suppression of EAE or the clinical stages of MS remain as interesting and exciting areas for future investigation.

REFERENCES

BOOKS AND REVIEWS

1. Bornstein, M. B. A tissue culture approach to demyelinative disorders. *Natl. Cancer Inst. Monogr.* 11:197, 1963.
2. Crain, S. M., Peterson, E. R., and Bornstein, M. B. Formation of Functional Interneuronal Connections Between Explants of Various Mammalian Central Nervous Tissues During Development *in Vitro.* In Wolstenholme, G. E. W., and O'Connor, M. (Eds.), *Growth of the Nervous System* (a Ciba Symposium). London: Churchill, 1968, pp. 13–31.
3. Gabrielson, A. E., and Good, R. A. Chemical Suppression of Adaptive Immunity. In Dixon, F. J., Jr., and Humphrey, J. H. (Eds.), *Advances in Immunology.* Vol. 6. New York: Academic, 1967, pp. 91–229.
4. Kies, M. W., and Alvord, E. C., Jr. *"Allergic" Encephalomyelitis.* Springfield: Thomas, 1959.
5. Norton, W. T., and Poduslo, S. E. Metachromatic Leucodystrophy: Chemically Abnormal Myelin and Cerebral Biopsy Studies of Three Siblings. In Ansell, G. B. (Ed.), *Variation in the Chemical Composition of the Nervous System.* Oxford: Pergamon, 1966, p. 82.
6. Paterson, P. Y. Experimental Autoimmune (Allergic) Encephalomyelitis. In Miescher, P. A., and Muller-Eberhard, H. J. (Eds.), *Textbook of Immunopathology.* Vol. 1. New York: Grune & Stratton, 1968, pp. 132–149.
7. Poser, C. M. Diseases of the Myelin Sheath. In Minckler, J. F. (Ed.), *Pathology of the Nervous System.* Vol. 1. New York: McGraw-Hill, 1968.
8. Uhr, J. W. Delayed hypersensitivity. *Physiol. Rev.* 46:359, 1966.

ORIGINAL SOURCES

9. Alvord, E. C., Jr., Shaw, Ch-M., Fahlberg, W. J., and Kies, M. W. An analysis of various types of inhibition of experimental "allergic" encephalomyelitis in the guinea pig. *Z. Immun.-Forsch.* 126:217, 1964.

10. Appel, S. H., and Bornstein, M. B. The application of tissue culture to the study of experimental allergic encephalomyelitis. II. Serum factors responsible for demyelination. *J. Exp. Med.* 119:303, 1964.
11. Baldwin, G. S., and Carnegie, P. R. Isolation and partial characterization of methylated arginines from the encephalitogenic basic protein of myelin. *Biochem. J.* 123:69, 1971.
12. Barron, K. D., and Bernsohn, J. Brain esterases and phosphatases in multiple sclerosis. *Ann. N.Y. Acad. Sci.* 122:369, 1965.
13. Berry, J. F., and Coonrad, J. D. Hydrolysis of nucleoside diphosphate esters in peripheral nerve during Wallerian degeneration. *J. Neurochem.* 14:245, 1967.
14. Bornstein, M. B., and Appel, S. H. The application of tissue culture to the study of experimental "allergic" encephalomyelitis. I. Patterns of demyelination. *J. Neuropathol. Exp. Neurol.* 20:141, 1961.
15. Bornstein, M. B., and Crain, S. M. Functional studies of cultured brain tissues as related to "demyelinative disorders." *Science* 148:1242, 1965.
16. Bornstein, M. B., and Iwanami, H. Experimental allergic encephalomyelitis: Demyelinating activity of serum and sensitized lymph node cells on cultured nerve tissues. *J. Neuropathol. Exp. Neurol.* 30:240, 1971.
17. Bornstein, M. B., and Murray, M. R. Serial observations on patterns of growth, myelination formation, maintenance and degeneration in cultures of newborn rat and kitten cerebellum. *J. Biophys. Biochem. Cytol.* 4:499, 1958.
18. Brostoff, S., and Eylar, E. H. Localization of methylated arginine in the A1 protein from myelin. *Proc. Natl. Acad. Sci. U.S.A.* 68:765, 1971.
19. Bunge, M. B., Bunge, R. P., and Ris, H. Ultrastructural study of remyelination in an experimental lesion in adult cat spinal cord. *J. Biophys. Biochem. Cytol.* 10:67, 1961.
20. Carels, G., and Cerf, J. A. Reversible depression of the reflex activity of the isolated spinal cord of the frog by the serum of subjects suffering from multiple sclerosis. *Rev. Neurol.* 115:242, 1966.
21. Carnegie, P. R. Amino acid sequence of the encephalitogenic basic protein from human myelin. *Biochem. J.* 123:57, 1971.
22. Crain, S. M. Development of "organotypic" bioelectric activities in central nervous tissues during maturation in culture. *Int. Rev. Neurobiol.* 9:1, 1966.
23. David, J. R., and Paterson, P. Y. In vitro demonstration of cellular sensitivity in allergic encephalomyelitis. *J. Exp. Med.* 122:1161, 1965.
24. Einstein, E. R., Dalal, K. B., and Csejtey, J. Increased protease activity and changes in basic proteins and lipids in multiple sclerosis plaques. *J. Neurol. Sci.* 11:109, 1970.
25. Eto, Y., Suzuki, K., and Suzuki, K. Globoid cell leukodystrophy (Krabbe's disease): Isolation of myelin with normal glycolipid composition. *J. Lipid Res.* 11:473, 1970.
26. Eylar, E. H. Amino acid sequence of the basic protein of the myelin membrane. *Proc. Natl. Acad. Sci. U.S.A.* 67:1425, 1970.

27. Eylar, E. H., Brostoff, S., Jackson, J., and Carter, H. Allergic encephalomyelitis in monkeys induced by a peptide from the A_1 protein. *Proc. Natl. Acad. Sci. U.S.A.* 69:617, 1972.
28. Eylar, E. H., and Hashim, G. A. Allergic encephalomyelitis: The structure of the encephalitogenic determinant. *Proc. Natl. Acad. Sci. U.S.A.* 61:644, 1968.
29. Eylar, E. H., Jackson, J., Rothenberg, B., and Brostoff, S. Suppression of the immune response: Reversal of the disease state with antigen in allergic encephalomyelitis. *Nature (Lond.)* 236:74, 1972.
30. Eylar, E. H., Westall, F. C., and Brostoff, S. Allergic encephalomyelitis. An encephalitogenic peptide derived from the basic protein of myelin. *J. Biol. Chem.* 246:3418, 1971.
31. Foote, J. L., Allen, R. J., and Agranoff, B. W. Fatty acids in esters and cerebrosides of human brain in phenylketonuria. *J. Lipid Res.* 6:518, 1965.
32. Kamoshita, S., Aron, A. M., Suzuki, K., and Suzuki, K. Infantile Niemann-Pick disease: A chemical study with isolation and characterization of membranous cytoplasmic bodies and myelin. *Am. J. Dis. Child.* 117:379, 1969.
33. Kamoshita, S., Rapin, I., Suzuki, K., and Suzuki, K. Spongy degeneration of the brain: A chemical study of two cases including isolation and characterization of myelin. *Neurology* 18:975, 1968.
34. Kibler, F. R., Shapira, R., McKneally, S., Jenkins, J., Selden, P., and Chou, F. Encephalitogenic protein: Structure. *Science* 164:577, 1969.
35. Klee, C. B., and Sokoloff, L. Changes in D (–)-β-hydroxylbutyric dehydrogenase activity during brain maturation in the rat. *J. Biol. Chem.* 242:3880, 1967.
36. Levine, S. Localization of allergic encephalomyelitis in lesions of cyanide encephalopathy. *J. Neuropathol. Exp. Neurol.* 19:239, 1960.
37. Levine, S., Hoenig, E. M., and Kies, M. W. Allergic encephalomyelitis: Passive transfer prevented by encephalitogen. *Science* 161:1155, 1968.
38. Lipton, M. M., and Freund, J. The transfer of experimental allergic encephalomyelitis in the rat by means of parabiosis. *J. Immunol.* 71:380, 1953.
39. Lumsden, C. E. The immunogenesis of the multiple sclerosis plaque. *Brain Res.* 28:365, 1971.
40. Lumsden, C. E., Robertson, D. M., and Blight, R. Chemical studies on experimental allergic encephalomyelitis. Peptide as the common denominator in all encephalitogenic "antigens." *J. Neurochem.* 13:127, 1966.
41. MacBrinn, M. C., and O'Brien, J. S. Lipid composition of the nervous system in Refsum's disease. *J. Lipid Res.* 9:552, 1968.
42. MacInnes, J. W., and Schlesinger, K. Effects of excess phenylalanine on in vitro and in vivo RNA and protein synthesis and polyribosome levels in brains of mice. *Brain Res.* 29:101, 1971.
43. McCaman, R. E., and Robins, E. Quantitative biochemical studies of Wallerian degeneration in the peripheral and central nervous system. II. Twelve enzymes. *J. Neurochem.* 5:32, 1959.
44. Mehl, E., and Jabzkewitz, H. Evidence for the genetic block in metachromatic leucodystrophy (ML). *Biochem. Biophys. Res. Commun.* 19:407, 1965.

45. Mezei, C. Cholesterol esters and hydrolytic cholesterol esterase during Wallerian degeneration. *J. Neurochem.* 17:1163, 1970.
46. Norton, W. T., Poduslo, S. E., and Suzuki, K. Subacute sclerosing leucoencephalitis. II. Chemical studies including abnormal myelin and an abnormal ganglioside pattern. *J. Neuropathol. Exp. Neurol.* 25:582, 1966.
46a. O'Brien, J. S., and Samson, E. L. Myelin membrane: A molecular abnormality. *Science* 150:1613, 1965.
47. Oldstone, M. B. S., and Dixon, F. J. Immunohistochemical study of allergic encephalomyelitis. *Am. J. Pathol.* 52:251, 1968.
48. Prensky, A. L., Carr, S., and Moser, H. W. Development of myelin in inherited disorders of amino acid metabolism. *Arch. Neurol.* 19:552, 1968.
49. Raine, C. S., and Bornstein, M. B. Experimental allergic encephalomyelitis: A light and electron microscope study of remyelination and "sclerosis" in vitro. *J. Neuropathol. Exp. Neurol.* 29:552, 1970.
50. Samorajski, T., Friede, R. L., and Reimer, P. R. Hypomyelination in the quaking mouse. A model for the analysis of disturbed myelin formation. *J. Neuropathol. Exp. Neurol.* 29:507, 1970.
51. Scheinberg, L. C., Kies, M. W., and Alvord, E. C., Jr. (Eds.). Research in demyelinating diseases. *Ann. N.Y. Acad. Sci.* 122:1, 1965.
52. Seil, R. J., Falk, G. A., and Kies, M. W. The in vitro demyelinating activity of sera from guinea pigs sensitized with whole CNS and with purified encephalitogen. *Exp. Neurol.* 22:545, 1968.
53. Shah, S. N., Peterson, N. A., and McKean, C. M. Cerebral lipid metabolism in experimental hyperphenylalaninaemia: Incorporation of ^{14}C-labeled glucose into total lipids. *J. Neurochem.* 17:279, 1970.
54. Steinberg, D., Avigan, J., Mize, C. E., Baxter, J. H., Cammermeyer, J., Fales, H. M., and Highet, P. F. Effect of dietary phytol and phytanic acid in animals. *J. Lipid Res.* 7:684, 1966.
55. Suzuki, K., Suzuki, K., and Kamoshita, S. Chemical pathology of GM_1 gangliosidosis (generalized gangliosidosis). *J. Neuropathol. Exp. Neurol.* 28:25, 1969.
56. Suzuki, Y., Tucker, S. H., Rorke, L. B., and Suzuki, K. Ultrastructural and biochemical studies of Schilder's disease. *J. Neuropathol. Exp. Neurol.* 29:405, 1970.
57. Turk, J. L. (Ed.). Delayed hypersensitivity. Specific cell-mediated immunity. *Br. Med. Bull.* 23:(No. 1), 1967.
58. Wenk, E. J., Levine, S., and Warren, B. Passive transfer of allergic encephalomyelitis with blood leukocytes. *Nature (Lond.)* 214:803, 1967.
59. Westall, F. C., Robinson, A. B., Caccam, J., Jackson, J., and Eylar, E. H. Essential chemical requirements for induction of allergic encephalomyelitis. *Nature (Lond.)* 229:22, 1971.
60. Winkler, G. F. In vitro demyelination of peripheral nerve induced with sensitized cells. *Ann. N.Y. Acad. Sci.* 122:287, 1965.
61. Wisniewski, H., and Morell, P. Quaking mouse: Ultrastructural evidence for arrest of myelinogenesis. *Brain Res.* 29:63, 1971.

62. Wisniewski, H., Prineas, J., and Raine, C. S. An ultrastructural study of experimental demyelination and remyelination. I. Acute experimental allergic encephalomyelitis in the PNS. *Lab. Invest.* 21:105, 1969.
63. Yonezawa, T., Ishihara, Y., and Matsuyama, H. Studies on experimental allergic peripheral neuritis. I. Demyelinating patterns studies in vitro. *J. Neuropathol. Exp. Neurol.* 27:453, 1968.

25

Vitamin and Nutritional Deficiencies

Pierre M. Dreyfus and
Stanley E. Geel

VITAMINS ARE INDISPENSABLE to the normal metabolic activity of the nervous system. A severe deficiency of these fundamental nutrients results in the improper function of a number of enzyme systems that are essential to the synthesis of basic constituents or to the removal of potentially toxic metabolites, thus interfering with normal development, growth, and function of the nervous system.

The most obvious cause for depletion of vitamins and other nutrients is an inadequate or improper diet. However, vitamin deficiency can occur even in the presence of an adequate diet when the organism is under stress or when the absorption of nutrients is limited. Vitamins, in order to be metabolically useful, must be converted into their active cofactor form; pathological states in which the mechanisms of conversion are faulty or incomplete have now been identified.

Among the various vitamins, some appear to be more essential than others. Despite increasing knowledge regarding the role of vitamins in general metabolism, the specific mechanisms by which a deficiency state affects the development and function of the nervous system remain essentially unknown.

Vitamin deprivation leads to several readily identifiable biochemical lesions, which invariably antedate clinical manifestations and histopathological alterations. In most instances, the correlation between the duration or the severity of these biochemical lesions and the irreversibility of symptoms or tissue changes has not been established, nor has it been possible to elucidate which of the various biochemical lesions are responsible for the neurological manifestations.

Knowledge concerning the neurochemical changes which attend nutritional disorders of the nervous system has been gleaned mainly from the experimentally induced deficiency of single vitamins without other associated variables or by the use of specific antivitamins. The results of these studies, for the most part, bear little or no resemblance to the naturally occurring disorders, which usually are the result of combined, multiple deficiencies. Any discussion of the neurochemistry of vitamin and other nutritional deficiencies relies, therefore, mainly on information derived from studies on experimental animals; speculations regarding human disease states are based on fragmentary clinical and biochemical observations.

An attempt will be made to summarize, rather than to review in detail, current knowledge of the effects of vitamin and nutritional deprivation on the nervous system.

THIAMINE

Thiamine, in its phosphorylated form (cocarboxylase, thiamine diphosphate, or thiamine pyrophosphate), acts as cofactor in two major enzyme systems. The first relates to glycolysis and involves the oxidative decarboxylation of pyruvic and α-ketoglutaric acid. The second concerns two transketolation steps of the phosphogluconate pathway (hexose monophosphate shunt), an alternate route of carbohydrate metabolism of importance in biosynthetic mechanisms (see Chap. 9):

ribose 5-phosphate + xylulose 5-phosphate
$\rightleftharpoons$ sedoheptulose 7-phosphate + glyceraldehyde 3-phosphate

xylulose 5-phosphate + erythrose 4-phosphate
$\rightleftharpoons$ fructose 6-phosphate + glyceraldehyde 3-phosphate

In addition to its coenzymatic function, thiamine appears to play an important role in nerve conduction and excitation [46]. It seems evident that the diphosphoric and triphosphoric esters of thiamine are involved in the process of nerve conduction and are essential to membrane permeability during nerve excitation [29]. Experimentally induced thiamine deficiency in animals bears certain clinical and pathological similarities to naturally occurring thiamine deficiency in man, as exemplified by the Wernicke-Korsakoff syndrome, an affliction of the central nervous system, and beriberi, a disease of the peripheral nervous system. The histopathological changes observed in animals and man assume a fairly constant topography, and although the afflicted regions or parts of the nervous system differ from species to species, the lesions appear to have a definite predilection for bilaterally symmetrical parts of the central nervous system and the distal parts of the peripheral nerves. Neurochemical data gathered to date have only partially elucidated some of the pathophysiological mechanisms which underlie this apparently selective vulnerability [20].

Brain thiamine in the normal rat and human brain is fairly evenly distributed, the content being slightly higher in the cerebellum than it is in the cerebral cortex, and considerably lower in white matter [23]. Observations made on rat brain during progressive depletion have shown that brain thiamine begins to fall during the first week of deprivation and that the greatest drop occurs during the second week. Total brain thiamine has to be reduced to less than 20 percent of normal before severe signs of deficiency (ataxia, loss of equilibrium, opisthotonus, loss of righting reflexes) and irreversible tissue changes (pannecrosis in the lateral pontine tegmentum) become manifest. Complete reversal of neurological manifestations is noted when brain thiamine returns to 30 percent of normal. These observations suggest that normal cerebral tissue has a substantial reserve of this vitamin. It has been demonstrated that during progressive depletion it is predominantly brain thiamine diphosphate that is decreased, while free thiamine and thiamine monophosphate remain fairly constant and thiamine triphosphate tends to increase. It is of interest that the loss of thiamine diphosphate is greater in the pons, the site of major pathological changes, than it is in any other part of the rat brain [39].

Studies of thiamine-dependent enzyme systems in the normal and deficient rat nervous system suggest that selective vulnerability and the character of histopathological changes may be attributed in part to the enzymatic topography of the normal nervous system and to the effect of vitamin deprivation on these enzyme systems.

In normal rat brain, the activity of pyruvic dehydrogenase (the enzyme responsible for the decarboxylation of pyruvate) is highest in areas richly endowed with neurons, such as the cerebellum and the cerebral cortex, and is considerably lower in the spinal cord (white matter). During progressive vitamin depletion, no appreciable decreases in activity can be measured when the animals are vitamin-deficient yet asymptomatic. Even at the most advanced stage of deficiency, when the animals exhibit severe neurological impairment, only the brain stem shows a 25 percent reduction in enzymatic activity. By contrast, even at the asymptomatic stage, organs such as the heart, liver, and kidney suffer a much sharper decrease in their ability to decarboxylate pyruvate. Lactate and pyruvate levels, measured in various parts of the rat nervous system during progressive depletion, generally tend to mirror the enzymatic defect [35]. Impaired pyruvate decarboxylation seems to have no effect on cerebral ATP or acetylcholine levels [35].

Normal brain transketolase activity is highest in the spinal cord (white matter) and brain stem (pons), whereas cellular aggregates, such as the cerebral cortex and the caudate nucleus, have the lowest enzymatic activity. During progressive vitamin depletion, the lateral pontine tegmentum, where the histological changes are most pronounced, exhibits a significant decrease of transketolase activity when compared to other parts of the brain. Severe signs and symptoms of deficiency and irreversible pathological changes become manifest when brain transketolase activity has been reduced by 58 percent. It appears that the defect in transketolation tends to reverse more slowly than does the abnormality of pyruvate metabolism.

It is generally accepted that the hexose monophosphate shunt plays an important role in synthetic mechanisms by virtue of NADPH and pentose production. Although the functional importance of this pathway in the metabolism of the nervous system has not been clearly defined, it has been postulated that it may be related to oligodendroglial metabolism [20]. In severe thiamine deficiency, decreased NADPH reduction could impair fatty acid and nucleic acid synthesis and the conversion of oxidized glutathione to its reduced form (GSH). Whereas no significant alterations in the profile of cerebral fatty acids have been demonstrated [21], GSH levels are reduced in the brain stem of symptomatic thiamine-deficient animals [35].

Experimental evidence suggests that thiamine-dependent metabolism differs in the developing and the mature rat nervous systems [21]. During fetal development, transketolase activity is half of that measured in adult brain. Within the first 10 days of life, the activity rises sharply to about twice the fetal level, reaching a plateau, or adult level, thereafter (Fig. 25-1). When the results are expressed in terms of the DNA or cellular content of the samples, a similar curve is obtained. The rise in brain transketolase activity during the first 10 days of life correlates with the rise of enzymes involved in oxidative phosphorylation and glycolysis. Transketolase activity may reflect an increase in glial cell duplication, proliferation, and migration. Animals born to thiamine-deprived mothers tend to be smaller and their growth rate is retarded. Cerebral transketolase activity of suckling rats is considerably more depressed than is that of their deficient mothers. This may be a reflection of enzyme immaturity or a difference in coenzyme binding in newborn rat brain rather than a selective reduction of the vitamin in neonatal brain.

Limited neurochemical observations have been made on the human nervous system. Estimation of transketolase activity in normal brain, obtained 6 hours after sudden death, has shown high transketolase activity in the mammillary body, a structure almost invariably affected in cases of Wernicke's disease, and in other structures of the brain stem and diencephalon that are not involved in this disease entity [20]. To date, it has not been possible to define the "biochemical lesion" in patients suffering from thiamine deficiency. However, evidence of a severe impairment of tissue transketolase activity has been obtained by simple biochemical measurements of blood. Determinations of pyruvate levels in blood before and after the administration of glucose, for many years the standard method of estimating thiamine deficiency, have proved to lack specificity. Other metabolic

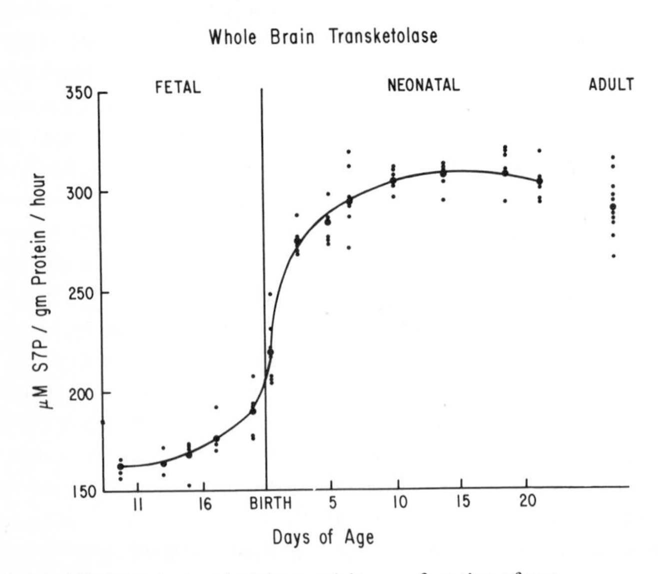

FIGURE 25-1. Whole brain transketolase activity as a function of age.

disorders can cause an elevation of pyruvate levels in blood. Determinations of transketolase levels in blood have been found to reflect, in a highly sensitive and specific manner, the state of thiamine nutrition in man [7]. Transketolase assays not only reveal the levels of available coenzyme; they also differentiate an acute state from a chronic state of deficiency by reflecting the levels of available apoenzyme. Blood transketolase assays provide clinical and biochemical evidence that at least some of the signs and symptoms of Wernicke's disease, particularly ophthalmoplegia, result from a specific lack of thiamine and that their prompt reversibility must be attributed to the presence of a "biochemical lesion" that antedates irreversible histopathological alterations [19].

Since Von Muralt [46] first proposed a dual role for thiamine in peripheral nerve, a number of important observations have been made which may have immediate relevance to neurological disease in man [29]. Experimental evidence has shown that electrical stimulation of peripheral nerve and spinal cord promotes the release of thiamine. Pyrithiamine, an antimetabolite of the vitamin that produces neurological symptoms of deficiency in experimental animals, has a profound and irreversible effect on the electrical activity of nerve preparations. This appears to be due to a displacement of the vitamin rather than to inhibition of thiamine-dependent enzyme systems. In the nervous system, thiamine phosphatases and phosphotransferases, which regulate the tissue levels of thiamine and its phosphoesters, appear to be localized in membranes rather than in axoplasms [29].

It is postulated that either thiamine diphosphate or triphosphate occupies a site on the nerve membrane and that cyclic dephosphorylation and rephosphorylation promote the passage of ions, probably sodium, across the membrane. Because there is rapid interconversion of these two thiamine esters, it is impossible to state which of the two is more important. It has been shown that, in the nervous system, the bulk of thiamine exists in the form of thiamine diphosphate and that thiamine triphosphate accounts for only 4 to 10 percent of total thiamine. This relatively small amount of triphosphoric ester may be of functional significance because it appears to be entirely lacking in the brains of patients who have succumbed to subacute necrotizing encephalopathy, a Wernicke-like disease of infants [11]. Furthermore, patients afflicted with this disease seem to elaborate a factor found in blood, urine, and cerebrospinal fluid that inhibits thiamine pyrophosphate-ATP phosphoryl transferase, the enzyme responsible for the synthesis of thiamine triphosphate in brain.

Although thiamine deficiency has been the subject of extensive investigations, large gaps exist in our knowledge concerning the sequence of events that ultimately leads to severe neurological dysfunction and irreversible histopathological changes.

VITAMIN B_6 (PYRIDOXINE)

Several forms of vitamin B_6 and its active phosphorylated coenzymes have been identified in mammalian tissue, including brain. The vitamin B_6 group bears the collective name of pyridoxine, and consists of pyridoxal, pyridoxol, and pyridoxamine. The predominant coenzyme forms of the vitamin in animal tissues are the phosphates of pyridoxal and pyridoxamine. The relative content and the rate of disappearance of these coenzymes during states of deficiency vary from species to species. Pyridoxal 5-phosphotransferase is responsible for the phosphorylation of the vitamin in brain and other organs. Isonicotinic acid hydrazide, the drug used in the treatment of tuberculosis, and 4-deoxypyridoxine, a vitamin B_6 antagonist, interfere with the phosphorylating enzyme and cause a decrease in the tissue levels of pyridoxal phosphate.

The dietary deprivation of pyridoxine has been shown to lower the seizure threshold in a variety of mammalian species, including man; the young appear to be more susceptible to convulsions than do the older and more mature. Pyridoxine deficiency in pigs is said to result in both convulsions and ataxia, the latter being caused by pathological changes in peripheral nerves, posterior root ganglia, and the posterior funiculi of the spinal cord (see also Chaps. 8 and 10).

The coenzyme forms of vitamin B_6 act as catalysts in a number of important enzymatic reactions related to the synthesis, catabolism, and transport of amino acids and the metabolism of glycogen and unsaturated fatty acids [47]. Many pyridoxine-dependent enzyme systems have been identified in the nervous system, yet knowledge concerning the biochemical pathology of vitamin B_6 deficiency continues to be limited.

Pyridoxine-dependent enzymes in the nervous system fall into two major categories: transaminases and L-amino acid decarboxylases. Some of these enzymes are involved in the γ-aminobutyric acid shunt, an alternate oxidative pathway restricted to nervous tissue, in which α-ketoglutaric acid is metabolized to succinate by way of glutamic and γ-aminobutyric acid. Vitamin B_6 deprivation leads to significant enzymatic depression in all tissues. The affinity of the coenzyme for its apoenzyme varies from enzyme to enzyme. Decarboxylases tend to have a lower affinity for coenzyme than do other enzymes; thus, the decarboxylases tend to be more readily affected than are the transaminases. Severe vitamin deprivation also results in a decrease in enzyme protein. The in vitro addition of excess pyridoxal phosphate to a vitamin-deficient enzyme preparation fails to restore complete activity; apoenzyme production stimulated by excess vitamin B_6 added to a normal tissue extract can be inhibited by puromycin. Thus, it would appear that pyridoxal phosphate regulates intracellular enzyme synthesis. It is generally believed that organs or cells with a high rate of protein turnover are most sensitive to pyridoxine depletion. Finally, pyridoxine has been shown to be required for cellular proliferation and for the synthesis of specific proteins involved in immunological reactions [1].

Of the various pyridoxine-dependent enzymes, two decarboxylases appear to be of particular importance to the integrity of neuronal function. The first, glutamic decarboxylase, is responsible for the production of the neuroinhibitor γ-aminobutyric acid from glutamic acid. Although generally believed to be restricted to neurons, this enzyme may also be active in other mammalian tissue. In rat CNS, enzymatic activity is highest in the hypothalamus and midbrain and lowest in the spinal cord (see Table 25-1). Although significant enzymatic activity can be demonstrated in human cerebral cortex (32 μ moles of glutamic acid decarboxylated per gram of protein per hour), none can be measured in white matter.

The second enzyme, 5-hydroxytryptophan decarboxylase, has been shown to be localized in nerve terminals and is involved in the synthesis of serotonin (5-hydroxytryptamine). The same enzyme may be involved in the decarboxylation of L-DOPA to dopamine (see also Chaps. 6 and 28). In the rat nervous system, the activity of this enzyme appears to be highest in the caudate nucleus, the hypothalamus, and the midbrain, which may be rich in serotoninergic terminals. The activity is surprisingly low in cerebral and cerebellar cortex (see Table 25-2).

When weanling rats are fed a diet deficient in vitamin B_6 for several weeks, severe depression of growth and acrodynia of paws, nose, ears, and tail ensue. The animals demonstrate unusual irritability; however, they rarely suffer convulsive seizures or motor weakness. This is in sharp contrast with observations of pyridoxine-deficient newborn rats, which frequently show seizures. When the activity of serum glutamic oxaloacetic transaminase reaches 50 percent of normal, glutamic decarboxylase and 5-hydroxytryptophan decarboxylase activities are sharply reduced. The loss of glutamic decarboxylase activity is 36 to 65 percent, while that of 5-hydroxytryptophan decarboxylase is 46 to 84 percent, depending upon the area of the nervous system (see Tables 25-1 and 25-2). Decreased enzymatic activity is even greater in pyridoxine-deficient neonatal animals which convulse, suggesting that there is a correlation between the seizure threshold and the level of activity of these two brain enzymes that are dependent on vitamin B_6 [51].

Evidence of vitamin B_6 deficiency in man can be obtained from a number of biochemical determinations in blood and urine. None is thought to reflect involvement of the nervous system specifically. Pyridoxal estimations in blood and

TABLE 25-1. Distribution of Glutamic Decarboxylase Activity in the Nervous System of Normal and Pyridoxine-Deficient Rats

Area Sampled	Micromoles Glutamic Acid Decarboxylated/Gram Protein/Hour	
	Normal[a]	Deficient[a]
Caudate nucleus	73.0	32.2
Cerebral cortex	74.3	34.0
Thalamus	96.7	33.8
Hypothalamus	101.2	47.2
Cerebellar hemisphere	65.8	42.2
Midbrain	100.7	42.8
Spinal cord	31.8	15.2

[a]Means of 7 animals. All determinations carried out in triplicate.
Source: Dreyfus, P. M., Meier, F., and York, C. Unpublished data, 1967.

TABLE 25-2. Distribution of 5-Hydroxytryptophan Decarboxylase Activity in the Nervous System of Normal and Pyridoxine-Deficient Rats

Area Sampled	Nanomoles ^{14}C-Serotonin Formed/ Gram Protein/Hour	
	Normal[a]	Deficient[a]
Caudate nucleus	4417.9	2390.1
Cerebral cortex	621.6	99.5
Thalamus	1319.3	428.0
Hypothalamus	4332.0	1767.2
Cerebellar hemisphere	474.7	165.2
Midbrain	4204.1	805.3
Spinal cord	1150.1	226.1
Liver	13,494.6	1722.2

[a]Means of 7 animals. All determinations carried out in triplicate.
Source: Dreyfus, P. M., Meier, F., and York, C. Unpublished data, 1967.

cerebrospinal fluid are of limited usefulness. Under normal circumstances, vitamin B_6 and its derivatives can be measured in the urine; none can be found during severe depletion, regardless of the patient's protein intake. Pyridoxine deficiency causes increased urinary excretion of the tryptophan metabolites, xanthurenic and kynurenic acids, after a high dose of tryptophan. A decreased 5-hydroxytryptophan decarboxylase activity results in decreased 5-hydroxyindoleacetic acid excretion in the urine, and faulty cysteine sulfinic acid or cysteic acid decarboxylase activity causes decreased taurine levels in urine. Vitamin B_6 deficiency affects cystathionine cleavage, with resultant cystathioninuria. Occasionally, reduced activity of serum glutamic oxaloacetic and glutamic pyruvic acid transaminases and oxaluria can be demonstrated.

In recent years, the state of pyridoxine dependency has been the subject of a number of investigations (see Chap. 22). This is a familial disorder of infants, characterized by seizures and a high daily vitamin B_6 requirement (10 mg per day), in which no specific metabolic defect has yet been demonstrated. Although the seizures respond promptly to the administration of pyridoxine and possibly to γ-aminobutyric acid, biochemical evidence of vitamin deficiency is lacking. An abnormality of the glutamic acid decarboxylase system has been postulated [40].

VITAMIN B_{12} AND FOLIC ACID

Although a great deal is known about the role of vitamin B_{12} in biochemical reactions, virtually nothing is known about its function in the nervous system. Biochemical studies on the nervous tissue of animals suffering from experimentally induced vitamin B_{12} deficiency have been difficult to interpret in view of the absence of clinical manifestations and histopathological changes in either the central or the peripheral nervous systems.

In man, however, the insufficient absorption of vitamin B_{12}, which may be caused by the absence of intrinsic factor, disease of the gastrointestinal tract (postgastrectomy, sprue, fish-tapeworm infestation), or vegetarianism, may result in subacute degeneration of the spinal cord, optic nerves, cerebral white matter, and peripheral nerves. The earliest visible change in the affected parts consists of swelling of individual myelinated nerve fibers in small foci. This is followed by coalescence of the lesions into larger, irregular, spongy, honeycomblike zones of demyelination. Fibers with the largest diameter seem to be predominantly affected and axis cylinders tend to be spared. Although demyelination appears to be the primary lesion, the possibility of initial involvement of axonal metabolism and axoplasmic flow has not been entirely excluded.

Only microorganisms that inhabit the mammalian gastrointestinal tract have the capacity to synthesize vitamin B_{12}. Although large amounts of the vitamin are manufactured in the rumen of some herbivorous animals, humans depend upon a dietary source, largely in the form of meat. The natural vitamin, cyanocobalamin, must be metabolized by the body to its active coenzyme form, 5-deoxyadenosylcobalamin. To date, only two reactions dependent on vitamin B_{12} have been demonstrated in human and other mammalian tissue (Table 25-3) [43].

The isomerization of L-methylmalonyl-CoA to succinyl-CoA requires the active coenzyme form of the vitamin, whereas the transmethylation of homocysteine to methionine probably depends upon another coenzyme form of the vitamin, methyl-B_{12}. The enzymes catalyzing these two reactions are L-methylmalonyl-CoA mutase and N^5-methyltetrahydrofolate homocysteine methyl transferase. Ribonucleotide reductase, an enzyme of great importance in the synthesis of DNA in microorganisms, has not as yet been measured in mammalian tissue, and virtually nothing is known about B_{12}-dependent enzyme systems involved in sulfhydryl reduction.

Vitamin B_{12} deficiency in humans and experimental animals results in decreased conversion of methylmalonyl-CoA to succinyl-CoA in the tissues, and the urinary excretion of methylmalonic acid ensues. It is not known to what extent reduced levels of vitamin B_{12} coenzyme and consequent reduction in methylmalonyl-CoA mutase activity are responsible for the neurological complications of vitamin B_{12} deficiency. The dietary restriction of vitamin B_{12} results in decreased coenzyme

TABLE 25-3. Vitamin B_{12}-Dependent Reactions in Mammalian Tissue

Reaction	Active Cofactor
Methylmalonyl-CoA mutase	B_{12} Coenzyme
Ribonucleotide reductase (?)	B_{12} Coenzyme
Methionine synthetase	Vitamin B_{12}
Sulfhydryl reduction	B_{12} Coenzyme (?)

levels in the liver, kidney, and brain of rats and miniature pigs [9], the percentage decrease being approximately the same in each organ. When rats or miniature pigs are fed a diet free of vitamin B_{12}, methylmalonuria is detectable after 4 to 6 weeks. The presence of this organic acid in the urine of animals and humans has been found to be both a specific and a sensitive index of vitamin B_{12} deficiency; it is, in fact, a more reliable indicator than are levels of B_{12} in serum [3, 22]. Methylmalonuria also occurs as the result of inherited defects of vitamin B_{12} metabolism [34]. Levels of methylmalonic acid in cerebrospinal fluid have been found to exceed those in plasma, suggesting that this acid may be elaborated by cerebral tissue as a consequence of faulty propionate metabolism in the brain [28].

Although rats kept on a deficient diet for 8 to 12 months appear clinically to be intact, miniature pigs show a significant reduction in food consumption and body weight, generalized weakness, and intermittent lethargy after 3 months. Brain, spinal cord, and peripheral nerves examined by standard light microscopic methods reveal no histopathological alterations [21].

It has been shown that in deficient liver, kidney, and brain, methylmalonyl-CoA mutase activity is reduced; methylmalonyl-CoA is hydrolyzed to methylmalonic acid rather than to succinyl-CoA. The rate of disappearance of methylmalonyl-CoA, however, is essentially the same in normal and deficient tissue [9]. These observations suggest that, in the deficient state, adaptive mechanisms exist that rid the organism of high levels of methylmalonyl-CoA.

Recently it has been shown that sural nerve obtained from patients with pernicious anemia incorporates ^{14}C-labeled propionate into abnormal fatty acids that have been identified as anteiso-C_{15} (branched-chain) and C_{17} (odd-chain) acids. The presence of these abnormal fatty acids may explain, in part, the structural changes observed in central and peripheral myelin [26].

The enzyme N^5-methyltetrahydrofolate homocysteine methyl transferase is involved in the synthesis of methionine from homocysteine. The remethylation of homocysteine to methionine is accomplished together with the regeneration of tetrahydrofolate from N^5-methyltetrahydrofolate. The reaction requires trace amounts of *S*-adenosylmethionine and a vitamin form of B_{12}, most likely methyl-B_{12}. The enzyme has been estimated in the liver and brain of both humans and rats [33]. In deficient rat liver and brain, enzymatic activity is drastically reduced (Table 25-4). The addition in vitro of excess vitamin B_{12} (cyanocobalamin) partially restores the activity of the enzyme in brain, but has no significant effect on liver enzyme activity.

The in vitro stimulation of enzymatic activity in normal and deficient brains suggests that this tissue contains excessive amounts of apoprotein. This may represent a protective mechanism against vitamin B_{12} deficiency. A substantial decrease in enzymatic activity may result in a critical reduction of methionine synthesis in neural tissue, which in humans could lead to neurological manifestations and histopathological changes in the nervous system.

Even though methionine is considered to be an essential dietary amino acid, there obviously exist metabolic pathways for its synthesis at the cellular level. The

TABLE 25-4. Activity of N^5-Methyltetrahydrofolate Homocysteine Methyl Transferase in Normal and Vitamin B_{12}-Deficient Rat Liver and Brain[a]

	Controls (6)		Deficient (6)	
	Plain	$+B_{12}$[b]	Plain	$+B_{12}$[b]
Liver	246.8 ± 8	248.2 ± 10	36.8 ± 5	48.8 ± 7
Brain	212.0 ± 10	341 ± 35	47.0 ± 3	97.3 ± 10

[a]Animals on deficient diet for 8 months. Results expressed as CPM/mg protein/hour ± SEM. Numbers in parentheses represent number of animals. All determinations were carried out in triplicate.

[b]Cyanocobalamin (0.3 μ mole) was added to the incubation mixture.

Source: Dreyfus, P. M., and Gross, C. Unpublished data, 1968.

brain and the spinal cord may depend upon these synthesizing mechanisms for their normal supply of methionine since the rate of penetration of most amino acids from blood into neural tissue is significantly slowed by the normal blood-brain barrier. Methionine, an important constituent of neural protein, may also be a constituent of proteolipids. The rate of its incorporation into protein is rapid; it is greatest in those regions of the nervous system that contain the highest density of cells, and it is more active in microsomes than in either nuclei or mitochondria. Profound changes in amino acid concentrations in brain affect the amino acid flux into protein as well as protein synthesis and degradation. It can be postulated that, in the nervous system, prolonged and severe vitamin B_{12} deprivation results in a critical reduction of methionine levels in tissue, which in turn may alter protein synthesis, turnover, and, conceivably, structure and function.

Abnormalities of sulfhydryl metabolism, as evidenced by depressed glutathione levels in blood and liver, have been observed in experimental and naturally occurring vitamin B_{12} deficiencies. Alterations in intracellular glutathione levels may lead to abnormal glycolysis and lipid metabolism. This biochemical lesion has not yet been demonstrated in neural tissue.

Vitamin B_{12} deficiency has been shown to cause hydrocephalus in newborn rats. It is believed that the anatomical defect is caused by stenosis of the aqueduct of Sylvius, associated with aplasia of the subcommissural organ, a special group of columnar ependymal cells in the roof of the aqueduct and the posterior part of the third ventricle [52]. It has been postulated that the congenital abnormality may be the result of faulty ribonucleic acid metabolism in the newborn brain. Levels of RNA per cell are decreased, whereas the amount of DNA remains normal [8].

Folic acid depletion occurs as a consequence of dietary deficiency, which is sometimes caused by increased needs during pregnancy or by gastrointestinal defects in absorption. The administration of anticonvulsant drugs, which act as competitors, may also lead to a state of deficiency. The neurological manifestations engendered by folic acid depletion alone are thought to be similar to those caused by the lack of vitamin B_{12} (see Chap. 22).

Folic acid and its derivatives are involved in the formation of methionine from homocysteine and in the synthesis of purines and pyrimidines and, hence, of RNA and DNA. Folate is therefore fundamental to normal cell division and growth. The active coenzyme form of folate contains four additional hydrogen atoms (tetrahydrofolate). Folic acid reductase permits the reduction of folate to tetrahydrofolate. Various derivatives of tetrahydrofolate have been identified, such as methyl-, methenyl-, methylene-, hydroxymethyl-, formyl-, and formiminotetrahydrofolate. In general, the various forms of folate are instrumental in the transfer of single carbon units. As yet, the specific role of folic acid in cerebral metabolism has not been elucidated. Interestingly, folate levels in cerebrospinal fluid are two to three times those in serum, even in states of folate deficiency [28].

The precipitation of neurological complications of vitamin B_{12} deficiency by the administration of folic acid remains essentially unexplained.

NIACIN AND PANTOTHENIC ACID

In the nervous system, as in other tissues, nicotinic acid, or niacin, is a constituent of the two coenzymes that transfer hydrogen or electrons. These coenzymes are nicotinamide adenine dinucleotide (NAD) and nicotinamide adenine dinucleotide phosphate (NADP). Both of these nucleotides are essential to a number of important enzymatic reactions involved in carbohydrate, fatty acid, and glutathione metabolism. The significance of these biochemical reactions has been covered in other chapters (see Chaps. 9 and 11).

Most mammalian cells synthesize nicotinic acid from tryptophan. It is not known whether this synthesis can also take place in the cells of the nervous system. In man, a deficiency of niacin is responsible for the neurological and psychological manifestations of pellagra: encephalopathy associated with signs of spinal cord and peripheral nerve involvement. The pathological changes observed in this disorder are known as central chromatolysis, or central neuritis, and consist of a characteristic degeneration of the large pyramidal cells (Betz cells) of the motor cortex.

It is not known how a deficiency of niacin brings about visible alterations of neuronal Nissl substance. While it has been demonstrated that niacin deficiency causes a reduction of cerebral NAD and NADP levels and decreased activity of the enzymes that depend upon these nucleotides, no specific neurochemical lesion has as yet been identified. It is well recognized that in patients afflicted with pellagra, NAD levels in red blood cells are reduced and the urinary excretion product of niacin, *N*-methylnicotinamide, is diminished. The administration of niacin to pellagrins and to deficient animals causes the nucleotide content of the blood and the tissues to rise temporarily above normal levels [36].

Pantothenic acid, a constituent of coenzyme A (CoA) is widely distributed in mammalian tissue. However, pantothenic acid differs from other vitamins in that it is not the active unit of CoA. In addition to pantothenic acid, CoA contains ribose, adenine, phosphoric acid, and β-mercaptoethylamine. The sulfhydryl group of the latter is the site that links acid and acetyl groups.

As part of CoA, pantothenic acid participates in a variety of biochemical reactions involved in fatty acid synthesis and oxidation, as well as in the metabolism of steroids and acetylcholine. Generally speaking, the reactions mediated by CoA fall into two categories: acetokinases and transacetylases [38]. The vitamin is also essential to the formation of certain amide and peptide linkages.

Relatively high concentrations of pantothenic acid exist in the brain, which does not readily yield its stores. Experimentally induced pantothenic acid deficiency in animals results in lesions of the peripheral nerves. In man, pantothenic acid deficiency causes numbness and tingling of hands and feet and occasionally a "burning-foot" syndrome. Biochemical changes that have been noted in such patients are characterized by the impaired ability to acetylate *p*-aminobenzoic acid and by a decline in blood levels of cholesterol and its esters. In addition, evidence of adrenal cortical hypofunction has been described [6]. In experimental animals, it has been noted that a deficiency of the vitamin results in deranged synthesis of acetylcholine, cholesterol, glucosamine, and galactosamine, in faulty fatty acid oxidation, and in reduced energy production. More specific information concerning the neurochemistry of pantothenic acid deficiency is lacking.

PROTEIN-CALORIE MALNUTRITION

Although the clinical manifestations of protein-calorie deficiency in early life are well documented, relatively little is known concerning the precise molecular changes that underlie this condition, particularly as it pertains to the central nervous system. The brain is highly susceptible, albeit less so than the body, to environmental influences during the period of morphological and functional differentiation; there may be serious impairment of psychological development, with ensuing disorders of behavior. Studies at the molecular level on the effects of undernutrition on the developing brain assume particular significance if one seeks to quantitate the extent of cerebral dysfunction, if any, and if one wishes to find means of correcting altered intellectual function.

It is clear from experimental studies in animals that a diet deficient in protein or calories alters the chemical composition of the brain during its development (see [17], [48], and [49] for reviews). The rat is a convenient species for such investigations because functional differentiation of the brain, which takes place in utero in most other animal species including man, occurs in rats during the neonatal period. In the rat, the period of ontogenesis between birth and 21 days of age is characterized by proliferation and migration of glial cells, outgrowth of axodendritic processes, rapid myelination, development of mature cerebellar and cerebral cytoarchitecture, and the maturation of adaptive enzyme patterns (see Chap. 14).

It has been demonstrated that the total cerebral DNA content is reduced in rats subjected to neonatal undernutrition. It has been postulated that this alteration reflects both the total number and the growth of neurons [50]. Prenatal restriction of dietary protein is said to reduce the number of neurons in the offspring [53, 54]. In the rat, neuronal cell division is confined to the prenatal period, so it is conceivable that malnutrition during pregnancy may lead to a permanent impairment of cerebral function. Neonatal dietary restriction has been reported to cause regional reductions in cerebral DNA content [4, 13, 50]. Parts of the brain, such as the cerebellum and the cerebral cortex, which develop later phylogenetically, are more sensitive to postnatal calorie and protein deficiency. A combined histological and microchemical study of the somatosensory cortex of undernourished suckling rats suggests that the altered DNA and ganglioside levels result from delayed glial cell proliferation and migration and the retarded growth of axodendritic processes [4]. Regional studies such as these are invaluable in detecting subtle changes that might otherwise be obscured by whole-brain analyses.

The abnormal histogenesis of glial cells in the central nervous system of undernourished rats probably results in defective myelin deposition. Nutritional inadequacy during a period coinciding with rapid myelination impairs the synthesis of sulfatides with a concomitant reduction in the galactocerebroside sulfokinase which catalyzes its synthesis [10]. The whole-brain concentration of other myelin lipids, such as cerebrosides, proteolipids, plasmalogens, and cholesterol, is most strikingly reduced during a similar period of undernutrition [5, 13, 27]. These findings correlate with the absence of stainable myelin [5]. Similarly, brain cerebroside and cholesterol levels are specifically reduced in pigs underfed for up to one year of age. This corresponds to the period of time during which these two constituents are being deposited at a maximal rate [14].

The activity of aldolase, creatinine phosphokinase, isocitric dehydrogenase, guanine deaminase, and acetylcholinesterase is said to be depressed in the brains of neonatally undernourished rats [30, 41, 45]. In mice exposed to protein and calorie malnutrition before and after birth, chromatographic profiles of brain extracts have revealed a significant reduction in soluble proteins [31]. It is logical to assume that a modification of protein metabolism must affect the morphology and functional activity of the developing central nervous system.

Of fundamental importance are the questions of reversibility of biochemical alterations induced by nutritional deficiency and the extent to which structural changes represent permanent functional impairment. A distinction must be made between those biochemical alterations that represent a temporary adaptive response and those that are caused by visible pathological changes which are not reversible by nutritional rehabilitation. The bulk of the evidence favors the view that undernutrition during a critical phase of brain development induces biochemical changes that are not reversible by subsequent feeding, even for prolonged periods. Implicit in these observations is the concept of a vulnerable period of brain development, which varies for different regions of the brain and for different species [16]. By contrast, dietary restriction later in life results in relatively slight and readily reversible changes in chemical patterns. Experimental undernutrition is commonly induced by rearing animals in large litters, by reducing the nursing time, by feeding a low-protein diet, or by a combination of several of these methods. It is obvious that the timing, duration, and extent of nutritional deficiency are important in the interpretation of experimental results and the subsequent effectiveness of rehabilitation.

It has been speculated that undernutrition during early life leads to behavioral changes. Although a variety of behavioral tests with experimental animals have suggested long-lasting alterations, it is not entirely clear whether learning capacity is impaired [2, 32]. The fact that brain neurotransmitters and other biochemical correlates of functional activity are suppressed in rats deprived of protein in the perinatal period does lend support to the idea that behavioral changes exist [41, 42]. On the other hand, knowledge concerning the chemistry and the intimate molecular mechanisms of learning and behavior is not sufficiently advanced to permit conclusions regarding the causal relationship between neurochemical abnormalities and brain function.

Neurochemical data on malnourished children are obviously scarce, and when available they are derived from a limited number of subjects about whom clinical information is scanty. Frequently these data are compared with inadequate controls. In humans, a period of rapid brain growth and cellular proliferation comparable to that in the rat extends from approximately 20 weeks of gestation to about the end of the first year of life [18]. Children malnourished during the first years of life possess smaller brains, with total and regional reductions in DNA content when compared with matched controls [50]. In agreement with animal studies, there is some indirect evidence to suggest that prenatal malnutrition leads to severe retardation of cell growth [50]. The deposition of myelin lipids appears to be arrested in undernourished infants [25] and the levels of gangliosides, particularly in the forebrain, are subnormal for a given brain weight [15].

As is the case in experimental animals, these abnormalities may be reflected in defective intellectual development [24]. Behavioral studies in children have persistently shown a reduction of mental capacity that is directly related to the age of onset of nutritional deprivation. Thus, children suffering from undernutrition before 6 months of age are more severely retarded and less likely to recover, even after prolonged treatment, than are children who experience malnutrition at a later age [12, 44].

The extrapolation of data from undernourished animals to similarly afflicted humans is hazardous and should be made with certain qualifications. Dietary restriction in experimental animals does not resemble the naturally occurring state in children, who frequently have suffered a prolonged deficit of several essential nutrients. Moreover, nutritional effects in human populations cannot be separated from the multitude of coexistent biological and sociological variables. A point frequently overlooked is the hormonal status of nutritionally deprived animals and humans. Diminished thyroid function and insulin secretion accompany undernutrition [37] and, conceivably, contribute to altered brain development.

A number of important findings have emerged from animal and human studies. There appears to be a definite correlation between the extent and persistence of cerebral dysfunction and the age of onset and duration of malnutrition. It seems well established that certain molecular indices of the structure and function of the central nervous system are irreversibly altered by protein and calorie malnutrition. Well-controlled studies are necessary before it can be established whether specific biochemical alterations attributable to nutritional deprivation are reflected in defective learning and intellectual development.

REFERENCES

ORIGINAL SOURCES AND SPECIAL TOPICS

1. Axelrod, A. E., and Trakatellis, A. C. Relationship of pyridoxine to immunological phenomena. *Vitam. Horm.* 22:591, 1964.
2. Barnes, R. H., Moore, A. U., and Pond, W. G. Behavioral abnormalities in young adult pigs caused by malnutrition in early life. *J. Nutr.* 100:149, 1970.
3. Barness, L. A. Vitamin B_{12} deficiency with emphasis on methylmalonic acid as a diagnostic aid. *Am. J. Clin. Nutr.* 20:573, 1967.
4. Bass, N. H., Netsky, M. G., and Young, E. Effect of neonatal malnutrition on developing cerebrum. I. Microchemical and histologic study of cellular differentiation in the rat. *Arch. Neurol.* 23:289, 1970.
5. Bass, N. H., Netsky, M. G., and Young, E. Effect of neonatal malnutrition on developing cerebrum. II. Microchemical and histologic study of myelin formation in the rat. *Arch. Neurol.* 23:303, 1970.
6. Bean, W. B., and Hodges, R. E. Pantothenic acid deficiency induced in human subjects. *Proc. Soc. Exp. Biol. Med.* 86:693, 1954.
7. Brin, M. Erythrocyte transketolase in early thiamine deficiency. *Ann. N.Y. Acad. Sci.* 98:528, 1962.
8. Bruemmer, J. H., O'Dell, B. L., and Hogan, A. G. Maternal vitamin B_{12} deficiency and nucleic acid content of tissues from infant rats. *Proc. Soc. Exp. Biol. Med.* 88:463, 1955.
9. Cardinale, G. J., Dreyfus, P. M., Auld, P., and Abeles, R. H. Experimental vitamin B_{12} deficiency. *Arch. Biochem. Biophys.* 131:92, 1969.
10. Chase, H. P., Dorsey, J., and McKhann, G. M. The effect of malnutrition on the synthesis of a myelin lipid. *Pediatrics* 40:551, 1967.
11. Cooper, J. R., Pincus, J. H., Itokawa, Y., and Piros, K. Experience with phosphoryl transferase inhibition in subacute necrotizing encephalomyelopathy. *N. Engl. J. Med.* 283:793, 1970.
12. Cravioto, J. Mental performance in school age children. *Am. J. Dis. Child.* 120:404, 1970.
13. Culley, W. J., and Lineberger, R. O. Effect of undernutrition on the size and composition of the rat brain. *J. Nutr.* 96:375, 1968.
14. Dickerson, J. W. T., and Dobbing, J. The effect of undernutrition early in life on the brain and spinal cord in pigs. *Proc. Nutr. Soc.* 26:5, 1967.
15. Dickerson, J. W. T., and Merat, A. Personal communication, 1971.

16. Dobbing, J. Vulnerable Periods in Developing Brain. In Davison, A. N., and Dobbing, J. (Eds.), *Applied Neurochemistry.* Philadelphia: Davis, 1968.
17. Dobbing, J. Undernutrition and the Developing Brain. In Himwich, W. A. (Ed.), *Developmental Neurobiology.* Springfield: Thomas, 1970.
18. Dobbing, J. Undernutrition and the developing brain. *Am. J. Dis. Child.* 120:411, 1970.
19. Dreyfus, P. M. Clinical application of blood transketolase determinations. *N. Engl. J. Med.* 267:596, 1962.
20. Dreyfus, P. M. Transketolase Activity in the Nervous System. In Wohlstenholme, G. E. W. (Ed.), *Thiamine Deficiency: Biochemical Lesions and Their Clinical Significance.* Boston: Little, Brown, 1967.
21. Dreyfus, P. M. Unpublished data, 1970.
22. Dreyfus, P. M., and Dubé, V. E. The rapid detection of methylmalonic acid in urine – a sensitive index of vitamin B_{12} deficiency. *Clin. Chim. Acta* 15:525, 1967.
23. Dreyfus, P. M., and Victor, M. Effects of thiamine deficiency on the central nervous system. *Am. J. Clin. Nutr.* 9:414, 1961.
24. Eichenwald, H. F., and Fry, P. C. Nutrition and learning. *Science* 163:644, 1969.
25. Fishman, M. A., Prensky, A. L., and Dodge, P. R. Low content of cerebral lipids in infants suffering from malnutrition. *Nature (Lond.)* 221:552, 1969.
26. Frenkel, E. P. Studies on mechanism of the neural lesion of pernicious anemia. The American Society for Clinical Investigation, 63rd Annual Meeting, 1971.
27. Geison, R. C., and Waisman, H. A. Effects of nutritional status on rat brain maturation as measured by lipid composition. *J. Nutr.* 100:315, 1970.
28. Girdwood, R. H. Abnormalities of vitamin B_{12} and folic acid metabolism – their influence on the nervous system. *Proc. Nutr. Soc.* 27:101, 1968.
29. Itokawa, Y., and Cooper, J. R. Ion movements and thiamine. II. The release of the vitamin from membrane fragments. *Biochim. Biophys. Acta* 196:274, 1970.
30. Kumar, S., and Sanger, K. C. S. Guanine deaminase in the developing rat brain: Effect of starvation. *J. Neurochem.* 17:1113, 1970.
31. Lee, C. J. Biosynthesis and characteristics of brain protein and ribonucleic acid in mice subjected to neonatal infection or undernutrition. *J. Biol. Chem.* 245:1998, 1970.
32. Levitsky, D. A., and Barnes, R. H. Effect of early malnutrition on the reaction of adult rats to adverse stimuli. *Nature (Lond.)* 225:468, 1970.
33. Levy, H. L., Mudd, S. H., Schulman, J. D., Dreyfus, P. M., and Abeles, R. H. A derangement in B_{12} metabolism associated with homocystinemia, cystathioninemia, hypomethioninemia and methylmalonic aciduria. *Am. J. Med.* 48:390, 1970.
34. Mahoney, M. J., and Rosenberg, L. E. Inherited defects of B_{12} metabolism. *Am. J. Med.* 48:584, 1970.
35. McCandless, D. W., and Schenker, S. Encephalopathy of thiamine deficiency: Studies of intracerebral mechanisms. *J. Clin. Invest.* 47:2268, 1968.

36. McIlwain, H. *Biochemistry and the Central Nervous System.* Boston: Little, Brown, 1966.
37. Milner, R. D. G. Malnutrition and the Endocrine System in Man. In Benson, G. K., and Phillips, J. G. (Eds.), *Hormones and the Environment.* New York: Cambridge University Press, 1970.
38. Novelli, G. D. Metabolic functions of pantothenic acid. *Physiol. Rev.* 33:525, 1953.
39. Pincus, J. H., and Grove, I. Distribution of thiamine phosphate esters in normal and thiamine deficient brain. *Exp. Neurol.* 28:477, 1970.
40. Scriver, C. R., and Whelan, D. T. Glutamic acid decarboxylase in mammalian tissue outside the central nervous system and its possible relevance to hereditary vitamin B_6 dependency with seizures. *Ann. N.Y. Acad. Sci.* 166:83, 1969.
41. Sereni, F., Principi, N., Perletti, L., and Sereni, L. P. Undernutrition and the developing rat brain. *Biol. Neonate.* 10:254, 1966.
42. Shoemaker, W. J., and Wurtman, R. J. Perinatal undernutrition: Accumulation of catecholamines in rat brain. *Science* 171:1017, 1971.
43. Stadtman, T. C. Vitamin B_{12}. *Science* 171:859, 1971.
44. Stoch, M. B., and Smythe, P. M. The effect of undernutrition on subsequent brain growth and intellectual development. *S. Afr. Med. J.* 41:1027, 1967.
45. Swaiman, K. F., Daleidan, J. M., and Wolfe, R. N. The effect of food deprivation on enzyme activity in developing brain. *J. Neurochem.* 17:1387, 1970.
46. Von Muralt, A. The role of thiamine in nervous excitation. *Exp. Cell Res.* 5:72, 1958.
47. Williams, M. A. Vitamin B_6 and amino acids – recent research in animals. *Vit. Horm.* 22:561, 1964.
48. Winick, M. Malnutrition and brain development. *J. Pediatr.* 74:667, 1969.
49. Winick, M. Nutrition and mental development. *Med. Clin. N. Am.* 54:1413, 1970.
50. Winick, M. Nutrition and nerve cell growth. *Fed. Proc.* 29:1510, 1970.
51. Wiss, O., and Weber, F. Biochemical pathology of vitamin B_6 deficiency. *Vit. Horm.* 22:495, 1964.
52. Woodard, J. C., and Newberne, P. M. The pathogenesis of hydrocephalus in newborn rats deficient in vitamin B_{12}. *J. Embryol. Exp. Morph.* 17:177, 1967.
53. Zamenhoff, S., Van Marthens, E., and Margolis, F. C. DNA (cell number) and protein in neonatal brain: Alteration by maternal dietary protein restriction. *Science* 160:322, 1968.
54. Zeman, F. J., and Stanbrough, E. C. Effect of maternal protein deficiency on cellular development in the fetal rat. *J. Nutr.* 99:274, 1969.

BOOKS AND REVIEWS

Dreyfus, P. M. Transketolase Activity in the Nervous System. In Wolstenholme, G. E. W. (Ed.), *Thiamine Deficiency; Biochemical Lesions and Their Clinical Significance.* Boston: Little, Brown, 1967.

Gyorgy, P., and Pearson, W. N. *The Vitamins.* New York: Academic, 1967.

McCance, R. A., and Widdowson, E. M. *Calorie Deficiencies and Protein Deficiencies.* Boston: Little, Brown, 1968.

Scrimshaw, N. S., and Gordon, J. E. *Malnutrition, Learning and Behavior.* Cambridge, Mass.: M.I.T. Press, 1968.

Smith, E. L. *Vitamin* B_{12}. London: Methuen, 1965.

Wagner, A. F., and Folkers, K. *Vitamins and Coenzymes.* New York: Interscience, 1964.

26

Cerebral Edema

Donald B. Tower

EDEMA is a frequent and important complication of trauma, vascular accidents, tumors, and various inflammatory or toxic states of the central nervous system, and it is often a major sequela of surgery on the brain and spinal cord. Clinical recognition of edema probably dates from the time of Hippocrates, but our understanding of the nature and therapy of the various types of edema is still very incomplete. The clinical picture of cerebral edema comprises general signs and symptoms of increased intracranial pressure (especially impaired consciousness), various and inconstant focal cerebral signs, and secondary complications, such as signs of cerebral herniation. These signs and symptoms may also characterize the primary, underlying pathological condition, so the diagnosis of edema as a complication and a critical evaluation of responses to therapy are often problematical. Moreover, the more objective means of ascertaining the presence and extent of the edema are seldom applicable to the acute clinical emergency. Most of the following discussion is based on studies in various animal species of experimentally induced forms of cerebral edema that resemble, but are not necessarily identical to, their respective clinical counterparts.

In the older literature, especially from Germany, a distinction was made between brain swelling *(Hirnschwellung),* or an increase in bulk of the brain with essentially no interstitial or intracellular fluid component, and brain edema *(Hirnödem).* Most investigators no longer consider this to be a useful or even a valid distinction. The two phenomena are usually viewed as different phases of the same process, and the two terms are often used interchangeably. In the biochemical literature, swelling is the term more commonly employed to denote imbibition of extra fluid by tissue samples.

The earlier concepts of cerebral edema, based on gross changes of brain volume and fluid content [10, 32], have been supplanted by the recognition that edema usually involves white matter rather selectively, although slices of cerebral cortex swell readily [6, 31]. Selective edema of subcortical white matter was demonstrated in the classic study on human brain tumors by Stewart-Wallace [26] and by more recent experimental studies by Aleu et al. [1] and Pappius and colleagues [23], among others. Much of the more recent work on the clinical, pathological, and experimental aspects of edema of the central nervous system was reviewed at a symposium held in Vienna in 1965 (for details beyond the scope of this chapter, the reader is referred to the published proceedings of that symposium [15]).

QUANTIFICATION

IN VIVO

The usual procedure for the definition of edema in vivo involves sampling the edematous area of brain (or spinal cord) and a control area (preferably the homologous contralateral area) and determining the respective percentage dry weights. This is most conveniently done by weighing each fresh tissue sample (wet weight) in a tared container, drying the sample at approximately 100°C to constant weight (usually within 12 to 24 hr), and reweighing to obtain the weight of the residue (dry weight). The percentage is simply calculated as:

$$(\text{wet weight}/\text{dry weight}) \times 100 = \%\ \text{dry weight (P)} \qquad (1)$$

The percentage of swelling or edema (or of shrinkage) can then be calculated by the formula of Elliott and Jasper [10]:

$$(P_{\text{control}} - P_{\text{exptl}})/P_{\text{exptl}} \times 100 = \%\ \text{swelling (or shrinkage)} \qquad (2)$$

If there is an appreciable content of solids (e.g., protein) in the edema fluid, a more accurate calculation is given by:

$$(P_{\text{control}} - P_{\text{exptl}})/(P_{\text{exptl}} - \text{P}_\text{F}) \times 100 = \%\ \text{swelling (or shrinkage)} \qquad (2a)$$

where P_F = percentage dry weight of the edema fluid. Elliott and Jasper [10] point out that it is misleading to report results in terms of percentage of water in the tissue, since a change in water content from 80 to 81 percent represents a swelling and hence an increase of weight of 5.3 percent. The foregoing method of calculation is usually necessary for studies in vivo because it is frequently difficult or impossible to obtain meaningful differences in tissue weights directly.

On the other hand, the foregoing procedures assume a uniform distribution of fluid within the respective normal and edematous samples. Such may not always be the case, especially when effects of therapeutic measures are being evaluated. In these cases, actual weights of control and experimental hemispheres should be obtained to supplement the dry-weight data. An example of this approach is the recent study by Pappius and McCann [24] on the effects of steroids on experimentally induced cerebral edema in cats (see pp. 548–549).

Additional insight into the nature of the edema fluid can be obtained by analyses of the tissue samples for proteins and electrolytes (Na^+, K^+, and Cl^-). Initially, such data will be obtained per unit of fresh (or wet) weight of tissue sample. In this form, the data may not be as meaningful as when expressed per unit of dry weight or as ratios (e.g., Na^+/K^+). In many cases, the edema fluid may be enriched in only one monovalent cation, with a consequent change in the Na^+/K^+ ratio. In such cases, the calculation in terms of dry weight will verify which ion remained unaffected by the influx of edema fluid.

IN VITRO

For slices of cerebral tissues studied in vitro, a different approach is required. Under most circumstances, once a slice has been cut, its dry weight (solids) will remain relatively constant. Hence, wet weights (W) must be taken initially and after the experimental period to assess the changes of fluid content within the tissue slice:

$$(W_{\text{final}} - W_{\text{initial}})/W_{\text{initial}} \times 100 = \%\ \text{swelling} \tag{3}$$

Of course, dry weights should also be obtained, whenever feasible, to check any significant change in the content of solids in such samples. Some investigators prefer to report their data for fluid content of slices as weight or volume of fluid per unit of dry weight. Differences between control and experimental samples can be appreciated readily, but unless the absolute values for wet and dry weights are reported, one may be unable to convert such data to percentage swelling.

TYPES OF CEREBRAL EDEMA

There is a distinct contrast between the edema of cerebral tissues encountered clinically or experimentally in vivo and that seen commonly in preparations in vitro. Swelling of cerebral tissues in vitro is characteristically of gray matter, whereas most types of edema in vivo are localized principally to white matter, usually in association with disruption of the blood-brain barrier. However, there may be some relationship between swelling in high K^+ media in vitro and cerebral edema in vivo.

EDEMA ASSOCIATED WITH TUMORS

Historically, the recognition of white matter as a site of predilection for many types of edema of neural tissues came from the studies by Stewart-Wallace [26] on the brains of patients with intracranial tumors. In these brains, the swelling of cerebral cortex was less than 5 percent, but in subcortical white matter homolateral or adjacent to the tumor, the swelling averaged 77 percent (Table 26-1). Several subsequent studies utilizing experimental models have confirmed this observation either morphologically or biochemically. The experimental models include implanted tumors,

TABLE 26-1. Types of Edema in Cerebral Tissues in Vivo

Species	Type of Lesion	Cerebral Cortex Dry Weight (%) Control	Cerebral Cortex Dry Weight (%) Exptl.	Cerebral Cortex Swelling (%)	Subcortical White Dry Weight (%) Control	Subcortical White Dry Weight (%) Exptl.	Subcortical White Swelling (%)
Man [26][a]	Tumor	17.4	16.9	< 5	30.7	17.4	77
Cat [23]	Freezing lesion	19.3	18.3	5.5	31.9	21.4	49
Rabbit [1, 13]	Triethyl tin	20.2	20.0	0	30.8	22.9	34
Cat [28]	Circulatory arrest	16.0	13.5	18.5[b]	29.6	29.0	0

[a]References to sources of data are given in brackets.
[b]For the monkey, the comparable value is 22.2% swelling [5].

the use of intracranial balloons, and the implantation of such hydrophilic materials as psillium seeds, polysaccharides, or proteins. Despite the focal and often variable nature of the edema in these preparations, there is good agreement among investigators that the edema fluid resembles a plasma exudate with a relatively high content of Na^+ and Cl^- and a demonstrable content of protein, notably albumin [13]. The presence of this protein exudate is indicative of the disruption of the blood-brain barrier.

FREEZING LESIONS

A similar type of edema is seen in the cerebral white matter of animals after the brain has been subjected to a localized freezing lesion. The technique involves application of a metal plate, cooled to –50°C or below, to the intact skull or directly to the surface of the brain. This produces a local necrotizing lesion that breaches the blood-brain barrier. In the 24 hr after placement of the lesion, edema fluid spreads from the white matter of the traumatized gyrus to involve adjacent white matter extensively, the rate of spread being a function primarily of the systolic blood pressure. The process seems to be most consistent with an extravascular migration of plasmalike fluid via extracellular routes to the white matter beneath and surrounding the lesion* [16]. Morphologically and biochemically, the freezing lesion closely resembles traumatic lesions of the brain, such as a stab wound, as well as the edema around tumors and foreign substances.

From the behavior of edema fluid around tumors or the freezing lesion, it seems that the interstitial spaces of cerebral gray matter are much less distensible than are those of cerebral white matter, so the exudates from the plasma that penetrate the breached blood-brain barrier tend to collect in the latter location.

Direct analyses of cerebral tissues from the control hemisphere and from the hemisphere in which the freezing lesion was placed demonstrate minimal swelling

*This form of edema has been termed vasogenic edema [14].

TABLE 26-2. Approximate Composition of Edema Fluids

Preparations	Swelling (%)	Na^+ (mM)	Cl^- (mM)	K^+ (mM)	Albumin (g/liter)
Rabbit: subcortical white matter; triethyl tin [1][a]	34	133	115	–[b]	0
Dog: subcortical white matter; freezing lesion [9]	43	124	87	59	19
Cat: cerebral cortex; slices incubated in 54 mM K^+ [4]	33	39	135	108	–

[a]References to sources of data are given in brackets.
[b]Per unit dry weight, the K^+ was identical to control values; hence the K^+ concentration in the edema fluid was presumably close to that of plasma (about 3 mM).

(about 5 percent) of cerebral cortex, but very pronounced swelling (about 50 percent) of subcortical white matter on the side of the lesion (see Table 26-1). The edema fluid consists principally of Na^+ and Cl^- plus an appreciable content of protein (Table 26-2). The composition of the edema fluid probably reflects some dilution and contamination with intracellular (nonedema) fluid during the isolation procedure [9], but it still primarily resembles a plasma exudate. An additional indication of plasma as the source for the protein in the edema fluid is provided by the selective uptake of radioactively labeled albumin into the edema fluid [13] in quantities that closely parallel the increases in volume and weight of the edematous hemisphere [24].

TRIETHYL TIN EDEMA

Both dialkyl and trialkyl tin compounds are fatally toxic for a variety of mammalian species, including man. However, only the trialkyl form is neurotoxic, producing edema of subcortical white matter that is distinct in several ways from the examples already described. Some swelling of astrocytes is seen in cerebral cortex, but the striking abnormality is an accumulation of fluid within the myelin sheaths of nerve fibers in the subcortical white matter [1]. The edema fluid is clearly intramyelinic, splitting the myelin lamellae at the interperiod line to form large blebs, or vacuoles. No swelling of gray matter can be demonstrated quantitatively, but the subcortical white matter exhibits swelling of some 34 percent (see Table 26-1).

The edema fluid is composed essentially of Na^+ and Cl^- (see Table 26-2), with no protein or other constituents that would be expected if the fluid were a plasma exudate. Evidence for the intactness of the blood-brain barrier is provided by studies with radioactively labeled albumin and vital dyes. In contrast to most forms of cerebral edema, in which the uptake of $^{24}Na^+$ by brain is normal or increased, the uptake of $^{24}Na^+$ in triethyl tin intoxication of the brain appears to be reduced, but further analyses have revealed that the apparent reduction is a reflection of the fact that a large fraction of the brain Na^+ (presumably that within the intramyelinic blebs) is no longer exchangeable. No evidence has been obtained for impairment of the activity of the Mg^{2+}-dependent, (Na^+, K^+)-activated ATPase under these conditions [13], and the precise mechanism of action of triethyl tin remains obscure. It is known that the earliest manifestation of toxicity is inhibition of oxidative phosphorylation in cerebral mitochondria. This inhibition apparently represents two effects: an inhibition of coupled phosphorylation (analogous to the effect of oligomycin) and an alteration of anion exchange across the mitochondrial membrane with intramitochondrial accumulation of Cl^- [27]. How these latter effects of triethyl tin may relate to the intramyelinic edema is not at all clear.

OUABAIN-INDUCED EDEMA

When ouabain (G-strophanthin) is introduced into the central nervous system by local application or by local perfusion, violent and usually fatal convulsions follow [19]. There is an associated leakage of K^+ from brain tissue into the perfusate and, by light microscopy, the cerebral cortex has the appearance of the classic status spongiosus, the underlying white matter being spared. Detailed examination of central nervous tissues by electron microscopy has not been reported, but Birks [3] has described perfusing the sympathetic ganglia of the cat with plasma containing a related glycoside, digoxin (5 μM), and the effects on the electron microscopic appearance of the ganglia after perfusion fixation. In ganglia perfused with plasma plus digoxin, the neurons exhibited significant swelling of the cytoplasm, as did the lumina of the rough endoplasmic reticulum and the nerve endings. The mitochondria appeared dense and shrunken. The Schwann cells appeared perfectly normal. If the ganglia were perfused with digoxin in a Cl^--free or low-Na^+ plasma, most or all of the abnormalities in the neurons were prevented.

These observations are consistent with the known effects of ouabain and related glycosides as specific inhibitors of Na^+ and K^+ transport, presumably mediated by binding to and inhibition of Mg^{2+}-dependent, (Na^+, K^+)-activated ATPase in neural membranes [25], and as mobilizers of cellular Ca^{2+} with subsequent accumulation in mitochondria [29]. The cellular locus of the swelling is also consistent with data from studies in vitro (see Table 26-5, below) that demonstrate swelling of slices of cerebral cortex incubated with ouabain, but no parallel increase in the inulin (extracellular) space of the slices, in association with the striking increase of intracellular Na^+ and Cl^- and depletion of intracellular K^+ [6, 29]. In contrast, sections of corpus callosum incubated with ouabain exhibit some extracellular swelling but no depletion of K^+, an indication that glial cells are relatively unaffected [29].

With this background, the changes in the distribution of cerebral fluids and electrolytes caused by ouabain (0.35 μM) when applied to the surface of the brain are more understandable (Table 26-3). The involved cerebral hemisphere exhibits about 20 percent swelling, with a considerable alteration of the Na^+/K^+ ratio

TABLE 26-3. Ouabain-Induced Edema in Rat Cerebral Hemispheres in Vivo

Preparation	Dry Weight (%)[a]		Swelling (%)	Na^+/K^+ Ratio[a]	
	Control H	Exptl. H		Control H	Exptl. H
Freezing lesion[b]	23.0	16.2	42	0.49	1.4
Ouabain (3.5 X 10^{-7} M)	22.6	18.6	21.5	0.46	1.1
Freezing lesion + ouabain	22.5	14.8	52	0.51	2.3

[a]Data are given for control hemispheres (H) and for hemispheres containing the lesion and/or to which ouabain was applied (exptl. H) [17, 18].

[b]The freezing lesions were produced by local application of vinyl ether sprayed onto the cortical surface. The results are comparable to those obtained by the application of cooled metal plates.

reflecting the depletion of tissue K^+ and its replacement by Na^+. The effect is much more pronounced if the ouabain is applied to a preexisting freezing lesion (Table 26-3), and striking focal seizures ensue.

CIRCULATORY ARREST

When the circulation to the brain is totally arrested for 15 to 30 min, edema of cerebral cortex develops to the extent of 18 to 22 percent, without any swelling of subcortical white matter (see Table 26-1). There is an associated leakage of K^+ from the cortical tissue and replacement by Na^+ without change in tissue Cl^- [28]. The edema cannot be attributed to hypoxia since respiratory arrest does not produce significant swelling of brain. The edema is clearly intracellular since the inulin and sucrose spaces decrease to one-half or one-third of normal after circulatory arrest (Table 26-4). It has been customary to dismiss such observations as postmortem artifacts [19], but these data may prove important in the problem of cerebrovascular occlusions (stroke).

Ames and colleagues [2, 8] have described the "no reflow" phenomenon in animals subjected to cerebral ischemia. When circulation was resumed, sizeable areas of brain failed to reperfuse, partly because of increased viscosity of blood and partly because of compression of capillary lumina caused by the swelling of perivascular glia and endothelial cells, with formation of blebs (endothelial ?) intruding into the capillary lumina. It has been speculated that the swelling of these cells may be caused by K^+ released by the anoxic tissue. Thus, this swelling would be similar to that seen with high-K^+ solutions in vitro. The investigators suggest that if these changes can be minimized or reversed (e.g., by increased blood perfusion pressure), the neurons may be able to resume normal function after periods of ischemia of 15 min or longer.

TABLE 26-4. Respiratory vs. Circulatory Arrest in Cerebral Gray Matter

Observations	Controls (%)	Arrest (%)	References
Respiratory arrest (5–7 min)[a]			
Cat cerebral cortex (circulation intact)			
Swelling, initial	–	3.8	[20]
Swelling after survival 1 hr	–	0	
Circulatory arrest (15–30 min)			
Cat cerebral cortex (not perfused)			
Swelling	–	18.5	[28]
Monkey cerebral cortex (perfused)[b]			
Swelling	0.7	22.2	[5]
After Cl^--free perfusate	–	5.8	
Dog caudate nucleus (perfused)[c]			
Sucrose space	18.1	6.6	[21]
Inulin space	12.7	5.4	

[a]Respiration was arrested until the blood pressure failed (5–7 min) and then was resumed with 100% oxygen.
[b]Perfusion over the surface of the cortex.
[c]Ventriculocisternal perfusion.

A practical corollary to the foregoing applies to samples of brain taken after any significant interval of circulatory arrest: the associated edema and shifts of electrolytes can seriously obscure the original state of the tissue prior to stopping the circulation [28].

COMPARATIVE AND ONTOGENETIC ASPECTS

Knowledge of edema in the central nervous system of nonmammalian species is relatively meager. The shark exemplifies how great the comparative differences may be; it is virtually impossible to produce edema in shark brain by some of the experimental procedures described above [16]. Presumably, the very different structural organization of the shark brain accounts for the failure to produce edema after placement of destructive surface lesions. Among the various mammalian species, there seems to be little qualitative or quantitative difference in the response of brain or spinal cord to various edema-producing procedures. However, the immature brain, exemplified by the neonatal kitten brain, is relatively resistant to production of edema either in vivo [16] or in vitro ([29, 30]; Table 26-5). Of the various body tissues that have been examined in vitro, slices of cerebral cortex exhibit much greater swelling under usual incubation conditions than slices of subcortical white matter, liver, renal cortex, or diaphragm [30].

SLICES OF BRAIN TISSUE INCUBATED IN VITRO

Several types of swelling characteristic of brain slices have been identified (Fig. 26-1; [6, 30, 31]).

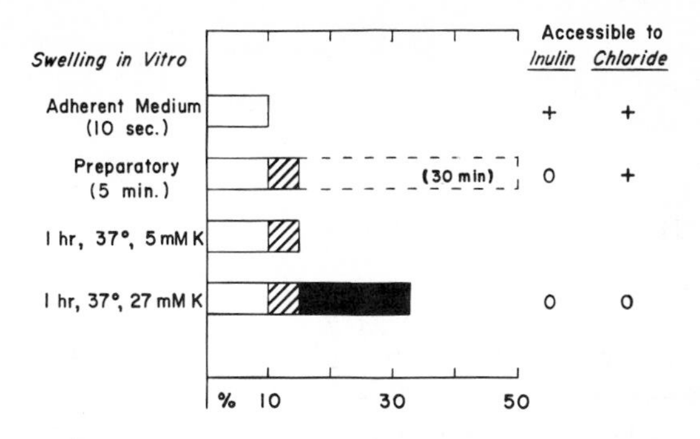

FIGURE 26-1. Types of swelling encountered in slices of cerebral cortex prepared and incubated in vitro. The values on the abscissa are percentages of initial fresh (wet) weight of slices. The slices were incubated or immersed in bicarbonate-buffered saline-glucose media containing 5 mM K^+ (normal) or 27 mM K^+ (high K^+). Incubations were for 1 hr at 37°C. The interpretations are based on studies by Varon and McIlwain [31], and the data are taken from the studies on cat cerebral cortex by Bourke and Tower [6]; see [6] for further details.

Inevitably, there is some adherent fluid from the incubation medium, amounting to 5 to 10 percent of the initial fresh weight of the tissue, that is probably associated primarily with the major cut surfaces of the slices. Swelling to this extent occurs when slices of cerebral cortex are dipped into isoosmotic media for as few as 10 sec, and it is clearly extracellular since it is accessible to both inulin and chloride.

A second type of swelling in cerebral cortical slices takes place during the immersion of slices at room temperature while the incubation flasks are being gassed. The extra fluid in the slices after immersion for 5 min totals about 15 percent of the initial fresh weight of the tissue. This "preparatory" swelling can progress under suboptimal conditions, so that after 30 min it may reach 50 percent of the initial fresh weight of the tissue. It takes place in a compartment inaccessible to inulin but accessible to chloride. This type of swelling accounts for the rather high values for slice swelling encountered in the earlier literature and in studies in which the preparation of slices is prolonged unduly. The foregoing types of swelling are clearly artifacts of *in vitro* conditions and probably have no direct relevance to *in vivo* forms of edema.

K^+- AND Cl^--DEPENDENT SWELLING

A third form of swelling that occurs in vitro may be related to intracellular swelling in vivo; this form is produced by elevated concentrations of K^+ and Cl^- in the incubation medium. This swelling is presumably intracellular because it takes place in a tissue compartment inaccessible to both inulin and Cl^- (Fig. 26-1). Superimposition of anoxia or inhibitory concentrations of ouabain that impair energy sources or op-

TABLE 26-5. Edema of Cerebral Tissues in Vitro

Preparations[a]	Swelling (%)	Na^+/K^+ Ratios
Adult cerebral cortex		
Controls	+ 32.7	0.75
Cl^--free media[b]	+ 14.4	0.58
Ouabain (10^{-5} M)	+ 45.1	2.75
Neonatal cerebral cortex		
Controls	− 1.2	1.05
Ouabain (10^{-5} M)	+ 19.6	3.1
Adult corpus callosum		
Controls	+ 6.7	1.2
Ouabain (10^{-5} M)	+ 15.3	1.85

[a]Slices of cat cerebral tissues were incubated for 1 hr at 37°C in 27 mM K^+ media.
[b]Isethionate (125 mM) was substituted for Cl^- [29].

eration of the Na^+-K^+ ion pumps depolarizes the cells, enhances the degree of edema, and results in gross shifts in distribution of electrolytes (see Table 26-5). Exposure of slices to inhibitory concentrations of ouabain is one of the few situations that cause slices of cerebral cortex of neonatal kitten to swell. The concomitant shift in the Na^+/K^+ ratio indicates that, despite the immaturity of these slices, the cation pumps must already be operative (see Table 26-5). For external concentrations of K^+ over about 20 mM, the swelling of slices is a linear function of increasing concentrations of K^+ in the incubation medium. This swelling occurs in a tissue compartment inaccessible to inulin, but is paralleled by an increase in fluid spaces accessible to Cl^-. The role of Cl^- is dramatically illustrated by the effect of substituting a less permeable anion, isethionate (2-hydroxyethanesulfonate), for Cl^-. Even at an external K^+ concentration of 125 mM, essentially no slice swelling occurs in the absence of Cl^- (Fig. 26-2).

Kinetic studies of these phenomena indicate that in incubated cerebral cortical slices there is a K^+-dependent, mediated transport of Cl^- into a cellular (astroglial ?) compartment and that the edema fluid in this compartment is composed primarily of K^+ and Cl^- ([4]; see Table 26-2). Presumably, these findings relate to the high-K^+ glial cells described by Kuffler and colleagues and for which they have proposed a K^+ "spatial buffering" function [22]. In their proposal the K^+ that is released, upon stimulation of neurons, into the immediate extraneuronal environment is removed by being taken up into adjacent astrocytes. Hence, the effects of accumulated external K^+ on neuronal membrane potentials and excitability are minimized.

Recent studies by Bourke et al. [5] have demonstrated that many of the foregoing phenomena can be reproduced in vivo. In monkeys fitted with a plastic perfusion dome in place of skull, the perfusion over the surface of the cerebral cortex of artificial cerebrospinal fluid containing various concentrations of K^+ or Cl^- or both reproduced qualitatively the same K^+- and Cl^--dependent edema as

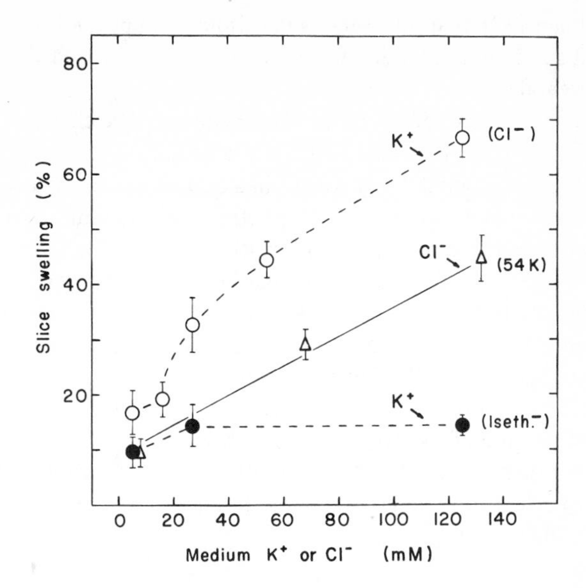

FIGURE 26-2. Swelling of incubated slices of cat cerebral cortex as a function of the concentration of K^+ (open circles) or of Cl^- (Δ) in the incubation medium. The effects of varying Cl^- in the medium were examined at a constant concentration of K^+ (54 mM); similarly, the effects of varying K^+ in the medium were examined at a constant concentration of Cl^- (125 mM). The effects of incubation in Cl^--free media (closed circles) were obtained by substituting isethionate (125 mM) for Cl^-. Slice swelling (ordinate) is the percentage of initial fresh (wet) weight of slices. All slices were incubated for 1 hr at 37°C in bicarbonate-buffered saline-glucose media with variations in composition as indicated. (From [4] and [6].)

depicted in Figure 26-2. Since in these experiments cerebral circulation was intact and only one cortical surface could be perfused, the quantitative changes were somewhat less, but edema of about 20 percent could be regularly elicited during perfusion with perfusates containing Cl^- and 100 mM K^+. Most of the edema could be prevented by perfusion with Cl^--free (isethionate) perfusates and could be reversed by switching to perfusates with normal (5 mM) K^+ concentrations. As already discussed, the relevance of these observations to edema associated with cerebrovascular ischemia or occlusions (stroke) should not be overlooked.

THERAPY

OSMOTICALLY ACTIVE AGENTS

Three principal approaches to the therapy of cerebral edema have been employed. Surgical intervention to decompress the swollen brain and relieve the increased intracranial pressure is an emergency, often a lifesaving, measure. It is obviously not a useful routine treatment, and in many cases it is only a temporary expedient. The use of osmotically active agents is an outgrowth of the classic observation of

Weed and McKibben [32] that intravenous injections of hypertonic (30 percent, w/v) NaCl would shrink normal brain, whereas intravenous distilled water would cause marked cerebral edema.

Several osmotically active agents are currently employed, such as urea [12], mannitol [33], and glycerol [7]. The first two are administered intravenously (urea: 0.5 to 1.5 g per kilogram body weight in a 30 percent w/v solution; mannitol: 1.5 to 2 g per kilogram body weight in a 15 to 20 percent w/v solution), and these will lower cerebrospinal fluid pressure significantly for 3 to 8 hr. Glycerol is administered orally or by gastric tube (0.5 to 0.7 g per kilogram body weight every 3 to 4 hr). Urea is eventually distributed throughout the total body water and, in high doses, can be toxic. It is also more difficult to prepare in sterile solutions and is not stable in solution. Mannitol is distributed only in extracellular fluid and crosses the blood-brain barrier with difficulty. It is relatively nontoxic and is excreted rapidly. Glycerol is useful because of its low toxicity and because it can be employed more easily in prolonged therapy. All these agents suffer from the disadvantage that they apparently do not affect the edema fluid per se but act to relieve signs and symptoms by dehydrating normal brain [23].

STEROIDS

The third type of therapy involves the use of adrenocortical steroids of the glucocorticoid type. The most widely used are dexamethasone (9α-fluoro-16α-methylprednisolone) and cortisone (17-hydroxy-11-dehydrocorticosterone). Dexamethasone, a synthetic fluorinated glucocorticoid, was chosen for its antiinflammatory potency and low salt-retaining activity [11]. In clinical practice, it is usually administered initially in a 10 mg dose intravenously, followed every 6 hr by a 4 mg dose intramuscularly until maximum response is obtained (usually within 18 to 24 hr) or for 2 to 4 days postoperatively. Oral maintenance doses of 1 to 3 mg three times daily may be continued for longer periods [11].

Despite widespread clinical evidence for the efficacy of dexamethasone and cortisone in relieving the signs and symptoms of cerebral edema in human patients, experimental verification has been difficult to obtain [24]. However, a recent study by Pappius and McCann [24] indicates that dexamethasone significantly diminishes the edema of cerebral hemispheres traumatized by a freezing lesion (Table 26-6). In these studies, cats were pretreated with dexamethasone for 48 hr before the freezing lesion was made in one hemisphere, and the animals were also injected with ^{131}I-serum albumin 48 hr before making the lesion. Differences in hemispheral weights and in content of ^{131}I-serum albumin between the control and traumatized hemispheres were measured at various times after making the freezing lesion. At 24 hr after the lesion, the differences between the two hemispheres were similar for control (untreated) and treated animals, but after 48 hr the treated animals exhibited significantly smaller differences in weight and in ^{131}I-serum albumin between control and traumatized hemispheres (Table 26-6). By 72 hr, when the edema had begun to resolve, there were no longer any significant differences between

TABLE 26-6. Effects of Pretreatment with Dexamethasone on Edema of Freezing Lesion in Cat Brain

Observations	Differences in Weight between Control and Edematous Hemisphere (g)[a] 24 hr after lesion	Differences in Weight between Control and Edematous Hemisphere (g)[a] 48 hr after lesion	Differences in Content of ^{131}I-Serum Albumin (ml serum equivalents)[a] 48 hr after lesion
Untreated controls	0.62 ± 0.14 (9)	0.85 ± 0.24 (15)	0.51 ± 0.15 (9)
Dexamethasone 2.5 mg/kg[b]	0.53[c] (8)	0.52 ± 0.19 (12)	0.36 ± 0.11 (12)
Significance (*P*) control vs. treated	n.s.	$P < 0.01$	$P < 0.05$

[a]The values represent means ± S.D. for the numbers of animals given in parentheses.
[b]Dexamethasone was given intraperitoneally in a dose of 1.25 mg per kilogram body weight twice daily for 48 hr prior to making the freezing lesion and was continued at the same dose schedule thereafter.
[c]The S.D. for this mean was not reported [24].

treated and control animals. Determinations of dry weights of subcortical white matter from the two groups of animals at 48 hr demonstrated no differences, the values being 24.2 ± 2.1 percent for untreated animals ($n = 11$) and 22.8 ± 0.4 percent for dexamethasone-treated animals ($n = 4$), in comparison with normal (untraumatized) white matter in which the dry weight averaged 32.3 ± 1.7 percent ($n = 8$).

Thus, there is an apparent discrepancy between the data for dry weights and the data for interhemispheric differences in total weight and radioactive albumin content. There was no evidence for shrinkage of normal brain, so the discrepancy could most readily be explained by assuming that the treatment with dexamethasone prevented spread of the edema in the period between 24 and 48 hr after the lesion was made. This effect can be calculated to correspond to limiting the edema to approximately one-third of the hemispheral white matter, in contrast to involvement of more than one-half of the hemispheral white matter in untreated animals.

Pappius and McCann [24] also reported that, in contrast to man, treatment of cats with cortisone was ineffective under the conditions described, apparently because of poor absorption after intraperitoneal administration. This is an important observation because it emphasizes the factor of species differences that may have complicated earlier attempts to examine therapeutic efficacy experimentally. From this brief review of the current status of therapy for cerebral edema, it is obvious that additional studies are needed to establish the efficacy and precise mechanisms of action of available therapeutic agents, as well as to develop more effective therapies.

CONCLUSION

Insofar as possible, discussion in this chapter has focused on edema or swelling of neural tissues. Only passing reference has been made to the blood-brain barrier, to the various fluid compartments within cerebral tissues, or to the electrolytes and other solutes in these fluids. And little cognizance has been taken of the fact that most of the edema-producing situations in vivo are commonly characterized by the occurrence of seizures.

Such aspects are discussed elsewhere in this volume, and those chapters should be read in conjunction with this one for a proper perspective of the normal and pathological factors responsible for the distribution of fluids and electrolytes in the central nervous system. The subject is still fraught with controversies, but it is one of the most important aspects of neurochemistry because the fluid environment surrounding neural membranes determines the viability and excitability of the neuronal and glial elements of the brain and spinal cord. The sensitivity of neural cells to even small changes in their environment accounts for the many mechanisms that operate to preserve the *milieu interne.* Such mechanisms produce or utilize the energy necessary for transporting a variety of solutes into or out of various compartments of the central nervous system, with concomitant appropriate shifts of water. The derangements of these processes that are associated with the various forms of edema promptly produce impaired consciousness, convulsions, coma, and often death – consequences that attest to the gravity of cerebral edema in clinical practice.

REFERENCES

1. Aleu, F. P., Katzman, R., and Terry, R. D. Structure and electrolyte analyses of cerebral edema induced by alkyl tin intoxication. *J. Neuropathol. Exp. Neurol.* 22:403, 1963. (The first definitive electron-microscopic study demonstrating the characteristic selective intramyelinic edema and evidence that the blood-brain barrier is not impaired. Confirms and extends the original light-microscopic and biochemical studies of Magee et al. [*J. Pathol.* 73:107, 1957].)
2. Ames, A., III, Wright, R. L., Kowada, M., Thurston, J. M., and Majno, G. Cerebral ischemia. II. The no-reflow phenomenon. *Am. J. Pathol.* 52:437, 1968. (See comments for the companion paper, reference 8.)
3. Birks, R. I. The effects of a cardiac glycoside on subcellular structures within nerve cells and their processes in sympathetic ganglia and skeletal muscle. *Can. J. Biochem.* 40:303, 1962. (An excellent electron microscopic study of the effects of perfusion of ganglia with plasma containing digoxin and of the prevention of the swelling of subcellular elements in the neurons by perfusion with Cl-free or low Na plasma. This is an important and unique study.)
4. Bourke, R. S. Studies of the development and subsequent reduction of swelling of mammalian cerebral cortex under isosmotic conditions *in vitro*. *Exp. Brain Res.* 8:232, 1969. (Together with the accompanying paper in the same journal

volume [p. 219], these studies demonstrate details of the K- and Cl-dependent edema in incubated slices of cerebral cortex, the kinetics of the mediated transport of Cl into astrocytes, and the composition of the edema fluid therein. Reference 5 reports comparable studies in vivo.)

5. Bourke, R. S., Nelson, K. M., Naumann, R. A., and Young, O. M. Studies on the production and subsequent reduction of swelling in primate cerebral cortex under isosmotic conditions *in vivo*. *Exp. Brain Res.* 10:427, 1970. (This paper reports the demonstration in vivo of the K- and Cl-dependent edema of cerebral cortex reported earlier [reference 4] for in vitro preparations. Monkeys were fitted with plastic domes over the cerebral hemispheres and the surface of the cortex was perfused with artificial CSF of various compositions. The observations are particularly relevant to problems of circulatory arrest and strokes.)
6. Bourke, R. S., and Tower, D. B. Fluid compartmentation and electrolytes of cat cerebral cortex *in vitro*. I. Swelling and solute distribution in mature cerebral cortex. *J. Neurochem.* 13:1071, 1966. (Confirmation and extension of earlier studies [reported in reference 31] on the distribution of fluids and electrolytes in cerebral tissues. Published with two accompanying papers, one on electrolytes [same journal volume, p. 1099] and one on ontogenetic and comparative aspects; see reference 30.)
7. Cantore, G., Guidetti, B., and Virno, M. Oral glycerol for the reduction of intracranial pressure. *J. Neurosurg.* 21:278, 1964. (First definitive report on efficacy of glycerol for treatment of cerebral edema.)
8. Chiang, J., Kowada, M., Ames, A., III, Wright, J. L., and Majno, G. Cerebral ischemia III. Vascular changes. *Am. J. Pathol.* 52:455, 1968. (Together with reference 2, describes the "no-reflow" phenomenon after experimentally induced cerebral ischemia and the associated pathological changes examined electron microscopically. These studies have much potential importance for the problems associated with cerebrovascular hypotension, transient ischemia and cerebrovascular occlusions. The therapeutic implications deserve attention.)
9. Clasen, R. A., Sky-Peck, H. H., Pandolfi, S., Laing, I., and Hass, G. M. The chemistry of isolated edema fluid in experimental cerebral injury. [In reference 15, p. 536.] (Review of experiences with edema resulting from freezing lesions produced through the intact skull, and details of biochemical characteristics, especially the protein content, of the edema fluid.)
10. Elliott, K. A. C., and Jasper, H. Measurement of experimentally induced brain swelling and shrinkage. *Am. J. Physiol.* 157:122, 1949. (The classic paper on the measurement and calculation of changes in the fluid content of cerebral tissues in vivo.)
11. Galicich, J. H., and French, L. Use of dexamethasone in the treatment of cerebral edema resulting from brain tumors and brain surgery. *Amer. Practit.* 12:169, 1961. (First definitive report on efficacy of dexamethasone for treatment of cerebral edema, with review of previous trials of steroid therapy.)
12. Javid, M. Urea – new use of an old agent. Reduction of intracranial and intraocular pressure. *Surg. Clin. North Am.* 38:907, 1958. (Review of experience

with hypertonic solutions of urea as therapy for cerebral edema, by the originator of the use of urea for this purpose.)

13. Katzman, R. Biochemical correlates of cerebral edema. [In reference 15, p. 461.] (Review of earlier work on cerebral edema produced by triethyl tin, implanted glial tumors, and implanted protein derivatives, with data on ultrastructure, nature of the edema fluid, and the involvement of the blood-brain barrier. The studies come from one of the leading groups in the field, especially for triethyl tin edema.)
14. Klatzo, I. Neuropathological aspects of brain edema. *J. Neuropathol. Exp. Neurol.* 26:1, 1967.
15. Klatzo, I., and Seitelberger, F. (Eds.). *Brain Edema.* New York: Springer-Verlag, 1967. (Publication of proceedings of a symposium on brain edema held in Vienna, Sept., 1965. This volume provides what is probably the best recent source of information on edema of neural tissues from clinical and experimental points of view, with numerous contributions on morphological and biochemical aspects. Therapy is not covered. In addition to chapters cited individually in this bibliography [references 9, 13, 16, 21, 23, and 28], the chapters in *Brain Edema* by Hoff and Jellinger [p. 3] on clinical aspects of cerebral edema, by Feigin [p. 128] on the human pathology, and by Long et al. [p. 419] on a review of various types of experimental cerebral edema should prove useful.)
16. Klatzo, I., Wisniewski, H., Steinwall, O., and Streicher, E. Dynamics of cold injury edema. [In reference 15, p. 554.] (Review of earlier work on edema resulting from freezing lesions produced directly on the surface of the brain, with special reference to the blood-brain barrier. Studies come from the originators of this procedure and encompass structural, biochemical, comparative, and ontogenetic aspects.)
17. Lewin, E. Focal "epileptogenic" lesions produced with ouabain in the rat. *Neurology* 19:310, 1969. (Abstract of studies on effects of topical ouabain, alone or after placement of a freezing lesion in cerebral cortex, on development of epileptiform activity and on the fluids and electrolytes of the experimental and control hemispheres. This paper provided date for Table 26-3 as supplement by unpublished data made available by the author.)
18. Lewin, E., Charles, G., and McCrimmon, A. Discharging cortical lesions produced by freezing: The effect of anticonvulsants on sodium-potassium-activated ATPase, sodium, and potassium in cortex. *Neurology* 19:565, 1969. (Data on the freezing lesion to supplement those in reference 17.)
19. Maccagnani, F., Bignami, A., and Palladini, G. Étude électroencéphalographique des effets de l'introduction intracrânienne d'ouabaïne chez le rat et le cobaye; mise en évidence d'un tracé périodique. *Rev. Neurol. (Paris)* 115:211, 1966. (One of the earliest contemporary reports on the clinical, EEG, and morphological effects of intracerebral ouabain. A good, detailed study extending observations originally made by Rizzolo [*Arch. Farmac. sper.* 50:16, 1929].)
20. Norris, J. W., and Pappius, H. M. Cerebral water and electrolytes. Effects of asphyxia, hypoxia and hypercapnia. *Arch. Neurol.* 23:248, 1970. (One of a

series of excellent studies from Pappius' laboratory that deals specifically with the effects of respiratory arrest in contradistinction to circulatory arrest. The paper contains a good review of the problem.)

21. Oppelt, W. W., and Rall, D. P. Brain extracellular space as measured by diffusion of various molecules into brain. [In reference 15, p. 333.] (Review of the studies on extracellular space of brain as measured by ventriculo-cisternal perfusion with inulin, dextran, and sucrose. The technique was developed in Rall's laboratory and represents one of the pioneering investigations that has contributed to current concepts of fluid distribution in the brain.)
22. Orkand, R. K., Nicholls, J. G., and Kuffler, S. W. Effect of nerve impulses on the membrane potential of glial cells in the central nervous system of amphibia. *J. Neurophysiol.* 29:788, 1966. (An important contribution from Kuffler's laboratory that proposes for the high K glia a "spatial buffering" function to remove excess K from the extracellular environs of neurons. This concept has been extended to mammalian CNS by Trachtenberg and Pollen [*Science* 167: 1248, 1970].)
23. Pappius, H. M. Biochemical studies on experimental brain edema. [In reference 15, p. 445.] (Review of many previous studies, particularly on the edema associated with freezing lesions, and the effects of intravenous urea thereon.)
24. Pappius, H. M., and McCann, W. P. Effects of steroids on cerebral edema in cats. *Arch. Neurol.* 20:207, 1969. (One of a series of excellent studies from Pappius' laboratory that demonstrates significant effects of dexamethasone on the development of cerebral edema after placement of a freezing lesion. The paper contains a good review of the status of steriod therapy in experimentally-induced edema.)
25. Siegel, G. J., and Albers, R. W. Nucleoside Triphosphate Phosphohydrolases. In Lajtha, A. (Ed.), *Handbook of Neurochemistry.* Vol. 4. New York: Plenum, 1970, p. 13. (An authoritative review of the enzymology of ATPases.)
26. Stewart-Wallace, A. M. Biochemical study of cerebral tissue, and changes in cerebral edema. *Brain* 62:426, 1939. (The classical observations localizing edema associated with brain tumors to cerebral white matter. A study from 30 years ago that has withstood the test of time.)
27. Stockdale, M., Dawson, A. P., and Selwyn, M. J. Effects of trialkyltin and triphenyltin compounds on mitochondrial respiration. *Eur. J. Biochem.* 15: 342, 1970. (The latest in a series of studies initiated by Aldridge and Cremer [*Biochem. J.* 61:406, 1955] on the earliest toxic effect of trialkyltin compounds – the interference with mitochondrial oxidative phosphorylation.)
28. Tower, D. B. Distribution of cerebral fluids and electrolytes *in vivo* and *in vitro.* [In reference 15, p. 303.] (A survey of factors affecting the distribution of fluids and electrolytes in the CNS based on in vitro studies reported in references 6 and 30, plus in vivo studies by Bourke et al. [*Am. J. Physiol.* 208:682, 1965]. In addition data on artifacts of fixation and on circulatory arrest are reported.)
29. Tower, D. B. Ouabain and the distribution of calcium and magnesium in cerebral tissues *in vitro. Exp. Brain Res.* 6:273, 1968. (Evidence for mobilization

by ouabain of tissue Ca and its sequestration in mitochondria; also data on comparative and ontogenetic effects of ouabain on neural tissues in vitro.)

30. Tower, D. B., and Bourke, R. S. Fluid compartmentation and electrolytes of cat cerebral cortex *in vitro*. III. Ontogenetic and comparative aspects. *J. Neurochem.* 13:1119, 1966. (Companion paper to reference 6, dealing primarily with in vitro studies on fluids and electrolytes in neonatal and developing brain and in body tissues other than cerebral cortex. Contains data not available elsewhere.)
31. Varon, S., and McIlwain, H. Fluid content and compartments in isolated cerebral tissues. *J. Neurochem.* 8:262, 1961. (A definitive study identifying principal factors affecting fluid distribution in incubated slices of cerebral cortex. Clarifies earlier studies from Elliott's laboratory [*Proc. Soc. Exp. Biol. Med.* 63:234, 1946; *Can. J. Biochem. Physiol.* 34:1007, 1956; 36:217, 1958] and extended subsequently by studies reported in references 6 and 30.)
32. Weed, L. H., and McKibben, P. S. Experimental alterations of brain bulk. *Am. J. Physiol.* 48:531, 1919. (Report of the first experimental demonstration of alterations in brain volume by intravenous injections of hyper- or hypo-tonic solutions. An important, classic study.)
33. Wise, B. L., and Chater, N. The value of hypertonic mannitol solution in decreasing brain mass and lowering cerebrospinal-fluid pressure. *J. Neurosurg.* 19:1038, 1962. (The first definitive report on the efficacy of hypertonic solutions of mannitol for treatment of cerebral edema.)

27

Neurotoxins

Maynard M. Cohen

EFFECTS OF NEUROTOXIC AGENTS

It has long been recognized that adverse effects of neural toxins result from industrial or accidental exposure, as well as from homicidal or suicidal attempts. Current environmental pollution now exposes the entire population to numerous toxic materials. Most toxins affecting the nervous system do so by interfering with biochemical systems, some through relatively specific mechanisms. The toxin may be attached to sites of metabolic action, blocking the active site from access to required compounds, or it may bear a chemical resemblance to certain compounds required for metabolic or physiological activity. When the toxic agent substitutes for the normally required compound, the metabolic or physiological activity is inhibited. A generalized toxic effect may be caused by processes such as protein denaturization or by alterations in acid-base balance or ionic equilibrium. Under some circumstances, there may be interference with biochemical systems, but added factors may be necessary to produce toxicity.

In addition to direct neural action, toxins may produce other effects contributing to morbidity and mortality. These include effects on cerebral blood vessels; physiological effects on cerebral centers, including depression of respiratory or circulatory centers or stimulation of neural activity; and effects on other organs and tissues.

The materials discussed in this chapter are those known to produce toxic neural manifestations in humans. Many other compounds are potentially toxic, but they are not considered here because either their effect on man is, as yet, of no serious consequence or their toxic action is not primarily on neural biochemical systems.

HEAVY METALS AND METALLOIDS

These compounds are protoplasmic poisons that exert specific effects on neural tissue by inactivating enzymes that contain sulfhydryl radicals. Oxidative metabolism is predominantly affected because it involves many such enzymes. The succinic dehydrogenase system is particularly sensitive to some of these compounds. Not all the metals and metalloids that produce neural intoxication have been tested specifically for toxic depression of oxidative metabolism. However, failure of glycolysis to proceed into oxidative metabolism has been demonstrated in the presence of many such toxins. This results in increasing blood concentration of pyruvic acid and lactic acid. Because heavy metal compounds exhibit differing affinities for sulfhydryl-containing enzymes, as well as different capacities to attack other organ systems, the symptoms of intoxication are not uniform.

ARSENIC

Arsenic is ubiquitous. Its use in insecticide sprays contaminates fruits and vegetables, and certain industrial processes also contribute heavily to its dissemination. The toxic effect of arsenic results from its binding sulfhydryl radicals of enzymes, thus inactivating them. α-Keto acid decarboxylases, such as pyruvate decarboxylase, are inactivated, and the result is an inhibition of oxidative metabolism. The general reaction of an arsenoxide or arsenite with adjacent sulfhydryl groups may be written:

$$R{-}As{=}O + (SH)_2Pr \longrightarrow R{-}As\begin{matrix} \diagup S \diagdown \\ \diagdown S \diagup \end{matrix}Pr + H_2O$$

where R is a substituent group and Pr is the enzyme protein.

Although arsenic intoxication usually results from oral ingestion, toxic symptoms may result from other routes. Arsenic is absorbed rapidly from parenteral sites of administration as well as from skin and mucous membranes. It rapidly leaves the bloodstream to be stored in liver, kidneys, intestines, spleen, lymph nodes, and bone. It is deposited in the hair within 2 weeks after administration and stays fixed in this site for years. It also remains within bone for extended periods. It is slowly eliminated in the urine and feces; excretion begins 2 to 8 hours after arsenic enters the body. A single dose may require 10 days before being completely excreted. After chronic arsenic administration, 70 days may be required before urinary levels return to normal. Renal excretion exceeding 0.1 mg of arsenic in 24 hr is usually abnormal. Toxic concentrations of arsenic may be present in the body even though urinary excretion levels are not elevated. When urinary concentrations are too low to be diagnostic, mobilization of tissue arsenic by a therapeutic regimen will increase concentrations to diagnostic levels. In long-standing intoxication, growing ends of the hair may be examined. Concentrations exceeding 0.1 mg per 100 g of hair indicates excessive ingestion or contamination [11].

Treatment for arsenic intoxication is based on providing compounds with sulfhydryl radicals which will unite with the arsenic and be excreted in urine subsequently. This can be accomplished by the administration of BAL (British antilewisit 2,3-dimercaptopropanol).

LEAD

Lead is one of the toxins known earliest to man. Evidence of lead in cosmetics was found in ancient Egyptian graves, and contamination of wines and syrup with lead has been suggested as contributing to the decline of Rome. Modern industrial culture results in the urban dweller ingesting from 100 to 2000 μg of lead daily, with additional inhalation of an average of 90 μg. Blood concentrations in urban dwellers usually reach 0.11 to 0.21 parts per million. Lead may be absorbed through the gastrointestinal tract, the skin, and, in vapor form, through the lungs. Large quantities are excreted in the feces, urine, and sweat.

Symptoms of intoxication appear quickly, particularly in industrial exposure, because the usual concentrations of lead in blood already border on the toxic. Individuals who use leaded paint and workers in gasoline industries, where tetraethyl lead is employed as an additive, are among the most frequently intoxicated. Pica, a craving often manifested by chewing paint that peels from cribs, walls, and other objects, is one of the leading causes of lead intoxication in children. In New York City, mass screening indicated that lead concentrations in blood reached potentially toxic levels in more than 5 percent of infants tested. When lead enters the bloodstream, it is distributed throughout the body and is carried chiefly to storage depots in cancellous bones. Lesser concentrations are present in the kidney, liver, spleen, bone marrow, and blood. The lead stored in bone is apparently innocuous, but lead in other organs represents the toxic portion. Lead stored in bone may be mobilized by debilitating disease or acute infectious conditions and liberated into the bloodstream, producing toxicity.

In addition to the involvement of enzymes concerned with oxidative metabolism, lead also interferes with porphyrin metabolism (Fig. 27-1).

Significantly increased concentrations of δ-aminolevulinic acid (ALA) in the urine appear to be caused partly by the induction of enzyme ALA synthetase. This enzyme, together with pyridoxal phosphate, leads to the formation of ALA from succinyl-CoA and glycine. ALA dehydrase is inhibited by lead, thus interfering with formation of porphobilinogen from ALA, and it is responsible for the major elevation of ALA concentration in urine. The conversion of coproporphyrinogen III to protoporphyrin-9 is mediated by the enzyme coproporphyrinogenase. This enzyme may be inhibited by lead, resulting in an increase in the urinary excretion of coproporphyrin III. The increase is relatively small in comparison with that of ALA, which is 25 to 50 times as great. Heme synthetase (ferrochelatase) also contains sulfhydryl groups. It has been suggested that inhibition of this enzyme results from lead intoxication, which diminishes the formation of heme from protoporphyrin-9.

The symptoms of the lead poisoning and those of acute intermittent porphyria are often strikingly similar. Many of the laboratory findings are also similar because of deranged porphyrin metabolism resulting from lead intoxication. The two conditions are distinguished biochemically by the greater increase in the urine, in acute porphyria, of uroporphyrin and porphobilinogen than of coproporphyrin or amino-

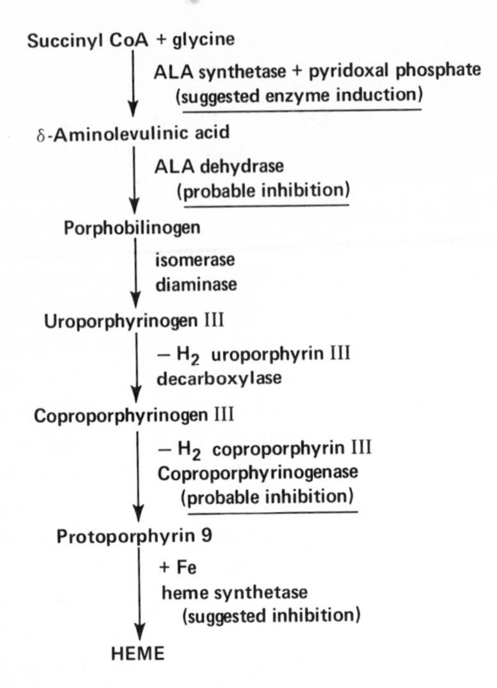

FIGURE 27-1. Probable and suggested sites at which lead interferes with porphyrin metabolism in the ultimate synthesis of heme from succinyl CoA and glycine.

levulinic acid. In lead poisoning, the opposite is true. Urinary coproporphyrin III is often increased up to 500 μg per 100 ml. The erythrocyte protoporphyrins, which normally amount to approximately 30 μg per 100 ml, may be increased up to 150 μg per 100 ml.

The chelating effect of calcium disodium EDTA is utilized in the treatment of lead poisoning. Chelation results in mobilization of lead, so symptoms may be exacerbated during treatment. The chelated form may become dissociated, thus increasing concentrations of toxic lead within the organism. Because large amounts of storage lead may be present, they are often incompletely mobilized during intoxication. As a result, additional chelation therapy for 1 to 6 months will still result in mobilization. This can be done by giving daily oral doses of penicillamine as the chelating agent [6, 13].

MANGANESE

Manganese may be absorbed through the lungs or the gastrointestinal tract. Intoxication usually results from a combination of inhaling manganese dust and swallowing particles in mining and industry. The highest incidence of intoxication is in Chilean manganese miners; toxic symptoms occur in approximately 15 percent of exposed mi

Although information on basic abnormalities of biochemical mechanisms is not available, some studies have employed injection of radioactive manganese in miners. The total body loss of the injected radioactive material occurs more rapidly in healthy miners than in those with manganese poisoning. It was suggested that tissue concentrations are greater in healthy miners than in those intoxicated by this heavy metal.

MERCURY

Mercury produces toxic effects on all biological enzyme systems [10]. In low concentrations, sulfhydryl radicals are bound, but higher concentrations of mercury also cause binding of other radicals. With exposure, mercury is rapidly taken up by the brain [7]. In experimental animals, approximately two-thirds of the total body uptake of mercury is found in the brain after inhalation of vapors. Mercury remains in the body for extensive periods, and has been found in the urine as long as 6 years after exposure has ceased. Urinary coproporphyrin levels are elevated in patients with mercury intoxication but, unlike lead intoxication, there is no correlation with concentrations of δ-aminolevulinic acid.

Acute mercury poisoning often follows accidental industrial or agricultural exposure. Particularly large numbers of poisonings result from contamination of sources of water supply, which causes toxic concentrations in edible seafood. Under some circumstances, fish and shellfish concentrate the mercury to form a protein complex compatible with continued life and function within these organisms. When shellfish are ingested by humans, mercury may be released and produce intoxication, as has been noted in the Minamata Bay epidemics in Japan. When mercury intoxication is treated with BAL, the metal is mobilized and excreted in the urine, but the cerebral concentrations appear to be unaffected by this treatment [8].

CYANIDE AND HYDROCYANIC ACID

Cyanide combines with many enzymes that contain ferric, cupric, or zinc ions, which results in enzyme inhibition. The cyanide anion also forms complexes with many proteins, such as heme, and may interfere with their physiological function. The acute action of cyanide on the nervous system is caused predominantly by its rapid combination with the ferric ion of cytochrome oxidase, thus inhibiting the utilization of molecular oxygen in the electron transport system. The administration of 5 mg NaCN per kilogram of body weight to experimental animals causes cytochrome oxidase activity to decrease to approximately half of normal values, with concomitant decrease in oxygen consumption by the tissues [1]. The compounds involved in glycolysis accumulate as a result of the failure of oxidative mechanisms. Cerebral lactate and phosphopyruvate concentrations increase, and lesser increases have been noted in hexose diphosphate and phosphoglycerate. These data, together with the concomitant decrease in cerebral glycogen concentrations, have been interpreted to indicate increased glycolysis under these conditions.

The principal function of cerebral oxidative metabolism is the production of high-energy phosphate, so a decrease in cerebral respiratory activity is reflected in the diminution of high-energy phosphate concentrations. Cyanide intoxication is accompanied by a significant decrease in cerebral phosphocreatine and ATP. An increase in ADP parallels the decrease in ATP; however, the increased inorganic phosphate concentrations are even greater than the decrease in phosphocreatine.

CHOLINESTERASE INHIBITORS

Substances that inhibit cholinesterase interfere with the enzymatic control of acetylcholine [2]. Toxicity results from use of compounds such as neostigmine, pyridostigmine, and similar materials in myasthenia therapy. More severe and dangerous forms of intoxication result from the alkylphosphates, which are employed as insecticides in farming and industry or as "nerve gas" [3]. Hexaethyltetraphosphate, tetraethylpyrophosphate, and diethyl-*p*-nitrophenylphosphate are the organic phosphorus insecticides most commonly used. Toxic symptoms as a result of cholinesterase inhibition may appear directly after exposure or may be delayed for 1 to 8 hr. The general structure of the alkylphosphate may be represented by

$$\begin{array}{l} R \quad O \\ \;\;\diagdown \; \| \\ \quad\;\; P{-}X \\ \;\;\diagup \\ R \end{array}$$

where R may be groups such as alkoxy, alkyl, acyl, or substituted ammonia groups, and X may be F, Cl, $O{-}C_6H_4NO_2$, or $O{-}PO(OR)_2$. The P–X bond, or compounds into which it may be converted, must be readily hydrolyzable.

Acetylcholinesterase has one anionic site which attracts the positively charged substrate and an esteratic site (as described in Chap. 7). When alkylphosphates are present, they combine with the enzyme at the esteratic site to form a dialkylphosphoryl compound. The cholinesterase activity in the brain is then decreased and the acetylcholine concentration increases.

Wilson [12] theorized that specific antidotes for alkylphosphate intoxication, such as pyridine-2-aldoxime methiodide (PAM) and other compounds, are effective because they have a quaternary structure to interact with the enzyme anionic site and a nucleophilic function to attract the phosphorus atom. The antidote reactivates acetylcholinesterase by withdrawing the toxic alkylphosphate from the enzyme. PAM also binds the free toxin, but in a slow reaction.

Triorthocresylphosphate is another anticholinesterase compound that has produced toxic neuropathy of epidemic proportions under several circumstances in which vessels subsequently used for cooking oil were contaminated. Earlier epidemics resulted from the ingestion of contaminated alcohol during the prohibition era. The peripheral neuropathy takes place some time after ingestion, when the anticholinesterase effect of the phosphate has dissipated. The toxic neuropathy is probably secondary to other reactions. With this intoxication, the so-called pseudo-cholinesterase, or butyrylcholinesterase, of plasma is inactivated to a greater extent than are the true acetylcholinesterases in red blood cells and the nervous system.

ISONICOTINIC ACID HYDRAZIDE

Symptoms of peripheral nerve involvement were found to result from isonicotinic acid hydrazide (isoniazid) administration shortly after initiation of its use in antitubercular therapy. The similarity of isoniazid neuropathy to those produced by vitamin B_6 deficiency or by deoxypyridoxine inhibition of pyridoxine led to the recognition that isoniazid produces its neurotoxic effect by interfering with normal pyridoxine action. Isoniazid combines with pyridoxine to form the isonicotinoyl-hydrazine derivative of pyridoxine, which is excreted in the urine. Pyridoxine-dependent reactions, such as decarboxylation of amino acids and transaminations, are thereby inhibited. Continued therapy with isoniazid is possible if it is combined with the administration of pyridoxine [4].

FURAN

Furan derivatives possess both antimicrobial and antineoplastic activity; they interfere with enzymatic processes necessary for bacterial or tumor growth. Three furans have been employed in therapy: nitrofurantoin, nitrofurazone, and furmethanol.

Nitrofurantoin is often used against *E. coli* in urinary-tract infections. When renal function is normal, antibacterial levels are achieved within the urine, but bactericidal or toxic levels are not reached in the blood. In renal dysfunction, the concentration of nitrofurantoin in blood may be sufficient to produce symptoms of neural involvement.

Nitrofurazone has been employed in the treatment of metastatic carcinoma and furmethanol has been employed against penicillin-resistant staphylococci. Because of the toxicity of the latter two furans, only nitrofurantoin is still used to any significant degree in therapy. Furan compounds interfere with energy metabolism in the tricarboxylic acid cycle by inhibiting formation of acetyl-CoA [9].

LATHYRISM

Ingestion of the chickpea *Lathyrus sativus* produces neurological disease. Cases of this disease are often found in humans and animals in areas of Europe and India where consumption of chickpeas is high. A number of potent neurotoxins have been isolated from the seeds of the chickpea, only one of which appears to be responsible for the neurological involvement. This compound is β-*N*-oxalyl-L-α,β-diaminopropionic acid [5]. When this material is administered to experimental animals, it produces convulsions that appear to result from chronic ammonia toxicity, which in turn could be caused by interference with the ammonia generating or fixing mechanisms in the brain. This mechanism has been postulated because administration of the toxin to experimental animals causes a striking increase in glutamine concentrations in the brain. The formation of glutamine from glutamic acid is the chief mechanism of detoxification of ammonia in the brain. A significant increase in the concentrations of free ammonia in the brain occurs at the onset of convulsions, with a subsequent decline to normal values. This is interpreted as detoxification through the formation of glutamine.

Another neurotoxin isolated from the chickpea, L-2,4-diaminobutyric acid, does not appear to produce nervous system damages by direct involvement with the biochemical mechanisms of the brain. The damage appears to be initiated in the liver, where a slight, chronic increase of blood ammonia concentration is produced.

REFERENCES

1. Albaum, H. G., Tepperman, J., and Bodansky, O. The *in vivo* inactivation by cyanide of cytochrome oxidase and its effect on glycolysis and on the high energy compounds in brain. *J. Biol. Chem.* 164:45, 1946.
2. Aldrich, F. D. Cholinesterase assays: Their usefulness in diagnosis of anticholinesterase intoxications. *Clin. Toxicol.* 2:445, 1969.
3. Aldridge, W. N., and Barnes, J. Neurotoxic side-effects of certain organophosphorus compounds. *Proc. Eur. Soc. Study Drug Toxicity* 8:162, 1967.
4. Aspinall, D. L. Multiple deficiency state associated with isoniazid therapy. *Br. Med. J.* 2:1177, 1964.
5. Cheema, P. S., Malathi, K., Padmanaban, G., and Sarma, P. S. The neurotoxicity of β-*N*-oxalyl-L-α,β-diaminopropionic acid, the neurotoxin from the pulse *Lathyrus sativus*. *Biochem. J.* 112:29, 1969.
6. Fromke, V. L., Lee, M. Y., and Watson, C. J. Porphyrin metabolism during versenate therapy in lead poisoning; intoxication from an unusual source. *Ann. Intern. Med.* 70:1007, 1969.
7. Magos, L. Mercury-blood interaction and mercury uptake by the brain after vapor exposure. *Environ. Res.* 1:323, 1967.
8. Magos, L. Effect of 2-3 dimercaptopropanol (BAL) on urinary excretion and brain content of mercury. *Br. J. Ind. Med.* 25:152, 1968.
9. Paul, M. F., Paul, H. D., Kopko, F., Bryson, M. J., and Harrington, C. Inhibition furacin of citrate formation in testis preparations. *J. Biol. Chem.* 206:491, 1954.
10. Webb, J. L. *Enzyme and Metabolic Inhibitors.* Vol. II. New York: Academic, 1966. (Chap. 7, Mercurials.)

11. Webb, J. L. *Enzyme and Metabolic Inhibitors.* Vol. III. New York: Academic, 1966. (Chap. 6, Arsenicals.)
12. Whitaker, J. A., Austin, W., and Nelson, J. D. Edathamil calcium disodium (versenate) diagnostic test for lead poisoning. *Pediatrics* 29:384, 1962.
13. Wilson, I. B. A specific antidote for nerve gas and insecticide (alkylphosphate) intoxication. *Neurology* 8:41 (Suppl. 1), 1958.

28

Parkinson's Disease and Other Disorders of the Basal Ganglia

Theodore L. Sourkes

SOME LARGE, ANATOMICALLY DISTINCT MASSES of gray matter lie at the base of the brain, and certain of these, namely the caudate nucleus, the putamen, and the globus pallidus, are collectively termed *basal ganglia.* These structures have many connections between one another and to and from other regions of the brain. Among the bodily functions that they regulate are the tone and posture of the limbs. This is achieved in conjunction with the so-called extrapyramidal system, i.e., the structures of the brain, apart from the cerebral cortex, which send efferent fibers to the spinal cord.

Regulatory functions of this type have been recognized since 1911, when S. A. K. Wilson demonstrated in a series of patients the relationship between lesions of the lenticular nucleus (putamen and globus pallidus, considered as a unit), on the one hand, and rigidity, tremor, and postural defects, on the other. Wilson's disease was thus regarded as the prototype of disorders of the basal ganglia.

Since that time, it has been widely recognized that the "shaking palsy," first described in 1817 by the London physician James Parkinson, stems from degenerative changes in the substantia nigra, a portion of the brain that has extensive connections with the striatum (i.e., the caudate nucleus and putamen). Besides the shaking, or tremor, of the limbs, patients with Parkinson's disease may also exhibit muscular rigidity, which leads to difficulties in walking, writing, speaking, and facial movements, as well as a lethargy, or loss of volitional movement, which may progress to akinesia.

Huntington's chorea is a third important disorder of basal ganglia, with apparently specific pathology of the caudate nucleus.

BIOCHEMICAL CHARACTERISTICS OF BASAL GANGLIA AND ASSOCIATED STRUCTURES

The basal ganglia present some interesting chemical characteristics. For example, they are especially sensitive to carbon monoxide and to manganese. This sensitivity probably plays a role in the neurological complications that accompany intoxication with these chemicals. In infants, the predilection of bilirubin for the cells of the basal ganglia may result in kernicterus, or "jaundice of the nuclei of the brain." Historically, kernicterus was the first basal ganglial disease described, and Wilson made use of the parallel in his classic study.

The globus pallidus contains much iron, a property it shares with other nuclei of the extrapyramidal system: the subthalamic nucleus of Luys, substantia nigra, the red nucleus (n. ruber), the dentate nucleus, and the inferior olive. In fact, these parts can be delineated by immersing slices of the brain in a solution of ammonium sulfide to stain for iron. It has been suggested that the metal is present in a ferritin-like molecular form.

The basal ganglia contain considerable amounts of acetylcholinesterase; this is particularly evident in the caudate nucleus. They are said also to be rich in gangliosides. The substantia nigra, the red nucleus, and the locus ceruleus each contains a distinctive pigment. The precursor of the pigmented material is thought to be DOPA (or dopamine), but the specific pathways of formation of the pigments are not known in any detail.

DOPAMINE AND BASAL GANGLIA

The biochemical characteristics that have thrown most light on the functions of the basal ganglia and certain associated structures in man and animals concern the distribution of monoamines, particularly dopamine. The biosynthetic relationships of this compound are illustrated in Figure 28-1. The starting point for its biosynthesis is tyrosine, which is present in the whole brain at a level of about 1.2 mg per 100 g fresh weight of tissue, or 70 μmoles per kilogram. The first step is catalyzed by tyrosine hydroxylase; it requires a pteridine cofactor. The product of this oxidation is DOPA (3,4-dihydroxyphenylalanine), an amino acid that is normally present only in the most minute concentrations in the body because it is so readily decarboxylated to dopamine (3,4-dihydroxyphenylethylamine). This is achieved through the action of the widely distributed enzyme DOPA decarboxylase, which also catalyzes amine formation from 5-HTP (5-hydroxytryptophan) and some other substrates. Hence, it is sometimes called 5-HTP decarboxylase or aromatic amino acid decarboxylase. Like other amino acid decarboxylases, it requires pyridoxal phosphate as its coenzyme. It is inhibited by many compounds, some of which are effective in vivo and have even been used therapeutically in certain diseases.

In certain neurons of the brain, the formation of dopamine is the terminal step in biosynthesis. One method by which this has been determined makes use of the specific fluorescence of derivatives of dopamine and other monoamines developed in situ by histochemical means. By pharmacological pretreatment of animals in a manner that specifically depletes the brain of one amine or another, areas rich in a particular monoamine can be differentiated by ultraviolet histofluorescence. The fiber connections passing from the substantia nigra to the caudate nucleus belong

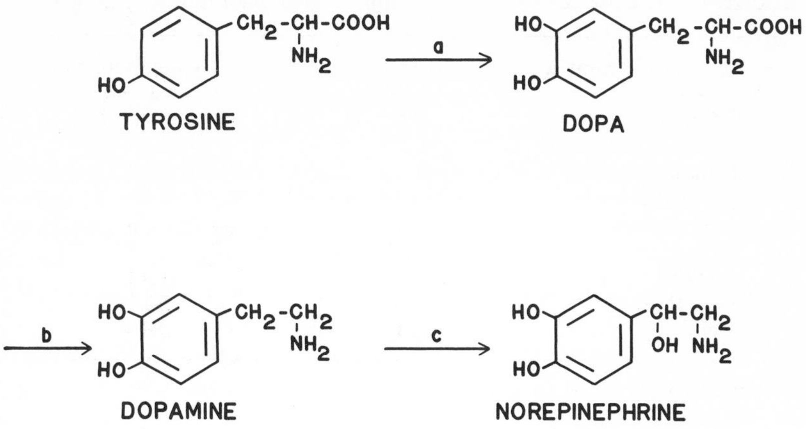

TYROSINE —a→ DOPA —b→ DOPAMINE —c→ NOREPINEPHRINE

FIGURE 28-1. Pathway of biosynthesis of catecholamines. The enzymes catalyzing each step are: *a,* tyrosine hydroxylase; *b,* DOPA decarboxylase; and *c,* dopamine β-oxidase.

to this group of neurons. They store dopamine in vesicles within the varicosities of the nerve terminations; the intravesicular concentration of the amine may rise to as high as 1 to 10 mg per gram fresh weight of nervous tissue, i.e., in the range of amine concentration in chromaffin tissue. Other neurons of the brain contain not only tyrosine hydroxylase and DOPA decarboxylase but also dopamine β-oxidase, which is located in the membrane of the storage vesicles. These are the norepinephrine-containing fibers. Under many circumstances, the biosynthesis of these monoamines is limited by the slow flux through the stage of tyrosine hydroxylase because of the low turnover rate of that enzyme.

The regional distribution of these compounds in the central nervous system is shown in Table 28-1. Dopamine and norepinephrine are irregularly distributed in the brain and spinal cord. Thus, dopamine is found in certain regions along with norepinephrine, but, unlike the latter compound, it is present in unusually high concentrations in some of the basal ganglia. On the other hand, norepinephrine, but not dopamine, is found in the spinal cord.

Neurochemical mapping of the brain, in conjunction with histofluorescence studies, has led directly to the concept of dopamine as a neurohumor acting at certain synapses within the basal ganglia and thus assisting in their neural function. This concept has drawn effectively upon our present knowledge of the function of norepinephrine as a transmitter substance at peripheral synapses served by post-ganglionic sympathetic fibers, and there is mounting evidence from many different branches of the neurosciences that dopamine plays such a role. Even though conclusive proof is still required, many investigators in the field have tacitly accepted dopamine into the family of neurohumors and speak of "dopaminergic" neurons.

The major metabolites of dopamine in the body are 3-*O*-methyldopamine

TABLE 28-1. Concentration of Catecholamines in Human Brain

Brain Region	Dopamine	Norepinephrine
Cerebral gray matter	0.02–0.17[a]	0–0.06[a]
Cerebral white matter	0.05	0
Striatum	3–8	0.06
Internal capsule	0.38	0.04
Red nucleus	1.17	0.23
Thalamus	0.30–0.46	0.05
Hypothalamus	1.12	1.11
Dentate nucleus	0.02	0–0.02
Medulla oblongata	0.17	0.14

[a]Micrograms per gram fresh weight of tissue.

(3-methoxytyramine) and HVA (homovanillic acid, 4-hydroxy-3-methoxyphenylacetic acid). These are depicted in Figure 28-2, which also shows the various alternative pathways of metabolism (in addition to the formation of norepinephrine). The amine and its *O*-methyl derivative are both subject to the action of monoamine oxidase (MAO), a flavoprotein that is present in the outer membrane of the mitochondria. Products of this oxidation include the aldehyde that corresponds to the amine substrate, hydrogen peroxide, and ammonia. Monoamines represent only a minor source of brain ammonia. Most of the aldehyde undergoes further dehydro-

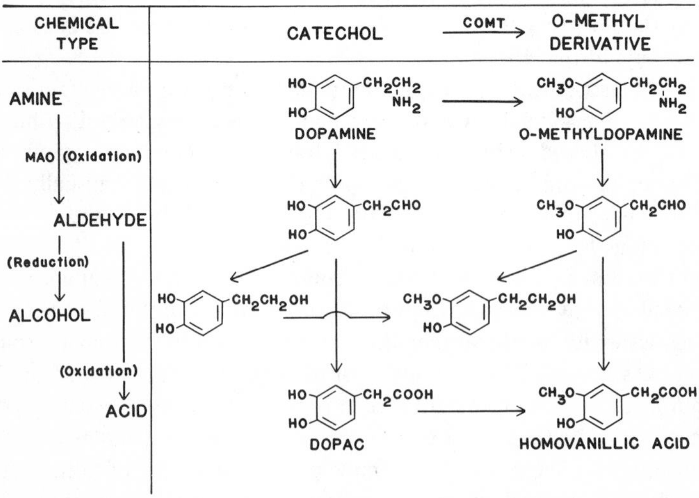

FIGURE 28-2. Alternative pathways of metabolism of dopamine. COMT, catechol-*O*-methyl transferase; MAO, monoamine oxidase; DOPAC, 3,4-dihydroxyphenylacetic acid.

genation to form DOPAC (3,4-dihydroxyphenylacetic acid) or HVA. DOPAC, like dopamine, is a substrate for catechol-*O*-methyl transferase (COMT), an enzyme which catalyzes the transfer of the methyl group from *S*-adenosylmethionine to an appropriate acceptor. This process appears to be very efficient in the brain since substantial amounts of HVA are normally present in the striatum, whereas only very small quantities of DOPAC can be found there. A portion of the aldehyde undergoes reduction, catalyzed by alcohol dehydrogenase or a similar enzyme and a reduced nicotinamide coenzyme. The alcoholic products which are formed are shown in Figure 28-2.

Biological activities have been sought for some of these metabolites of dopamine, so far with little success. Currently there is a growing interest in the possibility that the aldehyde derivatives of monoamines subserve a significant function.

Until the striking discrepancies in the regional distribution of the catecholamines were discovered (Table 28-1), dopamine was regarded as merely the precursor of norepinephrine. But the presence of dopamine in large amounts in localized areas of the brain, unaccompanied by comparable concentrations of norepinephrine, suggested a specific neural role for dopamine. As the dopamine-rich regions are pre-eminently the basal ganglia and some associated structures, apart from the thalamus and hypothalamus, it was natural to investigate diseases of the basal ganglia for evidence of deranged metabolism of catecholamines, in general, and dopamine, in particular. The search for such a faulty metabolism has been especially successful in Parkinson's disease, which now stands as the paradigm of cerebral diseases that involve dysfunction of monoamine-containing fibers, and in that respect it may serve as a model for the investigation of monoamine function in nervous and mental diseases.

CHEMICAL PATHOLOGY OF PARKINSON'S DISEASE

The evidence for disturbed metabolism of monoamines comes from the direct measurement of these substances in the brains of patients dying of Parkinson's disease. The regional concentrations of catecholamines are shown in Table 28-2. The three monoamines that have been measured are all present in lower concentrations than those observed in normal conditions, and the decrease of dopamine is greater than that of norepinephrine and serotonin. One of the first attempts to restore normal levels of brain monoamines was to administer MAO inhibitors to patients with Parkinson's disease. MAO inhibitors prevent the oxidative deamination of the amine (see Figure 28-2); hence, if its biosynthesis is proceeding normally, an excess of the compound could be expected to accumulate. Patients who were treated in the terminal stages of their illness with these inhibitors (some of which are used therapeutically in the treatment of mental depression) showed a significant increase in the brain concentrations of norepinephrine and serotonin but not of striatal dopamine. These results point clearly to some type of dysmetabolism of monoamines, particularly dopamine, in an important disease of the brain.

TABLE 28-2. Catecholamines and Homovanillic Acid (HVA) in Brain in Parkinson's Disease

Brain Region	Dopamine		HVA		Norepinephrine	
	Normal[a]	P.D.	Normal	P.D.	Normal	P.D.
Caudate nucleus	3.50[b]	0.32	1.87	0.34	0.07	0.03
Putamen	3.57	0.23	2.92	0.69	0.11	0.03
Globus pallidus	0.30	0.14	1.73	0.56	0.09	0.11
Substantia nigra	0.46	0.07	1.79	0.52	0.04	0.02
Thalamus	–	–	0.35	0.21	–	–
Internal capsule (anterior limb)	–	–	1.67	0.42	–	–

[a]Normal = nonneurological cases; P.D. = Parkinson's disease.
[b]Micrograms per gram wet weight of tissue.

However, neuropathological and necropsy findings by themselves cannot establish the precise biochemical defect that causes the loss of brain dopamine. Fortunately, one can prepare models in experimental animals for comparison with the clinical state in man. This has been especially successful with monkeys, in which a small lesion is placed stereotactically in the ventromedial tegmental area of the upper brain stem. In this way, it has been possible to induce in the animals various dyskinesias, including, significantly, tremor and hypokinesia of the limbs contralateral to the side of the lesion. Histological study of the brains of these monkeys indicates that fibers from the substantia nigra to the striatum are interrupted and undergo degenerative changes. Retrograde degeneration entails loss of staining characteristics of the nigra (i.e., loss of cell bodies), accompanied by significant decreases in the concentration of catecholamines in the striatum, without involving its serotonin content. The loss of striatal monoamines is attributed, then, to anterograde degeneration of the nigrostriatal fibers, leading to the disappearance not only of the neuronal structure but also of enzymes and metabolites.

Cats, rabbits, and rats have also been used effectively in the identification of neurochemical events that accompany specific morphological changes. However, the primates show the motor disturbances which permit neurological and neuropharmacological investigations that bear direct significance to the understanding of human diseases.

Examples of the loss of enzymes and metabolites from the striatum, i.e., from the nigrostriatal terminations there, are given in Tables 28-3 and 28-4. The latter table illustrates the progressive neurochemical changes accompanying the nigral degeneration as a result of experimental lesions: the greater the loss of nigral cell bodies and fibers, the greater the loss of DOPA decarboxylase and dopamine.

The decreased ability of the diseased nigra to produce dopamine in Parkinson's disease is also reflected in lower concentrations of HVA in the caudate nucleus and putamen. This can be reproduced experimentally by introducing lesions of the

TABLE 28-3. Monoamine Functions of Striatum of Animals with Unilateral Lesions of the Nigrostriatal Tract

Function	Species	No.	Intact Side	Operated Side
Concentration of:				
Dopamine[a]	Monkey	5	5.30	0.29
Homovanillic acid[a]	Monkey	4	12.50	6.10
Norepinephrine[a]	Monkey	5	0.30	0.15
Dopamine following injection of DL-DOPA (70 mg/kg)[a]	Cat	10	17.4	6.5
Homovanillic acid in cisternal fluid[b]	Cat	4	16.7[c]	6.8[d]
Dopamine following intraventricular injection of radioactive tyrosine[e]	Monkey	3	6500	750
Activity[f] of:				
Succinic dehydrogenase	Cat	6	100	87–110
Tyrosine hydroxylase	Cat	6	100	1–34
DOPA decarboxylase	Monkey	4	100	44
Monoamine oxidase	Monkey	4	100	89

[a]Micrograms per gram fresh weight.
[b]Nanograms per milliliter.
[c]Intact animals.
[d]Bilaterally operated cats.
[e]Counts per minute (as dopamine) per gram of caudate nucleus tissue.
[f]Relative activity.

TABLE 28-4. Relation of Neurochemical Changes in the Striatum to Histological Changes in the Substantia Nigra After Lesioning of the Nigrostriatal Tract

Cellularity of Substantia Nigra on the Side of the Lesion	DOPA Decarboxylase Activity[a]			Dopamine Concentration[b]		
	No.	Intact Side	Lesion Side	No.	Intact Side	Lesion Side
No cell loss (or nearly none)	6	75	63	16	4.79	4.12
Partial loss of cells	4	113	31	5	2.47	1.07
Complete or nearly complete loss of cells	6	99	0	13	3.35	0.27

[a]Nanomoles of dopamine formed per hour per 100 mg tissue.
[b]Micrograms per gram fresh weight of tissue.

nigrostriatal fibers in monkeys. The concentrations of the acidic metabolites of dopamine – HVA and DOPAC – then decline. Some HVA diffuses into the cerebrospinal fluid, particularly from the caudate nucleus that lies in the floor of the lateral ventricle, and samples of this fluid have easily detectable amounts of the acid. In patients undergoing brain surgery for the treatment of Parkinson's disease, the concentration of HVA is reduced far below the level found in patients who are being operated upon for reasons other than a disorder of basal ganglia. CSF taken at the suboccipital and lumbar levels has a much lower concentration of HVA than has the ventricular fluid; the sharp decrease is accounted for by the presence of an active transport mechanism that removes such metabolic acids as HVA. Nevertheless, the concentration of HVA in CSF obtained by a lumbar tap shows the relative deficiency of the acid in patients with Parkinson's disease. Just as in the case of the cerebral changes, the reduced concentration of HVA in spinal fluid can be produced experimentally in cats and monkeys that have received specific lesions of the nigrostriatal tract (Table 28-3).

The lowered HVA in CSF is not specific to Parkinson's disease. It is noted also in patients with inflammatory and other disorders of the central nervous system and in those suffering from mental depression. However, differentiation may be made by using probenecid (*p*-dipropylsulfamylbenzoic acid). This drug inhibits the mechanism that transports the metabolic acids across the CSF-blood barrier so that during its course of action, HVA and 5-hydroxyindoleacetic acid, derived from serotonin by enzymatic oxidation, increase in concentration in the CSF. However, parkinsonians tested with probenecid show little or no increase in HVA, although there is an increase in nonparkinsonians. Presumably the latter patients continue to form dopamine, and thus HVA, despite the low levels of the acid in the CSF; when the transport of HVA from the CSF is blocked, the acid can accumulate to some extent. This does not occur in Parkinson's disease because of the neuronal deficit and the consequent inability to produce dopamine, the precursor of HVA.

The concentration of HVA in CSF of the lateral ventricles is not necessarily low in all diseases with tremor. For example, in contrast to Parkinson's disease, attitudinal (heredofamilial) tremor is not accompanied by a drop in HVA. This may be because attitudinal tremor is cerebellar in origin; the neurochemical evidence indicates that the function of the dopaminergic nigrostriatal fibers is not significantly affected.

If an enzymatic defect in the brain were accompanied by a comparable one in the liver, the kidneys, or both, the abnormality might be more readily detectable and it could be studied conveniently during the life of the patient. This has motivated clinical-chemical studies of Parkinson's disease, including measurement of the urinary excretion of monoamines and some of their metabolites. Such studies have shown that the excretion of epinephrine and norepinephrine is normal, but the excretion of dopamine is subnormal, and may be especially low in those patients with the greatest degree of akinesia. The low values cannot be attributed, in our present state of knowledge, to variations in urinary volume, dietary factors, or a specific enzyme defect, nor can they be accounted for by a reduced turnover rate of cerebral dopamine. It has been suggested that the abnormality with respect to free urinary dopamine results from chronic treatment with antiparkinsonian drugs; one way to study this would be to examine newly detected cases of Parkinson's disease.

Other amines have been implicated in Parkinson's disease, but of the various reports describing abnormalities in urinary excretion, none has been substantiated.

The metabolism of exogenous amines has also been studied in Parkinson's disease. A simple load-test consists of giving patients 250 mg L-DOPA orally and subsequently measuring certain urinary products. Such tests show that, in comparison with healthy individuals, parkinsonian patients excrete subnormal amounts of dopamine. Other investigations with labeled dopamine indicate that these patients may convert relatively less of the amine to the norepinephrine fraction that is excreted unconjugated in the urine, with proportionately more going to other metabolites of dopamine. On the other hand, when labeled norepinephrine is infused into patients with Parkinson's disease, their pattern of excretion of urinary metabolites is similar to that of control subjects.

L-DOPA TREATMENT OF PARKINSON'S DISEASE

An important outgrowth of catecholamine studies has been the L-DOPA treatment of Parkinson's disease. In 1961, neuropharmacological studies in patients with the disease revealed that this amino acid exerts antirigidity and antiakinesic actions; that is, it influences the two most incapacitating symptoms in Parkinson's disease. Today it is said that L-DOPA provides about 60 percent improvement in at least 50 percent of the patients with Parkinson's disease; the treatment is considered the best one available. Other antiparkinsonian drugs have had a favorable action on the tremor, but not on the other symptoms.

It is generally assumed that the therapeutic role of L-DOPA lies in the ease with which it crosses the blood-brain barrier (in contrast to the exclusion of intravenously infused dopamine from the brain) and its ability to replenish the supply of dopamine at appropriate sites. The passage of L-DOPA into the parenchymal tissue of the brain entails its transfer from blood through the endothelial cells lining the capillaries. These cells, like many other peripheral cells, contain considerable amounts of DOPA decarboxylase, so that only a portion of a given amount of L-DOPA will eventually pass into the brain. Hence, large doses (up to 8 g per day) have been used. By giving the patient certain inhibitors of DOPA decarboxylase in doses that affect only the peripherally located enzyme, including that of the brain capillaries, it is possible to reduce the daily dose of L-DOPA to 2 to 3 g per day, with the same therapeutic response.

When L-DOPA has crossed the blood-brain barrier, it must be converted to dopamine, according to the commonly accepted hypothesis of its mode of action. This provides an apparent paradox, for if pathological processes have eliminated dopaminergic neurons, including their content of DOPA decarboxylase, how can the

L-DOPA be effective? Some have suggested that the amino acid or a nondecarboxylated product acts directly on dopamine-sensitive receptors in the striatum. Other hypotheses envisage formation of a Schiff's base between L-DOPA and its transaminated product, 3,4-dihydroxyphenylpyruvic acid, or analogously between L-DOPA and pyridoxal phosphate. In the latter cases, cyclization occurs with the formation of substituted tetrahydroisoquinolines, which, it is postulated, may be the biologically effective substances. A highly probable hypothesis is derived from the fact that the cells of the striatum receive connections from many fibers, including some that contain serotonin. Hence, L-DOPA could be acted upon by the decarboxylase within those neurons, i.e., by the enzyme that normally has 5-HTP as substrate. The dopamine formed would then be available in the presynaptic space, although its path of diffusion to sensitive neurons might now be longer than usual.

There are many other speculative hypotheses concerning the action of L-DOPA in Parkinson's disease, but the choice among them must await further research. Whatever the outcome, the use of L-DOPA in this disorder represents a successful biochemical treatment of an important cerebral disease.

HUNTINGTON'S CHOREA

Huntington's chorea has been of special interest to the neurochemist because it is genetically transmitted and involves pathology of the striatal tissue. Thus far, postmortem studies have not revealed any abnormalities in the catecholamine or HVA content of the brain. It is possible that serotonin metabolism is disordered, but investigations are incomplete. Moreover, urinary studies of this disease have not been especially revealing. It is important to keep in mind that the cell bodies in the striatum belong, for the most part, to cholinergic fibers. Hence, if there is a biochemical defect it may be in the formation, binding, or disposition of another neurohumoral agent, perhaps acetylcholine.

Extensive metabolic studies of Huntington's chorea have been carried out in a number of laboratories. Various blood lipids, serum enzymes, and urinary constituents, including trace metals and phenolic and indolic compounds, have been examined without detection of any characteristic defect. Patients with Huntington's chorea are said to have elevated levels of Mg^{2+} and Ca^{2+} in their erythrocytes, but normal levels of the two ions in serum. The plasma of fasting patients contains subnormal levels of proline, alanine, valine, isoleucine, leucine, and tyrosine, with similar, although less significant, changes in the CSF. These results may aid in the discovery of a specific biochemical error in Huntington's chorea.

WILSON'S DISEASE, OR HEPATOLENTICULAR DEGENERATION

Much of our knowledge of the metabolism of copper comes from studies of experimental and farm animals. In recent years, the interest in Wilson's disease, a disorder in which copper accumulates in a number of organs, has helped to stimulate research into the metabolism of this metal in man. Wilson's disease is a combined brain-liver disorder characterized by progressive rigidity, intention tremor, hepatic cirrhosis of the coarse type, and recurrent hepatitis. Renamed *hepatolenticular degeneration* in 1921, it is recognized as a familial disorder inherited in an autosomal recessive fashion. The frequency rate of the gene is estimated to be 1 in 500. Biochemically, hepatolenticular degeneration is characterized by low concentrations of copper and ceruloplasmin in the serum, elevated excretion of the metal in the urine, and deposition of excess copper in the brain, especially in the basal ganglia and in the liver and kidneys. In some cases, copper is deposited in the cornea, where it is reduced, forming the Kayser-Fleischer ring; this is then pathognomonic of the disease. In addition, there is a constant aminoaciduria, including excretion of some dipeptides, and sometimes abnormal excretion of monoamines and their metabolites. The amino acid levels in the plasma are normal, so the urinary findings may signify simply a renal defect caused by histotoxicity of copper.

NORMAL METABOLISM OF COPPER

Copper deficiency has been demonstrated in a number of mammalian species under laboratory or range conditions, but not unequivocally in the human, even in kwashiorkor or marasmus. However, it appears from many sources of evidence that the metal is an actual nutritional requirement for man. About 2 mg of copper suffices daily, but most diets supply more than this. The normal infant is born with a large liver store of copper, which provides temporary protection against the deficiency of the metal in milk (just as in the case of iron). Copper is needed for the utilization of iron in hemoglobin synthesis as well as for other functions. Conversely, iron seems to be required for the utilization of copper, so that deficiency of the one metal results in the accumulation of the other in liver or another storage organ. In some laboratory species, copper deficiency causes a microcytic hypochromic anemia. Farm animals maintained in regions where the soil (and therefore the pasturage) is deficient in copper may suffer from specific clinical disorders. In the sheep and goat, the primary symptom is anemia, but in others there is a neurological disturbance, such as "swayback" in cattle and "posterior paralysis" in swine. Myelination of nerve is delayed in lambs born of copper-deficient ewes.

The regulation of copper metabolism is attuned to the very small requirement of the metal in adult animals, so most of the copper ingested in the diet is excreted in the feces. A minute fraction of the dietary copper appears in the urine. Balance studies, however, disguise the fact that much copper is absorbed from the intestinal tract and reaches the liver in the portal circulation; a portion of it is utilized there, but the major part is excreted in the bile. Thus, there is an enterohepatic cycle for copper. The mechanism can be overwhelmed in experimental animals by repeated parenteral administration of copper, e.g., by daily intraperitoneal injection of solutions of copper sulfate. The metal then accumulates in the liver and can be found in many protein fractions. Regulation of the entry of copper into the brain is very strict, even under these conditions, so this method does not provide a model of Wilson's disease with its deposition of excess metal in the brain.

CUPROPROTEINS

Copper is an essential constituent of several important proteins, including cytochrome-c oxidase, dopamine β-oxidase, uricase (in mammals below the level of the primates), ceruloplasmin, and cytocuprein (also named cerebrocuprein, hepatocuprein, and erythrocuprein, depending on the tissue from which it is prepared). Copper is the prosthetic group of DOPA oxidase in the melanocytes of the skin and in certain pigmented regions of the brain. Indeed, there may be as much as 0.4 mg of copper per gram of dry weight in the subthalamic nucleus, with somewhat smaller amounts in the substantia nigra and dentate nucleus. However, all these supproproteins do not account for the total amount of copper in the body (100 to 150 mg in the human adult), and many other cuproproteins will probably be discovered.

Ceruloplasmin is a blue serum glycoprotein. Its molecular weight is about 155,000 and it contains eight atoms of copper, not all of which are equivalent in function. The divalent atoms of copper endow the protein with its blue color. The molecule displays some oxidase activity toward polyamines and polyphenols; phenylenediamine is often used as a substrate. Ceruloplasmin has a half-life in the serum of 54 hr. Its sialic acid residues seem to protect it from metabolic degradation since asialoceruloplasmin disappears from the circulation after its injection far more rapidly than ceruloplasmin.

Normally, the serum contains about 100 μg of copper per 100 ml, and most of this (95 percent) is present as ceruloplasmin. The concentration of this protein amounts to 20 to 40 mg per 100 ml. In hepatolenticular degeneration, serum copper may fall to one-half this value or less. In the past, investigators have sought a role for ceruloplasmin in copper metabolism. It does not seem to be involved in transport of the metal in the serum; this function is served by the small nonceruloplasmin fraction of serum, most of this copper being loosely bound to albumin. Recently it has been demonstrated that when ceruloplasmin is perfused through the isolated liver, there is a specific and rapid shift of transferrin from liver to the perfusate. This result is pertinent to the problem of the copper-iron interaction in metabolism.

A vexing question is that of the relationship between ceruloplasmin and hepatolenticular degeneration. The many investigations in this field have not yet provided a decisive answer, so we do not know what role the depressed level of ceruloplasmin plays, if any, in the pathogenesis of the disorder.

From time to time, a relationship between copper metabolism and schizophrenia has been claimed. For example, soon after the oxidase property of the protein became widely recognized, there were numerous studies attempting to relate the activity of serum ceruloplasmin to the psychosis, but there has been no valid evidence for such a relationship.

Some metal-chelating agents have been tested as therapeutic agents in Wilson's disease. BAL (British antilewisite) is effective in bringing about copper diuresis, but it must be given by intramuscular injection and has unpleasant side effects, which limit its use. EDTA has been tried, but it is not very effective in removing copper from the body. The most effective chelator that has been used is D-penicillamine, or 3,3-dimethylcysteine. It is given in doses up to about 3 g per day. Its effectiveness indicates that the cytotoxicity caused by copper can be reversed, even to the extent of bringing about remission of neurological and other symptoms of hepatolenticular degeneration.

CHRONIC MANGANESE POISONING

A small proportion of miners exposed to manganese dust develop "manganism." The disease is ushered in by self-limited psychiatric symptoms, followed by permanent neurological changes. The manifestations are those of extrapyramidal disease. Because they respond to treatment with L-DOPA, dopaminergic neurons of the brain are probably affected.

ACKNOWLEDGMENT

Research on the title subjects in the author's laboratory is supported by grants of the Medical Research Council of Canada.

REFERENCES

Calne, D. B., and Sandler, M. L-DOPA and Parkinsonism. *Nature* 226:21, 1970. (A review of the L-DOPA treatment of Parkinson's disease in relation to its pharmacological and biochemical aspects.)

Costa, E., Côté, L. J., and Yahr, M. D. (Eds.). *Biochemistry and Pharmacology of the Basal Ganglia.* New York: Raven, 1966. (The proceedings of a symposium at which the cholinergic and monoaminergic systems of the basal ganglia were considered.)

Cumings, J. N. Trace metals in the brain and in Wilson's disease. *J. Clin. Pathol.* 21:1, 1968.

Curzon, G. The biochemistry of dyskinesias. *Int. Rev. Neurobiol.* 10:323, 1967.

Dahlström, A., and Fuxe, K. Evidence for the existence of monoamine-containing neurons in the central nervous system. I. Demonstration of monoamines in the cell bodies of brain stem neurons. *Acta Physiol. Scand.* 62 (Suppl. 232):1, 1964.

Gillingham, F. J., and Donaldson, I. M. L. (Eds.). *Third Symposium on Parkinson's Disease.* Edinburgh: Livingstone, 1969. (The basic scientific aspects cover a substantial portion of these proceedings.)

Hornykiewicz, O. Dopamine (3-hydroxytyramine) and brain function. *Pharmacol. Rev.* 18:925, 1966.

Poirier, L. J., Sourkes, T. L., Bouvier, G., Boucher, R., and Carabin, S. Striatal amines, experimental tremor and the effect of harmaline in the monkey. *Brain* 89:37, 1966.

Sass-Kortsak, A. Copper metabolism. *Adv. Clin. Chem.* 8:1, 1965.

Sourkes, T. L. Cerebral and other diseases with disturbances of amine metabolism. *Progr. Brain Res.* 8:186, 1964. (An assessment of the problems involved in the biochemical investigation of diseases of the basal ganglia and other monoamine disorders.)

Sourkes, T. L., and Poirier, L. J. Neurochemical bases of tremor and other disorders of movement. *Can. Med. Assoc. J.* 94:53, 1966.

Walshe, J. M. The physiology of copper in man and its relation to Wilson's disease. *Brain* 90:149, 1967.

Wilson, S. A. K. Progressive lenticular degeneration: A familial nervous system disease associated with cirrhosis of the liver. *Brain* 34:295, 1912.

THREE

BEHAVIORAL NEUROCHEMISTRY

29

Psychopharmacology

Theodore L. Sourkes

PSYCHOPHARMACOLOGY is concerned with drugs which are active in the central nervous system and which influence behavior, perception, thought, and affect. In its earliest usage, the Greek word *psychopharmaka* was applied to drugs or measures with a healing action upon the psyche. In the context of modern therapeutics, the term psychopharmacology is ill chosen. As von Brücke and Hornykiewicz point out [10], it should really be reserved for placebos, because their actions seem to be mediated solely by mental functions.

The study of the biochemical effects of drugs made little progress until the discovery of the antibacterial effects of the sulfonamides. This was contemporaneous with the replacement of galenic pharmaceuticals by synthetic medicinal chemicals. The structural resemblance of sulfanilamide to *p*-aminobenzoic acid, a bacterial vitamin, led to the recognition of other comparable pairs of presumed antagonists. Antecedents can be found in Emil Fischer's concept of precise fit of enzyme and substrate, in Ehrlich's work on the relation of drug and cell receptor, and in the concept of competitive inhibition as first described by Quastel and Wooldridge. These concepts were further developed in their application to chemotherapy.

The antimetabolite hypothesis became a motive force in medicinal chemistry, pressing for the synthesis of new compounds to be tested pharmacologically and clinically, and it extended its influence into psychiatry. In recent years, the resemblance between the structures of the hallucinogen, mescaline, and the catecholamines has played a significant role in biochemical psychiatry by raising the question as to whether mescaline and drugs like it can antagonize the normal functions of catecholamines in the brain. Furthermore, although the psychological effects of the so-called phantastica had been known for many years, their significance for psychiatric research was largely dormant until the accidental discovery in 1943 of the remarkable actions of LSD-25. Minute amounts of this compound are able to bring about an acute psychological disturbance, and it was recognized that if an exogenous chemical of this kind could evoke protean changes in subjective experience, then an endogenous chemical, resulting from some metabolic disturbance, might conceivably be responsible for the development of psychoses.

In psychiatry, drugs affecting the concentration, synthesis, binding, or release of neurohumors or neurohumorlike substances in the brain are of the utmost interest because of the important role that some of these drugs play or have played in pharmacotherapy. Chlorpromazine's adrenolytic effects are thought to consist in preventing the attachment of the sympathetic neurohumor to a specific receptor site. Reserpine causes the release of amines from the special vesicles in which most of the intracellular content of these substances is stored; it also prevents their being taken up into the vesicles from the external medium. Monoamines are subject to the action of monoamine oxidase; this enzyme is inhibited by various compounds, some of which are valuable in the treatment of mental depression. Important findings of this kind have focused the attention of large numbers of neurobiologists on the role of brain amines in mental diseases.

The neurochemist who is oriented toward pharmacology seeks another type of information that is important in understanding the mode of action of drugs and their clinical uses: knowledge of the metabolism of the drugs themselves – their length of residence in the body, the chemical changes they undergo, and their mode and routes of excretion. Knowledge of drug metabolism may aid in understanding and even avoiding toxic and side actions of drugs.

An area of pharmacology in which neurochemistry aims to contribute is the classification of CNS-active drugs according to their modes of action. However, the present classification is necessarily based upon analysis by pharmacological and physiological techniques. Although the anatomical, biochemical, and physiological sites of action of many drugs in this category are being clarified, much of the terminology of psychopharmacology currently in use is subjective and is admittedly a compromise that ought not to last indefinitely.

The general pharmacological and clinical effects of psychopharmacological agents are outlined in Table 29-1. Detailed pharmacological information can be found in a number of useful monographs and multi-authored volumes, some of which are described in the references (see also [4] and [11]). Pertinent biochemical actions of some psychopharmacological agents are summarized in Table 29-2 and are discussed in more detail below.

NEUROLEPTICS, OR MAJOR TRANQUILIZERS

RESERPINE

The primary action of reserpine (Fig. 29-1) is on the storage and uptake of monoamines by nerve endings and synaptic vesicles. An interesting experiment has been carried out by injecting labeled reserpine into animals and then isolating the blood platelets. These organelles are rich in serotonin, among other constituents, and may thereby serve as a model for the study of the release and uptake of monoamines by vesicles of the nervous system. In this experiment, when the membrane of the platelet is separated from the soluble contents, most of the labeled material of the platelets is found to be associated with the membrane. Similar results are obtained even if the platelets are incubated in vitro with the labeled drug. This suggests that reserpine acts at the membrane by regulating movement of materials through it. When an animal is given a large dose of reserpine, the immediate effect on neurons is the loss of their content of monoamines. In addition, reserpine interferes with uptake of extravesicular monoamine, i.e., amine circulating in the blood, as well as amine temporarily in the neuronal cytoplasm or in the synaptic space after its release from the fiber. There is evidence that much of the amine is released intraneuronally under the influence of reserpine; in this case, it becomes subject to the action of mitochondrial monoamine oxidase and it is ultimately released from the nerve ending as an acidic compound (see Chap. 6).

The action of reserpine on brain monoamines is indiscriminate. Hence, this effect cannot by itself specify the amine whose loss from the brain is to be correlated with the sedative effect of the drug. A great many experiments have been carried out to establish which cerebral amine is affected by reserpine to bring about its tranquilizing action. Although much information useful to biochemical pharmacology has been acquired as a result of this effort, the issue has not yet been settled decisively. If the amines could pass the blood-brain barrier readily, they could be injected singly to determine which one is effective; but peripherally administered amines do not get into the parenchyma of the brain in appreciable amounts. Their amino acid precursors cross the blood-brain barrier much more easily, although an amino acid decarboxylase in the endothelial wall of the brain capillaries may limit the amount of amino acids getting into the parenchymal tissue. Within the brain, an amino acid such as DOPA or 5-hydroxytryptophan is readily decarboxylated, and the corresponding amine will be found not only in regions of the brain where it is an endogenous constituent, but also in others, for instance, in regions that normally do not contain that amine but possess the appropriate decarboxylase.

Experiments of this kind have shown that DOPA will temporarily reverse the sedation caused by reserpine, although 5-hydroxytryptophan will not. It has therefore been deduced that a catecholamine, but not serotonin, is the immediately effective arousal agent. Furthermore, under the experimental conditions, the β-oxidation of dopamine to norepinephrine is rate-limiting, so the injection of DOPA brings about large increases in the concentration of dopamine in many parts of the brain without significantly increasing the concentration of norepinephrine. This points to dopamine as being the amine with the antireserpine activity.

A synthetic analog of DOPA, 3,4-dihydroxyphenylserine, which peripherally gives rise on decarboxylation to norepinephrine, has also been tested in these brain experiments, but it does not seem to penetrate the blood-brain barrier in the way that DOPA can.

Some benzoquinolizines have a reserpinelike action pharmacologically, as well as with respect to storage of monoamines. These synthetic compounds have been studied with the purpose of selecting those that have desirable therapeutic features. One of them, tetrabenazine (see Fig. 29-1), has a much shorter-lasting action than does reserpine, but has the same type of activity qualitatively.

When reserpine is administered, it is distributed to many organs of the body

TABLE 29-1. Pharmacological Actions of Psychotropic Compounds[a]

Category	Typical Examples	Broad Pharmacological Effects	Broad Effects in Man	
			Clinical	Subjective
Neuroleptics, or major tranquilizers	Phenothiazines Butyrophenones Thioxanthenes Reserpine Benzoquinolizines	Decreased autonomic activity, especially of sympathetic branch; hence, antiadrenergic. Decreased motor activity. Decreased conditioned-avoidance reflex. No anesthesia.	Tranquilization of overactive psychotic syndromes. Indifference to environment. In high doses, extrapyramidal symptoms produced. Calming action. Antipsychotic effect.	Drowsiness, apathy, indifference to environment, but no effect on intellect. In high doses, extrapyramidal symptoms produced.
Anxiolytic sedatives or minor tranquilizers	Meprobamate Diazepoxides	Actions similar to those of the neuroleptics, but less potent. No anesthesia. Meprobamate in large doses causes paralysis.	Tranquilization, but without any antipsychotic effect. Sedation of anxious patients.	Muscular and mental relaxation. Less interest in environmental stimuli. Euphoria may occur.
	Barbiturates	As for other minor tranquilizers. Also, there is some blockade of parasympathetic functions. Reduced ability to undergo conditioning. In higher doses, ataxia and then anesthesia.	Tranquilization; hypnotic effect.	Sedation, sleepiness.

Antidepressants	Imipramine MAO inhibitors Pipradrol Amphetamine	Reduced autonomic activity, especially of cholinergic systems. Central sympathetic activation. Increased motor activity, with alerting behavior. Reduction of appetite and sleeping time.	Antidepressant action; stimulation; decreased fatigue. Anxiety, already present, may be increased. If psychotic symptoms are already present, they may be increased.	Reduced motor activity; feeling of relaxation. Increased mental energy, decreased feelings of fatigue, euphoria (but less of this with pipradrol). Greater awareness of the surroundings.
	Lithium salts	No specific effects known.	In manic-depressive psychosis there is control of the manic attack. On chronic use there may be prophylaxis against the clinical depression.	In appropriate patients: feeling of well-being.
Psychodysleptics, or hallucinogenic agents	Lysergide Mescaline Psilocybin Tetrahydrocannabinol	Increased motor activity. Variable autonomic effects. Catatonia.	Marked action on affect, perception, thought processes. Psychotic disturbance may be aggravated.	Autonomic phenomena (e.g., salivation, nausea), followed by psychotic reactions, with distortion of sensory perceptions, color visions, hallucinations.

[a]Source: based primarily on Tables 8 and 12 in [4] and Table 1 in [10].

TABLE 29-2. Biochemical Effects of Psychopharmacological Agents

Haloperidol:
Long-lasting blockade of receptors sensitive to norepinephrine and dopamine, i.e., with action at the postsynaptic site. Increased concentration of striatal HVA, probably by feedback-regulated overactivity of nigrostriatal fibers. May cause a temporary decrease in brain catecholamines.

Chlorpromazine:
Same as for haloperidol. Chlorpromazine causes an increased rate of synthesis of dopamine in brain from tyrosine. Inhibition of some flavin (yellow) enzymes.

Reserpine:
Interference with ATP- and Mg^{2+}-dependent reuptake of released monoamines from synaptic cleft, apparently by inhibition of Ca^{2+}, Mg^{2+}-stimulated ATPase of the vesicular membrane. This results in depletion of monoamines from peripheral sympathetic nerve endings and from central sites. Effects are long lasting; endogenous catecholamines and serotonin are released intraneuronally into the cytoplasm, where they become subject to action of MAO; temporary increase in formation of acidic metabolites (action of aldehyde dehydrogenase) is noted in brain, or of alcoholic metabolites by enzymic reduction of aldehyde product of MAO action. Effects of reserpinization are reversed temporarily by administration of DOPA.

Imipramine:
Interference with reuptake of catecholamines and serotonin from synaptic cleft.

MAO inhibitors:
Hydrazide type, tranylcypromine, and pargyline cause long-lasting, irreversible inhibition of mitochondrial MAO; monoamines are protected and increase in concentration (more pronounced for serotonin in brain than for catecholamines); some of conserved catecholamines may be diverted in increasing amounts into pathway of *O*-methylation. Harmaline and harmine are reversible, short-acting inhibitors of MAO. In large doses they may deplete dopamine from central stores.

Amphetamine:
Weak inhibitor of MAO; inhibits process of monoamine reuptake in peripheral nerve (perhaps also in CNS neurons); effective in releasing monoamines from storage vesicles into synaptic space. Stereotypies caused by amphetamine may be mediated by striatal dopaminergic mechanisms.

Lithium salts:
Stimulate the rate of turnover of norepinephrine and serotonin in the brain; effect is greater for serotonin than norepinephrine in chronic experiments. Inhibit the evoked release of monoamines from brain slices. Cause increase in total body water and extracellular water.

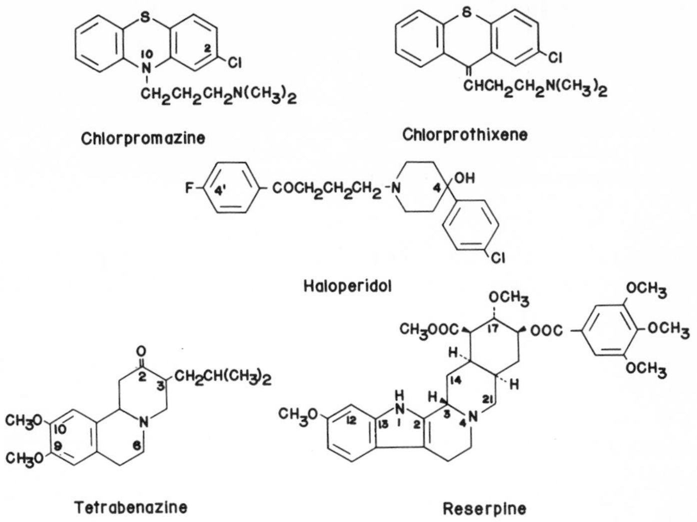

Chlorpromazine

Chlorprothixene

Haloperidol

Tetrabenazine

Reserpine

FIGURE 29-1. Some examples of neuroleptics.

without any preferential uptake by the brain. Initially, the viscera contain relatively high concentrations, but later the drug moves into adipose tissue. Careful experiments with radioactively labeled reserpine have demonstrated the presence of the drug in the brain during the long period over which it exerts its central depressive action. Reserpine contains two esterified groups and both of these may be hydrolyzed in the course of metabolism. Among the products are methyl reserpate, reserpic acid, 3,4,5-trimethoxybenzoic acid, and other compounds.

CHLORPROMAZINE

Pharmacologists have shown that chlorpromazine (see Fig. 29-1) acts at the postsynaptic receptor site, thereby blocking the action of transmitter substance released into the synapse. The blockade is long lasting and not reversed by excessive amounts of the transmitter. This type of action in the CNS is probably the basis of the antipsychotic action of phenothiazines, butyrophenones, and thioxanthenes.

Many of the major tranquilizers cause an accumulation of homovanillic acid (HVA) in dopamine-rich regions of the brain [1] without significantly affecting the dopamine concentration itself or that of serotonin and its metabolic product, 5-hydroxyindoleacetic acid (Table 29-3). This has been explained in the case of the striatum as follows: the afferent fibers from the substantia nigra are dopaminergic, and some of the striatal neurons with which they synapse provide a feedback influence on the nigra. If the striatal component of this nigro-strionigral loop is blocked – that is, if the pertinent striatal cells are no longer under the influence of dopamine because of neuroleptic blockade – feedback signals to the substantia

TABLE 29-3. Effects of Drugs on Homovanillic Acid Concentration of Brain

	Mouse Striatum				Rat Brain Stem				Rabbit Corpus Striatum	
Drug[a]	Dose mg/kg	HVA μg/g	Dose mg/kg	5-HIAA μg/g[b]	Dose mg/kg	HVA % of Control	Dose mg/kg	5-HIAA % of Control	Dose mg/kg	HVA μg/g
None (control)	0	0.31±0.01[c]	0	0.21±0.01	0	100	0	100	0	3.9±0.9[d]
Chlorpromazine	5	0.70±0.05	–	–	10	314±21[c]	20	119±3	–	–
Thioridazine	10	0.49	–	–	–	–	–	–	5	8.2
Chlorprothixene	10	1.28±0.08	–	–	10	300±17	20	125±4	10	8.0
Spiroperidol	0.1	0.66±0.04	0.5	0.24±0.03	–	–	–	–	0.5	8.3
Haloperidol	–	–	–	–	5	340±8	10	122±5	–	–
Reserpine	5	0.48±0.01	–	–	–	–	–	–	–	–
Tetrabenazine	50	0.73±0.06	–	–	–	–	–	–	–	–
Chlordiazepoxide	100	0.49±0.06	–	–	50	83±5	50	132±7	–	–

[a]Drugs were given 2 to 3.5 hr before death in most cases: i.p. for mouse and rat; i.v. for rabbit.
[b]Whole brain.
[c]Mean ± S.E.
[d]Mean ± S.D.
Source: Data for mouse from [8]; data for rat from [1]; data for rabbit from [7].

nigra result in the increased firing of the nigral fibers. The result is an excessive formation and release of dopamine without its actual concentration being affected at any given time; some of the excess accumulates as HVA. Hence, the ratio of HVA to dopamine increases above the resting levels [7].

The concentration of monoamines in the brain is the steady-state result of production and utilization. Kinetic studies are desirable to assess the action of neuroleptics on the rates of these two processes. This can be estimated by use of labeled precursors. For example, if radioactive tyrosine is infused into animals, some of it reaches the brain and is synthesized into catecholamines. In animals given chlorpromazine, a substantial increase in the amount of radioactivity derived from tyrosine can be accounted for as dopamine (Table 29-4 and [6]). This is not the case with norepinephrine, the concentration of which as measured by its radioactivity is only slightly elevated. In this type of experiment there is a small reduction (about 15 percent) in the concentration of endogenous tyrosine, which results in a modest increase in the specific activity of this amino acid in the brain. The apparently greater rate of utilization of tyrosine for the formation of dopamine indicates that chlorpromazine increases the activity of the neurons that synthesize dopamine; this has been postulated to be a consequence of its blockade of dopamine-sensitive receptors.

It is interesting that in Parkinson's disease, as well as in certain animal models in which nigrostriatal fibers have been interrupted, the ratio of HVA to dopamine is also increased. However, in these conditions the concentrations of both substances are subnormal.

Although no effect of neuroleptics on the concentration of catecholamines in the brain has been detected in acute experiments, repeated administration of chlorpromazine or haloperidol (see Fig. 29-1) decreases both striatal dopamine and cerebral norepinephrine. This result may be relevant to the long-term clinical usage of neuroleptics [8].

The major tranquilizers, when given in dosages that cause postsynaptic blockade, do not significantly affect oxidative metabolism in the brain. CNS depressants, such as barbiturates, reserpine, and chlorpromazine, have occasionally been reported to cause an increase in the levels of ATP and phosphocreatine of the brain, but, when special precautions are taken to prevent degradation of nucleotides during isolation from the brain, there proves to be no effect. On the other hand, chlorpromazine inhibits D-amino acid oxidase and other flavoproteins. It blocks the release of various tropic hormones of the anterior pituitary gland, but only at dosage levels in excess of those needed for neuroleptic action. Interestingly, some fibers are rich in dopamine that innervate the median eminence and apparently mediate higher nervous influences upon the hypophysis (see Chap. 17).

Chlorpromazine generally reduces the permeability of membranes. Thus, the drug prevents the release of acetylcholine from isolated synaptic vesicles that have been suspended in a hypotonic sucrose solution; it also inhibits uptake of acetylcholine by vesicles. Its net pharmacological action, however, involves a cholinergic blockade factor; in fact, chlorpromazine may potentiate the action of atropinelike compounds.

TABLE 29-4. Effects of Chlorpromazine and Pargyline on Rates of Formation of Catecholamines in Brain

		Precursor			Products		
						Norepinephrin	
Species[a]	Treat-ment	Compound	Conc. μg/g	S.A.[b] cpm/μg	Dopamine cpm/g	μg/g	cpm/g
Rat	Control	^{14}C-Tyrosine	–	6600±600[c]	615±30	–	183±1
	CPZ	^{14}C-Tyrosine	–	7500±500	1262±87	–	217±2
Guinea pig	Control	^{14}C-Tyrosine	17.7	1602	–	0.34	400
	MAOI	^{14}C-Tyrosine	17.2	1600	–	0.76	180
	Control	^{3}H-DOPA	–	6253	–	0.35	876
	MAOI	^{3}H-DOPA	–	6553	–	0.64	1826

[a]Rats were injected i.p. with chlorpromazine (CPZ), 15 mg/kg, 2 hr before killing them for brain measurements. Labeled tyrosine was infused i.v. for 25 min immediately preceding death [6]. Guinea pigs were injected with labeled tyrosine or DOPA 24 or 7 hr, respectively, after pargyline (MAOI) was given and were killed 1 hr later. The compounds were then measured in the brain stem [9].

[b]S.A. = specific activity.

[c]Mean ± S.E. Data for guinea pig are means of two or three animals.

The phenothiazine nucleus is a strong electron donor, and charge-transfer properties have been suggested as playing a role in the cellular actions of chlorpromazine and related compounds. Others have suggested that the molecule acts as a chelator of certain trace metals.

Chlorpromazine is readily absorbed from the gut, tissues (into which it may be injected), and blood; on the other hand, the release of the compound from most tissues is slow. Excretion by the kidney is rapid. The major portion of a single dose of chlorpromazine is excreted in the urine, but some is found in the feces. With chronic administration, there seems to be some accumulation in the tissues because cessation of treatment with chlorpromazine is followed by a period of many months during which small amounts of the compound or its metabolites are found in the urine. In the brain, which is easily penetrated by chlorpromazine, the drug is found in higher concentrations in the cerebral cortex and the basal ganglia than in other parts of the CNS.

In the course of tissue metabolism, chlorpromazine may undergo a variety of reactions. These include demethylation of the amine, oxidation of the ring to yield a hydroxyl group, conjugation of this hydroxyl with glucuronic acid, and oxidation to yield chlorpromazine sulfoxide and related metabolites. After long, continued use of this drug in large doses, as in the treatment of schizophrenia, a photosensitive metabolite may accumulate in the skin; exposure to sunlight converts this substance to a purple derivative, which lends a grayish or grayish-purple cast to the complexion Pigment may also be deposited in the cornea.

Chlorprothixene (see Fig. 29-1), a member of the thioxanthene family, is also oxidized to the sulfoxide, which represents a major metabolite of the drug. The compound disappears extremely rapidly from the blood after its administration.

ANXIOLYTICS, OR MINOR TRANQUILIZERS

BENZODIAZEPINES

The sedative, anticonvulsant, and taming activity of the benzodiazepines (Fig. 29-2), along with their relaxing effect on skeletal muscle, has given these drugs a prominent role as minor tranquilizers. Although their biochemical mode of action has not been specified as yet, their metabolism has been followed in a number of studies. Thus, a single dose of 15 mg of chlordiazepoxide, administered orally, provides a peak plasma concentration of about one μg per ml at 1 to 2 hr. The half-life of the radioactivity of the labeled compound in plasma is about 24 hr. Among the reactions that the molecule undergoes is a loss of the methylamino group and the formation of a lactam. The lactam ring is cleaved to yield a substituted glycine-*N*-oxide, which, like its precursor, is found in the urine. Half of the radioactivity of chlordiazepoxide is excreted in the urine in 3 to 3.5 days. In the case of diazepam, 10 mg taken orally provides a peak level in the blood of less than 0.1 μg per milliliter. Diazepam, which has 1-methyl-2-oxy substituents among others, undergoes 1-demethylation and 3-hydroxylation in metabolism, and these two reactions successively result in the formation of oxazepam, another benzodiazepine drug with a lactam ring as well. The hydroxy derivative of diazepam occurs in the urine as a glucuronide.

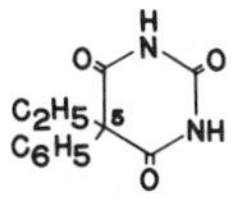

Phenobarbital

Chlordiazepoxide

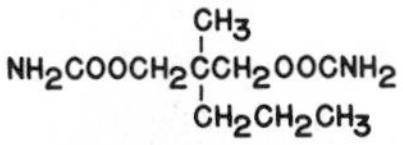

$NH_2COOCH_2C(CH_3)(CH_2CH_2CH_3)CH_2OOCNH_2$

Meprobamate

FIGURE 29-2. Examples of minor tranquilizers.

BARBITURATES

The barbiturates depress the metabolism of brain tissue in vitro to some extent; their effect is greater on electrically stimulated brain slices than on "silent" tissue. Moreover, some of these compounds inhibit the respiration of brain mitochondria, whose metabolism has been stimulated by the uncoupling agent, 2,4-dinitrophenol. Thiobarbiturates inhibit oxidative phosphorylation of the liver and brain mitochondria that metabolize pyruvate, and they have some action in uncoupling oxidative phosphorylation; the oxybarbiturates affect only the former process. The site of action of the oxybarbiturates appears to be in the respiratory chain between NAD and cytochrome c; this includes a flavoprotein long ago postulated to be inhibited by barbiturates.

One of the important parameters that pharmacologists use to classify the barbiturates is their duration of action. This is influenced, of course, by the molecular structure. Thus, the thiobarbiturates undergo metabolism more rapidly than does the oxybarbiturate series, and those with longer aliphatic side chains or an aromatic substituent are metabolized more slowly, presumably because of greater lipid solubility and longer residence in adipose tissue. These substituents are rendered more polar during metabolism through oxidations catalyzed by microsomal enzymes. In phenobarbital (see Fig. 29-2), a phenolic group is inserted; in pentobarbital, the alkyl side chain of the compound undergoes hydroxylation.

The administration of phenobarbital to rats causes increased activity of the microsomal oxidizing enzymes of the liver. In addition to this induction of "drug-metabolizing enzymes," there is an increase of a hemoprotein which may be a component of such oxidative systems.

ANTIDEPRESSANTS

MONOAMINE OXIDASE INHIBITORS

Monoamine oxidase (MAO) inhibitors as a class of antidepressant drugs have had great significance for biochemical pharmacology and for the theory of mental disease (Fig. 29-3). This significance remains, despite the dangers inherent in their use. The first member of this class was iproniazid (*N*-isonicotinyl-*N*'-isopropylhydrazide), which was originally tested as an antitubercular drug along with isoniazid (*N*-isonicotinylhydrazide). The elation observed in patients receiving iproniazid was later put to use in the treatment of mental depression. The reversal of depression was then related to the ability of the compound to inhibit MAO powerfully, and thereby increase the concentration of monoamines in the brain. Subsequently, chemists synthesized scores of compounds on the model of iproniazid in the search for new and useful inhibitors of MAO which might also be useful therapeutically.

For many years Zeller's classification into monoamine oxidase (acting on substrates like tyramine, isoamylamine, and epinephrine) and diamine oxidase, or histaminase, was used. As new types of amine oxidases have been discovered and as

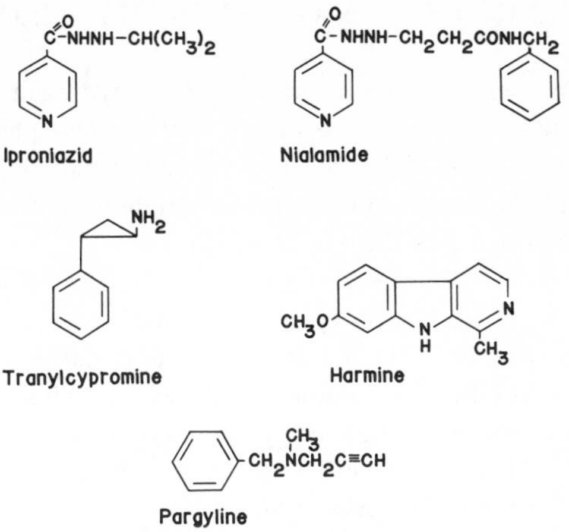

Iproniazid

Nialamide

Tranylcypromine

Harmine

Pargyline

FIGURE 29-3. Monoamine oxidase inhibitors.

the physical and chemical properties of these enzymes have become better known, it is now possible to classify them not merely by the spectrum of substrates whose oxidation they catalyze, but also on the basis of their respective prosthetic groups. For example, in the plasma of pig and some other species there is a pink amine oxidase, which requires pyridoxal phosphate and cupric ions for its activity; its function is unknown. A copper-containing amine oxidase, acting upon the ϵ-amino group of lysine, has been detected in birds and other species; it is needed for the formation of desmosine in the walls of arteries.

MAO is present in the outer membrane of mitochondria; it does not contain copper, but bears flavine adenine dinucleotide (FAD), probably in covalent linkage, as its coenzyme. The enzyme has been purified from the liver of several species and from bovine kidney and pig brain. The molecular weight of the mitochondrial enzyme in rat liver is approximately 150,000 (by ultracentrifugation measurements). This enzyme is especially sensitive to certain hydrazine derivatives, although not to hydrazine itself. It is also inhibited by quinacrine and by galactoflavin, an antimetabolite of riboflavin.

MAO has at least two functions. In the intestinal tract and liver, the enzyme probably serves in detoxification of certain amines formed by the gut flora. On absorption, these amines could exert pharmacological activity. In the nervous system, the enzyme serves to oxidize some of the serotonin, dopamine, and norepinephrine after their release into the synaptic space, thus terminating their action. This may also happen within the nerve fiber itself if the store of monoamines in the synaptic vesicles is released under the influence of reserpine: the "free" amine within the nerve ending can now be attacked as it diffuses toward the mitochondria.

MAO catalyzes the following reaction:

$$RCH_2CH_2NH_2 + O_2 + H_2O \longrightarrow RCH_2CHO + H_2O_2 + NH_3$$

The aldehyde undergoes oxidation by an aldehyde dehydrogenase to form the corresponding carboxylic acid, but some reduction may also take place through the action of an enzyme such as alcohol dehydrogenase (aldehyde reductase). The latter reaction is quantitatively important in the cerebral metabolism of norepinephrine, but less so for the amines that lack the β-hydroxyl group.

If an animal is treated with a MAO inhibitor, the concentrations of many amines increase in the tissues, including the brain. Thus, the rate of synthesis of serotonin in the brain is essentially unaffected, but, because degradation of the amine is slowed down, a net increase results in the amount of the amine. This phenomenon also occurs with the catecholamines, but the increases are more modest since two other factors enter here: (1) catecholamine biosynthesis is regulated by the norepinephrine feedback inhibition of tyrosine hydroxylase (Chap. 6 and [9]), and (2) dopamine and norepinephrine are subject to the action of catechol-*O*-methyl transferase (COMT) and the methylated products have less biological potency than do the parent amines. These effects are illustrated by the results in Table 29-4, in which norepinephrine formation from labeled tyrosine is reduced although the tissue concentration of the amine is increased. This is not true when labeled DOPA is used.

Animals treated with a MAO inhibitor become behaviorally alert and excited. This is attributed to the increase in concentration of monoamines, but which one of them is responsible is unknown. Indeed, it is not yet known if only one specific amine mediates the response or if several act in concert. Attempts to delimit the chemistry of the response have been made by injecting precursors of specific monoamines, e.g., DOPA, tryptophan, and 5-hydroxytryptophan, but as yet there are no definitive answers. The behavior elicited by a large dose of a MAO inhibitor is not observed if the animal has been treated previously with reserpine. On the other hand, if a MAO inhibitor is administered before reserpine, the action of the latter drug is potentiated. In this case, the released amines can exert their pharmacological activity at synapses for a protracted period, unmitigated by reuptake into the vesicles of the nerve ending or catabolism by MAO.

Some investigators have cautiously tested these and other amino acids in patients on MAO-inhibitor therapy, but again the results have not been clear-cut. The profound activation of autonomic systems caused by the combination of a MAO inhibitor with a monoamine precursor has precluded simultaneous use of the two types of agent in the treatment of depression. In fact, cardiovascular and other side effects of MAO inhibitors that are used therapeutically represent autonomic reactions mediated by endogenous amines. An important precaution in the clinical use of MAO inhibitors is the avoidance of foods prepared with cheese, especially of the aged varieties, and of certain types of wine, since these contain much tyramine formed through the action of the mold or yeast on tyrosine. Ordinarily, tyramine would be metabolized to *p*-hydroxyphenylacetic acid before it could pass into the general circulation, but, protected by the MAO inhibitor, the tyramine reaches sympathetic postganglionic nerve endings, where it displaces stored norepinephrine stoichiometrically. The excessive release of norepinephrine to receptor sites initiates the observed, and sometimes lethal, autonomic changes.

Paradoxically, MAO inhibitors cause a lowering of blood pressure. Several mechanisms for this have been suggested, based on events that may take place centrally or peripherally. One hypothesis suggests that the inhibitor allows the accumulation of abnormal amounts of amine metabolites that are normally not detectable; an example is *p*-hydroxyphenethanolamine, formed by β-oxidation of tyramine. At sympathetic nerve endings, this derivative would replace norepinephrine and then act as a substitute, or "false," transmitter, with little musculotropic action. Pargyline (see Fig. 29-3) is a MAO inhibitor that is particularly indicated in the treatment of hypertension.

Although the antidepressant action of MAO inhibitors is thought to reside in their ability to conserve monoamines and to increase the concentrations of these compounds in the brain, the drugs possess inhibitory actions on other enzymes as well. These inhibitions have not received the same thorough attention as has been given to MAO, but are nevertheless important from the toxicological point of view. For example, since the metabolism of ethanol is inhibited by MAO inhibitors, the effects of alcoholic beverages are potentiated. The MAO inhibitors also act upon the microsomal liver enzymes that catalyze the metabolic transformation, or detoxification, of other drugs. Thus, additional precautionary measures must be observed when a MAO inhibitor is used therapeutically. For example, this treatment must be stopped some days before a general anesthetic is to be administered.

The long-lasting action that most MAO inhibitors exert suggests that they form a covalent link with the enzyme. If so, the activity of the enzyme would be restored only some days after the drug was withdrawn, i.e., when new enzyme has been synthesized. This has been tested experimentally in rats that have been given iproniazid to inhibit MAO and measured at suitable time intervals to determine the ability of the animals to convert exogenous dopamine to urinary dihydroxyphenylacetic acid. This ability is restored to control levels in about 4 days.

There is evidence that iproniazid must undergo an oxidative reaction before it can act as an inhibitor; this may be a dealkylation, with loss of the *N*-isopropyl group. Tranylcypromine (see Fig. 29-3), a cyclopropylamine derivative, is thought to resemble the transient intermediate species formed by the substrate at the active site of the enzyme and, hence, to have a better fit to the enzyme. Pargyline possesses an acetylenic linkage and may act by still another mechanism. The related compound, chlorgyline, inhibits MAO in two successive stages distinguishable by two well-defined, successive ranges of inhibitory concentration.

Harmine (see Fig. 29-3) and harmaline (dihydroharmine) are unique among MAO inhibitors in that they occur naturally. Their inhibitory action is short lasting and reversible. Harmine, known at one time as telepathine, is considered to have some hallucinogenic activity. The mode of action of these β-carbolines is still being explored. Along with their *O*-demethylated derivatives, harmol and harmalol, and certain related compounds, they antagonize peripheral actions of serotonin, although only harmine and harmaline in this group inhibit MAO. Many analogs have been synthesized in the search for useful therapeutic agents.

The metabolism of hydrazide inhibitors of MAO involves the loss of one or another of the N-substituents. For example, iproniazid yields isonicotinylhydrazide, isopropylhydrazide, isonicotinic acid, and unchanged drug in the urine. Radioactive tracer studies indicate that the isopropyl label remains in tissues considerably longer than does the isonicotinyl label. Harmaline and harmine undergo metabolism by several pathways. One of them involves *O*-demethylation, with subsequent glucuronidation and sulfation of this phenolic type of derivative.

Experimental and clinical uses have been found for two hydrazine derivatives that inhibit DOPA decarboxylase strongly, both in vivo and in vitro, without affecting MAO and without providing antidepressant action. They protect easily degraded substrates, such as DOPA and 5-hydroxytryptophan, from the action of decarboxylase in the peripheral tissues. Thus, when the amino acid is administered, a larger portion of it is conserved for distribution to the brain, where the amine product is wanted. One of these drugs, *N*-seryl-*N*'-(2,3,4-trihydroxybenzyl)hydrazine (Ro4-4602), inhibits cerebral decarboxylase only if it is injected in amounts much higher than those required for action on the enzyme in the liver or kidneys; hence, a differential effect can be obtained by adjustment of the dosage. The other compound is the hydrazino analog of α-methylDOPA (MK-485 and MK-486); it does not pass the blood-brain barrier at all.

TRICYCLIC ANTIDEPRESSANTS

The tricyclic antidepressant compounds have structures based upon those of the phenothiazines (Fig. 29-4). In imipramine, the sulfur atom of promazine is replaced by an ethylene bridge, yielding a dibenzazepine; in amitriptyline, a carbon atom is also substituted for the nitrogen in the middle ring, giving a dibenzocycloheptadiene. The use of molecular models reveals that, despite the formal resemblance of imipramine and promazine, the two benzene rings of the former compound are twisted against one another in an unsymmetrical fashion.

Some of these drugs were originally tested in schizophrenic patients without success, but ultimately were found to have important applications in the treatment of mental depression. The tricyclic antidepressants do not inhibit MAO, but their therapeutic action has been related in another way to monoamine metabolism: they appear to interfere with the process of reuptake of norepinephrine and serotonin from the synaptic cleft back into the nerve ending and its vesicles. Such an action provides a net increase in the concentration of free amine (i.e., amine not bound in vesicles or to a macromolecule) in the synaptic cleft at any one time, with increased probabilities of stimulation of sensitive receptors on the postsynaptic neurons. There are some important differences between the tricyclic compounds in this respect; for example, imipramine and amitriptyline strongly inhibit reuptake of serotonin, whereas desipramine and nortriptyline (the corresponding *N*-monomethyl compounds) are less effective. The acute administration of imipramine to experimental animals causes a decreased turnover of serotonin in neurons; this could stem from the reduced rate of reuptake of the amine and its longer residence in the

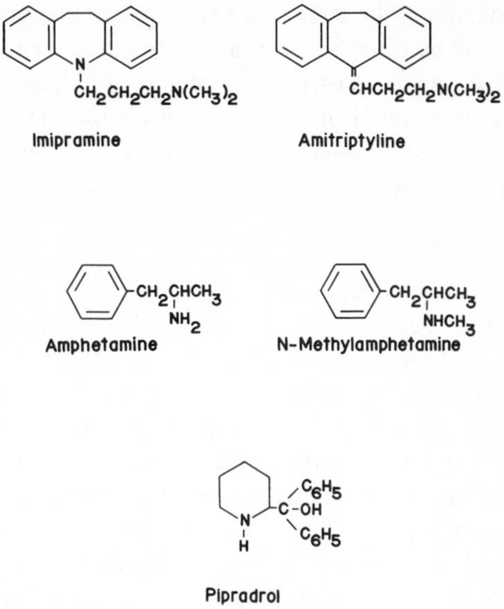

$CH_2CH_2CH_2N(CH_3)_2$

Imipramine

$CHCH_2CH_2N(CH_3)_2$

Amitriptyline

CH_2CHCH_3 NH_2

Amphetamine

CH_2CHCH_3 $NHCH_3$

N-Methylamphetamine

C_6H_5 C-OH C_6H_5 N H

Pipradrol

FIGURE 29-4. Some antidepressant drugs.

synaptic space. When the drug is given to animals chronically in order to approximate the therapeutic situation, the norepinephrine content of the brain decreases and the rate of metabolism of intracisternally injected (labeled) norepinephrine increases. The reduced rate of uptake of the amine under the influence of imipramine would render it more accessible to metabolic pathways. According to several hypotheses, the inhibition of monoamine reuptake plays an important role in the antidepressant action of imipramine and its congeners.

Imipramine is metabolized somewhat analogously to the phenothiazines: it undergoes dealkylation of the tertiary amine side chain, yielding desipramine as the first product, which also has antidepressant properties. Microsomal enzymes of the liver catalyze other oxidative reactions as well.

AMPHETAMINE

Amphetamine (see Fig. 29-4) and some chemically related compounds have been used as therapeutic agents for many years. After the Second World War, their illicit use as psychological stimulants and, more recently, as hallucinogenic agents expanded greatly. Amphetamine abuse can lead to serious disturbances of mental behavior. The resulting "amphetamine psychosis" may be virtually indistinguishable from paranoid schizophrenia and is often diagnosed correctly only when the patient is discovered to be taking amphetamine. Some amphetamine-dependent individuals exhibit peculiar chewing movements which are undoubtedly related to the stereotypy elicited in laboratory animals by large doses of amphetamine or by apomorphine and which may be mediated by dopamine-containing fibers of the CNS.

Amphetamine was recognized early as an inhibitor of MAO. It is not especially potent in this respect in comparison with the newer synthetic derivatives discussed above, but experiments with labeled amines show that it nevertheless has some inhibitory action on MAO in vivo. It is not itself a substrate since MAO acts only on ω-amines.

The euphoriant action of amphetamine is attributable not only to its ability to inhibit MAO but also to two other actions. It inhibits reuptake of norepinephrine into sympathetic nerve fibers, and thereby presumably causes retention of a greater supply of the amine in the synapse for stimulation of the postsynaptic fiber. Second, amphetamine can enter the nerve ending, penetrate into the amine storage vesicles, and replace the norepinephrine contained in them. This resembles the action of tyramine. The release of the amine is apparently directed outward, i.e., into the synapse. The vector of the releasing mechanism has been shown experimentally: after labeled norepinephrine is injected into an animal and time is allowed for the amine to be taken up into sympathetic nerve endings, administration of reserpine leads to a preponderance of acidic derivatives; the formation of these compounds is initiated by the action of MAO, an intraneuronal enzyme. On the other hand, injection of tyramine (or amphetamine) produces a preponderance of labeled amines, one fraction consisting of normetanephrine. This indicates that MAO has limited opportunity to act, but that COMT serves in the postrelease disposal of the amine; COMT occurs extraneuronally.

Despite their actions as CNS stimulants, both amphetamine and pipradrol have been used in the treatment of hyperkinetic children. Paradoxically, the drugs exert a quieting effect in this condition.

An interesting derivative of amphetamine is its *p*-chloro analog. This compound inhibits tryptophan hydroxylase, as does *p*-chlorophenylalanine. Either drug reduces the rate of formation of serotonin in the brain, with a resulting decrease in its serotonin concentration.

The excretion of amphetamine and related compounds is sharply dependent on the urinary pH. For example, a man taking about 10 mg of *N*-methylamphetamine (methamphetamine) orally, together with ammonium chloride, excretes 55 to 70 percent of the dose unchanged in the urine in 16 hr, and an additional 6 to 7 percent as amphetamine. If he takes sodium bicarbonate which renders the urine alkaline, only 0.6 to 2.0 percent of the dose appears in the urine in the amine fraction.

β-Phenylisopropylamines are metabolized along several pathways: *p*-hydroxylation, with or without *O*-glucuronidation; β-hydroxylation of the side chain; *N*-dealkylation if there is a substituent on the amino group; and oxidative deamination with formation of a ketone, the ketone being either reduced to the corresponding alcohol or further oxidized to an acid. Considerable species differences exist with regard to these metabolic pathways. In the dog, about one-third of a dose of *d*-amphetamine is excreted unchanged in the urine; another large portion appears as *p*-hydroxyamphetamine, some of which is conjugated. In man and rabbit, deamination is the predominant route. Some of the phenylacetone that is formed is converted to phenylisopropanol; another portion is oxidized to benzoic acid, most of which appears in the urine as hippuric acid. In the rat and cat, *d*-amphetamine is hydroxylated on both the ring and the side chain, forming *p*-hydroxynorephedrine. A small amount of this metabolite is formed in man via *p*-hydroxyamphetamine.

LITHIUM SALTS

The use of lithium salts offers considerable promise in the treatment of recurrent endogenous affective disorders. The lithium ion is therapeutically useful in the control of both mania and hypomania, and it provides prophylaxis not only against the manic but also the depressive episodes of phasic illness.

Lithium is a monovalent cation of the alkali metal group and is therefore related to sodium and potassium; it resembles magnesium and calcium in certain aspects of its metabolism. Its determination by flame photometry in the clinical laboratory facilitates therapeutic control. By this means, dosage can be adjusted to maintain a serum level of 0.6 to 0.8 meq per liter, which can be attained with a dosage of 25 to 50 meq of lithium per day.

Lithium salts are taken orally. There is good absorption from the gastrointestinal tract, with Li^+ passing from the blood into the tissues. Equilibration is attained quite quickly in liver, kidneys, and skin, and less rapidly in muscle and brain. The concentration of Li^+ in the cerebrospinal fluid reaches only about one-half the level in serum. Almost all the ingested Li^+ is excreted in urine.

Lithium salts appear to have some influence over the movements of norepinephrine at nerve endings and in synapses. There is evidence for a reduced availability of the amine at receptor sites as a result of Li^+ action, and this may stem from increased central release of norepinephrine at a rate that outstrips biosynthesis. However, the dynamics of monoamine metabolism during lithium treatment are still being worked out, especially in the case of chronic administration. In acute experiments, dopamine and serotonin are not affected.

Li^+ alters somewhat the distribution of Na^+, K^+, and water, but, with good therapeutic control of dosage, these changes are not serious.

PSYCHODYSLEPTICS, OR HALLUCINOGENIC AGENTS

The psychodysleptics represent a motley group of plant and laboratory products (Fig. 29-5). The remarkably potent *d*-LSD-25 is a synthetic derivative of a plant alkaloid, lysergic acid. The hallucinogens can be classified roughly into three major categories: (1) indolic derivatives, (2) phenylalkylamine derivatives, and (3) nonnitrogenous compounds. The first two groups have lent themselves readily to biochemical theories of hallucinogenic actions based on the concept of antagonism to cerebral serotonin and catecholamines, respectively. This antagonism is thought to interfere in some way with the as-yet-unknown role of the monoamines in mental

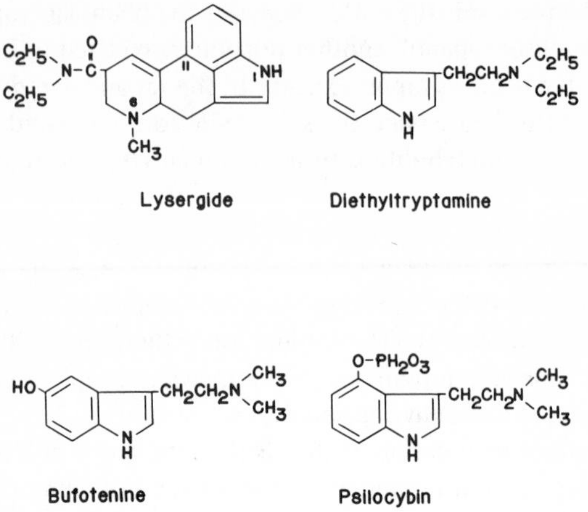

Lysergide

Diethyltryptamine

Bufotenine

Psilocybin

FIGURE 29-5. Indolic hallucinogens.

processes. Molecular orbital calculations have also been used to assess hallucinogenic potency in relation to chemical reactivity. However, our understanding of the action of these compounds at the cellular or molecular level is very limited at the present time.

INDOLIC HALLUCINOGENS

The simplest members of this category are the dialkyltryptamines, e.g., diethyltryptamine (see Fig. 29-5). Others are psilocybin, the *O*-phosphoryl ester of *N*-dimethyl-4-hydroxytryptamine, and *d*-LSD-25. Bufotenine (*N*-dimethyl-5-hydroxytryptamine) also is reputed to be hallucinogenic; it is found not only in plant sources but also in the poison gland of the toad. *d*-LSD-25, now known as lysergide, is one of four diastereoisomers of lysergic acid diethylamide, all of which have been tested for hallucinogenic activity. Only lysergide is active to any extent, and it is exceedingly potent. While exerting its psychodysleptic action, it is estimated to be present in brain to the extent of only a few picograms per gram of tissue.

In 1953 and 1954, several investigators, including D. W. Woolley, drew attention to the possible interaction of indolic alkaloids with serotonin receptors. According to this hypothesis, drugs such as lysergide would act in the brain by opposing the normal action of serotonin and thus cause a functional deficiency of serotonin at important CNS centers. This application of the antimetabolite hypothesis to psychodysleptics was soon recognized to be inadequate. For example, if lysergide is brominated to 2-bromolysergide, hallucinogenic activity is lost, but antagonism toward musculotropic actions of serotonin is retained. Thus, an antiserotonin action in one context does not necessarily insure that it will be manifested in another

There have been several attempts to relate the behavioral actions of lysergide in experimental animals to monoamine metabolism. An experiment that suggests the mediation of behavioral effects by serotonin depends upon the use of an unconditioned escape response in the rat [5]. The time from stimulus to escape is measured both for rats receiving various drugs, alone or in combination, and for control animals. Lysergide in such trials reduces the latent period sharply. If this effect is mediated by a monoamine, the administration of a specific inhibitor of the biosynthesis of this monoamine should abolish the action of lysergide. In the experiment described in Table 29-5 (Experiment A), two such inhibitors have been used: α-methyltyrosine, which inhibits catecholamine synthesis, did not alter the response brought about by lysergide, whereas *p*-chlorophenylalanine, which prevents the formation of serotonin, cancelled the effect of the hallucinogen. The ability of lysergide to produce strong central excitement along with a stereotyped pattern of behavior has been used similarly to measure the action of lysergide, as shown in Experiments B and C of Table 29-5. In these cases, interference with the formation of serotonin had no influence on the effects of lysergide. However, α-methyltyrosine, although itself causing some sedation, effectively antagonized the lysergide-induced excitement. This has also been observed in rabbits receiving small doses of lysergide intravenously; the animals exhibit dilation of the pupils, hyperventilation, increased motor activity, and hyperthermia. The administration of α-methyltyrosine also antagonizes the excitement caused by lysergide in this species, but has no influence on the hyperthermia [3]. These experiments indicate that the components of the pharmacological actions of lysergide may be mediated by different amines.

In other experiments, it has been shown that lysergide affects the metabolism of serotonin in the brain in a complex fashion, although its main known effect is to increase brain serotonin slightly and to decrease 5-hydroxyindoleacetic acid correspondingly. Studies of the distribution of serotonin in the subcellular fractions suggest that these changes probably take place within the neuron, i.e., without release of serotonin into the synaptic space, and that lysergide may cause increased binding of serotonin in the brain. This effect of lysergide also is seen with other indolic hallucinogens, and it is not due to inhibition of MAO. However, yohimbine, a nonhallucinogenic indolic alkaloid, has a similar action, so the specific relation of conservation of brain serotonin to the mental effects of drugs is not certain.

Lysergide has an interesting electrophysiological action in decreasing the rate of firing of cells of the midline raphe nuclei. These are structures that contain serotonin or a serotoninlike substance, according to histofluorescence examination.

There are large differences among species in the metabolism of lysergide. The biological half-life of the compound in the blood of the mouse and rat is 7 min, but in the monkey and cat it is 1.5 to 2 hr.

Lysergide is bound to plasma proteins when transported in the blood, but it is readily released into the brain and other organs. Its highest concentration is attained in these organs soon after injection, and it declines over the next few hours. The concentration of lysergide in the brain is always less than that in the blood.

Microsomal enzymes in the liver (but not in the brain) oxidize lysergide to its

TABLE 29-5. Antagonism of Effects of Lysergide by Inhibitors of Monoamine Biosynthesis

				Antagonist			
		Control		*p*-Chlorophenylalanine[c]		α-Methyltyrosine[d]	
Experiment[a]	+Lysergide[b]	−LSD	+LSD	−LSD	+LSD	−LSD	+LSD
A: Unconditioned escape response latency in rat (arbitrary units) [5]	1.0 (s.c.)	7.2	1.2	7.4	7.0	7.0	1.1
B: Aberrant behavior in rat, average score (arbitrary units) [2]	5.0 (s.c.)	0	25–30	No effect	25	Some sedation	8
C: Excited behavior in rabbit [3]	0.1 (i.v.)	Normal	Excitement	–	–	Some sedation	Some sedation[e]
D: Hyperthermia in rabbit (Change in rectal temp., °C) [3]	0.1 (i.v.)	0.2–0.3	1.5–1.8	–	–	0.3	2.0

[a] Reference numbers in brackets.
[b] Dose (+LSD) in mg/kg; route of administration in parentheses.
[c] Dose: 300 mg/kg intraperitoneally.
[d] Dose: 60 mg/kg intraperitoneally in A, 120 in B; 480 mg/kg during 24 hr in C and D.
[e] Same effect as α-methyltyrosine without lysergide.

biologically inactive 2-oxy derivative. Other products, more polar than lysergide itself, are formed. The bile serves as one route of excretion of the hallucinogen, but some of it is probably then reabsorbed from the intestine.

Some investigators claim that bufotenine is a constituent of the urine of schizophrenic patients, normal subjects, or both. However, these claims still require verification.

Psilocybin, the active principle of certain mushrooms found in the southern highlands of Mexico, is unusual in being an ester of phosphoric acid. Its nonhallucinogenic hydrolytic product is psilocin, the 4-hydroxy analog of bufotenine.

An interesting and novel pathway of metabolism of some indolic compounds is by way of 6-hydroxylation. This route accounts for two-fifths of a dose of *N,N'*-diethyltryptamine but only for about one-fifth of the dose of the dimethyl and dipropyl homologs.

MESCALINE

Mescaline, like lysergide, has been investigated for possible inhibitory action in many enzyme and respiratory systems in the as yet unsuccessful search for specific effects that might be related to its psychodysleptic action. Its structure (Fig. 29-6) suggests that it may act as an analog-inhibitor of some catecholamine system in the brain. One suggestion has been that, in schizophrenia, dopamine is doubly methylated to 3,4-dimethoxyphenethylamine. The resemblance of this hypothetical metabolic product of a catecholamine to mescaline, which is 3,4,5-trimethoxyphenethylamine, is clear enough, but there is no evidence for formation of the latter in the animal organism.

The dose of mescaline required to elicit hallucinations is much larger than that

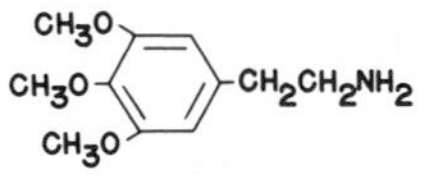

Mescaline

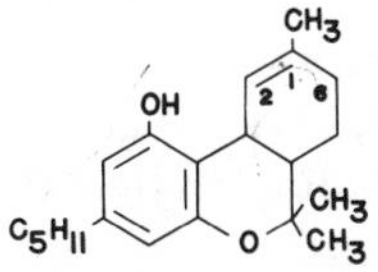

Δ^1-Tetrahydrocannabinol

FIGURE 29-6. Mescaline and tetrahydrocannabinol. The monoterpenoid numbering for the tetrahydrocannabinol has been used (see Mechoulam, 1970). The compound shown corresponds to Δ^9-THC in the formal numbering system.

needed for many of the other psychodysleptics. It is quite possible that some quantitatively minor metabolite of the drug is responsible for its effects.

Oral administration of labeled mescaline to humans shows that the compound is readily absorbed. It is also rapidly metabolized, for its half-life (as measured by urinary excretion of labeled products) is about 6 hr. Most of the tracer is excreted by the end of the second day. Among the metabolites detected are the carboxylic acid corresponding to mescaline, *N*-acetylmescaline, 3,4-dimethoxy-5-hydroxyphenethylamine, and the *N*-acetyl derivative of the last compound. Other metabolites remain to be characterized. In the rat, demethylation of the 4-methoxy group may also occur. The deamination of mescaline is sensitive to MAO inhibitors.

TETRAHYDROCANNABINOL

As must be evident from the discussion of the nitrogen-containing psychodysleptics, research on the mechanisms of action of hallucinogens is difficult enough when chemically pure agents are available, but it is much more difficult in the case of a galenic type of agent such as marijuana. Only recently has the chemistry of *Cannabis* been clarified with the identification of its most important psychodysleptic constituent as Δ^1-tetrahydrocannabinol (THC) (see Fig. 29-6). It is effective in man in an oral dose of 50 to 200 μg per kilogram body weight and, by inhalation (smoking), in a dose of 25 to 50 μg per kilogram. The isomeric Δ^3-THC was for many years the standard preparation for biological testing. It is said to be orally effective in a dose of about 2 mg per kilogram, but a dose of 0.4 mg per kilogram administered by smoking does not seem to have any action. Indeed, there is even some question about its activity altogether. Synhexyl, a synthetic compound, is the next higher homolog of Δ^3-THC (its side chain being C_6H_{13}); it is somewhat more active than Δ^3-THC by the oral route. The immediately active psychotomimetic substance may be a metabolite of THC.

The label of radioactive Δ^1-THC is eliminated slowly in the rat. Most of it passes into the feces and the remainder into the urine. By contrast, the compound is rapidly metabolized in the rabbit, the half-life of tritiated THC in blood being only 7 to 16 min, and the major route of excretion is in the urine. Whereas much of the urinary material in the rat tends to be nonpolar, the products are polar in the rabbit. The major metabolite in the rabbit is 7-hydroxy-Δ^1-THC, a compound that is also formed from THC by incubation with a liver enzyme; it, too, is biologically active. The isomer of Δ^1-THC that has a double bond between C-6 and C-1 (instead of between C-1 and C-2) is also metabolized in the liver to a pharmacologically active product, in this case 7-hydroxy-$\Delta^{1,6}$-THC.

ACKNOWLEDGMENT

Research in the author's laboratory in this field is supported by grants of the Medical Research Council of Canada.

REFERENCES

ORIGINAL SOURCES AND SPECIAL TOPICS

1. Da Prada, M., and Pletscher, A. On the mechanism of chlorpromazine-induced changes of cerebral homovanillic acid. *J. Pharm. Pharmacol.* 18:628, 1966.
2. Dixon, A. K. Evidence of catecholamine mediation in the 'aberrant' behavior induced by lysergic acid diethylamide (LSD) in the rat. *Experientia* 24:743, 1968.
3. Horita, A., and Hamilton, A. E. Lysergic acid diethylamide: Dissociation of its behavioral and hyperthermic actions by DL-alpha-methyl-p-tyrosine. *Science* 164:78, 1969.
4. Jacobsen, E. The comparative pharmacology of some psychotropic drugs. *Bull. W.H.O.* 21:411, 1959. (This is a useful monograph on psychopharmacological actions, testing in animals, and problems of drug classification.)
5. Knoll, J., Vizi, E. S., and Knoll, B. Pharmacological studies on para-bromomethamphetamine (V-111) and LSD. *Acta Physiol. Acad. Sci. Hung.* 37:151, 1970.
6. Nybäck, H., Sedvall, G., and Kopin, I. J. Accelerated synthesis of dopamine-C^{14} from tyrosine-C^{14} in rat brain after chlorpromazine. *Life Sci.* 6:2307, 1967.
7. Roos, B. E. Effects of certain tranquilizers on the level of homovanillic acid in the corpus striatum. *J. Pharm. Pharmacol.* 17:820, 1965.
8. Sharman, D. F. Changes in the metabolism of 3,4-dihydroxyphenylethylamine (dopamine) in the striatum of the mouse induced by drugs. *Br. J. Pharmacol.* 28:153, 1966.
9. Spector, S. Regulation of Norepinephrine Synthesis. In Efron, D. (Ed.), *Psychopharmacology, a Review of Progress, 1957–1967.* Washington, D.C.: Public Health Service Publication No. 1836, 1968.
10. Von Brücke, F. T., and Hornykiewicz, O. *The Pharmacology of Psychotherapeutic Drugs.* New York: Springer-Verlag, 1969. (E. B. Sigg has translated the authors' *Pharmakologie der Psychopharmaka.* This small book contains much of interest to the neurochemist. The reference in the opening paragraph of this chapter is in the German text only.)
11. World Health Organization. *Research in Psychopharmacology; Report of a WHO Scientific Group.* WHO Tech. Rep. Ser. No. 371, 1967. (The Scientific Group reviews the title subject and makes recommendations for advancing the field.)

BOOKS AND REVIEWS

Ban, T. A. *Psychopharmacology.* Baltimore: Williams and Wilkins, 1969.

Costa, E., and Garattini, S. *International Symposium on Amphetamines and Related Compounds.* New York: Raven Press, 1970.

Efron, D. (Ed.). *Psychopharmacology, a Review of Progress, 1957–1967.* Washington, D.C.: Public Health Service Publication No. 1836, 1968.

Mechoulam, R. Marihuana chemistry. *Science* 168:1159, 1970.
Schümann, H. J., and Kroneberg, G. (Eds.). *New Aspects of Storage and Release Mechanisms of Catecholamines.* (The Second Bayer Symposium.) New York: Springer-Verlag, 1970.
Usdin, E., and Efron, D. H. *Psychotropic Drugs and Related Compounds.* Washington, D.C.: Public Health Service Publication No. 1589, 1967.
World Health Organization. *Biochemistry of Mental Disorders; Report of a WHO Scientific Group.* WHO Tech. Rep. Ser. No. 427, 1969.

30

Biochemical Theories of Mental Disorders

Eli Robins and
Boyd K. Hartman

FIFTY OR SIXTY YEARS AGO, a substantial proportion of the beds in psychiatric hospitals was filled with patients suffering from one of two illnesses: general paralysis of the insane or pellagra. With the discovery that the former was an infectious disease (syphilis of the brain) and the latter a vitamin deficiency, these two psychiatric problems have become virtually nonexistent. Despite these triumphs of biological psychiatry, interest in biochemical approaches to psychiatric illness waned in the following decades. However, in the mid 1950s, with the advent of an effective "antipsychotic" medication, the phenothiazines, there was a major reawakening in biological psychiatry. The development of this group of drugs perhaps has done more to change psychiatry than has any other event. Only 15 years ago, straitjackets, restraints, padded cells, and cold baths were in general use in psychiatric care, but in most places they are now of little more than historical interest. The success of the phenothiazines intensified research into other drugs which might alter the course of psychiatric illness. A natural corollary was that if drugs could alter psychiatric illnesses, a thorough understanding of their pharmacological mechanisms might elucidate a biochemical basis for the illnesses.

The formulation of biochemical theories of psychiatric illness is largely dictated by available knowledge of neurochemistry and neuropharmacology. Thus, progress toward an understanding of a biochemical basis for psychiatric illness depends on understanding brain function. For example, it is no accident that many current biochemical theories in psychiatry are concerned with biogenic amine metabolism, for which extensive methodology has been developed, while acetylcholine has been largely neglected as a basis for such theories because of methodological difficulties in the study of its metabolism.

Given the relatively primitive state of neurochemistry at the present time, the theories to be discussed are more aptly classified as working hypotheses, which probably will require drastic revision as more basic information becomes available. The main reason for formulating such hypotheses is to give direction to research and suggest further experiments. The major disadvantage of these hypotheses is that they may be reiterated so frequently that they become more or less accepted in the absence of adequate evidence. Once this occurs, such hypotheses may actually do more harm than good by causing a premature channeling of efforts into a narrow area of essentially unrewarding research.

It is frequently thought that all biochemically oriented psychiatrists are committed to the belief that psychiatric problems result from biochemical abnormalities, such as a malfunctioning enzyme or circulating neurotoxin. This view has led to the categorization of psychiatric problems as either "organic" or "functional" illnesses. "Organic" in this sense refers to those illnesses in which a metabolic defect exists; "functional" illness is presumed to be due to psychological or environmental factors. The presupposition is that unless one assumes a "biochemical defect" to be the basis of a disorder, it is not profitable to examine that disorder biochemically. This restrictive view overlooks the basic function of the nervous system, which is to allow the organism to respond to events in the environment. Based on information presently available, the only way the nervous systen can carry out this function is through biochemical reactions. The stumbling block remains our ignorance about the biochemical basis of brain function. Once these processes are understood, the distinction between "organic" and "functional" psychiatric problems may disappear.

This chapter discusses three psychiatric disorders about which a substantial amount of biochemical data has been collected: affective disorder, schizophrenia, and anxiety neurosis. No attempt will be made to promote any particular hypothesis; at present the evidence is too inconclusive to warrant bias in one direction or another. Many problems arise in collecting reliable patient data; a high degree of variability in results is inevitable. This does not justify, however, lowering one's standards with regard to what constitutes acceptable evidence. The reverse seems indicated. Given the presence of difficulties in making reliable observations, one should be especially cautious of the tendency to overinterpret and overspeculate.

AFFECTIVE DISORDERS

Affective disorder and manic-depressive illness are essentially synonymous. Patients may have an excessively low mood (depression) or an excessively high mood (mania). Depressed patients usually have symptoms of sadness, hopelessness, low self-esteem, often accompanied by sleep and appetite disturbances, difficulty in concentrating, loss of interest, and slowing of thought and activity (psychomotor retardation). When severe, these symptoms are frequently associated with thoughts of suicide. On the other hand, patients with mania feel excessively elated and usually have high self-esteem. Such patients may speak rapidly, are overactive, and are easily distracted. As in depression, there may be appetite and sleep disturbances and difficulty in concentration.

There is evidence that affective illness is genetically transmitted. Diseases which are genetically transmitted presumably have a biochemical basis, so affective illness is of particular interest to the neurochemist.

Numerous systems have been devised for subclassifying affective disorders beyond the categories of mania and depression. None of these categories, however, has proved wholly satisfactory in producing homogeneous groups with regard to the natural history of the illness or response to treatment. Nevertheless, it is generally agreed that the criteria for patient selection are of high importance in obtaining reproducibility in any study of psychiatric patients. Thus, any report, regardless of the particular diagnostic nomenclature used, is of dubious value unless specific and clearly defined criteria of selection are given.

There are currently four main areas of interest with regard to possible biochemical changes in affective disorder. These are biogenic amines, steroid hormones, electrolyte and water distribution, and glucose metabolism. Two additional areas, cyclic AMP and tyrosine tolerance, will be discussed.

BIOGENIC AMINES

Norepinephrine (NE) is the peripheral neurotransmitter of the sympathetic nervous system that is associated with response to strongly emotional situations (fright, fight, flight). The interest in NE and other amines with regard to affective illness came about mainly from two observations made in patients.

The first involved antidepressant drugs such as MAO inhibitors (tranylcypromine, phenelzine) [162, 163] and tricyclic antidepressants (imipramine and amitriptyline). Both types of drugs affect central monoamines in animals.

The second observation involves drug-induced depressions. Reserpine, a drug used for hypertension, depletes brain monoamines in animals. It has been reported to cause depressions in a small proportion (about 15 percent) of patients treated with it.

Based on these observations, hypotheses have been formulated implicating the biogenic amines in affective disorders. Attempts have been made to extend and strengthen these hypotheses by various animal studies. It has been proposed, for example, that reserpine-sedated animals represent an experimental model of depression (reserpine was first used as a tranquilizer). The major problem with the "reserpine model" of depression is the arbitrary equating of sedation with depression. When reserpine is given to animals or humans, they become sedated. It is likely that the sedation in both animals and humans is analogous. Thus, experiments on reserpine-sedated animals would seem more likely to uncover the biochemical basis of the tranquilizing effect of reserpine than the cause of depression. Other drugs known to affect sedation-excitation level, such as amphetamines, cocaine, and morphine, also affect monoamine metabolism in animals, but the possible relation of these drugs to affective disorder is remote.

Another example in which animal studies have been used to strengthen proposed hypotheses concerning depression is the investigation of the mechanism of action of the tricyclic antidepressants. Their action is frequently attributed to the ability to block reuptake of NE into presynaptic neurons. However, other compounds inhibit reuptake of catecholamines, but have no antidepressant action. For example, chlorpromazine, which, if anything, tends to aggravate depression, also blocks reuptake of NE. Furthermore, the pharmacological effects of all of these drugs are much more complex than was originally believed. For example, imipramine, in addition to blocking reuptake of NE, has an anticholinergic effect and inhibits 3',5'-cyclic AMP phosphodiesterase. Thus, as more information becomes available, the relation of antidepressant activity to pharmacological mechanisms becomes increasingly ambiguous. Ultimately, the validity of a hypothesis concerning human affective illness can only be established with observations made on humans with the illness. The following discussion, therefore, emphasizes data concerned with human investigation.

Catecholamines

The catecholamines are generally (although not unanimously) considered to be excitatory transmitters. The "catecholamine hypothesis" [7, 28] is that in mania there is an excess, and in depression a deficiency, of catecholamines at functionally important CNS receptors. Techniques are not available for measuring active catecholamines at "functionally important" receptors in the brains of animals, not to mention humans with affective disorders, as would be required to test the proposed hypothesis directly. However, if such an excess or deficiency did exist in patients, it might be reflected indirectly by alterations in peripheral amines or metabolites.

Catecholamines and their metabolites are measurable in urine, but they are mostly of peripheral, not of central, origin. It is often assumed that central changes parallel peripheral alterations, although the validity of this assumption is unknown.

There have been numerous reports of alterations in unmetabolized urinary catecholamines in mania and depression, as well as in other mental disturbances. The most consistent finding is that catecholamines increase in manic states. This change, however, may be attributed to increased physical activity. Furthermore, because unchanged catecholamines represent only a very small and variable percentage of total catecholamine metabolites, more recent studies have included measurement of metabolic products in the hope of obtaining a more reliable index of total catecholamine metabolism. Normetanephrine (NMET) is currently assumed to be the metabolite most representative of noradrenergic activity. NMET therefore presumably would show the highest correlation with alterations in mood. Vanillylmandelic acid (VMA), the other major metabolite of NE, is presently thought to be formed by oxidation intracellularly and, presumably, does not act on receptor sites. Its relationship to mood is, therefore, not predicted by the catecholamine hypothesis.

Bunney et al. [28, 29] measured eight catecholamine metabolites, including dopamine, in 16 depressed patients, half of whom were psychotically depressed.

Urinary excretion of NE, epinephrine, and VMA were significantly increased in psychotic patients, compared with neurotically depressed patients or normal controls. Neither NMET nor any other metabolite was significantly different in any of the groups. The direction of change of NE and lack of change of NMET is contrary to that predicted by the catecholamine hypothesis.

Rosenblatt and Chanley [137] estimated the urinary excretion of NE metabolites by administering ^{3}H-NE. Amine-containing metabolites of NE (chiefly NMET) migrate electrophoretically (under acidic conditions), whereas oxidized metabolites (chiefly VMA) do not. The ratio of migrating radioactivity and nonmigrating radioactivity thus gives the ratio of amine-containing to oxidized metabolites (N/O). Generally higher urinary N/O ratios were found in patients with severe depression than in patients with less severe depression, other psychiatric diagnoses, or normal controls. These higher ratios were due primarily to a relative increase in NMET. This is contrary to the direction predicted by the catecholamine hypothesis. In the 2 patients treated with ECT (electroconvulsive therapy), the N/O ratio tended to return to normal levels. In contrast, imipramine and chlorpromazine tended to increase the N/O ratio, even in patients who improved on treatment. Because NE does not cross the blood-brain barrier, the N/O ratio probably reflects only the results of peripheral metabolism.

Schildkraut et al. [146, 147] studied 5 depressed patients prior to, during, and after treatment with imipramine. Urinary excretion of VMA and NMET were measured throughout the test. VMA fell in response to treatment, but returned to pretreatment levels after cessation of treatment. NMET, on the other hand, tended to increase with clinical response to treatment. Both a rise in NMET and a fall in VMA are consistent with the postulated pharmacological action of imipramine, which apparently blocks reuptake of NE, resulting in less oxidative metabolism by mitochondrial MAO. More important from the standpoint of the mechanisms of illness is that after treatment was stopped, the mean secretion rates of NMET generally did not return to the lower secretion rates observed when patients were depressed, and presumably represent normal levels. The changes in NMET are in the direction predicted by the catecholamine hypothesis.

Greenspan et al. [76] longitudinally studied 6 patients who had hypomania – an abnormal elevation of mood, but somewhat less severe than mania – as well as periods of normality (4 patients) and depression (2 patients) with regard to urinary excretion of five catecholamines and metabolites. In general, secretion rates tended to be higher in hypomania than in normal or depressed states. In this study, observed differences are also in the direction predicted by the catecholamine hypothesis.

Bunney et al. [30] reported mean urinary excretion rates of NE, epinephrine, and dopamine in 3 patients during "switches" from depression to mania. No significant differences were found in the metabolites measured.

Messiha et al. [108] measured urinary excretion of dopamine (both free and total) and VMA in 20 normal controls and compared these results with those in 7 manic and 6 depressive patients before and after lithium therapy. They found that both free and total dopamine were not only significantly increased in mania, but were also well outside normal limits. Lithium treatment returned excretion of dopamine to within normal limits. In depressed patients, dopamine excretion was also significantly elevated but was not altered by lithium treatment. Excretion of VMA was unaltered in both groups before and after treatment.

Interpreting data on urinary excretion products is difficult because of the impossibility of separating PNS from CNS effects. Moreover, even when changes are significant, they are usually small, and results differ from laboratory to laboratory. Additional studies are needed to establish a consistent change in urinary catecholamine excretion in any illness, with the possible exception of increased excretion of NE and epinephrine in manic patients. Given the difficulty of separating peripheral from central metabolites, it is dubious that urinary studies will confirm or disprove the role of catecholamines in affective disorders.

It has recently been reported that 3-methoxy-4-hydroxyphenylglycol (MHPG) may be a major metabolite of NE in the brain in some animals. If this is also true in humans, it may offer a way of deducing central adrenergic activity from data on urinary metabolites. Maas et al. [97] have reported that urinary MHPG was significantly decreased in 16 depressed patients as compared to 11 controls (MHPG in μg per milligram creatinine: depressed 0.88 ± 0.084, controls 1.16 ± 0.073, $P < 0.05$). This finding is consistent with the catecholamine hypothesis. No significant differences in NMET and metanephrine (MET) were observed in these two groups.

Analysis of brain tissue from autopsy and of cerebrospinal fluid (CSF) offers more direct means of obtaining information about brain catecholamine metabolism. Bourne et al. [24] measured NE in hindbrains of 21 suicide patients (most of whom presumably had affective disorders) and compared the results to 27 controls. No significant difference was found (controls: 439 ng per gram of tissue; suicides: 444 ng per gram). Pare et al. [120] were unable to show differences in brains of suicide cases and coronary death controls in the hypothalamus (NE) and in the caudate nucleus (dopamine).

Treatment of depressed patients with L-dihydroxyphenylalanine (L-DOPA) seems to provide a possible means of testing the catecholamine hypothesis. DOPA crosses the blood-brain barrier and presumably would correct the hypothesized deficiency of NE in affective disorders. Early trials with relatively small doses of DOPA were uniformly unsuccessful in alleviating depression. Recently, with the administration of higher doses of L-DOPA in treatment of Parkinson's disease, interest has been revived in the use of this drug for treating affective illness. It is too early to say whether higher doses will be more efficacious than lower doses in the treatment of depression.

In two studies [71, 72] where high doses of L-DOPA, with or without a peripheral DOPA decarboxylase inhibitor, were used, the proportion of patients who improved (25 to 33 percent) was not very different from that usually observed as improved in depressed placebo-treated control groups. In the first study, homovanillic acid (HVA), a metabolite of dopamine, was measured in CSF of patients before and during L-DOPA treatment. High elevations of HVA were found in two patients treated with high doses of L-DOPA, and moderate elevations were found in patients treated with lower doses of L-DOPA plus a peripheral DOPA decarboxylase inhibitor. In this study, where CSF metabolites were measured, no correlation was reported between elevation of CSF levels of HVA and clinical response. In another study in which L-DOPA was not given, Denker et al. [50] found no significant difference in CSF levels of HVA between depressed patients and controls.

Abnormal erythrocyte enzymes in affective disorders have also been cited. In one report [39], erythrocyte catechol-*O*-methyl transferase (COMT) was found to be significantly decreased in 36 depressed women compared to 19 controls. Histamine-*N*-methyl transferase was significantly increased in 11 depressed women versus 7 controls. No significant difference was found in comparisons of depressed men with controls. The low levels of erythrocyte COMT apparently did not change upon clinical recovery. The authors believed that these findings suggested "a possible sex-hormonal association." At present, the relationship between erythrocyte COMT activity and brain COMT activity is not known.

Table 30-1 summarizes the results of studies on urinary excretion of catecholamines. No study showed significantly lower catecholamines or metabolites in manic patients; elevated metabolites were frequently observed in both mania and depression. Adherents of the catecholamine hypothesis generally attribute the elevations in depression to a peripheral sympathetic response to agitation or stress. This argument is a double-edged sword, for, if true, it casts doubt on the original assumption that brain metabolism is reflected by peripheral catecholamine metabolism and urinary excretion of metabolites.

Table 30-1 shows evidence consistent with the catecholamine hypothesis, inconsistent with the hypothesis, or uncertain with regard to the hypothesis. The correlation rests on the assumption that urinary excretion of metabolites reflects some central catecholamine metabolism, even though these metabolites are primarily peripheral in origin. Moreover, the variant of the catecholamine hypothesis receiving most attention involves NE metabolism and implicates only three of the seven compounds listed in Table 30-1: NE itself, NMET, and MHPG. Of the other four, two are related to epinephrine (E and MET), and two (VMA and DA) have not been studied sufficiently to define their relationship to NE metabolism, especially in intact man. Nine studies are reported in Table 30-1, including 281 determinations of the seven constituents. In mania there were 22 determinations consistent with the catecholamine hypothesis and 6 inconsistent; in depression there were 21 consistent with the catecholamine hypothesis and 71 that were inconsistent. The remaining 161 determinations were uncertain with regard to the hypothesis. On the basis of these reports, the validity of the catecholamine hypothesis remains unproved.

In 1958, it was shown that excretion of both NE and epinephrine were higher in mania than in depression [155]. Dopamine excretion was unchanged. (These results could not be included in Table 30-1 because there were no baseline values reported for normals.)

Indoleamines

Although theories involving indoleamines in affective disorders generally have been less popular than those concerning catecholamines, there is a substantial number of studies that implicate altered metabolism of indoleamines in affective illnesses.

TABLE 30-1. Urinary Excretion of Catecholamines and Their Metabolites in Affective Disorders and Its Relation to the "Catecholamine Hypothesis"

Substance Measured[a]	Manic[b]		Depressed		
	Increased	Unchanged	Increased	Unchanged	Decreased
NE	+ (*N*=16; [18])	– (*N*=6; [76])	– (*N*=16; [29])	– (*N*=16; [18])	...
NMET	+ (*N*=6; [76])	...	– (*N*=7; [137])	– (*N*=16; [29]) – (*N*=16; [97])	+ (*N*=5; [146, 147])
MHPG	...	...	...	...	+ (*N*=16; [97])
VMA	...	0 (*N*=6; [76]) 0 (*N*=7; [108])	0 (*N*=16; [29])	0 (*N*=6; [108]) 0 (*N*=5; [146, 147])	...
E	0 (*N*=16; [18]) 0 (*N*=6; [76])	...	0 (*N*=16; [29])	0 (*N*=16; [18])	...
MET	...	0 (*N*=6; [76])	...	0 (*N*=16; [29]) 0 (*N*=16; [97])	...
DA	0 (*N*=7; [108])	...	0 (*N*=6; [108])	0 (*N*=16; [29])	...

[a]Data supporting "catecholamine hypothesis" indicated by +; uncertain, 0; and inconsistent with the hypothesis, –. *N* is the number of patients in the study; reference numbers in brackets.

NE = norepinephrine
NMET = normetanephrine
MHPG = 3-methoxy-4-hydroxyphenylglycol
VMA = vanillylmandelic acid
E = epinephrine
MET = metanephrine
DA = dopamine

[b]None of the substances measured was reported to decrease in manic patients.

Several studies show significant decreases in urinary or serum indoles in depression. Rodnight [6] reported decreased urinary excretion of serotonin (5-HT) and tryptamine in depressed patients ($P < 0.05$); he also found similar changes in schizophrenics. Van Praag [162, 163] found decreased 5-hydroxyindoleacetic acid (5-HIAA) – a major metabolite of 5-HT – in urine of depressed patients. Haskovec and Rysanek [78] found imipramine treatment resulted in significant decreases in urinary excretion of 5-HIAA from treatment days 9 through 27. Sarai and Kayano [145] measured serum 5-HT in several groups of patients and found significant decreases in depressed patients ($N = 40$) as compared to normal controls ($N = 34$) or to recovered depressives ($N = 24$) treated with imipramine. Serum levels in manic patients ($N = 10$) were significantly increased over control levels; schizophrenic patients ($N = 13$) did not show a significant difference. They found that imipramine increased serum 5-HT levels significantly but that electroconvulsive therapy (ECT) had no effect. Studies of serum and urinary levels of indoles suffer from the same lack as do studies of urinary excretion of catecholamines: the inability to distinguish central from peripheral metabolites.

In contrast to the catecholamines, however, there are several studies in which significant differences in indoleamines have been reported in CSF or brain.

In three studies, the brain stem concentration of indoles was found to differ in depressed suicides versus control groups (Table 30-2). Shaw et al. [149] found decreased 5-HT concentrations. Bourne et al. [24] found no change in 5-HT but did find decreased 5-HIAA concentrations. Pare et al. [120] studied indoleamines as well as catecholamines (see above) in brains of suicide victims and coronary controls. They were unable to show a significant difference in 5-HIAA in the brain stem. There were significantly lower levels of 5-HT in the brain stems of the 23 suicides studied compared with the 15 controls.

The analysis of CSF also offers a means of studying metabolites derived from brain. There is some evidence that metabolites of biogenic amines cannot exit directly from brain into blood because of the blood-brain barrier, but must first pass into CSF and then into the blood.

TABLE 30-2. 5-HT and 5-HIAA Concentration in Human Brain Stem: Depressed Suicides vs. Controls

	Shaw et al. [149]		Bourne et al. [24]			Pare et al. [120]	
	N	5-HT (ng/g)	*N*	5-HT (ng/g)	5-HIAA (ng/g)	*N*	5-HT (ng/g)
Depressed suicides	11	250	16	211	1271	23[a]	310
Controls	17[b]	307	15[c]	218	1698	15[c]	350
Significance		$P < 0.05$		N.S.	$P < 0.05$		$P < 0.05$

[a]All suicides (without regard to diagnosis).
[b]Accident victims and death from acute illness.
[c]Coronary deaths.

Several laboratories have reported differences of 5-HIAA levels in CSF of patients with affective disorders (Tables 30-3 and 30-4). Ashcroft and Sharman [16] showed significantly decreased levels ($P < 0.001$) of 5-HIAA in 9 depressed patients as compared with 10 controls. Denker et al. [50] reported significantly lower levels in both depressed and manic patients as compared with normal controls and hospitalized nonpsychiatric patients. Van Praag [160] attempted to show that decreased CSF 5-HIAA in depression was due to decreased synthesis by blocking the exit of 5-HIAA from CSF with probenecid, an agent that blocks transport of organic acids. Prior to probenecid treatment, he confirmed the lower 5-HIAA levels reported by Ashcroft et al. [15] in depressed versus control patients. After probenecid treatment, the mean difference was not significantly less in the depressed group (Table 30-3). Despite the lack of difference in the means, the depressed group included 7 of 14 patients whose 5-HIAA concentration in CSF rose by less than 20 ng per milliliter after probenecid, whereas in the control group only 1 of 11 increased by so small an amount. This difference was statistically significant ($P < 0.05$ by Fisher's exact probability test).

TABLE 30-3. 5-HIAA Concentration in Human CSF: Depressed vs. Controls

Type of Patient	Ashcroft and Sharman [16]		Denker et al.[a] [50]		Van Praag et al. [160] 5-HIAA[b] (ng/ml)			
	N	5-HIAA[b] (ng/ml)	*N*	5-HIAA[b] (ng/ml)	*N*	Before Probenecid	After Probenecid	Difference
Depressed	9	13 ± 3	14	11 ± 8	14	17 ± 17	39 ± 25	22 ± 20
Controls	10	32 ± 3	34	37 ± 11	11	40 ± 24	74 ± 26	35 ± 18
Significance		$P < 0.001$		$P < 0.01$		$P < 0.05$	$P < 0.05$	N.S.

[a]Denker also measured CSF from 6 manic patients: 5-HIAA, 12 ± 8, ng/ml, $P < 0.001$.
[b]Values ± S.D.

The most extensive study of indole metabolites was that of Ashcroft et al. [15] (Table 30-4), who have also pointed out some of the problems in interpreting such data. Lumbar CSF levels of hydroxyindole compounds were measured in 32 depressed patients and compared with levels in 21 "controls," consisting of 6 alcoholics, 7 neurological patients, and 8 psychiatric patients. They confirmed significantly lower CSF indole levels in depressed patients ($P < 0.001$) and found that in 10 patients who recovered, the levels returned to those of controls. They reported other data, however, which complicate CSF findings. In 7 "acute" schizophrenic patients (diagnostic criteria were not given), the levels were significantly less than controls. Chronic schizophrenics, however, were not significantly different from controls. Levels in hypomania were not different from controls. Treatment with MAO inhibitors resulted in CSF levels of indoleamines that were not significantly different from controls, even though the patients were depressed. And, finally, it was found that CSF samples drawn from the ventricles or from the lumbar area after injection of air for encephalography were considerably higher than samples drawn from the lumbar area in the standard manner. Their interpretation was that

TABLE 30-4. Concentrations of 5-Hydroxyindole Compounds in Human CSF

Type of Patient or Site of Sampling[a]	*N*	ng/ml ± S.D.	Significance Compared to Controls
Controls (neurological, psychiatric, and alcoholic)	21	19.1 ± 4	. . .
Depressed, imipramine and no treatment	32	10.3 ± 4	$P < 0.001$
Treated with MAO inhibitors (still depressed)	6	21.7 ± 3.5	N.S.
Depressed	10	11.9 ± 5	
After recovery		16.5 ± 4.2	$P < 0.02$[b]
Hypomanic	4	17.5 ± 8.6	N.S.
Acute schizophrenia	7	10.9 ± 2	$P < 0.01$
Chronic schizophrenia	7	16.4 ± 3	N.S.
Lumbar air encephalography	28	33.3 ± 11	$P < 0.001$
Ventricular drainage (patients had organic disease of CNS)	6	88.1 ± 23	$P < 0.001$

[a]Lumbar puncture unless otherwise noted.
[b]Significance of differences in depressed vs. recovered phase.
Source: [15].

there is a gradient of the concentration of metabolites from highest in ventricles to lowest in the lumbar area. This means that such variables as rate of mixing of fluid at various levels, which in part reflects physical activity, could affect concentrations at a given sampling site.

Precursor loading with tryptophan has been tried as a treatment for depression in an attempt to correct a hypothetical deficiency in indoleamines. The reports on the clinical use of these compounds generally have been somewhat more positive than on the use of DOPA for depression. Two reports [44, 46] have indicated that tryptophan and MAO inhibitors together have a greater antidepressant effect than either alone, and it has been claimed that this combination has a potency similar to tricyclic antidepressants [41] or ECT [46]. Carroll et al. [35] were unable to confirm any significant antidepressant activity for tryptophan alone when compared to ECT in a group of severely depressed patients.

p-Chloro-*N*-methylamphetamine has been reported by Van Praag [161] to have moderate antidepressant properties. Also, this drug reportedly influences overall indoleamine metabolism in animals, possibly by releasing 5-HT from storage sites, without influencing catecholamines. However, no significant differences in urinary excretion of 5-HT or 5-HIAA were observed in depressed patients treated with the drug before or after recovery, so the implications of these experiments with regard to amines and depression are not clear.

As a whole, the indoleamine data are somewhat more consistent than are the catecholamine data. The evidence tends to support a low level of 5-HIAA in the CSF of depressed patients. The probenecid study indicates this may be the result of decreased synthesis of 5-HT in some patients. At present, there is insufficient evidence to support a specific biochemical defect to account for these differences.

Dewhurst proposed a different system in which biogenic amines are divided into "excitor" amines – tryptamine and phenylethylamine – and "depressant" amines – epinephrine and NE. These amines act in turn on hypothetical excitant and depressant receptors. 5-HT is considered to be an intermediate amine which can act on both excitant or depressant receptors. The amines are grouped on the basis of their relative hydrophilic or lipophilic character (i.e., the presence of greater or lesser degree of hydroxyl substitution on phenylethylamine or indolylethylamine units). Dewhurst, although noting major flaws in the view that catecholamines are excitants, provided even fewer data to support the hypothesis he proposed, especially with regard to affective illness. His theory led to the use of methysergide, a 5-HT antagonist, in treatment of mania. Dewhurst believed that methysergide was a specific blocker of excitant receptors, but preliminary reports that methysergide is effective in treatment of mania [51, 79] have not been confirmed in several other laboratories [42, 57, 101].

STEROID HORMONES

Steroid hormones are known to affect mood and behavior. Mood changes sometimes occur during steroid therapy and Cushing's syndrome, and in women mood fluctuations are believed to be related to cyclic changes in steroid sex hormones during the menstrual cycle, pregnancy, and menopause. Steroids also increase in response to various stresses. These observations have led to attempts to relate steroid metabolism to affective disorders. Most often this consists of measuring steroids when the patient is ill and again upon recovery.

In eight studies (Table 30-5), prerecovery and postrecovery morning plasma cortisol levels (the major glucorticoid in humans) were measured. Generally, morning cortisol levels were higher when patients were ill than after recovery. This was found in seven of the eight studies, but in only one study was the difference significant [26]. The only study in which a significant elevation during illness was reported represented only 13 of 93 patients. In only one study [62], the mean of the morning plasma cortisol level in ill patients increased, but not significantly. In this latter study, however, plasma cortisol levels measured at other times during the day showed a higher level in ill than in recovered depressed patients. Sachar [141] noted that, although mean plasma cortisol levels were higher when patients were ill, in 3 of 10 patients there was a lower level during illness than upon recovery.

Table 30-6 summarizes five studies in which 24-hr adrenal cortical activity was estimated by measuring various urinary metabolites. Four classes of compounds were measured: 17-hydroxycorticosteroids (17-OHCS), 17-ketogenic steroids (17-KGS), 11-hydroxycorticosteroids (11-OHCS), and 17-ketosteroids (17-KS).

TABLE 30-5. Morning Plasma Cortisol Levels in Depressed Patients vs. Recovered Patients

Study	N	Mean Plasma Cortisol (μg/100 ml)		Significance	Method
		Depressed	Recovered		
Carroll et al. [34]	27	21.3	18.6	N.S.	Mattingly
Carroll [33]	16	19.4	16.8	N.S.	Mattingly
Fullerton [62]	9	12.5[a]	13.0[a]	N.S.	Mattingly
Butler and Besser [31]	3	27.3	12.7	No *P* given	Mattingly
Rimon et al. [134]	7	24.9[b]	22.5[b]	N.S.	Modified Porter-Silber
Sachar [141]	10	19.6	16.7	N.S.	Murphy
Bridges and Jones [26][c]	13	23.2	18.4	0.02	Zenker Bernstein
Gibbons [65]	8	17.6	10.6	N.S.	Peterson et al.

[a]Serum 17-OH corticosteroids given in μg/sq m.
[b]Plasma 17-OH corticosteroids.
[c]This study included hospitalized controls. The cortisol levels of these controls were higher than the depressed group both before and after recovery from depression.

TABLE 30-6. 24-Hour Urinary Excretion of Corticosteroids in Depressed vs. Recovered Patients

Study	Compound	N	Mean 24-hr Urinary Excretion (mg/24 hr)		P	Method
			Depressed	Recovered		
Fullerton et al. [62]	17-OHCS	9	4.7[a]	3.6[a]	N.S.	Porter-Silber
Sachar [141]	17-OHCS	20	5.7	4.8	0.05	Glenn, Nelson
	17-KGS	7[b]	7.6	7.2	N.S.	Rutherford, Nelson
Kurland [91]	17-OHCS	10	8.5	5.8	0.01	Porter-Silber
	17-KGS	10	39.1	23.6	0.01	Gibson, Norymberger
	17-KS	10	12.1	8.2	0.01	Norymberger
Gibbons [66]	17-OHCS	6	8.1	4.9	0.01	Peters et al.
	11-OHCS	6	0.223	0.135	0.01	Mattingly
	Cortisol[c]	6	23.0	14.4	0.05	Modification of Cope-Black
Gibbons [65]	CS[c]	6	3.5	2.4	0.02	isotope
	Cortisol	14	25.5	14.1	0.001	dilution

[a]17-OHCS mg/24 hr/sq m.
[b]These 7 patients are a subgroup contained in the 20 patients cited above.
[c]Values of cortisol and corticosterone (CS) *secretion,* mg/day, estimated by isotope dilution.

All determinations except that of 17-KS predominantly measure cortisol or its principal metabolites. The measurement of 17-KS includes androgens and their metabolites, whether of adrenal or gonadal origin. One investigator, Gibbons [65, 66], measured specific compounds by estimating total cortisol and corticosterone secretion during a 24-hr period, using an isotope dilution technique.

In contrast to studies measuring morning levels of cortisol in plasma, in which only one of eight showed a significant increase, four of five studies reported significant increases in 24-hr adrenal cortical activity during ill periods. The discrepancy may be explained on the basis that differences in plasma cortisol levels at only one isolated time are too small to reach significance, but that over an entire day, the total pattern becomes discernible. The relationship of these changes to the biochemistry of affective illness is unclear. Sachar [141], who investigated the largest number of patients for changes in urinary metabolites, noted that 7 of 20 patients had increases rather than the expected decreases in 24-hr 17-OHCS upon recovery, and there seemed to be an even greater decrease in steroid excretion in response to the first week of hospitalization than to recovery. The implication was that adaptation to the stress of hospitalization accounted for the greatest part of the differences observed in recovery. In a more recent study, Sachar et al. [142] used isotopes to measure cortisol secretion rates in 16 depressed patients before and after recovery. The higher rates of cortisol production during illness compared with those on recovery from depression correlated well (0.89, $P < 0.001$) with changes in scores on clinical rating tests for symptoms reflecting anxiety and emotional arousal. Other symptoms of depression did not correlate with cortisol production.

Gibbons [1] stated in a review, "Since even a modest increase in adrenal cortical activity is by no means invariable in depression, it is highly unlikely to be responsible for the affective disturbance and much more likely to be a physiological expression of the disturbance, a persistent variant of the usually transient increases in adrenocortical activity that accompany so many states of emotional arousal." This statement is supported by reports [4, 21] in which changes in steroid metabolism of similar magnitude are found in stressful situations. Furthermore, significant elevations in steroid levels have been reported in other psychiatric disorders – in schizophrenia [143] and in acutely disturbed patients in hospital emergency rooms [21] – suggesting that the changes are nonspecific.

Assuming that depressed patients have abnormally high corticosteroid levels, it has been postulated that depressed patients do not demonstrate corticosteroid suppression to dexamethasone, analogous to patients with Cushing's syndrome (hyperadrenocorticism). This possibility has been studied by three investigators. Butler and Besser [31] investigated three patients, two of whom did not suppress when ill; two of them, however, did not suppress when well, either. The other two reports [33, 34] describe 40 patients. Of these 40 patients, 18 followed the postulated pattern of nonsuppression when ill but suppression when well; the remaining 22 patients did not follow this pattern. In a control group of 22 normal individuals, 3 failed to suppress normally. The authors noted a high degree of correlation between nonsuppression and severity of depression.

With regard to dexamethasone suppression, there appears to be a lack of specificity for the affective state. In a recent report [22], 18 airmen in basic training who had anxiety symptoms presumably related to stress were studied in comparison with 30 airmen controls. The investigators found a significant lack of dexamethasone suppression in the patients with anxiety compared with the control group (mg 17-OHCS per 24-hr mg creatinine: patients 2.51, controls 1.46, $P < 0.05$). It would appear that nonspecific stress factors can affect dexamethasone suppression as well as plasma steroid levels and 24-hr urinary excretion of steroids.

Attempts have also been made to relate changes in steroid metabolism in affective disorders to (1) severity of illness, (2) the presumed cause (endogenous vs. reactive), (3) type of symptoms (psychotic vs. neurotic), and (4) periods of stress during a particular episode of illness. In general, the results indicate that steroid secretion is elevated during periods of stress and that the degree of elevation is related to the severity of stress, but that the effects are too nonspecific and variable to be of either diagnostic or prognostic value as a laboratory test in depression.

Several investigators have tried to link steroid metabolism with indoleamine metabolism through the demonstrated fact that steroids in sufficient amounts induce tryptophan pyrrolase. This rather comprehensive hypothesis is outlined in detail in a review by Curzon [49]. Increased steroids would result in the induction of tryptophan pyrrolase; increased activity of this enzyme would result in excessive metabolism of tryptophan by this route, leading to a deficiency of substrate for production of 5-HT, and hence, a lack of 5-HT. Lower levels of central indoleamines or metabolites in depression and the apparent, but disputed, effectiveness of tryptophan in the treatment of depression are consistent with the hypothesis. However, the slight and inconsistent elevations in steroid metabolism that have been found to occur in depression rule against this mechanism being of any major importance in the etiology of most depressions: such variations in the levels of steroids are not great enough to affect tryptophan pyrrolase sufficiently to lead to a 5-HT deficiency.

WATER AND ELECTROLYTE METABOLISM

Some of the earliest attempts to discover a metabolic abnormality in affective disorders were investigations of concentrations of electrolytes in serum or balance studies. In general, such investigations were negative when well controlled. More recent investigations, using radioisotopes to estimate electrolyte flux and distribution in various fluid spaces, are possibly more promising than the earlier studies.

Sodium and Water

Chiefly studied have been sodium and water metabolism, with direct measurements of total body water, extracellular water, exchangeable sodium, and sodium concentration in serum.

Total body water (TBW) is evaluated by measuring the dilution of administered tritiated water, which presumably equilibrates with TBW. Two studies of depressed patients before and after clinical improvement showed an increase in TBW with improvement. Coppen and Shaw [43] showed an increase from 35.1 liters to 36.3 liters ($P < 0.05$) in 23 patients who improved with treatment. Hullin et al. [85] observed an increase in TBW from 30.4 liters to 33.1 liters ($P < 0.01$) in eight depressed patients who improved with ECT. The change occurred during treatment and persisted afterward. Coppen et al. [47] could not show a significant change in TBW in manic patients with improvement (39.2 liters to 35.6 liters). Lithium ion was shown to increase TBW in 20 psychiatric patients [45] (before Li^+, 33.4 liters; after Li^+, 35.0 liters, $P < 0.001$).

Extracellular water (ECW) is estimated from the dilution of various substances which presumably do not significantly enter cells during the time of equilibration. The most common materials used are bromide ion and thiocyanate ion. An inherent problem in this method is that if treatment or illness alters the cellular permeability to these ions, their dilution at equilibrium will not accurately reflect the extracellular water. Coppen and Shaw [43] showed an increase in ECW (bromide space) from 15.2 liters to 15.7 liters ($P < 0.01$) in 23 depressed patients upon recovery. Hullin et al. [85] did not find a significant increase in the ECW (thiocyanate space) of eight depressed patients after improvement with ECT, although they measured a transient rise during treatment (12.9 liters to 14.5 liters, $P < 0.02$). Coppen et al. [47] found no difference in ECW in improved manics. Lithium treatment did not significantly alter the ECW in two studies [17, 45].

The exchangeable sodium (Na_E) is obtained by measuring the dilution of radioactive sodium by body sodium after equilibration (usually over a 24-hr period). Two studies have shown significant decreases in Na_E upon recovery. Gibbons [64] found a decrease from 2551 meq to 2343 meq ($P < 0.001$) upon improvement in depression. There was no significant difference in unimproved patients. Coppen and Shaw [43] were unable to confirm Gibbons' results in 23 depressed patients who recovered (2690 meq in depressed patients vs. 2590 meq in improved). Coppen et al. [47] showed that a decrease from 3267 meq to 2611 meq ($P < 0.02$) in manic patients correlated with improvement. Two studies showed nonsignificant differences in Na_E in patients before and after Li^+ treatment [17, 45].

In summary, these studies have shown the following:

1. Small but significant increases in TBW with improvement from depression (two studies) but not in mania (one study).
2. A small increase in ECW with improvement in depression (two studies), which was transient in one study, with no significant change in mania (one study).
3. Significant decrease in Na_E upon recovery from both depression and mania.

These data have stimulated speculation concerning possible mechanisms in affective disorders. For example, the increase in Na_E in depression and mania in the absence of an equivalent increase in total extracellular sodium could be due to an increase in residual sodium – mainly intracellular – possibly resulting from some alteration in the sodium transport system.

Regarding the effect of Li^+ on Na^+ and water metabolism, the only difference occurred in TBW, which was increased after Li^+ treatment. Baer et al. [17] found no significant changes in Na_E in patients who improved or failed to improve on Li^+ treatment.

Related Studies

Two groups have measured ^{24}Na transport from blood to CSF [36, 40]. Coppen reported that in 20 depressed patients the rate of transfer was less than normal, whereas after recovery the transfer rate was normal. Carroll et al. were unable to confirm these changes in a similar study; they pointed out that Coppen had only tested three patients in both depressed and recovery stages. The remaining patients were separate groups of depressed or recovered patients, whereas in the Carroll study, the rate of ^{24}Na transfer from blood to CSF was measured twice in the same 11 patients. They believed the differences observed by Coppen might be due to sampling error because of the high range of values obtained.

There have been a few reports on potassium, calcium, and magnesium metabolism in affective illnesses. Here again, positive reports are essentially balanced by negative reports, so additional studies are necessary to establish any trends.

GLUCOSE METABOLISM

Reports of abnormal glucose metabolism have generally shown a high incidence of decreased tolerance to glucose in patients with affective disorders.

Van der Velde and Gordon [159] found diabetic glucose tolerance tests (GTT) in 25 of 59 manic depressives. This proportion is twice that in a schizophrenic control group and many times higher than that usually reported for the general population. There was no correlation between whether patients were manic or depressed and the abnormality of the GTT.

Mueller et al. [109] found low glucose utilization (3 of 17 in diabetic range) in severe (psychotic) depressions but not in mild (neurotic) depressions. Increased insulin in blood correlated with a low rate of glucose utilization, indicating an abnormality in insulin sensitivity rather than inability to secrete insulin. Insulin sensitivity in severe depressions was measured directly in another study [110] that used the insulin tolerance test (ITT) and was found to be decreased in severe depressions. In several patients in this study, improvement of symptoms was correlated with a correction of abnormalities in glucose metabolism.

Aleksandrowicz [13] found that in a group of 52 depressed patients, those with both an abnormal GTT and ITT had the most severe symptoms of depression when compared to patients with only an abnormal ITT.

Van Praag and Leijnse [164] found that 14 depressed patients who responded to a hydrazine MAO inhibitor had, prior to treatment, higher blood glucose concentrations during fasting, lower glucose tolerance, and higher nonesterified fatty acids than did 11 nonresponders or 15 controls. During and after treatment, however, the three groups did not differ with respect to these tests.

Gildea, in an earlier study [67], measured oral and intravenous GTT in 30 manic depressive patients (6 manic and 18 retarded depressions and 6 agitated depressives). Six had an abnormal oral GTT but none had an abnormal intravenous GTT.

CYCLIC AMP

Two studies indicate that urinary excretion of 3',5'-cyclic adenosine monophosphate (cyclic AMP) may be increased in mania and decreased in depression. Abdulla and Hamadah measured urinary 24-hr cyclic AMP in 20 female psychiatric patients twice during their illness (selection criteria were unstated) [10]. The excretion of cyclic AMP was significantly lower in depressed patients before improvement than after improvement or in healthy controls (Table 30-7). Four depressed patients who did not improve showed a mean decrease in excretion of 34 percent.

Paul et al. [123] measured urinary excretion of cyclic AMP in 10 normal controls and 37 psychiatric patients (32 depressed and 5 manic). The mean excretion of cyclic AMP for the manic patients was increased, which was the only significant difference found in the study (Table 30-7).

Such studies are difficult to evaluate. Nearly one-half of the cyclic AMP in the urine originates in the kidney. Excretion is affected by physical activity, a factor that was not controlled in either study. Furthermore, the values reported by Abdulla and Hamadah [10] were much lower than those reported by Paul et al. [123], even though both used similar assays (enzymatic isotope displacement).

Abdulla and Hamadah proposed a mechanism of action of tricyclic antidepressants based on a disturbance in cyclic AMP metabolism. These antidepressants are competitive inhibitors of cyclic AMP phosphodiesterase. If depression were caused by decreased cyclic AMP, these drugs theoretically might correct the defect by preventing the breakdown of cyclic AMP. However, Ramsden [131] reported that millimolar concentrations of chlorpromazine inhibited phosphodiesterase activity

TABLE 30-7. Urinary Cyclic AMP Excretion in Affective Illness

Type of Patient	Abdulla and Hamadah [10]			Paul et al. [123]		
	N		Urinary Cyclic AMP (μM/24 hr)	*N*		Urinary Cyclic AMP (μM/24 hr)
Controls	18		2.282 ± 0.094	10		5.64 ± 0.68
Depressed	13	Ill	$0.532^{a,b}$ ± 0.078	25	Neurotic	6.70^{c} ± 0.37
	13	Improved	1.283^{b} ± 0.024	7	Psychotic	3.64^{c} ± 0.19
Manic	4		6.851^{c} ± 2.8	5		9.94^{d} ± 1.88

[a]Depressed (ill) vs. control, $P < 0.001$.
[b]Depressed (ill) vs. depressed (improved), $P < 0.001$.
[c]N.S. when compared with controls.
[d]Manic vs. controls, $P < 0.01$.

by 85 percent, as compared with imipramine which inhibited it by only 40 percent. Since chlorpromazine is more effective in mania (where cyclic AMP presumably is already elevated) than in depression, it may be incorrect to assign the antidepressant effect of imipramine to its ability to inhibit cyclic AMP phosphodiesterase.

TYROSINE TOLERANCE

Takahashi et al. [157] reported a decrease in tyrosine tolerance in 13 depressed patients and 5 manic patients as compared with 13 controls ($P < 0.005$) at 3 and 4 hr after oral tyrosine loading in both the depressed and manic patients. There was no significant difference between the patients with mania and those with depression. These investigators also found that tyrosine tolerance returned to normal after recovery, and that schizophrenic patients did not have a decreased tyrosine tolerance.

SCHIZOPHRENIA

Chronic schizophrenia is characterized by hallucinations, delusions, disorganized speech, a chronic course, and an insidious onset, usually prior to age 30. Long-term hospitalization is frequent. Similar symptoms occur in acute schizophrenia (also frequently called schizo-affective schizophrenia), but the patient recovers and the course tends to be episodic.

Early in the illness, clinical criteria may be inadequate to distinguish psychotic affective disorders from schizophrenia. In the absence of adequate criteria, the biochemical investigator should realize the difficulty in obtaining a homogeneous group of early schizophrenics. In contrast, there usually is agreement concerning the diagnosis of schizophrenia in patients who have been severely ill for at least 5 years.

Twin and adoption studies indicate the presence of a genetic factor in schizophrenia, which in turn suggests a biochemical abnormality. Psychotomimetic drugs (such as LSD and mescaline) have also provided support for the hypothesis that schizophrenia has a chemical cause. Nevertheless, a biochemical basis for schizophrenia has yet to be demonstrated. Recent attempts to correlate schizophrenia with possible biochemical abnormalities, primarily by analyses of body fluids, are reviewed in the following pages.

HISTORICAL NOTE

During the 1940s, a postulated lack of "adrenal-cortical responsivity" dominated the theories about a biochemical cause of schizophrenia [126]. By the late 1940s [122] and mid-1950s [20], it became evident that the adrenals showed normal responsivity in schizophrenia. Recently, normal responsivity again was observed [143]. This theory is mentioned chiefly as an example of how an incorrect hypothesis can so dominate thinking about a psychiatric illness that it hinders the pursuit of other chemical investigations and ideas.

SERUM: NORMALLY OCCURRING CONSTITUENTS

Small Molecular Weight Constituents

Serum or blood constituents have been reported as increased, decreased, or normal in schizophrenia. Generally, these results have not been consistent nor replicable. An extensive study before and after insulin treatment, for example, showed normal or virtually normal values of Na, K, Mg, Cl, glucose, lactate, pyruvate, phosphate, and diphosphoglycerate [73]. Reduced glutathione in blood also has been reported as normal or decreased, and as nonspecific for schizophrenia. Interest in this compound in relation to schizophrenia has virtually ceased. Liver function tests have been said to be abnormal in schizophrenia, but these findings are not replicable. For example, bromsulfalein retention, alkaline phosphatase, thymol turbidity, and cephalin flocculation are reported to be normal in schizophrenia [73].

Serum Enzymes

Several serum enzyme changes have been reported in schizophrenia, but attempts to replicate the studies have generally been unsuccessful.

Akerfeldt [12] reported that serum from schizophrenics oxidized *N,N'*-dimethyl-*p*-phenylenediamine (PPD) at a more rapid rate than did serum of normal controls (the so-called Akerfeldt test). The two most important factors influencing the oxidation of PPD were the serum levels of ceruloplasmin, a copper-containing protein with oxidase activity, and levels of serum ascorbic acid. All three parameters – oxidation rate of PPD, serum ceruloplasmin, and serum ascorbic acid – have been studied intensively and have been reported by various investigators to be significantly different or not significantly different in schizophrenic patients. Leach and Heath [95] attempted to correlate the adrenochrome hypothesis (see below) with the alleged increase in ceruloplasmin by showing that schizophrenic serum could oxidize epinephrine more rapidly than could normal serum. The concentrations of epinephrine used in their assay exceeded the physiological range. Cohen et al. [37] found that when epinephrine was used in concentrations closer to its physiological level, this type of oxidation was essentially negligible. The present consensus is that although differences in oxidative properties or levels of ceruloplasmin may on occasion be observed, they are not specific for schizophrenia (or other psychiatric illness), but probably represent nutritional or other nonspecific phenomena.

Other serum enzymes have been investigated but, again, attempts to confirm positive findings have been disappointing. Plasma cholinesterase was found increased in several psychiatric illnesses, including schizophrenia, but the change is now believed to be nonspecific [14, 129, 132, 136].

There is one report [148] of increased (Na^+, K^+)-ATPase in washed erythrocyte ghosts from schizophrenic patients, but, again, the study was not confirmed [121]. Antebi and King [14] investigated nine serum enzymes (including cholinesterase) from controls and schizophrenics, but none was consistently high or low in schizophrenic serum. Abnormal glucose 6-phosphate dehydrogenase activity was reported in schizophrenia, but this was not confirmed by two laboratories including the group that originally described the abnormality [25].

Changes in ATP-creatine phosphotransferase (creatine phosphokinase) in blood have been reported, but they are variable and nonspecific, occurring in some, but not all, cases of schizophrenia, manic-depressive illness, "periodic catatonia," and acute neurological illnesses. The increases in enzyme activity, when they occur, are large (5 to 50 times more than normal) but transient [107a]. This finding may lead to subclassifying schizophrenia on a biological basis. The enzyme in serum is from muscle, not brain. It is too early to tell whether the change is sufficiently specific to be useful in diagnosis. Many other illnesses and other factors (amount of activity and sleep disturbance, among others) that lead to an increase in this muscle-derived enzyme could vitiate its usefulness in helping to classify schizophrenia by a biological criterion.

Other Serum Proteins

Many reports of increased or decreased levels of various serum protein fractions in schizophrenia have appeared in the literature. Even statistically significant changes have been relatively small in magnitude and have not been found consistently in most patients with the illness. Furthermore, in nearly every instance, subsequent reports appeared in which other investigators were unable to replicate the results.

Various globulin fractions have been most frequently implicated. Fessel and co-workers [53, 56] reported increases in IgM (macroglobulin) fractions in schizophrenics when compared with controls. But, in another paper, the same group [54] reported that this elevation was not specific for schizophrenia and remained elevated during remissions [92]. Other workers have been unable to confirm any change in IgM [19, 70]. Some studies have reported differences in α-, β-, and γ-globulins [63, 130]. Jensen et al. [52, 87] found small differences in some serum protein fractions in some schizophrenic patients, but concluded they were probably nonspecific. Present data suggest that this latter view is correct.

SERUM: ABNORMAL CONSTITUENTS

Small Molecular Weight Constituents

The "adrenochrome hypothesis" about schizophrenia postulated abnormal metabolic pathways for epinephrine, resulting in the formation of adrenochrome and adrenolutin. The hypothesis appears to be untenable. Apparently there is no adrenochrome in the blood in schizophrenic patients [156]; in another study, 98 percent of an infused dose of ^{3}H-epinephrine could be accounted for without postulating abnormal metabolites [94]. Furthermore, the reported increased oxidation of epinephrine [95] was shown to result from dietary deficiency of ascorbic acid without a significant relationship to schizophrenia [102].

Serum Proteins

Lactate-Pyruvate Factor. Frohman and associates reported a difference between the plasma of schizophrenics and controls [59, 60]. The assay involved incubation of chicken erythrocytes with plasma of schizophrenic patients or with control plasma. The lactate-pyruvate ratio (L/P) was higher with schizophrenic plasma than with control plasma.

A more recent test [61] is one in which tryptophan, hydroxytryptophan, glutamic acid, alanine, and tyrosine are reported to be taken up more rapidly by chicken erythrocytes in the presence of schizophrenic plasma than in the presence of control plasma. Other amino acids tested were not differentially affected by the two plasmas.

Three groups attempted to replicate the L/P test results, but without success [27, 99, 140]. One group suggested that the apparent changes of the L/P ratio, which are paralleled by hemoglobin release (by erythrocyte lysis), are caused by reaction of a heterogenetic antigen of chicken erythrocyte plasma membrane with an antibody of human serum and complement [138]. Frohman and his group reported that an α_2-globulin, and not an antibody, is abnormal in schizophrenia, but it has not been isolated. Frohman also stated that the "pure" factor is not hemolytic and that hemolysis results from a β-globulin accompanying the α_2-globulin.

Two laboratories, using independent methods, tested the plasmas of the same subjects [139]. Excellent agreement between the methods of determination as measured by rank order of results was demonstrated, but the values were interpreted as a failure to confirm an elevated L/P ratio in schizophrenia [140]. It may remain of interest that 23 of 24 high L/P ratios were in schizophrenic sera, while only 19 of 42 of the lower measurements were in nonschizophrenic sera [140]. At this stage, pending purification and appropriate immunological studies, L/P ratio values cannot be used in diagnosis or in establishing a biological abnormality in schizophrenia.

Taraxein. In the late 1950s, an abnormal plasma factor in schizophrenia was reported by Heath et al., who named the factor "taraxein" [82, 83]. Early reports stated that when schizophrenic serum (or a fraction obtained by chloroform-ethanol precipitation) was injected intravenously into monkeys or humans, it produced a catatoniclike state, similar in some respects to such states observed in schizophrenia. Schizophrenics in remission reacted more strongly than did normal volunteers to these protein injections. In monkeys, fractions containing taraxein caused abnormal EEG spikes and slow-wave activity in septal nuclei and basal ganglia, as measured by implanted electrodes. In two other laboratories, however, attempts to replicate the behavioral changes in humans injected with taraxein were not successful [5, 150]. Monkeys were not tested in these studies.

This hypothetical protein originally was thought to be an α- or β-globulin, similar to ceruloplasmin. Subsequently, however, the active component of the "taraxein chloroform-ethanol" method of preparation was identified as a contaminant from the IgG (antibody) fraction [80, 81].

Taraxein is now thought by Heath to be an antibrain antibody and that schizophrenia is an autoimmune disease [80]. Globulin from schizophrenic sera reportedly binds to the neurons in schizophrenic patients, especially in the septal region and ventral part of the head of the caudate nucleus. Binding was much less when normal globulin was applied to schizophrenic brains in vitro or when schizophrenic globulin was applied to normal brains. Heath also reported endogenous binding of globulin to nuclei of neurons in schizophrenic brains but not in normal brains. Fluorescein-labeled antihuman globulin showed the presence of bound globulin.

Two main arguments were used to support a cause-effect relationship between the nuclear binding of schizophrenic globulin and the EEG changes in monkeys and behavioral changes in schizophrenic patients or normal patients given the schizophrenic serum. First, the globulin binding was localized predominantly in brain areas where EEG changes were found. Second, authentic antibrain antibodies, demonstrated by immunoprecipitation and prepared by inoculating sheep with brain extracts, produced EEG and behavioral changes similar to those observed with schizophrenic globulin.

Several points need clarifying. First, if the active factor is an antibody, it is not clear why it is so unstable; Heath states that serum fractions become inactive in 48 hr at 4°C. Second, it is unclear why fluorescent staining of human spleen did not occur when fluorescein-tagged antihuman globulin was administered. There are many globulin-producing cells in the spleen, and the serum globulin in the sinusoids would be expected to stain brightly. Inability to demonstrate such staining casts doubt on the specificity of the staining technique. Third, there is no direct evidence that the apparent binding of globulin to neuronal cells is an antigen-antibody reaction. It was stated that immunoprecipitates with brain homogenates did not occur, implying the "antibody" was not a precipitating antibody. However, attempts to demonstrate antigen-antibody interactions by other immunochemical methods, such as complement fixation or agglutination, were not reported. Schizophrenic brain tissue adsorbed the globulin that apparently was being detected by fluorescein-tagged antihuman globulin, but it was unclear whether the same results could occur with normal brain tissue.

Finally, if schizophrenia is an autoimmune disease, it will be the first reported to occur with apparent absence of inflammatory response or pathological change at the site of reaction. Severe inflammatory reaction and cellular destruction are the hallmarks of known diseases in which autoimmune reactions are believed involved. In schizophrenia, there are no observable pathological changes in the brain at the light or electron microscope level.

Heath was not the only investigator to attempt to demonstrate antibrain antibodies on the assumption that schizophrenia might be an autoimmune disease. Fessel [55], using agglutination of latex particles coated with brain protein or red blood cells from sheep, reported that sera from psychiatric patients (including schizophrenics) agglutinated more often than did normal sera. He attributed this agglutinating factor to a physicochemical change in sera rather than to a specific antibrain antibody because sera from psychiatric patients also caused agglutination of erythrocytes coated with other tissue extracts. Kuznetzova and Semanov [93] reported measuring a brain-specific antibody, using complement fixation with the sera from a group consisting of 26 percent schizophrenics, 30 percent patients with organic brain disease, 15 percent patients with collagen diseases, and 3 percent normal controls.

At present, no rigorous evidence exists for an autoimmune basis for schizophrenia.

"Rope-Climbing Test" Factor. The time it takes trained rats to climb a rope has been used as a bioassay for an abnormal serum factor in schizophrenia. Typically, 1 to 5 ml of schizophrenic serum or control serum is injected intraperitoneally per 100 g body weight. In "positive" experiments, administration of schizophrenic serum resulted in greater climbing time. Taraxein, prepared by Heath's method, increased climbing time in some studies [107], but not in others [9, 150]. Based on these conflicting observations, further attempts to purify the "active" protein in serum from schizophrenics were made. The two major groups working on the problem found the active protein to be either an α- or β-globulin, contrasting with Heath's identification of taraxein as a γ-globulin. One of these groups [144] later was unable to reproduce its earlier work. The other group [125] reported that they had partially purified (approximately 500-fold) active protein from the α_2-globulin fraction. The protein was extremely labile and present also in normal serum. However, the activity of a crude fraction was significantly greater in schizophrenics than in normal controls.

Abnormal Serum Antigens. Malis reported that sera from schizophrenics contained an antigen detectable by hemagglutination, complement fixation, and by the fact that the serum caused anaphylaxis in sensitized guinea pigs [3]. The guinea pigs were inoculated with schizophrenic serum and then desensitized with normal human serum. If the animals were then challenged with schizophrenic serum, anaphylaxis frequently occurred. No anaphylaxis occurred with normal serum. Malis speculated that the antigen responsible for anaphylaxis was of viral origin. These findings, using anaphylaxis as an assay, were confirmed by Haddad and Rabe [2], but could not be replicated in two reports from another laboratory [52, 87].

Another immunochemical method has been used in an attempt to demonstrate a unique protein in schizophrenia. Animals were inoculated with schizophrenic serum. The antiserum thus prepared was adsorbed with normal human serum. Schizophrenic serum was then tested against the adsorbed antiserum. The detection of an antigen-antibody reaction would indicate an antigen in schizophrenic serum. For the test to work, however, all antibodies to normal serum must be removed by adsorption. Noval and Mao reported [117] that a single precipitin band was obtained with 42 of 70 schizophrenic sera and in 20 percent of normal sera from 45 controls. The nature of the relatively specific antigen was not identified. Jensen et al. [52, 87], Rieder et al. [133], and Penell et al. [124] have reported negative results with this approach.

Recent advances in protein chemistry and immunochemistry ultimately may answer definitively whether there is a specific protein in the body fluids of schizophrenics; at present the question remains unresolved.

CEREBROSPINAL FLUID CONSTITUENTS

Several reports from one laboratory indicated that glycoprotein-bound neuraminic acid from CSF of schizophrenic patients was decreased compared to normal values and that the decrease was correlated with clinical severity of the disease [23, 32]. This could not be confirmed in several other laboratories, which employed either the method used in the original studies, or the more specific thiobarbituric acid method for neuraminic acid [75, 86, 111–115, 119, 135].

URINARY CONSTITUENTS

In 1962, Friedhoff and Van Winkle [58] reported that dimethoxyphenylethylamine (DMPEA) from the urine of schizophrenic patients occurred as a "pink spot" on chromatograms. This pink spot was absent in tests of normals. The report led to work by many investigators that can perhaps best be summarized by a quote from a recent review: "Thus schizophrenics do not differ from normals in their excretion of DMPEA" [154]. Creveling and Daly [48] showed that the pink spot was composed of multiple components of which DMPEA was present only in trace amounts. Watt and his co-workers [165] described the vicissitudes in attempting to isolate the pink spot and separate its components. Apparently, neither the pink spot nor DMPEA is unique to schizophrenia. This work illustrates the difficulties in clinical-chemical correlative research, and it also shows that it is possible to ascertain sources of error by replicative studies when the original work was carefully performed and described, as was the case in the Friedhoff and Van Winkle experiments. Formidable problems are posed in ascertaining diagnostic homogeneity and the influences of medication, diet, fluid intake, and so on.

There was a burst of interest in studying "aromaturia" (excretion of phenolic or hydroxyindole compounds) in the early 1950s, stimulated by the work of Armstrong in detecting urinary phenolic acids. Similar excitement was engendered by the discovery of the psychotomimetic effects of LSD, by Axelrod's work on catecholamine metabolism, by Rapport's work in purifying and characterizing 5-HT, by Woolley's suggestion that 5-HT may play a role in schizophrenia, and by Osmond and Smythies' theory of defective epinephrine metabolism in schizophrenia. The work on DMPEA was an extension of this interest. The McGeers described an excess of chromatographic spots, presumably of aromatic origin, in the urine of schizophrenics as compared with controls [103–106], but this has not been confirmed [11, 100].

Several papers suggested abnormalities in indole metabolism in schizophrenia, as evidenced by decreased excretion of HIAA after a tryptophan load. These findings have not been replicated [90].

SWEAT CONSTITUENTS

Schizophrenic patients frequently have been observed to have a peculiar body odor. Smith and Sines [152] reported that ether extracts of sweat from schizophrenics who had this odor could be distinguished (by trained rats and by personnel trained to detect and discriminate odors) from sweat extracts of normal controls and psychiatric patients without the odor. The odor was found in 69 percent of 85 chronic schizophrenics and 43 percent of 99 patients with other chronic psychiatric illness ($P < 0.001$). The chemical apparently responsible for the odor has been isolated from sweat and identified as *trans*-3-methyl-2-hexenoic acid [153]. The relationship of this compound to schizophrenia is not known. The suggestion that dermal contamination with *Pseudomonas aeruginosa* (an odor-producing bacteria) might be responsible was discounted in one report. The role of other bacteria has not been ruled out [151].

TRANSMETHYLATION HYPOTHESIS

A hypothesis concerning schizophrenia has been proposed that implicates an abnormality in transmethylation, which has been postulated to produce neurotoxins accounting for certain aspects of the illness (as suggested by Harley-Mason [118]). There is, however, no direct evidence of defective transmethylation.

A number of investigators have shown worsening or, in one instance, amelioration of schizophrenic symptoms when methionine and a monoamine oxidase inhibitor were given. A World Health Organization report emphasizes the ambiguities in interpreting such indirect evidence by stating: "Several investigators have reported that the administration of the major methyl donor, methionine, together with MAO inhibitors causes exacerbation in a minority of schizophrenic patients. The administration of other methyl donors has given similar results. However, the extent to which such exacerbations are toxic reactions superimposed on a schizophrenic psychosis is not clear" [8].

In the absence of direct evidence of defective transmethylation (for example, the demonstration of an abnormally methylated metabolite in schizophrenics), and in the presence of such unconvincing indirect evidence (reviewed in [89]), new methods of study will have to be found to demonstrate relationships between transmethylation and schizophrenia.

PERIODIC CATATONIA

Since R. Gjessing first described periodic catatonia, it has been thought to be a subtype of schizophrenia or at least related to schizophrenia. The hallmarks of the condition are periodicity, catatonia, and nitrogen retention. L. Gjessing, continuing his father's investigations into the syndrome [68, 69], reported that increased NMET, MET, and VMA excretion during exacerbations reverted to normal during remissions. Excretion of 5-HT, histamine, and tyramine were not increased [116].

In the reviewers' opinion, this syndrome has not been sufficiently studied clinically or chemically to justify its classification as a disease entity.

ANXIETY NEUROSIS

Anxiety neurosis is a chronic disorder estimated to affect 5 percent of adults. Common complaints are apprehension, shortness of breath, cardiac palpitation, and dizziness, which often occur together as "anxiety attacks." The syndrome has a strong familial predisposition and an onset prior to age 30.

Patients with anxiety neurosis have been shown to have higher blood lactate levels in response to exercise than do normal controls. This observation has been made in four laboratories [38, 84, 88, 96] and is a consistent finding, although it is not specific for the disorder (it is also reported in mitral stenosis and atrial septal defects).

Pitts and McClure [128] found that infusions of 0.5 M sodium lactate solution (10 ml per kilogram of body weight) can precipitate a syndrome indistinguishable from an anxiety attack in patients with anxiety neurosis (N=14). Identical infusions in normal controls (N=10) produced similar symptoms, although the number and severity of symptoms was less than in patients ($P < 0.001$). A glucose-saline control solution produced no apparent symptoms. This experiment was the first successful attempt to induce anxiety attacks in virtually all (13 of 14) of a series of patients with anxiety neurosis. Other means of inducing anxiety attacks – such as exercise, hypnosis, recall of traumatic life experiences, and plunging patients' arms into ice water – fail to induce such attacks in as high a proportion of patients.

These symptoms may be partially related to the ability of lactate to bind Ca^{2+} since addition of Ca^{2+} to the infusion medium substantially reduced the effect in both patients and controls. Grosz and Farmer [77] have challenged the proposed mechanism of calcium binding because of the low association constant of lactate and Ca^{2+} at the pH and ionic strength of blood. However, they were able to induce anxiety attacks by lactate infusion. Taylor et al. [158] also induced anxiety attacks with lactate infusion.

It has been reported that the β-adrenergic blocker, propranolol, is effective in reducing anxiety symptoms [74]. The drug has multiple pharmacological actions. In addition to blocking the β-receptor effects of the catecholamines (such as increased heart rate), it affects the presumed cyclic AMP-mediated effects of epinephrine on carbohydrate and lipid metabolism (e.g., it will block the lactademic effect of epinephrine).

On the basis of the increased blood lactate observed in patients with anxiety neurosis after exercise, the production of anxiety symptoms with lactate infusions, and the reported effectiveness of propranolol in treatment of anxiety symptoms, Pitts has hypothesized that the anxiety-neurotic patient is particularly sensitive to excess lactate production resulting from increased epinephrine production. The excessive sensitivity is theoretically coupled with "chronic overproduction of epinephrine, overactivity of the central nervous system, a defect in metabolism resulting in excess lactate production, a defect in calcium metabolism, or a combination of these conditions [127]. Further experimentation will be necessary to evaluate this hypothesis.

EPILOGUE

Studying the biochemical aspects of psychiatric illness is difficult. There are no animal models for any psychiatric illness. Extrapolating from animal data to human behavior is hazardous, at best. In vivo chemical studies in humans are largely limited to analysis of body fluids such as blood, urine, and CSF. Because of the blood-brain barrier and other factors, it is usually impossible to determine whether alterations in fluid constituents reflect changes in the CNS. As chemical assays become more sophisticated, technical expertise becomes more important in achieving reproducible results; many of the discrepancies reported in this review may arise simply from laboratory variability.

Finally, there are special problems inherent in any research involving human subjects. These include such variables as homogeneity of patient groups, patient cooperation, control of the experimental situation, ethical considerations, and cooperation with clinicians.

It was with full realization of these problems that this review was prepared. Although it was intended to be critical in the sense of impartially weighing the evidence, it was not intended to minimize the capabilities of the 100 or more investigators whose work was discussed. Anyone who works in this area knows the difficulties involved; inconsistencies are expected. What is objectionable are critiques that selectively assemble evidence to support particular theories and encourage the belief that we know more about psychiatric illness than we do.

Despite the skepticism implicit in this review, it is evident that a number of vigorous and resourceful investigators are working on biochemical aspects of psychiatric illness. The investigator who has the strength of his convictions (and the strength to lay them aside if they are not productive) may hit upon a basic biochemical defect in one of the major psychiatric illnesses, and perhaps has already done so. The reviewers may have overlooked such a finding only because insufficient time has elapsed for proof and replication. Not only would a basic biochemical discovery clarify the illness concerned but it would also encourage additional neurochemists to work in clinical areas.

REFERENCES

BOOKS AND REVIEWS

1. Gibbons, J. L. Biochemistry of Depressive Illness. In Coppen, A., and Walk, A. (Eds.), *Recent Developments in Affective Disorders. A Symposium.* Ashford, England: Headley, 1968.

2. Haddad, R. K., and Rabe, A. An Antigenic Abnormality in the Serum of Chronically Ill Schizophrenic Patients. In Heath, R. G. (Ed.), *Serological Fractions in Schizophrenia.* New York: Harper & Row, 1963.
3. Malis, G. Y. *Research on the Etiology of Schizophrenia.* New York: Consultants Bureau, 1961.
4. Mason, J. W. Psychological influences on the pituitary-adrenal cortical system. *Recent Progr. Horm. Res.* 15:345, 1959.
5. Robins, E., Lowe, I. P., and Smith, K. Attempts to Confirm the Presence of "Taraxein" in the Blood of Schizophrenic Patients. In Abramson, H. A. (Ed.), *Neuropharmacology.* New York: Josiah Macy, Jr. Foundation, 1959.
6. Rodnight, R. Body fluid indoles in mental illness. *Int. Rev. Neurobiol.* 3:251, 1961.
7. Schildkraut, J. J. The catecholamine hypothesis of affective disorders: A review of supporting evidence. *Am. J. Psychiatry* 122:509, 1965.
8. Scientific Group on Biological Research in Schizophrenia. *Biological Research in Schizophrenia. WHO,* Tech. Rep. Ser. No. 450, 1970.
9. Winter, C. A., Flataker, L., Boger, W. P., Smith, E. V. C., and Sanders, B. E. The Effects of Blood Serum and Serum Fractions from Schizophrenic Donors upon the Performance of Trained Rats. In Folch-Pi, J. (Ed.), *Chemical Pathology of the Nervous System.* New York: Pergamon, 1961.

ORIGINAL SOURCES

10. Abdulla, Y. H., and Hamadah, K. 3',5'-cyclic adenosine monophosphate in depression and mania. *Lancet* 1:378, 1970.
11. Acheson, R. M., Paul, R. M., and Tomlinson, R. V. Some constituents of the urine of normal and schizophrenic individuals. *Can. J. Biochem. Physiol.* 36:295, 1958.
12. Akerfeldt, S. Oxidation of N, N-dimethyl-*p*-phenylenediamine by serum from patients with mental disease. *Science* 125:117, 1957.
13. Aleksandrowicz, J. W. Insulin and glucose tolerance tests in depressive syndromes. *Pol. Med. J.* 7:759, 1968.
14. Antebi, R. N., and King, J. Serum enzyme activity in chronic schizophrenia. *J. Ment. Sci.* 108:75, 1962.
15. Ashcroft, G. W., Crawford, T. B. B., Eccleston, D., Sharman, D. F., MacDougall, E. G., Stanton, J. B., and Binns, J. K. 5-Hydroxyindole compounds in the cerebrospinal fluid of patients with psychiatric or neurological diseases. *Lancet* 2:1049, 1966.
16. Ashcroft, G. W., and Sharman, D. F. 5-Hydroxyindoles in human cerebrospinal fluids. *Nature (Lond.)* 186:1050, 1960.
17. Baer, L., Durell, J., Bunney, W., Levy, B., Murphy, D., Greenspan, K., and Cardon, P. Sodium balance and distribution in lithium carbonate therapy. *Arch. Gen. Psychiatry* 22:40, 1970.

18. Bergsman, A. The urinary excretion of adrenaline and noradrenaline in some mental diseases: A clinical and experimental study. *Acta Psychiat. Neurol. Scand.* 34:(Suppl. 133), 1959.
19. Bishop, M. P., Hollister, L. E., Gallant, D. M., and Heath, R. G. Ultracentrifugal serum proteins in schizophrenia. *Arch. Gen. Psychiatry* 15:337, 1966.
20. Bliss, E. L., Migeon, C. J., Branch, C. H. H., and Samuels, L. T. Adrenocortical function in schizophrenia. *Am. J. Psychiatry* 112:358, 1955.
21. Bliss, E. L., Migeon, C. J., Branch, C. H. H., and Samuels, L. T. Reaction of the adrenal cortex to emotional stress. *Psychosom. Med.* 18:56, 1956.
22. Blumenfield, M., Rose, L. I., Richmond, L. H., and Beering, S. C. Dexamethasone suppression in basic trainees under stress. *Arch. Gen. Psychiatry* 23: 299, 1970.
23. Bogoch, S., Belval, P. C., Dussik, K. T., and Conran, P. C. Psychological and biochemical syntheses occurring during recovery from psychosis. *Am. J. Psychiatry* 119:128, 1962.
24. Bourne, H. R., Bunney, W. E., Colburn, R. W., Davis, J. M., Davis, J. W., Shaw, D. M., and Coppen, A. J. Noradrenaline, 5-hydroxytryptamine, and 5-hydroxyindoleacetic acid in hindbrains of suicidal patients. *Lancet* 2:805, 1968.
25. Bowman, J. E., Brewer, G. J., Frisher, H., Carter, J. L., Eisenstein, R. B., and Bayrakci, C. A re-evaluation of the relationship between glucose-6-phosphate dehydrogenase deficiency and the behavioral manifestations of schizophrenia. *J. Lab. Clin. Med.* 65:222, 1965.
26. Bridges, P. K., and Jones, M. T. The diurnal rhythm of plasma cortisol concentration in depression. *Br. J. Psychiatry* 112:1257, 1966.
27. Brown, F. C. Blood factors in schizophrenia. *Arch. Gen. Psychiatry* 10:409, 1964.
28. Bunney, W. E., Jr., and Davis, J. M. Norepinephrine in depressive reactions. *Arch. Gen. Psychiatry* 13:483, 1965.
29. Bunney, W. E., Davis, J. M., Weil-Malherbe, H., and Smith, E. R. B. Biochemical changes in psychotic depression. *Arch. Gen. Psychiatry* 16:448, 1967.
30. Bunney, W. E., Murphy, D. C., Goodwin, F. K., and Borge, G. F. The switch process from depression to mania: Relationship to drugs which alter brain amines. *Lancet* 1:1022, 1970.
31. Butler, P. W. P., and Besser, G. M. Pituitary-adrenal function in severe depressive illness. *Lancet* 1:1234, 1968.
32. Campbell, R. J., Bogoch, S., Scolaro, M. J., and Belval, P. C. Cerebrospinal fluid glycoproteins in schizophrenia. *Am. J. Psychiatry* 123:952, 1967.
33. Carroll, B. J. Hypothalamic–pituitary function in depressive illness: Insensitivity to hypoglycaemia. *Br. Med. J.* 3:27, 1969.
34. Carroll, B. J., Martin, F. I. R., and Davies, B. Resistance to suppression by dexamethasone of plasma 11-OHCS levels in severe depressive illness. *Br. Med. J.* 3:285, 1968.
35. Carroll, B. J., Mowbrary, R. M., and Davies, B. Sequential comparison of L-tryptophan with ECT in severe depression. *Lancet* 1:967, 1970.

36. Carroll, B. J., Steven, L., Pope, R. A., and Davies, B. Sodium transfer from plasma to CSF in severe depressive illness. *Arch. Gen. Psychiatry* 21:77, 1969.
37. Cohen, G., Holland, B., and Goldenberg, M. The stability of epinephrine and arterenol (norepinephrine) in plasma and serum. *Arch. Neurol. Psychiatry* 80:484, 1958.
38. Cohen, M. E., and White, P. D. Life situations, emotions, and neurocirculatory asthenia (anxiety neurosis, neurasthenia, effort syndrome). *Ass. Res. Nerv. Ment. Dis. Proc.* 29:832, 1950.
39. Cohn, C. K., Dunner, D. L., and Axelrod, J. Reduced catechol-O-methyl transferase activity in red blood cells of women with primary affective disorder. *Science* 170:1323, 1970.
40. Coppen, A. Abnormality of the blood-cerebrospinal fluid barrier of patients suffering from a depressive illness. *J. Neurol. Neurosurg. Psychiatry* 23:156, 1960.
41. Coppen, A., and Noguera, R. L-Tryptophan in depression. *Lancet* 1:1111, 1970.
42. Coppen, A., Prange, A., Whybrow, P. C., Noguera, R., and Pare, C. M. B. Methysergide in mania. A controlled trial. *Lancet* 2:338, 1969.
43. Coppen, A., and Shaw, D. M. Mineral metabolism in melancholia. *Br. Med. J.* 2:1439, 1963.
44. Coppen, A., and Shaw, D. M. Potentiation of the antidepressive effect of a monoamine-oxidase inhibitor by tryptophan. *Lancet* 1:79, 1963.
45. Coppen, A., and Shaw, D. M. The distribution of electrolytes and water in patients after taking lithium carbonate. *Lancet* 2:805, 1967.
46. Coppen, A., Shaw, D. M., Herzberg, B., and Maggs, R. Tryptophan in the treatment of depression. *Lancet* 2:1178, 1967.
47. Coppen, A., Shaw, D. M., Malleson, A., and Costain, R. Mineral metabolism in mania. *Br. Med. J.* 1:71, 1966.
48. Creveling, C. R., and Daly, J. W. Identification of 3,4-dimethoxyphenethylamine from schizophrenic urine by mass spectrometry. *Nature (Lond.)* 216:190, 1967.
49. Curzon, G. Tryptophan pyrrolase – a biochemical factor in depressive illness? *Br. J. Psychiatry* 115:1367, 1969.
50. Denker, S. S., Malm, U., Roos, B. E., and Werdinius, B. Acid monoamine metabolites of cerebrospinal fluid in mental depression and mania. *J. Neurochem.* 13:1545, 1966.
51. Dewhurst, W. G. Methysergide in mania. *Nature (Lond.)* 219:506, 1968.
52. Faurbye, A., Lundberg, L., and Jensen, K. A. Studies on the antigen demonstrated by Malis in serum from schizophrenic patients. *Acta Pathol. Microbiol. Scand.* 61:633, 1964.
53. Fessel, W. J. Macroglobin elevations in functional mental illness. *Nature (Lond.)* 193:1005, 1962.
54. Fessel, W. J. Mental stress, blood proteins, and the hypothalamus. *Arch. Gen. Psychiatry* 7:427, 1962.

55. Fessel, W. J. The "antibrain" factors in psychiatric patients' sera. I. Further studies with a hemagglutination technique. *Arch. Gen. Psychiatry* 8:614, 1963.
56. Fessel, W. J., and Grunbaum, B. W. Electrophoretic and analytical ultracentrifuge studies in sera of psychotic patients: Elevation of gamma globulins and macroglobulins, and splitting of alpha globulins. *Ann. Intern. Med.* 54:1134, 1961.
57. Fieve, R. R., Platman, S. R., and Fleiss, J. L. A clinical trial of methysergide and lithium in mania. *Psychopharmacologia* 15:425, 1969.
58. Friedhoff, A. J., and Van Winkle, E. Isolation and characterization of a compound from the urine of schizophrenics. *Nature (Lond.)* 194:897, 1962.
59. Frohman, C. E., Czajkowski, N. P., Luby, E. D., Gottlieb, J. S., and Senf, R. Further evidence of a plasma factor in schizophrenia. *Arch. Gen. Psychiatry* 2:263, 1960.
60. Frohman, C. E., Latham, L. K., Warner, K. A., Brosius, C. O., Beckett, P. G. S., and Gottlieb, J. S. Motor activity in schizophrenia. *Arch. Gen. Psychiatry* 9:83, 1963.
61. Frohman, C. E., Warner, K. A., Barry, K. T., and Arthur, R. E. Amino acid transport and the plasma factor in schizophrenia. *Biol. Psychiatry* 1:201, 1969.
62. Fullerton, D. T., Wenzel, F. J., Lohrenz, F. N., and Fahs, E. D. Circadian rhythm of adrenal cortical activity in depression. I. A comparison of depressed patients with normal subjects. *Arch. Gen. Psychiatry* 19:674, 1968.
63. Gammack, D. B., and Hector, R. I. A study of serum proteins in acute schizophrenia. *Clin. Sci.* 28:469, 1965.
64. Gibbons, J. L. Total body sodium and potassium in depressive illness. *Clin. Sci.* 19:133, 1960.
65. Gibbons, J. L. Cortisol secretion rate in depressive illness. *Arch. Gen. Psychiatry* 10:572, 1964.
66. Gibbons, J. L. The secretion rate of corticosterone in depressive illness. *J. Psychosom. Res.* 10:263, 1966.
67. Gildea, E. F., McLean, V. L., and Man, E. B. Oral and intravenous dextrose tolerance curves of patients with manic-depressive psychosis. *Arch. Neurol. Psychiatry* 49:852, 1943.
68. Gjessing, L. R. Studies of periodic catatonia. I. Blood levels of protein-bound iodine and urinary excretion of vanillyl-mandelic acid in relation to clinical course. *J. Psychiatr. Res.* 2:123, 1964.
69. Gjessing, L. R. Studies of periodic catatonia. II. The urinary excretion of phenolic amines and acids with and without loads of different drugs. *J. Psychiatr. Res.* 2:149, 1964.
70. Goodman, M., Rosenblatt, M., Gottlieb, J. S., Miller, J., and Chen, C. H. Effect of age, sex, and schizophrenia on thyroid autoantibody production. *Arch. Gen. Psychiatry* 8:518, 1963.
71. Goodwin, F. K., Brodie, H. K. H., Murphy, D. L., and Bunney, W. E. Administration of a peripheral decarboxylase inhibitor with L-dopa to depressed patients. *Lancet* 1:908, 1970.

72. Goodwin, F. K., Murphy, D. L., Brodie, K. H., and Bunney, W. E., Jr. L-Dopa, catecholamines, and behavior: A clinical and biochemical study in depressed patients. *Biol. Psychiatry* 2:341, 1970.
73. Gottfried, S. P., and Willner, H. H. Blood chemistry of schizophrenic patients before, during and after insulin shock therapy. *Arch. Neurol. Psychiatry* 62:809, 1949.
74. Granville-Grossman, K. L., and Turner, P. The effect of propranolol on anxiety. *Lancet* 1:788, 1966.
75. Green, J. P., Atwood, R. P., and Freedman, D. X. Studies on neuraminic acid in the cerebrospinal fluid in schizophrenia. *Arch. Gen. Psychiatry* 12:90, 1965.
76. Greenspan, K., Schildkraut, J. J., Gordon, E. K., Levy, B., and Durell, J. Catecholamine metabolism in affective disorders. II. Norepinephrine, normetanephrine, epinephrine, metanephrine, and VMA excretion in hypomanic patients. *Arch. Gen. Psychiatry* 21:710, 1969.
77. Grosz, H. J., and Farmer, B. B. Blood lactate in the development of anxiety symptoms. A critical examination of Pitts and McClure's hypothesis and experimental study. *Arch. Gen. Psychiatry* 21:611, 1969.
78. Haskovec, L., and Rysanek, K. Excretion of 3-methoxy-4-hydroxymandelic acid and 5-hydroxyindoleacetic acid in depressed patients treated with imipramine. *J. Psychiatr. Res.* 5:213, 1967.
79. Haskovec, L., and Soucek, K. Trial of methysergide in mania. *Nature (Lond.)* 219:507, 1968.
80. Heath, R. G., and Krupp, I. M. Schizophrenia as an immunologic disorder. *Arch. Gen. Psychiatry* 16:1, 1967.
81. Heath, R. G., and Krupp, I. M. Schizophrenia as a specific biologic disease. *Am. J. Psychiatry* 124:1019, 1968.
82. Heath, R. G., Martens, S., Leach, B. E., Cohen, M., and Angel, C. Effect on behavior of humans with administration of taraxein. *Am. J. Psychiatry* 114:14, 1957.
83. Heath, R. G., Martens, S., Leach, B. E., Cohen, M., and Feigley, C. A. Behavioral changes in nonpsychotic volunteers following the administration of taraxein, the substance obtained from serum of schizophrenic patients. *Am. J. Psychiatry* 114:917, 1958.
84. Holmgren, A., and Strom, G. Vasoregulatory asthenia in a female athlete and da Costa's syndrome in a male athlete successfully treated by physical training. *Acta Med. Scand.* 164:113, 1959.
85. Hullin, R. P., Bailey, A. D., McDonald, B. R., Drainsfield, G. A., and Milni, H. B. Variations in body water during recovery from depression. *Br. J. Psychiatry* 113:573, 1967.
86. Jenner, F. A., Kerry, R. J., Fowler, D. B., and Graves, E. W. Bial's reaction for neuraminic acid in cerebrospinal fluid from schizophrenics. *J. Ment. Sci.* 108:822, 1962.
87. Jensen, K., Clausen, J., and Osterman, E. Serum and cerebrospinal fluid proteins in schizophrenia. *Acta Psychiatr. Scand.* 40:280, 1964.

88. Jones, M., and Mellersh, U. A comparison of the exercise response in anxiety states and normal controls. *Psychosom. Med.* 8:180, 1946.
89. Kety, S. S. Current biochemical approaches to schizophrenia. *N. Engl. J. Med.* 276:325, 1967.
90. Kopin, I. J. Tryptophan loading and excretion of 5-hydroxyindoleacetic acid in normal and schizophrenic subjects. *Science* 129:835, 1959.
91. Kurland, H. D. Steroid excretion in depressive disorders. *Arch. Gen. Psychiatry* 10:554, 1964.
92. Kurland, H. D., Fessel, W. J., and Cutler, R. P. Clinical aspects of a serologic study of psychoses. *Arch. Gen. Psychiatry* 10:262, 1964.
93. Kuznetzova, N. I., and Semenov, S. F. [Detection of anti-brain antibodies in the blood serum of patients with neuro-psychiatric diseases.] *Neuropatolil. Psikhiatr.* 61:869, 1961.
94. LaBrosse, E. H., Mann, J. D., and Kety, S. S. The physiological and psychological effects of intravenously administered epinephrine, and its metabolism, in normal and schizophrenic men. III. Metabolism of 7-H^3-epinephrine as determined in studies on blood and urine. *J. Psychiatr. Res.* 1:68, 1961.
95. Leach, B. E., and Heath, R. G. The in vitro oxidation of epinephrine in plasma. *Arch. Neurol. Psychiatry* 76:444, 1956.
96. Linko, E. Lactic acid response to muscular exercise in neurocirculatory asthenia. *Ann. Med. Intern. Fenn.* 39:161, 1950.
97. Maas, J. W., Fawcett, J., and Dekirmenjian, H. 3-Methoxy-4-hydroxy phenylglycol (MHPG) excretion in depressive states. *Arch. Gen. Psychiatry* 19:129, 19
98. Deleted in proof.
99. Mangoni, A., Belazs, R., and Coppen, A. J. The effect of plasma from schizophrenic patients on the chicken erythrocyte system. *Br. J. Psychiatry* 109: 231, 1963.
100. Mann, J. D., and LaBrosse, E. H. Urinary excretion of phenolic acids by normal and schizophrenic male patients. *Arch. Gen. Psychiatry* 1:547, 1959.
101. McCabe, M. S., Reich, T., and Winokur, G. Methysergide as a treatment for mania. *Am. J. Psychiatry* 127:354, 1970.
102. McDonald, R. K. Problems in biologic research in schizophrenia. *J. Chronic Dis.* 8:366, 1958.
103. McGeer, E. G., Brown, W. T., and McGeer, P. L. Aromatic metabolism in schizophrenia. II. Bidimensional urinary chromatograms. *J. Nerv. Ment. Dis.* 125:176, 1957.
104. McGeer, E. G., and McGeer, P. L. Physiological effects of schizophrenic body fluids. *J. Ment. Sci.* 105:1, 1959.
105. McGeer, P. L., McGeer, E. G., and Boulding, S. E. Relation of aromatic amino acids to excretory pattern of schizophrenics. *Science* 123:1078, 1956.
106. McGeer, P. L., McGeer, E. G., and Gibson, W. C. Aromatic excretory pattern of schizophrenics. *Science* 123:1029, 1956.
107. Mekler, L. B., Lapteva, N. N., Lozovskii, D. V., and Baliezina, T. I. [The chemical and biological properties of taraxein, the toxic protein of the blood

serum of schizophrenic patients.] *Dokl. Akad. Nauk. SSSR* 130:1148, 1960.
107a. Meltzer, H., Elkun, L., and Moline, R. A. Serum-enzyme changes in newly admitted psychiatric patients. *Arch. Gen. Psychiatry* 21:731, 1969.
108. Messiha, F. S., Agallianos, D., and Clower, C. Dopamine excretion in affective states and following Li_2CO_3 therapy. *Nature (Lond.)* 225:868, 1970.
109. Mueller, P. S., Heninger, G. R., and McDonald, R. K. Intravenous glucose tolerance test in depression. *Arch. Gen. Psychiatry* 21:470, 1969.
110. Mueller, P. S., Heninger, G. R., and McDonald, R. K. Insulin tolerance test in depression. *Arch. Gen. Psychiatry* 21:587, 1969.
111. Mukai, A., and Ejima, T. Studies on sialic acid (neuraminic acid) in cerebrospinal fluid. I. Total Bial-positive substance in cerebrospinal fluid in schizophrenic subjects. *Tohoku J. Exp. Med.* 77:120, 1962.
112. Mukai, A., and Ejima, T. Studies on sialic acid (neuraminic acid) in cerebrospinal fluid. II. Methodology of determination by direct Ehrlich reaction. *Tohoku J. Exp. Med.* 77:128, 1962.
113. Mukai, A., and Ejima, T. Studies on sialic acid (neuraminic acid) in cerebrospinal fluid. III. Methodology of determination by Bial's reaction. *Tohoku J. Exp. Med.* 77:223, 1962.
114. Mukai, A., and Ejima, T. Studies on sialic acid (neuraminic acid) in cerebrospinal fluid. IV. Methodology of determination by diphenylamine reaction. *Tohoku J. Exp. Med.* 77:230, 1962.
115. Mukai, A., and Ejima, T. Studies on sialic acid (neuraminic acid) in cerebrospinal fluid. V. Protein-bound sialic acid in cerebrospinal fluid in mental disorders. *Tohoku J. Exp. Med.* 77:343, 1962.
116. Nishimura, T., and Gjessing, L. R. Failure to detect 3,4-dimethoxyphenylethylamine and bufotenine in the urine from a case of periodic catatonia. *Nature (Lond.)* 206:963, 1965.
117. Noval, J. J., and Mao, T. S. S. Abnormal immunological reaction of schizophrenic serum. *Fed. Proc.* 25:560, 1966.
118. Osmond, H., and Smythies, J. Schizophrenia: A new approach. *J. Ment. Sci.* 98:309, 1952.
119. Papadpoulos, N. M., McLane, J. E., O'Doherty, D., and Hess, W. C. Cerebrospinal fluid neuraminic acid in Parkinsonism and schizophrenia. *J. Nerv. Ment. Dis.* 128:450, 1959.
120. Pare, C. M. B., Young, D. P. H., Price, K., and Stacey, R. S. 5-Hydroxytryptamine, noradrenaline, and dopamine in brain stem, hypothalamus, and caudate nucleus of controls and of patients committing suicide by coal-gas poisoning. *Lancet* 2:133, 1969.
121. Parker, J. C., and Hoffman, J. F. Failure to find increased sodium, potassium-ATPase in red cell ghosts of schizophrenics. *Nature (Lond.)* 201:823, 1964.
122. Parsons, E. H., Gildea, E. F., Ronzoni, E., and Hulbert, S. Z. Comparative lymphocytic and biochemical responses of patients with schizophrenia and affective disorders to electroshock, insulin shock, and epinephrine. *Am. J. Psychiatry* 105:573, 1949.

123. Paul, M. I., Ditzion, B. R., Pauk, G. L., and Janowsky, D. S. Urinary adenosine 3',5'-monophosphate excretion in affective disorders. *Am. J. Psychiatry* 126: 1493, 1970.
124. Pennell, R. B., Pawlus, C., Saravis, C. A., and Scrimshaw, G. Further characterization of a human plasma component which influences animal behavior. *Trans. N.Y. Acad. Sci. (Ser. II)* 28:47, 1965.
125. Pennell, R. B., and Saravis, C. A. A human factor inducing behavioral and electrophysiological changes in animals. I. Isolation and chemical nature of the agent. *Ann. N.Y. Acad. Sci.* 96:462, 1962.
126. Pincus, G., Hoagland, H., Freeman, H., Elmadjian, F., and Romanoff, L. P. A study of pituitary-adrenocortical function in normal and psychotic men. *Psychosom. Med.* 11:74, 1949.
127. Pitts, F. N., Jr. The biochemistry of anxiety. *Sci. Am.* 220:69, 1969.
128. Pitts, F. N., Jr., and McClure, J. N., Jr. Lactate metabolism in anxiety neurosis. *N. Engl. J. Med.* 277:1329, 1967.
129. Plum, C. M. Study of cholinesterase activity in nervous and mental disorders. *Clin. Chem.* 6:332, 1960.
130. Pospisilova, V., and Janik, A. The relationship of the blood serum protein fractions to the clinical picture in psychoses and psychotic states, determined by paper electrophoresis. *Rev. Czech. Med.* 4:29, 1958.
131. Ramsden, E. N. Cyclic A.M.P. in depression and mania. *Lancet* 1:108, 1970.
132. Richter, D., and Lee, M. Serum choline esterase and anxiety. *J. Ment. Sci.* 88:428, 1942.
133. Rieder, H. P., Ritzel, G., Spiegelberg, H., and Gnirss, F. [Serological tests for the detection of "taraxein."] *Experientia* 16:561, 1960.
134. Rimon, R., Salonen, S., and Pekkarinen, A. Antidepressive medication and diurnal variation of plasma 17-OHCS levels in depression. *J. Psychosom. Res.* 12:289, 1968.
135. Robins, E., Croninger, A. B., Smith, K., and Moody, A. C. Studies on N-acetyl neuraminic acid in the cerebrospinal fluid in schizophrenia. *Ann. N.Y. Acad. Sci.* 96:390, 1962.
136. Rose, L., Davies, D. A., and Lehmann, H. Serum-pseudocholinesterase in depression with notable anxiety. *Lancet* 2:563, 1965.
137. Rosenblatt, S., and Chanley, J. D. Differences in the metabolism of norepinephrine in depressions. *Arch. Gen. Psychiatry* 13:495, 1965.
138. Ryan, J. W., Brown, J. D., and Durell, J. Antibodies affecting metabolism of chicken erythrocytes: Examination of schizophrenic and other subjects. *Science* 151:1408, 1966.
139. Ryan, J. W., Brown, J. D., and Durell, J. Concordance between two methods of assaying the plasma effects on chicken erythrocyte metabolism. *J. Psychiatr. Res.* 6:45, 1968.
140. Ryan, J. W., Steinberg, H. R., Green, R., Brown, J. D., and Durell, J. J. Controlled study on effects of plasma of schizophrenic and non-schizophrenic psychiatric patients on chicken erythrocytes. *J. Psychiatr. Res.* 6:33, 1968.

141. Sachar, E. J. Corticosteroids in depressive illness. II. A longitudinal psychoendocrine study. *Arch. Gen. Psychiatry* 17:554, 1967.
142. Sachar, E. J., Hellman, L., Fukushima, D. K., and Gallagher, T. F. Cortisol production in depressive illness. A clinical and biochemical clarification. *Arch. Gen. Psychiatry* 23:289, 1970.
143. Sachar, E. J., Kanter, S. S., Buie, D., Engle, R., and Mehlman, R. Psychoendocrinology of ego disintegration. *Am. J. Psychiatry* 126:1067, 1970.
144. Sanders, B. E., Small, S. M., Ayers, W. J., Oh, Y. H., and Axelrod, S. Additional studies on plasma proteins obtained from schizophrenics and controls. *Trans. N.Y. Acad. Sci. (Ser. II)* 28:22, 1965.
145. Sarai, K., and Kayano, M. The level and diurnal rhythm of serum serotonin in manic-depressive patients. *Folia Psychiat. Neurol. Jap.* 22:271, 1968.
146. Schildkraut, J. J., Green, R., Gordon, E. K., and Durell, J. Catecholamine metabolism in affective disorders. I. Normetanephrine and VMA excretion in depressed patients treated with imipramine. *J. Psychiatr. Res.* 3:213, 1965.
147. Schildkraut, J. J., Green, R., Gordon, E. K., and Durell, J. Normetanephrine excretion and affective state in depressed patients treated with imipramine. *Am. J. Psychiatry* 123:690, 1966.
148. Seeman, P. M., and O'Brien, E. Sodium-potassium-activated adenosine triphosphatase in schizophrenic erythrocytes. *Nature (Lond.)* 200:263, 1963.
149. Shaw, D. M., Camps, F. E., Eccleston, E. G. 5-Hydroxytryptamine in the hindbrain of depressive suicides. *Br. J. Psychiatry* 113:1407, 1967.
150. Siegel, M., Niswander, G. D., Sachs, E., Jr., and Stavros, D. Taraxein, fact or artifact? *Am. J. Psychiatry* 115:819, 1959.
151. Skinner, K., Smith, K., and Rich, E. Bacteria and the "schizophrenic odor." *Am. J. Psychiatry* 121:64, 1964.
152. Smith, K., and Sines, J. O. Demonstration of a peculiar odor in the sweat of schizophrenic patients. *Arch. Gen. Psychiatry* 2:184, 1960.
153. Smith, K., Thompson, G. F., and Koster, H. D. Sweat in schizophrenic patients: Identification of the odorous substance. *Science* 166:398, 1969.
154. Smythies, J. R., and Antun, F. The biochemistry of psychosis. *Scot. Med. J.* 15:34, 1970.
155. Strom-Olsen, R., and Weil-Malherbe, H. Humoral changes in manic-depressive psychosis with particular reference to the excretion of catechol amines in urine. *J. Ment. Sci.* 104:696, 1958.
156. Szara, S., Axelrod, J., and Perlin, S. Is adrenochrome present in the blood? *Am. J. Psychiatry* 115:162, 1958.
157. Takahashi, R., Utena, Y., Machiyama, M., Kurihara, T., Nakamura, T., and Kanamura, H. Tyrosine metabolism in manic depressive illness. *Life Sci.* 7:1219, 1968.
158. Taylor, M. A., Volavka, J. V., and Fink, M. Anxiety precipitated by lactate. *N. Engl. J. Med.* 281:1429, 1969.
159. van der Velde, C. D., and Gordon, M. W. Manic-depressive illness, diabetes mellitus, and lithium carbonate. *Arch. Gen. Psychiatry* 21:478, 1969.

160. Van Praag, H. M., Korf, J., and Puite, J. 5-Hydroxyindoleacetic acid levels in the cerebrospinal fluid of depressive patients. *Nature (Lond.)* 225:1259, 1970.
161. Van Praag, H. M., Korf, J., Woudenberg, F. van, and Kits, T. P. Influencing the human indoleamine metabolism by means of a chlorinated amphetamine derivative with antidepressive action (p-chloro-N-methylamphetamine). *Psychopharmacologia* 13:145, 1968.
162. Van Praag, H. M., and Leijnse, B. The value of monoaminoxidase inhibitor as antidepressive. Principle I. *Psychopharmacologia* 4:1, 1963.
163. Van Praag, H. M., and Leijnse, B. The value of monoaminoxidase inhibitor as antidepressive. Principle II. *Psychopharmacologia* 4:98, 1963.
164. Van Praag, H. M., and Leijnse, B. Some aspects of the metabolism of glucose and of the non-esterified fatty acids in depressive patients. *Psychopharmacologia* 9:220, 1966.
165. Watt, J. A. G., Ashcroft, G. W., Daly, R. J., and Smythies, J. R. Urine volume and pink spots in schizophrenia and health. *Nature (Lond.)* 221:971, 1969.

31

Learning and Memory: Approaches to Correlating Behavioral and Biochemical Events

Bernard W. Agranoff

BIOLOGICAL BASIS OF MEMORY

There is a common belief among biologists that the detailed mechanisms of behavioral plasticity constitute a major remaining frontier in our understanding of living systems. Certainly it is at this time the most obscure frontier and, accordingly, hypotheses run rampant. While a vast literature on human and animal learning and memory exists, results among reported studies are not as readily comparable as are usual biological experiments. Factors such as the species and even the strain of animal used, the details of construction of experimental apparatus, the sequence of training, and the conditions of animal housing, appear to affect measured behavior drastically. Although there are many conventions for quantifying behavioral parameters, different methods may lead to different conclusions and no one method is more easily reconciled with known physiological facts than another. Nevertheless, such experiments have been directed at answering the question: What are the changes that take place in the brain during a few seconds of experience that will mediate the resulting altered response of an organism for long periods, even for a lifetime? That such changes are permanent and are resistant to subsequent electrical storms and silences of the brain point to a physicochemical alteration associated with learning, brought about by an underlying molecular process. The various investigations described in this chapter constitute attempts to establish, via chemical and behavioral means, the physiological changes which lead to the altered performance of an animal measured at various times after a training experience.

Major consideration in this chapter is given to correlation of neural metabolism with memory formation. One approach employs agents reported to inhibit or stimulate specific molecular processes. These agents are administered in association with training to test the hypothesis that one or another molecular process is critical in memory formation. Alternatively, in the absence of such agents, biochemical studies have been directed at gross or subtle changes in composition of the brain or on its selective incorporation of isotopically labeled precursors as a function of training.

CONSOLIDATION AND STAGES OF MEMORY FORMATION

Our present concept of memory formation and storage began with the observations in the last century that physical trauma to the human brain results in loss of memory [46]. Typically, a blow to the head may lead to a retrograde amnesia, i.e., loss of memory for events immediately preceding the trauma. The common interpretation of this phenomenon is that memory ordinarily forms after the actual experience, and the "fixation," or *consolidation,* of memory is blocked by the blow. A similar conclusion has been drawn from animal experiments in which electroconvulsive shock (ECS) administered after training is shown to result in poor subsequent performance, while the same treatment given sometime after training has no effect. In each instance, the growing insusceptibility is thought to reflect a physiological process of consolidation, whereby memory is converted from an unstable, or labile, form to a permanent form. This process has been reported by various investigators to take from seconds to as much as hours for completion.

Not all psychologists accept the consolidation concept, and alternate explanations for the developing strength of memory with passage of time following training have been offered. It is argued, for example, that the amnestic treatment, via ECS or some other means, is itself noxious. The subject learns that, when he performs correctly, he will be punished and he therefore does not exhibit the learned response after the treatment. The amnestic agent has been postulated to interfere with performance in other ways, for example by creating spurious electrical effects in the neuronal networks that mediate the new behavior. Both of these explanations argue that the new memory is not obliterated by the amnestic agent, but that some superimposed process prevents its retrieval. In the instance of a physical blow to the head, an argument against the consolidation hypothesis might state that fear or pain associated with the blow has repressed events closely associated in time with the accident. Whether interference is at the behavioral or electrical level, it is argued that memory of training very likely forms simultaneously and permanently with training experience.

In recent years, the consolidation hypothesis has gained considerable support from passive avoidance studies (see below), in which amnestic animals move in the direction that leads to the amnestic treatment [43]. If the treatment were aversive, the animals would not be expected to move in such a direction. Studies with antibiotic antimetabolites also give considerable support for the consolidation hypothesis. Animals may be given such agents prior to training. Normal acquisition is seen, but upon retesting some time later, there is a marked memory deficit. If the injection were to have the properties of an aversive unconditioned stimulus, it is not likely that it would, in addition, exert its effect if given prior to the conditioned stimulus (backward conditioning). As discussed below, these agents may not obliterate memory, but instead may block the formation of a more stable form from a labile form.

MEASUREMENT OF LEARNING AND MEMORY

LEARNING PARADIGMS

Perhaps the simplest kind of training we can easily measure is *habituation:* the decrease in some measured response seen with repeated stimulation [36]. The loss of a "startle" response after repeated tapping on an animal enclosure is an example of this phenomenon. Although it can be demonstrated in primitive organisms, habituation in higher animals is believed to involve neural networks and is not simply a reflection of muscle fatigue or receptor adaptation. Habituation has not been investigated extensively as a model for biochemical correlates of behavior because by its nature it is not long-lasting, and one cannot exclude the possibility that it is mediated exclusively via short-term (and possibly nonchemical) memory mechanisms. Augmentation of responses after repeated stimuli, or sensitization, similarly does not have sufficiently long-lasting properties to serve as a model for permanent alterations in the brain.

Most studies attempting to relate biochemical events to learning have employed conditioning paradigms. Commonly, a distinction is made between classical (type I) and instrumental (type II) conditioning. In classical conditioning, a neutral stimulus is conditioned by pairing it with an unconditioned stimulus. For example, an animal is exposed to a light flash (the conditioned stimulus, CS), followed by a mild punishing electrical shock (the unconditioned stimulus, UCS) which elicits a motor or autonomic response. After several pairings, the response, or one like it, is seen after the CS. In classical conditioning, the UCS presentation does not depend upon the animal's behavior. In instrumental conditioning, the response of the animal (such as pressing a bar) determines whether it will receive the UCS. Instead of a shock or other noxious stimulus, the UCS may be pleasant, such as food reward, for both type I and type II conditioning.

In general, use of a noxious UCS results in faster learning and perhaps more lasting memory than does positive reinforcement. The new behavior may be either active or passive. In a step-down task, a mouse learns that its natural tendency to step down from a small pedestal upon which it has been placed results in a mild electrical shock to the foot. It learns to remain on the pedestal (passive avoidance). Alternatively, an animal may be taught that shortly after a light or sound warning, it must leave the chamber in which it has been placed (e.g., by jumping over a hurdle) in order to avoid a punishing foot shock (active avoidance).

A further refinement in training techniques is *discrimination learning.* Animals learn to choose a given limb of an apparatus on the basis of position (right or left), illumination (light vs. dark), color, or geometrical pattern. Such tasks may or may not involve an avoidance component. For example, an animal may be placed in the starting box (the stem of a Y-maze) and be forced to make a choice of one limb of the maze by application of foot shock in the stem and the other arm of the maze. This is an escape-discrimination task. In the avoidance variant of this kind of problem, a warning signal is given prior to shock, and only those trials in which animals avoid the shock are considered valid. Discrimination tasks have the advantage of tending to minimize the effects of illness of an experimental subject. If a subject is sick after a drug treatment, it is not likely that the illness will influence its decision in a discrimination task.

In a nondiscriminative avoidance task, in which the rate of speed of responding may be the index of learning used, the animal's state of health could easily affect recorded scores. Various avoidance tasks, including "step-down," "step-through," and "shuttlebox," are such timed paradigms. If the animal does not respond within a fixed period, it does not receive a positive response score for the given trial. Nevertheless, these paradigms have several advantages over discrimination tasks. A single trial score for a timed paradigm may be highly significant. For example, in a particular situation, a naive animal will step down within 5 sec of being placed in an apparatus, whereas after foot-shock, it may take over 30 sec. In this way, measurement of one-trial learning is possible. A single discrimination task score, on the other hand, generally is not informative because on a random basis an animal will make the correct response 50 percent of the time, 25 percent of the time for two trials, and will achieve a criterion of three out of four in 12.5 percent of the sessions.

METHODS OF QUANTITATION

If we consider learning as a change in performance as a result of training, then memory is the demonstration of that new performance at later times. Loss of memory is variously referred to as extinction, forgetting, or amnesia, depending on the experimental history of a subject. When lost memory reappears, it is generally proposed that the subject had not lost memory but temporarily could not retrieve it.

A limitation to the study of biochemical correlates of behavior is the difficulty in quantifying the putative molecular and physiological factors that result in altered behavior. Molecular mechanisms have been inferred from bioassays in which the response of whole animals is measured. We do not, at present, have a simpler model for the measurement of long-term memory formation and must rely on total (usually overt) behavior.

In general, investigators tend to modify experimental conditions in the direction of maximizing an effect. They are confronted with several numbers after an experiment: these may include latency (the time between onset of stimulus and response), the tally of correct and incorrect responses for segments of a training session (first five trials, second five trials, and so forth), and various other criteria, which can be established before or even after an experiment (e.g., nine out of ten correct responses). From these, the investigator must measure the "amount" of memory. He might use raw scores (such as number of correct responses), "savings" (percentages of incorrect responses not seen in trained subjects as a function of number of errors made by untrained subjects), or a number of other scoring systems. None of these would seem, a priori, to be more valid than another, yet opposite conclusions can be drawn from a given set of data depending on which scoring method is used. Individual laboratories generally use only one particular method of calculation, so they are able to evaluate dependent variables, if only qualitatively. Often, however, results are not directly comparable to those of another laboratory because of a different scoring method. Given these problems, it is not surprising that conflicting reports occur.

Dosage and temporal gradients are often used in an attempt to demonstrate a partial impairment. In such instances, we are faced with the dilemma of whether some fraction of the specific components of a learned behavior are lost or whether there is a general impairment of all the components. For example, in experiments in which amnesia has been produced by ECS after training, the "lost" memory can be regained by means of a behavioral "reminder" [52]. If performance of the learned response involves a number of physiological units operating in "series," the impairment of any one could result in the loss of performance, and its introduction would permit performance of the learned response.

Despite such experimental difficulties, experiments designed to single out, by means of appropriate control experiments, the various components of behavior have proved useful. These inferred components do not yet correspond to anatomical or functional units of the brain. Hence reductionistic approaches, such as those outlined below, have been used with increasing frequency by neurobiologists in an attempt to correlate behavior with physiological parameters.

EXPERIMENTAL PREPARATIONS

WHOLE ANIMALS

For many years, the laboratory rat has dominated physiological psychology. Its hardiness under adverse laboratory conditions is somewhat balanced by the disadvantage that much interesting innate behavior may have been bred out of this docile beast. Differences in behavior among strains have long been recognized. Much variation from experiment to experiment and laboratory to laboratory has recently been attributed to diurnal variation and seasonal effects on behavior. Inbred strains of rats can vary from supplier to supplier, as can the conditions of rearing. Strains of mice that are highly homogeneous genetically are available, reducing at least this one source of variability in behavior. Mice are generally less expensive to purchase and to maintain than rats, and they effect additional savings in radioisotope costs by virtue of their small size. The importance of genetic factors is underscored by experiments in which opposite drug effects on learning have been found in mice of different strains [10].

Evidence for learning and memory has been sought in a wide variety of animals (and even some plants!). To say that any species is incapable of demonstrating learning or memory is probably unwise, since some new task can transform an animal from a "nonlearner" to a "learner." The following includes many of the better-known attempts at training among animal phylla.

Invertebrates

Touch and light pairing has been studied in protozoa, and well-documented examples of habituation have been reported in micrometazoa (see Bullock and

Eisenstein). Coelenterates have an elementary nervous system, and habituation has been reported, as well as learning. "Bait-shyness," the rejection of food previously coupled with an aversive stimulus, has been claimed for a wide variety of species, including the sea anemone. That planarians can learn is generally accepted, but whether they can retain what they have learned for long periods is less apparent. This uncertainty precludes, at least for the present, conclusions regarding more complex experiments, such as chemical transfer of memory. An old observation of Yerkes [60], who used a single annelid worm, reported its learning a right-left discrimination task. Further behavioral studies within this phyllum may be spurred by recent ultrastructural mapping of the relatively simple annelid nervous system. A rather rapid life cycle for some annelid species invites genetic investigation of the nervous system.

Echinoderms have been relatively little studied, although learning has been reported in the starfish. Molluscs have been investigated more extensively. The octopus can be trained to respond to various visual or tactile stimuli and to retain (for many hours) what it has learned [61]. Occasionally it has been speculated that long-term memory and its possible chemical basis may have first appeared phylogenetically in the vertebrates. The octopus would appear to be an exception or, perhaps, to disprove this idea. A great deal of interest in molluscs, particularly *Aplysia,* relates to study of reduced systems, discussed below.

Studies with arthropods include reduced systems such as the cockroach ganglion, discussed below, as well as of intact insects, such as learning among the social insects. A change in flying patterns of the honeybee can be considered a manifestation of plasticity of the nervous system. Experimentally, such social insects may be difficult to work with, particularly in regard to laboratory breeding and maintenance. Insects more suitable from this standpoint, such as *Drosophila,* do not have as complex behavior as do bees and ants. Nevertheless, responses such as phototaxis in *Drosophila* have been used as a basis for the genetic dissection of brain function [31]. Considering the size of the class Insecta, more ideal subjects for a combined behavioral-genetic study than are presently being used may yet be discovered. Readers fortunate enough to have seen a flea circus can attest to the potential behavioral skills of insects.

Vertebrates

Among the classes that make up the phyllum Vertebrata, amphibians and reptiles are generally unpopular for behavioral studies because of the difficulty, up to the present time, of easily demonstrating reproducible, stable learning changes. The turtle, however, has proved to be a useful behavioral subject.

Fish have been studied extensively both from the standpoint of learning and long-term memory formation. Because they are poikilotherms, temperature can readily be introduced as an experimental variable. Regeneration in the central nervous system of cold-blooded animals has made possible studies on specification and regrowth. A gynogenetic strain, *Poecilia formosa* [5], permits experiments with genetically homogeneous subjects. While most fish have highly developed visual systems (including color vision), many species, including the bullhead and salmon, also have highly developed olfactory systems which have been the subject of behavioral investigation. Conditioning of the lateral-line system has been reported in weakly electrical fish [9].

Birds have been the subject of behavioral and biochemical studies. The pigeon is a standard subject, especially for instrumental conditioning. The newly hatched chick is completely myelinated and capable of immediate behavioral responses [17]. The duckling has been studied extensively in imprinting and by behavioral phenomenon whereby lifelong patterns are specified during a critical posthatching period [30].

Although psychologists have reported learning in virtually every common species of mammal known, the destructive nature of biochemical analyses has limited correlative studies to inexpensive, small animals. Thus, very little biochemical information is available from studies of such mammals as cats, dogs, and monkeys. Changes in the concentration of metabolites and hormones in the blood in primates during learning and stress have been studied extensively, but will not be discussed here [39].

REDUCED SYSTEMS

If conditioning requires only that pairing of the CS and UCS eventually relate the CS to the unconditioned response, formal models may be constructed, using only a portion of the animal. The CS, UCS, and response may all be represented electrically and measured by means of electrodes in the central nervous system. It has been proposed that heterosynaptic facilitation in the *Aplysia* may serve as such a model [34]. The *Aplysia* has also been studied for a model habituation system, involving gill withdrawal after a mechanical stimulus [37]. This system is particularly attractive for study since the neuronal pathways that mediate the withdrawal response are known. In addition, this species has many large neurons which can be identified, impaled with electrodes, and injected. *Tritonia,* a related mollusc, exhibits a withdrawal response when a single known neuron is stimulated [59].

Neural systems have been elucidated in crayfish and lobster. In the latter, specific axons that mediate excitation and inhibition of muscle contraction have been identified, although behavioral modification of the limb has not yet been reported.

The cockroach has been extensively studied, using a model system described by Horridge and modified by Eisenstein and Cohen [26]. The thoracic ganglion of the headless cockroach mediates learning of a position habit. Separate electrodes are connected to a corresponding pair of legs. When the tip of a leg touches a saline bath beneath it, an electrical contact is made so that leg position (either in or out of the bath) is recorded. Under these conditions, the legs move independently. During the experiment, a mild punishing shock is administered to the experimental leg whenever it contacts the bath. During a 30-min trial, the experimental leg exhibits a developing avoidance of the bath. The paired, or "yoked," leg moves in and out of the bath, but is shocked whenever the experimental leg is extended. It receives as much shock as the positional leg, but not in relation to any position. No tendency for the paired leg to avoid the water bath is seen. The avoidance seen in the experimental leg appears, then, to be a specific position habit. Blocks of such learning by actinomycin D and by cycloheximide, inhibitors of RNA and protein synthesis, respectively, have been reported [29].

The reductionistic approach in a complex nervous system such as in that of a mammal might seem self-defeating. It appears, however, that in the highly evolved nervous system, specific structures – and, therefore, localized anatomical regions – may mediate specific components of the behavioral response. The striking effect of bilateral lesions of the hippocampus or of the mammilary bodies in man has been reported to lead to a specific memory deficit [53]. Such a phenomenon has not been demonstrated by lesioning in lower animals. Electrophysiological correlates of visual learning have been reported after visual and auditory units in the cat have been paired, using leg shock as the UCS [41]. Studies with brain explants and tissue culture, which tend to document the formation of new synapses, are being pursued actively at present. The in vitro demonstration of formation of new synaptic connections could be a valuable step forward in understanding brain plasticity.

AGENTS AFFECTING LEARNING AND MEMORY

Two kinds of arguments exist that point to an organic basis for memory. One can be inferred based on the administration of disruptive agents that have putative specific chemical or physical effects. The second is based on the claim that specific alterations in brain composition or isotopic incorporation are functions of a behavioral variable. By their nature, blocking agents are unphysiological, so that it is only by inference that they direct us to understanding brain mechanisms. They do, however, have the advantage of being usable in such complex systems as the whole animal, and, hopefully, can tell us at what point of the training session, in what region of the brain, and in which macromolecules putative changes may be expected to take place.

ABLATION

Lashley concluded, from a lifetime study of the rat, that no localization of learning or memory could be demonstrated by means of lesions, and generally rejected synaptic theories of memory. Current ideas attempting to resolve Lashley's findings with the prevalent connectionist views of memory are summarized as follows [18, 34

1. Lashley made inappropriate lesions across multiple functional areas, which were not well understood as differentiated structures at the time.
2. Infections may have created more brain damage than that due only to the surgical cut.

3. The mazes used involved too many sensory and motor variables. A variety of simpler tests might have shown behavior specifically correlated to lesions.
4. Lesions, particularly if given prior to training, are overcome by brain redundancy, i.e., an alternative pathway takes over.

Subsequent studies show that complex behaviors are, in general, highly sensitive to lesioning and that large lesions generally produce greater effects than do small ones.

SPREADING DEPRESSION

The topical application of concentrated KCl to the cortical surface causes cessation of electrical activity at the site of application and spreads over the hemicortex. Animals with spreading depression of one hemicortex can be trained and then will demonstrate learning at a later time when similarly treated, but not when KCl has been applied to the other hemicortex on retraining [14]. By varying the time after learning at which KCl is applied, it can be shown that within a few minutes of training there is an apparent transfer of the learned response to the opposite hemicortex. Inhibition of protein synthesis by KCl has been reported in rat, as well as in the goldfish. In the goldfish, KCl is a convulsant and also produces memory loss.

ELECTROCONVULSIVE SHOCK

ECS is the most commonly used experimental amnestic agent. Nevertheless, there is no uniform way in which it is applied. Electrodes are in some instances attached to ears, while in others, convulsions are produced transcorneally or through the skin. The amount of current, frequency of the pulse, and its duration are all varied, and there is no agreement as to whether the agent itself or the resulting convulsion is the amnestic effector. Amnesia has been reported after ECS was administered to sedated animals, although they do not convulse visibly. Some claim that there is a graded response to ECS, and others say that it is an "all-or-none" phenomenon. Use of gaseous convulsants, such as hexafluorodiethyl ether, would seem to be more easily regulated [17]. Other commonly used convulsant agents, in addition to electroconvulsive shock and KCl, include strychnine, pentylenetetrazol, and picrotoxin. Subconvulsant doses of picrotoxin and strychnine have been reported to enhance learning and memory [38]. When amphetamine, a stimulant, is given together with cycloheximide, an inhibitor of protein synthesis, the amnestic effect of cycloheximide is reportedly blocked, and memory is preserved [8].

GENETIC MANIPULATION

In 1940 Tryon reported [57] the results of breeding rats from a parent strain on the basis of their response to maze training. Two inbred populations, "maze-bright" and "maze-dull," were derived. Subsequent biochemical analyses for brain components and enzymes have not revealed striking differences. Further, the "brightness" is task-specific. The ideal subject for the geneticist would be a bacterium that possessed a demonstrable behavior. Adler has studied chemotaxis to sugars in bacteria, and various mutants have indeed been isolated, including some that have an apparent receptor defect and others with a "motor" defect, i.e., an inability to migrate in the direction of the chemical gradient [1].

TEMPERATURE

Ransmeier and Gerard [45] failed to demonstrate any interference with memory of a previously learned maze by suddenly lowering body temperatures in the hamster, although electrical activity of the brain was suppressed. It was subsequently demonstrated that hypothermia extended the period during which ECS could produce a performance deficit. Temperature experiments are more easily and rigorously performed in poikilotherms. Cooling goldfish after they have been trained in a shock-avoidance task prolongs the period during which memory of that task is susceptible to ECS [21]. Cooling also slows protein synthesis. Cooling trout for long periods after training does not appear to block memory [42]. Disruption of protein synthesis with blocking agents in the face of otherwise normal metabolism may somehow block memory formation, whereas a concerted slowing of all metabolism may simply delay the steps relevant to memory formation.

INHIBITORS OF PROTEIN SYNTHESIS

Flexner reported in 1964 [27] that bilateral temporal injections into the brains of etherized mice within 5 days after Y-maze training resulted in loss of memory retention. At later times, amnesia could be produced by injecting puromycin into six sites, including ventricular and parietal sites, in addition to the bitemporal injections. It was initially proposed that memory was localized in the temporal region and, after several days, spread throughout the brain. There appeared to be no time limit to the disruption resulting from six injections of puromycin. Because it is a selective inhibitor of protein synthesis, the suggestion was made that puromycin blocked the formation of protein required for memory function. On the basis of the amnesia produced by inhibition of protein synthesis throughout the brain (six sites), it was also proposed that protein synthesis was required for memory maintenance. Since these initial proposals, Flexner's findings and conclusions have been extended as follows [48, 49]:

1. Injection of saline into the bitemporal sites after long intervals appears to "restore" memory. This finding argues against a block of memory and for a chemical interference with retrieval of memory, relieved by saline.
2. In Flexner's experiments, a glutarimide antibiotic (acetoxycycloheximide, AXM) did not block memory. Because AXM is an even more potent inhibitor of protein synthesis than puromycin, it was concluded that some action of puromycin other than the block of protein synthesis was responsible for the amnesia. The

product of the reaction of puromycin with the polysomal complex, peptidyl puromycin, was proposed to be responsible for the behavioral effects of the drug. Support for this argument came from the purported block of the puromycin effect by AXM, which is also known to block the formation of peptidyl puromycin, in addition to blocking protein synthesis.

3. Puromycin was demonstrated to cause swelling in brain mitochondria. This effect was also thought to be mediated via peptidyl puromycin since protection was observed when AXM was simultaneously injected.

4. Most recently, it has been found that when a protein blocker is injected bitemporally into mice one day after training, the amnestic effects of this blocker can be negated by a number of drugs, including reserpine, L-DOPA, imipramine, tranylcypromine and D-amphetamine [48]. It is proposed that peptidyl puromycin is absorbed to adrenergic sites in the brain.

Memory in the goldfish has been shown to be blocked by antibiotics [2]. Fish were trained in a shuttlebox to swim over a barrier upon a light signal in order to avoid a punishing electrical shock administered through the water. During 30 trials, avoidances rose from an average of about 20 percent (trials 1 to 10) to about 40 percent for trials 11 to 20 and about 60 percent for trials 21 to 30. When fish were given 20 trials on the first day of an experiment, returned to home storage tanks, and retrained 3 days later, they still averaged about 60 percent response, demonstrating stable memory of what they had learned. Highly significant performance levels in comparison to untrained control groups are seen even 30 days after the day 1 training session. Findings with the fish are summarized as follows:

1. Puromycin (170 μg in 10 μl) can be injected easily into conscious goldfish by means of a 30-gauge needle and a Hamilton syringe. Protein synthesis is blocked by about 80 percent for nearly a day, but no gross neurological or behavioral abnormalities are seen.

2. If the drug is given immediately after training, no evidence of previous training is detected on the basis of response rates measured 3 days later, i.e., there is no memory of training. However, no effect resulted when the same amount of puromycin was injected into fish which had been trained but which had been returned to their home tanks for 1 hr prior to injection. Such fish showed response rates comparable to those of uninjected control subjects. An important difference between these studies and those of Flexner is the evidence that fish have a relatively short (1 hr) consolidation gradient. The demonstration of memory fixation without the production of unconsciousness by convulsants or anesthetics lends support for a consolidation theory of memory. Fish are functioning normally during the time the drug is active, so the hypothesis that gross electrical disruption of the brain is necessary for the production of amnesia is not supported.

3. Puromycin injected before training has little effect on acquisition of learning, although memory, judged by performance on day 4, is blocked.

4. Groups of fish tested at various times after training and immediately after the

injection of puromycin exhibit a slow decay of memory. Those tested immediately after training and injection of puromycin show normal response rates for trials 21 to 30. Other groups injected immediately after training and tested at various times thereafter exhibit a decrease in avoidance responses during the next 2 days. Thus, puromycin does not immediately obliterate memory, but rather initiates a process that takes several days for completion. Perhaps it weakens some physiological process, which then gives rise to defective intermediates that are subsequently destroyed.

5. AXM and cycloheximide produce effects similar to those of puromycin on memory in fish, in sharp contrast to the above findings in mice. Puromycin aminonucleoside (PAN), a derivative of puromycin that does not block protein synthesis, does not block memory in fish, nor does *O*-methyltyrosine, the amino acid moiety of puromycin [3]. In contrast, both puromycin and PAN potentiate seizures produced by pentylenetetrazol in the fish. The seizure-producing properties of puromycin would thus seem to be dissociated from its amnestic properties. Cycloheximides do not block the seizure potentiation via puromycin, nor do they alone potentiate the pentylenetetrazol seizures. Peptidyl puromycin thus does not appear to mediate the convulsant activity of puromycin or of PAN.

6. If fish are allowed to remain in the training apparatus after a training session, they remain susceptible to the amnestic action of the antimetabolites for several hours. This unexpected result suggests that presence of the fish in the training environment delays onset of the consolidation process. When trained fish are subjected to the training environment without trials on day 2, some mild amnestic effects of the antibiotics are seen when they are retrained 5 days later [22]. This latter result underscores the complexity of learning experiments and their interpretation. In experiments with rats, in which ECS is used as the amnestic agent, it can be shown that if ECS is administered after an additional training trial, which is given after consolidation is thought to be complete, then it can activate the amnestic effect [40]. As discussed below, such results may well be consequences of the complex nature of physiological factors leading to performance.

Experiments by Barondes and co-workers [8] have used paradigms very similar to those of Flexner, although the results are, in general, more in accord with those conducted with fish. Experiments employing pretrial or posttrial injections indicate a brief (15 to 30 min) consolidation time. Short-term memory in the mouse lasts about 3 hr. Both puromycin and cycloheximide derivatives produce amnesia. Cyclo heximides that do not block protein synthesis are not amnestic. Although differenc in the temporal effects of puromycin as demonstrated by the Flexner and Barondes experiments have not been systematically explored, it should be pointed out that Barondes used an escape discrimination procedure whereas Flexner employed avoidance discrimination.

The inhibitors of protein synthesis have provided several valuable clues for increa ing our understanding of memory formation. Short-term and long-term memory appear to be separable on the basis of sensitivity to these agents. A major question remains regarding the specificity of their action. Any generally toxic substance might produce a confusional state in such a way that acquisition is not affected, but permanent memory formation is blocked. At present, a clear-cut distinction between short-term and long-term memory is demonstrable only by the antimetabolite treatment method. Other drugs that have various neurotoxic effects have not been reported to provide evidence for this distinction. In man, several degenerative or traumatic conditions have been reported to produce a selective effect on formation of new permanent memories [58].

Flexner's studies of mice suggest that the effects of antimetabolites in this animal are localized in the entorhinal cortex and the hippocampus. At the molecular level, it is unclear whether protein inhibitors exert their amnestic effect via blocking the formation of some relatively prevalent brain protein at specific synaptic locations related to the new behavior, or whether, in addition, the proteins relevant to the block are chemically unique, like antibodies, for a particular neuronal pathway. Support for the existence of unique neuronal proteins is indirect. Specificity of regrowth of nerve bundles suggests the possibility of the presence of a high degree of chemical specificity within neurons [55]. Behaviorally, it has been claimed, but by no means accepted, that injection of specific proteins can alter behavior in recipient subjects. Further indirect evidence derives from a relatively high degree of hybridizable RNA found in brain as compared to other organs [12]. More direct proof would come from the identification of a change in brain protein patterns that is directly correlated with training.

If new protein is indeed formed at the synapse as a result of training, we should bear in mind the following:

1. Ribosomes are rare or absent at the presynaptic region. New protein in this region might come from some other source, e.g., presynaptic mitochondria, postsynaptic mitochondria, or postsynaptic ribosomes, the last being in relatively close proximity of the synapse. New presynaptic protein can migrate via a process of axonal flow, in which case the time for the new protein to reach the synapse is a function of the length of the presynaptic axon.
2. The formation of novel protein sequences, if it occurs as a function of training, is assumed to be ultimately under the regulation of DNA. As in the selective mechanism for antibody formation, experience might evoke a response from a vast repertoire that arose via the natural history of the species.
3. The hypothetical behavior-specific proteins should be found only in small quantities as the result of any one training experience. However, the presence of gross alterations in brain RNA or protein is evidence for correlation with some alteration that is not specifically related to the given behavior. A block in protein synthesis that results in amnesia does not necessarily involve a specific protein. Relatively prevalent proteins in specific loci may mediate the fixation process. Because short-term memory is labile, a signal may be required in the brain to secrete a "fixative" molecule that permeates the entire organ but fixes only the labile connections. The memory process may require the presence of a critical protein with

rapid turnover that catalyzes the formation of the fixative molecule. In this case, the protein itself does not contain information, and we might therefore see macroscopic changes in protein electrophoretic patterns or in radioactive labeling as a result of a training procedure.

Whether or not a hypothetical critical protein is task-specific, a further question arises as to whether the protein block that produces amnesia prevents formation of a new protein, as we have been considering up to this point, or whether the antibiotic inhibitor causes some rapidly turning over brain protein to fall below the effective level for normal memory formation. Although brain proteins in general turn over too slowly to support such a possibility, there are exceptions. For example, an isozyme of acetylcholinesterase in the rat brain appears to have a half-life of only 3 hr [19]. The concept of depletion of a critical protein predicts that a protein block given 2 to 3 hr prior to training would produce a greater deficit of memory than would injections given immediately after training, since the level of the critical protein would have reached an even lower level by the posttraining period. No evidence for this has yet been reported.

INHIBITION OF DNA AND RNA SYNTHESIS

Experiments in which amnesia is caused by protein inhibitors lead to the speculation that RNA blockers might also block memory. If enzyme induction mediates some aspect of the formation of long-term memory, a block at the transcriptional level (DNA-mediated synthesis of RNA) would be expected to produce a similar effect. If the protein blocker simply reduces the level of a rapidly turning over protein as discussed above, RNA blockers might or might not also cause reduction in a critical protein, depending on the stability of the appropriate messenger RNAs.

Experimentally, 8-azaguanine and actinomycin D have been used in vivo in learning experiments. These agents are more toxic than the protein inhibitors, and interpretation of experiments with them is often difficult. Although it was originally believed that actinomycin D did not block memory, recent studies tend to support a block of memory after injection of this agent [4, 56].

Although DNA may constitute the reservoir of behavioral information, current concepts of cell regulation do not imply that it turns over. DNA metabolism is not required for the synthesis of novel proteins in enzyme induction. Neurons are nondividing end cells, yet the brain is capable of enzyme induction. Other brain elements, notably glia, do divide. Memory theories involving glia have been proposed. Experimentally, arabinosyl cytosine, a DNA blocker, does not affect learning or memory in the goldfish [16].

ENZYME INHIBITORS

Deutsch has reported that diisopropylphosphorofluoridate (DFP), a blocker of acetylcholinesterase, interferes with memory in rats [23]. The kinetics of recovery have led to the conclusion that this agent primarily affects retrieval. An optimal level of brain acetylcholinesterase has been proposed, which is such that excessively high or low amounts result in poor behavioral performance. This optimal level is employed to explain the paradoxical findings that DFP administration leads to reduction in performance in the rat, whereas in animals that have "forgotten" the task, DFP leads to improved performance. It is difficult to relate these experiments to known brain mechanisms since the role of acetylcholine in the central nervous system is poorly understood. Actual brain levels of acetylcholine after administration of DFP in these experiments have not yet been reported.

ENVIRONMENTAL MANIPULATION

In animal memory experiments, increasing attention is being given to the animal's surroundings before and after training [20]. An example of interaction between the amnestic agent and the environment comes from *reinstatement* experiments. Reexposure to some elements of the training task can make rats susceptible to ECS amnesia at times after consolidation was believed to be complete [40]. The nature of these interactions is unknown but may be related in part to the general physiological state of the animal, such as stress factors.

BRAIN GROWTH

The simplest model for brain plasticity might require that the brain grows as a result of experience. Brain components should then have accretion properties. Their amount should increase with age and experience. New synaptic endings after stimulation have been reported [51], and increased brain length [7] and cortical thickness [24] have been described. Starvation in adults generally causes negligible effects; the brain is spared. However, in the growing animal this may not be true. It is clear that genetic abnormalities in man can produce defects during development that can be partially reversed by diet. Phenylalanine hydroxylase impairment (phenylketonuria) results in mental retardation in infants fed a normal diet. If a low phenylalanine diet is administered, the children develop relatively normally, and they are no longer very susceptible to brain damage by dietary phenylalanine. Thyroid deficiency during development results in defective brain growth, with reduction in dendritic spines in the cortex [25].

Administration of growth hormone to pregnant rats leads to progeny with permanently increased brain DNA [62]. Evidence that they are behaviorally superior is not convincing. Similarly, grafting a second brain into developing fish has been reported, although the behavioral significance is dubious [11].

ENHANCING AGENTS

Claims for the enhancement of learning via biochemically or pharmacologically active agents are cited below. Several investigators have claimed that when extracts of nucleic acid or proteins from brains of donor animals that have been trained are injected into recipients, the recipient animals exhibit significantly improved performance that is specific for the original task [44]. The conditions of injection, the species injected, and the task are sufficiently different among laboratories to prevent generalizations from the results.

BIOCHEMICAL CORRELATES OF EXCITATION, LEARNING, AND BEHAVIOR

The most direct confirmation of a role for macromolecules in behavior, inferred from the antimetabolite experiments, is the detection of changes in RNA or protein correlated with learning in the absence of the blocking agents. Such changes are indeed claimed, but they are not easily reconciled with the studies with blocking agents. For example, changes in labeling in the brain that are great enough to be detected with relatively crude methods, such as trichloroacetic acid precipitation, would seem too large to account for putative behavior-specific macromolecules. Alterations within a small number of neurons have been studied using special methods, such as those developed and employed by Hydén. His laboratory has reported [32] changes in total RNA and base ratios of isolated RNA of Dieters' nucleus cells after training of a balancing task. Freehand isolation of these large neurons and of the associated satellite glial cells was followed by microanalytical procedures. More recently, Hydén has examined electrophoretic patterns from the cytosol of cells in the rat hippocampus after training for handedness in an appetitive task [33]. Even with such relatively refined techniques, the alterations found seem rather gross compared to those anticipated from the learning of a single task.

Another technique for measuring regional alterations in the brain is autoradiography. Glassman reported [28] increased ^{3}H-uridine incorporation into macromolecular material, presumably RNA, in the diencephalon of mice after a training experience. Changes in protein labeling were not reported. Altman found increased incorporation of ^{3}H-leucine into protein in motor neurons of the rat lumbar cord after simple exercise [6].

There are many reports of alterations in total brain RNA and protein labeling. Such global changes tend to support association with some nonspecific correlate of training, exercise, or stress. Visual stimulation has been reported to cause changes in aggregation of brain polysomes, but not of liver polysomes. This change is interpreted as reflecting increased neuronal activity in general. On the other hand, an increase in the incorporation of labeled uridine into mouse brain has been reported to be specific for training of a shock-avoidance task [28]. Yoked control animals subjected to the same amount of light, buzzer, and shock as experimental animals, but which are not permitted to make the trained response, do not show the change in labeling. Overtrained animals, even when permitted to make the trained response, also do not show the increased labeling of RNA. It appears, then, that some changes in RNA metabolism take place with simple stimulation and that there are, in addition, conditions under which only training stimulates labeling. Changes in the distribution of incorporated radioactivity as a result of training have also been reported. When goldfish are injected with labeled orotic acid and RNA is isolated after training, an increase is found in the ratio of labeled uridine to labeled cytidine, compared to that seen in control animals [54].

Although it is difficult to draw any simple conclusion from these experiments, two sorts of issues can be delineated:

1. In none of the conditions in which RNA labeling is altered has a concomitant alteration in protein labeling been reported. Messenger, transfer, and ribosomal RNAs subserve protein synthesis, so a comparable, even greater, effect on protein labeling might have been anticipated.
2. In experiments in which increased labeling of RNA is observed, it is not certain that RNA synthesis is actually increased. An alternative explanation is a change in the size of the precursor pool. The injected radioisotopic precursor, usually a pyrimidine, must be phosphorylated intracellularly and converted to a triphosphate prior to incorporation into RNA. Efficacy of incorporation depends on the amount of endogenous intermediates present as well as upon the starting specific activity of the injected precursor. An increase in labeling might be due to reduction in the endogenous unlabeled precursor pool, rather than to an increase in RNA synthesis. The relevant intracellular pools are most likely intranuclear.

CLINICAL IMPLICATIONS

Nowhere in biology is the animal model less appropriate for studying human disease than in the area of higher brain function, yet many basic principles established in studies on animal learning and memory may carry over into practical biomedical problems. For example, the effect of electroconvulsive shock therapy (ECT) in the treatment of depression has not been explained adequately. We do not know if the amnesia associated with ECT is related to the therapeutic effect. It has been shown in animal studies that ECS causes alterations in catecholamine turnover in the brain [35], and catecholamines have been implicated in the depressive state [50]. The reinstatement phenomenon described in this chapter (p. 659) has prompted the suggestion that ECT effects could be enhanced by temporal proximity of the treatment with pretreatment clinical evocation of a patient's psychological problems by the therapist [47].

Numerous drugs have been proposed for the aid of defective learning and memory, mainly for the young and for the aged. Other than having indirect effects, such as depression of overactivity, increasing attention span, and mild stimulant effects, there has been no specific agent to gain acceptance as yet for the improvement of learning or memory. Drugs such as tricyanoaminopropene or magnesium pemoline, which have been reported to be effective in animals, have questionable biochemical, as well as behavioral, effects. In the aged, injection or ingestion of yeast RNA [15] has been reported to improve memory, but this work has been seriously challenged.

Specific amnestic defects in otherwise normal individuals have been reported as a result of tumors, surgery, or specific degenerative diseases such as Korsakoff's syndrome and Alzheimer's disease. Although the hippocampus appears to be involved in many of these conditions, the nature of its role, whether neural, humoral, or both, is unknown at present.

Finally, while relatively little is known about early dietary effects on brain function, animal experiments give us sufficient concern to recognize a potential problem in the malnourished infant, who may be destined for a life of restricted usefulness.

REFERENCES

ORIGINAL SOURCES AND SPECIAL TOPICS

1. Adler, J. Chemotaxis in bacteria. *Science* 153:708, 1966.
2. Agranoff, B. W. Agents That Block Memory. In Quarton, G. C., Melnechuk, T., and Schmitt, F. O. (Eds.), *The Neurosciences: A Study Program.* New York: Rockefeller University Press, 1967, pp. 756–764.
3. Agranoff, B. W. Protein Synthesis and Memory Formation. In Lajtha, A. (Ed.), *Protein Metabolism of the Nervous System.* New York: Plenum, 1970, pp. 533–543.
4. Agranoff, B. W., Davis, R. E., Casola, L., and Lim, R. Actinomycin D blocks memory formation of a shock-avoidance in the goldfish. *Science* 158:1600, 1967.
5. Agranoff, B. W., Davis, R. E., and Gossington, R. E. Esoteric fish. *Science* 171:230, 1971.
6. Altman, J. Differences in the utilization of tritiated leucine by single neurons in normal and exercised rats: An autoradiographic investigation with microdensity. *Nature (Lond.)* 199:777, 1963.
7. Altman, J., Wallace, R. B., Anderson, W. J., and Das, G. D. Behaviorally induced changes in length of cerebrum in rats. *Dev. Psychobiol.* 1:112, 1968.
8. Barondes, S. H. Cerebral protein synthesis inhibitors block long-term memory. *Int. Rev. Neurobiol.* 12:177, 1970.
9. Bennett, M. V. L. Neural Control of Electric Organs. In Ingle, D. (Ed.), *The Central Nervous System and Fish Behavior.* Chicago: University of Chicago Press, 1968, pp. 147–169.
10. Bovet, D., Bovet-Nitti, F., and Oliverio, A. Genetic aspects of learning and memory in mice. *Science* 163:139, 1969.
11. Bresler, D. E., and Bitterman, M. E. Learning in fish with transplanted brain tissue. *Science* 163:590, 1969.
12. Brown, I. R., and Church, R. B. RNA transcription from non-repetitive DNA in the mouse. *Biochem. Biophys. Res. Commun.* 42:850, 1971.
13. Bullock, T. H. Simple Systems for the Study of Learning Mechanisms. In Schmitt, F. O. (Ed.), *Neuroscience Research Symposium Summaries.* Vol. 2. Cambridge, Mass.: M.I.T. Press, 1967, pp. 203–327.

14. Bureš, J., and Burešova, O. The use of Leãos spreading depression in the study of interhemispheric transfer of memory traces. *J. Comp. Physiol. Psychol.* 53:558, 1960.
15. Cameron, D. E., and Solyom, L. Effect of RNA on memory. *Geriatrics* 16:74, 1961.
16. Casola, L., Lim, R., Davis, R. E., and Agranoff, B. W. Behavioral and biochemical effects of intracranial injection of cytosine arabinoside in goldfish. *Proc. Natl. Acad. Sci. U.S.A.* 60:1389, 1968.
17. Cherkin, A. Kinetics of memory consolidation: Role of amnesic treatment parameters. *Proc. Natl. Acad. Sci. U.S.A.* 63:1094, 1969.
18. Chow, K. L. Effects of Ablation. In Quarton, G. C., Melnechuk, T., and Schmitt, F. O. (Eds.), *The Neurosciences: A Study Program.* New York: Rockefeller University Press, 1967, pp. 705–713.
19. Davis, G. A., and Agranoff, B. W. Metabolic behaviour of isozymes of acetylcholinesterase. *Nature (Lond.)* 220:277, 1968.
20. Davis, R. E., and Agranoff, B. W. Stages of memory formation in goldfish: Evidence for an environmental trigger. *Proc. Natl. Acad. Sci. U.S.A.* 55:555, 1966.
21. Davis, R. E., Bright, P. J., and Agranoff, B. W. Effect of ECS and puromycin on memory in fish. *J. Comp. Physiol. Psychol.* 60:162, 1965.
22. Davis, R. E., and Klinger, P. D. Environmental control of amnesic effects of various agents in goldfish. *Physiol. Behav.* 4:269, 1969.
23. Deutsch, J. A., Hamburg, M. D., and Dahl, H. Anticholinesterase-induced amnesia and its temporal aspects. *Science* 151:221, 1966.
24. Diamond, M. C., Krech, D., and Rosenzweig, M. R. The effects of an enriched environment on the histology of the rat cerebral cortex. *J. Comp. Neurol.* 123:111, 1964.
25. Eayrs, J. T. Effects of Thyroid Hormones on Brain Differentiation. In *Brain-Thyroid Relationships.* Ciba Foundation Study Group No. 18, London: Churchill, 1964.
26. Eisenstein, E. M., and Cohen, M. J. Learning in an isolated prothoracic insect ganglion. *Anim. Behav.* 13:163, 1965.
27. Flexner, J. B., Flexner, L. B., and Stellar, E. Memory in mice as affected by intracerebral puromycin. *Science* 141:57, 1963.
28. Glassman, E. The biochemistry of learning: An evaluation of the role of RNA and protein. *Annu. Rev. Biochem.* 38:605, 1969.
29. Glassman, E., Henderson, A., Cordle, M., Moon, H. M., and Wilson, J. E. Effect of cycloheximide and actinomycin D on the behaviour of the headless cockroach. *Nature (Lond.)* 225:967, 1970.
30. Hess, E. H. Imprinting. *Science* 130:133, 1959.
31. Hotta, Y., and Benzer, S. Genetic dissection of the *Drosophila* nervous system by means of mosaics. *Proc. Natl. Acad. Sci. U.S.A.* 67:1156, 1970.
32. Hydén, H., and Lange, P. W. A differentiation in RNA response in neurons early and late during learning. *Proc. Natl. Acad. Sci. U.S.A.* 53:946, 1965.

33. Hydén, H., and Lange, P. W. S-100 protein: Correlation with behavior. *Proc. Natl. Acad. Sci. U.S.A.* 67:1959, 1970.
34. Kandel, E. R., and Spencer, W. A. Cellular neurophysiological approaches in the study of learning. *Physiol. Rev.* 48:65, 1968.
35. Kety, S. S., Jovoy, F., Thierry, A.-M., Julou, L., and Glowinski, J. A sustained effect of electroconvulsive shock on the turnover of norepinephrine in the central nervous system of the rat. *Proc. Natl. Acad. Sci. U.S.A.* 58:1249, 1967.
36. Kimble, G. A. In Hilgard, E., and Marquis, D. (Eds.), *Conditioning and Learning,* 2nd ed. New York: Appleton-Century-Crofts, 1961.
37. Kupferman, I., and Kandel, E. R. Neuronal controls of a behavioral response mediated by the abdominal ganglion of *Aplysia. Science* 164:847, 1969.
38. Mason, J. W. Organization of psychoendocrine mechanisms. *Psychosom. Med.* 30 (part 2):565, 1968.
39. McGaugh, J. L. Time-dependent processes in memory storage. *Science* 153: 1351, 1966.
40. Misanin, J. R., Miller, R. R., and Lewis, D. J. Retrograde amnesia produced by electroconvulsive shock after reactivation of a consolidated memory trace. *Science* 160:554, 1968.
41. Morrell, F. Electrical Signs of Sensory Coding. In Quarton, G. C., Melnechuk, T., and Schmitt, F. O. (Eds.), *The Neurosciences: A Study Program.* New York: Rockefeller University Press, 1967, pp. 452–469.
42. Neale, J. H., and Gray, I. Protein synthesis and retention of a conditioned response in rainbow trout as affected by temperature reduction. *Brain Res.* 26:159, 1971.
43. Pearlman, C. A., Sharpless, S. K., and Jarvik, M. E. Retrograde amnesia produced by anesthetic and convulsant agents. *J. Comp. Physiol. Psychol.* 54:109, 1961.
44. Quarton, G. C. The Enhancement of Learning by Drugs and the Transfer of Learning by Macromolecules. In Quarton, G. C., Melnechuk, T., and Schmitt, F. O. (Eds.), *The Neurosciences: A Study Program.* New York: Rockefeller University Press, 1967, pp. 744–755.
45. Ransmeier, R. E., and Gerard, R. W. Effects of temperature, convulsion and metabolic factors on rodent memory and EEG. *Am. J. Physiol.* 179:663, 1954.
46. Ribot, T. *Disorders of Memory.* London: Kegan, Paul, Trench and Co., 1885.
47. Robbins, M. J., and Meyers, D. R. Motivational control of retrograde amnesia. *J. Exp. Psychol.* 84:220, 1970.
48. Roberts, R. B., Flexner, J. B., and Flexner, L. B. Some evidence of the involvement of adrenergic sites in the memory trace. *Proc. Natl. Acad. Sci. U.S.A.* 66:310, 1970.
49. Roberts, R. B., and Flexner, L. B. The biochemical basis of long-term memory. *Q. Rev. Biophys.* 2:135, 1969.
50. Robins, E. (See Chapter 30, this volume.)
51. Schapiro, S., and Vukovich, K. R. Early experience effects upon cortical dendrites: A proposed model for development. *Science* 167:292, 1970.

52. Schneider, A. M., and Sherman, W. Amnesia: A function of the temporal relation of footshock to electroconvulsive shock. *Science* 159:219, 1968.
53. Scoville, W. B., and Milner, B. Loss of recent memory after bilateral hippocampal lesions. *J. Neurol. Neurosurg. Psychiatry* 20:11, 1957.
54. Shashoua, V. E. The Relation of RNA Metabolism in the Brain to Learning in the Goldfish. In Ingle, D. (Ed.), *The Central Nervous System and Fish Behavior.* Chicago: University of Chicago Press, 1968, pp. 203–213.
55. Sperry, R. N. Chemoaffinity in the orderly growth of nerve fiber patterns and connections. *Proc. Natl. Acad. Sci. U.S.A.* 50:703, 1963.
56. Squire, L. R., and Barondes, S. H. Actinomycin-D: Effects on memory at different times after training. *Nature (Lond.)* 225:649, 1970.
57. Tryon, R. C. Genetic differences in maze learning abilities on rats. *Yearbook Nat. Soc. Stud. Educ.* 39 (part 1):111, 1940.
58. Whitty, C. W. M., and Zangwill, O. L. *Amnesia.* London: Butterworths, 1966.
59. Willows, A. O. D. Behavioral Acts Elicited by Stimulation of Single Identifiable Nerve Cells. In Carlson, F. D. (Ed.), *Physiological and Biochemical Aspects of Nervous Integration.* Englewood Cliffs, N.J.: Prentice-Hall, 1968, pp. 217–243.
60. Yerkes, R. M. The intelligence of earthworms. *J. Anim. Behav.* 2:332, 1912.
61. Young, J. Z. Short and long memories in *octopus* and the influence of the vertical lobe system. *J. Exp. Biol.* 52:385, 1970.
62. Zamenhof, S., Mosley, J., and Schuller, E. Stimulation of the proliferation of cortical neurons by prenatal treatment with growth hormone. *Science* 152: 1396, 1966.

BOOKS AND REVIEWS

Agranoff, B. W. Effects of Antibiotics on Long-term Memory Formation in the Goldfish. In Honig, W. K., and James, P. H. R. (Eds.), *Animal Memory.* New York: Academic, 1971, pp. 243–258.

Barondes, S. H. Cerebral protein synthesis inhibitors block long-term memory. *Int. Rev. Neurobiol.* 12:177, 1970.

Bullock, T. H. Simple Systems for the Study of Learning Mechanisms. In Schmitt, F. O. (Ed.), *Neuroscience Research Symposium Summaries.* Vol. 2. Cambridge, Mass.: M.I.T. Press, 1967, pp. 203–327.

Eisenstein, E. M. The Use of Invertebrate Systems for Studies on the Bases of Learning and Memory. In Quarton, G. C., Melnechuk, T., and Schmitt, F. O. (Eds.)., *The Neurosciences: A Study Program.* New York: Rockefeller University Press, 1967, pp. 653–665.

Glassman, E. The biochemistry of learning: An evaluation of the role of RNA and protein. *Annu. Rev. Biochem.* 38:605, 1969.

John, E. R. *Mechanisms of Memory.* New York: Academic, 1967.

Kandel, E. R., and Spencer, W. A. Cellular neurophysiological approaches in the study of learning. *Physiol. Rev.* 48:65, 1968.

Whitty, C. W. M., and Zangwill, O. L. *Amnesia.* London: Butterworth, 1966.

Glossary

ACh	acetylcholine
AChE	acetylcholinesterase
ACTH	adrenocorticotropic hormone
ADP	adenosine diphosphate
ALA	δ-aminolevulinic acid
AMP	adenosine monophosphate (adenylic acid)
ATP	adenosine triphosphate
AXM	acetoxycycloheximide
BAL	British antilewisite (2,3-dimercaptopropanol)
BSA	bovine serum albumin
CBF	cerebral blood flow
CDP	cytidine diphosphate
ChAc	choline acetylase (choline acetyl transferase)
ChE	cholinesterase ("nonspecific" esterase)
CMP	cytidine monophosphate (cytidylic acid)
CMR	cerebral metabolic rate
CNS	central nervous system
CoA	coenzyme A
COMT	catechol-*O*-methyl transferase
CPZ	chlorpromazine
CRF	corticotropin releasing factor
CS	conditioned stimulus
CSF	cerebrospinal fluid
CTP	cytidine triphosphate
3',5'-cyclic AMP	cyclic 3',5'-adenosine monophosphate

DDT	1,1,1-trichloro-2,2-bis (*p*-chlorophenyl)-ethane
DEAE	diethylaminoethyl
DFP	diisopropylphosphorofluoridate
DMPEA	dimethoxyphenylethylamine
DON	6-diazo-5-oxo-L-norleucine
DOPA	3,4-dihydroxyphenylalanine
DOPAC	3,4-dihydroxyphenylacetic acid
dopamine	3,4-dihydroxyphenylethylamine
EAE	experimental allergic encephalomyelitis
EAN	experimental allergic neuritis
ECS	electroconvulsive shock
ECT	electroconvulsive shock therapy
ECW	extracellular water
EDTA	ethylenediaminetetraacetate
EGTA	ethyleneglycol-bis-(β-aminoethyl ether)-*N*,*N*-tetraacetic acid
EIM	excitability inducing material
ER	endoplasmic reticulum
E_R	resting potential
FAD	flavine adenine dinucleotide
FSH	follicle-stimulating hormone
FSH-RF	follicle-stimulating hormone releasing factor
GABA	γ-aminobutyric acid
GABA-T	γ-aminobutyric acid transaminase
GAD	glutamic acid decarboxylase
GH	growth hormone (somatotropin)
GH-RF	growth hormone releasing factor
GMP	guanosine monophosphate (guanylic acid)
GSH	glutathione (reduced form)
GTP	guanosine triphosphate
GTT	glucose tolerance test
HHH	hypothalamic hypophysiotropic hormone
5-HIAA	5-hydroxyindoleacetic acid
HIOMT	hydroxyindole-*O*-methyl transferase
HPr	histidine-containing protein

5-HT	5-hydroxytryptamine (serotonin)
5-HTP	5-hydroxytryptophan
HVA	homovanillic acid (4-hydroxy-3-methoxyphenylacetic acid)
ITT	insulin tolerance test
17-KGS	17-ketogenic steroid
17-KS	17-ketosteroid
LD	lethal dose
LH	luteinizing hormone
LH-RF	luteinizing hormone releasing factor
LMM	light meromyosin
LSD	*d*-lysergic acid diethylamide
MAO	monoamine oxidase
MeCh	acetyl-β-methylcholine (mecholyl)
MET	metanephrine
MHPG	3-methoxy-4-hydroxyphenylglycol
MLD	metachromatic leukodystrophy
MS	multiple sclerosis
MSH	melanocyte-stimulating hormone
MSH-IF	melanocyte-stimulating hormone inhibitory factor
NAD^+	nicotinamide adenine dinucleotide
NADH	reduced form of NAD^+
$NADP^+$	nicotinamide adenine dinucleotide phosphate
NADPH	reduced form of $NADP^+$
Na_E	exchangeable sodium
NANA	*N*-acetylneuraminic acid
NE	norepinephrine
NEM	*N*-ethylmaleimide
NGF	nerve growth factor
NMET	normetanephrine
NMR	nuclear magnetic resonance
11-OHCS	11-hydroxycorticosteroid
17-OHCS	17-hydroxycorticosteroid

OMP	orotidine monophosphate
PAM	pyridine-2-aldoxime methiodide
PAN	puromycin aminonucleoside
PAPS	3'-phosphoadenosine 5'-phosphosulfate
PEP	phosphoenolpyruvate
P_i	inorganic phosphate
PIF	prolactin inhibitory factor
PNMT	phenylethanolamine-*N*-methyl transferase
PNS	peripheral nervous system
PPD	*N*,*N*'-dimethyl-*p*-phenylenediamine
RER	rough endoplasmic reticulum
RF	releasing factor
SER	smooth endoplasmic reticulum
SME	stalk-median eminence tissue
SSA-D	succinic semialdehyde dehydrogenase
SSPE	subacute sclerosing panencephalitis
TBW	total body water
TCA	tricarboxylic acid; trichloroacetic acid
THC	Δ^1-tetrahydrocannabinol
TMB-4	*N*,*N*-trimethylene-bis-(pyridinium-4-aldoxime) dibromide
TRF	thyrotropin releasing factor
Tris	tris-(hydroxymethyl)-aminomethane
TSH	thyroid-stimulating hormone (thyrotropin)
TSP	total soluble protein
UCS	unconditioned stimulus
UDP	uridine diphosphate
UMP	uridine monophosphate (uridylic acid)
UTP	uridine triphosphate
VMA	vanillylmandelic acid

INDEX

Index